Handbook of Green Chemistry

Volume 4
Supercritical Solvents

Volume Edited by
Walter Leitner and Philip G. Jessop

Related Titles

Tanaka, K.

Solvent-free Organic Synthesis

Second, Completely Revised and Updated Edition

2009

ISBN: 978-3-527-32264-0

Lefler, J.

Principles and Applications of Supercritical Fluid Chromatography

2009

ISBN: 978-0-470-25884-2

Wasserscheid, P., Welton, T. (eds.)

Ionic Liquids in Synthesis

Second, Completely Revised and Enlarged Edition

2008

ISBN: 978-3-527-31239-9

Sheldon, R. A., Arends, I., Hanefeld, U.

Green Chemistry and Catalysis

2007

ISBN: 978-3-527-30715-9

Lindstrom, U. M. (ed.)

Organic Reactions in Water

Principles, Strategies and Applications

2008

ISBN: 978-1-4501-3890-1

Handbook of Green Chemistry

Volume 4
Supercritical Solvents

Volume Edited by Walter Leitner and Philip G. Jessop

WILEY-VCH

WILEY-VCH Verlag GmbH & Co. KGaA

The Editor

Prof. Dr. Paul T. Anastas
Yale University
Center for Green Chemistry & Green Engineering
225 Prospect Street
New Haven, CT 06520
USA

Volume Editors

Prof. Dr. Walter Leitner
RWTH Aachen
Institute for Technical Chemistry
Worringerweg 1
52074 Aachen
Germany

Prof. Dr. Philip G. Jessop
Department of Chemistry
Queen's University
90 Bader Lane
Kingston, ON K7L 3N6
Canada

Handbook of Green Chemistry – Green Solvents
Vol. 4: Supercritical Solvents
ISBN: 978-3-527-32590-0
Vol. 5: Reactions in Water
ISBN: 978-3-527-32591-7
Vol. 6: Ionic Liquids
ISBN: 978-3-527-32592-4

Set II (3 volumes):
ISBN: 978-3-527-31574-1

Handbook of Green Chemistry
Set (12 volumes):
ISBN: 978-3-527-31404-1

Library of Congress Card No.:
applied for

British Library Cataloguing-in-Publication Data
A catalogue record for this book is available from the British Library.

Bibliographic information published by the Deutsche Nationalbibliothek
The Deutsche Nationalbibliothek lists this publication in the Deutsche Nationalbibliografie; detailed bibliographic data are available on the Internet at http://dnb.d-nb.de.

Typesetting Thomson Digital, Noida, India
Printing Betz-Druck GmbH, Darmstadt
Binding Litges & Dopf GmbH, Heppenheim
Cover Design Adam-Design, Weinheim

Printed in the Federal Republic of Germany
Printed on acid-free paper

ISBN: 978-3-527-32590-0

Foreword

For several centuries, **Chemistry** has strongly contributed to a fast and almost unlimited trend of progress and innovation that have deeply modified and improved human life in all its aspects. But, presently, chemistry is also raising fears about its immediate and long-term impact on the environment, leading to a growing demand for development of "green chemistry" preserving the environment and natural non-renewable resources. Changing raw materials to renewable sources, using low energy-consumption processes, reprocessing all effluents and inventing new environment-friendly routes for the manufacture of more efficacious products are immense challenges that will condition the future of mankind.

In this context of sustainable development, Supercritical Fluids (SCF) and Gas-Expanded Liquids (GXL) are of rapidly-growing interest because either they are non-toxic and non-polluting solvents (like carbon dioxide or water) or they help one to avoid harmful intermediates through new processing routes. After two decades of development of new extraction/fractionation/purification processes using SCFs - mainly CO_2 - with about 250 industrial-scale plants now in operation around the world, other applications have been and will be at the centre of new developments for the present decade and the coming one:

- Manufacture of high-performance materials including pharmaceutical formulations, bio-medical devices and many specific polymeric, inorganic or composite materials, either by physical processes or chemical synthesis;
- New routes of chemical or biochemical synthesis, coupled with product purification;
- Innovative waste management and recycle.

It has to be understood that moving to SCF or GXL media for chemical synthesis shall not be considered as a "simple" substitution of "classical" organic solvents, but imposes a complete "reset" of knowledge of synthesis routes, reaction schemes and parameters. One main difference is related to the physico-chemical properties of these fluids that are both "tunable" solvents and separation agents. Some are also reactants at the same time. Because of these properties, reaction rate and selectivity are very different from those observed in liquid media, as well exemplified by

Handbook of Green Chemistry, Volume 4: Supercritical Solvents. Edited by Walter Leitner and Philip G. Jessop

ISBN: 978-3-527-32590-0

hydrogenation reactions over heterogeneous catalysts. Moreover, many new environmentally-friendly processes using CO_2 and water lead to innovative high-tech materials (especially nano-structured materials), biomass conversion and waste treatment such as, for example, PET-residues recycling by hydrothermal depolymerisation.

This is why this new edition based upon the 1999 book "Synthesis using Supercritical Fluids" but deeply revised and dealing with new areas, arrives at an optimal moment when scientists and engineers are facing the new challenges of sustainable development and demand for higher-performance products. In the fast-changing world of science, this update is a necessary tool offered to help the scientific community appreciate the opportunities presented by these fluids and to prepare chemists and engineers to incorporate these techniques in their process "tool-box".

March 2009 *Michel Perrut*

Contents

Foreword *V*
Preface *XV*
About the editors *XIX*
List of Contributors *XXI*

1 Introduction *1*
Philip Jessop and Walter Leitner
1.1 What is a Supercritical Fluid (SCF)? *1*
1.2 Practical Aspects of Reactions in Supercritical Fluids *4*
1.3 The Motivation for Use of SCFs in Modern Chemical Synthesis *6*
1.4 The History and Applications of SCFs *9*
1.4.1 The Discovery of SCFs and Their Use as Solvents *9*
1.4.2 Extraction and Chromatography in SCFs *14*
1.4.3 The History of Chemical Reactions in SCFs *16*
1.4.4 Industrial Use of SCFs as Reaction Media *20*
References *24*

2 High-pressure Methods and Equipment *31*
Nils Theyssen, Katherine Scovell, and Martyn Poliakoff
2.1 Introduction *31*
2.2 Infrastructure for High-pressure Experiments *32*
2.2.1 Location *32*
2.2.2 Gas Supply, Compression, and Purification *33*
2.3 High-pressure Reactors *34*
2.3.1 Materials of Construction for High-pressure Reactors *34*
2.3.1.1 Metal Components *34*
2.3.1.2 Sealing Materials *35*
2.3.2 Reactor Design *41*
2.3.2.1 General Considerations *41*
2.3.2.2 Pressure Vessels for Batch Processing *41*
2.3.2.3 Continuous Flow Reactors *44*
2.4 Auxiliary Equipment and Handling *45*

Handbook of Green Chemistry, Volume 4: Supercritical Solvents. Edited by Walter Leitner and Philip G. Jessop

ISBN: 978-3-527-32590-0

2.4.1 Tubes and Fittings *45*
2.4.2 Valves *47*
2.4.3 Pressure Transmitter and Manometer *47*
2.4.4 Reactor Heating and Temperature Control *48*
2.4.5 Stirrer Types *51*
2.4.6 Optical Windows *53*
2.4.7 Pressure Safety Valves and Bursting Discs *53*
2.4.8 Online Sampling *54*
2.4.9 Inline Spectroscopic Measurements *56*
2.4.10 Reactor Cleaning *57*
2.4.10.1 Cleaning with Organic Solvents *57*
2.4.10.2 Cleaning with Heated Solvents *58*
2.4.10.3 Cleaning with Acids *58*
2.4.10.4 Cleaning with a Combination of Organic Solvent and $scCO_2$ *58*
2.5 Dosage Under a High-pressure Regime *58*
2.5.1 Dosage of Gases *58*
2.5.1.1 Safety Warnings *58*
2.5.1.2 Dosing in Batch Processes *60*
2.5.1.3 Dosage for Continuous Flow Processes *61*
2.5.2 Dosage of Liquids *62*
2.5.3 Dosage of Solids *64*
2.6 Further Regulations and Control in Flow Systems *64*
2.6.1 Supply Pressure *64*
2.6.2 Reactor Pressure *65*
2.6.3 Flow and Total Volume Measurement of Depressurized Gas Streams *66*
2.7 Evaporation and Condensation *66*
2.7.1 Evaporation *66*
2.7.2 Condensation *66*
2.8 Complete Reactor Systems for Synthesis with SCFs *67*
2.8.1 Standard Batch Reactor System *67*
2.8.1.1 Essential equipment *67*
2.8.1.2 Brief Description of Work Steps *67*
2.8.2 Fully Automated System for Continuous Flow Operation *69*
2.9 Conclusion *73*
References *73*

3 Basic Physical Properties, Phase Behavior and Solubility *77*
Neil R. Foster, Frank P. Lucien, and Raffaella Mammucari
3.1 Introduction *77*
3.2 Basic Physical Properties of Supercritical Fluids *77*
3.3 Phase Behavior in High-Pressure Systems *86*
3.3.1 Types of Binary Phase Diagrams *86*
3.3.2 Asymmetric Binary Mixtures *88*
3.4 Factors Affecting Solubility in Supercritical Fluids *92*

3.4.1 The Supercritical Solvent *92*
3.4.2 Chemical Functionality of the Solute *94*
3.4.3 Temperature and Pressure Effects *96*
References *97*

4 Expanded Liquid Phases in Catalysis: Gas-expanded Liquids and Liquid–Supercritical Fluid Biphasic Systems *101*
Ulrich Hintermair, Walter Leitner, and Philip Jessop
4.1 A Practical Classification of Biphasic Systems Consisting of Liquids and Compressed Gases for Multiphase Catalysis *101*
4.2 Physical Properties of Expanded Liquid Phases *106*
4.2.1 Volumetric Expansion *106*
4.2.2 Density *109*
4.2.3 Viscosity *110*
4.2.4 Melting Point *111*
4.2.5 Interfacial Tension *112*
4.2.6 Diffusivity *114*
4.2.7 Polarity *115*
4.2.8 Gas Solubility *118*
4.3 Chemisorption of Gases in Liquids and their Use for Synthesis and Catalysis *120*
4.3.1 *In Situ* Generation of Acids and Temporary Protection Strategies *120*
4.3.2 Switchable Solvents and Catalyst Systems *124*
4.4 Using Gas-expanded Liquids for Catalysis *129*
4.4.1 Motivation and Potential Benefits *129*
4.4.2 Sequential Reaction–Separation Processes *130*
4.4.2.1 Tunable Precipitation and Crystallization *130*
4.4.2.2 Tunable Phase Separations *131*
4.4.2.3 Tunable Miscibility *134*
4.4.3 Hydrogenation Reactions *135*
4.4.4 Carbonylation Reactions *139*
4.4.5 Oxidation Reactions *143*
4.4.6 Miscellaneous *148*
4.5 Why Perform Liquid–SCF Biphasic Reactions? *150*
4.5.1 By Necessity (Unintentional Immiscibility) *151*
4.5.2 To Facilitate Post-Reaction Separation *152*
4.5.3 To Facilitate Product/Catalyst Separation in Continuous Flow Systems *154*
4.5.4 To Stabilize a Catalyst *155*
4.5.5 To Remove a Kinetic Product *156*
4.5.6 To Control the Concentration of Reagent or Product in the Reacting Phase *156*
4.5.7 To Permit Emulsion Polymerization *157*
4.5.8 To Create Templated Materials *158*
4.6 Biphasic Liquid–SCF Systems *159*

4.6.1 Solvent Selection *159*
4.6.2 Aqueous–SCF Biphasic Systems *159*
4.6.3 Ionic Liquid–SCF Biphasic Systems *163*
4.6.4 Polymer–SCF Biphasic Systems *167*
4.6.5 Liquid Product–SCF Biphasic Systems *171*
4.7 Biphasic Reactions in Emulsions *172*
4.7.1 Water-in-SCF Inverse Emulsions *172*
4.7.2 SCF-in-Water Emulsions *173*
4.7.3 Ionic Liquid-in-SCF Emulsions *173*
4.7.4 Applications of Emulsions *174*
References *175*

5 Synthetic Organic Chemistry in Supercritical Fluids *189*
Christopher M. Rayner, Paul M. Rose, and Douglas C. Barnes
5.1 Introduction *189*
5.2 Hydrogenation in Supercritical Fluids *190*
5.2.1 Asymmetric Hydrogenation and Related Reactions *198*
5.3 Hydroformylation and Related Reactions in Supercritical Fluids *202*
5.4 Oxidation Reactions in Supercritical Fluids *205*
5.5 Palladium-mediated Coupling Reactions in Supercritical Fluids *208*
5.6 Miscellaneous Catalytic Reactions in Supercritical Fluids *214*
5.6.1 Metal-catalyzed Processes *214*
5.6.2 Base-catalyzed Processes *218*
5.6.3 Acid-Catalyzed Processes *219*
5.7 Cycloaddition Reactions in Supercritical Fluids *221*
5.8 Photochemical Reactions in Supercritical Fluids *224*
5.9 Radical Reactions in Supercritical Fluids *228*
5.10 Biotransformations in Supercritical Fluids *229*
5.11 Conclusion *234*
References *235*

6 Heterogeneous Catalysis *243*
Roger Gläser
6.1 Introduction and Scope *243*
6.2 General Aspects of Heterogeneous Catalysis in SCFs and GXLs *244*
6.2.1 Utilization of SCFs in Heterogeneous Catalysis *245*
6.2.1.1 General Considerations *245*
6.2.1.2 Rate Enhancement *247*
6.2.1.3 Selectivity Tuning *248*
6.2.1.4 Lifetime/Stability Enhancement *250*
6.2.1.5 Reactor and Process Design *251*
6.2.2 Utilization of GXLs in Heterogeneous Catalysis *251*
6.3 Selected Examples of Heterogeneously Catalyzed Conversions in SCFs and GXLs *252*
6.3.1 Conversions in SCFs *252*

6.3.1.1 Hydrogenations *256*
6.3.1.2 Fischer–Tropsch Synthesis *260*
6.3.1.3 Hydroformylations *262*
6.3.1.4 Oxidations *263*
6.3.1.5 Alkylations *266*
6.3.1.6 Isomerizations *269*
6.3.1.7 Miscellaneous *270*
6.3.2 Conversions in GXLs *271*
6.4 Outlook *273*
References *274*

7 Enzymatic Catalysis *281*
Pedro Lozano, Teresa De Diego, and José L. Iborra
7.1 Enzymes in Non-aqueous Environments *281*
7.2 Supercritical Fluids for Enzyme Catalysis *283*
7.3 Enzymatic Reactions in Supercritical Fluids *285*
7.4 Reaction Parameters in Supercritical Biocatalysis *289*
7.5 Stabilized Enzymes for Supercritical Biocatalysis *292*
7.6 Enzymatic Catalysis in IL–scCO_2 Biphasic Systems *294*
7.7 Future Trends *298*
References *298*

8 Polymerization in Supercritical Carbon Dioxide *303*
Uwe Beginn
8.1 General Aspects *303*
8.1.1 Introduction and Scope *303*
8.1.2 Supercritical Fluids *304*
8.1.3 Solubility of Macromolecules in scCO_2 *306*
8.1.4 Stabilizer Design for Dispersion Polymerizations *310*
8.1.5 Limitations of Polymer Preparation in scCO_2 *314*
8.2 Polymerization in scCO_2 *315*
8.2.1 Radical Polymerization in scCO_2 *315*
8.2.1.1 Side-chain Fluoropolymers *317*
8.2.1.2 Fluoroolefin Polymers *319*
8.2.1.3 Poly(Methyl Methacrylate) *326*
8.2.1.4 Polystyrene *332*
8.2.1.5 Other Vinyl Monomers *335*
8.2.2 Metal-catalyzed Polymerizations *340*
8.2.2.1 Polyolefins *340*
8.2.2.2 Other Metal-catalyzed Polymerizations *342*
8.2.3 Ionic Chain Polymerizations *346*
8.2.3.1 Cationic Polymerizations *346*
8.2.3.2 Coordinative Anionic Polymerization *348*
8.3 Conclusion *352*
References *353*

9 **Synthesis of Nanomaterials** *369*
Zhimin Liu and Buxing Han
9.1 Introduction *369*
9.2 Metal and Semiconductor Nanocrystals *369*
9.2.1 Direct Synthesis of Nanocrystals in SCFs *369*
9.2.1.1 Synthesis in $scCO_2$ *370*
9.2.1.2 Synthesis in Supercritical Organic Solvents *371*
9.2.1.3 Synthesis in Supercritical Water (scH_2O) *373*
9.2.2 Synthesis of Nanomaterials in SCF-based Microemulsions *374*
9.2.2.1 Water-in-Supercritical Alkane Microemulsion *374*
9.2.2.2 Water-in-$scCO_2$ Microemulsions *375*
9.2.2.3 Recovery of Nanoparticles from Reverse Micelles Using $scCO_2$ *377*
9.3 Metal Oxide Nanoparticles *377*
9.3.1 Supercritical Hydrothermal Synthesis *377*
9.3.2 Direct Sol–Gel Synthesis in $scCO_2$ *380*
9.3.3 Synthesis Using Water-in-CO_2 Microemulsions *382*
9.4 Carbon Nanomaterials *383*
9.4.1 Carbon Nanotubes (CNTs) *383*
9.4.2 Carbon Nanocages *385*
9.5 Nanocomposites *385*
9.5.1 Synthesis of Polymer-based Composites *386*
9.5.2 Decoration of Nanoparticles on Carbon Nanotubes *388*
9.5.3 Deposition of Nanoparticles on Porous Supports *391*
9.5.4 Some Other Nanocomposites *393*
9.6 Conclusion *393*
References *394*

10 **Photochemical and Photo-induced Reactions in Supercritical Fluid Solvents** *399*
James M. Tanko
10.1 Introduction *399*
10.1.1 "Solvent" Properties of Supercritical Fluids *399*
10.1.2 Scope of This Chapter *400*
10.1.3 Experimental Considerations *400*
10.2 Photochemical Reactions in Supercritical Fluid Solvents *403*
10.2.1 Geometric Isomerization *403*
10.2.2 Photodimerization *403*
10.2.3 Carbonyl Photochemistry *405*
10.2.4 Photosensitization and Photo-induced Electron Transfer *409*
10.2.5 Photo-oxidation Reactions *410*
10.3 Photo-initiated Radical Chain Reactions in Supercritical Fluid Solvents *410*
10.3.1 Free Radical Brominations of Alkyl Aromatics in Supercritical Carbon Dioxide *410*

10.3.2 Free Radical Chlorination of Alkanes in Supercritical Fluid Solvents *411*
10.4 Conclusion *414*
References *415*

11 Electrochemical Reactions *419*
Patricia Ann Mabrouk
11.1 Introduction *419*
11.2 Electrochemical Methods *419*
11.3 Analytes *420*
11.4 Electrolytes *421*
11.5 Electrochemical Cell and Supercritical Fluid Delivery System *421*
11.6 Electrodes *422*
11.6.1 Working Electrode *422*
11.6.2 Reference Electrode *422*
11.7 Solvents *423*
11.7.1 Supercritical Carbon Dioxide *424*
11.7.1.1 Electrode Modification *425*
11.7.1.2 Hydrophobic Electrolytes *426*
11.7.1.3 Water-in-Carbon Dioxide Microemulsions *426*
11.7.2 Hydrofluorocarbon Supercritical Solvents *426*
11.8 Applications *429*
11.8.1 Electrochemical Synthesis in Supercritical Solvents *429*
11.8.2 Electrochemical Detection in Supercritical Solvents *429*
11.9 Conclusion and Outlook *431*
References *431*

12 Coupling Reactions and Separation in Tunable Fluids: Phase Transfer-Catalysis and Acid-catalyzed Reactions *435*
Pamela Pollet, Jason P. Hallett, Charles A. Eckert, and Charles L. Liotta
12.1 Introduction *435*
12.2 Phase Transfer Catalysis *435*
12.2.1 Background *435*
12.2.2 Phase Transfer Catalysis Quaternary Ammonium Salt-catalyzed Reactions *436*
12.2.3 PTC Separation and Recycling Using CO_2 *437*
12.3 Near-critical Water *438*
12.3.1 Definition *438*
12.3.2 Properties *439*
12.3.3 Friedel–Crafts Chemistry in NCW *442*
12.4 Alkylcarbonic Acids *448*
12.4.1 Probing Alkylcarbonic Acids – Alkylcarbonic Acids with Diazodiphenylmethane (DDM) *448*
12.4.2 Reactions Using Alkylcarbonic Acids *451*
12.4.2.1 Ketal Formation *451*

12.4.2.2 Formation of Diazonium Salts *452*
12.5 Conclusion *453*
References *454*

13 Chemistry in Near- and Supercritical Water *457*
Andrea Kruse and G. Herbert Vogel
13.1 Introduction *457*
13.2 Properties *457*
13.3 Synthesis Reactions *459*
13.3.1 Hydrations *460*
13.3.2 Hydrolysis *460*
13.3.2.1 Esters *460*
13.3.2.2 Ethers *461*
13.3.2.3 Amides *461*
13.3.3 Dehydrations *461*
13.3.4 Condensations *462*
13.3.5 Diels–Alder Reactions *462*
13.3.6 Rearrangements *463*
13.3.7 Partial Oxidations *464*
13.3.8 Reductions *464*
13.3.9 Organometallic Reactions *465*
13.4 Biomass Conversion *465*
13.4.1 Platform Chemicals *465*
13.4.1.1 Carbohydrates *465*
13.4.1.2 Lignin *467*
13.4.1.3 Proteins *467*
13.4.2 Oil, Gases, Coke *468*
13.5 Supercritical Water Oxidation (SCWO) *470*
13.6 Inorganic Compounds in NSCW *471*
13.6.1 Particle Formation *471*
13.6.2 Corrosion *471*
13.6.3 Unwanted Salt Precipitation and Salt Plugging *472*
13.6.4 Poisoning of Heterogeneous Catalysts *472*
13.7 Conclusion *472*
13.8 Future Trends *473*
References *473*

Index *477*

Preface

Reactions under supercritical conditions have been used for industrial production on various scales for most of the 20th century, but the current intense academic interest in the science and applications of supercritical fluids (SCFs) dates from the mid 1980's (Figure 1) and the application of SCFs in the chemical synthesis of organic molecules or materials became a "hot topic" starting in the early 1990's. Processes involving SCFs can be conducted in a fully homogeneous monophasic fluid or in biphasic systems. Biphasic conditions can involve a supercritical or subcritical gas as the upper phase and a gas-expanded liquid (GXL) below. The optimum situation is often a delicate balance of thermodynamic and kinetic boundaries for a given transformation. This book is intended to introduce the reader to the wide range of opportunities provided by the various synthetic methodologies developed so far for synthesis in SCFs and GXLs.

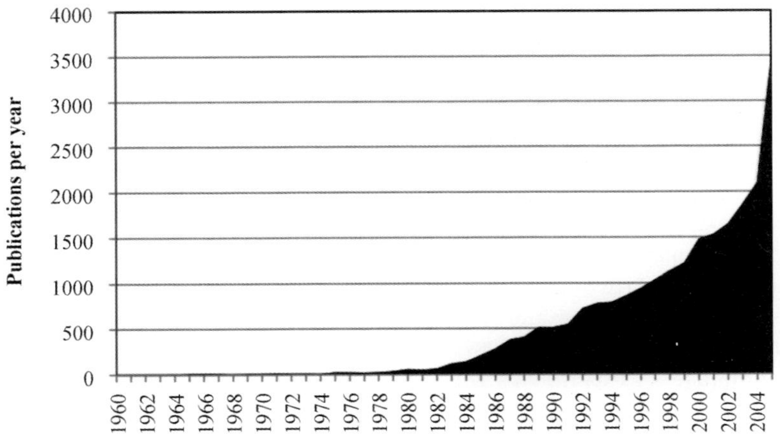

Figure 1 Publications on the topic of supercritical fluids per year (data mined from the Chemical Abstracts Service).

Handbook of Green Chemistry, Volume 4: Supercritical Solvents. Edited by Walter Leitner and Philip G. Jessop

ISBN: 978-3-527-32590-0

Applications of SCFs include their use as solvents for extractions, as eluents for chromatography, and as media for chemical reactions. All of these are worthy topics for extensive scientific and technical discussion, and in fact have been topics of books in the past. We decided that a satisfactory coverage of all three topics would not be possible in a single monograph of a reasonable size, and therefore we chose to cover only one. While extractions such as decaffeination of coffee and chromatography such as the supercritical CO_2-based preparative chromatography used in the pharmaceutical industry are great examples of the environmental and economic benefits of SCFs, we focus here on chemical synthesis where the fluid is not only used as a mass separation agent, but also directly affects the molecular transformation.

Supercritical fluids and gas-expanded liquids may be alternatives to liquid solvents, but they are neither simple nor simply replacements of solvents. The experimental chemist could not modify a written synthetic method by simply crossing out the word "benzene" and replacing it with the words "supercritical carbon dioxide". Many other modifications to the procedure would be necessary, not only because of the need for pressurized equipment but also because of the inferior solvent strength of many SCFs. On the other hand, additional degrees of freedom in the reaction parameters emerge from the high compressibility of SCFs, allowing density to be introduced as an important variable. At the same time, mass transfer can be greatly enhanced in the presence of SCFs. Selective separation and compartmentalization of elementary processes in multiphase systems offer another parameter that can be exploited especially in catalytic processes. These are only some of the reasons why the result of a chemical synthesis can sometimes be dramatically changed, often for the better, by this solvent switch. If such beneficial effects can be combined with the benign nature of many SCFs such as CO_2 or H_2O, they can contribute to the development of more sustainable chemical processes, explaining why SCFs and GXLs are often referred to as "Green Solvents".

It is only fair to say that we are still far away from a detailed understanding of all the effects of using SCFs and GXLs. More basic research will be needed before we learn how to exploit the benefits in the most efficient way. In the meantime, it is our hope that the chemist or engineer considering using one of these fluids as a medium for a reaction will turn to this volume to find out both what has been done, how to do it, and, more importantly, what new and innovative directions are yet to be taken.

At this point, we must offer a safety warning and disclaimer. Supercritical fluids are used at high pressures and in some cases at elevated temperatures. The chemist contemplating their use must become acquainted with the safety precautions appropriate for experiments with high pressures and temperatures. Some SCFs also have reactive hazards. The safety considerations mentioned in Chapters 1 and 2 are meant neither to be comprehensive nor to substitute for a proper investigation by every researcher of the risks and appropriate precautions for a contemplated experiment.

We have selected the chapter topics to guide the reader through the process of planning and carrying out chemical syntheses in SCFs and GXLs. The subjects include a brief overview of the historical development and current use, as well as a description of equipment, methods, and phase behaviour considerations. The properties of biphasic conditions and gas-expanded liquids are spelled out in chapter 4, and all these themes are elaborated upon in the largest part of the book which is devoted to various types of chemical reactions involving SCFs and GXLs as solvents and/or reactants. The emphasis is on synthetic reactions, rather than reactions tested for the purpose of investigating near-critical phenomena.

This book represents a partial update of our 1999 book on "Chemical Synthesis Using Supercritical Fluids", but most of the chapters are entirely new and the selection of topics is not the same. We therefore encourage readers, if they want more information, to look up the 1999 book.

The contributors to the present volume, all leading experts in the field, have given us a wide view of the types and methods of chemistry being performed in supercritical fluids and expanded liquids. Many of the techniques that the reader will find described in these pages have been laboriously developed by these contributors and their colleagues. We gratefully thank all of the contributors for agreeing to take time out from their research schedules to write chapters for this volume.

We also thank the following people and institutions for providing us with information or photographic material on the historical aspects and the industrial use of SCFs: Dr. J. Abeln (Forschungszentrum Karlsruhe), Dr. U. Budde (Schering AG), Dr. H.-E. Gasche (Bayer AG), Dr. P. Møller (Poul Møller Consultancy), Dr. T. Muto (Idemitsu Petrochemical), Prof. G. Ourisson and the Académie des Sciences, Dr. A. Rehren (Degussa AG), M.-C. Thooris (Ecole Polytechnique Palasieau) and representatives of Eco Waste Technology and General Atomics.

Special thanks are due to Dr. Markus Höslcher at ITMC, RWTH Aachen, and Drs. Elke Maase and Lesley Belfit at Wiley-VCH for their competent help and collaboration in producing this book. Furthermore, we wish to express our sincere thanks to all the members of our research groups, for their talents and their enthusiasm, which make our research efforts devoted to SCFs and GXLs so much fun.

Finally, and most importantly, we dedicate our own contribution to this book to our wives and families, for all their love and understanding throughout the years and especially during the preparation of this volume.

February 2009 *Philip Jessop and Walter Leitner*

About the editors

Series Editor

Paul T. Anastas joined Yale University as Professor and serves as the Director of the Center for Green Chemistry and Green Engineering there. From 2004–2006, Paul was the Director of the Green Chemistry Institute in Washington, D.C. Until June 2004 he served as Assistant Director for Environment at the White House Office of Science and Technology Policy where his responsibilities included a wide range of environmental science issues including furthering international public-private cooperation in areas of Science for Sustainability such as Green Chemistry. In 1991, he established the industry-government-university partnership Green Chemistry Program, which was expanded to include basic research, and the Presidential Green Chemistry Challenge Awards. He has published and edited several books in the field of Green Chemistry and developed the 12 Principles of Green Chemistry.

Volume Editors

Philip Jessop is the Canada Research Chair of Green Chemistry at Queen's University in Kingston, Ontario, Canada. After his Ph.D. (Inorganic Chemistry, UBC, 1991) and a postdoctoral appointment at the University of Toronto, he took a contract research position in the Research Development Corp. of Japan under the supervision of Ryoji Noyori, investigating reactions in supercritical CO_2. As a professor at the University of California-Davis (1996–2003) and then at Queen's University, he has studied green solvents, the conversion of waste CO_2 to useful products, and aspects of H_2 chemistry. He has presented popular chemistry shows to thousands of members of the public.

Handbook of Green Chemistry, Volume 4: Supercritical Solvents. Edited by Walter Leitner and Philip G. Jessop

ISBN: 978-3-527-32590-0

Distinctions include the Canadian Catalysis Lectureship Award (2004), a Canada Research Chair (2003 to present), and the NSERC Polanyi Award (2008). He has chaired the 2007 *CHEMRAWN and ICCDU Conference on Greenhouse Gases*, will chair the 2010 *3rd International IUPAC Conference on Green Chemistry*, and serves as Technical Director of GreenCentre Canada.

Walter Leitner was born in 1963. He obtained his Ph.D. with Prof. Henri Brunner at Regensburg University in 1989 and was a Postdoctoral Fellow with Prof. John M. Brown at the University of Oxford. After research within the Max-Planck-Society under the mentorship of Profs. Eckhard Dinjus (Jena) and Manfred T. Reetz (Mülheim), he was appointed Chair of Technical Chemistry and Petrochemistry at RWTH Aachen University in 2002 as successor to Prof. Willi Keim. Walter Leitner is External Scientific Member of the Max-Planck-Institut für Kohlenforschung and Scientific Director of CAT, the joint Catalysis Research Center of RWTH Aachen and the Bayer Company.

His research interests are the molecular and reaction engineering principles of catalysis as a fundamental science and key technology for Green Chemistry. In particular, this includes the development and synthetic application of organometallic catalysts and the use of alternative reaction media, especially supercritical carbon dioxide, in multiphase catalysis. Walter Leitner has published more than 170 publications in this field and co-edited among others the first edition of "Synthesis using Supercritical Fluids" and the handbook on "Multiphase Homogeneous Catalysis". Since 2004, he serves as the Scientific Editor of the Journal "Green Chemistry" published by the Royal Society of Chemistry. The research of his team has been recognized with several awards including the Gerhard-Hess-Award of the German Science Foundation (1997), the Otto-Roelen-Medal of Dechema (2001), and the Wöhler-Award of the German Chemical Society (2009).

List of Contributors

Douglas C. Barnes
University of Leeds
School of Chemistry
Leeds LS2 9JT
UK

Uwe Beginn
University of Osnabrück
Institute for Chemistry
Organic Materials Chemistry
Barbarastrasse 7
49076 Osnabrück
Germany

Teresa De Diego
Universidad de Murcia
Facultad de Química
Departamento de Bioquímica y Biología
Molecular "B" e Inmunología
P.O. Box 4021
30 100 Murcia
Spain

Charles A. Eckert
Georgia Institute of Technology
School of Chemical and Biomolecular
Engineering
School of Chemistry and Biochemistry
Specialty Separations Center
Atlanta, GA 30332
USA

Neil R. Foster
The University of New South Wales
School of Chemical Sciences and
Engineering
Sydney 2052
Australia

Roger Gläser
University of Leipzig
Institute of Chemical Technology
Linnéstrasse 3
04103 Leipzig
Germany

Jason P. Hallett
Georgia Institute of Technology
School of Chemical and Biomolecular
Engineering
School of Chemistry and Biochemistry
Specialty Separations Center
Atlanta, GA 30332
USA

Buxing Han
Chinese Academy of Sciences
Institute of Chemistry
Beijing National Laboratory for
Molecular Sciences
Beijing 100080
China

Handbook of Green Chemistry, Volume 4: Supercritical Solvents. Edited by Walter Leitner and Philip G. Jessop

ISBN: 978-3-527-32590-0

Ulrich Hintermair
RWTH Aachen
Institut für Technische Chemie
Worringerweg 1
52074 Aachen
Germany

José L. Iborra
Universidad de Murcia
Facultad de Química
Departamento de Bioquímica y Biología
Molecular "B" e Inmunología
P.O. Box 4021
30 100 Murcia
Spain

Philip G. Jessop
Queen's University
Department of Chemistry
90 Bader Lane
Kingston
ON K7L 3N6
Canada

Andrea Kruse
Technische Universität Darmstadt
Ernst-Berl-Institut für Technische und
Makromoleculare Chemie
Technische Chemie I
Petersenstrasse 20
64287 Darmstadt
Germany

Walter Leitner
RWTH Aachen
Institut für Technische Chemie
Worringerweg 1
52074 Aachen
Germany

Charles L. Liotta
Georgia Institute of Technology
School of Chemical and Biomolecular
Engineering
School of Chemistry and Biochemistry
Specialty Separations Center
Atlanta, GA 30332
USA

Zhemin Liu
Chinese Academy of Sciences
Institute of Chemistry
Beijing National Laboratory for
Molecular Sciences
Beijing 100080
China

Pedro Lozano
Universidad de Murcia
Facultad de Química
Departamento de Bioquímica y Biología
Molecular "B" e Inmunología
P.O. Box 4021
30 100 Murcia
Spain

Frank P. Lucien
The University of New South Wales
School of Chemical Sciences and
Engineering
Sydney 2052
Australia

Patricia Ann Mabrouk
Northeastern University
Department of Chemistry and
Chemical Biology
360 Huntington Avenue
Boston, MA 02115
USA

Raffaella Mammucari
The University of New South Wales
School of Chemical Sciences and
Engineering
Sydney 2052
Australia

Martyn Poliakoff
University of Nottingham
Department of Chemistry
Nottingham NG7 2RD
UK

Pamela Pollet
Georgia Institute of Technology
School of Chemical and Biomolecular
Engineering
School of Chemistry and Biochemistry
Specialty Separations Center
Atlanta, GA 30332
USA

Christopher M. Rayner
University of Leeds
School of Chemistry
Leeds LS2 9JT
UK

Paul M. Rose
University of Leeds
School of Chemistry
Leeds LS2 9JT
UK

Katherine Scovell
University of Nottingham
Department of Chemistry
Nottingham NG7 2RD
UK

James M. Tanko
Virginia Polytechnic Institute and
State University
Department of Chemistry
Blacksburg, VA 24061
USA

Nils Theyssen
Max-Planck-Institut für
Kohlenforschung
Kaiser-Wilhelm-Platz 1
45470 Mülheim
Germany

G. Herbert Vogel
Technische Universität Darmstadt
Ernst-Berl-Institut für Technische und
Makromolekulare Chemie
Technische Chemie I
Petersenstrasse 20
64287 Darmstadt
Germany

1
Introduction

Philip Jessop and Walter Leitner

1.1
What is a Supercritical Fluid (SCF)?

A supercritical fluid is a compound, mixture, or element above its critical pressure (p_c) and critical temperature (T_c) but below the pressure required to condense it into a solid (Figure 1.1). This definition is modified from that of IUPAC [1], which unfortunately omits the clause concerning condensation into a solid. That the melting curve extends over the supercritical region [2–4] is often forgotten even though the pressure is not always impractically high. For example, the minimum pressure required to solidify supercritical CO_2 is only 570 Mpa [5].

The conditions under which SCFs are investigated are often described in terms of "reduced temperature" (T_r) and "reduced pressure" (p_r), defined as the actual values of T and p divided by T_c and p_c, respectively (Equations 1.1 and 1.2). The "law of corresponding states" as introduced by van der Waals [6] implies that compounds behave similarly under the same values of the reduced variables. This allows valuable comparisons of different compounds under various conditions, but deviations can be substantial in close proximity to the critical point.

$$T_r = T/T_c \tag{1.1}$$

$$p_r = p/p_c \tag{1.2}$$

The properties of SCFs are frequently described as being intermediate between those of a gas and a liquid. This Janus-faced nature of SCFs arises from the fact that the gaseous and liquid phases merge together and become indistinguishable at the critical point. Figure 1.2 shows how the meniscus between the phases disappears upon reaching the critical point for CO_2. Not all properties of SCFs are intermediate between those of gases and liquids; compressibility and heat capacity, for example, are significantly higher near the critical point than they are in liquids or gases (or even in the supercritical state further from the critical point). Although the properties of a compound may change drastically with pressure near the critical point, most of them show no discontinuity. The changes start gradually, rather than with a sudden onset, when the conditions approach the critical point.

Handbook of Green Chemistry, Volume 4: Supercritical Solvents. Edited by Walter Leitner and Philip G. Jessop

ISBN: 978-3-527-32590-0

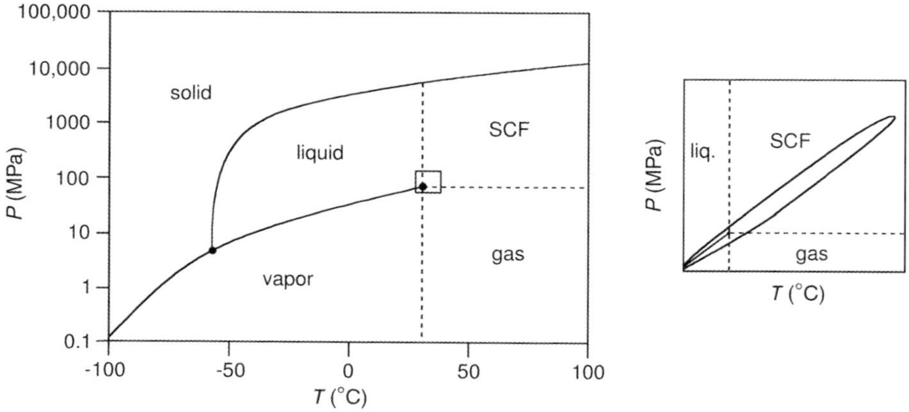

Figure 1.1 The phase diagram of CO_2 [192, 193]. The critical and triple points are shown as filled circles. The inset (with a linear pressure scale) shows an expanded view of the area around the critical point; the tear-shaped contour indicates the compressible region.

It is common to refer to the somewhat ill-defined region where such changes are noticeable as the "near-critical" region. Technically, the "near-critical region" extends all around the critical point, but the expression is commonly used to refer to the non-supercritical section only. The very similar expression "compressible region" refers to the area around the critical point in which the compressibility is significantly greater than would be predicted from the ideal gas law. In fact, the compressibility at the critical point itself approaches infinity, and hence the speed of sound in the fluid reaches a minimum; a method for the determination of critical data of mixtures based on this phenomenon has been devised [7]. Although a significant portion of the compressible region lies inside the SCF section of the phase diagram, there is also overlap with the liquid and vapor regions as well (Figure 1.1, inset). Thus, even liquids have significant compressibility near the critical point, although they are virtually incompressible at

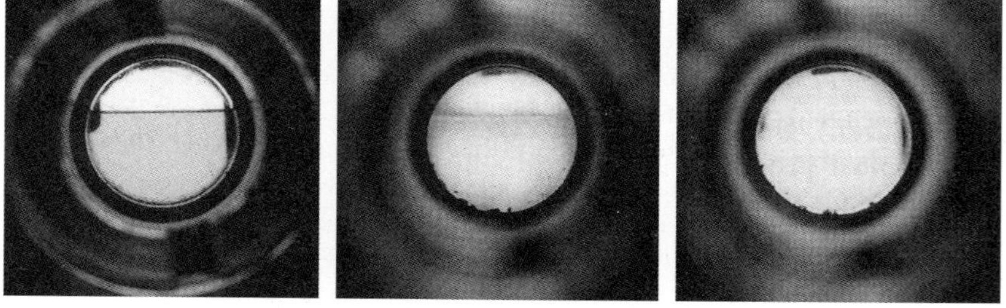

Figure 1.2 The meniscus separating liquid and gaseous CO_2 disappears when the critical point is reached by heating liquid CO_2 in a closed vessel. A small amount of a highly CO_2-soluble and brightly colored metal complex [194] was added for better contrast.

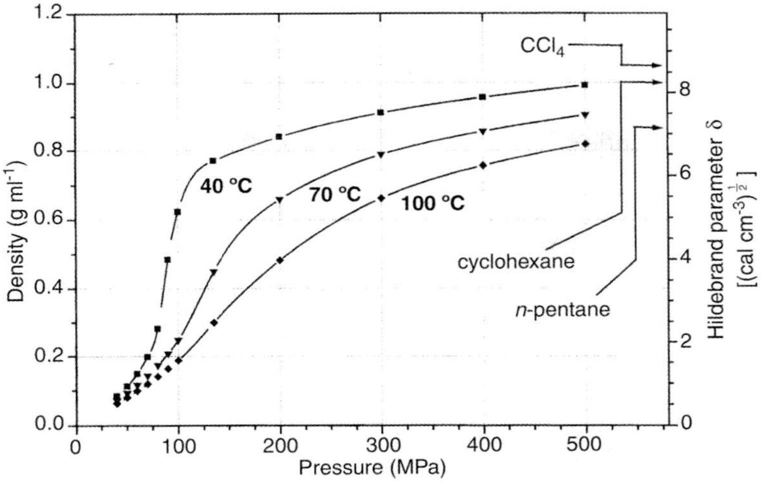

Figure 1.3 The density and the solvent power (as expressed by the Hildebrand parameter) of $scCO_2$ as a function of temperature and pressure [8, 9].

$T_r \ll 1$. Liquid phases at temperatures below, but not too far below, T_c are called "subcritical liquids", whereas "subcritical gases" are those at pressures below p_c.

When working with an SCF, it is valuable to refer to a plot of the dependence of density (d) on pressure and temperature, as presented for supercritical CO_2 ($scCO_2$) in Figure 1.3. Note that the density changes sharply but continuously with pressure in the compressible region, illustrating the properties outlined above. At higher pressures, the density changes occur more gradually. The critical density d_c (i.e. d at T_c and p_c) is the mean value of the densities of the gas phase and the liquid phase and amounts to 0.466 g ml^{-1} for CO_2. The reduced density is defined in analogy with the other reduced variables (Equation 1.3). The density data shown in Figure 1.3 correspond to the bulk density of the medium, but density fluctuations lead to microscopic areas of decreased and increased local densities in SCFs ("local density augmentation"). Because of the very large compressibility, these density fluctuations are most pronounced very near to the critical point. If the fluctuations are of the same order of magnitude as the wavelength of visible light, scattering of the light leads to critical opalescence, which may be apparent as a clouding or coloration of the SCF and can also be used to determine the critical point.

$$d_r = d/d_c \tag{1.3}$$

Many solvent properties are directly related to bulk density and will therefore have a pressure dependence similar to that shown in Figure 1.3. The best known example is the continuous variation in "solvent power" over a fairly wide range, which provides the basis for the technical use of SCFs in highly selective natural product extractions. The solvent power is a rather ill-defined property, but there have been experimental approaches to devise scales for liquid solvents. One of the most successful attempts was put forward by Hildebrand and Scott [8]; the so-called Hildebrand parameter for solvent power was found to be directly proportional to the density of SCFs [9], as shown in Figure 1.3 for CO_2. Some typical organic solvents are marked on the Hildebrand scale for comparison to give some indication of the tunability of the solvent power of CO_2. It should be apparent from the diagram that in order for a SCF to have significant solvating ability, it must usually have a d_r of >1. Note, however, that the concentration of a solute in a compressed gas or SCF does not depend on solvation only, but also involves volatility as an important parameter. Changing the solvating ability of an SCF will have different effects on the solubility of individual solutes.

The possibility of using SCFs as "tunable solvents" not only for supercritical fluid extraction (SFE) but also for chemical reactions is one of the many interesting features associated with their application in modern synthesis. Before we discuss the many potential benefits in detail (Section 1.3), it seems appropriate to give a brief introduction to some practical aspects of the use of SCFs on a laboratory scale.

1.2
Practical Aspects of Reactions in Supercritical Fluids

Considerations when selecting a SCF to serve as a reaction solvent include critical temperature, solubilizing power, inertness, safety, environmental impact, and cost. An attractive scenario is to use one of the reagents as the supercritical solvent; many reactions in SCFs in industry use a reagent as the solvent. Inorganic and organic compounds which are frequently used as SCFs are listed, together with leading references for their volumetric behavior, in tables in Chapter 2. Because most organic syntheses are performed between room temperature and 120 °C, SCFs with critical temperatures around room temperature are most commonly used; these are CO_2, ethane, ethylene, fluoroform, nitrous oxide, and the partially fluorinated methanes and ethanes. For many reasons, including inertness, safety, impact, and cost, $scCO_2$ is the most popular of these.

The purity of the SCF is an important consideration in the planning of a synthesis. Low concentrations of impurities can have noticeable effects on the volumetric and phase behavior of SCFs. For example, helium can be present in commercial CO_2 because it is sometimes added as a "head-gas" to ensure nearly complete delivery of the cylinder contents and this has been found to affect the use of $scCO_2$ as a solvent for analytical and preparative purposes [10–12]. The He head-gas is unnecessary if a cooled pump is used for CO_2 delivery. Purity can also have an effect on the cost of the SCF. For some materials, especially the C_2 or higher hydrocarbons, the price is highly dependent on the purity and very high purities are prohibitively expensive.

Specialized equipment is required for experiments with supercritical fluids, as described in more detail in Chapter 2, mainly because of the requirements to work at elevated pressures and/or temperatures. The physical and chemical properties of the SCFs can sometimes present hazards to the experimentalist [13]. All researchers in the field should search the literature for information concerning the hazards of the materials with which they are working. The following information is presented as a brief overview only, and should not be considered a comprehensive review on the subject.

All SCFs are compressed gases, and therefore contain a great deal of potential energy, which can be released upon catastrophic failure of the equipment. As safety regulations vary from country to country and also depend on the size of the reactor and the maximum applied pressure, we can only give some general advice here. All new equipment should be pre-tested by filling with an incompressible liquid (water, oil, hexane) under pressures of approximately 1.5–2 times the maximum operating pressure. One of the simplest and most effective safety rules when working with SCFs is to avoid direct exposure of the operator to the pressurized vessel, for example by using strong polycarbonate or Lexan shields or similar optically transparent safety equipment. The use of angled mirrors or video equipment also allows visual inspection without direct exposure.

Other than the large potential energy due to compression, most of the hazards of SCFs are related to the chemical reactivity of the gas itself. Several SCFs such as scH_2O, scHCl and the other acids corrode standard stainless-steel reaction equipment, which could result in catastrophic failure. Also, organic acids such as acetic or formic acid dissolved in scCO_2 can have significant corrosive properties. Explosive deflagration or decomposition is common with acetylene even at subcritical pressures [14]. Perfluoroethylene (scC_2F_4) will explode at pressures above 0.27 MPa unless inhibitor is added [15]. Runaway polymerization can be a concern when polymerizable SCFs such as scC_2H_4 are used [16]. The polymerization can be initiated by free radicals, O_2, or metal catalysts, including even stainless-steel components. The initiation of radical oxidations involving O_2 as terminal oxidant can, however, also be used for selective oxidation processes [17]. SCFs, like most other compounds, can be incompatible with some other materials; examples of incompatible combinations which have exploded or violently reacted in the past include NH_3–ethylene oxide, HBr–Fe_2O_3, HCl–Al, C_2H_2–Cu and many others [15].

Flammable SCFs include all of the hydrocarbons plus others such as scCH_3OH, scCH_3OCH_3, scNH_3, and ethylenediamine (en). Silane (scSiH_4) is particularly dangerous because it could autoignite upon leaking from the vessel, independent of any spark source [18]. The only commonly used oxidizing SCF is scN_2O. Mixing significant quantities of combustible material with scN_2O has led to at least two explosions in the past. One of these occurred when ethanol (9 vol.%) was used as a cosolvent in scN_2O [19, 20]. and the other occurred when 1 g of roasted coffee was exposed to scN_2O in a 2.5 ml extraction vessel [21].

Most common SCFs have a comparatively low level of acute toxicity [22], but the high local concentrations which may result from use under high pressures require appropriate safety considerations such as sufficient ventilation. Irritant poisons include en, NH_3 and the acidic SCFs (HCl, HBr, and HI) [15]. SCFs such as scCHF_3 and scN_2O (laughing gas) are known to act as narcotics when at high concentrations [23]. Carcinogenic SCFs such as benzene should be replaced by other SCFs with similar properties wherever possible.

Toxic compounds dissolved in SCFs can be spread throughout the laboratory if the pressurized solutions are vented outside a fume hood. The same action can result in contamination of stock chemicals in the laboratory [24]. Those SCFs with particularly high T_cs (e.g. scH_2O) could cause thermal burns to operators if a leak were to occur. Chemical burns could result from leaks if irritants or acids are used as SCFs or are dissolved therein. These chemical risks, and the procedures for avoiding them, should be familiar to the practicing chemist utilizing SCFs on a laboratory or technical scale.

1.3
The Motivation for Use of SCFs in Modern Chemical Synthesis

Why use SCFs as solvents for chemical reactions? There are numerous advantages associated with the use of SCFs in chemical synthesis, all of which are based on the unique combination of properties of either the materials themselves or the supercritical state. Different types of reactions may benefit particularly from a specific property, and these sometimes spectacular effects will be discussed in detail in the individual chapters of this book. Here, we try to summarize briefly the various potential improvements that can be expected if SCFs are employed as solvents for synthetically useful chemical reactions. The advantages fall into four general categories; environmental benefits, health and safety benefits, process benefits, and chemical benefits (Table 1.1).

Table 1.1 Advantages of using SCFs rather than liquids as reaction media.

Category	Advantage	Which SCFs
Environment	Do not contribute to smog	Many
	Do not damage ozone layer	Many
	Non-VOC	CO_2, H_2O
	No acute eco-toxicity	CO_2, H_2O
	No liquid wastes	CO_2 and other volatile SCFs
	Available as a recycled waste	CO_2
Health/safety	Non-carcinogenic	Most (but not C_6H_6)
	Non-toxic	Most (but not HCl, HBr, HI, NH_3)
	Non-flammable	CO_2, N_2O, H_2O, Xe, Kr, CHF_3
Process	No solvent residues	CO_2 and other volatile SCFs
	Facile separation of products	CO_2 and other volatile SCFs
	Process integration and intensification	All
	High diffusion rates	All
	Low viscosity	All
	Adjustable solvent power	All
	Adjustable density	All
	Inexpensive	CO_2, H_2O, NH_3, Ar, hydrocarbons
Chemical	High miscibility with gases	All
	Variable dielectric constant	SCFs having dipole moments
	High compressibility	All
	Local density augmentation	All
	High diffusion rates	All
	Altered cage strength	All

Environmental benefits are most often cited for processes with $scCO_2$ or scH_2O as the solvents. Although CO_2 and many of the other SCFs are greenhouse gases, the use of CO_2 as an industrial solvent would still be of benefit to the environment because it would utilize only already existing CO_2 and allow the replacement of environmentally far more damaging liquid organic solvents. Processes involving CO_2 as a solvent provide an opportunity for the recycling of waste CO_2. Also, reactions which result in the fixation of the $scCO_2$ would consume a small amount of waste CO_2 and decrease our dependence on fossil fuels as sources of carbon-containing molecules. At present, most recovered CO_2 is generated as a byproduct of ammonia and hydrogen production, but efforts to sequester CO_2 from flue gases of power plants are rapidly increasing [25].

In terms of greenhouse gas emission, the energetic balance of two alternative processes will most likely be more significant than the material balance of CO_2. This balance is not always straightforward; for example, energy would be saved during removal of the solvent by releasing CO_2 instead of distillation of an organic solvent. On the other hand, the compression of CO_2 is energy costly. Nevertheless, even for products with fairly low solubility in $scCO_2$, integrated reaction–extraction processes have been shown to compare well with conventional techniques in certain cases [26]. It is often forgotten in these comparisons that a process operating with compressed CO_2 would not alternate between the pressure required in the supercritical mixture and a full expansion to ambient pressure; reducing the pressure to any value that allows separation of the product is sufficient and the required pressures may well be close to or even above the p_c of pure CO_2. Recompression is usually achieved by cooling the gas to a liquid and then repumping the liquid, a process which is considerably less energy intensive than recompressing the gas directly. Furthermore, temperature rather than pressure may be the variable of choice for many separation processes. In the decaffeination process, the caffeine is isolated by extraction with water from the CO_2 stream in a countercurrent flow column rather than by precipitation through release of pressure. Therefore, comparisons of energy consumption require sufficiently detailed information on potential technical solutions for the process under scrutiny in order to avoid decisions based on prejudice.

Health and safety benefits include the fact that the most important SCFs, $scCO_2$ and scH_2O, are non-carcinogenic, non-toxic, non-mutagenic, non-flammable and thermodynamically stable. Very few traditional liquid solvents fit this description. While the high critical temperature of supercritical water prevents it from being used as a solvent for some organic syntheses, it has great potential for applications that require elevated temperatures (see Chapter 13).

Process benefits derived from the physical properties of SCFs, such as high diffusivity, low viscosity, and intermediate density, make SCFs particularly suitable for continuous-flow processes. The high flow rates and fast reactions often encountered with SCFs allow the design of high-throughput reactions in relatively small-scale reactors. The engineering solution for up-scaling of reactions in SCFs which have been tested in batch or semi-batch reactors on a laboratory scale will in most cases not involve an increase in the size of the reaction vessel, but rather the design of a continuous-flow system with high space–time yields. In other words, the actual technical solution of a successful synthesis in SCFs will most likely bear more resemblance to a gas-phase process, whereas the exploratory test phase can be closely similar to the screening of liquid-phase reactions.

One of the most obvious advantages of SCFs for chemical synthesis is their adjustable solvating power. The considerable body of knowledge concerning extraction and solubility in SCFs can be brought to bear to solve engineering problems in the separation and purification steps of industrial chemical processes in SCFs. For example, the extractive properties of SCFs may be exploited to separate products from by-products or to recover homogeneous catalysts. The design of integrated SCF-based processes replacing wasteful and time-consuming work-up and separation schemes seems highly attractive, but has been met piecemeal at best. An elegant application of this concept is the *in situ* regeneration of heterogeneous catalysts by SCF extraction of wax or tar byproducts that would block pores and active sites in gas-phase reactions. Some control of the molecular weights of growing polymer chains is also possible by control of the precipitation. It might also be possible to isolate intermediate reaction products by selective precipitation or extraction and prevent them from further reacting or decomposing.

The volatility of many SCFs allows complete removal from the product without the need for costly or energy-consuming drying processes. Thus, solvent residues in products can be avoided; this is particularly valuable in the preparation of cosmetics, pharmaceuticals, food additives, and materials for use in electronics.

In addition to making processes cleaner and more efficient, the use of SCFs as solvents can also have beneficial effects directly on chemical reactions. Many reactions that can be performed in SCFs occur also in liquid solvents, but there is a considerable number of examples where the use of an SCF as the solvent increases the rate of the reaction over that which would be observed in a liquid medium. Several of the unique properties of SCFs can cause such a change in rate, and it is not always easy to identify the most important contribution. For example, extremely rapid reactions which are diffusion controlled or altered by solvent cage effects can be more rapid in SCFs because of the higher diffusivity and weaker cage effects. Local solute–solute "clustering" can lead to high local concentrations of reagents, increasing the rates of reactions which are performed near the critical point of the mixture. Another often cited advantageous property of SCFs is their miscibility with other gases, which can lead to high rates of reactions if the kinetics are first order or higher in the concentration of the dissolved gas. The presence of a single homogeneous phase is especially important for catalytic reactions that would be operating under mass transport limitations under two-phase liquid–gas conditions. Similarly, reactions which involve mass transfer between a liquid phase and a solid phase such as a polymer or zeolite are often mass transport limited. The rates of such reactions can be greatly enhanced under supercritical conditions. The same arguments predict that diffusionally limited reactions involving suspended enzymes in liquid organic solvents would be faster in an SCF. Optimization of the rate of a reaction in an SCF can be achieved by varying the choice and concentration of cosolvent, a technique known as "cosolvent tuning". Finally, the SCF may take part directly in the chemical reaction, for example as one of the reagents (hydrolysis, CO_2 fixation) or by modifying substrates or catalytically active sites.

Other chemical benefits can be related to selectivity changes. Any of the above factors known to affect rates could affect selectivity by altering the rates of competing reactions. In addition, the pressure dependence of typical solvent parameters such as the dielectric constant of some SCFs may cause a considerable tuning effect on the selectivity of enzymatic and homogeneous catalysis. Reactions with large and different volumes of activation will show a distinct rate dependence upon pressure in the compressible region. All these factors may affect the chemo-, regio-, and stereoselectivity of chemical reactions, and open additional degrees of freedom for the optimization of synthetic processes.

It is possible to become too caught up in the excitement over environmentally benign processes, rate increases, and other potential benefits of reactions in SCFs. If the reaction can be performed anywhere close to adequately in a relatively benign liquid solvent, then there is little motivation for a switch to SCFs, because their use is still considered to be expensive. On the other hand, in cases where chemical, process, or environmental benefits can be obtained, industrial use of supercritical conditions is economically feasible and often already a reality (Section 1.4.4). Although the costs of the implementation of high-pressure equipment and of operating an SCF process are arguably higher than using an existing reaction vessel, a more detailed analysis must include the costs of all steps of the process, including work-up and waste treatment, and may well lead to different results. As mentioned above, it is also important to consider possible engineering solutions which are completely different from the initial screening procedure and may lead to much lower equipment and operating costs than anticipated. Furthermore, one should bear in mind that small-scale operating units for SCF reactions can be highly flexible and may allow the equipment to be switched from one process to another without long downtimes.

Nevertheless, it seems fair to conclude that new applications of chemistry in SCFs will be most likely in the synthesis of high-value fine chemicals which are given directly to the customer rather than for commodities which are used as intermediates for further downstream processing. One key aspect which is often neglected in this context is marketing; it might still be cheaper to produce decaffeinated coffee using CH_2Cl_2 as a solvent, but it is very unlikely that the customer would knowingly accept this product. We suspect that the market would also react favorably to food preservatives, pharmaceuticals or cosmetics which were produced using "natural carbonic acid" ($scCO_2$) instead of organic solvents.

1.4 The History and Applications of SCFs

1.4.1 The Discovery of SCFs and Their Use as Solvents

The interest in reactions in SCFs has increased over the last 10–15 years because the special properties of SCFs make them particularly attractive solvents for modern synthetic chemistry as outlined above. We should be aware of the fact, however, that

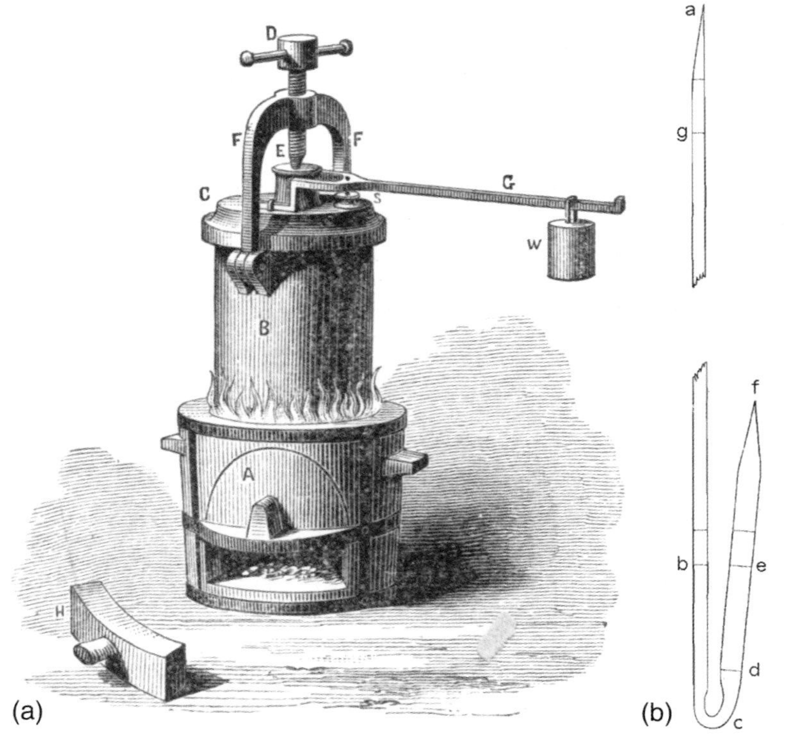

Figure 1.4 (a) The "digester" made by Denys Papin in 1680 [195]. The first vessel of de LaTour was a small vessel of this type. (b) The glass tube design used by de LaTour to observe the transition from a liquid to a supercritical fluid. Mercury was introduced into section *bcde* and ether was put into *ef*. The ends *a* and *f* were then sealed and the tube heated over a fire. At the moment when the ether was transformed into a vapor, the level *b* of the mercury had climbed to point *g*. The pressure was calculated from the distances to be 3.7–3.8 MPa [27].

the idea of using SCFs as reaction media has been emerging ever since the discovery of this "peculiar state of matter" early in the nineteenth century by Baron Charles Cagniard de LaTour, an experimental physicist in France [27].

The experiments which led to the discovery of the critical point were prompted by the research of Denys Papin in England in 1680. He designed a high-pressure vessel, his "digester" (Figure 1.4a), and used it to prove that the boiling of water could be suppressed by the action of pressure. He demonstrated a practical application for the raised boiling temperature of water by cooking a meal in his digester for King Charles II [28]. In France, Baron Cagniard de LaTour speculated that this suppression of boiling must have a limit, and his risky experiments to test this theory in 1822 proved the existence of the critical point [27].

> *I introduced into a small Papin's digester, built from the end of a thick-walled gun barrel, a certain quantity of alcohol at 36 degrees and a marble or sphere of flint; the liquid occupied nearly a third of*

the interior capacity of the apparatus. Having observed the kind of noise that the marble produced upon my making it roll in the barrel at first cold, and then heating little by little over a fire, I arrived at a point where the marble seemed to bounce at each collision, as if the liquid no longer existed inside the barrel.

The same effect was observed with "l'ether sulfurique" (diethyl ether) and petroleum ether, but not with water because of the high critical temperature of water. He called this new state of matter "l'état particulier". Despite the popular myth, it is unlikely that the Baron used a cannon for these experiments (the Baron's original French wording "canon de fusil" means "gun barrel" rather than "cannon"; he also specifies that the digester he made was "small"). Later, with the use of sealed glass tubes (Figure 1.4b), he was able to observe the transition, describing it in the following manner [27]:

The liquid, after approaching double its original volume, completely disappeared, and was converted into a vapor so transparent that the tube appeared entirely empty.

He refined his methods to allow for the determination of critical temperatures and pressures [29]. His values for diethyl ether and carbon disulfide are within 15 °C and 0.4 MPa of the values accepted today.

The Baron's timing could not have been better. Physicists at the time were working on trying to liquefy gases by pressurization and/or cooling, a field of research started by Gaspard Monge late in the previous century [30]. de LaTour's discovery of the critical point showed that the success of attempts to liquefy gases by pressure would depend entirely on the temperature being lower than the critical temperature. Michael Faraday [31] liquefied a number of gases, including CO_2, in experiments begun in 1823 at the request of his supervisor Sir Humphrey Davy [32]. Faraday also observed the transition of liquid N_2O to supercritical and recognized that this represented the same transition as observed by de LaTour [33].

According to Faraday [30], supercritical CO_2 had probably first been prepared in 1797 by Count Rumford [34] during his experiments on the detonation of gunpowder in a chamber that could expand as needed but would prevent the escape of the produced gases. Ironically, the Count, unlike the Baron, *did* use a cannon, albeit only as a weight (Figure 1.5a). The detonation caused the production of hot gases but the high pressure caused the gases to take only a tiny fraction of the volume that one would have expected of gases, because, according to Faraday, the CO_2 was supercritical.

Initial observations of the critical point of CO_2 were made in 1835 by Thilorier [35], who noted that a glass tube containing both liquid and gaseous CO_2 would contain only one phase if it were warmed above 30 °C; the pressure at this point was 7.4 MPa. Cailletet, recognizing the importance of combining low temperatures with pressure, was able to liquefy gases with lower critical temperatures, including oxygen. He also succeeded in measuring the critical temperature and pressure of water by using a steel pressure apparatus calibrated by a 300 m tall column (manometer) of mercury attached to the Eiffel Tower in Paris (Figure 1.5b) [36].

The nature of the supercritical state and the significance of the critical point were debated by Michael Faraday, Dimitrii Mendeleev, Thomas Andrews, and others [37].

Figure 1.5 Early equipment for experiments with supercritical fluids. (a) Rumford's equipment. Gunpowder was detonated in a small barrel C which opened upwards but contained a moveable plug blocking the egress of any gases. A cannon was used as a weight upon the plug to inhibit the plug's movement upwards. (b) Part of Cailletet's manometer in the Eiffel Tower. It was made of 4.5 mm i.d. steel, 300 m high, and had small glass manometers connected via valves placed every 3 m so that the exact position of the top of the mercury column could be observed. An assistant used a mobile telephone to call down the measurements to the experimentalist far below [196]. Part (b) courtesy of Bibliothèque Nationale de France.

Andrews (Figure 1.6) introduced the expression "critical point" [38] and described the true nature of the supercritical state in his thorough study of carbon dioxide [39, 40].

> *Carbonic acid at 35.5° and under 108 atmospheres of pressure, stands nearly midway between the gas and the liquid; and we have no valid grounds for assigning it to the one form of matter any more than to the other .... The gaseous and liquid states are only distant stages of the same condition of matter, and are capable of passing into one another by a process of continuous change [38].*

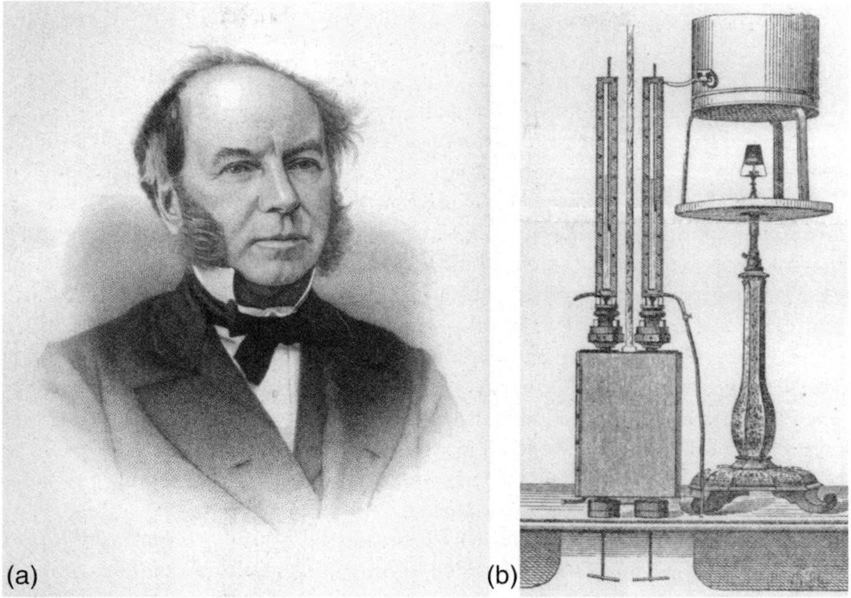

Figure 1.6 (a) Thomas Andrews (1813–1885) [197] and (b) his high-pressure equipment. A glass tube containing CO_2 was sealed at the upper end and had a "plug" of liquid mercury in the open lower end. The tube was placed within a copper tube containing water, which was compressed with a steel screw to pressures up to 40 MPa [38].

Only 4 years after these remarkable experiments, van der Waals wrote his PhD thesis [6] at the University of Leiden on "Die Kontinuität des flüssigen und gasförmigen Zustands" (on the continuity of the liquid and gaseous states), where he developed his equation of state for non-ideal gases. His theory also provided a first qualitative explanation for the critical phenomena of gases at temperatures around or above T_c, but it failed for $T < T_c$. In 1875, Maxwell introduced the equal area construction which also allowed the inclusion of subcritical temperatures [41]. Our current understanding of the behavior of gases is still mainly based on their work.

The sometimes spectacular phenomenon of critical opalescence also attracted the interest of many physicists. Altschul proposed in 1892 that opalescence be used as a method for pinpointing the critical point [42]. The relationship between density fluctuation and light scattering formed the basis of theoretical work by von Smoluchowski [43] and Einstein [44, 45], with important contributions devoted directly to critical opalescence from Ornstein and Zernike [46].

Discovery of the high-pressure limit of the supercritical state (i.e. the solid–SCF boundary in Figure 1.1) was first claimed by Tammann in 1903 [47], who showed that supercritical "phosphonium chloride" (PH_4Cl) could be condensed into a solid by the action of pressure, but the SCF later turned out to be a mixture of PH_3 and HCl [48].

The first true example came 11 years later, when Percy Bridgman showed that supercritical CO_2 could be solidified [5].

While the early phase behavior and critical point studies were mainly devoted to pure gases [49–52], the possibility of making gaseous solutions by dissolving solids in compressed gases soon became of interest. Thilorier described liquid CO_2 as being miscible in all proportions with alcohol, ether, naphtha, CS_2, and turpentine [35]. Gore showed in 1861 that liquid CO_2 also has the ability to dissolve solids such as camphor, naphthalene, and iodine. However, the failure of other solids, including charcoal, aluminum metal, and iron sulfate, to dissolve led Gore to remark that liquid CO_2 was "a very feeble solvent of substances in general" [53]. Hannay and Hogarth undertook the first systematic solubility studies on SCFs in 1879 [54, 55], demonstrating that chlorophyll, KI, and other solids were soluble in supercritical alcohol. They extended the work to sulfur and other reagents in supercritical carbon disulfide. Their astute observations included the comments that "a liquid has its critical point raised by the solution in it of a solid" [55] and that a very finely divided solid could be produced by rapidly reducing the pressure of a supercritical solution. The latter phenomenon is the basis for the modern-day RESS process.

Paul Villard, a physicist better known for his later discovery of gamma rays [56], was working on phase behavior in 1896 when he found that iodine, alkanes, stearic acid, and camphor could be dissolved in supercritical ethylene (scC_2H_4) [57, 58]. He noticed that the action of the scC_2H_4 caused the camphor to melt before it dissolved, even though the temperature was not being raised (cf. the PGSS process). Studies of solubilities and phase behavior continued through the twentieth century, and the results are described in a number of reviews [59–64]. More details on our current understanding of solubility data and phase behavior can be found in Chapter 3.

1.4.2
Extraction and Chromatography in SCFs

Starting from the early 1920s, practical applications of compressed gases were recognized for extraction and purification. Auerbach in Berlin filed several international patents on a process utilizing liquid CO_2 in the period between 1926–1928 [65, 66]. SCFs were investigated especially in the rapidly growing petrochemical industry. Shell Oil patented various processes utilizing "para-critical" (i.e. near-critical) hydrocarbons and other gases including CO_2 [67–69]. In the mid-1950s, several technical solutions for SFE processes for coal, mineral oil, and wool were proposed by Zhuze [70]. In 1955, Todd and Elgin pointed out the analogy of SCF–solid and SCF–liquid separation with conventional extraction [71].

A major technical breakthrough came with the development of natural product extraction using $scCO_2$ by Kurt Zosel (Figure 1.7a) at the Max Planck Institut für Kohlenforschung in the early 1960s [72]. His interest in SCFs originated from accidental observations made during the preparation of higher olefins from ethylene in the presence of aluminum trialkyls (Aufbau reaction). Different product compositions were obtained when the high-pressure reactors were vented at temperatures either above or below T_c. He realized that this was related to scC_2H_2 extraction of

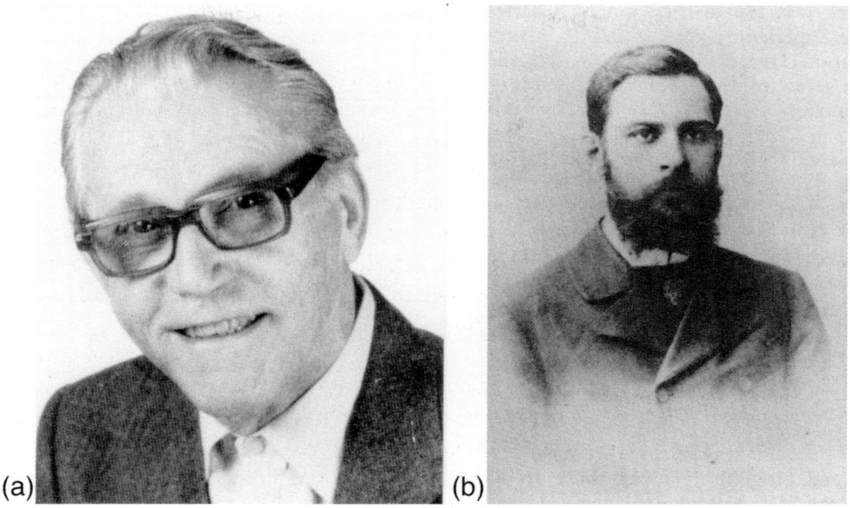

Figure 1.7 (a) Kurt Zosel (1913–1989). Reproduced with permission from the Photoarchiv MPI für Kohlenforschung, Mülheim/Ruhr. (b) Vladimir N. Ipatiev (1867–1952) in 1897, 6 years before he started his high-pressure work [105].

components from the mixture at $T > T_c$. He started to study SCFs more systematically, facilitated by the excellent knowledge of high-pressure experimentation at the Institute's workshop. He recognized the importance of both volatility and polarity for the solubility of various components in $scCO_2$ and termed the separation process "destraktion", composed from the German "destillation" and "extraktion", to describe both contributions. Unfortunately, this useful terminology was not adopted by others, probably because of its unfavorable transfer into English.

His enthusiasm for $scCO_2$ as a separation agent was initially seen with some skepticism by the Institute's Director and Nobel Laureate Karl Ziegler, but it soon became apparent that extraction with toxicologically benign "carbonic acid" was highly attractive for the food industry. The first application to the extraction of caffeine from green coffee beans was already implemented on a technical scale in the 1970s by the German company Hag and later transferred to the United States by General Foods. Today, more than 100 000 t of decaffeinated coffee are produced per year world-wide using this technology. The "destraktion" of hops aroma from hops flowers on industrial scale soon followed and is still state of the art for a significant part of the supply to the brewing industry.

Another modern application of SCFs came about with a breakthrough in the early 1960s; supercritical fluid chromatography (SFC). Klesper *et al.* showed in 1962 [73] that supercritical fluids could be used as eluents for chromatographic separations. While they were interested particularly in separating nickel porphyrin complexes, the technique has become widely accepted for analytical and preparative separations in research and development. Pharmaceutical companies in particular have recently adopted chiral preparative SFC with $scCO_2$ as eluent for the preparation of

Table 1.2 Industrial applications of supercritical fluids.

Process	References
Extraction:	[77, 81, 85–88]
of caffeine from coffee and tea	[81]
of pharmaceuticals, neutraceuticals	[77]
of flavors from hops, spices, herbs	[77, 81]
of deasphalted oil from crude oil residuum	[89, 90]
Chromatography:	[78, 91]
of polyunsaturated fatty acids	[85]
of pharmaceuticals, enantiomers	[92]
Fractionation (of aromas, polymers, lubricants)	[93]
Impregnation:	
of biocides into wood	[77, 94]
of dyes into textiles	[77, 95]
Atomization of paint (UNICARB process)	[96]
Micronization	[83, 97, 98]
Power generation (coal-fired scH_2O boilers)	[99]
Viscosity reduction (e.g. of oil in reservoirs)	[100]
Foaming (of polymers)	[76]
Drying (e.g. of aerogels)	[101]
Reactions	See Section 1.4.4

pre-production quantities of pharmaceuticals, thereby greatly reducing their need for organic solvents.

Supercritical fluids are now used in a wide variety of applications in many industries (Table 1.2). While non-reactive applications are not discussed in this book, there are many reviews and monographs available on supercritical fluids in general [74–77] and specifically for chromatography [78, 79], extractions [75, 79–81], cleaning [82], drug product development [83], and polymer synthesis [84].

1.4.3
The History of Chemical Reactions in SCFs

When was the first use of a SCF as a medium for a reaction? If we exclude the fact that Nature has been doing chemistry in scH_2O in the Earth's crust for a very long time [102], the first reactions observed in SCFs appear in the nineteenth century. It was again Baron Cagniard de LaTour who noticed as early as 1822 that near-critical water seemed particularly reactive. He observed that it was "capable of decomposing glass by dissolving its alkali" and speculated that "one could possibly obtain some other interesting results for chemistry, in multiplying the applications of this process of decomposition". Indeed, most of the studies of reactions in SCFs during the next few decades primarily concerned H_2O at temperatures above T_c.

Pioneering studies were carried out by Gabriel-Auguste Daubrée, a geologist who held the Chair of Mineralogy and Geology at Strasbourg [103]. In 1857, he reported tests of the reactions of scH_2O with glass and with minerals at 400 °C. Daubrée put the water and mineral in a glass tube, and put the sealed tube in water in an iron outer tube. The whole assembly was then heated. Thus, the glass would not break because it would experience roughly equal pressures from both inside and out. The pressures were, in his words, "enormous", but as they were not measured, we know only that the water was supercritical in terms of temperature (ignoring the effect of dissolved minerals on T_c). Hot spring water containing potassium silicate, when heated in this manner, deposited quartz. Reaction of this water with kaolin produced crystals of feldspar. Wood was transformed to a material resembling anthracite under similar conditions [104].

Early research on hydrothermal reactions was also carried out by Charles Friedel, an instructor at the École Normale Supérieure [105], the same institution where Paul Villard received his education. Over many years, Charles, and later his son Georges Friedel, published a series of papers on the reaction of minerals in H_2O at temperatures and presumably also pressures far above critical, starting with the preparation of quartz in 1879 [106, 107]. Charles Friedel is more famous for his discovery of the Friedel–Crafts alkylation of aromatics. The works of de LaTour, Daubrée and the Friedels together ensured that the first experimental reactions of and in SCFs were those in H_2O, as reviewed in great detail by Morey and Niggli [108].

Reactions in SCFs other than water started, so far as we have been able to determine, with Hannay and Hogarth's [55, 109] solubility experiments in 1879. They found that a blue liquid ammonia solution of elemental sodium underwent a reaction when heated past the critical temperature of the ammonia, producing a white solid and hydrogen gas. With hindsight, we would formulate the reaction as shown in Equation 1.4. Just as unexpected was their observation that solid zinc oxide reacted with $scCCl_4$ at 300 °C without dissolving [56]. The products were zinc chloride and an unidentified gas. Elemental arsenic seemed to dissolve in $scCS_2$, but when the pressure was released, reddish yellow crystals were precipitated. Hannay and Hogarth believed these to be arsenic sulfide arising from a reaction with the SCF. Villard also observed reactions during his measurements of the solubility of solids in SCFs in 1898 [58, 59]. When he tried to dissolve iodine in scC_2H_4 at 17 °C and 30 MPa, he found that once dissolved, the iodine slowly reacted with the ethylene (Equation 1.5). Over a period of an hour or two, the deep violet color of the solution faded. The solid white product was isolated in crystalline form by pressure reduction.

$$Na_{(s)} \xrightarrow[<132\,°C]{liqNH_3} Na^+_{(NH_3)} + e^-_{(NH_3)} \xrightarrow[>132\,°C]{} NaNH_{2(s)} + \tfrac{1}{2}H_{2(g)} \qquad (1.4)$$

$$H_2C{=}CH_2 + I_2 \xrightarrow[\substack{17\,°C \\ 30\,MPa}]{scC_2H_4} H_2IC{-}CH_2I \qquad (1.5)$$

A wide variety of high-pressure reactions were described by Vladimir Ipatiev (Figure 1.7b) [105], a Professor at the Chemistry Laboratory of the Michail Artillery Academy in St Petersburg, in articles from 1904 to 1914 (Scheme 1.1). Many of these reactions were at supercritical conditions, although not all of his reports accurately specified the pressures at which the reactions were performed. We now take for

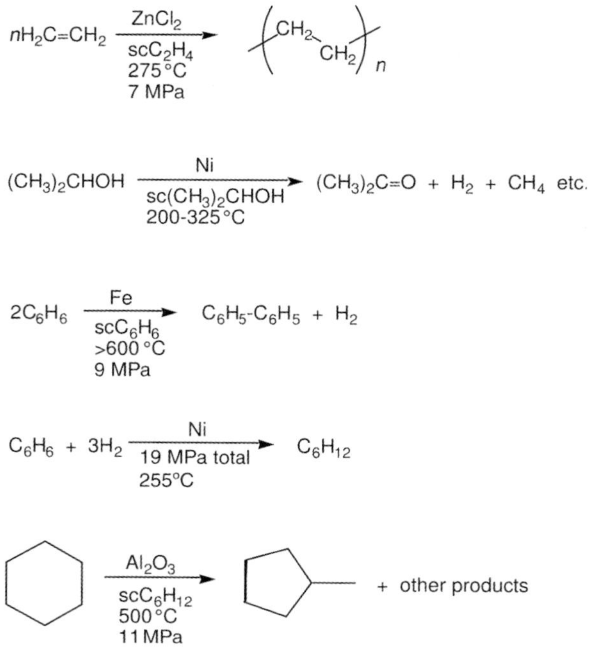

Scheme 1.1 Some of the reactions of supercritical fluids or mixtures studied by Ipatiev.

granted the high pressure "bomb" or autoclave, which he was the first to develop [110]. Among his significant discoveries was the first observation that heating and pressurizing scC_2H_4 caused it to be non-catalytically oligomerized into liquid alkanes and cycloalkanes of a range of molecular weights up to cyclotetradecane [111–113]. He also noticed that with $ZnCl_2$ or $AlCl_3$ as presumably heterogeneous and homogeneous catalysts, respectively, oligomerization occurred at much lower temperatures [113].

Other heterogeneously catalyzed supercritical phase reactions reported by Ipatiev included the dehydrogenation of 2-propanol [114], the dehydration of 2-butanol [115], the dehydrogenative coupling of benzene to diphenyl [116], and the isomerization of cyclohexane to methylcyclopentane and other products [117]. Most of his work was dedicated to the development of heterogeneously catalyzed hydrogenation reactions such as the Ni-catalyzed hydrogenation of benzene in what was probably a supercritical mixture of benzene and hydrogen [116]. He revolutionized heterogeneous catalysis [110], introducing not only the use of high pressures but also multicomponent catalysts such as Ni_2O_3/Al_2O_3. In all, 550 articles and patents were written or co-written by Ipatiev [110].

Several other chemists were active in the field of supercritical fluid reactions at the beginning of the twentieth century. For example, Briner's group studied the decomposition and reactivity of supercritical fluids such as scNO and scCO (e.g. Equation 1.6) [118–121]. He also investigated the system N_2–H_2 and reported that N_2 and H_2 do not combine at room temperature and 90 MPa [121]. At the Chemical

Institute of the University of Berlin, Arthur Stähler studied the reactions of alkyl halides such as chloroethane with $scNH_3$ (Equation 1.7) [122]. In the early 1940s, Patat at the University of Innsbruck studied the hydrolysis of aniline under supercritical conditions (Equation 1.8) [123].

$$2NO + 2HCl \xrightarrow[\text{RT, 30 MPa}]{\text{scNO}} NOCl + H_2O + \tfrac{1}{2}Cl_2 + \tfrac{1}{2}N_2 \qquad (1.6)$$

$$\underset{}{EtCl} + NH_3 \xrightarrow[\text{220 °C, 22 MPa}]{scNH_3} \underset{80\%}{[EtNH_3]Cl} + \underset{15\%}{[Et_2NH_2]Cl} + \underset{5\%}{[Et_3NH]Cl} \qquad (1.7)$$

$$PhNH_2 + H_2O \xrightarrow[\text{69 MPa, 440 °C}]{[NH_4][H_2PO_4]} PhOH + NH_3 \qquad (1.8)$$

Early mechanistic work on reactions in SCFs includes a kinetic study of the bimolecular decomposition of supercritical HI in 1928 [124]. In the following two decades, a number of researchers tried to address the question of the effect of near-critical temperatures on reaction rates, although not all of them took into account pressure and density effects [125–128]. In one of the more careful studies, Holder and Maass [126] in 1938 described tests of the rate of the reaction between HCl and propene in liquid and supercritical mixtures at different temperatures. The density within a series of experiments was kept constant by adjusting the pressure. The rates of reaction passed through a minimum at the critical temperature for three of the four chosen densities with $d_r > 1$. The minimum was most marked, in percentage terms, when d_r was close to 1. In 1946, Toriumi [128] reported that the rate of uncatalyzed oxidation of NH_3 by oxygen reached a maximum at the critical temperature. Similarly, the rate of the Pt-catalyzed oxidation of SO_2 by oxygen was found to be highest at the T_c of SO_2. Rapid diffusion at the critical temperature was offered as one possible explanation. The themes of rate changes near the critical point and the effectiveness of SCFs as solvents for heterogeneous catalysis (Chapter 6) have fascinated researchers ever since and have been the subject of many more recent studies.

Chemical fixation of CO_2 under supercritical conditions was attempted for the first time in the middle of the twentieth century in the polymerization field. In a 1949 patent, Sargent suggested that CO_2 could be incorporated into polyethylene to give a more waxy product. This was achieved by copolymerizing scC_2H_4 and $scCO_2$ at 72–88 °C and 100 MPa in the presence of a free-radical initiator such as benzoyl peroxide. The resulting polymers were reported to incorporate one CO_2 for every 29 ethylene units [129]. This contrasts with more recent studies, which have demonstrated the use of $scCO_2$ as an "innocent" solvent for free radical polymerizations (see Chapter 8). In 1951, Buckley and Ray [130] patented the preparation of polymeric ureas by the treatment of carbamates with $scCO_2$ at 200 °C and 50 MPa, but it is not clear to what extent any of the carbamates dissolved in the supercritical phase. Stevens patented in 1966 the copolymerization of CO_2 and ethylene oxide to form polycarbonates; in several of his examples the reaction mixtures were clearly at supercritical conditions [131].

There are many more reports and patents of reactions in SCFs than the selected examples mentioned here, especially in the period 1910–1945. Since 1945, the literature has become so extensive that it is not possible to review it in this short chapter, but the reader will find detailed coverage of individual aspects in the other chapters of this book. Further information on the historical background of reactions in SCFs is also given in the reviews written by Subramaniam and McHugh (covering the years 1945–1985) [132], Savage *et al.* (covering 1986–1994) [133], and Scholsky (covering polymerization from 1940 to 1990) [134]. The reader is also referred to other monographs on reactions in SCFs [135–137].

1.4.4
Industrial Use of SCFs as Reaction Media

Of the industrial applications of chemistry in SCFs, the first and the most famous is arguably the synthesis of ammonia, although some might not include this as an example of a reaction in an SCF because the operating temperature is much higher than the critical temperature. High-pressure studies of the reaction between H_2 and N_2 to form ammonia started in 1901 with Le Chatelier's patent on the uncatalyzed process at up to 10 MPa [138]. Fritz Haber later found heterogeneous catalysts such as osmium for this process, but using pressures as high as 17.7 MPa [139]. The chemical and technical problems associated with the ammonia formation and the drastic conditions required were solved by Haber and Bosch. These studies led to the commercialization of the iron-catalyzed process by BASF (Equation 1.9), with operations of the first plant starting in September 1913 at Oppau, Germany [138, 140, 141].

$$N_2 + 3H_2 \xrightarrow[\substack{10\text{–}35\ \text{MPa} \\ 400\text{–}530\,^\circ\text{C}}]{\text{Fe}} 2NH_3 \tag{1.9}$$

Modern plants for ammonia production by the Haber–Bosch process [140] operate at 10–35 MPa and 400–530 °C, well above both T_c and p_c for the reaction mixture. Most notably, the largest plants use the highest pressures because this allows a greater throughput per unit volume of vessel. The vessels themselves can be enormous (Figure 1.8a). The actual ammonia synthesis step may not be the only part of the plant which operates at supercritical conditions. The production of the H_2 in the Shell and Texaco ammonia processes is performed by the partial oxidation of hydrocarbons by O_2 at 1250 °C and pressures up to 8 MPa, followed by the water gas shift reaction. The conditions of this partial oxidation step may well be supercritical depending on the composition of the hydrocarbon mixture and the hydrocarbon:O_2 ratio.

In 1913, the same year as the first commercial ammonia process was launched, BASF began initial tests of a synthesis of methanol from pressurized H_2–CO mixtures, leading to the first plant for the production of synthetic methanol (Equation 1.10) at Leuna in Germany in 1923 [142, 143]. Patart in France later claimed prior invention [144, 145].

Figure 1.8 Early industrial chemistry at supercritical conditions. (a) Reactors, with man-sized bolts, for the synthesis of NH_3 from N_2 and H_2 at ICI's 1920 plant in Billingham, England [198]. (b) A 9 l high-pressure reactor in ICI's 1937 pilot plant for the production of polyethylene at up to 90 MPa and 200 °C [199].

$$CO + 2H_2 \xrightarrow[\substack{25\text{–}30\ \mathrm{MPa} \\ 320\text{–}450\,^\circ\mathrm{C}}]{\mathrm{Zn/Cr\ oxide}} CH_3OH \tag{1.10}$$

The high pressure of the process was required because of the poor activity of the catalyst, but this problem was alleviated by the introduction of better catalysts by ICI and Lurgi in the 1960s and 1970s [143]. The current process for the synthesis of methanol, as operated by ICI in England [143], requires a CO–CO_2–H_2 feed mixture and a pressure of between 5 and 10 MPa, which may or may not be supercritical, depending on the composition and pressure. The growing interest in methanol as a transportable liquid form of carbon stemming from coal or gas[146, 147] leads to the construction of "MegaMethanol" plants using mainly Lurgi technology, which operates in a similar pressure and temperature regime. Methanol can also be obtained from only CO_2 and hydrogen; Mitsui Chemicals has announced a demonstration plant to operate in Japan [148]. Again, the reaction mixture may well be supercritical, but not enough process details are available in the open literature to make a definite conclusion.

Some other industrial processes are or have been carried out under supercritical conditions even though the fact may not have been generally recognized. For example, McKee and Parker pointed out in 1928 that some oil cracking processes performed industrially at the time occurred above the critical temperature of the reaction mixtures [149]. These processes have since been replaced by lower temperature catalytic cracking methods. Other historical industrial processes which probably involved supercritical phases include the synthesis of melamine [150] from dicyanodiamide, N_2, and NH_3, and the alkylation of aniline by methanol [150].

The oxidation of light alkanes by air or O_2 at supercritical temperatures and pressures was explored by Standard Oil in the mid-1920s [151]. Experiments were performed in the laboratory and then at a semi-commercial plant level. The primary products were alcohols. For example, the oxidation of pentane was performed at supercritical conditions (240 °C, around 20 MPa and a few mole percent of O_2) and produced primarily C_2–C_3 alcohols and acids. However, the oxidation of heptane was performed at subcritical temperatures (225 °C) and produced primarily C_6–C_7 alcohols. The change in selectivity was attributed to either the difference in phase or more likely the difference in temperature. Other commercial processes for the formation of alcohol denaturants or formaldehyde were reported in the same decade [152, 153], but it is unclear whether those reactions were operated at supercritical pressures. Modern processes involving alkane oxidation are heterogeneously catalyzed at subcritical pressures [154].

The polymerization of ethene attracted a great deal of commercial interest two decades after Ipatiev's pioneering studies. Perrin and co-workers, in a 1937 patent for ICI, reported a dramatic improvement in the method [155–157]. If a higher pressure of around 200 MPa and rigorous temperature control (~170 °C) were used, solid products with molecular weights around 3500 could be obtained. Failure to control the temperature resulted in explosive production of carbon and hydrogen. Further, they found that if oxygen was present in the scC_2H_4, then solid products could be obtained at only 50 MPa. The fortuitous introduction of oxygen in their experiment was due to a leaky vessel! Their 9 l vessel is shown in Figure 1.8b.

ICI built a plant in Cheshire which produced polyethylene, primarily as an insulator for cables which were badly needed for the wartime radar system. To increase production, the secrets of the method were passed on to American companies [158]. Immediately after the War, DuPont patented refinements to the process. Krase, an investigator at DuPont, found that stepwise reduction of the pressure of the scC_2H_4 allowed the fractionation of the polyethylene by selective precipitation [159, 160]. The system was further developed and the relevant phase behavior published by Ehrlich in a series of papers in the 1960s and 1970s [161–164], including a major review on the subject [165]. The high-pressure synthesis of low-density polyethylene (LDPE) as currently practiced [166] involves the polymerization of scC_2H_4 at 100–300 MPa and at 80–300 °C. Oxygen is still one of the initiators employed, others being benzoyl peroxide and azobisisobutyronitrile (AIBN). Approximately 3.2×10^{-6} t per annum of LDPE are produced in the United States alone [167].

Supercritical or near-critical water has found technical applications for hydrothermal syntheses, as discussed in detail in Chapter 13. Another more recent industrial application of chemical reactions in SCFs is the oxidative destruction of chemical wastes in scH_2O (SCWO, supercritical water oxidation). A detailed coverage of the large and prolific field of SCWO is outside the scope of this book. The extensive pilot plant activity, primarily by MODAR (now General Atomics) and Eco Waste Technologies, has been summarized by Schmieder *et al.* [168]. The first commercial plant was opened by Huntsman Chemical in collaboration with Eco Waste Technologies.

The industrial hydration of ethene to ethanol occurs when a mixture of ethene and water is fed through the reactor at 300 °C and 70 MPa. The conditions are kept above the dew point of the ethene–water mixture because condensed water would deactivate the catalyst, typically phosphoric acid on a silica support. This process is carried out by Shell, BP, Erdöl-Chemie and Hibernia-Chemie [169, 170], with the first plant being opened by Shell in 1948 [171].

Several processes have been commercialized that use biphasic catalysis, in which there is an aqueous catalyst-bearing phase and a supercritical reactant phase. These processes include methanol carbonylation and the hydration of propene and butene, as described in Chapter 4.

Synthesis of fluoropolymers has been performed in scCO_2 by DuPont in a $40 million plant that started up in 2000 [172]. The supercritical process was invented by DeSimone and co-workers at the University of North Carolina [173–175] and was further developed by DuPont. Compared with the old water-based process, the scCO_2 process has the advantage of not requiring fluorinated surfactants.

The highly successful applications of SCFs in the production of polymers and bulk chemicals illustrate that the technical challenges associated with reactions under high pressure and high temperature can be overcome even on a very large scale. Despite this encouraging precedence, the industrial application of SCFs to the production of fine chemicals, pharmaceuticals, or other more specialized products is still in its infancy. The following examples illustrate the current industrial interest in fine chemical synthesis in SCFs.

The reduction of 1,4-androstanediene-3,17-dione (AAD) as an industrial route to estradione using supercritical tetralin as both the solvent and the hydrogen donor was investigated at Schering in cooperation with the University of Göttingen [176, 177]. The reaction kinetics were studied between 350 and 600 °C and 50 and 30 MPa. A pilot plant with a possible throughput of up to 30 l h^{-1} was studied successfully over 1 year with continuous processes run up to 4 days at 575 °C and pressures up to 10 MPa with contact times of less than 1 s.

Heterogeneously catalyzed hydrogenation processes involving gaseous hydrogen have been extensively studied for industrial applications. Hoffman La Roche (now DSM Vitamins) investigated the technical feasibility of a variety of such processes [178], including the heterogeneously catalyzed hydrogenation of vitamin precursors in a continuous-flow process in scCO_2 on a pilot-plant scale. The reactor vessel has an internal volume of 40 l and an output of 800 t per annum [179, 180]. Flowcharts comparing batch and continuous-flow processes have been published [181].

Longstanding research efforts at Degussa (now Evonik Degussa) have revealed that the hardening (hydrogenation of the C=C double bonds) of edible oils and fatty acids and their esters can be achieved with high efficiency and selectivity in scCO_2 using commercially available supported fixed-bed Pd catalysts [182–185]. Up to 15 times higher space–time yields were achieved and three times higher catalyst productivities resulted from the extended lifetime of the catalyst in scCO_2 compared with reactions in a trickle-bed process. Based on this technology, Thomas Swan Ltd in cooperation with the University of Nottingham developed a miniature flow-through system which allows the hydrogenation of large amounts of unsaturated substrates in surprisingly small reactors [186, 187]. This has led to the construction of a multipurpose plant for catalysis in $scCO_2$ at Consett in the United Kingdom. For typical hydrogenation conditions as operated on an industrial scale in this plant, later studies revealed that the reactions most likely occur under expanded liquid rather than single-phase supercritical conditions [188].

Finally, the synthesis of energetic materials has been suggested to be performed in liquid or supercritical CO_2. The Indian Head Division of the U.S. Naval Surface Warfare Center built a pilot plant with a 100 l vessel for the synthesis of such energetics as MTV (magnesium–Teflon–Viton) and poly-3-nitratomethyl-3-methyloxetane [189–191].

Some of the advantages associated with the use of SCFs for chemical reactions have been known for a very long time, ever since chemists started to investigate these media. Nevertheless, it is only now that we are starting to appreciate fully all of the benefits and to integrate them together with the development of new synthetic methodologies into the design process of green chemistry. The basic principles of carrying out synthetic processes in SCFs are now, to a significant extent, laid out, although we are still far from a complete understanding of all molecular and physicochemical interactions in such complex reactive systems. At the same time, SCFs are currently being used in a number of processes in industry, including several chemical syntheses. The challenge that lies ahead is to build upon these achievements and to target specifically more and more processes that would benefit from the application of SCF technology. If the right applications are identified, then supercritical fluids will further contribute to the greening of chemical processes.

References

1 McNaught, A.D. (1997) *Compendium of Chemical Terminology, IUPAC Recommendations*, 2nd edn, Blackwell Science, Oxford.

2 Bridgeman, P.W. (1952) *The Physics of High Pressure*, G. Bell, London.

3 Bridgeman, P.W. (1934) *Physical Review*, **46**, 930.

4 Scholsky, K.M. (1989) *Journal of Chemical Education*, **66**, 989.

5 Bridgman, P.W. (1914) *Physical Review*, **3**, 126.

6 (a) van der Waals, J.D. (1873) Die Kontinuität des Flüssigen und Gasförmigen Zustands, PhD Dissertation, University of Leiden, (b) for an English translation, see Threlfall, R. and Adair, J. (1888) *Physical Memoirs*, vol. **1**, Part 3, Taylor and Francis, London.

7 Popov, V.K., Banister, J.A., Bagratashvili, V.N., Howdle, S.M. and Poliakoff, M. (1994) *Journal of Supercritical Fluids*, **7**, 69.

8 Hildebrand, J.H. and Scott, R.L. (1950) *The Solubility of Nonelectrolytes*, 3rd edn, Reinhold, New York.

9 Giddings, J.C., Myers, M.N., McLaren, L. and Keller, R.A. (1968) *Science*, **162**, 67.

10 Kordikowski, A., Robertson, D.G. and Poliakoff, M. (1996) *Analytical Chemistry*, **68**, 4436.

11 Roth, M. (1998) *Analytical Chemistry*, **70**, 2104.
12 Porter, N.L., Richter, B.E., Bornhop, D.J., Later, D.W. and Beyerlein, F.H. (1987) *Journal of High Resolution Chromatography & Chromatography Communications*, **10**, 477.
13 Clavier, J.Y. and Perrut, M. (1996) *High Pressure Chemical Engineering: Proceedings of the 3rd International Symposium on High Pressure Chemical Engineering, Zurich, Switzerland, 7–9 October 1996* (eds P.R. von Rohr and C. Trepp), Elsevier, Amsterdam, p. 627.
14 Urben, P.G. (1995) *Bretherick's Handbook of Reactive Chemical Hazards*, 5th edn, Butterworth Heinemann, Oxford.
15 Lewis, R.J. Sr. (1992) *Sax's Dangerous Properties of Industrial Materials*, 8th edn, Van Nostrand, New York.
16 Albert, J. and Luft, G. (1998) *Chemical Engineering and Processing*, **37**, 55.
17 Theyssen, N., Hou, Z. and Leitner, W. (2006) *Chemistry - A European Journal*, **12**, 3401.
18 (1976) *Encyclopedie des Gaz*, L'Air Liquide/Elsevier, Amsterdam.
19 Hansen, B.N., Hybertson, B.M., Barkley, R.M. and Sievers, R.E. (1992) *Chemistry of Materials*, **4**, 749.
20 Sievers, R.E. and Hansen, B. (1991) *Chemical & Engineering News*, **69**, (29), 2.
21 Raynie, D.E. (1993) *Analytical Chemistry*, **65**, 3127.
22 Kühn, R. and Birett, K. (1990) *Merkblätter Gefährliche Arbeitsstoffe*, 9th edn, vol. 8, Ecomed, Landsberg/Lech.
23 Budavari, S., O'Neil, M.J., Smith, A., Heckelman, P.E. and Kinneary, J.F. (1996) *Merck Index*, 12th edn, Merck, Whitehouse Station, NJ.
24 Fang, Y. and Chau, Y.K. (1995) *Applied Organometallic Chemistry*, **9**, 365.
25 Pierantozzi, R. (1993) *Encyclopedia of Chemical Technology* 4th edn,vol. 5 (eds J.I. Kroschwitz and M. Howe-Grant), John Wiley & Sons, Inc., New York, p. 35.
26 Kuehne, E., Witkamp, G.-J. and Peters, C.J. (2008) *Green Chemistry*, **10**, 929.
27 Cagniard de LaTour, C. (1822) *Journal de Physique et Le Radium*, **21**, 127.
28 Asimov, I. (1982) *Asimov's Biographical Encyclopedia of Science and Technology*, 2nd edn, Doubleday, Garden City, NJ.
29 Cagniard de LaTour, C. (1823) *Journal de Physique et Le Radium*, **22**, 410.
30 Faraday, M. (1824) *Quarterly Journal of Science*, **16**, 229.
31 Faraday, M. (1823) *Philosophical Transactions of the Royal Society of London*, **113**, 189.
32 Davy, H. (1823) *Philosophical Transactions of the Royal Society of London*, **113**, 164.
33 Goudaroulis, Y. (1995) *Review of History and Political Science*, **48**, 353.
34 Count Rumford, B. (1797) *Philosophical Transactions of the Royal Society of London*, **87**, 222.
35 Thilorier, M. (1835) *Annales de Chimie et de Physique*, **60**, 427.
36 Cailletet, L.P. and Colardean, E. (1891) *Comptes Rendus de l'Academie des Sciences Paris*, **112**, 1170.
37 Rowlinson, J.S. (1969) *Nature*, **224**, 541.
38 Andrews, T. (1869) *Philosophical Transactions of the Royal Society of London*, **159**, 575.
39 Andrews, T. (1870) *Annales de Chimie et de Physique*, **21**, 208.
40 Andrews, T. (1876) *Proceedings of the Royal Society of London*, **24**, 455.
41 Niven, W. (ed.) (1927) *The Scientific Papers of James Clerk Maxwell*, Herrmann, Paris.
42 Altschul, M. (1892) *Zeitschrift fur Physikalische Chemie-International Journal of Research in Physical Chemistry & Chemical Physics* **11**, 577.
43 von Smoluchowski, M. (1906) *Annals of Physics*, **21**, 756.
44 Einstein, A. (1905) *Annals of Physics*, **17**, 549.
45 Einstein, A. (1906) *Annals of Physics*, **19**, 371.
46 Ornstein, L.S. and Zernike, F. (1914) *Proc. Sect. Sci. K. Med. Akad. Wet.*, **17**, 793.
47 Tammann, G. (1903) *Kristallisieren und Schmelzen*, E. Barth, Leipzig.
48 Briner, E. (1907) *J. Chim. Phys. Phys.-Chim. Biol.*, **4**, 476.

49 Estreicher, T. and Schaerr, A.A. (1913) *Z. Komprim. Flüss. Gase*, **15**, 161.

50 Kobe, K.A. and Lynn, R.E. (1953) *Chemical Reviews*, **52**, 117.

51 Kudchadker, A.P., Alani, G.H. and Zwolinski, B.J. (1968) *Chemical Reviews*, **68**, 659.

52 Mathews, J.F. (1972) *Chemical Reviews*, **72**, 71.

53 Gore, G. (1861) *Philosophical Transactions of the Royal Society of London, Ser. A*, **151**, 83.

54 Hannay, J.B. and Hogarth, J. (1879) *Proceedings of the Royal Society of London, Section B*, **29**, 324.

55 Hannay, J.B. and Hogarth, J. (1880) *Chemical News*, **41**, 103.

56 Dostrovsky, S. (1976) in *Dictionary of Scientific Biography*, vol. XIV (ed C.C. Gillispie), Charles Scribner's Sons, New York, p. 31.

57 Villard, P. (1896) *Seances Soc. Fr. Phys.*, 234.

58 Villard, P. (1898) *Chemical News*, **78**, 297.

59 Bartle, K.D., Clifford, A.A., Jafar, S.A. and Shilstone, G.F. (1991) *Journal of Physical and Chemical Reference Data*, **20**, 713.

60 Dohrn, R. and Brunner, G. (1995) *Fluid Phase Equilibria*, **106**, 213.

61 Brennecke, J.F. and Eckert, C.A. (1989) *AIChE Journal*, **35**, 1409.

62 Kiran, E. and Levelt Sengers, J.M.H. (1994) *NATO ASI series*, **273**, 796.

63 Sadus, R.J. (1992) *High Pressure Phase Behaviour of Multicomponent Fluid Mixtures*, Elsevier, Amsterdam.

64 Streett, W.B. (1983) in *Chemical Engineering at Supercritical Fluid Conditions* (eds M.E. Paulaitis, J.M.L. Penninger, R.D. Gray, Jr. and P. Davidson), Ann Arbor Science, Ann Arbor, MI, p. 3.

65 Auerbach, E.B. (1926) British Patent 277, 946.

66 Auerbach, E.B. (1931) US Patent 1,805,751.

67 Pilat, S. and Godlewicz, M. (1940) to Shell Development Co., US Patent 2,188,013.

68 Schaafsma, A. (1941) US Patent 2,252,864.

69 Schaafsma, A. (1938) US Patent 2,118,454.

70 Zhuze, T.P. (1960) *Petroleum (London)*, **23**, 298.

71 Todd, D.B. and Elgin, T.C. (1955) *AICHE Journal*, **1**, 20.

72 Zosel, K. (1978) *Angewandte Chemie (International Edition in English)*, **17**, 702.

73 Klesper, E., Corwin, A.H. and Turner, D.A. (1962) *The Journal of Organic Chemistry*, **27**, 700.

74 Arai, Y. (2001) *Supercritical Fluids*, Springer, New York.

75 Gopalan, A.S. (2003) *Supercritical Carbon Dioxide: Separations and Processes*, Oxford University Press, Oxford.

76 Beckman, E.J. (2004) *Journal of Supercritical Fluids*, **28**, 121.

77 Theyssen, N. (2005) in *Multiphase Homogeneous Catalysis* (eds B. Cornils, W.A. Herrmann, D. Vogt, I. Horvath, H. Olivier-Bourbigon, W. Leitner and S. Mecking), Wiley-VCH Verlag GmbH, Weinheim, p. 630.

78 Anton, K. and Berger, C. (1998) *Supercritical Fluid Chromatography with Packed Columns: Techniques and Applications*, Chromatographic Science Series, vol. 75, Marcel Dekker, New York.

79 Caude, M.H. and Thiebaut, D. (1999) *Practical Supercritical Fluid Chromatography and Extractions*, Harwood Academic Publishers, Amsterdam.

80 McHugh, M. and Krukonis, V. (1994) *Supercritical Fluid Extraction*, 2nd edn, Butterworth-Heinemann, Boston, MA.

81 Taylor, L.T. (1996) *Supercritical Fluid Extraction*, Wiley-Interscience, New York.

82 McHardy, J. (1998) *Supercritical Fluid Cleaning*, William Andrew Publishing, Norwich, NY.

83 York, P., Kompella, U.B. and Shekunov, B. (2004) *Supercritical Fluid Technology for Drug Development*, Marcel Dekker.

84 Kemmere, M.F. and Meyer, T. (2005) *Supercritical Carbon Dioxide in Polymer Reaction Engineering*, Wiley-VCH Verlag GmbH, Weinheim.

85 Perrut, M. (2000) *Industrial & Engineering Chemistry Research*, **39**, 4531.

86 Raventós, M., Duarte, S. and Alarcón, R. (2002) *Food Science and Technology International*, **8**, 269.

87 Martinez, J.L. (2007) *Supercritical Fluid Extraction of Nutraceuticals and Bioactive Compounds*, CRC Press, Boca Raton, FL.

88 Mukhopadhyay, M. (2000) *Natural Extracts Using Supercritical Carbon Dioxide*, CRC Press, Boca Raton, FL.

89 Gearhart, J.A. (1980) *Hydrocarbon Processing*, **59**, (5), 150.

90 (1996) *Hydrocarbon Processing*, **75**, (11), 105.

91 Saito, M., Yamauchi, Y. and Okuyama, T. (1994) *Fractionation by Packed-Column SFC and SFE: Principles and Applications*, VCH, New York, p. 276.

92 Majewski, W., Valery, E. and Ludemann-Hombourger, O. (2005) *Journal of Liquid Chromatography & Related Technologies*, **28**, 1233.

93 Perrut, M. (2004) *Ion Exchange Solvent Extr.*, **17**, 1.

94 Acda, M.N., Morrell, J.J. and Levien, K.L. (2001) *Wood Science and Technology*, **35**, 127.

95 Montero, G.A., Smith, C.B., Hendrix, W.A. and Butcher, D.L. (2000) *Industrial & Engineering Chemistry Research*, **39**, 4806.

96 Lewis, J., Argyropoulos, J.N. and Nielson, K.A. (2000) *Met. Finish.*, **98**, 254.

97 Jung, J. and Perrut, M. (2001) *The Journal of Supercritical Fluids*, **20**, 179.

98 Yeo, S.-D. and Kiran, E. (2005) *The Journal of Supercritical Fluids*, **34**, 287.

99 Akagawa, K. (1999) in *Steam Power Engineering* (ed. S. Ishigai), Cambridge University Press, Cambridge, p. 204.

100 Jessop, P.G. and Subramaniam, B. (2007) *Chemical Reviews*, **107**, 2666.

101 Brinker, C.J. and Scherer, G.W. (1990) *Sol–Gel Science*, Academic Press, Boston, MA.

102 Ayers, J.C. and Watson, E.B. (1991) *Philosophical Transactions of the Royal Society of London, Ser. A*, **335**, 365.

103 Chorley, R.J. (1971) in *Dictionary of Scientific Biography*, vol. III (ed. C.C. Gillispie) Charles Scribner's Sons, New York, p. 586.

104 Daubrée, (1857) *Ann. Mines* [5], **12**, 289.

105 Ipatieff, V.N. (1946) *The Life of a Chemist*, Stanford University Press, Stanford, CA.

106 Friedel, C. and Sarasin, E. (1879) *Bull. Soc. Min.*, **2**, 113.

107 Friedel, C. and Sarasin, E. (1883) *Comptes Rendus de l'Academie des Sciences Paris*, **97**, 290.

108 Morey, G.W. and Niggli, P. (1913) *Journal of the American Chemical Society*, **35**, 1086.

109 Hannay, J.B. (1879–1880) *Proceedings of the Royal Society of London*, **30**, 484.

110 Kuznetsov, V.I. (1973) in *Dictionary of Scientific Biography*, vol. VII (ed. C.C. Gillispie), Charles Scribner's Sons, New York, p. 21.

111 Ipatiev, V. (1906) *J. Russ. Phys. Chem. Soc.*, **38**, 63.

112 Ipatiev, V. (1911) *J. Russ. Phys. Chem. Soc.*, **43**, 1420.

113 Ipatiev, V. and Rutala, O. (1913) *Chemische Berichte*, **46**, 1748.

114 Ipatiev, V. (1906) *J. Russ. Phys. Chem. Soc.*, **38**, 1180.

115 Ipatiev, V. and Zdzitovetsky, B.S. (1907) *J. Russ. Phys. Chem. Soc.*, **39**, 897.

116 Ipatiev, V. (1907) *Chemische Berichte*, **40**, 1270.

117 Ipatiev, V. and Dovgelevich, N. (1911) *J. Russ. Phys. Chem. Soc.*, **43**, 1431.

118 Briner, E. and Boubnoff, N. (1913) *Journal of Chemical Physics*, **11**, 597.

119 Briner, E. and Wroczynski, A. (1909) *Comptes Rendus de l'Academie des Sciences Paris*, **148**, 1518.

120 Briner, E. and Wroczynski, A. (1909) *Comptes Rendus de l'Academie des Sciences Paris*, **149**, 1372.

121 Briner, E. and Wroczynski, A. (1910) *Comptes Rendus de l'Academie des Sciences Paris*, **150**, 1324.

122 Stähler, A. (1914) *Chemische Berichte*, **47**, 909.

123 Patat, F. (1945) *Monatshefte fur Chemie*, 77, 352.

124 Kistiakowsky, G.B. (1928) *Journal of the American Chemical Society*, **50**, 2315.

125 Sutherland, H.S. and Maass, O. (1931) *Canadian Journal of Research* , **5**, 48.

126 Holder, C.H. and Maass, O. (1938) *Canadian Journal of Research* , **16B**, 453.

127 Emschwiller, G. (1936) *Comptes Rendus de l'Academie des Sciences Paris*, **203**, 1070.

128 Toriumi, T. (1946) *J. Soc. Chem. Ind. Jpn.*, **49**, 1.

129 Sargent, D.E. (1949) to E. I. DuPont de Nemours, US Patent 2,462,680.

130 Buckley, G.D. and Ray, N.H. (1951) US Patent 2,550,767.

131 Stevens, H.C. (1966) to Pittsburgh Plate Glass Co., US Patent 3,248,415.

132 Subramaniam, B. and McHugh, M.A. (1986) *Industrial and Engineering Chemistry Process Design and Development*, **25**, 1.

133 Savage, P.E., Gopalan, S., Mizan, T.I., Martino, C.J. and Brock, E.E. (1995) *AICHE Journal*, **41**, 1723.

134 Scholsky, K.M. (1993) *The Journal of Supercritical Fluids*, **6**, 103.

135 Ikariya, T. (1998) *Principles and Developments of Supercritical Fluids Reactions*, CMC, Tokyo.

136 DeSimone, J.M. and Tumas, W. (2003) *Green Chemistry using Liquid and Supercritical Carbon Dioxide*, Oxford University Press, New York.

137 Brunner, G. (2004) *Supercritical Fluids as Solvents and Reaction Media*, Elsevier, Amsterdam.

138 Tamaru, K. (1991) in *Catalytic Ammonia Synthesis* (ed. J.R. Jennings), Plenum Press, New York, p. 1.

139 Haber, F. (1910) *Z. Elektrochem.*, **16**, 244.

140 Hooper, C.W. (1991) in *Catalytic Ammonia Synthesis*, (ed. J.R. Jennings) Plenum Press, New York, p. 253.

141 Shankster, H. (1937) in *Thorpe's Dictionary of Applied Chemistry*, 4th edn, vol. 1, (eds J.F. Thorpe and M.A. Whiteley), Longman's Green, London, p. 326.

142 Huebner, D.W. (1947) in *Thorpe's Dictionary of Applied Chemistry* 4th edn, vol. VIII (ed. M.A. Whiteley), Longman's Green, London, p. 326.

143 English, A., Rovner, J. and Davies, S. (1995) in *Kirk–Othmer Encyclopedia of Chemical Technology*, 4th edn, vol. 16 (ed. M. Howe-Grant), John Wiley & Sons, Inc., New York, p. 537.

144 Lormand, C. (1925) *Industrial and Engineering Chemistry*, **17**, 430.

145 Patart, G. (1925) *Industrial and Engineering Chemistry*, **17**, 859.

146 Asinger, F. (1986) *Methanol: Chemie- und Energierohstoff*, Springer, Berlin.

147 Olah, G.A., Goeppert, A. and Prakash, G.K.S. (2006) *Beyond Oil and Gas: the Methanol Economy*, Wiley-VCH Verlag GmbH, Weinheim.

148 (2008) *Focus Catal.*, (11), 4.

149 McKee, R.H. and Parker, H.H. (1928) *Industrial and Engineering Chemistry*, **20**, 1169.

150 Bandel, G., Böcker, E., Henninger, D. and Petri, H. (1972) in *Chemische Technologie, Band 4, Organische Technologie II* (eds K. Winnacker and L. Küchler), Carl Hanser Verlag, Munich, p. 124.

151 Wiezevich, P.J. and Frolich, P.K. (1934) *Industrial and Engineering Chemistry*, **26**, 267.

152 Carman, F.J. and Chilton, T.H. (1929) US Patent 1,697,106.

153 Bitler, W.P. and James, J.H. (1928) *Chem. Met. Eng.*, **35**, 156.

154 Franz, G. and Sheldon, R.A. (1991) in *Ullmann's Encyclopedia of Industrial Chemistry*, vol. A18 (eds B. Elvers S. Hawkins and G. Schulz), VCH Verlag GmbH, Weinheim, p. 261.

155 Fawcett, E.W., Gibson, R.O., Perrin, M.W., Paton, J.G. and Williams, E.G. (1937) to ICI, British Patent 471,590.

156 Fawcett, E.W., Gibson, R.O. and Perrin, M.W. (1939) to ICI, US Patent 2,153,553.

157 Perrin, M.W., Paton, J.G. and Williams, E.G. (1940) to ICI, US Patent 2,188,465.

158 Kaufman, M. (1963) *The First Century of Plastics*, Plastics Institute, London.

159 Krase, N.W. (1945) to E. I. Du Pont de Nemours, US Patent 2,388,160.

160 Krase, N.W. and Lawrence, A.E. (1946) to E. I. DuPont de Nemours, US Patent 2,396,791.

161 Ehrlich, P. (1965) *J. Polym. Sci., Part A*, **3**, 131.

162 Ehrlich, P. and Fariss, R.H. (1969) *The Journal of Physical Chemistry*, **73**, 1164.

163 Ehrlich, P. (1971) *Journal of Macromolecular Science-Chemistry*, **A5**, 1259.

164 Takahashi, T. and Ehrlich, P. (1982) *Macromolecules*, **15**, 714.

165 Ehrlich, P. and Mortimer, G.A. (1970) *Advances in Polymer Science*, **7**, 386.

166 Brydson, J.A. (1995) *Plastics Materials*, 6th edn, Butterworth Heinemann, Oxford.

167 McCoy, M., Reisch, M.S., Tullo, A.H., Trembley, J.-F. and Voith, M. (2009) *Chemical & Engineering News*, **87**(27), 29.

168 Schmieder, H., Dahmen, N., Schön, J. and Wiegand, G. (1997) in *Chemistry Under Extreme or Non-classical Conditions* (eds R. van Eldik and C.D. Hubbard), John Wiley & Sons, Inc., New York, p. 273.

169 Bestian, H., Friedrich, H.-J. and Horn, O. (1972) in (eds K. Winnacker and L. Küchler), *Chemische Technologie, Band 4, Organische Technologie II*, Carl Hanser Verlag, Munich, p. 1.

170 Kosaric, N., Farkas, A., Sahm, H., Bringer-Meyer, S., Goebel, O. and Mayer, D. (1987) in *Ullmann's Encyclopedia of Industrial Chemistry*, 5th edn, vol. A9, (eds W. Gerhartz, Y.S. Yamamoto, L. Kaudy, J.F. Rounsaville and G. Schulz), VCH Verlag GmbH, Weinheim, p. 587.

171 Asinger, F. (1968) *Mono-olefins Chemistry and Technology*, English Edition, Pergamon Press, Oxford.

172 (2002) *Industrial Lubrication and Tribology*, **54**, 188.

173 DeSimone, J.M., Guan, Z. and Elsbernd, C.S. (1992) *Science*, **257**, 945.

174 Romack, T.J., DeSimone, J.M. and Treat, T.A. (1995) *Macromolecules*, **28**, 8429.

175 Davidson, T.A. and DeSimone, J.M. (1999) in *Chemical Synthesis Using Supercritical Fluids*, (eds P.G. Jessop and W. Leitner), Wiley-VCH Verlag GmbH, Weinheim, p. 297.

176 Buback, M. (1994) in *Supercritical Fluids Fundamentals for Application* (eds E. Kiran and J.M.H.L. Levelt Sengers,), Kluwer, Dordrecht, p. 481.

177 Hanke, A. (1995) *Anwendungsperspektiven von Überkritischen Medien*, DECHEMA, Frankfurt am Main.

178 Jansen, M. and Rehren, C. (1998) European Patent 0 841 314 A1.

179 Poliakoff, M. and Howdle, S. (1995) *Chemistry in Britain*, 118.

180 *Roche Magazine*, (1992) 2.

181 Pickel, K.H. and Steiner, K. (1994) in *3rd International Symposium on Supercritical Fluids*, vol. 3, International Society for the Advancement of Supercritical Fluids, Strasbourg, p. 25.

182 Tacke, T., Wieland, S. and Panster, P. (1996) *Process Technol. Proc.* **12**, 17.

183 Tacke, T. (1995) *Chemie–Anlagen + Verfahren*, **28**(11), 48.

184 Tacke, T., Wieland, S. and Panster, P. (1997) in 4th International Symposium on Supercritical Fluids, Sendai, Japan p. 511.

185 Tacke, T., Wieland, S., Panster, P., Bankmann, M., Brand, R. and Mägerlein, H. (1998) US Patent 5,734,070.

186 Poliakoff, M., Swan, T.M., Tacke, T., Hitzler, M.G., Ross, S.K. and Wieland, S. (1997) Patent, WO 97138955.

187 Licence, P. and Poliakoff, M. (2005) in *Multiphase Homogeneous Catalysis* (eds B. Cornils, W.A. Herrmann, I.T. Horváth, W. Leitner, S. Mecking, H. Olivier-Bourbigou and D. Vogt), Wiley-VCH Verlag GmbH, Weinheim 7, p. 734.

188 Licence, P., Ke, J., Sokolova, M., Ross, S.K. and Poliakoff, M. (2003) *Green Chemistry*, **5**, 99.

189 Farncomb, R.E. and Nauflett, G.W. (1997) *Waste Management*, **17**, 123.

190 Farncomb, R.E. and Nauflett, G.W. (1998) presented at the International ICT Conference Karlsruhe 30 June–3 July.

191 Stern, A.G., Kenar, J.A., Trivedi, N.J., Koppes, W.M., Farncomb, R.E., Turner, S., Bomberger, D.C., Penwell, P., Manser, G.E., Spas, S.E. and Nahlovsky, B.D. (2004) in *Defense Industries: Science and Technology Related to Security*, (eds P.C. Branco, H. Schubert and J. Campos), Kluwer, Dordrecht, p. 141.

192 Angus, S., Armstrong, B. and de Reuck, K.M. (1976) International Theromodynamic Tables of the Fluid State: Carbon Dioxide, IUPAC, Pergamon Press, Oxford.

193 Bridgeman, P.W. (1914) *Physical Review*, **3**, 126.

194 Kainz, S., Koch, D., Baumann, W. and Leitner, W. (1997) *Angewandte Chemie (International Edition in English)*, **36**, 1628.

195 Thurston, R.H. (1895) *A History of the Growth of the Steam Engine*, 5th edn, Kegan, Paul, Trench, Trübner, London.

196 Figuier, L. and Gautier, É. (1892) *Année Sci. Ind.*, **35**, 88.

197 Andrews, T. (1889) *The Scientific Papers of the Late Thomas Andrews with a Memoir by P G Tait and A. Crum Brown*, Macmillan, London.

198 Hardie, D.W.F. and Pratt, J.D. (1966) *A History of the Modern British Chemical Industry*, Pergamon Press, Oxford.

199 Wilson, G.D. (1994) in *The Development of Plastics* (eds S.T. I. Mossman and P.J. T. Morris) Royal Society of Chemistry, Cambridge, p. 70.

2
High-pressure Methods and Equipment

Nils Theyssen, Katherine Scovell, and Martyn Poliakoff

2.1
Introduction

Carrying out reactions at pressures far above those conventionally used, as is necessary in practically all studies that use supercritical fluids (SCFs), requires a specialized knowledge of how experiments should be performed in such a way that they are both meaningful (reproducible) and inherently safe. To reach this goal, the equipment used and the methods applied differ substantially from those usually used by experimentalists. Hence the aim of this chapter is to give guidelines for beginners in this field, rather than trying to formulate a comprehensive overview of high-pressure technology, which would, on its own, easily fill a complete textbook. Nevertheless, it is hoped that some sections of the chapter will contribute to the knowledge of those chemists who have been working in the field for many years, either as a recapitulation or as a first introduction to methods that have not been of importance to them in the past, but may be of interest in upcoming projects.

References are given in cases where more detailed information might be helpful. In this regard, the authors would like to apologize for the frequent citation of web pages, which may be edited regularly. However, this rather unusual practice is only done for cases in which the Internet turned out to be a superior information source to that which can be found in scientific journals (English or German; searched using *ISI Web of Knowledge*). Hence the reasoning behind the decision to include these references originates not from the preferences of the authors but from the technical nature of this particular topic. It should be emphasized that in showing equipment or citing web pages from one manufacturer or another, no preference for a particular product is being expressed. It should also be noted that although the information has been collected to the best of our knowledge, no responsibility is taken by the authors for the correctness or safety of any information given in the present chapter. Safety regulations for installation and handling of high-pressure equipment can vary from country to country and even from institution to institution, and it is essential for the experimentalist to become familiar with and obey these regulations.

Handbook of Green Chemistry, Volume 4: Supercritical Solvents. Edited by Walter Leitner and Philip G. Jessop

ISBN: 978-3-527-32590-0

2.2 Infrastructure for High-pressure Experiments

When a reaction is carried out under supercritical conditions, high pressures, sometimes in excess of 40 MPa, are required. The potential danger of such conditions means that full safety precautions are essential when carrying out experiments.

This section outlines the instrumentation in a high-pressure laboratory (in process and production engineering, high pressure is generally understood as 10–400 MPa) satisfying such safety demands. As a result of the increased safety requirements, the potential risks, and hence increased costs, must be weighed up carefully against the benefit of a particular experiment in a careful manner. Also, there may be local legislature or codes of practice which need to be adhered to.

2.2.1 Location

When studying high-pressure reactions, there must be a guarantee that, in case of a leakage or even the rupture of the pressure vessels, there is no chance of injury. High-pressure cells built of ferroconcrete (wall thickness about 30 cm), sand-filled steel plates with a thickness of 6–7 mm on each side, or massive steel plates with a wall-thickness of about 10–20 mm, which are all specially designed for use with high pressures, can provide a safe environment. Additionally, each pressure chamber should be equipped with a bursting wall, ensuring a fast pressure discharge, in order to reduce shock waves and to restrict the strain from the protective walls [1–3]. It is also worth considering fully the possible impact of noise and the protection that may be necessary (Table 2.1).

Inspection windows made of pressure- and impact-resistant glass should be installed in the cells to allow for continuous observation from a safe position.

Table 2.1 Relationship between the peak overpressure of an explosion, the sound level, and physical effects [4].

Peak over-pressure (kPa)	Sound level (dB)	Physical effects
0.32	144	Average human threshold of pain
1.38	157	Window damage
6.89	170	Rocket engine, knocks personnel over
34.5	185	Damage to reinforced concrete buildings, threshold for eardrum rupture
100 (1 bar)	194	Threshold for lung damage
130		Damage to reinforced concrete blast-resistant structures
240		Threshold for fatalities
450		Fatalities 99% probable

Shut-off devices for electricity and gas supply must be installed in the observation area. To avoid the build-up of an explosive or toxic atmosphere in case of a leakage, each pressure box should be equipped with an extractor which can extract gases via both an upper (light gases) and a lower (heavy gases or vapors) port. The power should at least guarantee a 20-fold air renewal per hour. The installation of a gas warning system together with automatic shut-off devices for gas supply is highly recommended. Moreover, it is recommended to equip high-pressure chambers with a data acquisition system so that temperature and pressure profiles can easily be obtained. In addition to the monitoring of the reaction data, an easy and effective control for compliance with the maximum operating parameters and the determination of the operating time of a given reactor is possible simultaneously.

Depending on the individual experiment and the specific legislation regulations of the individual country or state (in Germany a risk assessment has to be conducted in advance; this has to be done by the person who is responsible for the operational safety in this particular area), small-scale high-pressure experiments can also be performed in the fumehood of a normal chemical laboratory. In many cases, this will be the preferred setup for exploratory research and fundamental investigations. However, if this is done, the following conditions must be met:

- The components of the system that are under pressure must be checked for corrosion before an experiment.
- No corrosive chemicals should be used in these experiments.
- The reactor is equipped with a pressure relief valve, a rupture disk, or jackscrews (which elongate and open the vessel in case of overpressure).
- The experiments cannot generate sudden and potentially explosive increases in temperature or pressure.
- The experiment does not exceed the maximum safe operating parameters (especially temperature and pressure) of the system.
- A blast shield should be installed in front of the reactor.

It is recommended that specialized high-pressure equipment, meeting the criteria described above, is used for all scientific experiments with compressed gases. It must also be considered that full knowledge of the thermodynamic and kinetic properties of novel reactions are not always available in exploratory research. Literature data should be acquired from related processes to estimate safe operation parameters in conjunction with the above criteria when planning such experiments using high pressure.

2.2.2
Gas Supply, Compression, and Purification

When using compressed gases, it is important to consider the supply of the gas, its compression and any purification that may be required. The following components can be used in combination to provide an efficient large-scale gas supply for use with high-pressure vessels:

1. gas discharging devices with automatic switching for redundant gas cylinders, which should be easily accessible by an outer entrance (preferably with a double door) to facilitate the exchange of the gas cylinders
2. diaphragm or reciprocating compressors (electrically or pneumatically driven) to ensure maximum utilization of the contents gas cylinders
3. double contact manometers to maintain the pressure level automatically
4. buffer vessels connected to ring line systems to keep the pressure more constant.

The initial expense of the central unit described above is, at least for regular usage, compensated for by its benefits. It can reduce the number of cylinders required, which not only results in a safer system, but can also reduce the costs of leasing the cylinders, which is substantial. The standard charge per day and cylinder is about US$1. Moreover, another safety feature of such a system is that it allows the convenient installation of a gas warning device and automatic shut-off valve.

However, for beginners in the field, who usually do not want to invest several tens of thousands of dollars before performing the first experiments, a single SCF gas cylinder and a screw pump or a high-performance liquid chromatographic (HPLC) pump with a coolable pump head are sufficient.

The purification of the gases is also an important factor to consider. Impurities, such as moisture, oxygen, carbon dioxide, carbon monoxide, hydrocarbons, and sulfur compounds, can usually be removed by commercially available gas purifying cartridges. Many of these are fully functional at room temperature; however, some require heating to enable them to work. It should also be considered, in the context of SCFs, that many purifiers are only rated for pressures $\leq$1 MPa.

2.3 High-pressure Reactors

2.3.1 Materials of Construction for High-pressure Reactors

2.3.1.1 Metal Components

A synthetic chemist is normally used to performing reactions in borosilicate glass apparatus. However, such apparatus is impractical for high-pressure reactions due to its low tensile strength (6 N mm^{-2}). Hence metal vessels are required [5]. Metal is more sensitive to chemical attack than glass, hence corrosion needs to be considered [6]. For this reason, a glass liner can be inserted into a metal reactor to protect the metal during a reaction in the gas–liquid phase. However, it is difficult to line every part of a metal system. Such preventive measures usually fail when using SCF technology, as corrosive compounds can dissolve in the SCF phase, and they reach the most distant regions of a reactor. Therefore, one has to analyze whether the pressure reactor, seals, and associated accessories (manometers, valves, capillaries etc.) comply with the expected chemical (corrosion) and physical (pressure, temperature) strains. Details concerning material instabilities are listed in corrosion tables, which can usually be obtained from steel distributors and the chemical literature (see the references given in Table 2.2). If data on a corrosive chemical cannot be found from a manufacturer, it is best to test the vessel material by exposing it to the substrate for 24 h. The corrosion can then be determined gravimetrically and evaluated by a technical expert before its suitability for the task is decided.

Components made from metals such as stainless steel can also suffer from so-called "stress corrosion cracking." This happens when corrosion occurs preferentially along the boundaries between the grains in the metal. The result is that the corrosion can penetrate deep into the metal very rapidly, often in the form of pinholes which traverse the whole thickness of a pipe or vessel. The speed of stress corrosion cracking makes it potentially more dangerous than other forms of corrosion since it can lead to complete and unexpected failure of components. With stainless steel, it occurs particularly in the presence of halogens and at high temperatures. Therefore, great care should be taken when working with free halogens or with compounds such as halogenated organic compounds which could decompose to halogens or hydrogen halides.

If corrosive gases, such as ammonia, boron trifluoride or chloride, vinyl chloride or bromide and also other halocarbons, di- or trimethylamine, ethylene oxide, phosgene, hydrogen sulfide, nitrogen monoxide, or nitrogen dioxide are used, or potentially formed in the reaction, the use of brass valves or copper capillaries is not appropriate. In these cases, instruments made of steel, and possibly seals made of lead, PTFE, or gold, should be used.

When selecting materials for high-pressure components, special attention should be paid to high strength, good isotropy, homogeneity, toughness, and quality. Homogeneity is achieved by the smelting method and isotropy is achieved by forging the slugs as much as possible from all sides. An analysis of a number of types of steel is shown in Table 2.2.

2.3.1.2 **Sealing Materials**

Nowadays a wide variety of sealing materials are available, and all of them have advantages and disadvantages. In general, the vessel will be leak free if the seal pressure rating is higher than the inner pressure of the vessel. These are described in Table 2.3.

The simple O-ring design (Figure 2.1a) is suitable for pressures up to 30 MPa. Similar seal designs with an optimized support for the O-ring (or a soft metal flat gasket) allow significantly higher pressure ratings (several hundred MPa, depending on the material; Figure 2.1b).

In high-pressure chemical plants, metal seals are usually necessary. Here, flat sealing rings and metallic O-rings are seldom used. Details of metal seal types which are used for industrial high-pressure processes can be found elsewhere [20]. In any case, the sealing surfaces of the autoclave have to be cleaned (metal-on-metal-seals have additionally to be oiled slightly) before the reactor is closed. The question of cleaning is particularly important in reaction chemistry where particles of product (e.g. polymer) can easily foul sealing surfaces or screw threads.

Table 2.2 Overview of selected metal materials which are accredited for high-pressure reactors.

Material number (AISI) Symbol *New designation*	Density ($g\,ml^{-1}$)	Maximum application temperature (°C)	Surface polishing	Machinability	Tensile strength, R_m ($N\,mm^{-2}$) Yield point, $Rp_{0.2}$ ($N\,mm^{-2}$)	
1.4306 (304 L)	7.90	200	++	+	460–680 (640)	
X 12 Cr 13	**Resistant against:** nitric acid (up to high temperatures and concentrations) and many organic acids, intergranular attack				20 °C	170
X10Cr13					100 °C	147
	Unstable against: chloride-containing media, sulfuric acid				200 °C	118
1.4429 (316 LN)	7.98	400	+	0	580–800 (680)	
X 2 CrNiMoN 17 13 3	**Resistant against:** intercrystalline corrosion (increased chemical stability)				20 °C	275
X2CrNiMoN17-13-3	**Unstable against:** seawater				200 °C	175
					300 °C	140
1.4571 (316 Ti)	7.98	400	–	+	500–700 (600)	
X 6 CrNiMoTi 17 12 2 (V4A)	**Resistant against:** most organic acids, phosphoric acid, dilute sulfuric acid, sodium hydroxide solution, ammonia, steam				20 °C	215
X6CrNiMoTi17-12-2					100 °C	185
	Unstable against: halogen-containing media, hot alkali metal hydroxides, concentrated nitric acid, concentrated sulfuric acid, hot hydrogen sulfide, 30% hydrogen peroxide				200 °C	167
					300 °C	145
					500 °C	120
2.4617 [7]	9.22	500	–	–	745–1000	
NiMo28 (Hastelloy B-2)	Very good resistance in reducing media, e.g. hydrochloric, sulfuric, hydrofluoric and phosphoric acid. Well resistant against chloride-induced stress corrosion cracking				20 °C	407
					204 °C	361
	Unstable against: oxidizing compounds, ferric or cupric salts				316 °C	336
					427 °C	319

2.4600 [8]	9.22	700	–	–	885	
NiMo29Cr (Hastelloy B-3)	Same excellent resistance to hydrochloric acid and other strongly reducing chemicals as B-2 alloy, but with significantly better thermal stability at higher temperature, fabricability and stress corrosion cracking resistance				20 °C	400
					204 °C	330
					427 °C	285
					649 °C	290
2.4610 [9]	8.64	550	–	–	700–900	
NiMo16Cr16Ti (Hastelloy C-4)	**High resistance to corrosion**. Virtually the same corrosion resistance as alloy C-276. Good resistance against highly oxidizing salts, pitting, stress corrosion cracking and intercrystalline corrosion. For pressure vessels with high requirements concerning durability in the range between –196 and 400 °C **Unstable against:** concentrated hydrochloric acid				20 °C	335
					93 °C	301
					204 °C	264
					316 °C	247
					427 °C	236
					538 °C	205
2.4602 [10]	8,69	700	–	–	800	
NiCr21Mo14W (Hastelloy C-22)	**High resistance to corrosion**. Better overall corrosion resistance in oxidizing corrosives than C-4, C-276 and 625 alloys. Outstanding resistance to localized corrosion and excellent resistance to stress corrosion cracking				20 °C	372
					204 °C	283
					427 °C	241
					760 °C	214
2.4819 [11]	8.89	550	–	–	700–1000	
NiMo16Cr15W (Hastelloy C-276)	**High resistance to corrosion**. Versatile, corrosion-resistant alloy. Very good resistance to reducing and mildly oxidizing corrosives. Excellent stress corrosion cracking resistance with very good resistance to localized attack **Unstable against:** alkali metal hydroxides at higher temperatures				20 °C	365
					204 °C	263
					316 °C	235
					427 °C	235
					538 °C	226
2.4675 [12]	8.50	650	–	–	740	
NiCr23Mo16Cu (Hastelloy C-2000)	**High resistance to corrosion.** Most versatile, corrosion-resistant alloy with excellent resistance to uniform corrosion in oxidizing or reducing environments. Excellent resistance to stress corrosion cracking and superior resistance to localized corrosion compared with C-276 alloy				20 °C	350
					204 °C	283
					427 °C	216
					538 °C	214
					649 °C	209

(Continued)

Table 2.2 *(Continued)*

Material number (AISI) Symbol *New designation*	Density ($g\,ml^{-1}$)	Maximum application temperature (°C)	Surface polishing	Machinability	Tensile strength, R_m ($N\,mm^{-2}$) / Yield point, $Rp_{0.2}$ ($N\,mm^{-2}$)	
2.4643 [13]	8.22	550	–	–	700	
NiCr33Mo8 (Hastelloy G-35)	**High resistance to corrosion.** Excellent resistance to corrosion in highly oxidizing media and acidic chloride environments. Outstanding corrosion resistance to oxidizing acids, alkalis, and chloride-containing media. Especially suited for oxidations in scH_2O with chloride-containing systems				20 °C	330
					93 °C	313
					204 °C	248
					427 °C	215
					649 °C	184
2.4816 [14, 15]	8.43	550	+	–	500–750	
NiCr15Fe (Inconel 600)	Very good resistance in many oxidizing and reducing media even at high temperatures **Unstable against:** strong oxidizing solutions such as hot, concentrated nitric acid, less hydroxide resistant than Inconel 625				20 °C	240
					200 °C	230
					427 °C	203
					649 °C	183
2.4856 [11]	8.44	850	–	–	690–900	
NiCr22Mo9Nb (Inconel 625)	**Resistant against:** chloride ions, stress corrosion cracking, pitting, crevice and high-temperature corrosion, and also carburization. Durable in oxidizing media. Outstanding strength and toughness in the temperature range from cryogenic to 800 °C **Unstable against:** alkali metal hydroxides at higher temperatures				20 °C	496
					204 °C	429
					760 °C	381
					871 °C	241
					982 °C	75
3.7065 [15]	4.51	400	+	++	370–550	
Titanium (grade 4)	**Resistant against:** chloride-containing or oxidizing media. Stable against wet chlorine gas, chlorine dioxide, nitric and other acids **Unstable against:** sulfuric acid, sodium hydroxide, and (most notably) hydrofluoric acid				20 °C	550
					205 °C	250
					315 °C	165
					425 °C	145

Table 2.3 Overview of sealing materials used for high-pressure purposes [16].

Sealing material	Price[a]	Low temperature	High temperature	Compression set[b]	Wear/abrasion[b]
Perbunane (nitrile–butadiene rubber, NBR) Copolymer of butadiene and acrylonitrile	A	−35 °C	120 °C	2	2
	Description: excellent resistance to petroleum-based oils and fuels, water and alcohols. Nitrile also has good resistance to acids and bases, except those with a strong oxidizing effect **Limitations:** avoid highly polar solvents (acetone, methyl ethyl ketone, etc.) and direct exposure to ozone and sunlight				scCO_2 2 scH_2O 5 scNH_3 5 scMeOH 5 scC_3H_8 2
Viton (fluorinated rubber, FKM or FPM) Copolymer of vinylidene fluoride and hexafluoropropylene	D	−25 °C	205 °C	1	2
	Description: excellent resistance to petroleum products and solvents. Very good high-temperature performance **Limitations:** avoid polar solvents, amines, anhydrous ammonia, hydrazine, and hot acids				scCO_2 5 scH_2O 5 scNH_3 5 scMeOH 5 scC_3H_8 1
Teflon (polytetrafluoroethylene, PTFE)	D	−155 °C	230 °C	5	2
	Description: probably the most frequently used sealing material. The gas permeability is low. Has the tendency to flow at higher temperatures (>100 °C). A variety of composite material exists which fulfill the demands for different applications. Due to the low elasticity, processes with cooling intervals are critical **Resistant against:** nearly all organic and inorganic chemicals **Unstable against:** elementary fluorine under pressure or at high temperatures, fluorohalogen compounds and alkali metal melts				scCO_2 2 scH_2O 5 scNH_3 1 scMeOH 4 scC_3H_8 1

(Continued)

Table 2.3 (*Continued*)

Sealing material	Price[a]	Low temperature	High temperature	Compression set[b]	Wear/abrasion[b]	
PEEK (polyether-ether-ketone)	D	−200 °C	250 °C	5	2	
	Description: not as inert as Teflon but much lower tendency to flow at higher temperatures (hardly any softening). Due to the low elasticity, processes with cooling intervals are critical **Resistant against:** most chemicals **Unstable against:** concentrated sulfuric and nitric acid, some halocarbons (e.g. dichloromethane)				$scCO_2$ scH_2O $scNH_3$ scMeOH scC_3H_8	1 5 1 2 1
Kalrez, Simriz, Chemraz, Parafluor (perfluorinated rubber) Co-polymer of tetrafluoroethylene and perfluorovinyl ether.	H	Up to −20 °C	Up to 325 °C	2	2	
	Description: excellent resistance to almost all chemicals. Excellent outgassing performance in vacuum environments. Probably the best, but also the most expensive elastomer. A variety of composite materials for special applications exists **Limitations:** avoid low molecular weight, fully halogenated fluids and molten alkali metals. Especially strong oxidizing acids may cause some swelling. Helium permeability is slightly higher than that of fluoroelastomer compounds. Specific Simriz compounds provide better low-temperature performance and amine resistance				$scCO_2$ scH_2O $scNH_3$ scMeOH scC_3H_8	1 5 1 1 1
Self-energizing metal O-rings [17–19]	J	−250 °C	1100 °C	1	2	
	Description: the application of metal seals is expedient if a combination of high temperatures, high pressures, and/or swelling solvents such as ethers, halocarbons, aldehydes, and ketones are used. Their possible catalytic activity is detrimental **Design:** the seal is a spring-actuated, pressure-assisted sealing device consisting of an adapted metal (or polymer) jacket partially encapsulating a corrosion-resistant metal spring energizer. The spring forces the jacket lips against the gland walls, a process that is assisted by the system pressure				$scCO_2$ scH_2O $scNH_3$ scMeOH scC_3H_8	1 1 1 1 1

[a] A = low; J = high.
[b] 1 = Recommended; 2 = satisfactory; 3 = poor; 4 = marginal; 5 = not recommended.

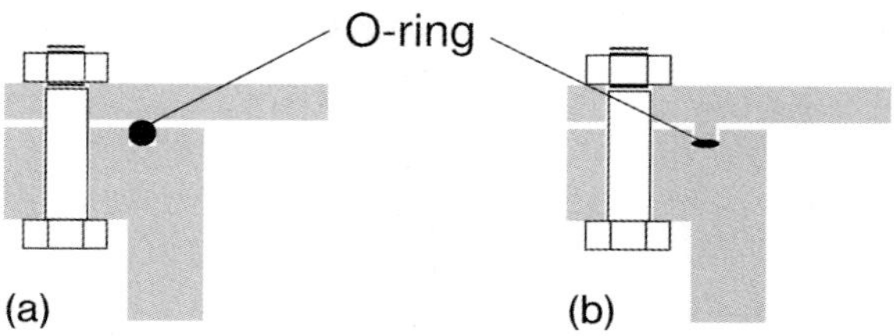

Figure 2.1 Schematic diagrams of different O-ring seal design.

2.3.2 Reactor Design

2.3.2.1 General Considerations

In addition to ascertaining a suitable reactor material, the following points should be considered when designing a high-pressure reactor:

- The maximum pressure at which the reactor will be operated and, equally important, the maximum pressure that could be reached in the event of malfunction.
- The temperature range at which the reactor will be run, including the method of heating and control, the possible rate of change, and the possibility of heat removal.
- The volume and shape of the inner chamber.
- The type of stirrer to be used (design and shape).
- The operating mode of the system (batch or continuous).
- Any safety precautions or features required in the system, for example, safety valve, rupture disk, jackscrews, limit value monitoring, and cut-off.
- Will it be necessary to open the vessel rapidly, and how high a priority is easy cleaning?

A standard high-pressure reactor typically has a cylindrical design, but there are often modifications, some more significant than others. From an engineering point of view, the calculation of the necessary wall thickness of these reactors is a fundamental issue to consider. This value depends not only on the operating temperature and pressure, material (individual yield point), and inner diameter, but also on the length, the kind of base (planar or curved), and any additional fixtures on the reactor. Hence, if a new reactor is being constructed, it is advisable to use approved computer software to calculate the appropriate specification for legal and safety requirements. Often insurance companies will insist on this.

The closure head of an autoclave is usually best fixed by a screwed flanged joint, which is the optimal compromise between easy handling and robustness in sealing properties. Other connections to the inner part of a reactor are preferentially made using *National Pipe Thread* fittings ("NPT fittings"; see Section 2.4.1) and a sealing material (e.g. Teflon band).

2.3.2.2 Pressure Vessels for Batch Processing

Stirred Tank Reactors (STRs) When conducting an experiment in batch mode, reactants are added to the vessel (sometimes at different rates) and are held under controlled conditions until the desired end-point for the reaction is reached. These reactions are often lengthy and, at the end, product isolation is required. This can be achieved by a variety of separation processes such as filtration, extraction, distillation, and adsorption.

A simple but well-established design for a batch reactor consists of a main body, a closure head, two valves, a manometer, one or more temperature probes and magnetic stirring. Such a setup is often heated and stirred with a hot-plate stirrer using metal top frames for better heat input (see Figure 2.3c). The aspect ratio of the tank (height/diameter, H/D) is normally a compromise between mixing efficiency (best if $H/D = 1$) and heat transfer capability (best if $H/D \rightarrow \infty$). If H/D is >1.5, a second stirrer is recommended [21]. The basic design of the STR can be modified and extended to allow for visual inspection, dosing, online sampling, or inline spectroscopy. As these extensions can be applied also to other reactor designs, they will be discussed in Sections 2.4.52.4.9.

Variable-volume View Cells When SFCs are used as a medium, the phase behavior (single- or multi-phase) often plays a central role in terms of selectivity and/or overall efficiency. A crucial parameter to measure, so as to learn about the phase behavior, is the fluid density, which can be changed *in situ* by a variable-volume view cell [22–24]. Two thick-walled, large-aperture sapphire windows (one stationary and one mobile) allow the experimentalist to observe the phase behavior accurately. The thickness of the window also helps to avoid a temperature differential to the metal cell, which may cause unwanted condensation or reflux. In order to produce an optically clear image, backlighting with an adjustable and diffuse light source is required. In addition, it is very convenient to enhance the observation by using a camera to record the activity in the cell. The core of the phase analyzer system is the high-pressure variable-volume view cell. It consists of three major parts:

1. A cell body, which contains a stirrer, an electric heating system, service ports for filling, a pressure sensor, and temperature probes.
2. A holder for the static sapphire window.
3. A hydraulic unit with a sliding carriage, which allows the position of the mobile sapphire window to be changed to increase or reduce the cell volume in a controlled manner. By attaching a capacitive position encoder to the moving window, a continuous readout of the actual volume can be obtained after calibration.

Reactors for High-pressure NMR Investigations There are various designs for coupling inline spectroscopic techniques with high-pressure reactors. Owing to the high magnetic fields used in NMR spectroscopy, a special cell design is required in this case. Accordingly, a large variety of different high-pressure cells for NMR investigations have been developed, some of which can be used readily with standard spectrometers and the more common SCFs, such as supercritical carbon dioxide ($scCO_2$) (see below). However, there are also cells available with maximum operating conditions of up to 10 GPa (diamond anvil cell, DAC) [25] or 700 °C (titanium alloy

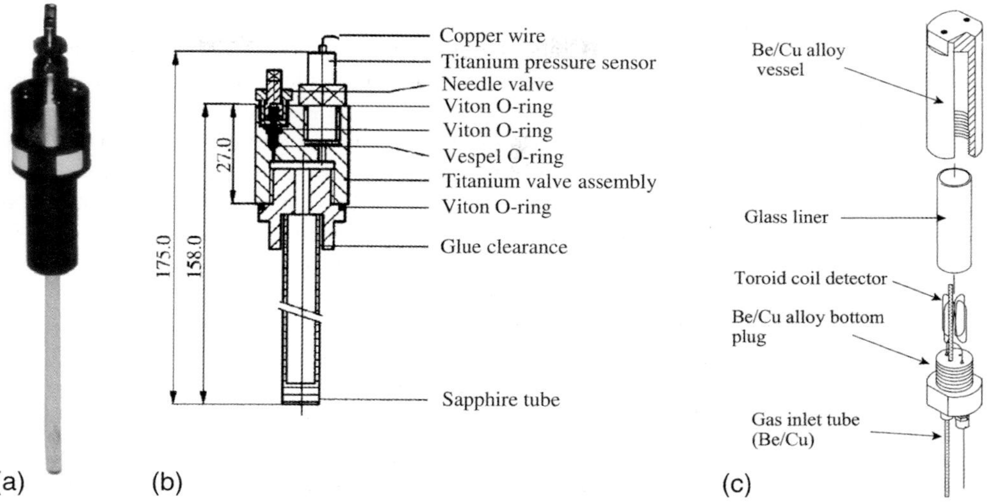

Figure 2.2 (a) Photograph and schematic diagram of an NMR high-pressure cell and (b) a diagram of an NMR high-pressure probe. (a) Reproduced with permission from [32], copyright John Wiley & Sons, Ltd. (b) Reproduced with permission from [29], copyright Elsevier.

probe) [26]. In principle, two basic approaches, referred to as the *high-pressure probe technique* and the *high-pressure cell technique*, have been utilized in high-pressure NMR studies (Figure 2.2) [27, 28].

A complete high-pressure probe contains both the sample tube and an appropriately designed RF circuit, which is used for the irradiation of the resonance frequency. The most common materials used to make these are copper–beryllium and titanium alloys. In general, they have higher sensitivity than high-pressure cells owing to the high filling factor of the coils, especially if a toroidal design is employed where the high-pressure vessel itself acts as a toroidal cavity [29]. Additionally, this approach allows stirring/agitation [30], which is difficult when using the high-pressure cell technique. The main disadvantage of high-pressure probes is the technical and experimental complexity associated with the manufacture and implementation, such as wide-bore magnets and probe tuning instrumentation. These factors preclude the use of such probes for a routine NMR investigation in most high-pressure laboratories.

In contrast, high-pressure cells fit within the probeheads of standard commercial NMR spectrometers. The most frequently used design is a single-crystal sapphire NMR cell with titanium alloy heads which was first developed by Roe in 1985 [31]. Although the temperature and pressure ranges are limited to about 100 °C and 25 MPa, their capability to work in modern spectrometers with higher frequencies (500 MHz and above) and in spinning mode (to average out inhomogeneities of the magnetic field) offers state-of-the-art performance in terms of resolution. Assemblies in which needle valves and pressure sensors (made from materials such as titanium with low magnetic susceptibilities) are incorporated, can only be operated in the non-spinning mode [32]. However, it is possible to provide easy filling and depressurization, and continuous pressure measurements are also possible, allowing further information to be gained on chemical kinetics (e.g. consumption of a gaseous reactant or the determination of the activation volumes). Apart from sapphire, high-strength organic polymers such as PEEK have been used for cell construction [33]. Pioneering work in this field was done by Merbach and co-workers using polyimide [34]. A new setup was developed which is dedicated to the construction of high-pressure probes to fit in commercial narrow-bore magnets [35].

2.3.2.3 Continuous Flow Reactors

The scale-up of SCF reactions will in most cases involve continuous flow operation, as simple scale-up of batch processes in high-pressure technology is problematic due to limitations in wall thickness, in addition to the weight and cost of the equipment. In fact, the design of flow systems holds many opportunities even for laboratory-scale operation in modern synthetic chemistry. Hence, when the reactions are rapid enough (minutes or less) and the required mass transport can be achieved in a reliable manner, continuous operation is preferred. In continuous flow processing, substrates are fed continuously into the reactor. The product mixture, containing the desired product and by- and side-products in addition to unconverted starting material, leaves the reactor at the same rate as the input stream, thus maintaining a constant flow under steady-state conditions [36].

SFCs, as a mobile phase, can be used with a reaction setup which resembles a typical gas-phase process. Owing to the solvent properties of an SCF, this favorable situation can be reached at temperatures that are far beyond the boiling points of the individual substrates. Thus, highly selective transformations are possible under these mild conditions, in a quasi-solvent-free, continuous flow mode [37, 38]. In addition, carrying out a reaction in continuous flow is beneficial when higher production capacities are needed. Other advantages of continuous processing include:

- A small flow reactor has a much larger surface area-to-volume ratio and will permit much better heat control. The heat removal capability of a cylindrical reactors is related to the wall area πHD (H = height, D = diameter), whereas the heat generation area is proportional to the reactor volume $\pi HD^2/4$. The ratio of the two values ($4/D$) shows that halving the diameter increases the maximum reaction heat removal by a factor of two. This point is of particular importance when (highly) exothermic reactions, such as hydrogenation or oxidation, are carried out.
- The mixing efficiency often decreases significantly in larger vessels, which may result in a loss of selectivity for a batch-wise scaling up, because the reduced mass transport counteracts the chemical equilibria and reduced heat transport creating localized hot spots.
- The quality of products from different runs is normally more uniform.
- Continuous reactors simplify reaction optimization because they permit the use of online monitoring and, hence, real-time optimization. Spectroscopic monitoring (see Section 2.4.9) is particularly useful in this context.

- The reactor can be used in a multi-pass mode with the process stream being recycled through the reactor to increase the yield. In some cases, products and by- and side-products can be separated after each pass, which permits both longer reaction times and reduced by-product formation arising from unwanted consecutive reactions.

Currently, three different types of reactor are available for performing continuous flow reactions using SCFs:

1. continuously stirred tank reactors (CSTRs) [21, 39]
2. plug flow reactors (PFRs) [39, 40]
3. the more recently developed microstructured reactors (MSRs) [41, 42].

Which reactor type should be use for which type of reaction depends on many factors [43, 44]. A proper selection of reactor type should be based on the optimum residence time distribution and the rate of reaction, especially if the reactions are other than first order [45]. Characterisations of the basic reactor types are summarized in Table 2.4. In Section 2.8, we will describe a fully automated apparatus for continuous flow operation in more detail.

2.4 Auxiliary Equipment and Handling

2.4.1 Tubes and Fittings

In general, all of the materials mentioned in Section 2.3.1.1 (and many more) are suitable for the manufacture of high-pressure tubes. A broad range of different sized tubing is commercially available. Rolled tubes are normally made of a softer material than cold drawn tubing and should be avoided where possible.

Tubes can be cut with either a hacksaw or a tube cutter. Tube cutters throw a burr into the ID of the tubing whereas a hacksaw will burr both the ID and OD of the tube. All burrs must be removed after cutting a tube with a deburring tool and a smooth file. Any shards of metal must be removed from the end of the tubing as they can cause fittings to leak, or damage other components in a system. A detailed manual for tubing and tube fitting handling can be found elsewhere [55, 56].

If (a) the seamless tubes are operated mainly under steady stress conditions, (b) the ratio of diameters corresponds to $\mathrm{OD}/\mathrm{ID} \leq 1.7$ and (c) the operating temperature is below 120 °C, the required wall thickness for a given pressure load can be estimated according the following approximation [57]:

$$s \approx \frac{\mathrm{OD} \times p \times 2.25}{R_\mathrm{m}} \tag{2.1}$$

where s = approximately required wall thickness in mm, OD = outer diameter in mm, p = pressure in MPa, and R_m = tensile strength in $\mathrm{N\,mm^{-2}}$ (see Table 2.2).

Table 2.4 Basic reactor types for SCF applications.

Reactor type		Applications (with special emphasis for SCF technology)
STR	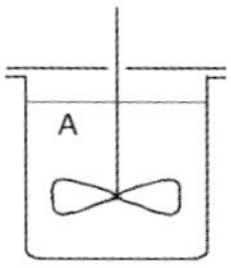	• Single- and multi-phase reactions [46] • Useful for preliminary screening • Easy use of *in situ* spectroscopy [28, 47, 48] and calorimetric measurements [49]
CSTR	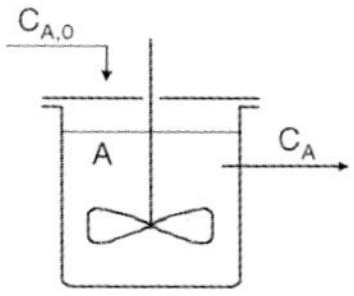	• Homogeneous reactions • Multiphase catalysis with a stationary, non-volatile liquid [50, 51]
PFR	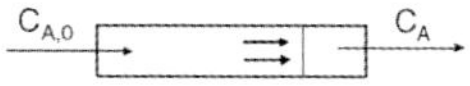	• Homogeneous reactions • Fixed-bed technology with a homogeneous SCF-phase [38, 52, 53]
CSTR cascade	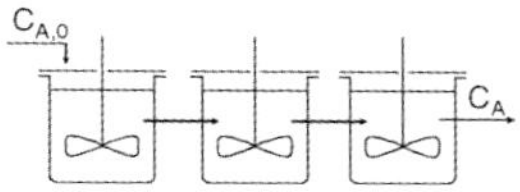	• Optimization of conversion and yield
MSR	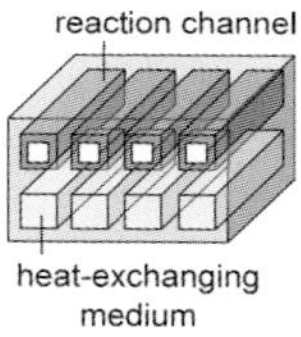	• Fast reactions which need to be quenched rapidly (e.g. radical propagation) [54] • Highly exothermic or endothermic reactions • Reactions in the explosive regime • Fixed-bed technology

The following considerations are important in installation procedures:

- The metal tubing material must be softer than or at least of the same hardness as the fitting material. For example, stainless-steel tubing should not be used with brass fittings.
- Wall thickness should always be checked against the suggested minimum and maximum wall thickness limitations.
- Surface finish is very important for proper sealing. Tubing with any kind of depression, scratch, raised portion, or other surface defect or may cause a seal to fail.
- Tubing that has a distorted cross-section will not fit through fitting nuts easily. Ferrules and tubing should never be forced into the fitting.

In order to perform tubing–vessel connections, the simplest possibility is the use of NPT fittings. The sealing is achieved by screwing the taper of the connector (male) and the vessel (female) together. Additionally, a sealant (e.g. Teflon band, two wraps) is placed between the threads. The tube is fixed to the connector by a bite-type fitting joint. Such a connection is suitable for pressures up to 50 MPa. For higher pressures, the tubing can be threaded into additional connectors. Depending on the applied pressure, the connection of two tubes is done in a manner identical with that for normal connections. If bored-through fittings are used (e.g. for inserting capillaries the reactor), one should be aware that the pressure rating will be reduced.

There is a bewildering selection of tube fittings available from different manufacturers to satisfy the varying demands of customers. The most important forms of fittings are crosses, elbows (90° and 45°), straights, tees, end caps, plugs, reducers, filter units, and quick-connects. Nearly all parts are available in a wide variety of sizes and with male, female, or weld connectors. For pressures of up to about 40 MPa (depending on temperature, wall thickness and OD of the tube), the use of nut–ferrule systems is recommended, as these connections are easy to install, disconnect, and retighten and are also comparatively inexpensive. Higher pressures require welded connections or screw couplings of special geometry. When installing and handling any fitting, one has to comply with the following safety precautions:

- Do not depressurize a system by loosening a fitting.
- Do not tighten fittings when the system is pressurized.
- Make sure that the tubing rests firmly on the shoulder of the tube fitting before tightening the nut.
- Fittings are not interchangeable. Do not mix fitting components from different manufacturers.
- Do not mix imperial and metric fittings in a single apparatus.
- Avoid unnecessary disassembly of unused fittings

In general, it is advised to be meticulous in the assembly of fitting. 'Rough and ready' installations should be disregarded. To avoid situations where it might be tempting to ignore the above stipulations, one should always have enough spare parts available.

2.4.2 Valves

There is a sufficiently large variety of commercially available valve designs that almost any requirement can be met concerning working conditions (size, pressure range, operating temperature, corrosion resistance etc.) and purpose (on/off valve, flow rate adjustment). Most valves can be actuated automatically by electric, pneumatic, or hydraulic devices. Table 2.5 gives a short overview.

2.4.3 Pressure Transmitter and Manometer

When measuring the relatively high pressures required for SCFs, relative pressure instruments should be used in preference to absolute pressure monitors, which are

Table 2.5 Overview of common valves for SCF handling.

Valve type	Characteristics	Available pressure rating (MPa)
Ball valve	"On/off"	10–60
Plug valve	"On/off", simple design	5–20
Metering valve	"Flow rate adjustment"	5–15
Needle valve	"On/off"- or "flow rate adjustment" (depending on the stem tip design)	10–60; special designs: up to 1500
Pressure relief valve	Protect the vessel and its components from overpressure ("proportional" or "on/off")	1–200 (desired value is exactly adjusted by the manufacturer)
Check valve	Ensures flow in one direction only	0.1–1000
Diaphragm valve	"On/off"- or "flow rate adjustment" (depending on the design), all-metal containment, packless	5–30

of more use for vacuum pressure measurements. Two types of pressure sensor types are commonly used, Bourdon tube pressure gauges and electronic pressure transmitters. Both have advantages and disadvantages, which are listed in Table 2.6.

2.4.4
Reactor Heating and Temperature Control

Autoclaves on the scale used for academic research are almost exclusively heated by electric devices. Smaller reactors can be heated by using a heating plate/magnetic stirrer. It is important to consider whether the contact is large enough for the necessary heat exchange (e.g. by manufacture of adapted reactor top frames, see Figure 2.3c). Arrangements where heating coils are inserted in the reactor walls (Figure 2.3a, left reactor) or placed directly around the reaction zone (Figure 2.3a) are more convenient and allow a more uniform heat distribution with smaller temperature hysteresis, which is of particular importance for kinetic investigations.

Modern compact controllers enable a high quality of control to be reached as they allow both extensive manual optimization and highly sophisticated autotuning (self-optimization) procedures. Moreover, such compact controllers and microprocessor-controlled digital indicators usually include limit comparators. These allow (with the use of suitable cabling) a relay-based switch-off of the heat input when the maximum operating parameters (both temperature *and* pressure) are exceeded. Therefore, the operational safety of the device is significantly improved. However, one should keep in mind that autoclaves with inserted heating coils are not usually suitable for cooling procedures or leak testing by immersion.

To control and to monitor the temperature of a reactor, thermocouples or resistance temperature detectors (RTDs) are used almost exclusively (Table 2.7). RTDs, despite being more expensive, are slowly replacing thermocouples in many industrial applications below 600 °C as they are more accurate. Normally, the

Table 2.6 Characterization of common pressure sensors for SCF handling.

Feature	Pressure transmitter[a]		Bourdon tube pressure gauge[b]	

Measuring principle	The affecting pressure causes a deflection of the metallic membrane. A resistance strain gauge, which is superimposed on an isolated part of the membrane, is thereby elongated. Hence the electric resistance is increased, which causes a rise in the measured voltage. The voltage signal is translated into a standard electric signal (0–10 V or 4–20 mA)		The media flow into a closed coiled tube with an oval cross-section. The affecting pressure will tend to uncoil, because the pressure acts to bring the coil into a round cross-section. The resulting movement is transferred mechanically into a circular motion, which causes a deflection of the needle	
Pressure range (relative)	++	Typically: up to 60 MPa relative; up to 1500 MPa can be obtained	++	Typically: up to 60 MPa, up to 700 MPa can be obtained
Temperature range	0	Typically: −30 to 100 °C; some designs allow up to 200 °C (but at lower pressure)	+	Typically: −30 to 100 °C; some designs allow up to 250 °C
	The expense of a special high-temperature pressure instrument can often be avoided by using a temperature stand-off (a short length of tubing) to isolate the instrument from the high-temperature pressure media. For example 15 cm of a 6 mm capillary is sufficient to reduce the temperature level from 450 to 100 °C. However, a calibration then needs to be done to allow an inference of the pressure value in the hot area (reactor)			
Monitoring	++	Can easily be done by a continuous recording of the electric output signal (shorter response time with current output)	−/+	Can only be done with special versions (e.g. with a rotary position sensor, see right-hand photograph)
Clearness of display	++	Small changes (trends!) can be easily observed	0	Small changes or exact values cannot be read out in a reliable manner

(Continued)

Table 2.6 (*Continued*)

Feature	Pressure transmitter[a]		Bourdon tube pressure gauge[b]	

Robustness	0	More sensitive to mechanical shock, mechanical vibrations and overstepping of the measurement range	+	Some of them are not suited for external vacuum environments (e.g. glove-box handling)
Independence	–	Transmitter needs especially configured display units with power supply	++	No read-out unit needed
Chemical inertness	+		+	
	Instruments can be produced of highly corrosion-resistant steels but the Bourdon gauge can easily be plugged (e.g. by polymer particles)			
Signal transmission	++	Electric (standard), even battery-powered radio transmitters are commercially available	–/+	Can only be done with special versions (see above)
Price	–	Typically: €400 per unit	++	Typically: €80 per unit
Accuracy	++	~0.5% of full-scale	+	~1% of full-scale

[a]Photographs © Jumo GmbH & Co KG.
[b]Photographs © WIKA Alexander Wiegand GmbH & Co. KG.

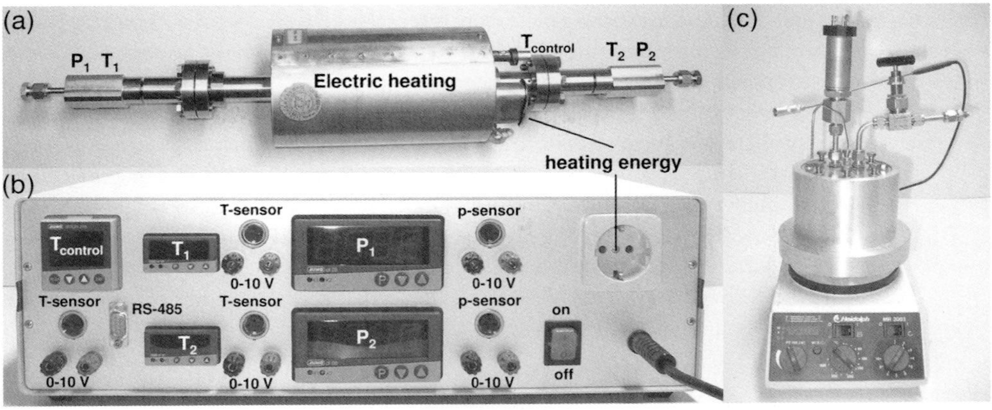

Figure 2.3 Plug flow reactor (PFR) with an outer electric heating unit (a), temperature and pressure control unit for the PFR (b), and stirred-tank reactor setup which is heated in a metallic top frame of a magnetic stirrer (c).

connection to the reactor is done with an NPT fitting. In cases where there is a long distance between the sensor and transducer, one should always ensure that there is sufficient amplification and/or correction of the measurement signal.

2.4.5 Stirrer Types

As austenitic stainless steels are not magnetic, it is sufficient in some cases (e.g. in small reactors) to place a simple magnetic stirring bar inside the autoclave which is

Table 2.7 Overview of common temperature sensors for SCF handling.

Type	Material	Temperature range (°C)	Sensitivity	Accuracy
J	Fe–Cu/Ni	−40 to +760	50 μV °C^{-1}	Class 1: dT = ±1.5 between −40 and 375 °C ±0.004T between 375 and 750 °C Class 2: dT = ±2.5 between −40 and 333 °C ±0.0075T between 333 and 750 °C
K	Ni–Cr/Ni	−90 to +1200	41 μV °C^{-1}	Class 1: dT = ±1.5 between −40 and 375 °C ±0.004T between 375 and 1000 °C Class 2: dT = ±2.5 between −40 and 333 °C ±0.0075T between 333 and 1200 °C
Pt100	Pt	Class A: −30 to +350 Class B: −70 to +500	0.385 Ω °C^{-1}	Class A: dT = ±(0.15 °C + 0.002T) Class B: dT = ±(0.30 °C + 0.005T)

rotated by a standard stirring plate or an external mechanical stirrer. In the latter case, the autoclave must be equipped with a rotatable iron core on the bottom plate. Such an arrangement allows higher agitator power.

For larger volumes, very high viscosities, and in particular for reactions containing a solid phase, more vigorous stirring is often necessary. In principle, packless magnetic drives and stirrers with self-sealing packing glands can be used. The magnetic drive offers higher achievable working pressures because the inner rotor is completely enclosed within a non rotating housing with fixed seals (Figure 2.4) [58, 59].

The particular stirrer design should comply with the phase system, the viscosity, and the desired flow pattern of the medium. Details addressing these parameters can be found in [60].

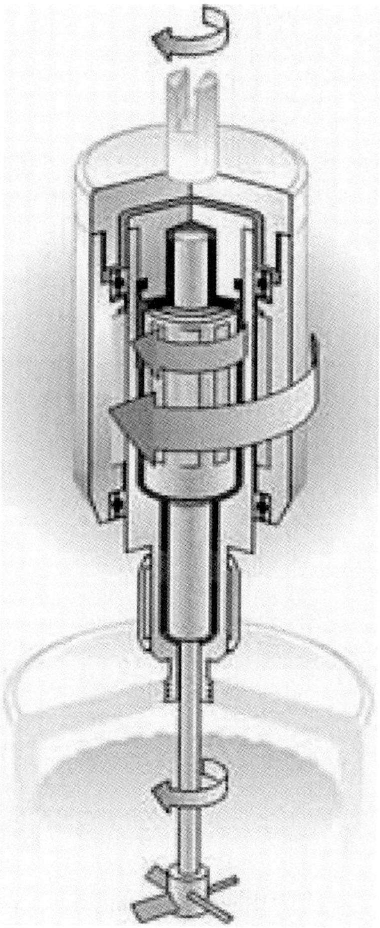

Figure 2.4 Schematic diagram of a stirrer with a packless magnetic drive. Copyright Parr Instrument Company, USA.

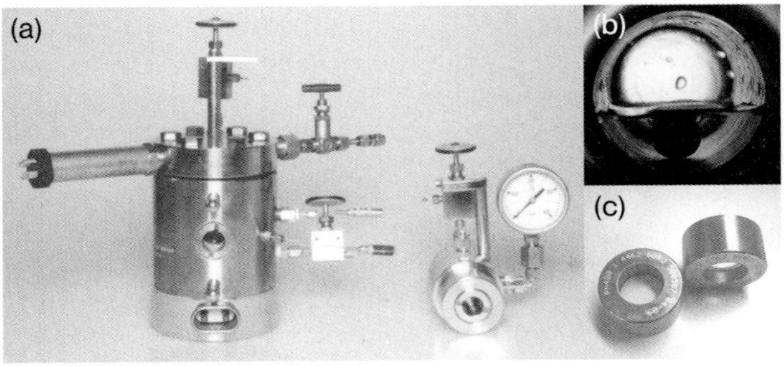

Figure 2.5 Window-equipped autoclaves (a), view through the windows of the autoclave (b), and metal fused high-pressure windows (c).

2.4.6
Optical Windows

As the knowledge of the phase behavior in SCF technology is often crucial for the effectiveness of the chemical transformation under study, the installation of two windows on opposite sides of the reactor walls, at least in stirred tank reactors, is highly recommended (Figure 2.5). Whereas such a setup was for long time possible only with expensive synthetic sapphires, thick-walled glass windows made of borosilicate are now commercially available. These allow pressures of up to 20 MPa. Metal fused versions are even rated to 40 MPa [61].

2.4.7
Pressure Safety Valves and Bursting Discs

Due to both the high pressure level and the rapid increase of pressure with temperature curves of SCFs around or above their critical density, the use of pressure relief devices in a system is highly recommended. In general, jackscrews should be used for connecting the main body of the reactor with a closure head. However, they normally have very rough dimensions and for this reason are unreliable in their response behavior. Therefore, small batch reactors should also be equipped with bursting disks, which are available for pressures from 1 to 800 MPa. The most widespread installation system is probably the clamping of a conical disk, with a nominal width of 0.25 in (6.35 mm), in a conical seat. For safe use, the maximum operating pressure of the reactor should not exceed 80% of the nominal pressure of the bursting disk. The advantage of bursting disks is their simplicity (low failure rate and costs). The main downside, however, is the fact that, when they rupture, the complete reactor content is vented into the environment.

Whenever possible (mandatory in the case of toxic or combustible loadings) the released substances should be discharged into a closed exhaust system. The same holds for pressure safety valves (PSVs), also called pressure relief valves (PRVs), which ensure automatic release when the pressure exceeds a preset limit. In comparison with a bursting disk, a PSV does not cause a complete depressurization, but released pressure until the set pressure level is reached again. In principle, PSVs work in a similar manner to back-pressure regulators (see Section 2.6.2). For academic research, a PSV is cost-effective when using larger autoclaves (≥1 l) or continuous flow setups. In order to ensure an error-free function, a sufficiently large difference in working pressure and the set pressure of the PSV should be allowed (approximately 10% difference for working pressures up to 100 MPa). Normally, the setting of the PSV is done by the manufacturer in the presence of an authorized inspector to obtain a valid insurance certificate. The PSV should be mounted vertically and vibrations must be avoided. As the tightness of the valve seat is not perfect, a bursting disk is often implemented upstream of the PSV when they are used in industrial processes. Such an arrangement requires pressure monitoring of the volume between the bursting disk and PSV so that one can detect a rupture of the bursting disk.

2.4.8 Online Sampling

By exploiting the "gas-like" properties of SCFs, one can to perform online gas chromatographic (GC) sampling of the supercritical phase in which substrates and products are dissolved. To achieve a representative transfer from the highly compressed SCF phase to the GC column (at nearly ambient pressure), an automatic sampling and injection system is necessary. As shown in Figure 2.6, this usually consists of two two-way pneumatic valves (connected in series) which are controlled by programmable timers or the GC software [62]. The six-port valve contains ports for the reactor, the gas chromatograph, a sample loop of adjusted volume (several microliters, depending on the concentration), the carrier gas, and the pneumatic three-port valve. To counteract partial condensation of dissolved substances, the valve should be heated by a temperature-controlled surrounding metal device (~60–90 °C). The three-port valve allows controlled purging of the injection system before the sampling is performed. This ensures that eventual residues from the last sampling are removed.

Shortly before the sampling, the three-port valve is opened (purging), then it is closed and the sample loop is filled with fresh material (Figure 2.6, position A). At this point, the six-port valve switches position and the carrier gas enters, transporting the contents of the sample loop in to the GC injector (position B). Finally, the valve switches back to position A, whereby the initial state resumed.

When sampling from batch reactors, it is important that they should be large enough to tolerate a substantial reduction in the amount of compound in the reactor, which comes about as a result of the obligatory purging. A consequence of the associated pressure reduction is that the solvent power of the SCF is reduced. This could interfere with the reaction significantly as a result of (additional) condensation effects. Such a disadvantage is completely eliminated when the sampling is done from a flow system. Here the three-port valve is no longer necessary as the substrate flows through the six-port valve, which switches from time to time for sampling.

The sampling of SCFs differs from the sampling of conventional solvents and one should be aware that substances which are dissolved in the SCF phase condense fairly

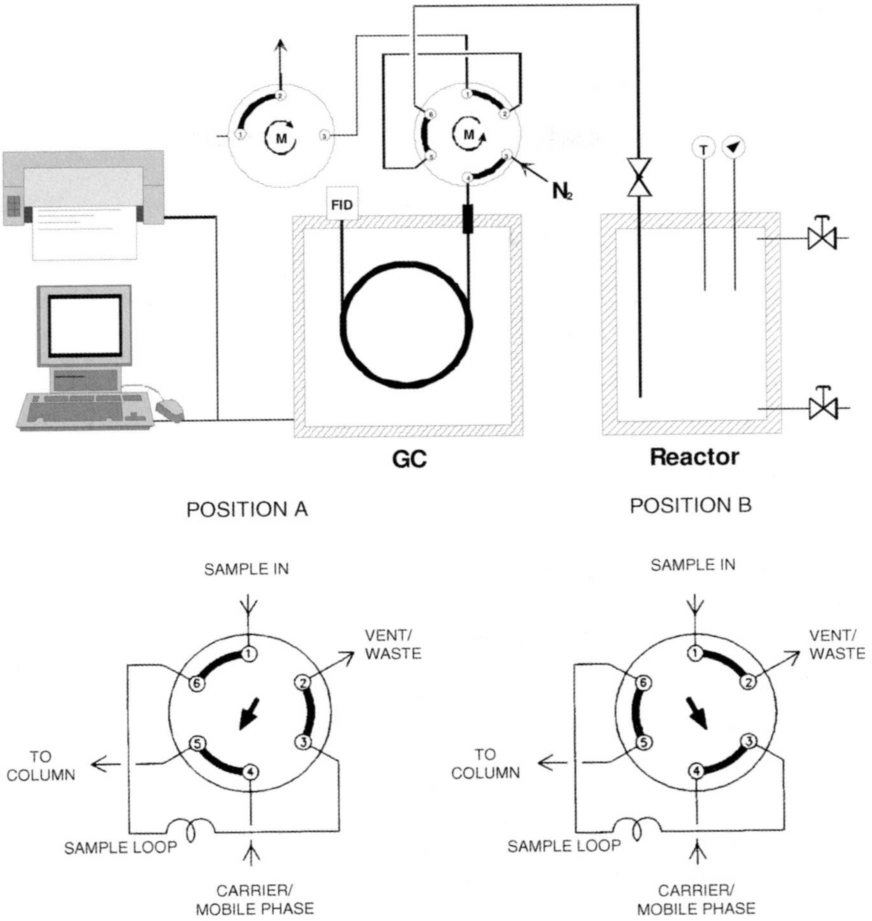

Figure 2.6 A possible setup for online GC sampling from an STR together with a schematic view of the two possible positions of the six-port sampling valve. Position A: purging (open three-port valve) and resting position (closed three-port valve); Position B: sampling (closed three-port valve). Copyright VICI AG International.

easily on contact with the carrier gas, which is at near ambient pressures, as a result of depressurization. This results in either substantial tailing or even complete failure of the transfer if solids are precipitating. For that reason, this methodology is restricted to volatile (preferentially low-polarity) organic compounds having eight or fewer heavy atoms. Such a restriction can be overcome by connection of a reactor to a an HPLC or supercritical fluid chromatography unit using sampling devices similar to those described here. Similar combinations have been reported for extraction processes [63].

2.4.9
Inline Spectroscopic Measurements

In contrast to online GC analysis, IR analysis is not only a non-invasive method, but is also a truly *in situ* method. The IR probe used is in direct contact with the reactor contents at the place of reaction and is not necessarily in the form of a bypass or a flow cell as with online GC measurement (inline versus online monitoring). When considering the determination of reaction kinetics, the advantage of *in situ* IR is the improved time resolution (less than 1 min to obtain a full spectrum in good quality versus several minutes for a chromatogram). Reactions can be examined if at least one of the stretching frequencies of the reactant (decrease) or of the product (increase) changes. For monitoring under high pressure, it is often convenient to use the attenuated total reflection mode (ATR, mid-IR range) for the measurement. Here a beam of IR light is passed through the ATR crystal in such a way that it reflects at least once off the internal surface in contact with the sample (Figure 2.7b) The refractive index of the crystal must be significantly greater than that of the sample, or total internal reflection will not occur, that is, the light will be transmitted rather than internally reflected in the crystal. The reflection forms an evanescent wave which extends into the sample by one half wavelength of the light. The beam is then focused with a detector after it exits the crystal (Figure 2.7b). As with all Fourier transform (FT) IR measurements, an IR background has to be collected. In the case of SCFs, it is

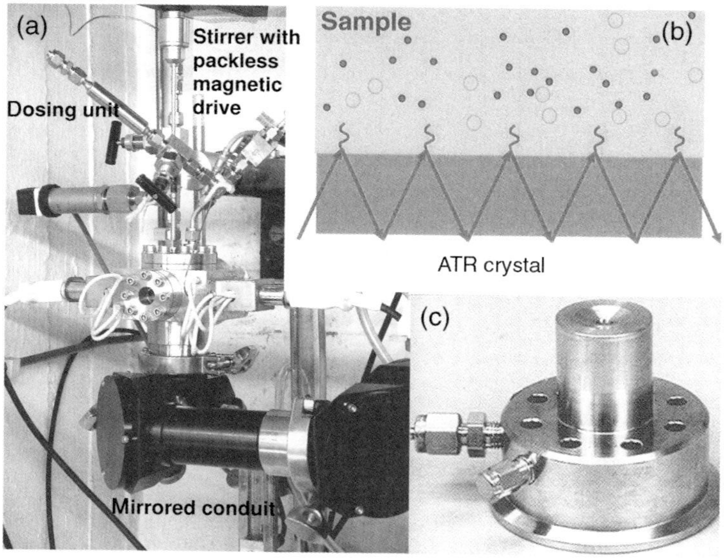

Figure 2.7 (a) Setup for high-pressure ATR–FTIR inline measurements with the ATR probe at the bottom of the reactor (a rotation of the reactor allows a measurement from the top); (b) IR beam reflection in the ATR crystal; (c) high-pressure ATR probe. Copyright Mettler-Toledo Autochem.

beneficial, in most cases, to do this after filling the reactor with the SCF (dosing of the SCF at the location of the IR measurement is not possible in most cases, at least not for batch rectors, which are filled by gravimetric dosing) to avoid negative bands [28, 64–66]. That being said, due to the short penetration depth, the detection limit is reduced in comparison with traditional FTIR measurements. Therefore, depending on the strength of the IR chromophores in the probed compounds, concentrations <0.5 mol % are often difficult to monitor by ATR in an accurate manner.

Interference from dispersed solids is minimal in ATR-FTIR as constant contact between the probe and the sample is necessary (for measurement of solids under reaction conditions, see [67]). Problems may arise if an additional liquid phase is present; this tends to wet the sensor preferentially, since the liquid phase has a higher surface tension than the SCF (as in the case of a water phase).

In most high-pressure applications, for high pressure either silicon ($n = 3.43$) or diamond ($n = 2.42$) ATR crystals are used as an internal reflection elements of high refractive index. Although silicon is not as hard and chemically resistant as diamond, it has the great advantage of being relatively transparent across the whole range 4000–650 cm^{-1}. Diamond absorbs strongly between 1900 and 2200 cm^{-1} and has weak performance up to 2600 cm^{-1} which precludes, for instance, the detection of metal carbonyls, free carbon monoxide, several triple bonds and cumulative double bonds or some non-classical metal hydrides. The absorption of CO_2 masks the following regions of the spectrum: 3800–3500 cm^{-1}, 2600–2100 cm^{-1}, and the region below 800 cm^{-1}. The connection between the spectrometer can be realized either with rigid, mirrored conduits (full mid-IR range), as shown in Figure 2.7a, or flexible fiber conduits (easy to handle but with a limited range and normally lower sensitivity).

2.4.10 Reactor Cleaning

An autoclave must be cleaned thoroughly after each use to obtain reproducible results that are of scientific value. Owing to the gas-like transport properties of SCFs, great care has to be taken to reach all parts of a reactor setup. Blank experiments between individual runs are recommended if catalytically active materials are used or generated in the SCF phase. The cleaning methodology has to be appropriate to the type of contamination. A list of different cleaning methods is given below.

2.4.10.1 Cleaning with Organic Solvents

First one should try to remove contaminants with conventional organic solvents. For toxicity reasons, the use of ethanol, acetone, and also, particularly for polymeric components, toluene is preferred. It is important to ensure that the *whole* autoclave is cleaned, including bore holes, valves, temperature sensor, manometer, capillaries, and connecting pieces. For narrow components, the use of a pipe-cleaner may be helpful. For the cleaning of lopsided tight holes (e.g. as is the case for many manometers) one can feed in a thin polyethylene tube (possibly extracted) to the bottom of the hole. The solvent can then be inserted through this tube, thereby allowing complete wetting. After thorough flushing of the system, the reactor is dried with the help of tissues and compressed air. The autoclave can then be reassembled and dried in a drying oven for at least 30 min at 60 °C. Then the still warm autoclave is evacuated to remove all solvent residues. $scCO_2$ can often be helpful for removing the final traces of solvent. Note that chlorinated solvents can cause corrosion and even failure in stainless-steel systems. Therefore, whenever possible they should be avoided.

2.4.10.2 Cleaning with Heated Solvents

It might be helpful to fill the autoclave with about 50% of a suitable solvent (e.g. toluene) and to heat the closed autoclave under vigorous stirring to about 120 °C, especially for the displacement of persistent polymeric components. After cooling the reactor, the solvent is removed and the cleaning is continued with the solvent procedure described above.

2.4.10.3 Cleaning with Acids

In addition to the application of standard organic solvents, the use of acids for cleaning purposes is often required. For this purpose, acetic acid is particularly suitable. In some cases a half-concentrated methanolic hydrochloric acid solution is even more effective. Due to the corrosiveness of this mixture, the residence time has to be short (<5 min). In the case of persistent discolorations/encrustations, either a commercial scouring agent can be used or the acid mixture can be combined with Celite as an abrasive. In general, after an acid or an abrasive has been used (also bases when required), one has to flush thoroughly with copious amounts of water, which is then removed with the solvent procedure described above.

2.4.10.4 Cleaning with a Combination of Organic Solvent and $scCO_2$

$scCO_2$ has the advantage of being distributed uniformly through the whole reactor volume, which allows it to reach parts of the reactor that are difficult to access with conventional solvents. Its high diffusibility allows for excellent penetration of solid contaminants. In combination with conventional solvents (e.g. acetic acid for metal-containing compounds, toluene or dichloromethane for polymers), a very high cleaning power can be achieved. After depressurization, the solvent procedure should then be applied. If an acid or base is used, one has to flush with copious amounts of water before use.

2.5 Dosage Under a High-pressure Regime

2.5.1 Dosage of Gases

2.5.1.1 Safety Warnings

Pressure Rise Due to Temperature Increase Before gaseous compounds are added to a closed and heated system, one should have a rough knowledge about the resulting pressure–temperature curves. This is especially important in the case of SCFs, which

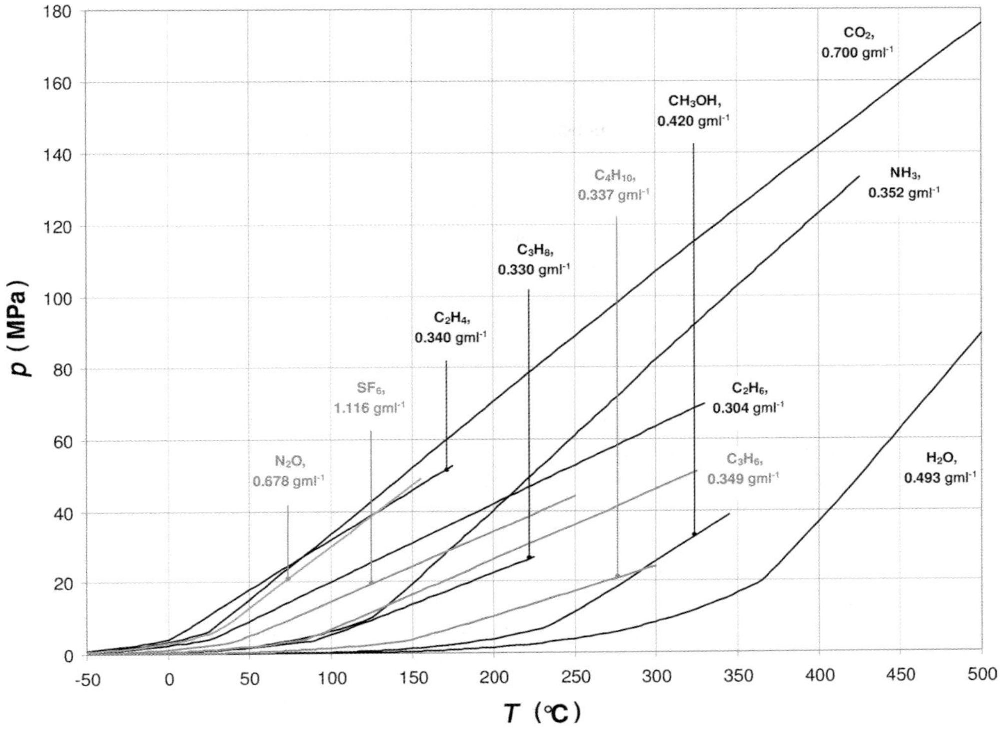

Figure 2.8 Calculated pressure–temperature curves for frequently used SCFs (density = $1.5\rho_{crit.}$) [68].

are often dosed as a liquid, because a small increase in temperature might cause a drastic pressure rise (Figure 2.8). Here the applied densities are normally substantially higher compared with the multitude of reactions, including a compressed gas phase. Hence, in other words, it is absolutely essential to understand these systems fully before performing a reaction for the first time. A pressure–temperature curve calculation (and a large variety of other thermophysical properties of fluid systems) can easily be done using the excellent website of the National Institute of Standards and Technology (NIST), which considers individual equations of the state of 75 different fluids [68].

Explosion Prevention in Oxygen-containing Systems Because of the complete miscibility of SCFs and gases, SCFs are frequently used for processes in which a gaseous reactant is part of a transformation. The inherent inertness towards oxygen of carbon dioxide or water makes them the favored reaction media for oxidation reactions. However, one should be aware that even with these SCFs, high oxygen concentrations can promote explosions if combustible compounds are present. Although detailed studies concerning this particular problem are not available for most cases, one is well advised only to use oxygen concentrations below the limiting oxygen concentration (LOC). This value, which is tabulated for many volatile combustible com-

Table 2.8 LOC values of flammable compounds in $scCO_2$ as a reaction medium.

Flammable compound	Temperature (°C)	LOC in CO_2 (mol%)	Flammable compound	Temperature (°C)	LOC in CO_2 (mol%)
Benzene	100	11.8	*n*-Butane	20	12.0
Isobutane	20	13.1	Cyclopropane	20	12.0
Carbon monoxide	20	4.6	Ethane	20	11.7
Ethylene	20	10.5	*n*-Hexane	100	11.6
n-Heptane	100	10.9	Hydrogen	20	5.2
Methane	20	13.7	*n*-Pentane	20	11.0
Propane	20	12.6	Propylene	20	12.6
Toluene	100	12.9	Xylene	100	13.1

pounds, describes the maximum oxygen concentration as an arbitrary mixture of the flammable material and an inert gas (such as CO_2, N_2, Ar, or H_2O) at which explosion does not occur. It must be taken into to consideration, however, that these values are stated for ambient pressure and room temperature. Nevertheless, they are certainly a good starting point for planning experiments. In practice, when using $scCO_2$ as a reaction medium, an additional safety factor of 0.8 should be applied to the values given in Table 2.8. If several combustible compounds are present, the one with the lowest LOC value is the most significant [69].

Finally, it must be stressed that nitrous oxide (N_2O) is potentially a very powerful oxidizing agent. There have been disastrous explosions with scN_2O and even small quantities of organic compounds [70].

2.5.1.2 Dosing in Batch Processes

Gravimetric Dosage The most frequently used SCFs have critical temperatures above room temperature. Hence they condense if they are compressed beyond their vapor pressure (at ambient temperature), making a dosage by pressure adjustment impossible. Therefore, a gravimetric dosing of the SCF is applied for batch processes in most cases. It must be considered, when dosing in this manner, that the weighing scales used to measure this value may struggle to work efficiently (in terms of weighing performance, accuracy, and stabilization time) due to the weight of the average pressure vessel. To ensure reproducible loadings of the SCF, the following points should be considered:

- The dosing valve should be installed in a fixed position (the valve of the reactor should not be used for this purpose).
- The connection between the dosing valve and the reactor should be made of a short, thin, and flexible (strainless) capillary.
- The weight of the unconnected autoclave should be measured before and after loading.

In order to trace leaks and to exclude possible loss of material during the experiment, an additional weight check before depressurization is highly recommended.

Dosage by Pressure Adjustment In general, for batch processes, and for those temperature ranges where condensation of gaseous compounds is not a problem, dosing of gases by pressure adjustment is possible. If one is planning to dose the gas directly into the loaded reactor, one should be aware of possible interactions between the reactor content and the added gas phase which might distort the actual value significantly. Such a dosing methodology is also problematic if only small amounts of a gaseous compound are to be added which influence the total pressure to only a marginal extent. To be on the safe side, one can use an additional storage vessel with known volume and temperature. The dosing is then performed by transferring gas out of the additional storage vessel; the necessary pressure drop is calculated in advance, for example using the following virial equation including the second virial coefficient:

$$p_2 = \frac{RTB}{V^2}\left[\left(n + \sqrt{\frac{1}{4}\left(\frac{V}{B}\right)^2 + \frac{p_1 V^2}{RTB}}\right)^2 - \frac{1}{4}\left(\frac{V}{B}\right)^2\right]$$

where n = required amount of gas (mol), R = gas constant = 83.145 ml bar K^{-1} mol^{-1}, T = temperature of the storage vessel (K), B = second virial coefficient of the individual gas as a function of temperature (ml mol^{-1}), V = volume of the storage vessel including the tube connection to the reactor (ml), p_1 = initial pressure of the storage vessel (bar), and p_2 = desired final pressure of the storage vessel (bar).

Reproducible dosing can be expected for pressure differences ≥ 1.0 MPa (the size of the storage vessel has to be adjusted, accordingly) in combination with a pressure monitor with a resolution of 10 kPa.

2.5.1.3 Dosage for Continuous Flow Processes

Dosing from a Storage Vessel The methods for gas dosing described above cannot be applied when using continuous flow processes. In some cases, one can use larger heatable storage vessels in which the SCF and some (or all) substrates (except the catalyst) are mixed in advance. With the help of an adapted pressure reducer, one is able to feed a flow reactor for a limited period. This period depends on (a) the size of the storage vessel, (b) the required pressure level, and (c) the desired flow rate. However, such a setup restricts flexibility, in terms of optimization procedures, and has longer run times.

Dosage by a Mass Flow Controller A mass flow controller (MFC) is a device used to measure and control the flow of gases. All mass flow controllers have an inlet port, an outlet port, a mass flow sensor, and a proportional control valve. The MFC is fitted with a closed loop control system which is given an input signal by the operator (or an external circuit/computer). This is compared with the value from the mass flow sensor and the proportional valve is automatically adjusted accordingly to achieve the required flow. Two different measurement principles are currently applied to flow controllers in the majority of cases, thermal mass flow controllers and those based on the Coriolis force.

A thermal mass flow meter is based around a thermal sensor. The thermal sensor consists of a small-bore tube with two resistance thermometer elements wound around the outside of the tube. The sensor tube is heated by an electric heating element which is placed in the middle of the sensors [71]. The medium is heated to a defined temperature which is measured by comparing the values at temperature sensors before (background value at laminar flow) and after (response value). The height of the response value decreases with increase in flow rate. In modern controllers ΔT is kept constant and the mass flow is calculated from the necessary heat input. Such a measurement setup avoids overheating in the case of small mass flows [72, 73].

Further properties are as follows: available flow ranges: 0.02 $ml_n\,min^{-1}$–5000 $l_n\,min^{-1}$; accuracy (including linearity): ≤1% full-scale; setting time (controller): <2 s; pressure rating: up to 70 MPa; medium temperature: −10 to 70 °C (typical), up to 200 °C; independent of medium density; dependent on isobaric heat capacity (not suitable for SCFs close to T_c).

Thermal mass flow controllers for liquids are, in principle, fairly similar. Here a sensor bypass is often not installed. Instead, the complete flow is heated and measured. Although a higher energy input is necessary, such a design has the advantage that unwanted retention phenomena are avoided and the controller is insensitive to different mounting positions [74]. One should be aware that bubble formation by the heating procedure will adversely affect accuracy and reproducibility and could cause malfunction of the MFC. Thus, their application for SCFs with low critical temperatures (such as carbon dioxide, ethane, nitrous oxide, fluoroform, and sulfur hexafluoride) or conventional solvents with low boiling points is not recommended [75].

The Coriolis mass flow controller is based on Coriolis forces which are generated in oscillating systems when a liquid or a gas moves away from or towards an axis of oscillation. A Coriolis measuring system is of symmetrical design and consists of one or two measuring tubes, either straight or curved. A twisting of the system is induced which depends on the flow rate of the medium and the mass of the tube (which is a function of its loading), its oscillating frequency, and its spring constant. This twisting can be measured. An explanation of the measuring principle in more detail is beyond the scope of this chapter, but the websites cited contain helpful animations to help explain the phenomenon [76]. Operation of Coriolis flow meters is almost independent of the physical properties of the medium (temperature, density, viscosity, phase behavior, pressure fluctuations), which makes them well suited for handling SCFs in the vicinity of the T_c. The main disadvantage of Coriolis flow meters is their high initial setup costs and the difficulty in cleaning them.

Further properties are as follows: available flow ranges: 0.03 $kg\,h^{-1}$–1500 $t\,h^{-1}$; accuracy (including linearity): ≤0.2% full-scale; setting time (controller): <2 s; pressure rating: 10 MPa typical, up to 90 MPa available; medium temperature: 0–70 °C (typical), 250–400 °C available.

2.5.2 Dosage of Liquids

The simplest way to dose a liquid substrate into a pressurized autoclave is by using a dosing unit of adjustable size, connected to the reactor by rigid tubing. Such a unit is shown in Figure 2.7a. The contents of the dosing unit are separated from the reactor by a valve, which is open at a certain time to start the addition by diffusion. If a spontaneous addition is desired, one should fill the dosing unit with an inert gas cushion at a higher pressure level than the reactor contents at the reaction temperature. This methodology has the advantage that the dosing is very reliable (gravimetric metering) and is nearly under isobaric conditions. However, it has low variability and this approach is not suitable for flow systems. Sample injection valves are a possible alternative. In these, the content of a defined inner volume is discharged by a piston (actuated by a handle) [77].

Alternatively, the dosing of fluids into a supercritical environment is mostly done with two types of metering pumps: HPLC pumps and high-pressure syringe pumps. Both pump types can also be used for SCF supply in continuous flow setups on a small scale. Table 2.9 compares their properties.

Table 2.9 Comparison of reciprocating and syringe pumps.

Property	HPLC pump (reciprocating pump)	Syringe pump
Working principle	A motor moves a small reciprocal sapphire piston in a correspondingly shaped chamber in the pump head of about 10 ml volume (analytical scale). When the piston is moving outwards it sucks liquid into the cylinder through an inlet check valve in the pump. When the piston moves forward, the liquid is compressed and flows through the outlet check valve. The pump head can be heated or cooled by an external temperature control jacket (liquid heat transfer medium)	A metallic cylinder of about 30–300 ml is loaded and discharged by a sealed piston which is moved by a stepper motor. A distributing valve and two microswitches permit the pump to be purged and refilled automatically. The pump cylinder can be thermostated by an external temperature control jacket (liquid heat transfer medium). A dual pump, which has two syringes (alternating loading and discharging), allows a fully continuous supply
Maximum pressure (MPa)	Up to 60	Up to 140
Temperature range (°C)	0–40, pump head can be cooled	0–40 (typical), up to 200
Flow rates	0.01–200 ml min^{-1}	0.01 µl min^{-1}–400 ml min^{-1}
Flow accuracy (%)	$<\pm1$	0.3–0.5
Pressure accuracy	Pressure control requires an extra module	0.5% full-scale
Flow characteristic	Pulses are reduced by a pulse damper	Pulseless flow
Operating modes	Constant flow rate, constant volume, time-based sequence	Constant flow rate, constant pressure, time-based sequence
Continuous flow	Possible	A dual pump is required
Media	Liquids and SCFs	Liquids and SCFs
Price (€)	~3000	~12 000 (dual pump ~20 000)

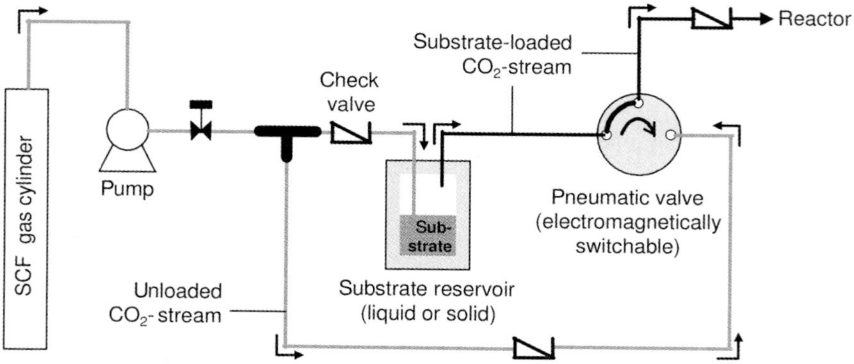

Figure 2.9 Possible setup for dosage of a solid compound into an SCF flow system.

2.5.3
Dosage of Solids

For some reactions, the addition of a solid key component (e.g. catalyst) into the otherwise loaded and equilibrated reactor is desirable. For discontinuous applications, specially designed ball valves or addition devices are commercially available [78]. In some cases, a solid-containing sealed glass ampoule which is broken by pressure or upon agitation might also be of help.

The continuous dosing of solids into a high-pressure system is a severe problem, especially on a small scale since the necessary mass transport into a sealed environment can be realized only with considerable effort and cost. One possibility might be the use of an automatic sluice ensemble working in an alternating manner. Another approach could be the generation of a suspension of the solid substance and the SCF (in its liquid form) which is transported into the reactor by a slurry pump [79]. A further solution to that problem assumes a sufficient solubility of the solid compound in the SCF; then the solid can be loaded in a substrate reservoir through which the SCF passes. With the help of an automatic switchable three-port valve, the desired concentration in the reactor can be adjusted after a calibration procedure (Figure 2.9) [80].

2.6
Further Regulations and Control in Flow Systems

2.6.1
Supply Pressure

The regulation of the gas supply pressure is best done with the help of a pressure regulator which is a normally open valve. It is installed at the start of a system or before pressure-sensitive equipment to regulate or reduce undesirably high inlet pressures (e.g. from a gas cylinder, compressor, ring line, or reservoir autoclave).

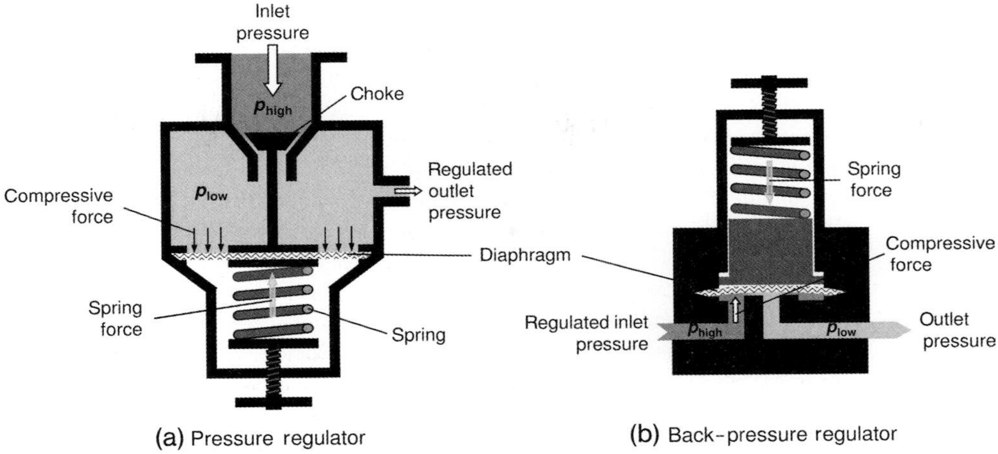

Figure 2.10 Schematic diagrams of a pressure and a back-pressure regulator.

The principle by which it works is as follows. The outlet pressure places a force on a diaphragm, which is connected to an adjustable spring on the bottom and works against its spring force. The top of the diaphragm is connected to the choke of the valve which separates the upstream from the downstream pressure. If the pressure on the downstream side becomes too high, the diaphragm flexes and the choke closes (Figure 2.10a).

2.6.2 Reactor Pressure

For performing continuous flow reactions with SCFs, reliable control of the pressure level inside the reactor is essential. The most elegant way of doing this is probably by the use of a back-pressure regulator, which is installed at the end of the piping system to provide an obstruction to unhindered flow and thereby to regulate the back-pressure.

In principle, these regulators are similar in operation to precision pressure relief valves. When the inlet pressure increases, an opposing force is generated which acts on the diaphragm and attached poppet against the set pressure load in the dome. When the inlet pressure level reaches the preset pressure, the poppet is gradually lifted from its seat. A consequent decrease in upstream pressure is experienced as fluid flows out faster than the inlet pressure can re-supply. With decreasing upstream pressure, the pressure-loaded dome starts to move the poppet towards its closed position, thus maintaining the desired set pressure level within a narrow band (Figure 2.10b).

Automated versions are also commercially available but are more expensive. One should be aware that a back-pressure regulator tends to pulse at low flow rates. Therefore, one should know the minimum flow coefficient of the regulator, k_V. To prevent damage to the sealing surfaces, it is recommended to install a filter before the flow reaches the back-pressure regulator.

2.6.3
Flow and Total Volume Measurement of Depressurized Gas Streams

Especially in simple flow setups, where the use of expensive mass flow controller appears inappropriate, the total gas flow of uncondensed compounds (e.g. the SCF together with gaseous reactants or products) can be easily measured at the exit of the reactor (after the individual decompression and condensation unit) by a variable-area flowmeter or highly accurate (and much more expensive) laboratory meter (or wet meter).

Variable-area flowmeters are very popular as they are comparatively cheap, easy to install and to handle, maintenance free, and independent in their operation. Depending on the shape of the floating body, different measuring ranges are commercially available starting from 0.1–1 $L_n\,h^{-1}$ and going up to 200–2000 $L_n\,h^{-1}$. Their relative accuracy lies typically between $\pm1\%$ and $\pm5\%$ of the final value. Nowadays, adapted designs are also available, which are based on thermal mass flow. These battery-driven units offer not only digital convenience but also higher accuracy, alarm functions, and a totalizer [81].

Laboratory meters, which are also called wet meters, are predominantly designed for highly accurate volume measurements. Operated by the difference between the inlet and outlet pressures, which causes a liquid-embedded drum with propeller shaped blades to rotate, this drum is connected to a counter unit. Both the actual flow (how fast the watch hand is rotating) and the total volume passing through (the number of rotations) can be monitored. Depending on the application, different scales are available ranging from 1 to 100 L_n gas displacements per rotation [82].

2.7
Evaporation and Condensation

2.7.1
Evaporation

In most processes using an SCF as the reaction medium, the fluid is stored in its liquid form which allows for its controlled evaporation in flow systems at a certain stage. The same holds for condensed reactants mixed with the fluid. Nowadays, specially designed evaporation units are commercially available. These allow both efficient mixing and controlled evaporation. The units are fed with predefined liquid (and maybe gas) flows by using MFCs or metering pumps.

2.7.2
Condensation

In nearly all cases, and independently of the mode of processing, some form of separation of the product mixture from the SCF used (and maybe other gaseous compounds present) is necessary. To obtain high recovery rates of the product mixture, one requires both a controlled depressurization and efficient cooling.

For batch experiments, different types of needle valves can be used to control the expansion, whereas back-pressure regulators are the preferred option for flow applications. The expansion is governed by the Joule–Thomson effect, which causes a temperature decrease for most SCFs. Hence heating of the expansion unit is often necessary to avoid plugging due to the formation of solid particles.

The decrease in density of the SCF on expansion cause a drastic reduction in solvent power; dissolved compounds with a comparatively lower vapor pressure condense as solids or liquids. This effect is used to isolate the product from the SCF (which returns to the gas phase). Sometimes, the expansion of the SCF is carried out in two stages; the first stage, down to 1 or 2 MPa, is often sufficient to separate a liquid product efficiently. The pressure is then reduced to ambient after the liquid product has been removed and the second, more volatile, product can be isolated.

Sometimes, to ensure high recovery of the product, the expanded gas may be cooled further to complete condensation. For this purpose, heat exchangers of various designs can be used [83]. In academia, cold traps made of borosilicate glass which ensure intensive heat transport between a cooling medium and the passing gas stream by spiraled internals or superstructures can be used. In more problematic cases, several cold traps can be connected in a sequential manner. Ethanol and acetone–dry-ice baths or powerful refrigerating machines function as a cooling source.

2.8 Complete Reactor Systems for Synthesis with SCFs

2.8.1 Standard Batch Reactor System

This section gives an overview of the essential equipment that is needed to perform chemical reactions in SCFs as reaction media. Subsequently the single steps are listed briefly.

2.8.1.1 Essential equipment

The following are required: a high-pressure reactor with one (or better two) needle valves, a manometer, a bursting disc or jackscrews and two thermocouples (inner and wall temperature), a reactor enclosure made of steel plates and/or ferroconcrete with a polycarbonate window, a heating plate/magnetic stirrer, a gas cylinder, an HPLC pump equipped with a coolable head and a needle valve at the delivery side, a cryostat, a balance for dosing the SCF into the reactor, a cold trap, a valve heating device, a bubbler, capillaries and a waste-gas line.

2.8.1.2 Brief Description of Work Steps

Before high-pressure experiments are performed, an instruction from a supervisor is mandatory!

Experimental Planning

- Evaluation of corrosion risks (Section 2.3.1)
- Calculation of substance amounts and pressure at the reaction temperature (Section 2.5.1.1; it is crucial to know how much reactor volume is already occupied by solid and/or liquid substrates).
- Required setup: possible need for a special reactor configuration (e.g. glass inlet, additional capillaries).

Preliminary Work Concerning the Reactor

- Reactor cleaning (Section 2.4.10).
- Reactor assembly (the use of a torque spanner is recommended).
- Tightness control (preferentially with hydrogen).
- Drying at about 60 °C for 30 min → evacuation of the still warm autoclave.

Dosing of Condensed Substances into the Reactor (at Ambient Pressure)

- Dosing of solid and/or liquid substances at ambient pressure and temperature. For loading of air-sensitive compounds: switch three times between evacuation and argon treatment (solid compounds are best filled in a glove-box; liquids are best transferred under an argon counterflow).

SCF Dosing (Section 2.5.1.2)

- Start pump head cooling ($T \approx 5\,°C$ for carbon dioxide).
- Reactor balancing.
- Gas filling: connection of the closed reactor (still placed on the balance) with the feed line → opening of the (firmly mounted!) valve of the HPLC pump at the delivery side and flushing (three times).
- Start gas compression → close the delivery valve → tare weight → open the reactor valve → regulated addition with the delivery valve of the HPLC pump until the balance shows the desired mass → close the reactor valve → disconnect the feed line.
- Reactor balancing.
- Leak testing by immersion, leak detector spray or an electronic leakage tester.

Performing the Reaction

- Transfer the autoclave to the experimental station (reactor enclosure, Section 2.2.1).
- Connect the thermocouples with compact controllers and/or digital indicators (Section 2.4.4) → start the stirring and temperature regulation → possibly addition of a further substance into the equilibrated reactor (Section 2.5).

Work-up

- Depressurization: reactor cool-down to ambient temperature → connect the reactor with a cold trap via a flexible Teflon or polyethylene tube. The cold trap is placed in a dry-ice bath (temperature not below −55 °C for carbon dioxide). The exit port of the cold trap is connected with a bubbler, which is connected to the waste-gas line. The exit valve of the reactor is heated with the help of an enclosing electric heating device or a heat gun (much less energy efficient!). The valve of the reactor is opened in a very regulated manner so that two to eight gas bubbles pass the bubbler per second (Section 2.7.2).

- Completion of product mixture recovery is done by addition of one to three fractions of a suitable solvent into the depressurized autoclave. Each fraction is stirred inside the closed autoclave and subsequently collected.
- The combined fractions (including the one from the cold trap) are analyzed or treated further.

2.8.2
Fully Automated System for Continuous Flow Operation

This section describes how the technique described above can be used to produce a fully automated system for the study of continuous flow reactions in $scCO_2$ based on online sampling with GC as the analytical tool.

The automated supercritical flow reactors [84] utilize HPLC pumps to deliver liquids into the reactor at set volumetric flow rates. Gases are dosed into the reactor using Rheodyne dosage units. All the materials are then mixed in a heated chamber before reaching the reactor. The reactor is made from a stainless steel-pipe and is generally filled with a solid catalyst, resulting in a fixed-bed catalyst setup. The pressure is controlled at the back-pressure regulator (BPR). The automated system also incorporates an additional online sampling loop that samples into a gas chromatograph, which allows the reaction to be monitored in real time. This gas chromatograph takes a very small sample of the reaction stream, providing a momentary observation of the reaction products at the current reactor conditions. By using programmable controllers for all the reaction parameters, the automated system can be used to study systematically the effects of the different variables on the reaction outcome.

The following features are all automated in the system:

- Fully automated liquid delivery is achieved by the utilization of programmable HPLC pumps to deliver both CO_2 and liquid substrates/solvents.
- Fully automated gas delivery is possible by using an in-house designed Rheodyne dosage unit. This unit is directly controlled by the computer via an interface.
- Commercially available programmable temperature control units are used to control external reactor temperatures.
- Automated pressure control is achieved by using programmable back-pressure regulators.

The automated analytical features include the following:

- A picolog TC-08 eight-channel data logger to monitor the internal and external temperatures of the reactor by attaching k-type thermocouples. It can also be

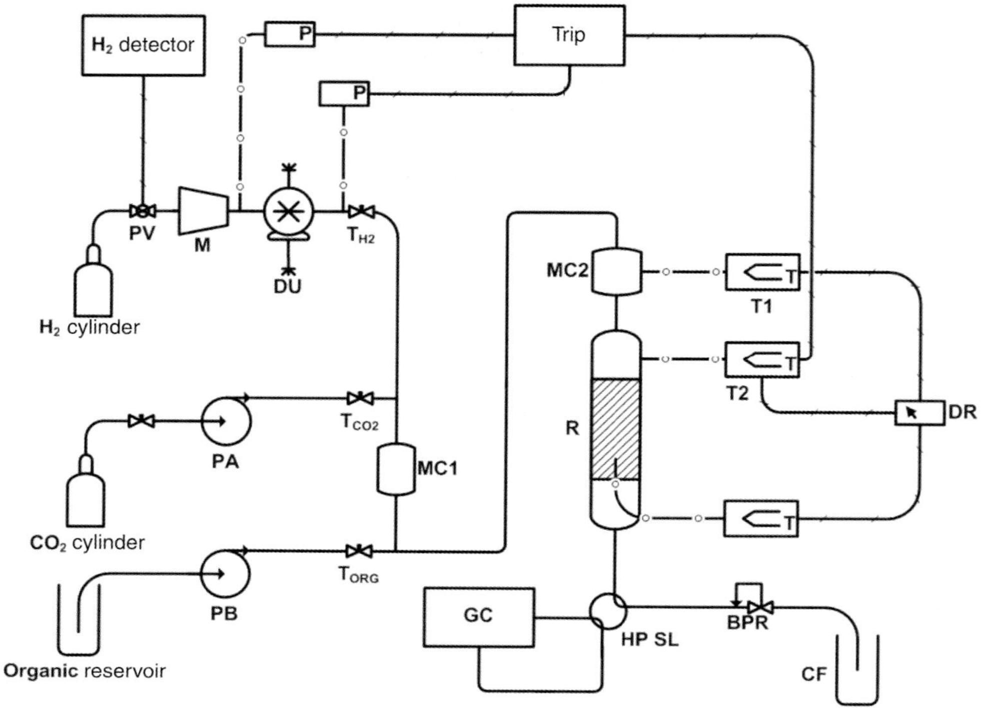

Figure 2.11 Detailed process schematic of the supercritical hydrogenation automated reactor. CO_2 is fed into pump PA (Jasco PU-1580-CO_2) where it is refrigerated at ca $-10\,^{\circ}C$ to ensure that it remains in a liquid state for pumping. Hydrogen is pressurized by a Pickel NWA CU 105 Maximator (M), which allows pressures greater than that of the system pressure to be achieved. This over-pressurized H_2 is then dosed by an in-house modified Rheodyne dosage unit (DU). Organic substrates are delivered via the HPLC pump, PB (Jasco PU980 HPLC pump). The three feeds outlined above can be isolated at taps T_{CO_2}, T_{H_2} and T_{ORG}. The CO_2 and H_2 are both initially mixed in a 2 ml mixing chamber filled with glass beads (MC1). This combined feed is then mixed with the organic substrate from the PB in a second mixing chamber (MC2). The feed is then introduced into the reactor (R). Catalyst is contained within the reactor by frits attached to each end of the reactor and a plug of glass-wool which is replaced each time the catalyst is changed. This arrangement prevents any catalyst particles from passing into the tubing beyond the reactor causing blockages or catalyst loss. Both MC2 and R are heated by a pair of aluminum heating blocks each containing a heating cartridge. The temperature of the aluminum blocks is controlled by Eurotherm 2216e programmable heating controllers connected by k-type thermocouples (T1 + T2). The temperature of the heating block is also monitored by the data recorder (DR), which is also connected thermocouples attached to the heating blocks and at the internal outlet of the reactor. The data recorder can also be connected to the BPR and pumps for continuous pressure monitoring. The product stream then passes through the high-pressure sample loop (HP SL). Small aliquots of the product stream can be introduced directly into the carrier gas stream of a gas chromatograph (Shimadzu GLC-17A), passing on to the column for product composition analysis. The pressure of the system is controlled by a programmable back-pressure regulator (BPR) (Jasco BP-1580-81). Following depressurization, liquid materials are collected in a conical flask (CF) and gases are released into a fumehood. For safe operation, a trip monitors the temperatures and pressures of the system to ensure that the apparatus is controlled within safe operating

expanded to monitor flow rates and pressures by directly connecting to the system's HPLC pumps in experiments where these factors are varied.

- A high-pressure four-port sample loop connected downstream of the reactor before expansion occurs at the BPR. This sample loop is connected to the carrier line of a dedicated gas chromatograph and is triggered at set intervals to inject small-volume, high-pressure aliquots of the product stream directly for GC analysis.
- Extensive safety features to ensure that the system can be left to run continuously over several hours/days with automatic safety cut-outs in the event that a system parameter goes outside safe operating limits.

The major advantage of an automated reactor is that the apparatus can be left unattended for long periods of time while collecting data throughout. More traditional supercritical flow equipment does not permit such operations, and requires samples to be taken manually by the experimentalist for analysis. In addition, operating variables need to be changed much more abruptly, thereby necessitating long equilibration times and sample collecting periods. In contrast, by careful planning and programming of the automated equipment, reaction variables can be slowly and automatically varied and analysis of the reaction stream can be performed via the high-pressure sample loop without the need for human input. This leads to a significant increase in the productivity of a given $scCO_2$ continuous flow reactor when it is fully automated, in addition to being less labor intensive for the experimenter.

The following guidelines show the practices which should be observed when using the automated systems described above.

1. The Jasco PU 1580 CO_2 pump should remain on with the Peltier cooling unit switched to the cooling position (the switch is on the back of the pump). The CO_2 cylinder should remain open and the valve (T_{CO_2} in Figure 2.11) should be left open and only closed when changing the cylinder. These procedures avoid sudden changes in the temperature and pressure of the chilled pump, which can shorten the lifetime of check valves.
2. Regular testing of the system should be conducted. This includes the following important checks:
   –Ensure that the Jasco pumps are working at the desired volumetric flow rate by pumping organic substrate directly into a measuring cylinder.

parameters. Further, a hydrogen detector (ANALOX 1300/2) is fitted with a powered valve (PV) to shut off the hydrogen supply in the event of hydrogen being released into the laboratory. The components are connected via 1/16 in stainless-steel Swagelok fittings and pipework. The reactor is mounted on a 1/4 in NPT cross-piece (MC2) with necessary adapters to accommodate thermocouples and standard 1/16 in piping connections and is filled with glass beads to act as a pre-heater. The other end of the reactor is attached to a 1/4 in NPT T-piece which enables an internal k-type thermocouple to be positioned at the outlet of the reactor.

–The Jasco pumps check valves should be tested regularly by closing the taps immediately downstream of the pumps (T_{ORG} and T_{CO_2} for the organic and CO_2 HPLC pumps, respectively) and observing the increase in pressure.
–The system should be tested for leaks by pressurizing the equipment with CO_2 at operating pressures using a Swagelok Snoop liquid leak detector.
–Ensure gas dosage systems are working correctly by checking the correct dosage volumes are being delivered by dosing gases directly into an inverted water-filled container and measuring the water displaced.
–Check that pressure transducers are reading correctly and recalibrate if necessary.

An example of the application of the system is detailed in this section. The dehydrogenation of 1-vinylcyclohex-4-ene to form a mixture of ethylcyclohexane and ethylbenzene was studied [85]. When a study of the effect of temperature was carried out, it was found that, within a very small temperature range, the conversion of 1-vinylcyclohex-4-ene had gone from 0 to 100%. The results observed and shown in Figure 2.12 would have been very difficult to obtain using a manual system due to the number of data points needed to locate such a rapid change and the resulting time it would take to carry out the experiment. The automated system also allowed the temperature to be decreased, to show that the results obtained were reproducible for decreasing temperatures. Again, these experiments would have been impractical on a manual system due to the large amount of data required to produce these results.

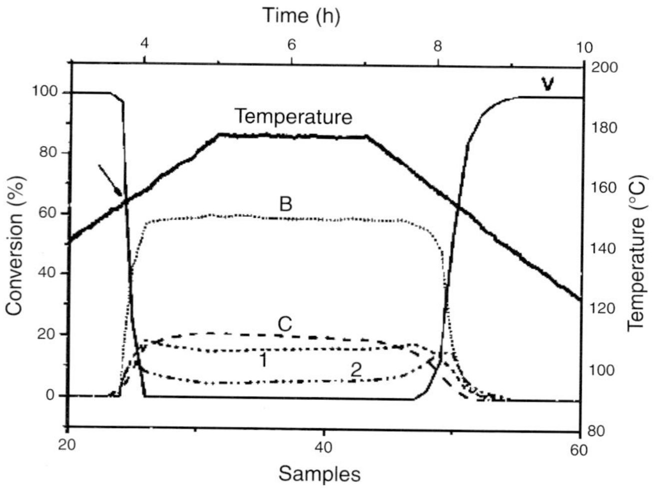

Figure 2.12 Plot of the product distribution of ethylbenzene (B) and ethylcyclohexane (C), observed while the catalyst bed is heated (bold line) between 140 and 180 °C in the dehydrogenation of 1-vinylcyclohex-4-ene (V). Note the arrowed autothermal ignition point at around 150 °C and the presence of intermediates 1-ethylcyclohex-4-ene (1) and 1-ethylcyclohexa-4,6-diene (2). Flow rates: CO_2, 1 ml min^{-1}: V, 0.2 ml min^{-1}. The product composition was sampled every 15 min over the 10 h period. Reproduced with permission from [86], copyright Wiley-VCH Verlag GmbH.

2.9 Conclusion

It is hoped that this chapter will contribute to the increased usage of SCF both in academia and in industrial research. We are aware of the fact that the handling of stainless-steel apparatus and its measurement and control technology might be strange at the beginning – at least for chemists who are used to working mainly with standard glassware. However, even after 1–2 months of regular usage, almost all students who have passed through our high-pressure laboratories had become fairly familiar with the handling of high-pressure experiments. It is not an exaggeration to assert that most of them have really sensed lots of fun working in this field. Another barrier might be the high investment costs at the outset, and of course one cannot deny that. However, one should take into account that an HPLC pump (which has to be equipped with a coolable head), a magnetic stirrer and a suitable balance are already present in many synthetic laboratories. The remaining initial costs for a single high-pressure autoclave with valves, thermocouples, a manometer and a reactor enclosure will often not exceed US$5000. In other words: give SCFs a chance in your laboratory!

References

1 Franko-Filipasic, B.R. and Michaelson, R.C. (1984) *Chemical Engineering Progress*, **80**, 64–69.

2 Livingston, E.H. (1984) *Chemical Engineering Progress*, **80**, 70–75.

3 Meyn, I. Ausführung und Ausrüstung von Hochdruck-Forschungslaboratorien, company literature of Andreas Hofer Hochdrucktechnik GmbH, year of publication unknown.

4 Penninger, J.M.L. and Okazaki, J.K. (1980) *Chemical Engineering Progress*, **76**, 65–71.

5 Gräfen, H. (2005) *Ullmann's Encyclopedia of Industrial Chemistry, Construction Materials in Chemical Industry*, Wiley-VCH Verlag GmbH, Weinheim.

6 Gräfen, H., Horn, E.-M., Schlecker, H. and Schindler, H. (2005) *Ullmann's Encyclopedia of Industrial Chemistry, Corrosion*, Wiley-VCH Verlag GmbH, Weinheim.

7 http://www.hightempmetals.com/techdata/hitempHastBdata.php.

8 http://www.haynesintl.com/pdf/h2104.pdf.

9 http://www.haynesintl.com/pdf/h2007.pdf.

10 http://www.haynesintl.com/pdf/h2019.pdf.

11 http://www.hightempmetals.com/techdata/hitempHastC276data.php.

12 http://www.haynesintl.com/pdf/h2118.pdf.

13 http://www.haynesintl.com/pdf/h2121.pdf.

14 http://www.hightempmetals.com/techdata/hitempInconel600data.php.

15 http://www.pxprecimet.ch/upload/alliages/de/Recapitulatif_alliages.pdf.

16 http://www.marcorubber.com/material_chart.htm.

17 http://www.helicoflex.com/.

18 http://www.seals.saint-gobain.com.

19 http://www.sealco.co.uk/index.htm.

20 Vetter, G. and Karl, E. (2005) *Ullmann's Encyclopedia of Industrial Chemistry, High-pressure Technology*, Wiley-VCH Verlag GmbH, Weinheim.

21 Middleton, J.C. and Carpenter, K.J. (2005) *Ullmann's Encyclopedia of Industrial*

Chemistry, Stirred-tank and Loop Reactors, Wiley-VCH Verlag GmbH, Weinheim.

22 Licence, P., Dellar, M.P., Wilson, R.G.M., Fields, P.A., Litchfield, D., Woods, H.M., Poliakoff, M. and Howdle, S.M. (2004) *Review of Scientific Instruments,* **75**, 3233–3236.

23 Hoffmann, M.M. and Salter, J.D. (2004) *Journal of Chemical Education,* **81**, 411–413.

24 Byun, H.-S. and Choi, T.-H. (2004) *Korean Journal of Chemical Engineering,* **21**, 1032–1035.

25 Lee, S.H., Luszynski, K., Norberg, R.E. and Conradi, M.S. (1987) *Review of Scientific Instruments,* **58**, 415.

26 de Fries, T.H. and Jonas, J. (1979) *Journal of Magnetic Resonance,* **35**, 111.

27 Rathke, J.W., Klingler, R.J. Gerald, R.E. II, Fremgen, D.E., Woelk, K., Gaemers, S. and Elsevier, C.J. (1999) in *Chemical Synthesis Using Supercritical Fluids* (eds P.G. Jessop and W. Leitner), Wiley-VCH Verlag GmbH, Weinheim, pp. 65–194.

28 Grunwaldt, J.-D., Wandeler, R. and Baiker, A. (2003) *Catalysis Reviews,* **45**, 1–96.

29 Kramarz, K.W., Klingler, R.J., Fremgen, D.E. and Rathke, J.W. (1999) *Catalysis Today,* **49**, 339–352.

30 Rathke, J.W., Klingler, R.J., Gerald, R.E. II, Kramarz, K.W. and Woelk, K. (1997) *Progress in Nuclear Magnetic Resonance Spectroscopy,* **30**, 209–253.

31 Roe, D.C. (1989) *Journal of Magnetic Resonance,* **85**, 150.

32 Gaemers, S., Luyten, H., Ernsting, J.M. and Elsevier, C.J. (1999) *Magnetic Resonance in Chemistry,* **37**, 25–30.

33 Wallen, S.L., Schoenbachler, L.K., Dawson, E.D. and Blatchford, M.A. (2000) *Analytical Chemistry,* **72**, 4230–4234.

34 Vanni, H., Earl, W.L. and Merbach, A.E. (1978) *Journal of Magnetic Resonance,* **29**, 11–19.

35 Zahl, A., Igel, P., Weller, M. and van Eldik, R. (2004) *Review of Scientific Instruments,* **75**, 3152–3157.

36 Anderson, N.G. (2001) *Organic Process Research & Development,* **5**, 613–621.

37 Hyde, J.R., Licence, P., Carter, D. and Poliakoff, M. (2001) *Applied Catalysis A,* **222**, 119–131.

38 Hou, Z., Theyssen, N. and Leitner, W. (2007) *Green Chemistry,* **9**, 127–132.

39 Henkel, K.-D. (2005) *Ullmann's Encyclopedia of Industrial Chemistry, Reactor Types and Their Industrial Applications,* Wiley-VCH Verlag GmbH, Weinheim.

40 Cresswell, D., Gough, A. and Milne, G. (2005) *Ullmann's Encyclopedia of Industrial Chemistry, Tubular Reactors,* Wiley-VCH Verlag GmbH, Weinheim.

41 Kiwi-Minsker, L. and Renken, A. (2005) *Catalysis Today,* **110**, 2–14.

42 Watts, P. and Haswell, S.J. (2005) *Chemical Society Reviews,* **34**, 235–246.

43 Schmidt, L.D. (1998) *The Engineering of Chemical Reactions,* Oxford University Press, Oxford.

44 Fogler, H.S. (2005) *Elements of Chemical Reaction Engineering,* 4th edn, Prentice-Hall, Englewood Cliffs, NJ.

45 Westerterp, K.R. and Wijngaarden, R.J. (2005) *Ullmann's Encyclopedia of Industrial Chemistry, Principles of Chemical Reaction Engineering,* Wiley-VCH Verlag GmbH, Weinheim.

46 Theyssen, N., Hou, Z. and Leitner, W. (2006) *Chemistry - A European Journal,* **12**, 3401–3409.

47 Grunwaldt, J.-D., Ramin, M., Rohr, M., Michailovski, A., Patzke, G.R. and Baiker, A. 2005 *Review of Scientific Instruments,* **76**, 054104/1–054104/7.

48 Grunwaldt, J.-D., Caravati, M. and Baiker, A. (2006) *The Journal of Physical Chemistry. B,* **110**, 25587.

49 Minnich, C.B., Küpper, L., Liauw, M.A. and Greiner, L. (2007) *Catalysis Today,* **126**, 191–195.

50 Webb, P.B., Sellin, M.F., Kunene, T.E., Williamson, S., Slawin, A.M.Z. and Cole-Hamilton, D.J. (2003) *Journal of the American Chemical Society,* **125**, 15577–15588.

51 Hou, Z., Theyssen, N., Brinkmann, A. and Leitner, W. (2005) *Angewandte Chemie*

(International Edition in English), **44**, 1346–1349.

52 Elbashir, N.O. and Roberts, C.B. (2005) *Industrial & Engineering Chemistry Research*, **44**, 505–521.

53 Licence, P., Ke, J., Sokolova, M., Ross, S.K. and Poliakoff, M. (2003) *Green Chemistry*, **5**, 99–104.

54 Ikushima, Y., Hatakeda, K., Sato, M., Sato, O. and Arai, M. 2002 *Chemical Communications*, 2208–2209.

55 Callahahn, F.J. (1998) *Swagelok Tube Fitter's Manual*, Swagelok Marketing Co., Solon, OH.

56 http://www.parker.com/ICD/cat/english/4200-B4.pdf.

57 Bierweth, W. (1997) *Tabellenbuch Chemietechnik*, Europa Lehrmittel, Haan.

58 http://www.parrinst.com/.

59 http://www.premex-reactorag.ch/e/.

60 Zlokarnik, M. (2005) *Ullmann's Encyclopedia of Industrial Chemistry, Stirring*, Wiley-VCH Verlag GmbH, Weinheim.

61 http://www.metaglas.com.

62 Koch, D. and Leitner, W. (1998) *Journal of the American Chemical Society*, **120**, 13398–13404.

63 Sato, K., Sasaki, S.S., Goda, Y., Yamada, T., Nunomura, O., Ishikawa, K. and Maitani, T. (1999) *Journal of Agricultural and Food Chemistry*, **47**, 4665–4668.

64 Hind, A.R., Bhargava, S.K. and McKinnon, A. (2001) *Advances in Colloid and Interface Science*, **93**, 91–114.

65 http://las.perkinelmer.com/content/TechnicalInfo/TCH_FTIRATR.pdf.

66 Minnich, C.B., Buskens, P., Steffens, H.C., Baeuerlein, P.S., Butvina, L.N., Kuepper, L., Leitner, W., Liauw, M.A. and Greiner, L. (2007) *Organic Process Research & Development*, **11**, 94–97.

67 Buergi, T. and Baiker, A. (2006) *Advances in Catalysis*, **50**, 227–283.

68 http://webbook.nist.gov/chemistry/fluid/.

69 Hoppe, T. and Jaeger, N. (2005) *Process Safety Progress*, **24**, 266–272.

70 Raynie, D.E. (1993) *Analytical Chemistry*, **65**, 3127–3128.

71 http://en.wikipedia.org/wiki/Mass_flow_controller.

72 Hoffman, J. (1998) *Taschenbuch der Messtechnik*, Carl Hanser Verlag, Munich, pp. 186–187.

73 http://www.emg.ing.tu-bs.de/pdf/MNG/7_Durchflussmessung.pdf.

74 http://www.bronkhorst-maettig.de/files/downloads/brochures/folder-liqui-flow.pdf.

75 http://www.advanced-energy.com/upload/File/White_Papers/SL-MFCFUND-270-01.pdf.

76 (a) http://en.wikipedia.org/wiki/Mass_flow_meter; (b) http://www.edn.com/contents/images/305490.pdf; (c) http://www.efunda.com/DesignStandards/sensors/flowmeters/flowmeter_cor.cfm;(d) http://www.rheonik.com/c.php/englisch/Products/Product_Overview/product_overview.rsys; (e) http://www.krohne-mar.com/Measuring_Principle_Coriolis_Mass_Flowmeters_en.730.0.html.

77 (a) Hutchison, J.W. *ISA Handbook of Control Valves*, 2nd edn, Instrument Society of America, Pittburgh, PA; (b) http://www.highpressure.com/pdfs/section/ps.pdf.

78 http://www.parrinst.com/default.cfm?page_id=22.

79 Method for suspending and introducing solid matter in a high pressure process. Patent WO/2005/090667, 2005.

80 Method for carrying out continuous ring closing metathesis in compressed carbon dioxide, Patent WO/2006/075021, 2006.

81 http://www.voegtlin.com.

82 http://www.actaris.com.

83 Shaw, R.K. and Mueller, A.C. (2005) *Ullmann's Encyclopedia of Industrial Chemistry, Heat Exchange*, Wiley-VCH Verlag GmbH, Weinheim.

84 Walsh, B., Hyde, J.R., Licence, P. and Poliakoff, M. (2005) *Green Chemistry*, **7**, 456–463.

85 Hitzler, M.G. and Poliakoff, M. 1997 *Chemical Communications*, 1667–1668.

86 Hyde, J.R., Walsh, B. and Poliakoff, M. (2005) *Angewandte Chemie (International Edition in English)*, **44**, 7588–7591.

3
Basic Physical Properties, Phase Behavior and Solubility

Neil R. Foster, Frank P. Lucien, and Raffaella Mammucari

3.1
Introduction

The ability of supercritical fluids (SCFs) to act as solvents has been known for well over a century [1], yet the most significant developments in the application of this technology have taken place in the last few decades. Among the most recent applications, SCFs have been employed as media for chemical reactions. As a reaction medium, the SCF may either participate directly in the reaction or simply act as a solvent for the various chemical species. The physical properties of SCFs are highly dependent on pressure and temperature, which makes it possible to fine tune the reaction environment. These characteristics are unique to SCFs and provide the potential to tune the reaction environment in order to optimize reaction rate and selectivity.

In view of their importance to process design, fundamental aspects of SCF technology are described in this chapter. Basic physical properties of SCFs are presented initially, followed by consideration of phase behavior in high-pressure systems. Lastly, the factors affecting the solubility of components in SCFs are described.

3.2
Basic Physical Properties of Supercritical Fluids

A fluid is defined as being supercritical if it is maintained at conditions above its critical temperature and pressure (T_c and p_c). The critical point, as defined by T_c and p_c, marks the end of the liquid–vapor coexistence curve in the phase diagram for a pure substance, as shown in Figure 3.1. A supercritical region originating from the critical point can also be identified in this phase diagram. In general, only a single homogeneous phase can exist in the supercritical region, irrespective of the temperature and pressure. However, in some cases, solidification may occur at very high pressure. Changes in phase from liquid to vapor are accompanied by abrupt changes in physical properties, but in the supercritical region properties may be varied continuously by

Handbook of Green Chemistry, Volume 4: Supercritical Solvents. Edited by Walter Leitner and Philip G. Jessop

ISBN: 978-3-527-32590-0

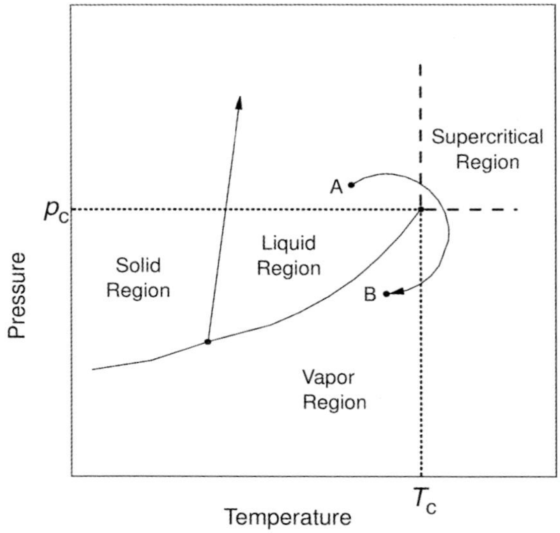

Figure 3.1 Phase diagram for a pure substance.

manipulating the temperature and pressure. It is possible to construct the path shown in Figure 3.1, where the progression from a liquid state to a gas state, and vice versa, proceeds without any visual signs of boiling or condensation.

The properties of SCFs vary over a wide range depending on the temperature and pressure, but are generally intermediate to those of gases and liquids. Selected physical properties of SCFs are shown in Table 3.1 along with typical values for gases and liquids.

The properties of SCFs are very sensitive to small changes in temperature and pressure in the vicinity of the critical point. This applies especially to density, as shown in Figure 3.2, which represents the projection of the liquid–vapor coexistence curve in the density–pressure plane. Pressure and temperature are represented in terms of their reduced values, which are simply the actual values divided by the critical values ($p_r = p/p_c$, $T_r = T/T_c$). The left portion of the diagram reveals the two-phase envelope and the equilibrium between liquid and vapor phases is indicated by a vertical tie-line. The supercritical region in this view of the phase diagram is defined by the region to the right of the $T_r = 1.0$ isotherm and for $p_r > 1.0$.

In the immediate vicinity of the critical point, the density of supercritical CO_2 is around $0.4\,g\,ml^{-1}$. For reduced pressures greater than 2, it can be seen that the density of supercritical CO_2 is comparable to that for liquid CO_2. It is this liquid-like

Table 3.1 Comparison of the physical properties of gases, liquids, and SCFs [2].

Property	Gas	SCF	Liquid
Density ($g\,cm^{-3}$)	10^{-3}	0.3	1
Viscosity (Pa s)	10^{-5}	10^{-4}	10^{-3}
Diffusivity (cm^2 s)	0.1	10^{-3}	5×10^{-6}

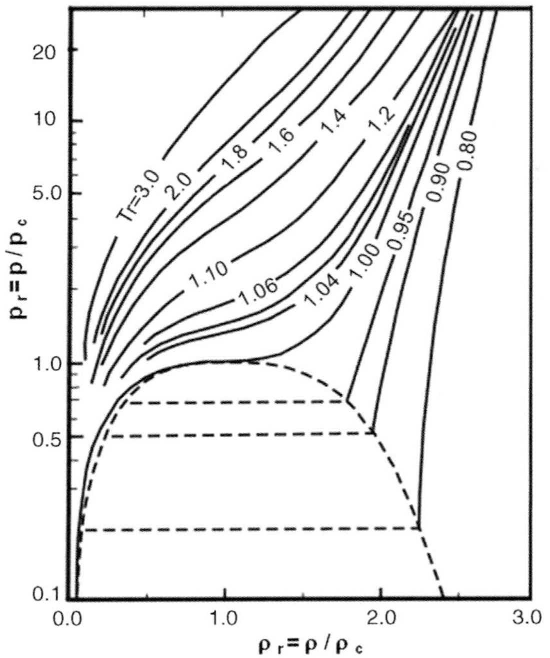

Figure 3.2 Variation of the reduced density of CO_2 in the vicinity of its critical point [2].

density which enables many materials to be solubilized to a level which is several orders of magnitude greater than that predicted by ideal gas considerations. Since density is a measure of the solvating power of a SCF, temperature and pressure can be used as variables to control the solubility and separation of a solute. For example, at $T_r = 1.1$ in Figure 3.2, decreasing the reduced pressure from 3 to 1 lowers the density by around 80%. The conditions of temperature and pressure of most interest in SCF processes are usually bounded by reduced temperatures between 1.0 and 1.2 and reduced pressures greater than 1.

The diffusion coefficient (or diffusivity) and viscosity represent transport properties which affect rates of mass transfer. In general, these properties are at least an order of magnitude higher and lower, respectively, compared with liquid solvents. This means that the diffusion of a species through a SCF medium will occur at a faster rate than in a liquid solvent, which implies that a solid will dissolve more rapidly in a SCF. In addition, a SCF will be more efficient at penetrating a microporous solid structure. However, this does not necessarily mean that mass transfer limitations will be absent in a SCF process. For example, in the extraction of a solute from a liquid to a SCF phase, the resistance to diffusion in the liquid phase will probably control the overall rate of mass transfer. Stirring will therefore continue to be an important factor in such systems.

The self-diffusivity of CO_2 is presented in Figure 3.3 along with the range of diffusivities for solutes in organic liquids. The self-diffusivity of CO_2 is about one to two orders of magnitude higher than the diffusivity of solutes in normal liquids and is comparable to the diffusion coefficient in CO_2 of molecules having a similar size [3].

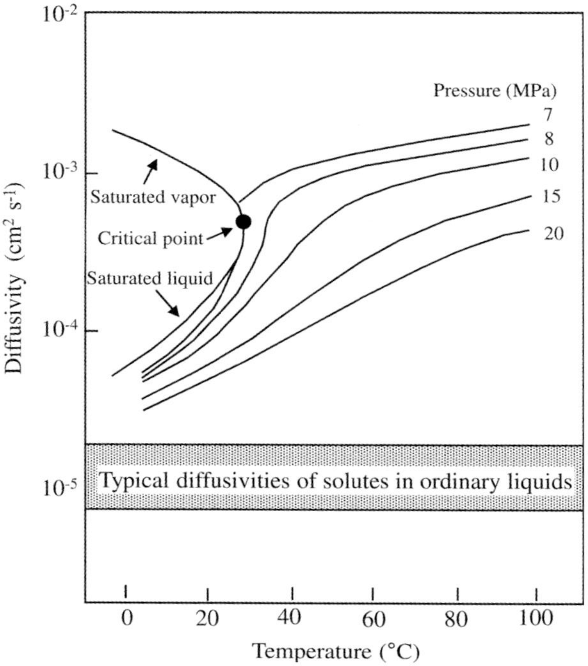

Figure 3.3 Self-diffusivity of CO_2 [3].

The diffusion coefficient varies with both temperature and pressure and is strongly influenced by density and viscosity [4]. Density and viscosity both increase with pressure (Figure 3.4) with a corresponding decrease in the diffusion coefficient. The effect is less pronounced at higher pressure because density becomes less sensitive to pressure. The diffusion coefficient generally increases with temperature at constant pressure. However, at constant density, temperature appears to have a minimal effect.

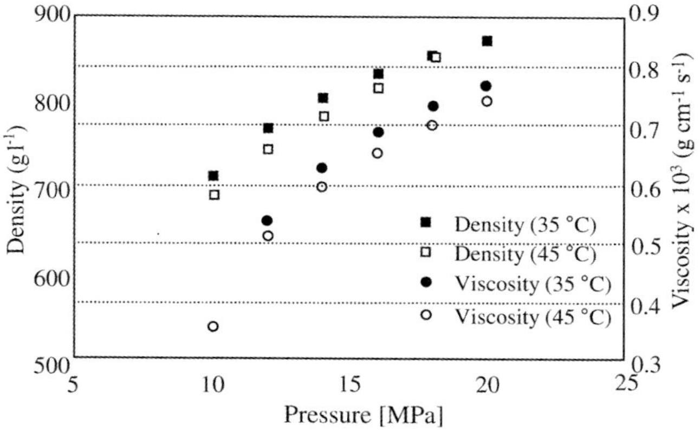

Figure 3.4 Effect of pressure on the density and viscosity of CO_2 [5].

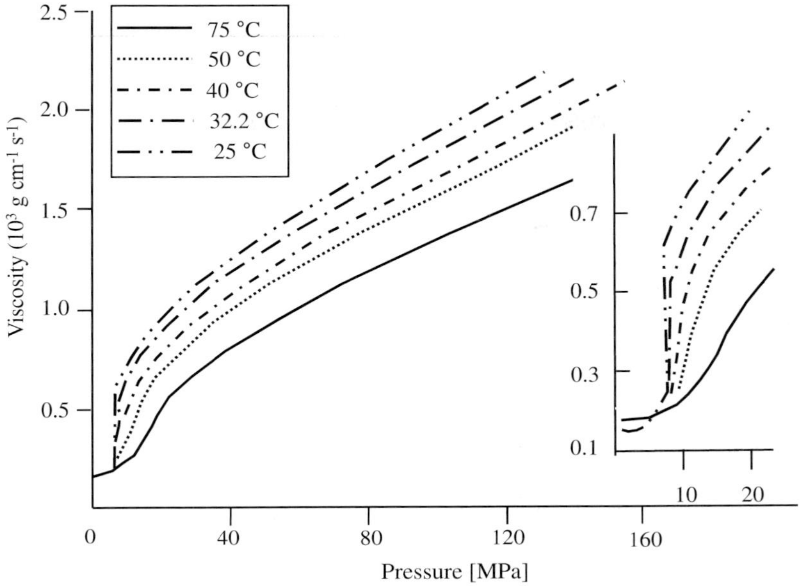

Figure 3.5 Viscosity of CO_2 versus pressure [6].

Temperature has a dual effect on viscosity. Increments of temperature at constant pressure reduce the density of the fluid and increase the intermolecular distance, which tends to facilitate the relative movement of molecules. An increase in temperature, however, also corresponds to higher kinetic energy and collision rates of the molecules, which hinder the movement of molecules. The variation of viscosity with pressure at different temperatures is represented in Figure 3.5 for CO_2.

Studies of the viscosity of several pure fluids, including CO_2, xenon, ethane, ethylene and nitrogen, have shown that an increase in viscosity occurs in the proximity of the critical point [5, 7]. The viscosity of SCF multi-component systems has also been studied. Systems containing CO_2 and hydrocarbons, or CO_2 and organic solvents, are of interest for applications such as natural gas processing [8] and the processing of a wide range of materials as the addition of small amounts of organic cosolvents can enhance the solvating power and the selectivity of SCFs [9]. In most cases, the viscosity of a SCF increases with the addition of modifiers. The extent of the variation depends on the molecular size, polarity, shape, and concentration of the modifier. Viscosity increments can be graded within a homologous group of modifiers (i.e. low molecular weight alcohols) based on the molecular weight. Cosolvent polarity, the possibility of establishing hydrogen bonds and the concentration of the modifier also generally increase the viscosities of multi-component CO_2 mixtures [9]. The effect of modifiers on the viscosity of CO_2 is presented in Figure 3.6.

The dielectric constant (ε) is closely related to properties such as solubility and reaction rate constant. In SCFs, ε has a dependence on pressure which parallels that of density (Figure 3.7a). The relationship between ε and reduced density is presented in Figure 3.7b, which shows how the variability of ε depends on the nature of the fluid.

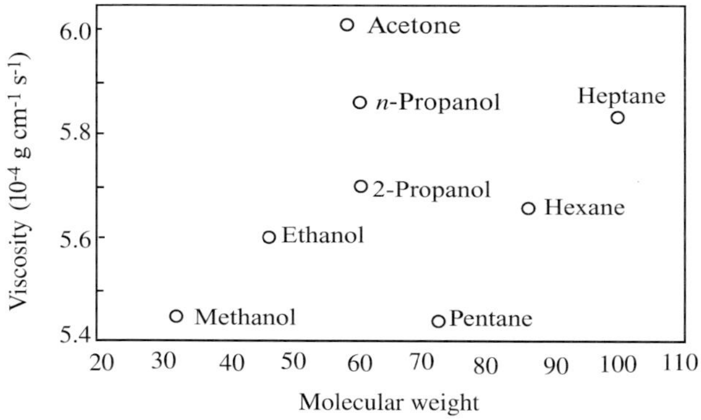

Figure 3.6 Viscosity of 2 mol% cosolvent–CO_2 systems at 45 °C and 12 MPa [9].

The dielectric constant of supercritical CO_2 changes minimally over a wide range of values of reduced density. The ε value for supercritical CO_2 diverges slightly from about 1.5, which indicates a substantially non-polar solvent. The dielectric constant of water, by comparison, exhibits an increase of more than 200% when the reduced pressure changes from 1 to 2.

Fluids with an ε which varies substantially with the operating conditions have potential application as tunable solvents. Figure 3.8 presents a comparison between the dielectric constant of water as a function of temperature (estimated at 100 MPa) [15, 16] and the dielectric constant of conventional organic solvents at room conditions (20 °C and 0.1 MPa) [16]. Practical applications that have developed from the tunability of the dielectric constant of water include extraction [17, 18] and chromatography [19].

In systems in which the dielectric constant is relatively unaffected by variations in temperature and pressure, tunability can be achieved through the addition of

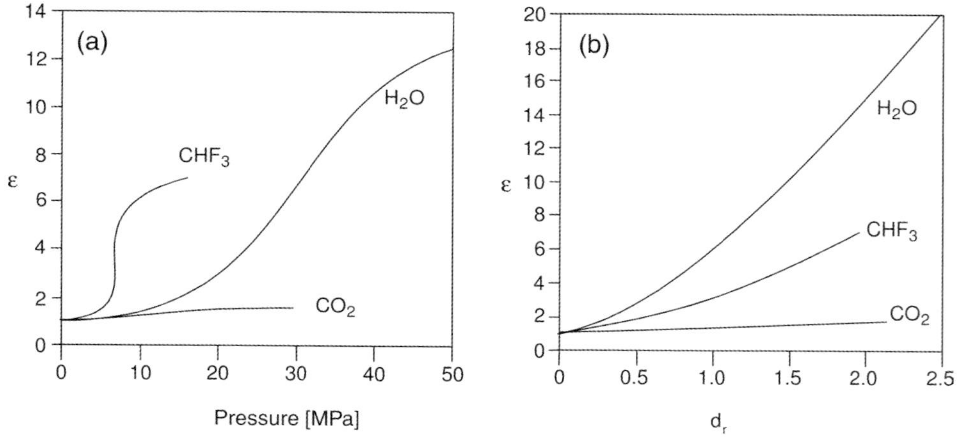

Figure 3.7 The dielectric constant of H_2O (400 °C, T_r = 1.04) [10, 11], CO_2 (40 °C, T_r = 1.03), [12] and CHF_3 (30 °C, T_r = 1.01), [13, 14] as functions of (a) pressure and (b) reduced density.

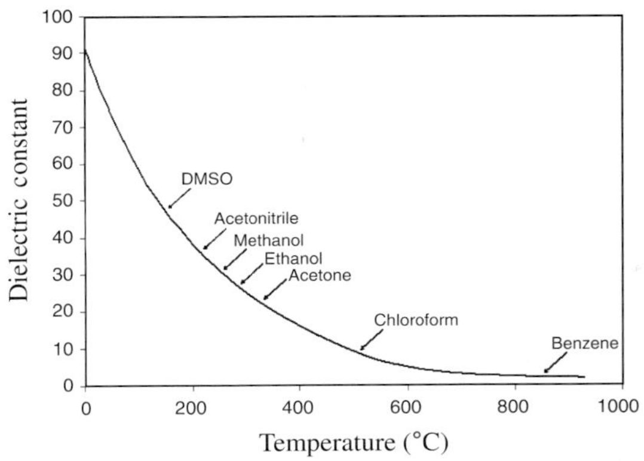

Figure 3.8 Dielectric constant of water as a function of temperature at 1000 bar. Comparison to organic solvent properties at room conditions [16].

modifiers. The effects of modifiers on the dielectric constant of SCFs can be illustrated by the ethane and CO_2 systems modified with methanol. As can be seen in Figure 3.9, the dielectric constants of ethane and CO_2 increases with the methanol content. The dielectric constant for both systems approaches that of pure methanol at high methanol mole fractions and the values for each of the pure ethane and CO_2 at low methanol mole fractions.

However, ethane and CO_2 have similar pure-component dielectric properties, it can be observed from Figure 3.9 that the dielectric constant of the CO_2–methanol system is higher than the dielectric constant of the ethane–methanol system at any of the tested conditions [20]. The phenomenon highlights how the effect of modifiers on the dielectric properties of SCFs depends on specific interactions. In the CO_2–methanol system, the effect of methanol on the dielectric properties can be related to the formation of CO_2–methanol complexes and to CO_2–methanol intermolecular interactions resulting from the quadrupole moment of the CO_2 molecules. Ethane is not expected to exhibit either of the two properties [20].

Other parameters affecting the dielectric constant of supercritical systems including a modifier are temperature and density. In the ethane–methanol and

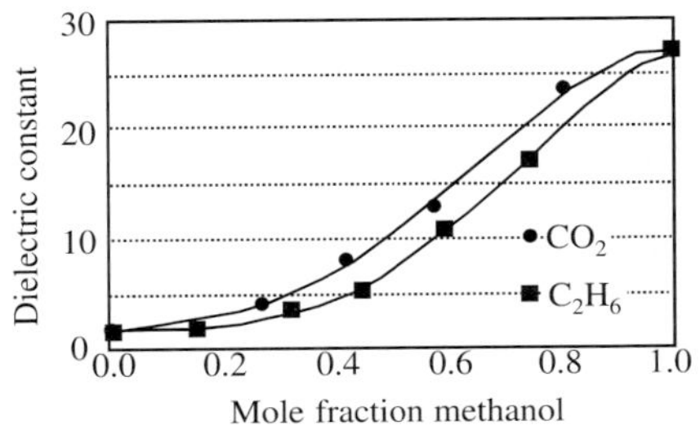

Figure 3.9 Dielectric constant of CO_2–methanol (●) and ethane–methanol (■) mixtures as a function of mole fraction at 17.5 MPa and 65 °C [20].

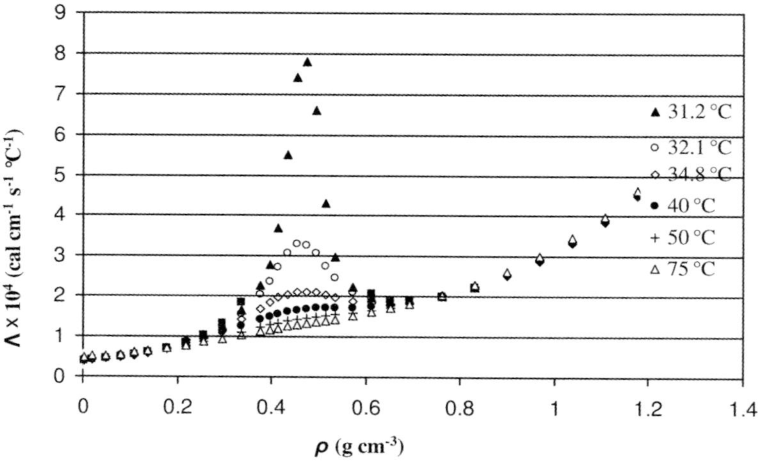

Figure 3.10 Thermal conductivity of CO_2 versus density [21].

CO_2–methanol systems, the dielectric constants increases with temperature reductions and density increments [20].

Knowing the thermal conductivity of SCFs is essential in the design of process equipment such as heat exchangers for the recovery of thermal energy in SCF extraction or fractionation processes. The thermal conductivity of fluids changes dramatically during phase changes and in the proximity of the critical point. The thermal conductivity of supercritical CO_2 as a function of density is presented in Figure 3.10. Thermal conductivity is higher in proximity to the critical density of CO_2 and the critical temperature [21]. Similarly, ammonia, ethene, ethane, propane, *n*-butane, isobutane, pentane, sulfur hexafluoride, and CF_3Cl presented enhancements of thermal conductivity in proximity of the critical point [22]. Generally, the thermal conductivity of pure fluids exhibits a strong and finite enhancement near the vapor–liquid critical points (plait points), behavior that has been confirmed for binary SCF mixtures [23].

The changes in thermal conductivity of CO_2 with density follow a similar trend to that of the specific heat. In particular, experimental observations have shown that the variation of the thermal conductivity with viscosity is in a fixed ratio with the variation of the specific heat. The similarities observed suggest that the variations of thermal conductivity and other phenomena are based on similar mechanisms, specifically the formation and breaking up of molecular clusters because of temperature gradients [21].

Supercritical fluids have a number of distinct advantages over conventional liquid solvents. The adjustable solvent strength and favorable transport properties have already been mentioned, and it is these features which really differentiate SCFs from liquid solvents. Most SCFs are low molecular weight gases which have relatively low critical temperatures. Operations may therefore be carried out at moderate temperatures, which is desirable in the recovery of thermally labile materials. Perhaps the most important advantage offered by SCFs is that after the release of pressure, components are left virtually free of residual supercritical solvent.

The majority of studies involving SCFs has focused on four fluids: CO_2, ethylene, ethane, and water. The first three fluids all have critical temperatures below 35 °C. Carbon dioxide is by far the most widely used SCF and has several advantages over the hydrocarbons in that it is non-toxic, non-flammable and readily available in high purity. Carbon dioxide also exhibits low miscibility with water and is a moderately good solvent for many low to medium molecular weight organics. Reported solubilities for non-polar solids and liquids in CO_2 range from 0.1 to 10 mol%. Marginally higher solubilities are obtained in ethane and ethylene.

Supercritical water (SCW) has a very high critical temperature and pressure in comparison with the other commonly used SCFs. Nonetheless, SCW has been investigated for more than a decade as a medium for the oxidation of organic wastes [24]. The properties of SCW are different to those of ordinary water. In the vicinity of the critical point, SCW behaves like a moderately polar organic liquid. The dielectric constant is reduced to the point where organic materials are readily soluble whereas the solubility of inorganic species is greatly reduced. These solvation characteristics make SCW an ideal medium for the oxidation of organics. The oxidation process leads to the formation of simple and non-toxic compounds such as water and CO_2. The addition of a caustic solution to the process promotes the conversion of heteroatoms such as Cl, F, P, and S into simple salts which precipitate from the reactor effluent.

Lists of inorganic and organic compounds which are commonly used as SCFs are presented in Tables 3.2 and 3.3. The compounds are listed with their critical parameters, physical properties, and molecular weights. References for their volumetric behavior are given in the last column. It is evident from the data reported that SCFs usually require the use of pressure in excess of at least 4 MPa. It is interesting that linear hydrocarbons generally have a critical pressure below 5 MPa and a critical temperature which increases with molecular weight. In addition, substances which are capable of hydrogen bonding require relatively high critical temperatures and pressures.

Table 3.2 Selected inorganic supercritical fluids.

SCF	Symbol	T_c (°C)	p_c (MPa)	d_c (g ml^{-1})	MW	μ (D)[a]	Ref.
Ammonia	NH_3	132.4	11.32	0.235	17.03	1.47	[25–27]
Argon	Ar	−122.5	4.86	0.531	39.95	0	[28, 29]
Carbon dioxide	CO_2	31.1	7.38	0.466	44.01	0	[30, 31]
Hydrogen bromide	HBr	90.0	8.55	n.a.	80.91	0.82	[32]
Hydrogen chloride	HCl	51.5	8.26	0.42	36.46	1.08	[32–34]
Hydrogen iodide	HI	150.7	8.3	n.a.	127.9	0.44	[32]
Krypton	Kr	−63.76	5.49	0.912	83.80	0	[28]
Nitrous oxide	N_2O	36.4	7.25	0.453	44.01	0.167	[35]
Sulfur hexafluoride	SF_6	45.5	3.76	0.737	146.1	0	[36–39]
Water	H_2O	374.0	22.06	0.322	18.02	1.85	[40–42]
Xenon	Xe	16.6	5.83	1.099	131.3	0	[28, 43]

[a]Ref. 16.

Table 3.3 Selected organic supercritical fluids.

SCF	Symbol	T_c (°C)	p_c (MPa)	d_c (g ml^{-1})	MW	μ (D)[a]	Ref.
Benzene	C_6H_6	289.5	4.92	0.300	78.11	0	[44, 45]
n-Butane	C_4H_{10}	152.0	3.80	0.228	58.12	<0.05	[46, 47]
Difluoromethane	CH_2F_2	78.1	5.78	0.424	52.02	1.97	[48, 49]
Dimethyl ether	C_2H_6O	126.9	5.4	0.242	46.07	1.30	[50]
Ethane	C_2H_6	32.2	4.87	0.207	30.07	0	[30, 46, 51, 52]
Ethene	C_2H_4	9.2	5.04	0.214	28.05	0	[53–55]
Ethylenediamine	$C_2H_8N_2$	320	6.28	0.29	60.10	1.99	[56]
Fluoroform	CHF_3	25.9	4.82	0.525	70.01	1.65	[57, 58]
n-Hexane	C_6H_{14}	234.5	3.03	0.234	86.18	n.a.	[46]
Isobutane	C_4H_{10}	134.7	3.64	0.224	58.12	0.132	[47]
Methane	CH_4	−82.6	4.60	0.163	16.04	0	[59, 60]
Methanol	CH_4O	239.5	8.08	0.273	32.04	1.70	[61–63]
n-Pentane	C_5H_{12}	196.6	3.37	0.232	72.15	n.a.	[46]
Propane	C_3H_8	96.7	4.25	0.220	44.10	0.084	[46, 47, 64]
Propene	C_3H_6	91.8	4.60	0.228	42.08	0.366	[47, 55, 65]

[a]Ref. [16].

In some applications, the pressure required for a SCF process may result in prohibitively high capital investment for process equipment. However, it is possible to increase the solvent power of a primary SCF with the addition of small amounts of cosolvents, such as methanol or acetone. The significance of this is that lower operating pressures (and temperatures) are made possible. The selectivity of the SCF for a particular component may also be improved with the addition of a cosolvent.

3.3 Phase Behavior in High-Pressure Systems

The development of SCF processes involves considering of the phase behavior of multicomponent systems. The influence of pressure and temperature on phase behavior in such systems is complex. For example, it is possible to have multiple phases, such as liquid–liquid–vapor or solid–liquid–vapor equilibria, present in the system. In many cases, the operation of a SCF process under multi-phase conditions may be undesirable and so phase behavior should first be investigated. The limiting case of equilibrium between two components (binary systems) provides a convenient starting point in the understanding of multicomponent phase behavior.

3.3.1 Types of Binary Phase Diagrams

The phase behavior of most binary systems can be described by nine types of phase diagrams which can be predicted qualitatively with the van der Waals equation of state [66]. The nine types of phase diagrams may be further grouped into five major classes designated I–V. A sixth class exists for some aqueous systems but it is much less common than the other classes and is not considered here. The phase diagrams are based on the projection of three-phase lines and mixture critical lines from pressure–temperature-composition (p–T–x) space onto the p–T plane. A detailed discussion on the three-dimensional features of the various classes of phase diagrams was given by McHugh and Krukonis [67] and Streett [68]. It should be noted that some differences exist between the classification systems of these two authors. The system referred to here is that described by Streett [68].

The simplest class of binary phase diagram is class I, as shown in Figure 3.11. The component with the lower critical temperature is designated component 1. The solid

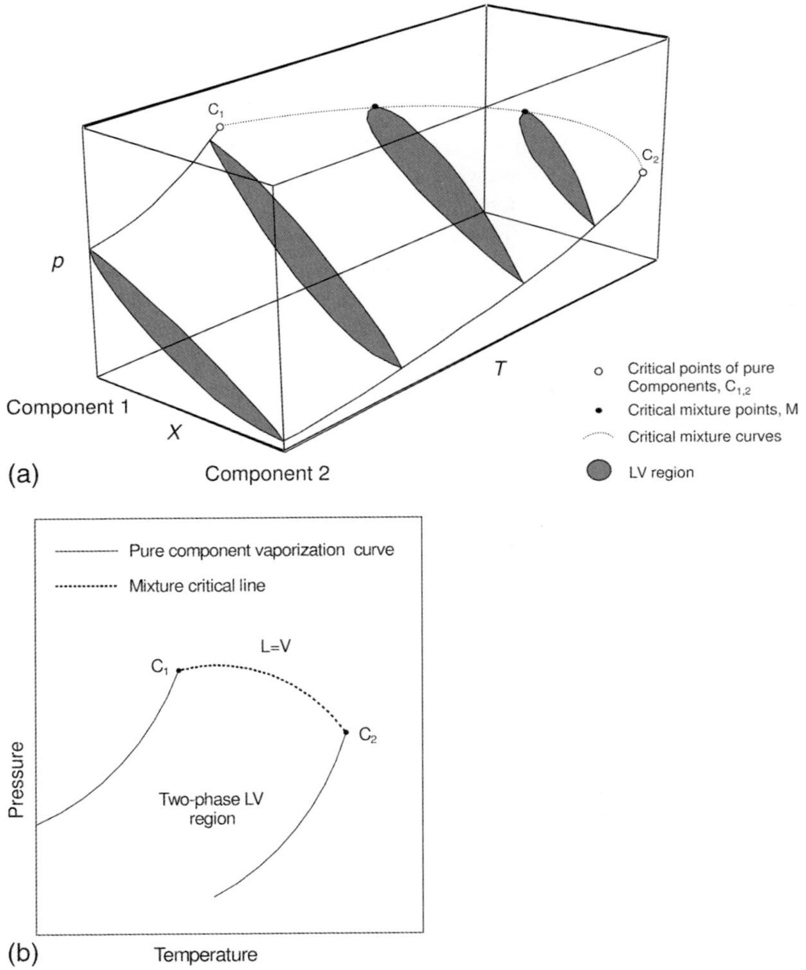

Figure 3.11 Class I binary phase diagram: (a) three-dimensional representation in p–T–x space; (b) p–T projection.

Table 3.4 Examples of the five classes of phase diagrams [68].

Class	System
I	Ar–Kr, N_2–O_2
II	CO_2–*n*-octane, CO_2–2-hexanol
III	Ethane–methanol, CO_2–H_2O
IV	CO_2–nitrobenzene
V	Ethane–ethanol, ethane–*n*-propanol

lines represent the pure component liquid–vapor coexistence curves, which terminate at the pure component critical points (C_1 and C_2). The feature of importance in this phase diagram is that the mixture critical line (dashed line) is continuous between the two critical points. The mixture critical line represents the locus of critical points for all mixtures of the two components. The area bounded by the solid and dashed lines represents the two-phase, liquid–vapor (LV) region. The mixture critical line is denoted L = V since it represents the merging of liquid and vapor phases. The mixture critical line shown in Figure 3.11 does not necessarily exhibit a maximum and can take on a variety of shapes.

The distinctions between the various classes of phase diagrams are based mainly on the behavior of the mixture critical line. The class II phase diagram, like class I, also has a continuous mixture critical line between the pure component critical points. In classes IV and V, the mixture critical line originating from C_2 exhibits a maximum in pressure and then intersects an additional three-phase, liquid–liquid–vapor (LLV) line. In the class III phase diagram, the mixture critical line which originates from C_2 rises to very high pressures and can take on a number of different shapes. Examples of the different shapes of mixture critical lines in class III systems were given by Alwani and Schneider [69]. Some typical examples for each of the five classes of phase diagrams are given in Table 3.4.

3.3.2 Asymmetric Binary Mixtures

Phase diagrams for binary systems consisting of components with widely separated critical temperatures are considered to be a subset of class III systems [68]. This situation often arises in supercritical extraction operations in which relatively nonvolatile components are extracted. The general features of the class III phase diagram are shown in Figure 3.12. The mixture critical line originating from C_2 (L = V) may pass through an inflection point before rising to higher pressures. These critical lines usually end at an intersection with a three-phase, solid–liquid–vapor (SLV) line (not shown). The dashed-dotted line represents a three-phase, LLV line, which terminates at a critical end-point (open triangle). The critical end-point also marks the termination of another mixture critical line (L = V) originating from C_1. The points M_1, M_2 and M_3 are located on the respective mixture critical lines or three-phase LLV line for the given temperature T_1.

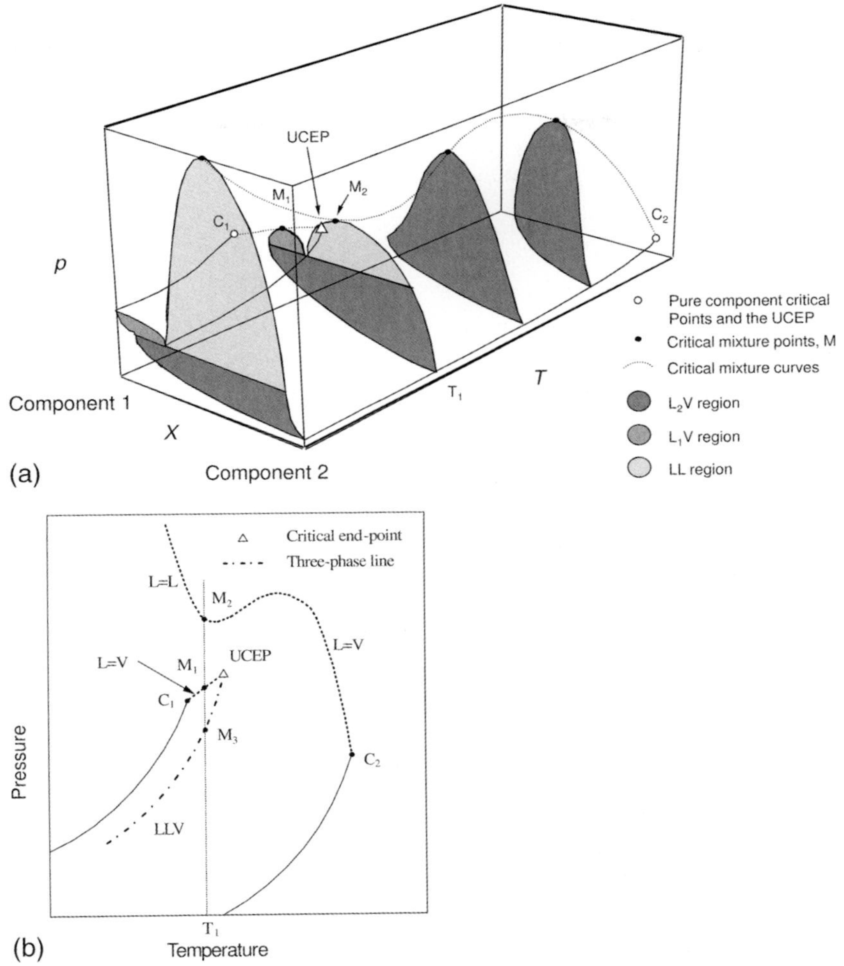

Figure 3.12 Class III binary phase diagram: (a) three-dimensional representation in *p–T–x* space; (b) *p–T* projection.

Critical end-points represent the conditions at which two of three coexisting phases merge and become identical. Both lower and upper critical end-points are possible depending on whether the end-point occurs at a high-temperature branch of a three-phase line (UCEP) or a low-temperature branch (LCEP). At the UCEP in Figure 3.12, a liquid and vapor phase critically merge into a single vapor phase in the presence of another liquid phase.

The presence of the LLV line in Figure 3.12 causes the upper mixture critical line to pass continuously from L = V to L = L in the vicinity of the UCEP. At a temperature T_1, between C_1 and the UCEP, the p–x diagram has the characteristics shown in Figure 3.13. As pressure is increased from p_1 to p_2, a single vapor phase splits into a two-phase LV region and eventually a three-phase LLV point is reached. The

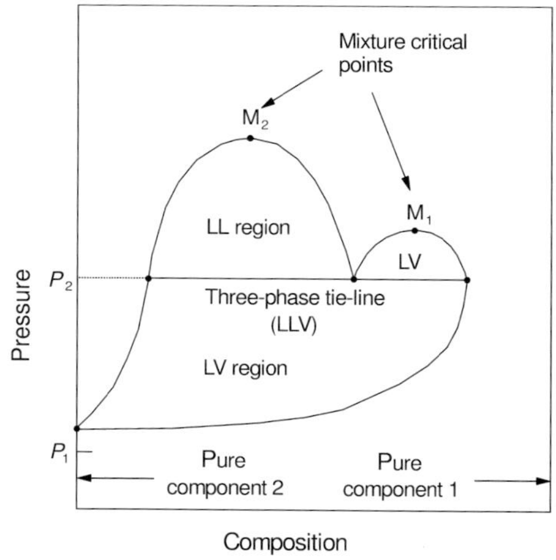

Figure 3.13 *p–x* diagram for a class III binary mixture at a temperature which intersects the three-phase LLV line.

horizontal line at p_2 is a tie line which connects the three coexisting phases at a fixed temperature and pressure. It is the p–T projection of the tie-line which appears as one point (M_3) on the three-phase LLV line in Figure 3.12b.

At pressures higher than p_2 there are two distinct two-phase regions in the form of closed domes. The stationary point on each closed dome represents the critical point of the two-phase mixture. The critical point of the LL envelope at the left (M_2) represents a point on the mixture critical line which originates from C_2. The other critical point (M_1) appears on the mixture critical line from C_1.

In a binary mixture consisting of a solid and a SCF, the critical temperatures of the pure components are normally far apart. The critical temperature of the supercritical solvent, in particular, lies below the triple point of the solid and there is no common range of temperature where the pure components both exist as liquids. As a consequence, the three-phase LLV line in Figure 3.12 becomes a SLV line which terminates at a LCEP as shown in Figure 3.14. A second, higher temperature branch of a SLV line originates from the triple point of the solid. This SLV line, which was not shown in Figure 3.12, intersects the L = V mixture critical line from C_2 at an UCEP. At both the LCEP and UCEP, a liquid and vapor phase critically merge into a single fluid phase in the presence of a solid phase.

An interesting feature of Figure 3.14 is that the higher temperature branch of the SLV line lies at lower temperatures than the solid–liquid coexistence curve of the pure solid. This indicates that melting point depression of the solid can occur in the presence of a SCF. As an example of melting point depression, consider the three-phase SLV line for the binary octacosane–ethane system [70]. The normal melting point of octacosane occurs at 64.5 °C whereas the UCEP occurs at 38.9 °C and 8.6 MPa. For temperatures between the UCEP and the normal melting point of the solid, the phase behavior is modified by the presence of the SLV line as shown in Figure 3.15.

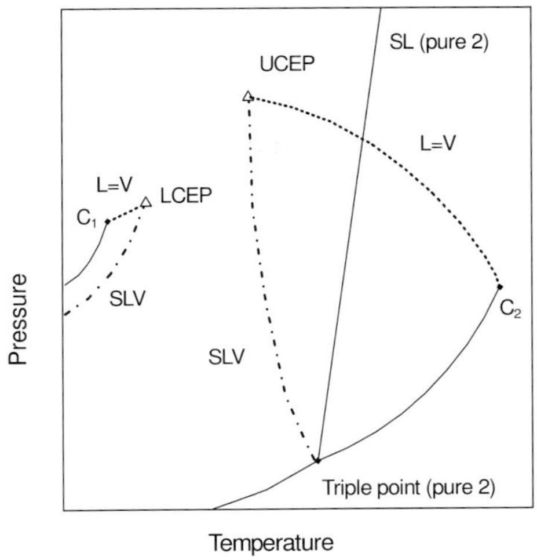

Figure 3.14 Class III phase diagram for an asymmetric binary mixture consisting of a solid and a SCF.

As pressure is increased from p_1 to p_2, a single vapor phase splits into a two-phase SV region until the SLV line is reached, as indicated by the horizontal tie-line at pressure p_2. For pressures greater than p_2, two regions of two-phase behavior appear. When the overall composition of the system is less than x_L, a region of solid–vapor equilibria exists up to very high pressures. If the overall composition of the system is greater than x_L, LV equilibria are established up to the stationary point at pressure p_3,

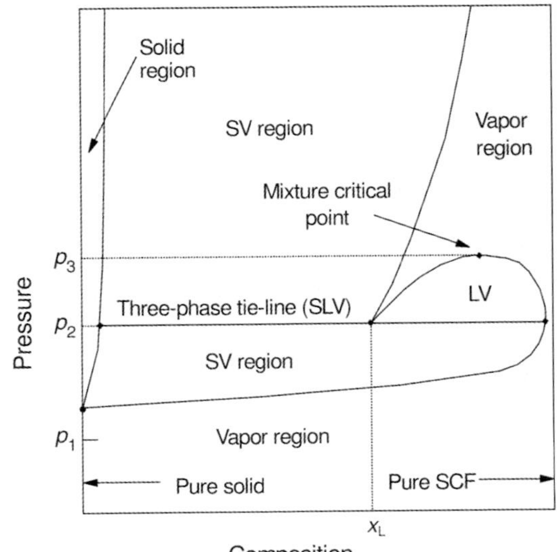

Figure 3.15 *p–x* diagram for an asymmetric binary mixture at a temperature which intersects the three-phase SLV line.

which represents the critical point of the two-phase mixture. This point lies on the mixture critical line connecting the UCEP and C_2 (Figure 3.14).

The solids of interest in SCF processes usually have relatively low vapor pressures, which results in a LCEP which is located close to the critical point of the SCF. The region between the critical point of the SCF and the UCEP generally defines the temperature range where unconstrained solid–vapor equilibria exist, that is, liquid phase formation is absent. However, operating at temperatures below the UCEP does not always ensure unconstrained solid–SCF equilibria because the SLV line can exhibit a temperature minimum. At low pressures the SLV line begins with a negative slope, passes through a temperature minimum, and then continues with a positive slope to the UCEP. Octacosane exhibits this type of behavior under the influence of CO_2 [70].

The complexity of the phase diagrams presented clearly demonstrates that phase behavior under high-pressure conditions can vary markedly. This is particularly important for chemical synthesis in SCFs since the number and types of phases have a direct impact on the progress of reaction. A knowledge of phase behavior is therefore essential for the proper interpretation of experimental data.

3.4
Factors Affecting Solubility in Supercritical Fluids

The solubility of compounds in SCFs has perhaps been the most extensively investigated area of SCF research. Solubility data indicate how well the SCF performs as a solvent for a particular solute and are an important starting point in the consideration of potential process applications. Although the solubility of a single solute in an SCF is not necessarily the same as that obtained in a multicomponent system [71], binary solubility data are nonetheless useful for estimating the selectivity of an SCF for a particular solute. The solubility of a component is mainly influenced by its chemical functionality, the nature of the SCF solvent, and the operating conditions.

The discussion which follows is centered on the solubility of solids, although similar principles apply for the case of liquids. Referring back to Figure 3.14, the region between the LCEP and the UCEP defines the temperature range where unconstrained solid–vapor equilibria exist with respect to pressure. Solid solubility data available in the literature are normally measured in this range of temperature. Furthermore, a check is usually made to determine the maximum temperature at which data can be measured before the onset of melting. As stated previously, the high-temperature branch of the SLV line may exhibit a minimum with respect to temperature, and this has the effect of reducing the temperature range for unconstrained solid–vapor equilibria.

3.4.1
The Supercritical Solvent

The solubility (y) of a material in an SCF is usually expressed in terms of the overall mole fraction of the solute in the SCF phase. The ability of SCFs to dissolve many substances arises from the highly non-ideal behavior of pure SCFs. The solubility of a component, as predicted by the ideal gas law, decreases asymptotically with increasing pressure because the solubility is simply the ratio of the vapor pressure (p^{sat}) to the total pressure (p). Under supercritical conditions, however, the solubility is enhanced by several orders of magnitude above that predicted by the ideal gas law.

The solubility enhancement of a component, particularly in the vicinity of the critical point, is driven primarily by the augmentation in the density of the SCF. The extent of solubility enhancement which occurs in the SCF phase is usually expressed in terms of an enhancement factor (E), which is defined as the actual solubility divided by the solubility predicted from the vapor pressure of the solute and ideal gas considerations:

$$E = \frac{yp}{p^{sat}} \tag{3.1}$$

The enhancement factor measures the extent of solubility in excess of that generated from the vapor pressure of the pure solid. This provides a convenient way of comparing the solubilities of solids with different vapor pressures. Since the actual solubility of a solid is heavily influenced by the SCF density, the enhancement factor displays a similar dependence on density. Values of the enhancement factor typically vary between 10^3 and 10^6, although enhancement factors as high as 10^{10} have been reported for some systems [72].

The solubility of a given solute also depends on the type of SCF, as shown in Table 3.5. Fluoroform has the highest mass density under the conditions shown but displays the lowest affinity for naphthalene. The variation in the solubility of naphthalene in different SCFs therefore suggests that there are varying degrees of intermolecular interaction between the solid and the SCF. The different levels of intermolecular interaction can be explained in terms of solvent polarity.

The overall effect of solvent polarity on the solubility of naphthalene follows the same general solubility rule in liquid extractions that 'like dissolves like'. Naphthalene is a non-polar solid and is most soluble in supercritical ethane. Carbon dioxide behaves as a non-polar solvent but less so because of its quadrupole moment [74]. Fluoroform is the most polar solvent because of the electron-withdrawing capability of the fluorine atoms. In general, non-polar solvents such as ethylene and ethane are the preferred solvents for aromatic hydrocarbons.

For polar solids, the effect of solvent polarity is not as straightforward. Although it is true that non-polar SCFs exhibit lower affinities for polar solutes, the maximum solubility of a polar solute does not necessarily occur in the most polar SCF. For

Table 3.5 Solubility of naphthalene in various SCFs at 20 MPa and 45 °C [73].

Solvent	Solubility (mole fraction)	Density (g cm^{-3})
Ethane	4.75×10^{-2}	0.39
Carbon dioxide	2.42×10^{-2}	0.81
Fluoroform	1.17×10^{-2}	0.94

example, CO_2 is a better solvent for benzoic acid than fluoroform [75]. Polar SCFs may exhibit greater potential for polar molecules when they contain functional groups that increase the level of intermolecular interaction with the solvent. Fluoroform, for example, is as good a solvent as CO_2 for 2-aminofluorene, probably as a result of hydrogen bonding between the amino group and the acidic proton in fluoroform [75]. In the case of 2-cyanonaphthalene, intermolecular interactions between the solute and fluoroform lead to much higher solubilities than those obtained with CO_2 [76].

3.4.2
Chemical Functionality of the Solute

The difference between solid solubilities in a given SCF depends mainly on the solid vapor pressure and intermolecular interactions between the solvent and solute. The individual solubilities of solids can vary greatly, although most values are well below 10 mol%. The differences between enhancement factors, however, are less pronounced [77, 78], which suggests that the vapor pressure of the solid exerts the primary influence on solubility. Intermolecular interactions between the solvent and solute depend on the types of functional groups present in their chemical structures. In general, the intermolecular interactions are dominated by dispersion forces, which accounts for the similar values of the enhancement factor for many solids.

It was previously described how the degree of intermolecular interaction can increase when a polar SCF is used for polar solutes. As another example, consider the solubilities of naphthalene and 1,4-naphthoquinone in fluoroform shown in Table 3.6. The solubility of naphthalene is much higher than that of 1,4-naphthoquinone under similar conditions, mainly because of the difference in their vapor pressures. With the effect of vapor pressure removed, the enhancement factor for 1,4-naphthoquinone is substantially greater than that for naphthalene. This suggests that a higher degree of intermolecular interaction occurs between fluoroform and 1,4-naphthoquinone. The additional interaction between solvent and solute may be associated with hydrogen bonding between the acidic proton in fluoroform and the carbonyl groups in 1,4-naphthoquinone [73].

The vapor pressure of a solid and the intermolecular interactions that it is capable of are ultimately determined by its chemical structure. Differences between the solubilities of solids can also be explained in terms of structural features which limit or enhance solubility. Beginning with a parent compound, the addition of a

Table 3.6 Comparison of the enhancement factors for naphthalene and 1,4-naphthoquinone in supercritical fluoroform at 20 MPa and 45 °C [73].

Solute	Solubility (mole fraction)	Enhancement factor
Naphthalene	1.17×10^{-2}	3380
1,4-Naphthoquinone	3.11×10^{-3}	92500

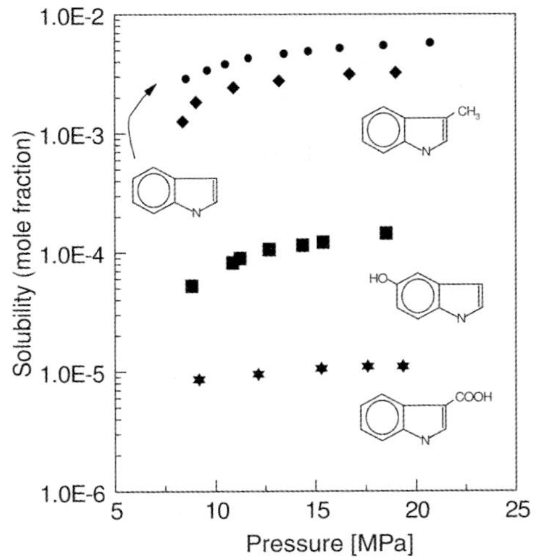

Figure 3.16 Dampening effect of substituents on the solubility of indole derivatives at 35 °C [84–86].

functional group generally has the effect of reducing solubility. This applies especially to hydroxyl (–OH) and carboxyl –COOH) functional groups as described by Stahl *et al.* [79].

The carboxyl functional group has a greater dampening effect on solubility than the hydroxyl group, as shown in Figure 3.16. Beginning with indole as the parent compound, the solubility decreases by around 1.5 orders of magnitude with the addition of the hydroxyl group (5-hydroxyindole). The addition of the carboxyl group to indole (indole-3-carboxylic acid) decreases the solubility by a much larger amount of around 2.5 orders of magnitude. For comparison, the addition of a methyl group to indole (skatole) causes a relatively minor solubility reduction of less than 0.5 orders of magnitude.

A detailed study on the effect of structural features on solubility was given by Dandge *et al.* [80]. This study considers structural features such as chain length, branching, and the number of rings in addition to the position and types of substituents. Structure–solubility relationships using CO_2 as a solvent are illustrated for several classes of compounds, including hydrocarbons, alcohols, phenols, aldehydes, ethers, esters, amines, and nitro compounds. *n*-Alkanes, for example, are completely miscible with CO_2 when the carbon number is below 12. Solubility decreases rapidly beyond a carbon number of 12. For a given carbon number, the solubility of acyclic hydrocarbons is favored by unsaturation; for example, 1-octadecene is more soluble than *n*-octadecane. Branching leads to more favorable solubilities compared with *n*-alkanes. The maximum limit of complete miscibility for branched alkanes also occurs at a much higher carbon number, between 19 and 30.

Although the degree of intermolecular interaction can increase when a polar SCF is used for polar solutes, the overall solubility of polar solutes is normally much lower than that for non-polar solutes. Organic salts will exhibit a wide range of solubility values in SCFs depending primarily on vapor pressure. Inorganic salts exhibit very low solubility in SCFs and in most cases may be considered as insoluble. Metal ions are insoluble in SCFs because of the charge neutralization requirement but if they are bound to organic ligands, they may become soluble to an appreciable extent. A good example of this is the complexation of metal ions with tri-*n*-butyl phosphate (TBP) and di(2-ethylhexyl)phosphoric acid [81].

The choice of the organic chelating agent impacts significantly on solubility. For example, diethyl dithiocarbamate (DDC) forms stable complexes with a wide range of metals and such complexes are soluble in supercritical CO_2 to varying extents. Fluorination of the ligands, as in bis(trifluoroethyl) dithiocarbamate (FDDC), can increase the solubility of the metal complexes by several orders of magnitude [82]. It is interesting that fluorination of an organic structure is a useful way of increasing its solubility in supercritical CO_2. As another example, the technique has been used as a means of increasing the solubility of hydrocarbon polymers [83].

3.4.3
Temperature and Pressure Effects

Solubility data are typically presented in the form of solubility isotherms as shown for the naphthalene–CO_2 system in Figure 3.17. The characteristic feature to note in this diagram is that solubility increases rapidly at lower pressures, generally near the critical pressure of the SCF, whereas at higher pressures the increase in solubility is less pronounced. The behavior of the solubility isotherm in this way simply reflects the density changes which are occurring in the SCF.

The effect of temperature on solubility is more complex and involves consideration of both the solute vapor pressure and the density of the SCF. The solubility isotherms shown in Figure 3.17 are typical of most solid–SCF systems in that they intersect

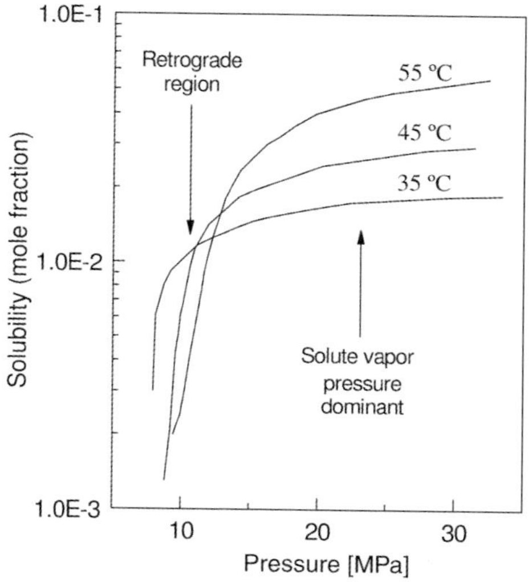

Figure 3.17 Solubility of naphthalene in supercritical CO_2 [87].

within a narrow range of pressure. For any two isotherms, the point of intersection, or crossover pressure, represents a change in the temperature dependence of solubility.

The region of pressure below the crossover pressure is known as the retrograde region. In this range of pressure, solubility decreases with increase in temperature because the density of the SCF falls sharply. The decrease in density is sufficient to overcome any increases in solute vapor pressure which would normally lead to an increase in solubility. Above the crossover pressure, the decrease in solvent density is less sensitive to temperature and so solubility increases with temperature because the vapor pressure effect becomes dominant.

Solubility data provide important information on achieving separation between the solute and the SCF. The solubility data shown in Figure 3.17 suggest that there are two ways in which this separation might be achieved. The first separation scheme, which is perhaps the most conventional procedure, involves an isothermal decrease in pressure. This would preferably be accomplished in the vicinity of the critical pressure of the SCF where solubility is sensitive to changes in pressure. The reduction in pressure need not proceed completely to atmospheric pressure since an order of magnitude reduction in solubility is easily accomplished in the steeply rising portion of the solubility isotherm.

The second separation scheme involves an increase in temperature under isobaric conditions. This can be achieved in the retrograde region, and again an order of magnitude reduction in solubility can occur with a modest increase in temperature. This separation method is probably more favorable than the first method in terms of energy consumption because the first method involves a significant recompression step after separation. However, a separation based on a temperature increase requires much more specific information on the solubility behavior of the solute. In contrast, a separation based on a pressure reduction mainly involves consideration of the critical pressure of the solvent. A near 100% separation can be achieved once the pressure is reduced to below the critical pressure.

Solubility data also suggest potential operating conditions for an extraction process. Referring again to Figure 3.17, an extraction at 20 MPa requires a lower solvent-to-feed ratio than extraction at 10 MPa. This advantage may be offset by the higher capital costs and a larger recompression duty. A separation at 20 MPa based on a temperature decrease involves a significant reduction in solubility but the overall separation efficiency remains low in comparison with a temperature increase at 10 MPa in the retrograde region. This example demonstrates that higher pressures favor the extraction step whereas lower pressures, particularly in the retrograde region, favor the separation step. The actual choice for the operating pressure is therefore a compromise between process yield (or solvent-to-feed ratio) and operating cost.

References

1 Hannay, J.B. and Hogarth, J. (1879) *Proceedings of the Royal Society of London*, **29**, 324.

2 Johnston, K., (1984) in *Kirk–Othmer Encyclopedia of Chemical Technology*, 3rd edn, Suppl. Vol. (eds M. Grayson and

D.E. Eckroth), John Wiley & Sons, Inc., New York, p. 872.
3 McHugh, M.A. and Krukonis, V.J. (1986) Introduction, in *Supercritical Fluid Extraction: Principles and Practice* (ed H. Brenner), Butterworth-Heinemann, Boston, MA. p. 15.
4 Liong, K.K., Wells, P.A. and Foster, N.R. (1991) *Industrial & Engineering Chemistry Research*, **30**, 1329–1335.
5 Angus, S., Armstrong, B. and De Reuck, K.M. (1976) *International Thermodynamic Tables of the Fluid State: Carbon Dioxide*, IUPAC, Pergamon Press, Oxford.
6 Michels, A., Botzen, A. and Schuurman, W. (1957) *Physica*, **23**, 95–102.
7 Iwasaki, H. and Takahashi, M. (1981) *Journal of Chemical Physics*, **74**, 1930–1943.
8 Diller, D.E., Van Poolen, L.J. and Dos Santos, F.V. (1988) *Journal of Chemical and Engineering Data*, **33**, 460–464.
9 Tilly, K.D., Foster, N.R., MacNaughton, S.J. and Tomasko, D.L. (1994) *Industrial & Engineering Chemistry Research*, **33**, 681–688.
10 Fernandez, D.P., Mulev, Y., Goodwin, A.R.H. and Sengers, J.M.H.L. (1995) *Journal of Physical and Chemical Reference Data*, **24**, 33–69.
11 Uematsu, M. and Franck, E.U. (1980) *Journal of Physical and Chemical Reference Data*, **9**, 1291–1306.
12 Kita, T., Uosaki, Y. and Moriyoshi, T. (1994) in *High Pressure Liquids and Solutions* (eds Y. Taniguchi, M. Senoo and K. Hara), Elsevier, Amsterdam (*Curr. Jpn. Mater. Res.*, **13**, 181–198).
13 Downing, R.C. (1988) *Fluorocarbon Refrigerants Handbook*, Prentice-Hall, Englewood Cliffs, NJ.
14 Makita, T., Kubota, H., Tanaka, Y. and Kashiwagi, H. (1977) *Reito*, **52**, 543–551.
15 Fernandez, D.P., Goodwin, A.R.H., Lemmon, E.W., Levelt Sengers, J.M.H. and Williams, R.C. (1997) *Journal of Physical and Chemical Reference Data*, **26**, 1125–1166.
16 Weast, R.C. (ed.) (1982) *CRC Handbook of Chemistry and Physics*, CRC Press, Boca Raton, FL.
17 Ogunsola, O.M. and Berkowitz, N. (1995) *Fuel Processing Technology*, **45**, 95–107.
18 Rodil, R. and Popp, P. (2006) *Journal of Chromatography A*, **1124**, 82–90.
19 Smith, R.M. (2006) *Analytical and Bioanalytical Chemistry*, **385**, 419–421.
20 Roskar, V., Dombro, R.A., Prentice, G.A., Westgate, C.R. and McHugh, M.A. (1992) *Fluid Phase Equilibria*, **77**, 241–259.
21 Michels, A., Sengers, J.V. and Van der Gulik, P.S. (1962) *Physica*, **28**, 1216–1237.
22 Le Neindre, B., Garrabos, Y. and Tufeu, R. (1984) *Berichte der Bunsen-Gesellschaft-Physical Chemistry Chemical Physics*, **88**, 916–920.
23 Sakonidou, E.P., van den Berg, H.R., ten Seldam, C.A. and Sengers, J.V. (1998) *Journal of Chemical Physics*, **109**, 717–736.
24 Shaw, R.W., Brill, T.B., Clifford, A.A., Eckert, C.A. and Franck, E.U. (1991) *Chemical & Engineering News*, **69**, 26–39.
25 Vargaftik, N.B., Vinogradov, Y.R. and Yargin, V.S. (1996) *Handbook of Physical Properties of Liquids and Gases*, Begell House, New York.
26 Harlow, A., Wiegand, G. and Franck, E.U. (1997) *Berichte der Bunsen-Gesellschaft-Physical Chemistry Chemical Physics*, **101**, 1461–1465.
27 Haar, L. and Gallagher, J.S. (1978) *Journal of Physical and Chemical Reference Data*, **7**, 635–677.
28 Rabinovich, V.A. and Selover, T.B. Jr. (eds) (1988) *Thermodynamic Properties of Neon, Argon, Krypton, and Xenon*, Hemisphere, Washington, DC.
29 Stewart, R.B. and Jacobsen, R.T. (1988) *Journal of Physical and Chemical Reference Data*, **18**, 639–798.
30 Augus, S., Armstrong, B. and de Reuck, K.M.E. (eds) (1976) *International Thermodynamic Tables on the Fluid State – 3: Carbon Dioxide*, IUPAC, Perghamon Press, Oxford.
31 Span, R. and Wagner, W. (1996) *Journal of Physical and Chemical Reference Data*, **25**, 1509–1596.
32 Mathews, J.F. (1972) *Chemical Reviews*, **72**, 71–100.

33 Horvath, A.L. (1973) *Process Technology*, **18**, 67–69.

34 Mangold, K. and Franck, E.U. (1962) *Berichte der Bunsen-Gesellschaft-Physical Chemistry Chemical Physics*, **66**, 260–266.

35 Couch, E.J., Hirth, L.J. and Kobe, K.A. (1961) *Journal of Chemical and Engineering Data*, **6**, 229–237.

36 Berg, J. and Wagner, Z. (1990) *Fluid Phase Equilibria*, **54**, 35–45.

37 Pöhler, H. and Kiran, E. (1997) *Journal of Chemical and Engineering Data*, **42**, 389–394.

38 Gokmenoglu, Z., Xiong, Y. and Kiran, E. (1996) *Journal of Chemical and Engineering Data*, **41**, 354–360.

39 Mears, W.H., Rosenthal, E. and Sinka, J.V. (1969) *The Journal of Physical Chemistry*, **73**, 2254–2261.

40 Kell, G.S. (1972) Thermodynamics and transport properties of fluid water, in *Water, a Comprehensive Treatise* (ed. F. Franks), Plenum Press, New York. pp. 363–412.

41 Levelt Sengers, J.M.H., Kamgar-Parsi, B., Balfour, F.W. and Sengers, J.V. (1983) *Journal of Physical and Chemical Reference Data*, **12**, 1–28.

42 Sato, H., Watanabe, K., Levelt Sengers, J.M.H., Gallagher, J.S., Hill, P.G., Straub, J. and Wagner, W. (1991) *Journal of Physical and Chemical Reference Data*, **12**, 1023–1044.

43 Šifner, O. and Klomfar, J. (1994) *Journal of Physical and Chemical Reference Data*, **23**, 63–152.

44 Deul, R., Rosenzweig, S. and Franck, E.U. (1991) *Berichte der Bunsen-Gesellschaft-Physical Chemistry Chemical Physics*, **95**, 515–519.

45 Cheremisinoff, P.N. and Morresi, A.C. (1979) *Benzene. Basic and Hazardous Properties*, Marcel Dekker, New York.

46 Ambrose, D. and Tsonopoulos, C. (1995) *Journal of Chemical and Engineering Data*, **40**, 531–546.

47 Younglove, B.A. and Ely, J.F. (1987) *Journal of Physical and Chemical Reference Data*, **16**, 577–798.

48 Tillner-Roth, R. and Yokozeki, A. (1997) *Journal of Physical and Chemical Reference Data*, **26**, 1273–1328.

49 Malbrunot, P.F., Meunier, P.A., Scatena, G.M., Mears, W.H., Murphy, K.P. and Sinka, J.V. (1968) *Journal of Chemical and Engineering Data*, **13**, 16–21.

50 Kudchadker, A.P., Alani, G.H. and Zwolinski, B.J. (1968) *Chemical Reviews*, **68**, 659–735.

51 Friend, D.G., Ingham, H. and Ely, J.F. (1991) *Journal of Physical and Chemical Reference Data*, **20**, 275–347.

52 Sychev, V.V., Vassarman, A.A., Kozlov, A.D., Zagoruchenko, V.A., Spiridov, G.A., Tsymarny, V.A. and Selover, T.B. Jr. (eds) (1987) *Thermodynamic Properties of Ethane*, Hemisphere, Washington, DC.

53 Jahangiri, M., Jacobsen, R.T., Stewart, R.B. and McCarty, R.D. (1986) *Journal of Physical and Chemical Reference Data*, **15**, 593–734.

54 Synchev, V.V., Vasserman, A.A., Golovsky, E.A., Kozlov, A.D., Spiridinov, G.A., Tsymarny, V.A., and Selover, T.B. Jr. (1987) *Thermodynamic Properties of Ethylene*, Hemisphere, Washington, DC.

55 Tsonopoulos, C. and Ambrose, D. (1996) *Journal of Chemical and Engineering Data*, **41**, 645–656.

56 Reid, R.C., Prausnitz, J.M. and Poling, B.E. (1987) *The Properties of Gases and Liquids*, McGraw-Hill, New York.

57 Aizpiri, A.G., Rey, A., Davila, J., Rubio, R.G., Zollweg, J.A. and Streett, W.B. (1991) *The Journal of Physical Chemistry*, **95**, 3351–3357.

58 Altunin, V.A., Geller, V.Z., Petrov, E.K., Rasskazov, D.C., Spiridinov, G.A., and Selover, T.B. Jr. (eds) (1987) *Thermophysical Properties of Freons. Methane Series, Part 1*, Hemisphere, Washington, DC.

59 Setzmann, U. and Wagner, W. (1991) *Journal of Physical and Chemical Reference Data*, **20**, 1061–1155.

60 Wadner, W. and de Reuck, K.M. (eds) (1996) *International Thermodynamic Tables*

of the Fluid State – 13: Methane, IUPAC, Pergamon Press, Oxford.

61 Goodwin, R.D. (1987) *Journal of Physical and Chemical Reference Data*, **16**, 799–892.

62 Gude, M. and Teja, A.S. (1995) *Journal of Chemical and Engineering Data*, **40**, 1025–1036.

63 Wagner, W. and de Reuck, K.M. (eds) (1993) *International Thermodynamic Tables of the Fluid State – 12: Methanol*, IUPAC, Pergamon Press. Oxford.

64 Sychev, V.V., Vasserman, A.A., Kozlov, A.D., Tsymarny, V.A. and Selover, T.B. Jr. (eds) (1991) *Thermodynamic Properties of Propane*, Hemisphere, Washington, DC.

65 Angus, B.A.S. and de Reuck, K.M. (1980) *International Thermodynamic Tables of the Fluid State – 7: Propylene*, IUPAC Pergamon Press, Oxford.

66 van Konynenburg, P.H. and Scott, R.L. (1980) *Philosophical Transactions of the Royal Society of London*, **298**, 495–540.

67 McHugh, M.A. and Krukonis, V.J. (1986) *Supercritical Fluid Extraction: Principles and Practice* (ed H. Brenner), Butterworth-Heinemann, Boston, MA. Chapter 3, p. 512.

68 Streett, W.B. (1983) in *Chemical Engineering at Supercritical Conditions* (eds M.E. Paulaitis, J.M.L. Penninger, R.D. Gray and P.E. Davidson), Ann Arbor Science, Ann Arbor, MI. Chapter 1.

69 Alwani, Z. and Schneider, G.M. (1976) *Berichte der Bunsen-Gesellschaft-Physical Chemistry Chemical Physics*, **80**, 1310–1315.

70 McHugh, M.A. and Yogan, T.J. (1984) *Journal of Chemical and Engineering Data*, **29**, 112–115.

71 Lucien, F.P. and Foster, N.R. (1996) *Industrial & Engineering Chemistry Research*, **35**, 4686–4699.

72 Liphard, K.G. and Schneider, G.M. (1975) *Journal of Chemical Thermodynamics*, **7**, 805–814.

73 Schmitt, W.J. and Reid, R.C. (1985) in *Supercritical Fluid Technology*, (eds J.M.L. Penninger, M. Radosz, M.A. McHugh and V.J. Krukonis), Elsevier, Amsterdam.

74 Prausnitz, J.M., Lichtenthaler, R.N. and Gomez de Azevedo, E. (1986) in *Molecular Thermodynamics of Fluid-phase Equilibria*, 2nd edn, Prentice-Hall, Englewood Cliffs, NJ. Chapter 4, p. 600.

75 Schmitt, W.J. and Reid, R.C. (1986) *Journal of Chemical and Engineering Data*, **31**, 204–212.

76 Nakatani, T., Ohgaki, K. and Katayama, T. (1991) *Industrial & Engineering Chemistry Research*, **30**, 1362–1366.

77 Brennecke, J.F. and Eckert, C.A. (1989) *AICHE Journal*, **35**, 1409–1427.

78 Dobbs, J.M. and Johnston, K.P. (1987) *Industrial & Engineering Chemistry Research*, **26**, 1476–1482.

79 Stahl, E., Schilz, W., Schuetz, E. and Willing, E. (1978) *Angewandte Chemie-International Edition*, **90**, 778–785; *Angewandte Chemie International Edition in English*, **1978**, **17**, 731–738.

80 Dandge, D.K., Heller, J.P. and Wilson, K.V. (1985) *Industrial & Engineering Chemistry Poduct Research and Development*, **24**, 162–166.

81 Dehghani, F., Wells, T., Cotton, N.J. and Foster, N.R. (1996) *Journal of Supercritical Fluids*, **9**, 263–272.

82 Laintz, K.E., Wai, C.M., Yonker, C.R. and Smith, R.D. (1991) *Journal of Supercritical Fluids*, **4**, 194–198.

83 Betts, D., Johnson, T., Anderson, C. and DeSimone, J.M. (1997) *Polym. Prepr., ACS Div. Polym. Chem.*, **38**, 760–761.

84 Nakatani, T., Ohgaki, K. and Katayama, T. (1989) *Journal of Supercritical Fluids*, **2**, 9–14.

85 Sako, S., Ohgaki, K. and Katayama, T. (1988) *Journal of Supercritical Fluids*, **1**, 1–6.

86 Sako, S., Shibata, K., Ohgaki, K. and Katayama, T. (1989) *Journal of Supercritical Fluids*, **2**, 3–8.

87 Tsekhanskaya, Y.V., Iomtev, M.B. and Mushkina, E.V. (1964) *Zhurnal Fizicheskoi Khimii*, **38**, 2166–2171.

4
Expanded Liquid Phases in Catalysis: Gas-expanded Liquids and Liquid–Supercritical Fluid Biphasic Systems

Ulrich Hintermair, Walter Leitner, and Philip Jessop

4.1
A Practical Classification of Biphasic Systems Consisting of Liquids and Compressed Gases for Multiphase Catalysis

A rigorous classification of biphasic systems of liquids and compressed gases, according to the thermodynamic principles of multi-component mixture phase behavior, is often perceived as too abstract from a practical point of view. This has led to several attempts to develop more pragmatic definitions in the literature, which take into account the potential application and also the observed physico-chemical property changes [1–3]. In this chapter, we have aimed at giving a unified description of these phenomena that allows the practicing chemist to identify the most promising system for a given synthetic (reaction and separation) process.

Initially, the interest in supercritical fluids (SCFs) for catalysis and reaction engineering was based mainly on their nature as homogeneous, single-phase reaction media, providing highly diffusive solvents which are completely miscible with gases [4]. It became apparent, however, that multiphasic systems comprising a liquid phase and a compressed supercritical or subcritical gas phase can provide benefits similar to those of single-phase SCFs and create at the same time new opportunities for integration of reaction and separation [5]. The liquid phases under these conditions can be referred to as expanded liquid phases (ELPs). The term "expanded liquid phase" is chosen in order to highlight the fact that an ELP consists of more than one component and that the expansion is a result of their combination only, whereas the term "expanded liquid" or "expanded fluid" has been used to describe a liquefied metal that thermally expands upon further heating. ELPs often show properties very different to those of the same liquids in the absence of compressed gas. These property changes often occur even at pressures and temperatures well below the critical parameters of the biphasic mixture or even of the pure gas phase. Hence they may allow the use of simpler equipment, leading to lower investment costs, and lower the entrance barrier for practical application.

Handbook of Green Chemistry, Volume 4: Supercritical Solvents. Edited by Walter Leitner and Philip G. Jessop

ISBN: 978-3-527-32590-0

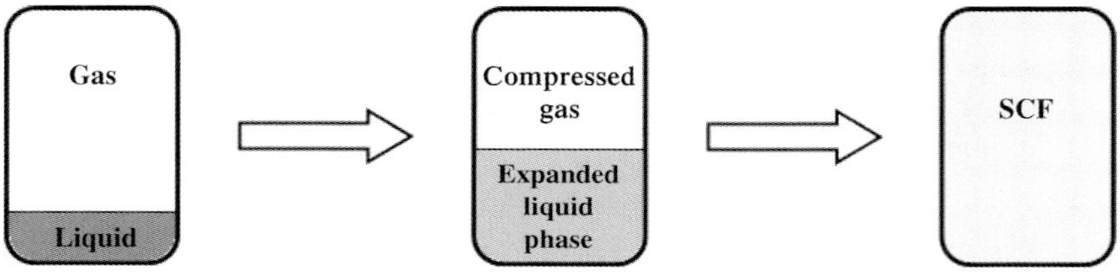

Figure 4.1 Phase behavior of a binary liquid–gas system with increasing gas density at $T \geq T_c$. The compressed gas can be at sub- or supercritical conditions for the pure gas. The transition into the homogeneous SCF requires temperatures and pressures beyond the mixture critical point.

The typical behavior of a binary mixture of a liquid and a gas under increasing gas density is depicted schematically in Figure 4.1. The increase in gas density results from an increase in system pressure, which can be achieved experimentally either by decreasing the volume of the cell (constant mole fractions) or more typically by increasing the gas pressure (increasing gas mole fraction). Three main stages can be distinguished in this scenario.

When the liquid is in contact with the gas at moderate pressure, dissolution of the gas into the liquid takes place. In this two-phase system, the low gas concentration in the condensed phase is determined by the solubility properties of the liquid and the low liquid concentration in the gas phase is mainly determined by its vapor pressure. This situation is encountered in typical homogeneous catalysis with organic solvents and gaseous reagents under moderate conditions in terms of temperature and pressure (left-hand diagram in Figure 4.1). The dissolved gas does not alter the physical properties of the liquid phase significantly, nor does the liquid affect the vapor phase very much. In such ideally diluted systems, Henry's law linearly describes gas solubility as a function of its partial pressure. When the density of the gas is increased, more gas dissolves into the condensed phase, causing a volumetric expansion, the magnitude of which depends on the nature of the liquid and the gas (middle diagram in Figure 4.1). This condensed phase is then called an expanded liquid phase (ELP). At such elevated pressures, the fugacity of the gas differs significantly from its partial pressure and the mole fraction of the gas in the liquid phase no longer follows the linear extrapolation of Henry's law. If the gas is kept above its critical temperature, a further increase in pressure cannot lead to condensation, but still renders the densities of the gas and liquid phase more and more similar. This is most pronounced around the critical pressure of the gas, where its density strongly increases. When both phases have equal density *and* compositions (identical chemical potential), the mixture critical point is reached and a single supercritical phase is formed. This transition may not be accessible for mixtures where the components differ greatly in their thermodynamic properties, as for example with supercritical carbon dioxide ($scCO_2$) and polymers or ionic compounds [6]. Based on this phenomenological description of the phase behavior, the general definition of an ELP can be given as follows:

An expanded liquid phase is the condensed phase of a mixture of liquids and gases at conditions below the mixture critical point exhibiting a measurable volume expansion of the liquid phase compared with the pure liquids in the absence of the gases.

This definition deliberately excludes low levels of gas dissolution which do not much affect the physical properties of the liquid phase. Liquids which have a sufficient amount of gas dissolved to exhibit volume expansion under isothermal conditions differ from non-expanded phases, and this chapter specifically deals with such situations. Conditions in a regime either to evaporate the condensed phase or to condense the vapor phase are also excluded by the above definition. However, it should be noted that condensing a gas into a liquid may create a mixture with properties similar to those of an ELP but can be better described as a liquid dilution by a low-viscosity solvent. So-called enhanced fluidity liquids (EFL) (liquids with pressurized gas dissolved *without* a separate vapor phase), which are used, for example, in chromatographic separations and extractions [7], also fall into the definition of an ELP.

The transition of an ELP to a single supercritical phase as depicted in Figure 4.1 requires complete mutual dissolution in an equilibrated system. If the system is either not in equilibrium or if the SCF is oversaturated with the liquid solute, a two-phase system consisting of an ELP and an SCF will be obtained. This situation corresponds to the interesting class of liquid–supercritical systems for multiphase catalysis, which will be described in detail in Sections 4.5 and 4.6. The SCF in such systems has significant solvent power, making the situation reminiscent of conventional liquid–liquid biphasic catalysis [8]. The physico-chemical properties of the ELP, however, differ from those of the original liquid phase. These changes have their onset at pressures and temperatures well below the critical point of the pure gas phase. In order to distinguish ELPs under a subcritical vapor phase from the expanded liquid–SCF biphasic systems, the term gas-expanded liquid (GXL) has been introduced [GXLs obtained through pressurization with CO_2 have also been called CO_2-expanded liquids (CXLs)].

The mutual solubility properties of liquids and gases, and hence their behavior as ELPs, span a wide range and can be classified into three groups:

- *Class I* systems comprise liquid–gas combinations in which the liquid expands measurably but by less than 10% of the original volume even at higher pressures. The expansion is small because the liquid dissolves the gas poorly. Even though the physical properties are not changed in the sense of an ELP, chemical properties of the liquid (such as acidity or polarity/ionicity) might be affected by chemisorption of the gas. The most prominent example of such a behavior is water and CO_2. As Class I liquids are usually also poorly soluble in the respective SCF of the gas, they can provide useful systems for liquid–SCF biphasic catalysis. Genuine ELPs containing Class I liquids in limited amounts can be created in multi-component systems, as exemplified for water in CO_2-expanded MeCN–alkene mixtures [9] or CO_2-expanded dimethyl sulfoxide (DMSO) [10].
- *Class II* systems comprise liquid–gas combinations in which the liquid dissolves large amounts of the gas at elevated pressure, leading to significant deviations from

Henry's law. They exhibit expansions that can be several times larger than their initial volume and consequently undergo significant changes in virtually every physical property. Most organic liquids which are volatile at atmospheric pressure are in this class and form the bulk of GXLs. Fluorinated organic molecules also readily form Class II systems, especially with non-polar compressible gases such as CO_2.

- *Class III* systems comprise liquid–gas combinations in which the liquid dissolves the gas moderately, but the gas uptake is limited owing to strong intermolecular forces or a high molecular weight of the liquid. Hence the observed expansion is considerably smaller than with Class II systems (typically 10–30% of the original volume). The amount of gas dissolved still leads to significant changes in certain physical properties. Conversely, the solubility of the liquids in the gas and its SCF is usually immeasurably small due to their low volatility. Class III systems include ionic liquids (ILs), polymers, and crude oil in combination with compressible gases such as CO_2, ethane, and ethylene. They make the most interesting systems for liquid–SCF biphasic catalysis, as will be seen in Section 4.6.

Figure 4.2 qualitatively illustrates the mutual gas–liquid solubilities and also the degree of expansion for Class I–III combinations, and provides some prototypical examples for each class.

As each of the stages GXL, ELP–supercritical, and single-phase SCF has distinct physico-chemical properties and thus particular advantages and disadvantages for multiphase reaction engineering, knowledge of the phase behavior is crucial both for the fundamental understanding and for the practical use of these systems [11]. The first extensive measurements of the solubility of CO_2 in organic solvents at high pressure were made as early as 1911 in the laboratories of Gustav Tammann at the University of Göttingen, Germany. They revealed that the dissolution of CO_2 caused the liquid phase to expand considerably [12, 13] and showed that volumetric expansion depends on the liquid used (data listed here for 4.9 MPa and 35 °C):

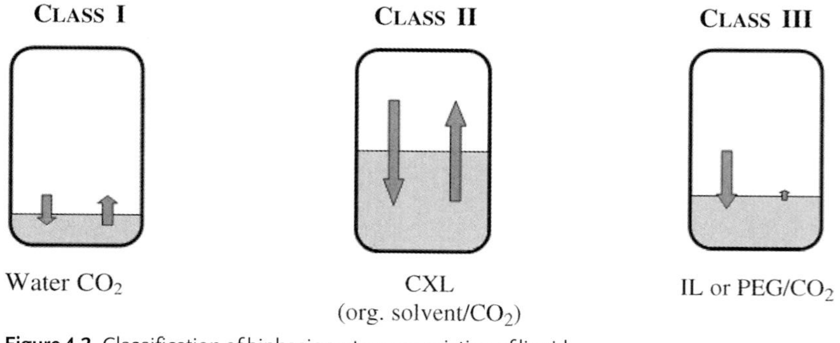

Figure 4.2 Classification of biphasic systems consisting of liquids and compressed gases and selected prototypical examples (gas: upper phase, white; liquid: lower phase, gray). The size of the arrow qualitatively indicates mutual solubility.

EtOAc ≈	diethyl ether >	nitrobenzene	toluene >	propanol >	ethanol >>	water
256	247	130	128	81	70	3.5%

Today, a large number of vapor–liquid equilibrium (VLE) data for binary and even ternary systems are available in the literature [14]. Figure 4.3 shows an isotherm of the Class II system MeOH–CO_2 as an example. Inside the phase envelope an ELP is in coexistence with a vapor phase, which might be subcritical (in which case the liquid is a GXL) or supercritical (ELP–SCF). Above the bubble point curve there is only one liquid phase [the so-called enhanced fluidity liquid (EFL)] whereas to the right of the dew point curve there is only a vapor phase. Bubble and dew point curves meet at the mixture critical point where the liquid and the vapor phase converge into a single homogeneous supercritical phase. However, not all binary systems have mixture critical points. As the bubble point curve represents the solubility of the gas in the liquid, Class III combinations have a bubble point curve converging to infinite pressures at the saturation limit of the maximum gas mole fraction. The dew point curve being the solubility of the liquid in the gas does not exist for Class III combinations.

Following the bubble point curve from the beginning in Figure 4.3, the deviation from Henry's law is evident on going to higher gas mole fractions. In Figure 4.3, decreasing the cell volume is represented by a vertical isopleth, whereas increasing the gas pressure corresponds to a diagonal movement from lower left to the upper right.

As is suggested by Figure 4.1, ELPs have properties that are intermediate between those of conventional condensed solvents and those of single-phase SCFs. They are by no means imperfect supercritical systems, which failed to be monophasic, but

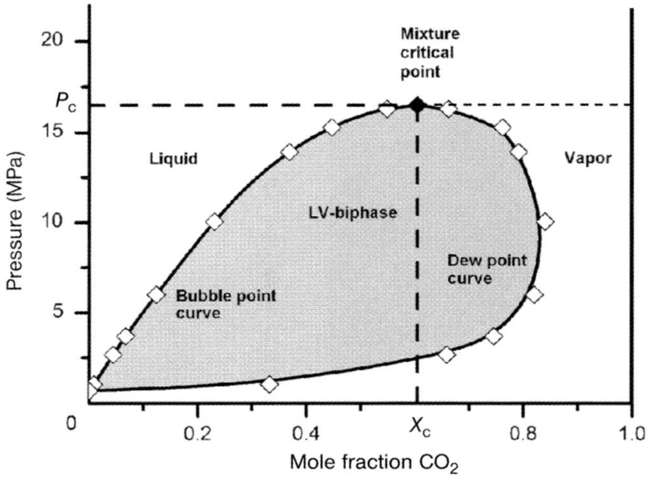

Figure 4.3 Vapor–liquid equilibrium (VLE) data for a Class II liquid–gas binary system exemplified for methanol + CO_2 at 121 °C [15].

constitute solvent systems that are tunable often in an even wider range than single-phase SCFs. These adjustable properties make them attractive for application to sustainable processes, based on the principles of green chemistry and engineering, including materials processing, separation techniques, chemical synthesis, and catalysis. In the following sections we will briefly highlight important physical properties of ELPs which are of relevance to application in synthesis and catalysis. Note that other properties not discussed here also undergo changes, and might be important for other techniques using ELPs. Some of these (including speed of sound, spectral shifts or refractive indices) can even be directly correlated to the expansion behavior [16–18], and thus may serve as useful process probes in closed large-scale installations. For coverage of applications beyond the scope of this chapter, we refer the reader to a recent review [2] and the Proceedings of the ACS Symposium on GXLs in 2007 [19]. In the remaining parts of the chapter, we will discuss selected reactive systems, which emphasize the beneficial nature of EPLs as "tunable solvents". For convenience, GXLs and liquid–SCF systems will be treated separately, even though the expansion of the liquid phase in liquid–SCF systems often contributes to the effectiveness of the system.

4.2
Physical Properties of Expanded Liquid Phases

4.2.1
Volumetric Expansion

According to the classification given in Section 4.1, the solubility of a gas in a liquid defines the extent to which an ELP is formed under given reaction conditions. Gases with a potential to induce significant expansions are near-critical, compressible gases such as ethane [20, 21], ethylene [20, 22], propane [23], CHF_3 [24], CCl_2F_2 [25], $CClF_3$ [25], N_2O [26] (because of its strong oxidizing power, N_2O should never be used with flammable or combustible liquids), SF_6 [25], xenon [27], and CO_2 [2]. Gases far beyond their critical temperature (e.g. H_2, CO, O_2, N_2, CH_4, and He at room temperature) are poorly soluble in most liquids and do not expand them significantly upon pressurization. As an expansion gas, CO_2 has been the most extensively used until now owing to its high solubility in many liquids and its high molar density therein (see below), in combination with its chemical reactivity, low toxicity, non-flammability, and low cost. From the large body of data available for this gas, some general conclusions about the expansion phenomena can be drawn.

Volumetric expansion of Class II liquids increases with pressure in a non-linear manner, with only moderate expansion occurring below 3 MPa (Figure 4.4a). Class III liquids typically have a more linear response. Comparing the volumetric expansion ($\Delta V/V$) during CO_2 uptake, Class II and III liquids behave fairly similarly when the expansion is plotted against wt% CO_2 (Figure 4.4b). In fact, the ionic liquid [bmim][BF_4] (bmim = 3-butyl-1-methylimidazolium) and the liquid polymers poly(ethylene glycol) (PEG) and poly(propylene glycol) (PPG) expand more than MeCN

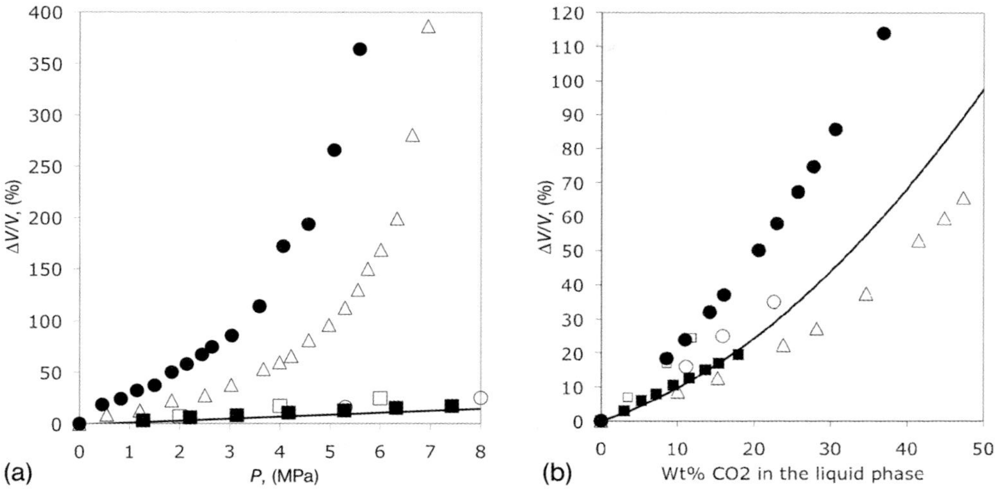

Figure 4.4 Expansion of solvents as a function of (a) CO_2 pressure and (b) wt% dissolved CO_2 at 40 °C for Class II liquids ethyl acetate (●) [20] and MeCN (△) [20] and Class III liquids [bmim][BF_4] (■, interpolated) [28], crude oil (line, at 43 °C) [29], PPG (□) [30], and PEG (○) [30].

at the same wt% CO_2. It is therefore the limited ability of Class III liquids to dissolve a high mass fraction of gas which limits the volumetric expansion. However, as these liquids have higher molecular weights than Class II liquids, they dissolve CO_2 fairly well on a mol% basis. Table 4.1 contrasts the wt% CO_2 and the mol% CO_2 for all three classes.

The smaller volume expansion of Class III liquids at similar mol% CO_2 cannot be explained by smaller molar volumes of CO_2 in these liquids. The average molar

Table 4.1 Different solvents and their relative volume expansion ($\Delta V/V$) during CO_2 uptake at 40 °C [2].

Class	Solvent	p (MPa)	Volume expansion (%)	CO_2 (wt%)	CO_2 (mol%)	CO_2 molar volume[a] (l mol^{-1})	Ref.
I	H_2O	7.0	n.a.[d]	4.8	2	n.a.	[31]
II	MeCN	6.9	387	83	82	46	[20]
	1,4-Dioxane	6.9	954	80	89	100	[20]
	DMF	6.9	281	52	65	119	[20]
III	[bmim][BF_4][b]	7.0	17	15	47	39	[28]
	PEG-400	8.0	25	16	63	53	[30]
	PPG-2700[c]	6.0	25	12	89	83	[30]

[a] Calculated from the increase in volume and moles of CO_2 from data in this table.
[b] Interpolated from the literature data.
[c] At 35 °C.
[d] Not available.

volumes of CO_2 in the expanded liquids were calculated as $\Delta V/n_{CO_2}$, where n_{CO_2} is the number of moles of CO_2 dissolved, from the data in Table 4.1. The values obtained in that manner are in accordance with data from the literature [32]. Obviously, the range of molar volumes in Class II systems overlaps the range in Class III systems. In particular, comparison of MeCN and PPG reveals that the lower volumetric expansion of PPG is not a result of a lower molar volume for CO_2 in PPG. Rather, the explanation is that at comparable mole fractions of CO_2 (e.g. 85%), the expanded MeCN contains a very large amount of CO_2 (109 mol of CO_2, assuming an initial volume of MeCN of 1 l) whereas the expanded PPG contains a very small amount of CO_2 (only 2 mol). Thus the difference between Classes II and III is caused by the different amount (i.e. number of moles or weight) of CO_2 dissolved rather than a difference in volume increase per mole of CO_2.

Within Class II systems, there is little variation in the molar density of a certain gas in different liquids. For example, CO_2 has a partial molar density of 1.0–1.1 g ml^{-1} in Class II organic liquids [33], and its mass uptake is therefore governed by the density of the organic solvent only [34]. Expansions of different Class II liquids as a function of mole fraction are hence very similar for a given gas (Figure 4.5) [20].

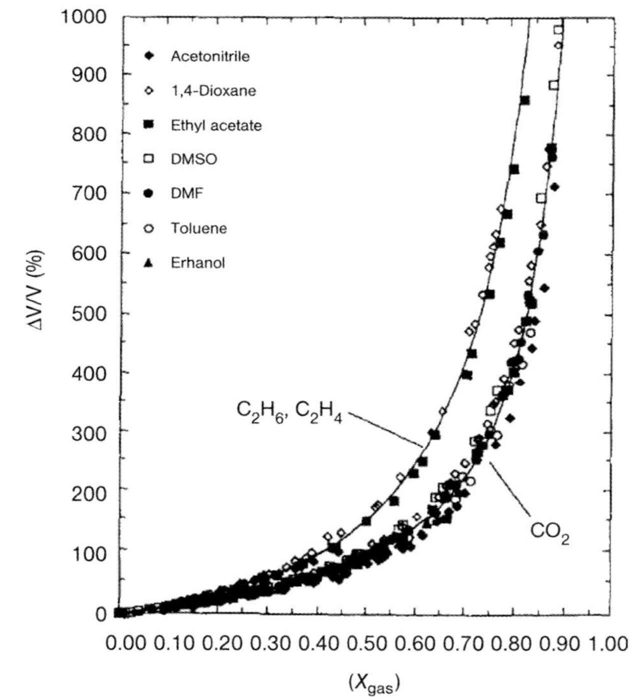

Figure 4.5 Volume expansion of different Class II liquids as functions of ethane, ethylene, and CO_2 mole fraction in the liquid phase at 40 °C (Reprinted with permission from Elsevier [20]).

4.2.2
Density

The sight of a liquid expanding in the presence of a compressed gas might intuitively suggest that density of the condensed phase should decrease due to the increase in volume. A more detailed analysis reveals, however, that the density of the ELP is not simply a linear function of the densities of the pure components. Figure 4.6 compares the ELP density using ethylene and CO_2 to expand different Class II liquids. It can be seen that the uptake of CO_2 does not change or even lead to a measurable increase in density over a wide range of composition. This is generally true for systems which have a strong negative volume of mixing and where the molar mass of the gas is similar to or even higher than that of the liquid. Examples of such combinations include CO_2 in light organic solvents such as MeCN, DMF, ethyl acetate, and short-chain alcohols [20, 35]. Fluorinated Class II liquids, however, form ELPs with steadily decreasing density when pressurized with CO_2 because of their high initial density ($>1.5\ \mathrm{g\,ml^{-1}}$) [36].

The high density of the ELP can be a significant advantage for the practical application of CO_2-expanded liquid phases as they retain high solvent power while replacing a substantial volume of potentially hazardous organic solvent by more benign CO_2. Moving closer to the mixture critical point leads to significant density reductions, allowing the initiation of selective separation processes coupled directly with a reaction in an ELP. Examples of this strategy will be discussed in more detail in Section 4.4.1.

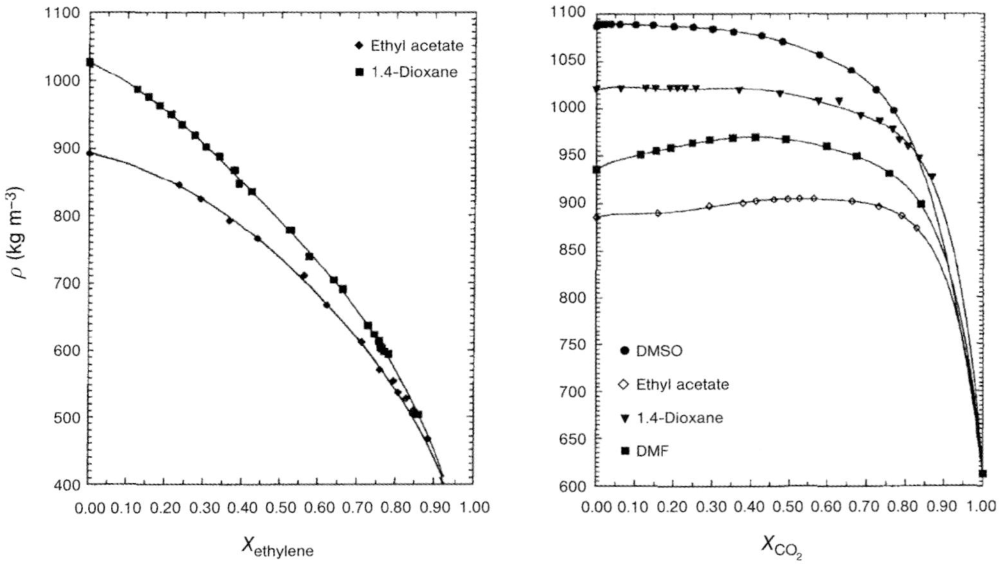

Figure 4.6 ELP densities of different Class II liquids as function of (a) ethylene and (b) CO_2 mole fraction at 30 °C (Reprinted with permission from Elsevier [20]).

4.2.3
Viscosity

In contrast to density, the viscosity of an ELP is always strongly reduced in comparison with the pure liquid. The reduction in viscosity of an ELP usually correlates with the mole fraction of dissolved gas and no threshold is detected upon contact of the liquid with the gas (Figure 4.7). For methanol and ethanol, a linear decrease in viscosity up to 0.5 mole fraction CO_2 (corresponding to about 100% increase in volume) was measured. At higher gas contents and volumetric expansions, viscosities converge to a maximum reduction of about 80% [37, 38]. Aprotic solvents such as acetone showed a smaller reduction in viscosity (maximum 60%) with a linear dependence on CO_2 content over the entire composition [39]. Highly non-polar, fluorinated hydrocarbons exhibit viscosity reductions which linearly correlate down to 85% at high gas contents [36].

Comparing Figures 4.6 and 4.7, it should become apparent that ELPs do not follow Enskog's theory, which correlates viscosities with density [40]. In the regime of constant or even slightly increased density (up to 0.6 mole fraction CO_2) viscosity is already greatly reduced. This is mainly due to the very low viscosity of the pure gaseous components even in a highly compressed state. For example, liquefied CO_2 (7 MPa at 27 °C) has a viscosity of 0.054 cP, which is roughly 20 times less than that of liquid acetone at the same temperature [41]. Consequently, the viscosity of the liquid phase is also greatly reduced in contact with $scCO_2$ or when liquid CO_2 is condensed into it. As the viscosity of a fluid is inversely correlated with important process parameters such as heat and mass transport, the significant decrease in viscosity at high density and solvent power constitutes a highly attractive property of Class II ELPs for reactive multiphase applications.

As Class III systems comprise dense, non-volatile, and rather viscous liquids, the effects on viscosity are typically even more pronounced. The reductions achieved with only moderate CO_2 pressures are very high on an absolute scale. The idea of lowering the viscosity of crude oil by dissolving gases in the oil has been known since 1895 [42–45] and industrially exploited for enhanced oil recovery processes.

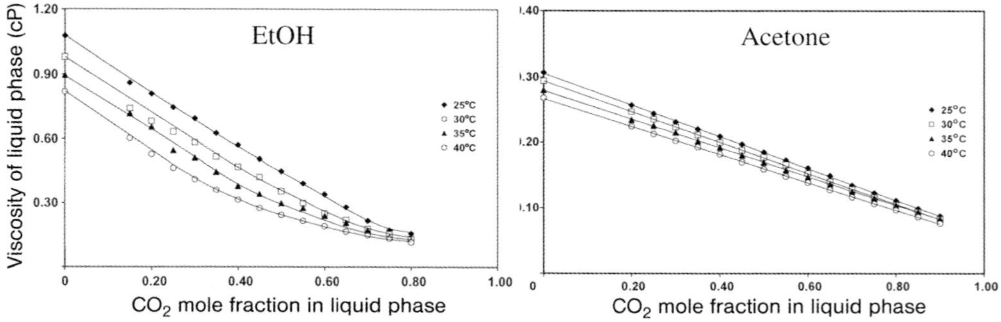

Figure 4.7 Viscosities of Class II ELPs as a function of CO_2 mole fraction at various temperatures (Reprinted with permission from Elsevier [38, 39]).

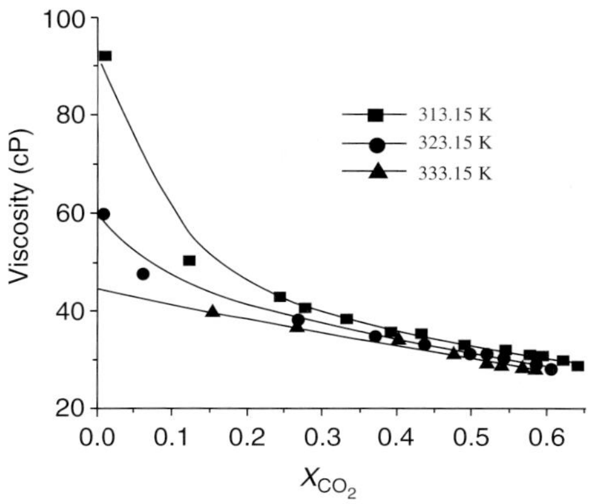

Figure 4.8 Viscosity of [bmim][PF_6] as a function of CO_2 mole fraction at various temperatures [48].

Compressed CO_2 is also applied to allow polymer processing at milder temperatures [46]. For example, molten high molecular weight poly(ethylene glycol) (PEG 6000) undergoes a reduction in viscosity from 820 to 200 cP when pressurized with 15 MPa of CO_2 at 80 °C [47]. As discussed in detail in Sections 4.4 and 4.5, this effect can be exploited favorably in multiphase catalysis using PEG–scCO_2 systems.

Ionic liquids are also Class III liquids that benefit greatly from the viscosity reduction with compressed gases. Viscosities of typical standard ILs are often two to three orders of magnitude higher than those of organic solvents (e.g. 160 cP for [bmim][PF_6] at room temperature), which can limit their application in chemistry and chemical engineering. As can be seen in Figure 4.8, [bmim][PF_6] exhibits strong non-linear viscosity reductions of up to 65% with a pronounced initial reduction of about 50% within the first 0.2 mole fraction (2.0 MPa) of CO_2 [48]. Spectroscopic studies predict a maximum reduction of 80% for the same system [49]. Thus, compressed CO_2 must not only be considered as an SCF in combination with ILs, but can also be envisaged for process improvement in applications of ILs at much lower pressures.

4.2.4 Melting Point

The dissolution of a solute in a liquid typically causes a melting point depression and the same holds for compressed gases which dissolve in ELPs. This is not restricted to lowering the melting point of compounds that are already liquids at ambient conditions, but can also result in a gas-induced melting of solids. Owing to the noticeable solubility of gases in solids, the ELP can actually be formed by melting point depression as part of the physico-chemical changes in the presence of the compressed gas. As early as 1898, Paul Villard reported on "solutions of solids and liquids in gases" where he found that camphor liquefies below its normal melting temperature upon contact with compressed ethylene [26]. This phenomenon has

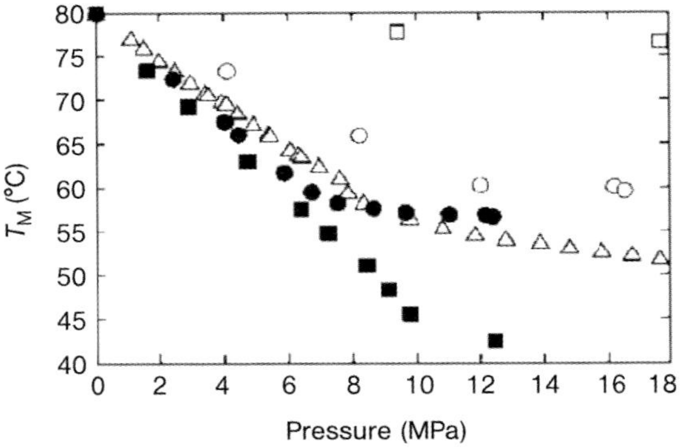

Figure 4.9 Melting point of naphthalene as a function of pressure of CO_2 (○), ethane (●), ethylene (△), methane (□), and xenon (■) (Reprinted with permission from American Chemical Society [2]).

since been observed for many organic compounds [50], and also for polymers [51–53], organic/inorganic salts (to become ionic liquids) [54, 55] and lipids [56].

In general, induced phase change is best seen in Class III systems. Melting point depressions of solutes usually go along with boiling point elevations of the gas and can be explained on a thermodynamic level as the interplay of the solid–liquid and solid–liquid–vapor equilibria of the pure components [6]. Hence there is no direct quantitative correlation between solubility and melting point depression, but high gas solubility is generally a prerequisite for the ability to lower significantly the melting point of the respective solid. If the gas is very poorly soluble, melting point elevations are observed because most organic solids expand upon melting and thus hydrostatic pressure (as applied by a non-dissolving gas) inhibits the phase transition. Examples include naphthalene + helium [57], but these systems are beyond the scope of this chapter. Examples of melting point depressions of naphthalene with different gases are shown in Figure 4.9.

In most cases, melting point depressions are moderate (<30 K) and reach a limit at the upper critical end-point of the mixture. If binary systems do not have upper critical end-points, larger melting point depressions can occur (e.g. 40 K for menthol + ethylene) [58]. The largest effects known in the literature are melting point depressions of organic salts with CO_2 to become CO_2-induced ionic liquids (Figure 4.10). Certain tetraalkylammonium and -phosphonium salts exhibit melting point depressions reaching up to 120 K for example in the case of $[(^n\mathrm{Bu})_4\mathrm{N}][\mathrm{BF}_4]$ [55]. This allows the generation of ELPs for multiphase catalysis employing compounds that would not be considered as ILs under conventional conditions.

4.2.5
Interfacial Tension

Interfacial tension strongly influences transport properties across the gas–liquid interface. It determines bubble and droplet sizes, which are important characteristics

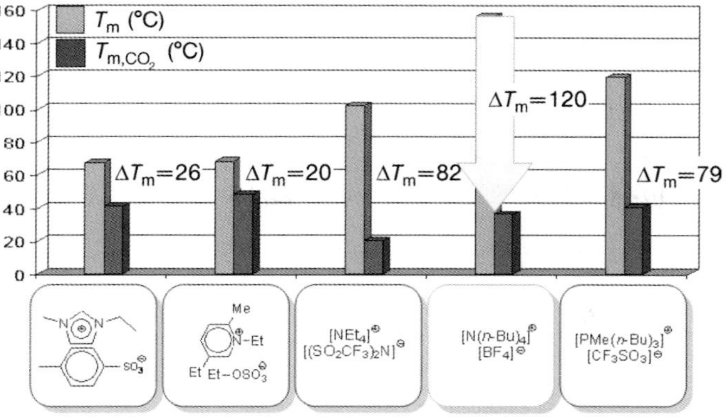

Figure 4.10 Generation of CO_2-induced ionic liquids: melting point depression of selected salts in the presence of compressed CO_2 (15 MPa) [55].

for emulsions. Similarly, the surface tension of a liquid determines the wetting behavior of solid surfaces, which is especially important in heterogeneous catalysis with high surface-area porous particles. For compressible gases, a largely temperature-independent correlation between gas density and interfacial tension of the liquid phase has been found for many systems, including Class II and even hardly expanded Class I combinations (mainly water) [59–61]. Pressurizing ethanol with CO_2 causes reductions in interfacial tension of up to 90% and even in the subcritical regime of a GXL more than 50% reductions have been obtained (Figure 4.11). Very

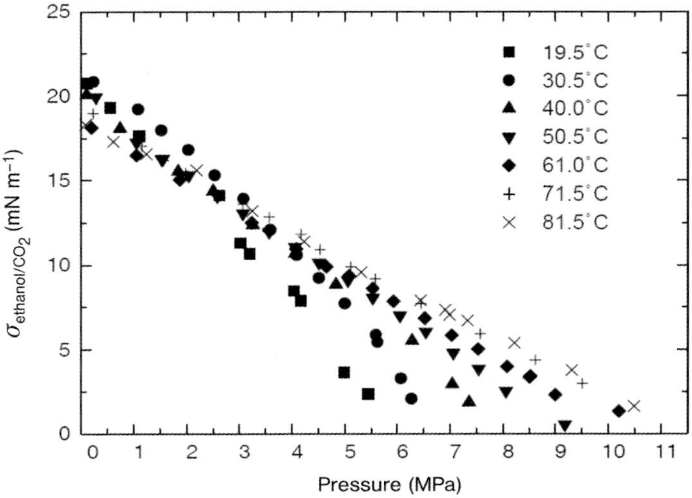

Figure 4.11 Interfacial tension of ethanol in contact with compressed CO_2 at various temperatures (Reprinted with permission from Elsevier [60]).

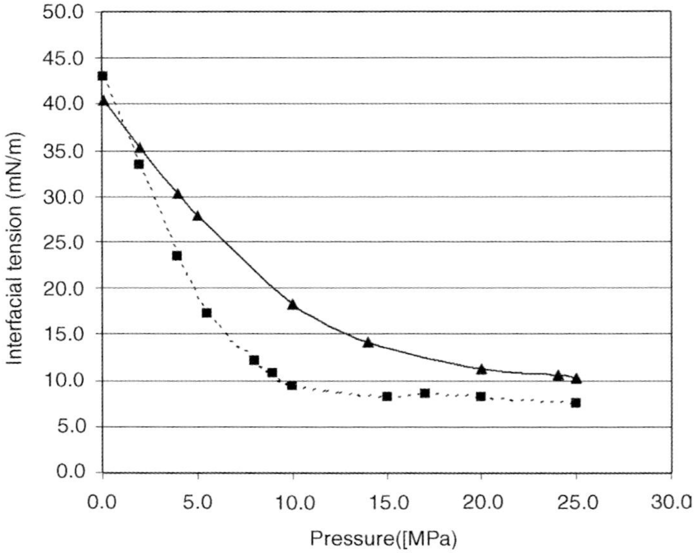

Figure 4.12 Interfacial tension of [bmim][PF_6] in contact with CO_2 at 40 °C (■) and 80 °C (▲) (Reprinted with permission from Elsevier [63]).

similar results were obtained with ethane as the expansion gas, whereas no correlation with density and lower reductions in interfacial tensions were found with non-compressible N_2 [60]. Very large reductions have been reported when non-polar hydrocarbons were pressurized with CO_2: the interfacial tension of *n*-butane decreased by up to two orders of magnitude at 8.0 MPa CO_2 [62].

Very recently ionic liquids were investigated with respect to interfacial tension changes upon contact with compressed gases (Figure 4.12) [63]. Both [bmim][PF_6] and [bupy][BF_4] showed reductions in interfacial tension comparable to those for other Class III compounds such as lubricant oils. A decrease of 75% was achieved at 20 MPa, which is comparable to the reduction for water but slightly lower than for Class II liquids. By application of 5.0 MPa CO_2 at 40 °C, interfacial tensions comparable to those for organic Class II solvents under atmospheric pressures could be achieved.

As the interfacial tension is related to the diffusion resistance of a gas into a liquid phase, the observed reductions should have beneficial impacts on the kinetics of gas intake into ILs or other viscous liquids. Together with increased liquid diffusivity (see below), it can help to overcome transport limitations for fast catalytic reactions with gaseous substrates. This demonstrates again the potential of using even moderate CO_2 pressures for process intensification based on ELP systems.

4.2.6 Diffusivity

According to the Stokes–Einstein equation [64] and other often used derivations such as that of Wilke and Chang [65], the diffusion coefficient of any compound in

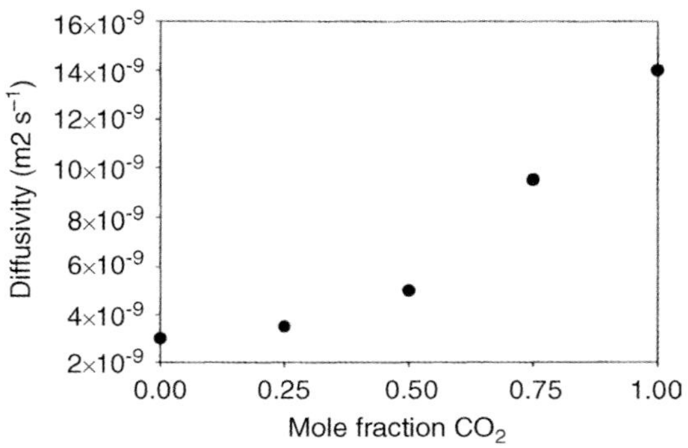

Figure 4.13 Diffusion coefficients of benzene in methanol at 40 °C as a function of CO_2 mole fraction at 15 MPa [67].

a liquid phase is inversely proportional to the viscosity of that liquid. Therefore, the decrease in viscosity upon expansion directly correlates with enhanced diffusivity. Several Class II systems have been examined and 3–4-fold enhancements of diffusivity were measured for aromatic probe molecules in CO_2-expanded methanol [66–68]. These enhancements correlate well with the respective viscosity measurements (Figure 4.13) [37].

As outlined in Section 4.2.3, viscosity reductions are most pronounced for Class III ELPs. Consequently, very high diffusion enhancements are expected for these systems. However, only a limited set of data is currently available, as only gas diffusivities have been measured in ILs and polymers so far. Extrapolations from these data should be treated with caution, as non-ideal behavior beyond the Stokes–Einstein equation has been found for heavier gases in ILs [69]. A lower than expected dependence of gas diffusivity on viscosity was determined, in line with the finding that diffusion-controlled reaction rates in imidazolium ILs were higher than predicted by the Stokes–Einstein equation [70].

It should be noted that self-diffusion of a liquid is not only related to mass transfer, but is also directly proportional to convective heat dissipation. This indicates that improved thermal equilibration can also be expected in ELPs, which may be of importance for both exothermic (heat removal) and endothermic (heat input) reactions. Although there is no direct evidence yet, this may be part of the beneficial characteristics of hydrogenation reactions over heterogeneous catalysts, where hot spots may be avoided under ELP conditions.

4.2.7 Polarity

A gas physically dissolved in a liquid interferes with the intermolecular interactions of the liquid molecules. Therefore, an ELP has different polarity and hydrogen bonding properties than the pure liquid in the absence of the gas. The change in polarity can be represented graphically similarly to other property changes in a polarity–composition diagram: the polarity of the ELP varies between the polarities of the pure components. The fact that in most combinations the pure components differ significantly in polarity allows wide tuning of the resulting ELP. Because non-polar gases are usually combined with polar liquids to create Class II ELPs, polarity decreases with increasing mole fraction of the gas in the liquid phase. This does not mean, however, that ELPs are generally more polar than SCFs: cosolvents can increase the polarity of many SCFs considerably [71] and a large number of polar substances form SCFs, including chlorofluoromethanes [72], fluoroethanes [73], ethanol [74], ammonia [73], and water [75]. CO_2 has a considerable quadrupole moment, which allows interactions with medium-polarity substances [76], but the fact that it has no dipole moment means that it decreases the polarity of most ELPs.

ELPs formed from medium- to high-polarity liquids with CO_2 have considerably higher polarities than $scCO_2$-based media. Solvent polarity is usually quantified either macroscopically via the bulk dielectric constant (relative permittivity) or locally using various solvatochromic probe molecules. These are well-established methods for the characterization of conventional organic solvents [77]. Whereas most probe molecules provide information on only one specific solvent–solute interaction, Kamlet and Taft developed a useful set of three complementary solvatochromic parameters that characterize solvents with regard to hydrogen bond acidity (α) [78], hydrogen-bond basicity (β) [79] and polarity and polarizability (π^*) [80]. These are referenced parameters calculated from spectroscopic shifts of a series of solvatochromic probe dyes and typically vary between 0 and 1, although a few solvents fall just outside that range. Pure $scCO_2$ at 40 °C has π^* values between –0.05 and 0.05 [81] and β values between −0.08 and +0.06 [82], varying as functions of pressure, which makes it a rather non-polar solvent with little hydrogen bonding ability. As a consequence, Class II ELPs consisting of polar organic solvents undergo dramatic changes when pressurized with CO_2.

The polarities of methanol and acetone are greatly reduced as CO_2 dissolves (Figure 4.14), whereas their hydrogen bonding abilities are not much affected by the gas: essentially no change in α and, only above 5 MPa, a small decrease in β are

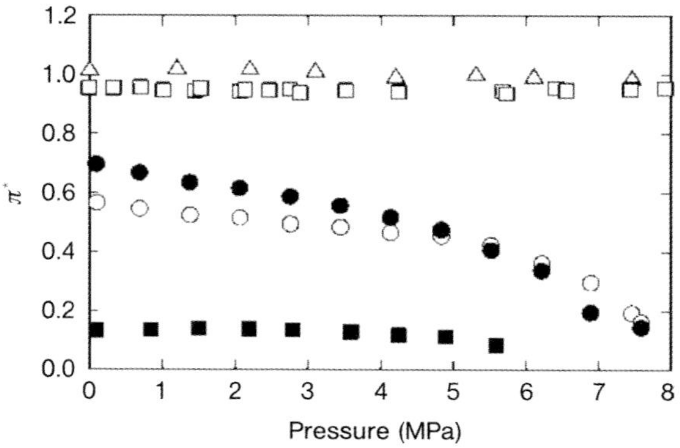

Figure 4.14 Polarity parameter (π^*) of acetone (•), methanol (○), [bmim][BF_4] (△), [bmim][PF_6] (□), and ethoxynonafluorobutane (■) as a function of CO_2 pressure at 40 °C (Reprinted with permission from American Chemical Society [2]).

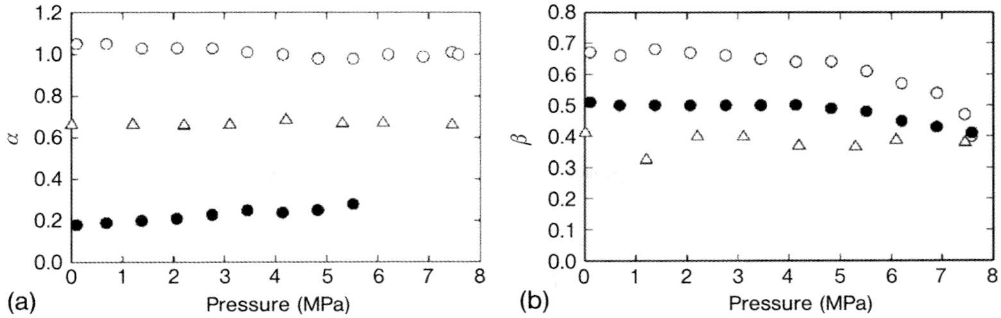

Figure 4.15 Hydrogen bond-accepting (α) and hydrogen bond-donating (β) abilities of acetone (•), methanol (○), and [bmim][BF_4] (△) as a function of CO_2 pressure at 40 °C (Reprinted with permission from American Chemical Society [2]).

observed (Figure 4.15b). The polarity of very low-polarity Class II liquids such as fluorinated hydrocarbons barely changes as non-polar CO_2 dissolves.

A direct comparison of the relative permittivities of different organic solvents under CO_2 pressure provides a practical overview of the polarities of Class II ELP systems (Figure 4.16) [83]. As the data obtained were collected at constant pressure, the different solubility of CO_2 in the various solvents is not accounted for. Nevertheless, the results demonstrate qualitatively the fact that the reduction in polarity of the resulting ELP depends on the difference in polarity of the pure components.

Ionic liquids show no measurable changes in polarity at similar pressures (Figure 4.14), although the difference in polarity between the liquid and the gas is fairly large [74, 84]. However, as shown in Table 4.1, the mole fraction of CO_2 in this IL is lower than in Class II solvents at the same pressure. Using solvatochromic probe

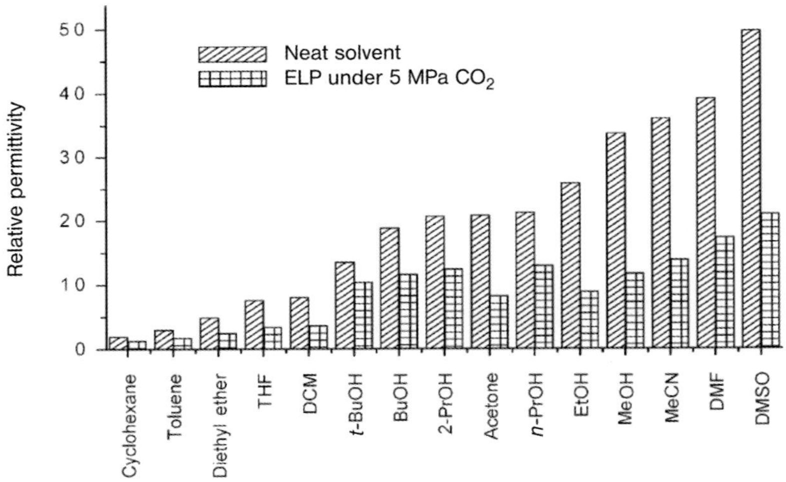

Figure 4.16 Relative polarities of organic solvents and their ELPs obtained under 5 MPa CO_2 at 25 °C (Reprinted with permission from Royal Society of Chemistry [83]).

molecules, it has been shown that no significant reduction in polarity can be observed even at pressures up to 15 MPa [85]. Other Class III solvents, such as PEG-1000 and PPG-3500, do show a small but steady decrease in polarity as measured by the solvatochromic dye Nile Red [86]. The strong intermolecular interactions especially in the case of ILs and the limited gas uptake of Class III systems in general have been held responsible for the comparably small changes in polarity.

4.2.8
Gas Solubility

The tunable variation of the polarity of ELPs as described above opens up a wide field of possible separation techniques based on reversible changes in mutual solubility, which will be discussed in Section 4.3.1. The presence of a compressible gas in a liquid also affects the solubility of other non-compressible gases in the resulting ELP. This is intuitive as ELPs can be pictured as intermediates between condensed molecular solvents and SCFs, which are completely miscible with permanent gases. Class II systems show enhanced solubility of non-compressible gases as a function of increasing density of the expanding gas as measured for H_2, O_2, and CO in polar CO_2-expanded liquids including methanol, acetonitrile, and acetone [87–89]. The enhancement factors (ratio of solubility in neat solvent to that in the respective ELP at identical partial pressure/fugacity) reach values of up to 2.8 as a function of CO_2 density at temperatures between 25 and 40 °C with little difference between the reactive gases. The nature of the compressible gas seems to have an influence on the solubility enhancements as higher H_2 solubility enhancements were found in 2-propanol expanded with propane as compared with CO_2 [90], albeit the measurements were carried out at different temperatures.

Figure 4.17 compares the solubility of H_2 in a non-polar organic solvent with its solubility in the ELP of the same solvent at 60 °C. Here, enhancement factors of up to

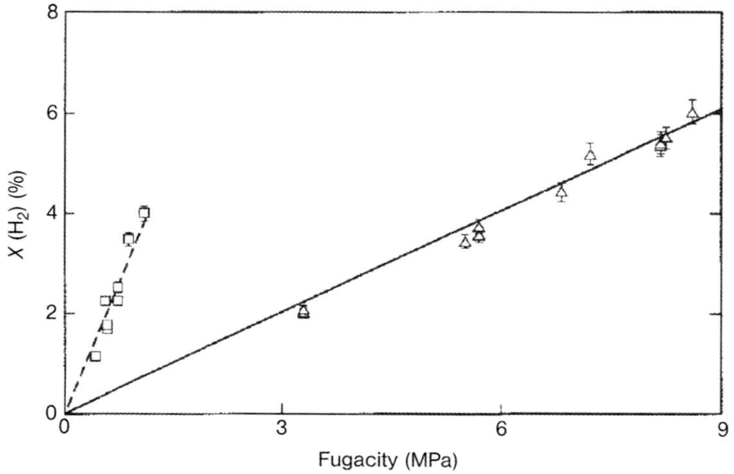

Figure 4.17 H_2 (mol%) in 1-octene at 60 °C versus H_2 fugacity for pure H_2 (△) and a 1:1 mixture of H_2 and CO_2 (□) [87].

4 were found even at relatively mild pressures. Note that both the solubility of H_2 in the neat organic solvent (a Class I system) and its initial solubility in the GXL ($\leq$1 MPa) are in the linear regime of Henry's law. On comparing solubilities at the same total pressure ($P_{CO_2} + P_{H_2}$), however, it turns out that the solubility in CO_2-expanded liquids is not increased beyond the solubility achieved with the pure gas at the same total pressure. The enhancement at constant partial pressure is important for the understanding of the reactivity in multiphase systems based on ELPs. Furthermore, it allows a substantial part of a reactant gas to be replaced by an inert component without decreasing the reactant gas concentration in the liquid phase where a catalyst may reside. This aspect constitutes a considerable safety benefit, especially for oxidation reactions with molecular oxygen performed in CO_2-expanded liquids.

Class III systems also exhibit pronounced effects of gas solubilities upon expansion with a compressible gas. O_2 was found to experience enhancement factors in the ionic liquid [hmim][NTf$_2$] of up to 5 even at pressures below 1 MPa [91]. Most interestingly, even an absolute enhancement (solubility of pure gas compared with mixed gas at identical total pressure) can be achieved in such systems, meaning that diluting the gas phase leads to an increase of the concentration in the liquid phase. The solubility of CH_4 in this IL was found to be enhanced by a factor of 2.6 in the same study. For hydrogen solubility, high-pressure ^{1}H NMR spectroscopy can be used to measure H_2 contents in liquids also in the presence of expanding gases [92]. If corrected for the equilibrium between *ortho*- and *para*-H_2 [93], the technique allows direct access to molar ratios independent of volume changes. In Figure 4.18, the increase in H_2 concentration in an IL at a constant partial pressure of 3.0 MPa H_2

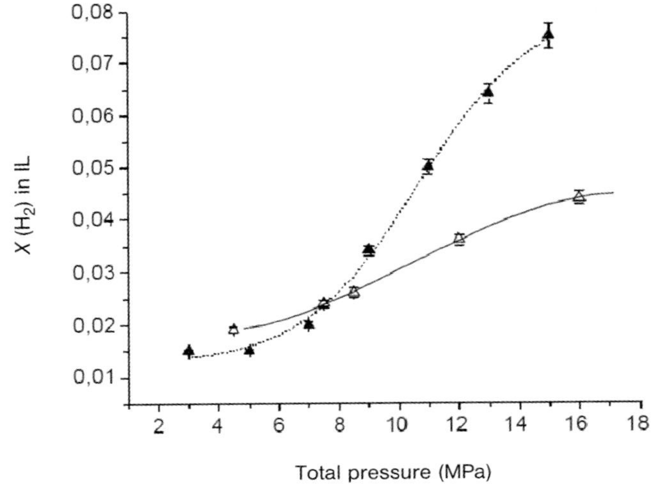

Figure 4.18 H_2 mole fraction in [bmim][NTf$_2$] at a constant partial pressure of H_2 = 3.0 MPa as a function of CO_2 (▲) and C_2H_4 (△) pressure added at 22 °C [92].

caused by adding compressed CO_2 or ethylene is shown. With CO_2 at 15 MPa, enhancement factors of up to 4.5 are achieved for H_2 solubility in [bmim][NTf_2]. Note that the enhancement through CO_2 is most pronounced after its liquefaction point of 6 MPa (at 22 °C), whereas the lower enhancement through ethylene is caused by a supercritical fluid ($T_c = 9\,°C$, $p_c = 5$ MPa) in this case.

Remarkably, the solubility enhancement of gases does not correlate with any other trends such as volumetric expansion or polarity change either for Class II or Class III systems. It appears that a synergy between the expansion gas and the permanent gas is responsible for the enhancements, whereby the highly soluble compressible gas acts like a carrier for the less soluble gas. This concurs with the finding that the solubility of the compressible gas in the liquid decreases as the solubility of the non-compressible gas increases when a mixture of both is applied [91]. Clearly, more data on mixed gas solubilities, especially for Class III ELPs, are needed in order to understand fully the underlying principles. However, the increased solubility together with the increased diffusivity due to reduced viscosity leads to an enhancement of both the kinetics and the thermodynamics of gas availability in ELPs at the same time. Selected examples in Section 4.4 will highlight the exploitation of these effects for multiphase catalysis.

4.3
Chemisorption of Gases in Liquids and their Use for Synthesis and Catalysis

The properties of ELPs as discussed in the previous section result from physical interactions between liquid and gas molecules such as dipole interactions, induced dipole interactions, and dispersion forces (physisorption). Gas solubility as a macroscopic phenomenon may also comprise chemisorption [94] caused by chemical reactivity between the components involving bond formation or electron transfer. These processes can also be reversible, just like the weaker physical interactions. Whereas pure physisorption leads to a gradual change with pressure (linear in the range of Henry's law), chemisorption is typically characterized by a step change directly upon contact of the two components. The chemical changes quickly reach saturation at reaction equilibrium, which is usually not shifted by higher pressures [94]. To distinguish these more abrupt changes from pressure-tunable variations of ELPs and SCFs, the term switchable solvents [95] has been introduced to highlight the fact that in the case of chemical reactivity the gas serves as a trigger rather than a modulator as with physical dissolution.

4.3.1
In Situ Generation of Acids and Temporary Protection Strategies

Despite its common description as being "non-reactive", CO_2 readily interacts even with fairly weak bases to generate relatively stable Brønsted acids under equilibrium conditions. The most important reactions are the additions of water, alcohols, or amines as shown in Scheme 4.1. These reactions happen to these compounds not

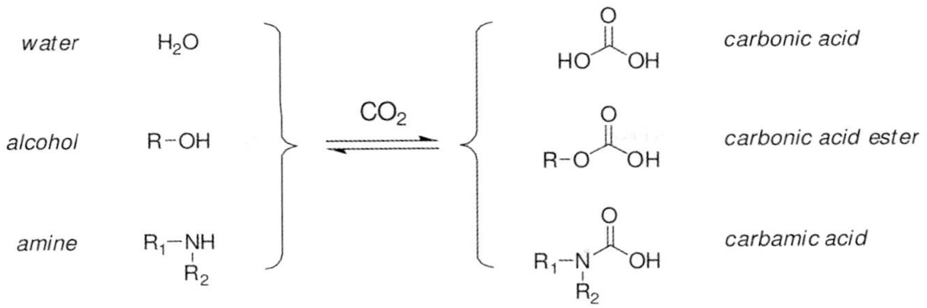

Scheme 4.1 Primary equilibrium reactions of water, alcohols, and amines with CO_2; further equilibria are not shown (R = alkyl, aryl, H).

only when they are serving as solvents, but also when they are solutes in any other solvent under CO_2 pressure. As these *in situ* acid formations are completely reversible by removal of CO_2, they can be utilized as self-neutralizing acids. This is a distinct advantage over permanent acids, which need to be neutralized in downstream processes creating large amounts of salt waste. On the other hand, the acids may also negatively impact on catalysts or reaction pathways in the process under scrutiny. A straightforward test of whether this reactivity of CO_2 interferes with a reaction conducted in a multiphase system is a direct comparison with a non-reactive compressible gas such as ethane, argon, or SF_6 [96].

As all the reactions shown in Scheme 4.1 involve Brønsted acid transformations, the equilibrium position can be probed via the pH of the liquid. As stronger acids are being formed upon introduction of CO_2, a decrease in pH can be observed that exhibits the sharp pressure dependence typical of chemisorption. For instance, pure water at 60 °C under 1.0 MPa CO_2 has a pH of 3.6, which only reduces further to 3.2 at 9.0 MPa [97]. Depending on the pH and ionic strength, there are further acid–base equilibria to the bicarbonate and carbonate ion in solution. Thus, addition of the corresponding base of the *in situ*-generated acid can buffer pH values to almost neutral levels, but high buffer concentrations in the range 0.4–1 mol l^{-1} are needed at higher pressures [97].

The formation of carbonic acid esters is manifested by, for example, incorporation of solvent alcohols into ethers and carbonates from diazodiphenylmethane, which proceeds only when the alcohol is pressurized with CO_2 [98]. Similarly, pressurizing a 1:1 mixture of water and methanol with 2.0 MPa CO_2 at 75 °C allowed effective hydrolysis of β-pinene without the need for extra acid and at even higher selectivity compared with conventional systems. By varying the water to alcohol ratio and comparison with other non-protic expanded solvents of similar polarity, it was concluded that the carbonic acid ester of the alcohol was the promoting component in this reaction [99].

A strong increase in reaction rate in the presence of CO_2 has been reported for the transesterification of glycerol monostearate with methanol to give methyl stearate, a potential biofuel. Full conversion was achieved within 5 h at 60 °C when the neat mixture of reactants was expanded with 6.5 MPa CO_2, whereas 35 h were needed without CO_2 [100]. Although the accelerating effects of CO_2 were explained solely on the basis of physical effects such as enhanced miscibility and diffusivity, it seems plausible that carbonic ester formation leading to increased acidity in the medium also helps to catalyze the transesterification reaction. Similarly, the esterification of acetic acid with ethanol was found to be enhanced in presence of CO_2. In addition to accelerated rates due to catalytic effects by increased acidity, the equilibrium position could also be shifted towards the products by adding 5.9 MPa CO_2 [101]. It was concluded that preferential extraction of the product ester enhanced the equilibrium conversion.

Nitroarenes can be selectively reduced to their respective hydroxylamines using zinc [102] or iron [103] when the reaction is carried out in water under 0.1 MPa CO_2. The *in situ* formation of carbonic acid makes the addition of NH_4Cl unnecessary, thus removing the need for downstream neutralization.

The *in situ* generation of acids with CO_2 has also been exploited for the oxybromination of phenols and anilines in H_2O–CO_2 media (Scheme 4.2). The reaction uses NaBr–H_2O_2 as the bromine source and proceeds without any metal catalyst or additional acid in the presence of CO_2 [104]. In this case, the CO_2 plays a dual function as the formation of peroxocarbonic acids from H_2O_2 in CO_2-expanded water increases the acidity and also the oxidation ability (see Section 4.4.5). Thus, conversion could be increased from 32% in neat water to 91% by addition of CO_2 under otherwise identical conditions.

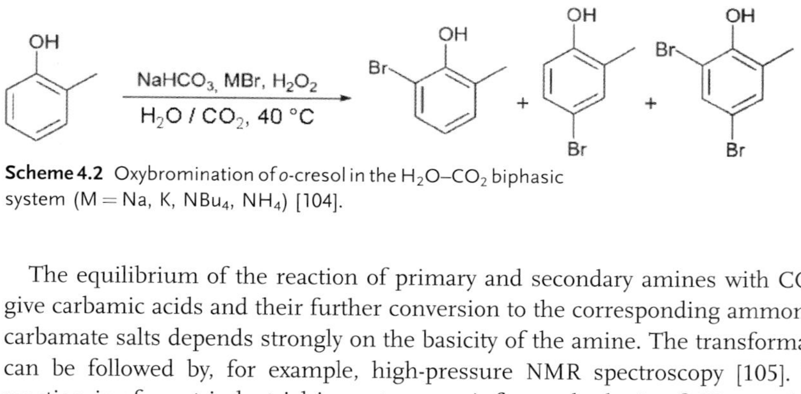

Scheme 4.2 Oxybromination of *o*-cresol in the H_2O–CO_2 biphasic system (M = Na, K, NBu_4, NH_4) [104].

The equilibrium of the reaction of primary and secondary amines with CO_2 to give carbamic acids and their further conversion to the corresponding ammonium carbamate salts depends strongly on the basicity of the amine. The transformation can be followed by, for example, high-pressure NMR spectroscopy [105]. This reaction is of great industrial importance as it forms the basis of CO_2 scrubbing processes used on a large scale in hydrogen production and discussed also for post-combustion CCS (carbon dioxide capture and storage) technologies in power plants [106, 107].

The reversible reactivity of CO_2 with amines can also be used to direct the selectivity of reactions by acting as a temporary protecting group for N–H groups. An example using this effect in a complex catalytic cascade reaction is the intramolecular rhodium-catalyzed hydroaminomethylation of allylic secondary amines under hydroformylation conditions (Scheme 4.3) [108]. In organic solvents, the free amine reacts to form exclusively cyclic amides as products. When CO_2 is applied to

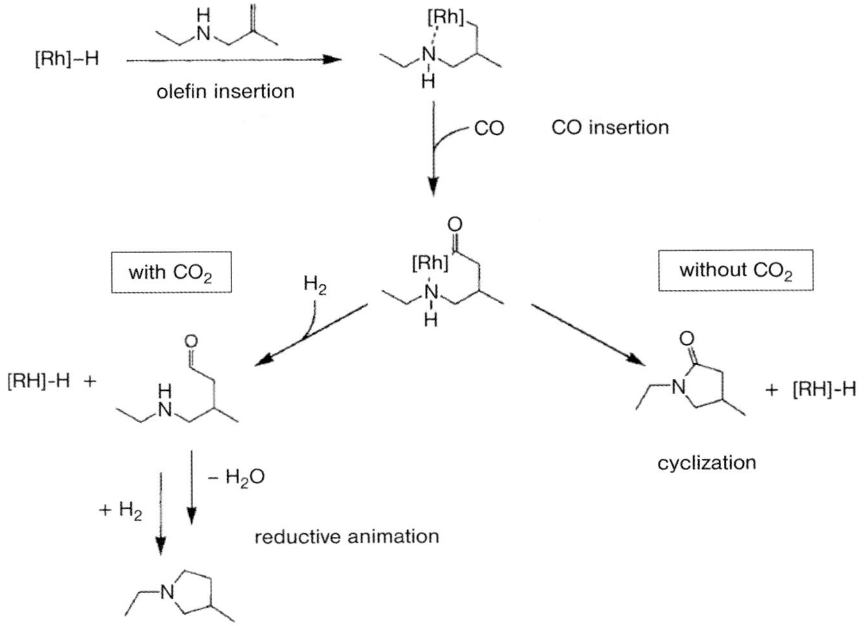

Scheme 4.3 Selectivity control in the hydroaminomethylation reaction of allylic amines through temporary CO_2 interaction. The Rh–acyl intermediate of the free amine cyclizes intramolecularly to give amides whereas under CO_2 pressure cyclization is suppressed and hydroformylation occurs followed by reductive amination to yield pyrrolidines [108].

the catalytic system, the amine is temporarily converted into the carbamate and the corresponding pyrrolidines are formed as the main products.

Similarly, the temporary *in situ* protection of amines with CO_2 has been applied to alkene metathesis of amine-containing substrates in $scCO_2$. Whereas classical Grubbs catalysts cannot be used with basic amines under conventional conditions because of catalyst deactivation [109], the reaction can be efficiently carried out in the presence of CO_2 [110]. Owing to the reversible interaction of CO_2 and the N–H function, the free amines are isolated after standard workup, saving two reaction steps (protection and deprotection) in the reaction sequence.

Another example of selectivity control by application of CO_2 has been reported for the heterogeneously catalyzed hydrogenation of aryl nitriles (Scheme 4.4). If no CO_2 is added, the imine intermediates react with the free amines to give secondary and tertiary amines and no primary amines are obtained [111]. When the reaction is conducted in ethanol expanded with 3.0 MPa CO_2, the primary amines can be isolated as the corresponding ammonium carbamate salts.

The reactivity of CO_2 with amines can also be used for further derivatization to urethanes. Amines can be converted to methyl carbamates in the presence of a methylating agent and CO_2 with low levels of direct *N*-methylation [112]. The reaction

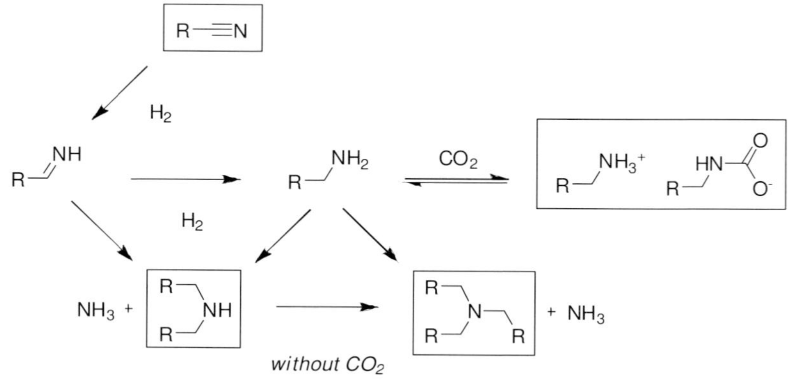

Scheme 4.4 The formation of secondary and tertiary amines during the hydrogenation of nitriles is suppressed when CO_2 is used to protect the primary amines [111].

proceeds both in $scCO_2$ and in an ELP of the neat substrates, but with a maximum conversion in the latter around 5.0 MPa of CO_2. This strategy has been used for Pictet–Spengler cyclizations to synthesize quinolines. Whereas the substrates do not react in the presence of CO_2 due to inhibition of the amine, the reaction does proceed selectively when dimethyl carbonate is added [113]. In this case, alkylation of the carbamate allows for further reactivity of a primary amine in CO_2-containing media [114].

4.3.2
Switchable Solvents and Catalyst Systems

The reactions of CO_2 shown in Scheme 4.1 and their consecutive proton transfer reactions do not just allow interaction with substrates or reagents, but can also be used to alter the nature of a reaction medium and thus change the molecular environment of a chemical process. If a stoichiometric amount of a potentially acidic compound (typically alcohol or amine) and a suitable base are exposed to CO_2, the acid–base equilibria create salts of the corresponding carbamates or carbonates. Given that these salts exhibit low melting points, this reaction will generate a highly polar ionic liquid from a medium-polarity organic solvent [95]. Remarkably, the only step needed for the transformation is application of a low (often only atmospheric) pressure of CO_2. Gentle heating to 50 °C and/or stripping CO_2 out by an inert gas at atmospheric pressure lead to a complete reversal of the reaction recreating the non-ionic organic solvent.

This principle was successfully demonstrated for equimolar mixtures of amidines and alcohols where the latter react with CO_2 to give the corresponding carbonic acid ester, which is deprotonated by the amidine to give amidinium alkylcarbonate ILs (Figure 4.19) [115]. Further, it has been extended to amidines combined with chiral alkyl esters [116] or alcohols [117] derived from natural amino acids affording chiral ILs. By solvatochromic measurements, it has been demonstrated that the change in

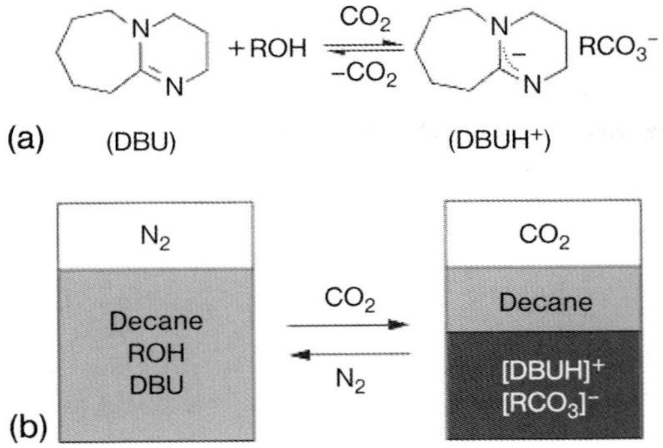

Figure 4.19 (a) Example of a switchable solvent system: protonation of the amidine DBU (1,8-diazabicyclo [5.4.0]undec-7-ene) by a reversibly *in situ*-generated carbonic acid ester from an alcohol and CO_2. (b) Exposure of the non-polar (light gray) amidine–alcohol mixture to CO_2 for 1 h at room temperature results in the formation of the corresponding polar IL (dark gray). If a non-polar liquid such as decane is present in the mixture, it separates from the resulting IL, qualitatively indicating the polarity change during the switch. N_2 reverses the process [95].

polarity of alcohol–base mixtures to protic ILs corresponds to the difference of chloroform and dimethylformamide. In addition to amidines as N-bases, guanidines such as butyltetramethylguanidine (TMBG) have also be shown to react reversibly with CO_2 and thus proved useful for generating switchable solvents. The length of the alkyl chain of the alcohol is crucial for the melting point of the resulting salts. Whereas water, methanol, and ethanol afford solids with 1,8-diazabicyclo[5.4.0] undec-7-ene (DBU) and CO_2, longer chain alcohols (propanol to decanol) create room temperature ILs [115]. On the basis of thermodynamic analyses, such mixtures have been suggested as CO_2 capture agents because of their high gravimetric CO_2 uptake and convenient handling. As they remain liquid even at high CO_2 uptake, they can more easily be recycled and thus provide potential advantages over aqueous ammonia or monoethanolamine (MEA) solutions currently used in industry [118].

For the system TMBG–methanol, conductivity measurements have nicely shown the reversible formation of ionic species upon switching between reactive CO_2 and inert N_2 (Figure 4.20) [119]. By switching from an organic solvent to an ionic liquid, not only are conductivity, acidity, and polarity affected, but redox potentials, heat capacity, viscosity, and related transport properties may also differ for the two stages.

Switchable systems based on only one component to act both as the acid generator and the base are also known, as demonstrated for pure secondary dialkylamines. Here, a molecule of free amine deprotonates the carbamic acid formed from the same amine and CO_2. This system has the practical advantage that it does not depend on accurate mixing of two components. Furthermore, the system has a lower polarity in the non-ionic form as compared with alcohol–base mixtures, thus exhibiting an even larger change upon switching [120]. Another example of such solvents is primary amines functionalized with siloxanes. Usually short-chain primary amines are volatile and thus elusive as solvents in their non-ionic state, but incorporating a

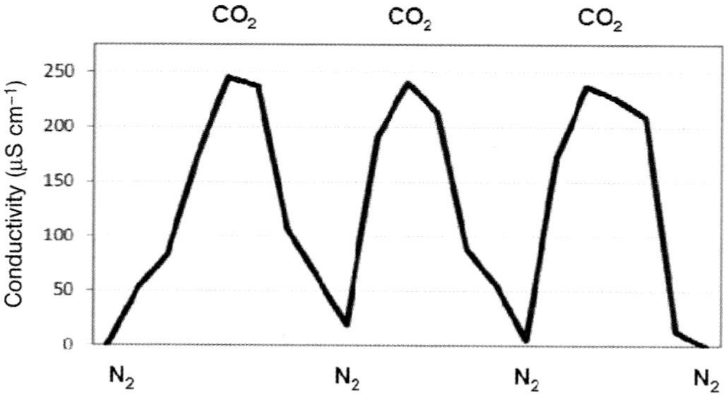

Figure 4.20 Conductivity of a 1:1 TMBG–methanol mixture in chloroform upon bubbling CO_2 and N_2 sequentially through the solution (Reprinted with permission from Elsevier [119]).

siloxane moiety into the molecule makes them available as switchable liquids [121]. The sharp change in solubility properties has been demonstrated in sequential crude oil separation. Due to the reactivity of the siloxane group, however, applications are limited to anhydrous media.

Amine functionalities incorporated into permanent ILs exhibit the same reactivity with CO_2 as in organic media. This has been used to switch the basicity of imidazolium ILs bearing a primary amine function by application of CO_2 pressure. The reaction can also be repeatedly reversed by stripping out CO_2 with N_2 [122]. Again, such systems have been suggested as alternative composite materials for CCS processes [123].

Switchable solvent systems are not restricted to acid–base equilibria based on CO_2 reactions. The principle of stepwise and reversible changes of solvent properties has been demonstrated also for the cycloaddition of 1,3-pentadiene with SO_2 to give piperylene sulfone [124]. The sulfone has solvent properties similar to those of DMSO whereas the diene after removal of gaseous SO_2 is non-polar. Eventually, the diene itself is also fairly volatile, allowing complete removal of the polar solvent and reversible recombination with SO_2 in a separate reactor for reuse.

The control over solvent polarity and solubility offered by switchable solvents has been utilized to design novel separation schemes coupled with chemical reactions. For example, the radical-initiated polymerization of styrene proceeds smoothly under homogeneous conditions in the low-polarity form of the switchable solvent DBU–1-propanol. Once the polymerization is complete, the addition of CO_2 triggers the switch of the solvent to the polar form, which in turn causes the precipitation of the product. The polystyrene is filtered off and the solvent switched back to its neutral form with N_2 and reused for another reaction sequence (Figure 4.21). Four cycles have been demonstrated with one batch of solvent with 97% overall yield of isolated product [115].

The Claisen–Schmidt condensation of 2-butanone with benzaldehyde can be carried out in neat TMBG, which plays the dual role of a base catalyst and solvent. After the reaction, octane, methanol, and CO_2 are added, which results in the

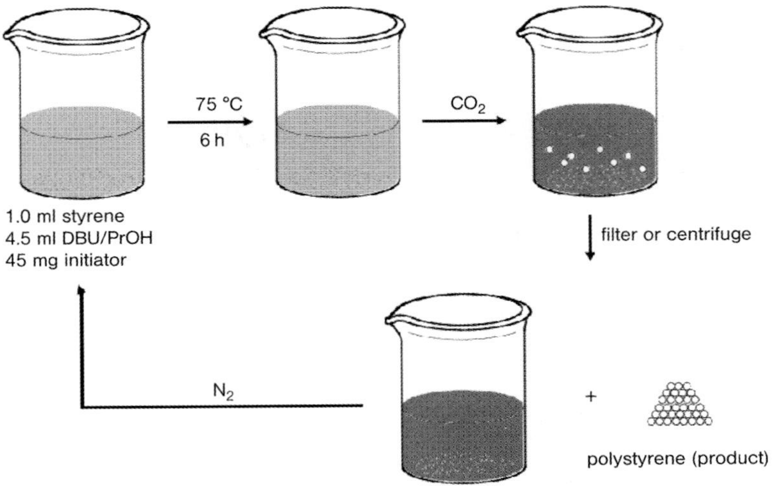

Figure 4.21 The polymerization of styrene in the low-polarity form of a switchable solvent (light gray) followed by switching to high polarity (dark gray) allows isolation of the precipitated polymer and recycling of the switchable solvent (Reprinted with permission from American Chemical Society [115]).

formation of a biphasic system consisting of octane and the ionic form of the switchable solvent TMBG–MeOH. The octane phase containing the products can now easily be decanted. If the reaction mixture is freed from the byproduct water prior to switching, the solvent system can be reused. Similarly, cyanosilylations of ketones and Michael additions have been demonstrated using the non-ionic form of switchable solvent systems both as solvents and as base catalysts [119].

The first demonstration of a metal-catalyzed process taking advantage of switchable solvents was reported for the copolymerization of epoxides with CO_2 catalyzed by cobalt complexes with salen-type ligands [120]. The homogeneous reaction is performed in neat substrate expanded with 3.5 MPa of CO_2. After the reaction, butylethylamine is added to the mixture, which in the presence of CO_2 generates an ionic solvent from which the product precipitates while the catalyst remains dissolved. The method makes it possible to avoid any additional solvents (conventionally, CH_2Cl_2 and methanol are used for reaction and precipitation). Both the catalyst and the switchable solvent can be reused at least one more time.

Pd-catalyzed Heck reactions and related C–C coupling strategies are very important tools in modern organic synthesis. Several examples of Heck reactions in ionic liquids have been reported, but the halide salts formed as by-products in the catalytic cycle accumulate in the ionic solvent and ultimately limit the repeated use of the catalytic system. A switchable IL system offers the possibility of sequentially separating the organic products, the salt by-products, and the catalyst. This has been demonstrated for the coupling of bromobenzene and styrene in the ionic form of a DBU–hexanol solvent system: the non-polar product can be extracted with heptane from the ionic liquid after the reaction. Switching the solvent to its neutral form cause

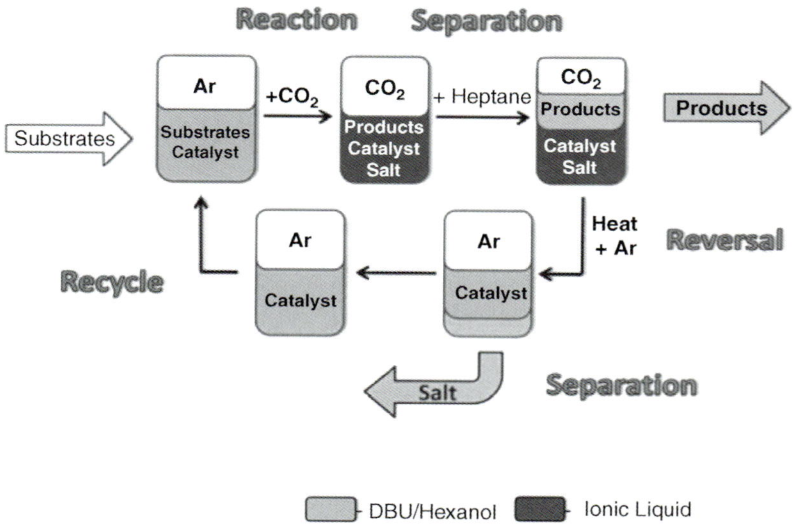

Figure 4.22 Recycling scheme for the Heck reaction in a switchable solvent system (Reprinted with permission from Elsevier [119]).

the salts to precipitate while the catalyst remains dissolved for reuse (Figure 4.22) [119]. In addition, the switchable medium provided intrinsic basicity during the reaction step by equilibrium amounts of free DBU, rendering the addition of an external base obsolete. Only the stoichiometric amounts of DBU removed as salt would have to be re-added during an optimized recycling procedure.

The concept of switching from neutral to ionic by reaction with CO_2 has also been used for solutes [125] and surfactants [126]. A very elegant application of the concept for homogeneous catalysis is the design of phase-switchable catalysts. It has been shown that the solubility properties of rhodium complexes bearing triarylphosphine

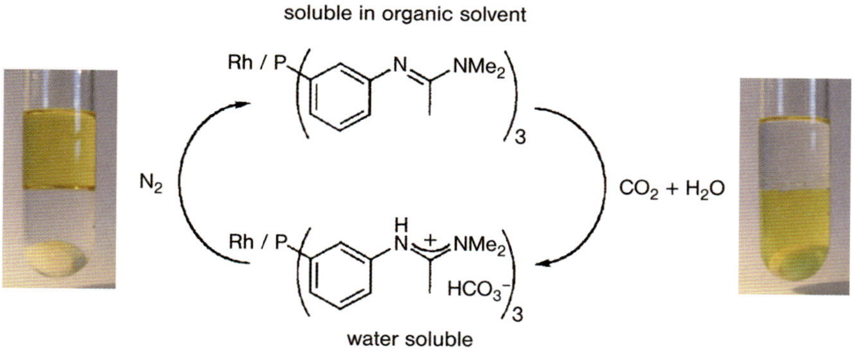

Figure 4.23 Phase-switchable homogeneous catalyst system based on an amidine-functionalized phosphine ligand. The yellow color indicates the presence of the rhodium complexes either in toluene (top) or water (bottom) [127].

ligands with amidine groups can be switched between hydrophilic and hydrophobic by alternate application of CO_2 and N_2 (Figure 4.23). The reversible transformation of the ligand from neutral to its bicarbonate salt is responsible for the phase transition. The complexes are active catalysts for hydroformylation in both stages and therefore can be used in either phase to convert substrates of various polarities [127]. The catalytic reaction occurs homogeneously in the phase with the suitable polarity for the substrate whereby the catalyst is controlled to be in the same phase by the choice of conditions. After reaction, the catalyst is switched to the form which is insoluble in the product phase, the latter removed, and the catalyst recycled for the next transformation.

4.4 Using Gas-expanded Liquids for Catalysis

4.4.1 Motivation and Potential Benefits

As mentioned in the outline of this chapter, the practical advantage of gas-expanded liquids as compared with SCFs is the lower requirement in terms of pressure, which is strongly connected with energy input and equipment investment. Furthermore, concentrations are typically higher in ELPs than in SCFs due to higher solvation power allowing smaller reaction volumes. Compared with systems using solely condensed organic solvents, the efficiency of many synthetic processes can also be greatly enhanced by expanding a liquid solvent with a compressible gas. This can result either from integrated reaction and separation steps or by inherent reaction advantages. Both cases will exemplarily be addressed in the following sections.

In addition, the use of benign compressible expanding gases such as CO_2 can also result in considerable ecological and safety benefits. First, the amount of organic solvent is greatly reduced as a considerable volume is replaced by the compressible gas. In some cases, solvents can even be completely omitted and the reaction conducted in neat expanded substrate. Second, for reactions using gaseous reagents such as H_2, O_2, or CO, the compressible gas also allows for a reduction in the amount of reactive gas in the system while maintaining high concentrations in the liquid phase. Thus, dilution of gaseous reagents with expanding gases such as CO_2 is typically more effective than dilution with inert N_2, for instance [128]. Furthermore, as the heat capacity of a compressible gas is exceptionally high around its critical point, additional safety is gained when highly exothermic reactions are conducted in the presence of such a gas just below its critical point [129]. Overall, these effects result in increased intrinsic process safety with reduced environmental impact.

Owing to the significant volume changes occurring in most ELP systems, care has to be taken when using typical evaluation criteria for catalytic reactions such as conversions at fixed reaction times and turnover frequencies (TOFs), especially when comparing a given reaction in a conventional solvent with the same system in an ELP and an SCF. This is due to the fact that conversions and TOFs do not take into account concentration changes. Consequently, these data underestimate increases in reactions rate constants or even suggest deceleration due to dilution effects.

When conversions at a given time or TOF are interpreted in terms of intrinsic kinetics, they have to be compared at equal concentrations and thus at similar reaction volumes. In practical terms, it would often be most useful to refer to space–time yields for direct comparison of different systems, but this analysis is still rarely carried out.

4.4.2 Sequential Reaction–Separation Processes

4.4.2.1 Tunable Precipitation and Crystallization

The pressure-dependent change in polarity and solvation power of Class II ELPs has long been used to separate homogeneous mixtures and solutions. In 1838, Mitchell noted that dissolving compressed CO_2 in an ethanolic solution of shellac caused the precipitation of a white powder, and releasing the CO_2 caused the shellac to be redissolved into the ethanol [130]. In the 1930s, researchers at Shell and De Bataafsche Petroleum used methane, CO_2, propane, and butane to precipitate heavy fractions such as asphaltenes from oil or solutions of oil in solvents [131–133]. This phenomenon was later developed into a gas-induced crystallization technique, the so-called GAS (gas–antisolvent) process [134–137]. By knowing the saturation limits of a solute in an ELP over pressure, mild crystallizations without solvent inclusions can be achieved. As the kinetics of nucleation and particle growth are driven by the gas dissolution only, the morphology of the resulting crystals can be varied by the rate of pressurization.

More recently, the isolation of useful chemicals including vanillin, syringaldehyde, and syringol from mixed waste biomass has been demonstrated using tunable precipitation of the different components from methanol by CO_2 pressure [138]. In addition to purification, selective precipitation can also be integrated into reactive systems, where it may help to increase yields by shifting the reaction equilibrium towards completion. This has been demonstrated for the synthesis of an anti-inflammatory copper complex which was obtained in high yields and high purity using CO_2 to expand a DMF solution of the reactants under optimized conditions [139].

Selective precipitation has been used also to recover homogeneous catalysts after the reaction. Cobalt complexes precipitate from CO_2-expanded acetonitrile at certain CO_2 contents, making it possible to run an oxidation reaction homogeneously in an ELP and recover the catalyst from the same solvent by simply adding more CO_2 [140]. Similarly, rhodium complexes for homogeneous hydroformylation can be selectively precipitated after the catalysis by knowing their respective saturation limits as a function of CO_2 content in the ELP [87]. Whereas in acetone the PPh_3-based Rh catalyst remained soluble even at higher pressures, the neat mixture of substrates and products allowed the precipitation of the catalyst around 9 MPa (Figure 4.24).

Crystallizations of solutes from ionic liquids are also possible upon pressurization with CO_2, even though this represents a Class III system [141]. In view of the very low polarity changes during the limited gas uptake, it can be assumed that the dissolved gas affects the solute more than the solvent causing an interruption of the solute–solvent interactions. This procedure may reconstitute an alternative to extraction

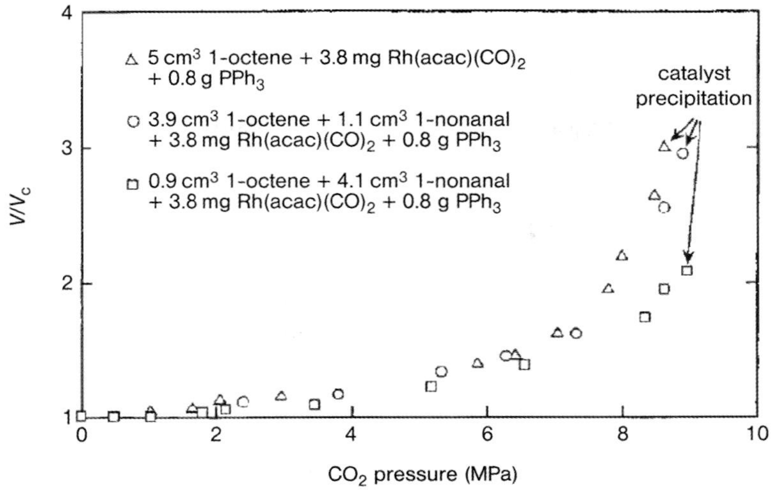

Figure 4.24 Expansion behavior of neat substrate with catalyst (△) and substrate–product mixtures with catalyst (○ and □) at 50 °C versus CO_2 pressure. At the end-point of maximum expansion, precipitation of the catalyst occurs [87].

when the solute has limited solubility in the SCF used. Remarkably, analogous to GAS processes from organic solvents, no solvent inclusions were found for crystals of a phenylalanine methyl ester grown by antisolvent precipitation from [bmim][BF_4] [141].

4.4.2.2 **Tunable Phase Separations**

Dissolution of compressible gases into solutions can also induce phase separations of otherwise miscible solvents, provided that one of the liquid components has a considerably higher ability to dissolve the gas than the other. Such combinations typically consist of mixtures of Class I and Class II liquids, which separate upon formation of an ELP of the Class II liquid. As a prototypical example, Baker and Anderson showed in 1957 that CO_2 pressure makes ethanol and water become immiscible at room temperature [142]. Such ternary mixtures may exhibit complex phase behaviors, which can even lead to four-phase equilibria [143]. Under optimized conditions, however, this can be exploited for the mild recovery of biomolecules from aqueous solutions [144].

In catalysis, so-called OATS (organic–aqueous tunable solvents) make it possible to run reactions homogeneously in mixed aqueous–organic media and effect post-reaction separation by pressurization. As an example, the biocatalytic conversion of hydrophobic ketones was successfully demonstrated in mixed dimethyl ether–water solvent [145]. The enzyme and the cofactor could be recovered in the aqueous phase and recycled after induced phase split through application of 3.0 MPa CO_2. Figure 4.25 exemplifies this behavior for a homogeneous water–THF mixture where the dye stands for a hydrophilic catalyst. This combination has been used successfully

Figure 4.25 Water–THF mixture containing a hydrophilic dye (dark) at room temperature and atmospheric pressure (a). Application of 3.0 MPa CO_2 results in phase separation into an acidic water phase containing the dye and an ELP of THF with CO_2 (b) [145].

for the homogeneous Rh-catalyzed hydroformylation of higher alkenes, which usually are very poorly soluble in pure water, strongly limiting the reaction by mass transfer. With 30 vol.% THF as cosolvent, the solubility of 1-octene is greatly enhanced, and by adding 3.2 MPa CO_2 to the homogeneous solution after the catalysis effective separation of the ionic catalyst in the aqueous phase from the product-rich organic phase can be achieved [146].

Phase separation of mixtures of two Class II liquids upon expansion with CO_2 has also been observed in certain cases. This appears counterintuitive when compared with the case of Class I–II separation. As CO_2 affects the physical properties of both components in a similar way, no change in mutual solubility would be expected. However, the two Class II liquids may have different gas–liquid mixing enthalpies, e.g. methanol + CO_2 versus toluene + CO_2 [147]. If the difference in the mixing enthalpies of the gas in each separate liquid is larger than the excess mixing enthalpy of the pure liquids, separation of the liquid phases occurs in the ternary system because one solvent prefers to dissolve the gas instead of the other liquid component. This is the thermodynamic force for all CO_2 - induced phase separations including those interpreted on the basis of polarity changes. A simplified overview of various combinations of organic solvents, which can be phase split upon pressurization with CO_2 is shown in Figure 4.26. Note that phase behavior and hence mutual miscibility are always also a function of composition. The changes shown in Figure 4.26 were obtained with mixtures of equal liquid volumes at one constant CO_2 pressure. Therefore, different behavior might be obtained for mixtures of varying compositions.

Gas-induced phase splitting has also been used to achieve separations of homogeneous mixtures of Class II and Class III liquids. Especially separations of high-boiling polar organic compounds from ILs are difficult to achieve by conventional techniques such as extraction or distillation. Application of moderate CO_2 pressures to alcohol–IL solutions results in phase separation due to the formation of an ELP of the alcohol. Importantly, no contaminations of IL were detected in the organic

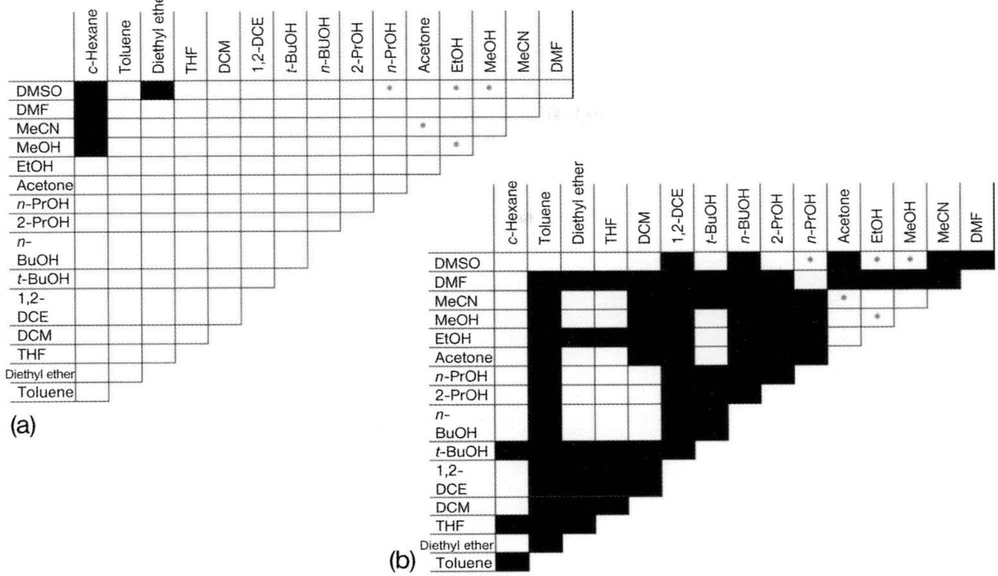

Figure 4.26 Phase behaviors of organic solvent mixtures of equal volume at room temperature and ambient pressure (a) and under 5.0 MPa CO_2 (b). Black boxes indicate two-phase systems; asterisks highlight combinations that are not affected by CO_2 under the conditions applied (Reprinted with permission from Royal Society of Chemistry [147]).

ELP [148]. Similarly to OATS, this effect has been exploited for product recovery after homogeneous catalysis in ILs [149].

Eventually, Class I and Class III mixtures such as water and ILs have also been examined and found to exhibit reversible phase separations through CO_2 pressure [150]. Both hydrophobic and hydrophilic ILs formed two liquid phases at pressures below 5 MPa even though CO_2 is not very soluble either in ILs or in water. The mixing thermodynamics of such combinations have not been addressed yet, but the mechanism of separation appears not to be purely physical. Ethane as an inert, non-polar expansion gas was not capable of separating the two components, indicating that a chemical interaction might play a role in some way. Furthermore, the segregation achieved with CO_2 was only observed in certain concentration regimes and also not in a useful range for effective separation; both phases still contained considerable amounts of the second component. Especially hydrophilic ILs remained at nearly 10 mol% in the water phase. Initially, the authors suggested this procedure as an alternative to distillation for the purpose of purification of water from ionic contaminants, but it has also been used in reactive systems by others. Using the esterification of acetic acid with ethanol as model reaction, it has been shown that the equilibrium conversion of the reaction in [bmim][HSO_4] depends on the CO_2 pressure applied to the mixture because of induced phase changes [151]. Depending on the number and nature of the phases present, the distribution coefficients and thus local concentrations of the substrates and products differ, thereby directly affecting the reaction equilibrium.

4.4.2.3 **Tunable Miscibility**

Not only separation but also mixing can be induced by applying pressure of a compressible gas. This can be achieved for mixtures where both components form a common ELP with the respective gas but are immiscible as pure liquids. In addition to the few combinations in Figure 4.26, one example is fluorinated hydrocarbons and organic solvents, which reversibly become monophasic upon pressurization with CO_2 [152]. In such systems, the decrease in polarity, or more specifically the increase in "fluorophilicity" of the more polar component, with increase in CO_2 concentration is used to explain the observed mixing.

This has been exploited for conducting homogeneous catalysis in CO_2-controlled binary solvents as an alternative to thermally induced fluorous–organic miscibility systems [153]. The epoxidation of cyclohexene with fluorinated cobalt catalysts and the hydrogenation of allyl alcohol with fluorinated dendrimer-stabilized Pd nanoparticles proceeded at 50–70% higher rates in a monophasic organic–fluorous solvent system under CO_2 compared with the same mixture under biphasic conditions in the absence of CO_2 [152]. Both reactions benefited from an increase in catalyst accessibility and also higher gas availability in the monophasic ELP.

Not only liquids but also highly fluorinated solid materials that are insoluble in non-fluorinated organic Class II solvents become temporarily soluble during pressurization with CO_2 [154]. This has been used to shuttle a fluorinated catalyst

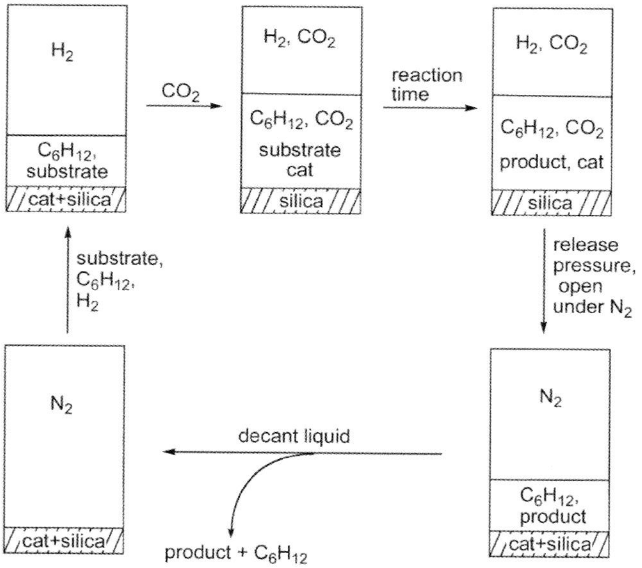

Figure 4.27 Homogeneous hydrogenation and catalyst recycling using tunable catalyst solubility in an ELP of cyclohexane [155].

from a fluorinated support material into an ELP of an organic solvent for homogeneous catalysis. As a model reaction, the hydrogenation of styrene with a fluorinated version of Wilkinson's catalyst was performed in cyclohexane (Figure 4.27). The medium becomes reversibly fluorophilic upon expansion with CO_2 to dissolve the catalyst from a fluorinated support on which the catalyst readsorbs after release of CO_2 pressure [155]. Five cycles with only a slight decrease in activity were demonstrated using the same batch of catalyst and support.

4.4.3 Hydrogenation Reactions

Reductions using molecular hydrogen can be efficiently carried out in ELPs with both heterogeneous and homogeneous catalysts. Various examples exist in which a low-temperature, heterogeneously catalyzed hydrogenation reaction occurs faster in an ELP than in a single-phase SCF [156–161]. The continuous selective hydrogenation of isophorone over heterogeneous palladium catalysts in compressed CO_2, which was commercialized by Thomas Swan (Consett, UK) in 2002, was also revealed to operate in an ELP rather than in a single-phase SCF under optimized reaction conditions [162]. The enhanced performance of heterogeneous catalysts in ELPs has mainly been attributed to higher local substrate concentrations combined with good hydrogen availability [163]. As mentioned in Section 4.2.6, the enhanced heat dissipation compared with non-expanded liquids and the higher heat capacity compared with SCFs might be especially advantageous for the exothermic surface reaction. In high-temperature hydrogenations with metallic catalysts, care has to be taken when using CO_2 as expansion gas as reverse water gas shift reactions and consecutive Fischer–Tropsch reactions have been reported as possible catalyst deactivation mechanisms [164].

Not only the activity but also the selectivity of a heterogeneous hydrogenation can be influenced by the phase behavior. For example, the selectivity of the continuous hydrogenation of limonene over platinum catalysts could be tuned by the pressure of CO_2. As the ratio of H_2 to substrate is different in the single-phase SFC and the ELP, it was possible to adjust the reaction outcome by CO_2 density: whereas under supercritical conditions complete reduction was observed, the mono-hydrogenation product could be obtained with nearly 80% selectivity when the reaction took place in an ELP of the neat reactants (Scheme 4.5) [165].

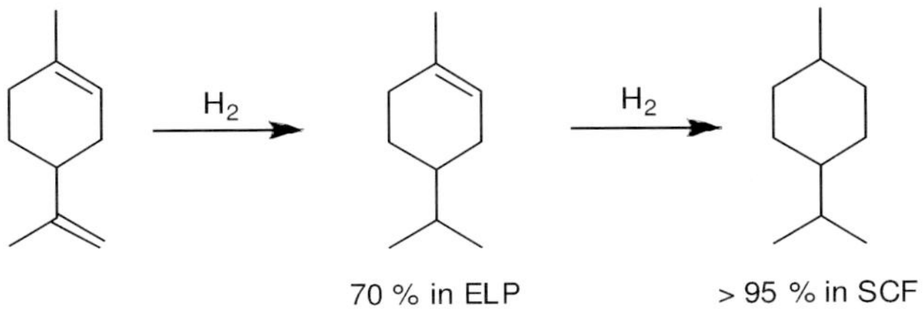

Scheme 4.5 Selectivity control and product yields of the sequential hydrogenation of limonene with heterogeneous Pt catalysts at 50 °C [165].

The heterogeneously catalyzed continuous asymmetric hydrogenation of ethyl pyruvate has been investigated with ethane and CO_2 as expansion gas and SCF [21]. Even under conditions above the critical points of the pure expansion gases, an ELP of the neat substrate was observed, highlighting the importance of mixture critical points. In certain temperature and pressure regimes even two liquid phases were observed in equilibrium with a gas phase. The amount of H_2 present in the mixture played a crucial role in this phase behavior and similar anti-solvent effects of H_2 on the phase behavior were also observed in the hydrogenation of limonene [166]. The interpretation of the observed changes in activity and selectivity in terms of phase behavior was complicated by the complex catalytic system involving adsorption–desorption processes of the chiral modifier. Nevertheless, the range for optimal conditions to afford very rapid and selective hydrogenation could be identified in the ELP region [167], whereby ethane generally gave both higher activities and enantioselectivities than CO_2 [21].

A strong influence of the nature of the expansion gas on turnover rates was also observed during the homogeneous hydrogenation of CO_2 with ruthenium complexes in methanol (Scheme 4.6) [24]. With only the reagent gases 770 catalytic turnovers per hour were observed, which decreased to $160\,h^{-1}$ when the solution was expanded with 4.0 MPa ethane. Dilution of the reaction with hexane gave rise to a similar reduction in conversion, indicating that a decrease in the polarity of the medium is detrimental to the reaction. Indeed, expansion of the methanolic reaction mixture with polar CHF_3 (also 4.0 MPa) increased the TOF to $910\,h^{-1}$.

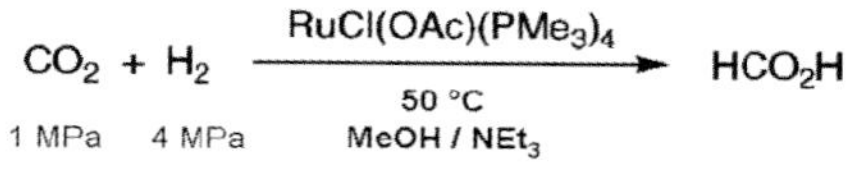

Scheme 4.6 Homogeneous hydrogenation of CO_2 in methanol [24].

The solventless ruthenium-catalyzed hydrogenation of cinnamaldehyde was also shown to be strongly affected by the phase behavior with compressed CO_2 [168]. Conversions were low in the non-expanded neat substrate and pressurization with N_2 did not improve the reaction. In a single-phase SCF with CO_2, higher activities were achieved, which could be improved further by running the reaction in a ELP (Table 4.2). However, as different catalyst and substrate concentrations existed, the

Table 4.2 Solventless hydrogenation of cinnamaldehyde with $RuCl_3$–perfluoro-TPP catalyst at 50 °C [168].

Reaction phase	Gases	Conversion after 2 h (%)	Selectivity to unsaturated alcohol (%)
Liquid substrate	H_2	11	88
Liquid substrate	$H_2 + N_2$	12	88
Single-phase SCF	$H_2 + CO_2$	22	25
Expanded liquid substrate	$H_2 + CO_2$	54	98

conversions are not indicative of the intrinsic reaction rate. This point was addressed later in a more extensive study [169]. The selectivity to cinnamyl alcohol was also highest for the ELP system.

The same reaction was compared in organic–water and CO_2–water biphasic systems using the sulfonated version of the TPP ligand to immobilize the ruthenium catalyst in the aqueous phase. At 40 °C, conversion in the toluene–water system reached 11% with 92% selectivity to the unsaturated alcohol within 2 h, whereas the $scCO_2$–water system led to 38% conversion at 99% selectivity [170]. Changing the metal to rhodium or palladium caused a complete reversal in selectivity to give exclusively the saturated aldehyde at similar conversions under identical conditions [170].

The effect of melting point depression was exploited in the solventless hydrogenation of oleic acid to stearic acid, where the formation of a solid product inhibited complete conversion because the heterogeneous platinum catalyst was progressively blocked. Thus, the conventional reaction did not exceed 95% conversion even after 25 h at 50 °C, whereas quantitative conversion was achieved within 4 h when 6.0 MPa CO_2 was added to create an ELP that remained liquid even at high conversion [171]. Melting point depression through the presence of a compressible gas was also used in homogeneous catalysis for the solventless hydrogenation of 2-vinylnaphthalene with Wilkinson's catalyst. By application of 5.6 MPa of CO_2, the substrate was liquefied at 36 °C, i.e. significantly below its normal melting temperature, and the reaction proceeded more than 10 times faster than in the absence of the compressible gas [171].

Similarly, solid PEG-1000 melts under 17 MPa of CO_2 at 40 °C and has been used as solvent in the biphasic hydrogenation of styrene with Wilkinson's catalyst [172]. The re-solidified polymer encapsulates the rhodium complex yielding an air-stable solid (Section 4.6.4) [86]. The CO_2-induced melting of ammonium salts was used to design an IL-type reaction system where the catalyst phase is liquid under reaction conditions, but solid during storage or recycling. A cationic rhodium complex embedded in $[R_4N][BF_4]$ could be used consecutively in reaction sequences of hydrogenation, hydroboration, and hydroformylation (Figure 4.28) [55]. A similar strategy has been developed for the preparation of rhodium nanoparticles and their use in selective hydrogenation in CO_2-induced ILs [173].

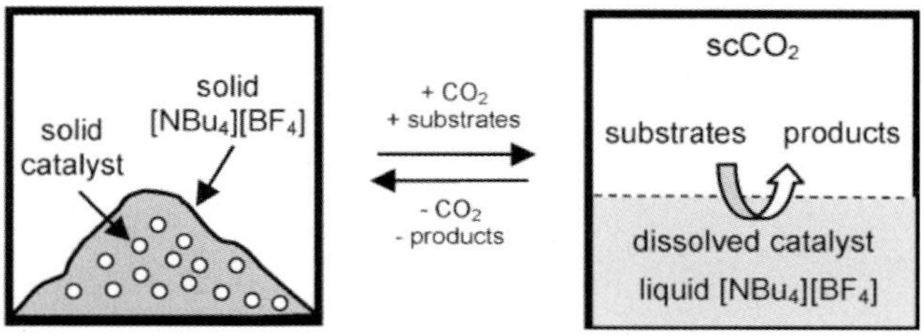

Figure 4.28 CO_2-induced liquefaction of organic salts for IL–CO_2 homogeneous catalysis. Reactions can be carried out at 55 °C, i.e. more than 100 °C below the regular melting point of the salt when 10 MPa CO_2 is added [55].

Table 4.3 Asymmetric hydrogenation of atropic and tiglic acid with Ru–BINAP complexes in [bmim][PF_6] in the presence and absence of CO_2 as expansion gas [175].

Substrate	H_2 pressure (MPa)	CO_2 pressure (MPa)	*ee* (%)
Tiglic acid	0.5	0	93
	0.5	7.0	79
Atropic acid	5.0	0	32
	5.0	5.0	57
	10.0	0	49

The asymmetric C=C hydrogenation of a drug precursor to produce the non-steroidal anti-inflammatory drug (NSAID) naproxen was conducted in CO_2-expanded methanol and compared with supercritical conditions and non-expanded methanol. Both reaction rates and enantioselectivity of the Ru–BINAP–NEt_3 catalytic system were found to be decreased by the presence of CO_2 [174]. The data do not allow a definite conclusion to be drawn about the influence of hydrogen solubility or acid–base equilibria between methanol, triethylamine, and CO_2 as possible reasons for the observation.

A successful demonstration of how the enhanced gas availability in an ELP can be used to tune selectivity is the enantioselective C=C hydrogenation of functionalized alkenes with chiral ruthenium catalysts in ionic liquids. Substrates which are hydrogenated with high enantioselectivity at low local H_2 concentrations (e.g. tiglic acid) were shown to proceed with high enantiomeric excess (*ee*) in neat ILs, whereas substrates requiring high H_2 concentrations for high selectivity (e.g. atropic acid) run more selectively in CO_2-expanded ILs (Table 4.3) [175].

Exploiting the increased availability of H_2 in CO_2-expanded ILs, the rate of the enantioselective hydrogenation of imines with chiral iridium complexes (Scheme 4.7) could be greatly enhanced by expansion of the IL with CO_2. Whereas in the absence of CO_2 only 3% conversion was detected in [emim][NTf_2] after 22 h under 3.0 MPa H_2, complete conversion took place when 10 MPa CO_2 was added under otherwise identical conditions [176]. In order to achieve similar activities without CO_2, the pressure of pure H_2 had to be raised to over 10 MPa, demonstrating

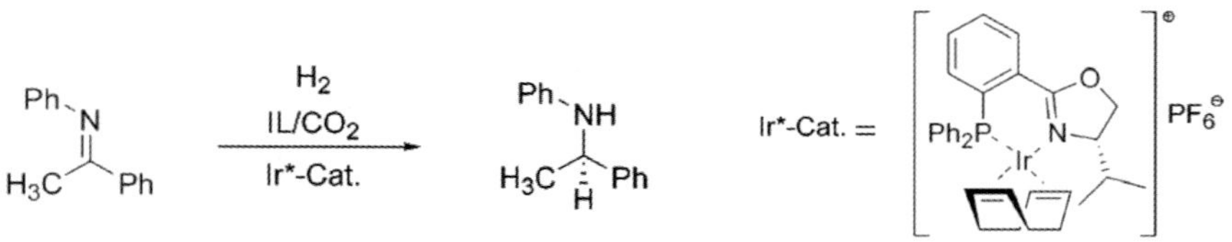

Scheme 4.7 Enantioselective hydrogenation of *N*-(1-phenylethylidene)aniline in the IL–CO_2 biphasic system using chiral (phosphanodihydroxooxazole)iridium complexes [176].

well the beneficial effect of an inert expansion gas for homogeneous catalysis with permanent gases.

4.4.4 Carbonylation Reactions

The hydroformylation of alkenes with syngas (H_2–CO) is the largest scale industrial application of homogenous catalysis. Catalyst separation and recycling are very important parts of the overall process schemes, and the reaction is often limited by mass transfer of gaseous or liquid reagents under multiphase conditions. Therefore, homogeneous catalytic hydroformylation has often served as a model reaction to test novel solvent systems including SCFs and ELPs.

The low-temperature hydroformylation of 1-octene (Scheme 4.8) with unmodified rhodium–carbonyl catalysts has been shown to proceed more than four times more rapidly in CO_2-expanded acetone than in neat acetone. On varying the CO_2 content in the media from 0 to 100%, it was observed that at about 75% CO_2 the reaction proceeded best in an organic ELP even when compared with liquid CO_2. The balance between syngas availability and octene dilution was used to explain this observation, but other effects such as solvent polarity might also play a role. Within the range of pressures and concentrations studied, the kinetics of the reaction in the ELP followed the trends known for the reaction in non-expanded liquids. Additionally, the regioselectivity towards linear and branched aldehydes was not affected by CO_2 [177].

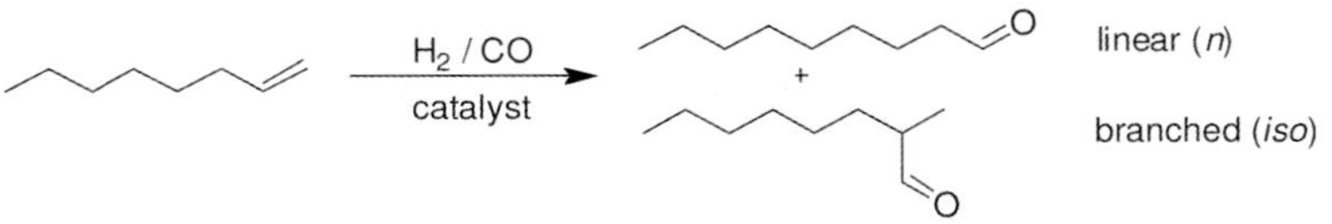

Scheme 4.8 Hydroformylation reaction of 1-octene yielding regioisomers of nonanal.

The same reaction system was later optimized further using different phosphine-modified catalysts, which allowed for higher regio- and chemoselectivities. Eventually a set of conditions was found which proved competitive in terms of productivity to industrial processes while operating under even milder conditions [178]. Product separation and catalyst recycling could also be achieved by pressure-tunable precipitation of the catalyst (Figure 4.24) [87].

The asymmetric hydroformylation of styrene with chiral Rh–BINAPHOS complexes in CO_2 was found to provide significant enantioselectivities at CO_2 densities close to the critical density under conditions involving an ELP [179]. Both reaction rate and selectivity decreased on passing into the SCF regime, because the complex with the chiral ligand became insoluble in the medium. With a fluorinated version of the same ligand, efficient asymmetric hydroformylation was later also achieved under single-phase conditions in $scCO_2$ [180].

Even though multi-component reactions with compressible gases may be started under single-phase conditions, the change in composition over time due to conversion of the substrates might lead to phase changes of the mixtures. This effect is

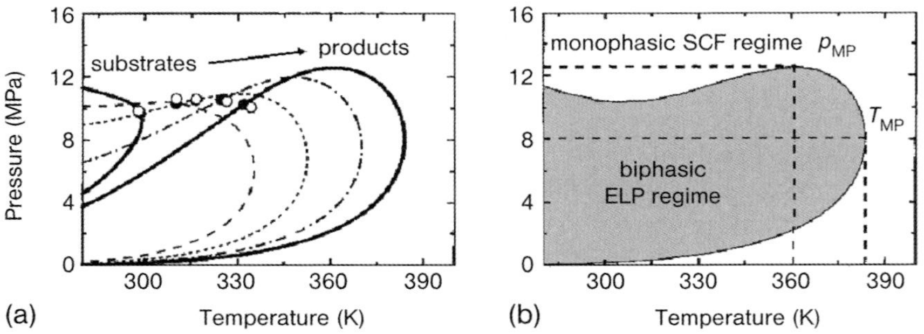

Figure 4.29 Phase diagrams for the six-component mixture CO_2, H_2, CO, propene, and butyraldehydes (*n*- and *iso*-) of various composition. (a) Calculated phase boundaries corresponding to different conversions of 0, 30, 49, 75, and 100% with calculated (●) and experimental (○) mixture critical points. (b) Global reaction phase envelope with T and p above which the mixture remains monophasic (gaseous for $T > T_{MP}$, supercritical for $p > p_{MP}$) irrespective of conversion (Reprinted with permission from American Chemical Society [181]).

generally more likely to occur in systems with high substrate concentrations and specifically for reactions in which the products and substrates differ either in volatility or polarity. The latter is significant to hydroformylation as aldehydes and alkenes, respectively, form mixtures with very different mixture critical points with a given compressible gas. This effect has been studied for the hydroformylation of propene in compressed CO_2. Phase behavior studies of mixtures of CO_2, H_2, CO, propene, and butyraldehydes with various compositions were carried out to map a reaction mixture phase envelope for the system (Figure 4.29) [181].

From Figure 4.29, it can be seen that there are regimes of T and p in which an initially monophasic reaction mixture for the hydroformylation of propene in CO_2 will split into two phases as a result of catalytic conversion (e.g. 8.0 MPa at 50 °C). Similar results have been observed for the synthesis of dimethyl carbonate from CO_2 and methanol [182]. Furthermore, it could be shown that higher total pressures are needed in batch mode to maintain single-phase conditions throughout the reaction than in continuous flow mode because pressure equilibration can take place along the continuous reactor. The phase behavior of various compositions has also been investigated for the same system without CO as a model for the hydrogenation of propene. The variation in the mixture critical points showed the opposite trend: the critical conditions decreased due to increased volatility of the products compared with the substrates [183]. The difference in solubility between substrate and product in a compressible gas can be sufficiently large to take advantage of this effect for the development of "solventless" continuous flow processes (see Section 4.6.5).

In the continuous hydroformylation of 1-octene using a molecular rhodium catalyst in a supported ionic liquid phase (SILP) on mesoporous silica, it was observed that the reaction rate depended strongly on the phase behavior of the mobile phase. First, the performance of the SILP-catalyst system was optimized by variation of IL film thickness, substrate flow, and syngas to substrate ratio using

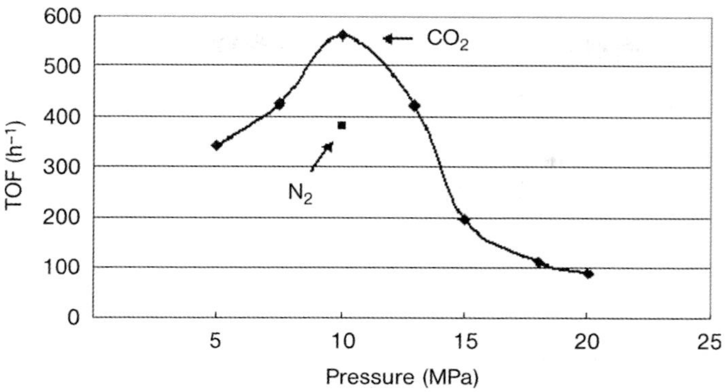

Figure 4.30 Rates of 1-octene hydroformylation using Rh–[pmim][$Ph_2P(3\text{-}C_6H_4SO_3)$] in [omim][$NTf_2$] supported on dehydroxylated silica gel in continuous flow mode at 100 °C as a function of pressure [185].

statistical design of the experiment [184]. For the best parameters found, phase behavior studies were then undertaken by visual observation of the mixture at different pressures. As can be seen from Figure 4.30, the reaction rate increases as the CO_2 pressure approaches the mixture critical pressure of 10.6 MPa. After a single phase has been reached at $p > 11$ MPa, the rate decreases again.

This behavior was again interpreted on the basis of local substrate concentrations. As the expansion gas is introduced, the rate is enhanced because syngas availability in the catalytically active SILP increases. At monophasic conditions, however, the concentration of 1-octene in the SILP decreases because it partitions into the CO_2 phase and thus the rate decreases again with increasing solvent power of the mobile phase. Non-compressible N_2 is not capable of expanding the substrate in the IL on the support and therefore has no significant effect on the rate.

The retention of the molecular catalyst in the SILP is also affected by the phase behavior of the mobile phase. Rhodium and IL leaching values are unacceptably high at low pressures because the ELP of the substrate–product shows a noticeable solubility for the IL and the complex. As the polarity of the ELP is reduced in the vicinity of the mixture critical point, catalyst leaching drops to levels below 0.5 ppm because the solvent power of the mobile phase for both IL and catalyst decreases. With N_2, very high leaching was observed at any pressure, substantiating further that the use of CO_2 is essential to form an ELP of substrates and products. Under optimized conditions, 40 h of continuous flow catalysis were achieved with a constant TOF of 550 h^{-1} at rhodium leaching as low as 0.2 ppm [185].

Other types of heterogenized catalyst systems have also shown improved performance in ELP systems. Higher activities have been reported for the Rh-catalyzed hydroformylation of 1-hexene with silica-tethered PPh_2Et as ligand using either CO_2-expanded toluene or $scCO_2$ as compared with the reaction with the catalyst suspended in non-expanded toluene [186]. Most notably, the situation was different when the free complex was used as homogeneous catalyst in solution. Here the

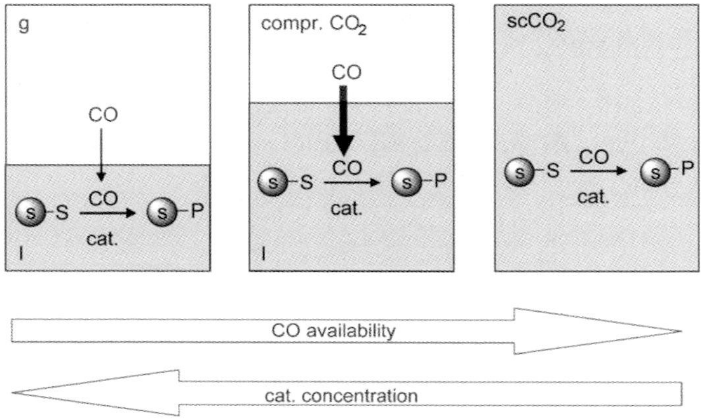

Figure 4.31 Solid-phase organic synthesis (SPOS) in carbonylation catalysis with CO_2. With no expansion gas reaction rates are low due to low CO availability in the reactive phase. The rates increase when the medium is pressurized with CO_2 (g = gas; l = liquid) [187].

highest rates were achieved in non-expanded toluene, followed by CO_2-expanded toluene, with very low rates in $scCO_2$. The different behavior in the supercritical phase reflects the limited solubility of the organometallic complex in $scCO_2$. For the ELP, it seems that dilution is the dominant factor with the homogeneous catalyst, whereas increased gas availability and mass transfer prevail for the heterogenized system.

The beneficial mass transfer properties are not restricted to heterogenized catalysts, but apply similarly to transformations of solid-phase bound substrates. Solid-phase organic synthesis is a widely used technique of increasing importance, but reactions with pressurized gases are rarely used because of insufficient rates resulting from poor mixing. This limitation can be ameliorated by performing the reactions in ELPs or under supercritical conditions (Figure 4.31), as demonstrated recently for the hydroformylation and the Pauson–Khand reaction (Scheme 4.9) as versatile and prototypical carbonylation reactions [187].

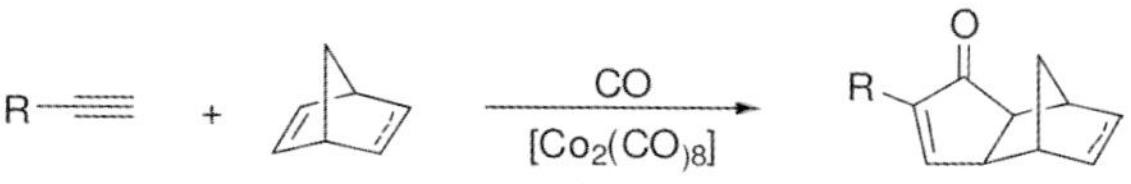

Scheme 4.9 Cobalt-catalyzed [2 + 2 + 1] cycloaddition of alkynes, alkenes, and CO (Pauson–Khand reaction). For solid-phase synthesis, R was a trityl polystyrene resin connected via an alkoxy spacer. Treatment with trifluoroacetic acid detached the alcoholic products from the solid after the catalysis.

Carbonylations of the Pauson–Khand type have also been performed with soluble substrates in ELPs. Both alkynes and strained alkenes were successfully reacted in carbonylative couplings with allyl halides using a catalyst system consisting of $NiBr_2$,

NaI, and iron powder in CO_2-expanded acetone [188]. Most notably, the selectivity of monocarbonylation versus bicarbonylation was reported to depend strongly on the CO concentration in the catalytic phase, which can be adjusted through the pressure of expanding gas. Whereas under 4.8 MPa CO_2 a product ratio of 98:2 for the bicarbonylated product was obtained, the monocarbonylated product was observed exclusively using 5.3 MPa CO_2 under otherwise identical conditions. Similar trends were found for carbonylative couplings of alkynes with allyl halides using the same system [188].

4.4.5
Oxidation Reactions

Homogeneous aerobic oxidation catalysis benefits from gas-expanded media in many respects. Combustible organic solvents and O_2 form explosive mixtures in wide composition ranges, making the use of inert dilution gases inevitable for practical application, particularly at elevated temperatures. Oxidations in inert SCFs have proven successful but are limited by substrate solubility, require high pressures (20 to >30 MPa), and demand functionalized catalysts soluble in the SFC. Furthermore, homogeneous catalysts, which are more easily over-oxidized than heterogeneous catalysts, may exhibit enhanced lifetimes in ELPs because of higher substrate to O_2 ratios compared with SCFs. The combination of safety advantages, enhanced productivity, and reduced amounts of hazardous chemicals is a prime example of sustainable process intensification for homogeneous catalysis using ELPs. Heterogeneous systems also sometimes benefit from biphasic ELP conditions, but there are also examples that operate better under monophasic conditions (either gaseous or supercritical) than in an ELP [189]. Overall, for heterogeneous oxidation catalysts there seems to be no general rule and each system has to be carefully analyzed for phase behavior effects.

The rate of the aerobic oxidation of cyclohexane to cyclohexanol and cyclohexanone with fluorinated iron–porphyrin complexes in the presence of an aldehyde as sacrificial co-reductant (Mukaiyama conditions) has been studied as function of CO_2 pressure added [190]. It was found that the reaction gave the highest conversions when the CO_2 pressure was just below the critical point of the mixture forming an ELP of the substrates (Figure 4.32). Although the authors did not specifically interpret the effect of phase behavior on the reaction, the enhanced O_2 availability combined with high substrate concentrations in the ELP presumably causes this trend. Experiments without the iron complex gave no conversion, ruling out the influence of catalytic wall effects [191] in the range of temperature studied, although this was later shown to occur at higher temperatures [192].

The industrially important aerobic oxidation of *p*-xylene using cobalt complexes and various promoters to produce terephthalic acid is usually carried out in acetic acid. Promoting effects of moderate pressures of CO_2 (0.4–2.0 MPa) on a multicomponent catalyst system consisting of cobalt, manganese, alkali metal, HBr, and nickel have been reported for such oxidations at temperatures between 150 and 195 °C [193]. After investigating different catalyst compositions and analyzing

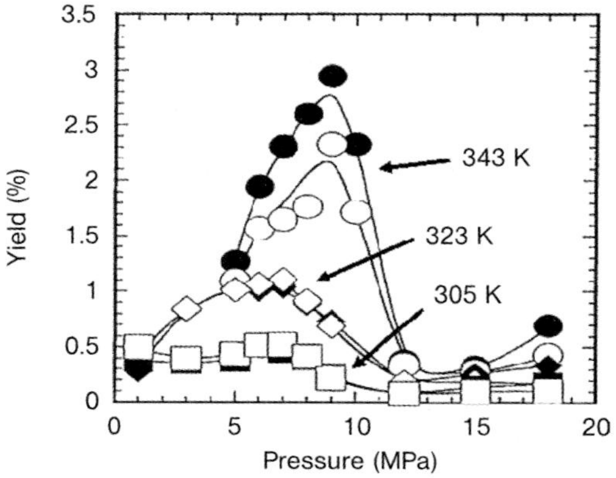

Figure 4.32 Yields of cyclohexanol (open symbols) and cyclohexanone (filled symbols) after 1 h from the homogeneous oxidation of cyclohexane with 1.0 MPa O_2 and acetaldehyde using fluorinated iron–porphyrins under CO_2 as function of pressure at various temperatures (Reprinted with permission from the Chemical Society of Japan [190]).

product distributions, it was concluded that CO_2 acts as co-oxidant in the reaction, forming more reactive peroxocarbonates, especially in combination with nickel [194]. Physical effects such as enhanced miscibility in the ELP were also discussed. In addition to higher yields (48% with CO_2 compared with 37% without CO_2), the purity of the crude product was also found to be higher when CO_2 was added to the reaction. Similar observations were made for the aerobic oxidation of other alkylaromatics such as ethylbenzene [195] and methylanisoles [196] with the same multi-component catalyst.

Very high rate enhancements have been found for the homogeneous oxidation of phenols to quinones with Schiff base cobalt catalysts (Scheme 4.10). With [*N,N*′-bis (3,5-di-*tert*-butylsalicylidene)-1,2-cyclohexanediimino(2)]cobalt(II) [Co(salen*)] as the catalyst, the reaction proceeds up to 100 times faster in CO_2-expanded MeCN than in non-expanded MeCN or $scCO_2$ [140]. Whereas a single-phase SCF of the reaction mixture requires pressures >20.0 MPa, the ELP can be created with only 6.0 MPa, allowing for an 80 vol.% replacement of the organic solvent. Higher CO_2 pressures caused anti-solvent precipitation of the transition metal complex for product separation and catalyst recycling (see Section 4.4.2.1).

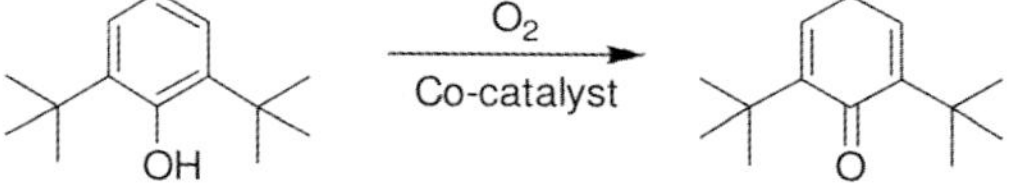

Scheme 4.10 Oxidation of 2,6-di-*tert*-butylphenol (DTBP) to 2,6-di-*tert*-butyl-1,4-benzoquinone (DTBQ) with molecular oxygen using cobalt complexes.

The increase in rate compared with neat MeCN could be correlated with increasing O_2 contents in the ELP caused by the presence of compressed CO_2. Compared with $scCO_2$, it was suggested that the polar solvent molecules stabilize the active cobalt complex in solution, thus lowering activation energies. This is in line with the finding that the influence of temperature on the reaction rate was different in $scCO_2$ and in MeCN, but virtually independent from CO_2 in the liquid phase. A decrease in the polarity of the ELP at higher CO_2 contents also proved detrimental to the rate [197]. The selectivities of the reaction in $scCO_2$ and in the ELP were comparable, with no signs of MeCN oxidation in the latter.

In the scope of the study, iron porphyrins were also tested in the oxidation of cyclohexene to its epoxide. In non-expanded MeCN the radical reaction involves an induction period of up to 16 h, which is ascribed to the build-up of a critical radical concentration required to ignite the reaction in absence of a sacrificial co-oxidant. In CO_2-expanded MeCN (6.2 MPa at 50 °C), product formation starts after 4 h. However, after longer times the product profiles converge. Comparing the fluorinated version of the catalyst in the ELP with its application in $scCO_2$, seven-fold rate enhancements were achieved at 80 °C, providing another example of a reaction running more efficiently in an ELP compared with both an SCF and a conventional organic solvent.

Steel-promoted radical oxidations using aldehydes as sacrificial co-reductants could be applied to selective oxidations of cycloalkanes and arylalkanes in CO_2-expanded liquids at mild temperatures. Mechanistic studies suggested that the aldehyde serves as an H-atom donor for peroxo and oxo radicals and as a reductant for the hydroperoxo intermediates [198]. Similar intermediates may be involved in oxidations of other substrates under supercritical conditions using the same oxidation system [191, 199, 200].

The overall reaction shown in Scheme 4.11 is based on coupled autoxidation processes of the aldehyde and the hydrocarbon substrate. To achieve appreciable hydrocarbon conversion, the reaction has to be run under kinetic control because in a free radical chain the oxidation of the aldehyde is thermodynamically favored. This was realized by slow diffusion of the aldehyde into the reaction media. Conducting the reaction under biphasic conditions with an ELP of the substrates instead of single-phase $scCO_2$ additionally improved hydrocarbon conversion because of the different partitioning of aldehyde (mainly in CO_2 phase) and hydrocarbon (mainly in ELP).

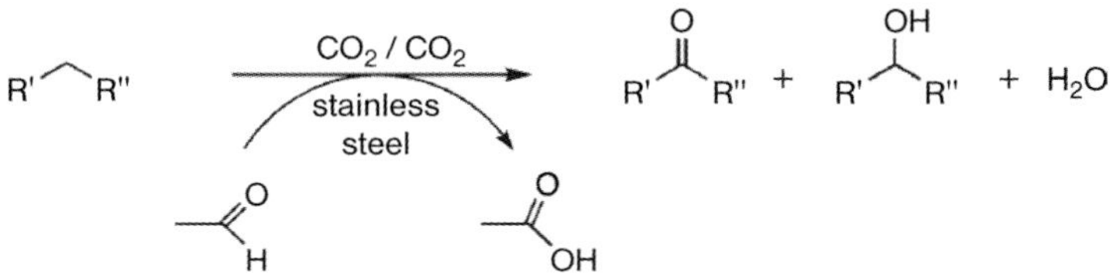

Scheme 4.11 Steel-promoted radical oxidation of hydrocarbons in compressed CO_2 with acetaldehyde as sacrificial co-reductant [198].

A 34% conversion of cumene was achieved with 66% selectivity to 2-phenyl-2-propanol within 27 h at temperatures as low as 52 °C. CO_2-induced melting was also exploited for the oxidation of cyclododecane in its ELP at 45 °C, which is 15 °C below its normal melting point [198].

The aerobic oxidation of cinnamyl alcohol to cinnamaldehyde over alumina-supported palladium catalysts has been optimized by variation of O_2 content and CO_2 pressure. A strong influence of the inert gas on conversion was found in continuous flow mode with small amounts of toluene as co-solvent (Figure 4.33) [201]. From phase behavior measurements using *in-situ* ATR-JR spectroscopy, the decrease in activity above 13 MPa could be correlated with the transition of an ELP to a monophasic SCF. It was concluded that gas-liquid mass transfer was not limiting in this case and higher local substrate concentrations caused the higher activities in the ELP. High activities were also obtained at medium pressures (45% conversion at 5 MPa) but conversions were <10% below 2 MPa. The increased selectivity to the unsaturated aldehyde at higher CO_2 densities was correlated with suppressed attack at the C=C of the substrate due to its preferential partitioning into the CO_2 phase.

A similar behavior was reported for the continuous aerobic oxidation of secondary octanols to the corresponding ketones catalyzed by palladium on alumina under compressed CO_2 [202] and also for the aerobic oxidation of benzyl alcohol to benzaldehyde catalyzed by metal oxide-supported gold colloids in batch mode with CO_2 [203], both systems exhibiting the highest conversions in ELPs at temperatures between 80 and 100 °C.

The use of compressed CO_2 has also been suggested for the direct synthesis of H_2O_2 from H_2 and O_2 because it could lead to solvent reduction, increased process safety, and enhanced productivity [204]. The *in situ* generation of H_2O_2 has been used for the synthesis of propylene oxide from propene, H_2, and O_2 with heterogeneous

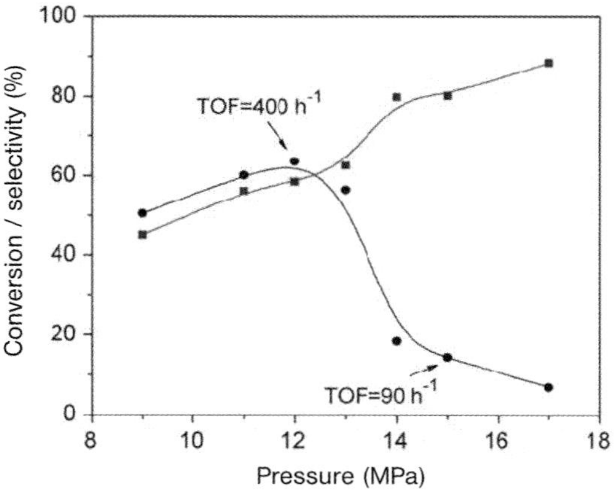

Figure 4.33 Conversion (●) and selectivity (■) of the oxidation of cinnamyl alcohol with compressed CO_2 as function of pressure at 80 °C (Reprinted with permission from Elsevier [201]).

palladium catalysts in CO_2. In batch experiments at 45 °C, the effect of CO_2 versus N_2 and addition of methanol or water was investigated, yielding the highest selectivity to propylene oxide (94%) in a CO_2-expanded liquid phase of the product [204]. Similarly, homogeneous palladium catalysts have been employed for the direct synthesis of H_2O_2 in H_2O–CO_2 biphasic media and used for efficient epoxidation reactions of alkenes in the same solvent system [205]. Enzymatic oxidations, which suffer from inhibition by high concentrations of H_2O_2, have also been successfully coupled with the *in situ* generation using compressed CO_2 [206]. The enhanced activity of H_2O_2 in aqueous solutions under CO_2 pressure has been connected with the formation of peroxocarbonic acid species (Scheme 4.12) (see also Section 4.3.1) [207].

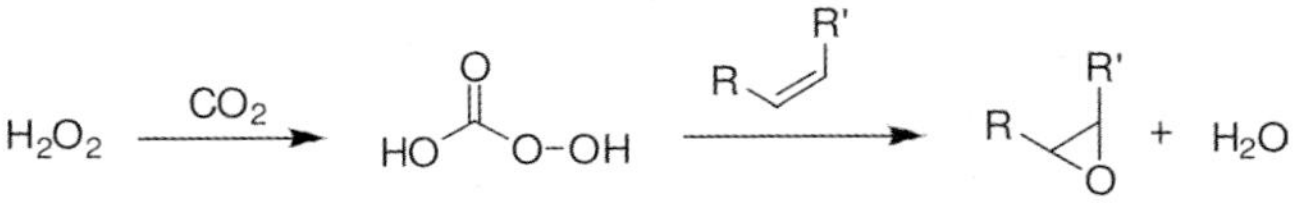

Scheme 4.12 Formation of peroxocarbonic acid or peroxocarbonates during oxidation processes with H_2O_2 in the presence of CO_2, exemplified by the epoxidation of alkenes. Depending on the conditions, the diols may be isolated from ring opening of the primary products [208].

The heterogeneously catalyzed oxidation of styrene to acetophenone with aqueous H_2O_2 over Au–Pd catalysts on metal oxides has been optimized using CO_2 pressure. After catalyst optimization, an ideal balance of both conversion and selectivity was found at around 9 MPa CO_2 at 120 °C, where an ELP of the substrate is formed. Whereas conversion increased with CO_2 addition up to 4.0 MPa and then remained constant, the chemoselectivity to acetophenone showed a distinct maximum of 87% around 9 MPa because over-oxidation increased at higher pressures [209].

The homogeneous epoxidation of various alkenes with H_2O_2 could be greatly enhanced by CO_2 expansion of aqueous MeCN reaction media. Cyclohexene, styrene, and other cyclic alkenes were epoxidized at high selectivities exceeding 85%. After careful evaluation of the maximum CO_2 pressures applicable without inducing a phase split of the ELP of MeCN from the aqueous H_2O_2, the influence of various bases on the system was examined and ELP was compared with SCF conditions [9]. In addition to perxocarbonate formation (see Scheme 4.12), the roles of both the solvent and the bases as oxygen transfer agents were discussed as reasons for the enhanced activity. Physical solvent effects were also suspected to influence the reaction network: THF expanded with CO_2 gave rise to only half the activity compared with CO_2-expanded MeCN. However, expanding gases other than CO_2 were not tested [9].

A very recent study investigated the feasibility of the homogeneous production of ethylene oxide using methyltrioxorhenium (MTO) as catalyst in aqueous methanol containing H_2O_2 [22]. Phase behavior modeling and flammability analysis of the mixtures allowed the identification of a broad range of pressures, temperatures, and compositions for safe and efficient production. The first batchwise test reactions showed promising results: over 95% selectivity was achieved at 48% conversion within 6 h at 40 °C and 5.0 MPa ethylene.

4.4.6 Miscellaneous

A number of stoichiometric organic reactions have been conducted in ELPs taking advantage of their tunable properties in different ways. Examples include Diels–Alder reactions [210] and nucleophilic substitutions [211]. Here, however, we concentrate on catalytic reactions and discuss some more examples that do not fall into the categories in Sections 4.4.3–4.4.5.

Alkene metathesis with Grubbs' catalyst in a neat substrate has been studied in the presence of compressed CO_2 and found to depend both kinetically and thermodynamically on the phase behavior of the mixture. The ethenolysis of oleic acid esters to give 1-decene and alkyl decenoates (Scheme 4.13) is interesting as two useful chemicals are produced from a biomass-derived substrate [212].

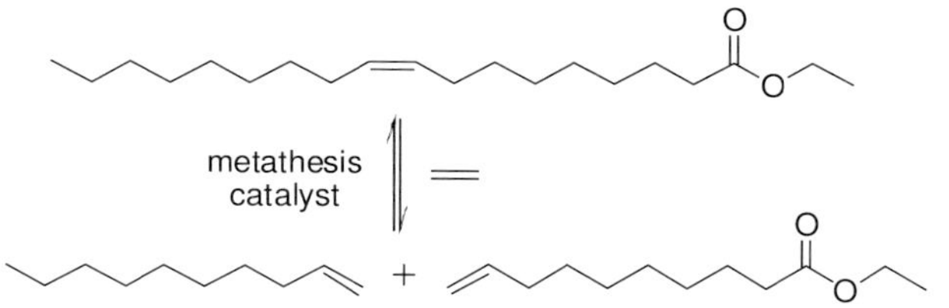

Scheme 4.13 Cross-metathesis of ethylene and ethyl oleate to give 1-decene and ethyl decenoate.

The phase behavior of mixtures of various compositions were measured and calculated and catalytic reactions were investigated in ELPs of neat substrate (Figure 4.34). Without CO_2 the reaction is fast, but equilibrium stops below 80% conversion. Expansion of the reaction medium with CO_2 allows for equilibrium conversions of over 90%, probably because of product extraction from the catalyst phase by CO_2. At higher CO_2 densities the reaction becomes slower again because the substrate starts to partition into the CO_2 phase.

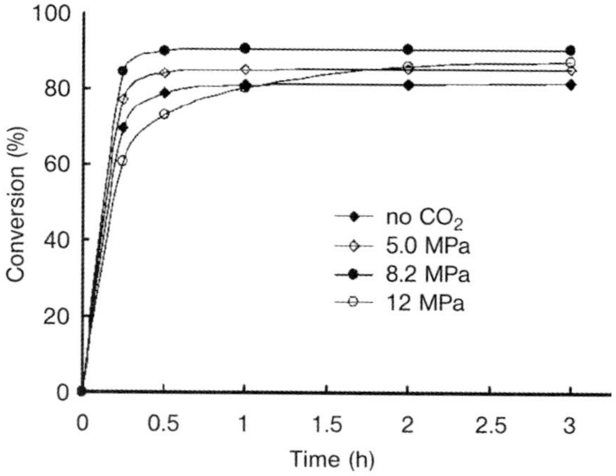

Figure 4.34 Ethyl oleate conversion using first-generation Grubbs catalyst at 35 °C and 1.5 MPa ethylene with added CO_2 (Reprinted with permission from American Chemical Society [212]).

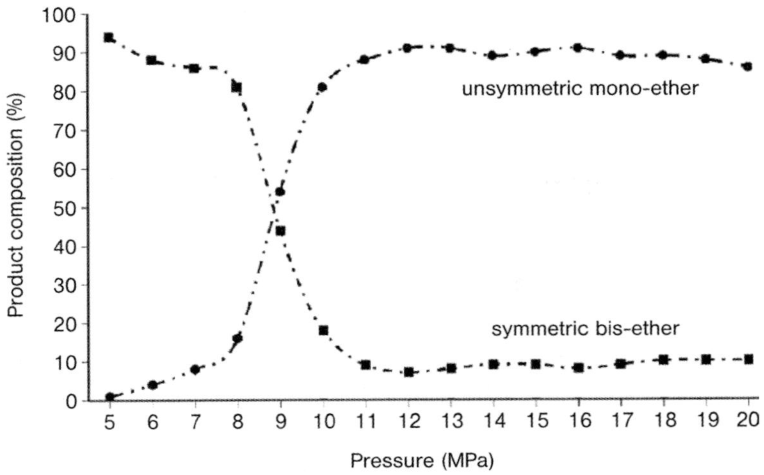

Figure 4.35 Product distribution of the solid acid-catalyzed etherification of 1,6-hexanediol with methanol at 150 °C as a function of CO_2 pressure (Reprinted with permission from American Chemical Society [213]).

The selectivity of the continuous acid-catalyzed etherification of symmetrical diols shows a very pronounced dependence on the CO_2 pressure (Figure 4.35). Phase behavior studies showed that the marked change of the ratio of mono- to bis-etherification corresponds to the transition of an ELP into a single-phase SCF at higher pressures. The interpretation of the authors was based on a difference in local concentrations of the reagents in the different states influencing the selectivity of the consecutive condensation reaction [213].

The homogenously palladium-catalyzed coupling reaction of two butadiene molecules with one molecule of CO_2 to give a δ-lactone is another example of a phase equilibrium-controlled reaction system. After careful evaluation of the phase diagram of the mixture and optimization of reaction parameters, experiments under distinctively different phase behavior were carried out. Using a non-CO_2-soluble catalyst, the highest conversions were achieved in the biphasic region where the reaction proceeded in an ELP [214].

Electrocatalysis has also been shown to benefit from ELPs. Benzyl chloride was efficiently electrocarboxylated to phenylacetic acid in CO_2-expanded DMF. The optimum working conditions were found to be 6.0 MPa CO_2 at 40 °C using stainless-steel electrodes with ammonium ILs as supporting electrolytes. Higher CO_2 contents gave poorer results because of the lowered electric conductivity of the media. The enhanced mass transport at the electrodes caused by CO_2 dissolution was used to explain the superior performance of the ELP system over non-expanded solutions [215]. However, applications are limited to substrates with reduction potentials higher than that of CO_2 in this case.

Catalytic polymerizations also benefit from the presence of an expanding gas because of reduced viscosities. The radical chain transfer during the polymerization

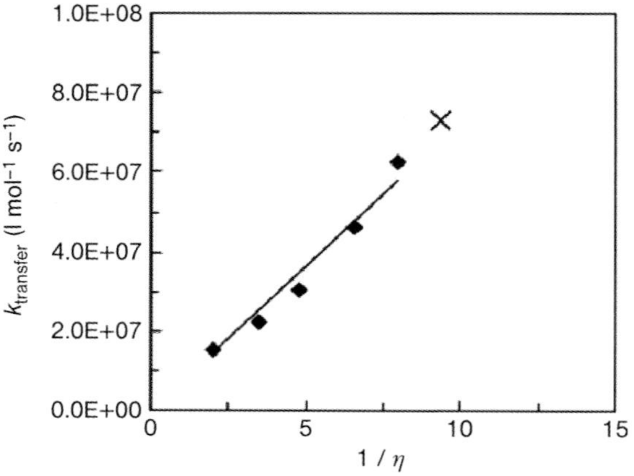

Figure 4.36 Relation between the chain transfer rate coefficient of the catalytic polymerization of MMA and the viscosity of CO_2-expanded MMA (◆) at 50 °C for 0.1, 2.0, 3.5, 5.0, and 6.0 MPa CO_2 (Reprinted with permission from Elsevier [216]); × = under monophasic (supercritical) conditions with 9.5 MPa CO_2 [217].

of methyl methacrylate (MMA) catalyzed by a homogeneous cobalt–oxime complex is a diffusion-controlled reaction both in neat monomer and in organic solution. When pure MMA was expanded with 6.0 MPa CO_2 at 50 °C, the rate of polymerization was increased by a factor of 4, and the enhancement could be directly related to the viscosity of the reaction media (Figure 4.36) [216].

Biocatalysis using non-conventional reaction media such as ILs and SCFs is an area of growing interest [218]. Despite the fact that ELPs can offer aqueous environments with tunable properties under mild conditions, so far they have rarely been explored for biocatalytic processes. One example is the use of immobilized lipases for the ethanolysis of triglycerides from milk fat. The highest conversions and selectivities to short-chain fatty esters were obtained with the commercial product Novozym 435 in CO_2-expanded neat ethanol–fat mixtures [219].

4.5
Why Perform Liquid–SCF Biphasic Reactions?

There are many possible motivations for designing a reaction to be conducted under conditions where two fluid phases ("solvents") are present. This section describes each of these motivations, and gives examples of reactions that have been performed in a liquid–SCF biphasic mixture.

Traditionally, reactions in biphasic fluid mixtures have been done with liquid–liquid mixtures [8], with the main goal of achieving an efficient separation of catalyst and products. However, liquid–SCF biphasic mixture can have additional important advantages that do not exist in a liquid–liquid system. These result mainly from the very unsymmetrical phase diagrams of many liquid–SCF systems, leading to a high or at least noticeable solubility of the SCF in the liquid phase, with marginal or negligible solubility of the liquid phase in the SCF. Such a situation is generally not observed for liquid–liquid systems, where mutual partial miscibilities always exist and lead to cross-contamination of both phases. The situation in a liquid–SCF system is more complex, but offers the following advantages for reactions in biphasic systems (see Section 4.2 for details):

1. The presence of dissolved CO_2 (or other SCF) in the liquid phase increases the solubility of reagent gases in the liquid.
2. It also reduces the viscosity of the liquid phase and accelerates mass transfer between the two phases.
3. The solubility of CO_2 or other SCFs can substantially lower the melting point of materials, making them available as liquid phases at temperatures significantly below their regular melting points.
4. The SCF has gas-like transport properties and can be readily used as a mobile phase under continuous flow operation.
5. Many SCFs are extremely poor at dissolving certain liquid solvents (water, liquid polymers, and ionic liquids, for example), which decreases the level of contamination of product by solvent if the SCF is the product-containing phase.
6. The SCF is easily removed from dissolved material, offering possibilities to simplify downstream processing if the SCF is the product-containing phase.
7. The SCF can also be used as a stationary phase, providing additional opportunities for continuous operation.

The following section outlines each of the major motivations for performing reactions in liquid–SCF biphasic media, with examples.

4.5.1
By Necessity (Unintentional Immiscibility)

If one fluid reagent happens to be immiscible in the other reagent or in the best solvent for the reaction, then finding a solvent or conditions that render them miscible may be difficult or may not result in practical advantages. This is the most common reason for the use of biphasic conditions, even though it is intellectually the least satisfying. There are several industrial processes where liquid–SCF systems are or were present without specifically addressing or exploiting the biphasic nature of the reactive systems.

Industrial production of 2-propanol by propene hydration, for example, is performed in a supercritical propene–liquid water biphasic mixture because the high-pressure propene has poor solubility in water, not because the supercritical state of the propene confers particular advantages (Figure 4.37). The catalyst and conditions differ from one company to another; over tungsten oxide catalysts at 270 °C and 25 MPa {$d_r = 1.1$ (calculated for pure propene from the temperature and partial pressure of propene, after subtracting the vapor pressure of water [219])} at ICI [220, 221], over a supported phosphoric acid catalyst at 240–260 °C and ≤6.5 MPa ($d_r \leq 0.14$) at VEBA [222], and over a cation-exchange resin at 6–10 MPa and 130–150 °C

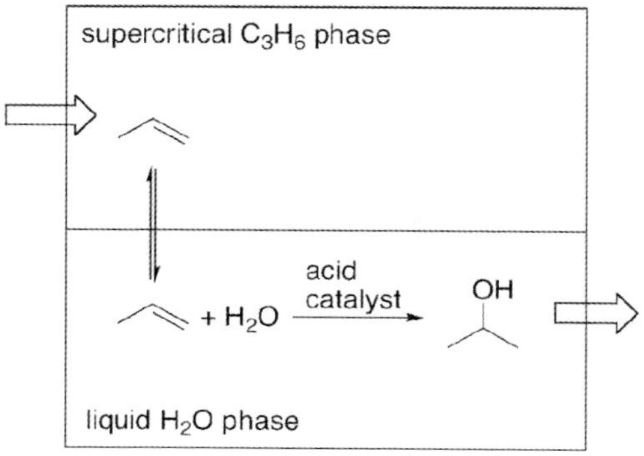

Figure 4.37 Biphasic catalysis during the direct hydration of propene. Supercritical propene is fed into the reactor and the product 2-propanol is recovered as an aqueous solution. Arrows indicate flow of starting material and product.

($d_r = 0.5$–1.2) at Deutsche Texaco [223]. In contrast, the Tokuyama Soda process [224, 225] uses a homogeneous catalyst, $[Si(W_3O_{10})_4]^{4-}$, dissolved in the aqueous liquid phase under a supercritical propene phase at 240–270 °C and 15–20 MPa ($d_r = 0.6$–0.9 at 250 °C). The product 2-propanol remains in the aqueous phase.

Methanol carbonylation for the formation of acetic acid was also performed in a biphasic mixture consisting of an aqueous phase containing dissolved cobalt iodide and methanol and a supercritical phase containing CO and more methanol at 250 °C and 76 MPa ($d_r = 1.6$ for CO). BASF's plant at Ludwigshafen, Germany, was the first to open, coming on-line in 1963 [226–228]. The development of the rhodium-catalyzed Monsanto process in 1968 [229] lowered the pressure and temperature requirements to 150–200 °C and 3.3–6.6 MPa [226], so that although the CO phase is still technically supercritical, d_r is now far below 1. Similar low pressures are applied in the most recent Cativa technology based on Ir catalysts [230].

4.5.2
To Facilitate Post-Reaction Separation

The separation of product from catalyst is a significant problem which hinders the wider application of homogeneous catalysis. Even in uncatalyzed reactions, the separation of product from unreacted reagents requires extra unit operations in the process and consumes energy. Designing reactions to facilitate post-reaction separation can result in significant energy savings and the ability to recover and recycle precious homogeneous catalysts.

Biphasic catalysis is a process in which there are two immiscible fluid phases; one preferentially dissolves the catalyst and the other preferentially dissolves the product (Figure 4.38). Because the product-containing fluid is typically removed from the vessel while the catalyst-bearing phase is typically left in the vessel to be used again, we shall use the terms "mobile phase" and "stationary phase" to describe the product-bearing and catalyst-bearing phases, respectively.

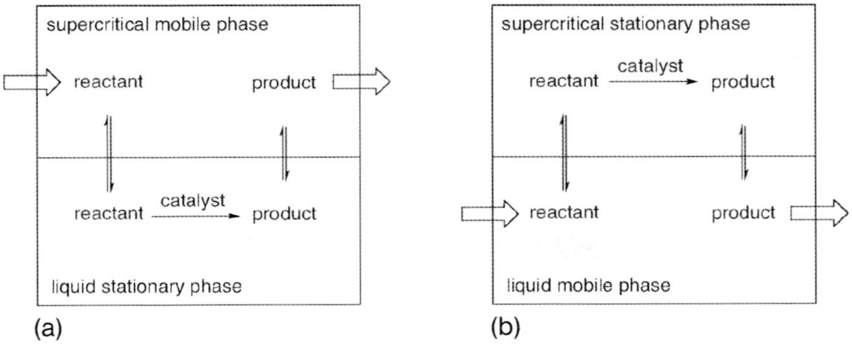

Figure 4.38 Biphasic catalysis in normal (a) or inverted (b) configurations. For batch processes, the product is removed in the "mobile" phase, while the reactant(s) can enter the system in either phase. In continuous flow systems, the reactant(s) must enter along with the mobile phase (shown with block arrows).

Typically, the mobile and stationary phases consist of two mutually immiscible solvents, either two liquids or a liquid and an SCF. However, if the product is a fluid and is immiscible with the stationary phase, then the product itself can be the mobile phase. When one of the two phases is an SCF, then conventionally that is the mobile phase because many homogeneous catalysts are more soluble in liquids than in SCFs. However, if the product of a reaction is particularly hydrophilic and therefore unlikely to dissolve preferentially in an SCF, then it may be better to use the "inverted" configuration (Figure 4.38b), in which the catalyst resides in the supercritical fluid and the product is removed as an aqueous solution. For example, Hâncu and Beckman [205, 231] reported that the reaction of H_2 and O_2 in $scCO_2$ could be catalyzed by a CO_2-soluble catalyst, $PdCl_2(PAr_3)_2$ (Ar = C_6H_4-*p*-CF_3 or C_6H_4-*p*-$CH_2CH_2C_6F_{13}$). The product H_2O_2 dissolved in the aqueous phase while the catalyst remained in the SCF. Leitner and co-workers [5, 232] described the hydroformylation of a water-soluble alkene in a biphasic H_2O–$scCO_2$ system (Scheme 4.14), where the aqueous solution of the product was drained out, leaving the catalyst-containing $scCO_2$ solution in the pressurized vessel. Fresh aqueous solution of substrate was then injected.

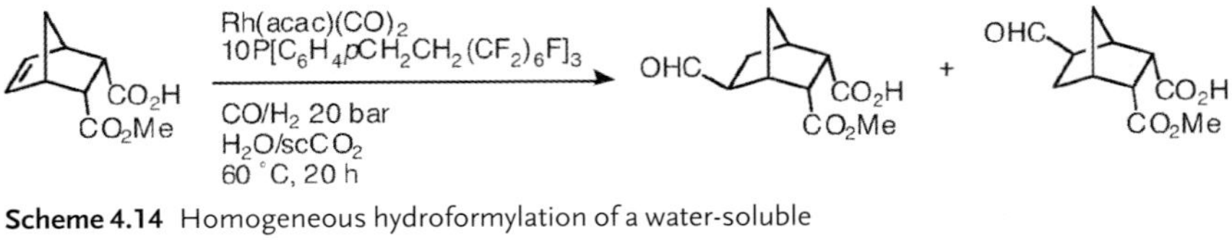

Scheme 4.14 Homogeneous hydroformylation of a water-soluble substrate in an inverted H_2O–CO_2 system.

4.5.3
To Facilitate Product/Catalyst Separation in Continuous Flow Systems

In some cases, the use of biphasic media makes it possible to use continuous flow reactors, rather than batch reactors, with homogeneous catalysts. In such systems, one fluid (the stationary phase) dissolves the homogeneous catalyst and remains in the reactor vessel while another fluid (the mobile phase) carries starting material into the vessel. The starting material then dissolves in the stationary phase and undergoes the reaction. The product is then extracted into the mobile phase and exits the reactor. In normal systems of this type, the stationary phase is water, an ionic liquid or a liquid polymer and the mobile phase is an SCF. In "inverted" systems, the stationary phase is the SCF (Figure 4.38).

The best industrial example of liquid–SCF biphasic catalysis is the hydration of supercritical butene. 2-Butanol for eventual dehydrogenation to 2-butanone [methyl ethyl ketone (MEK)] is prepared by hydration of butene in an Idemitsu Petrochemical plant (Figures 4.39 and 4.40) in Tokoyama, Japan, by a direct hydration process. Supercritical butene (a mixture of 1- and 2-butene) is bubbled through liquid water at low pH, 200 °C, and 20 MPa ($d_r = 1.8$ for 1-butene [233]). The hydration takes place in the aqueous phase, catalyzed by a dissolved heteropoly acid, following which the 2-butanol dissolves preferentially in the scC_4H_8 [234]. The supercritical phase then exits the reactor from the top, is cooled to the liquid state, and is then separated into 2-butanol and unreacted butene. The butene is recycled. The Idemitsu Petrochemical process also demonstrates that a homogeneous catalyst can be used in a continuous flow system if the catalyst is dissolved in a solvent that is immiscible with the product-bearing phase. Annual production of 2-butanone is 40 000 metric tons, and the plant has been in production since 1985 [235, 236].

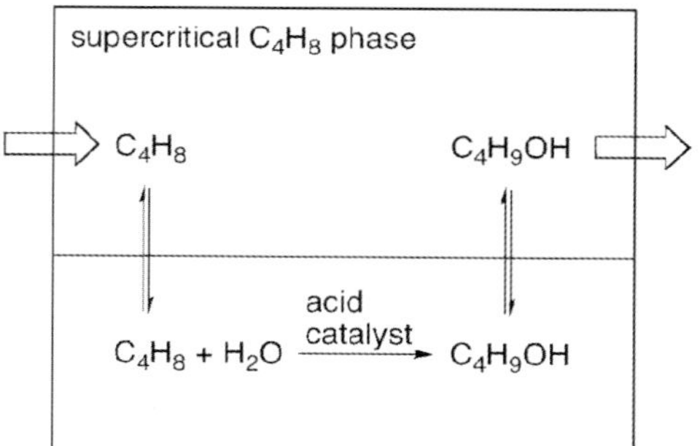

Figure 4.39 Biphasic catalysis during the hydration of 2-butene in the Idemitsu Petrochemical process. Supercritical butene is fed into the reactor (actually from below) and a solution of 2-butanol in scC_4H_8 exits the vessel. Arrows indicate flow of starting material and product.

Figure 4.40 The Idemitsu Petrochemical plant for the hydration of supercritical butene to butanol, for eventual conversion to 2-butanone (MEK). Of the three tallest towers, that on the right (with the framework around it) is the hydration reactor, and that on the left is the MEK fractionation column. Photograph courtesy of Idemitsu Petrochemical.

The use of the supercritical alkene reagent as a solvent with which to extract the alcohol product could also be possible in the hydration of propene [237], although this is not currently practiced.

4.5.4 To Stabilize a Catalyst

Many catalysts are unstable in the absence of a substrate or reactant, and therefore decompose at the end of a batch reaction once the substrate has been consumed. In such a situation, a continuous flow process is highly desirable because the catalyst is always being exposed to fresh substrate. The nickel-catalyzed hydrovinylation of styrene provides an illustrative example of such a system (Scheme 4.15) [238]. Because the chiral Wilke catalyst decomposes in the absence of substrate, catalyst recycling after batch reactions was futile. Using an IL–CO_2 system under continuous

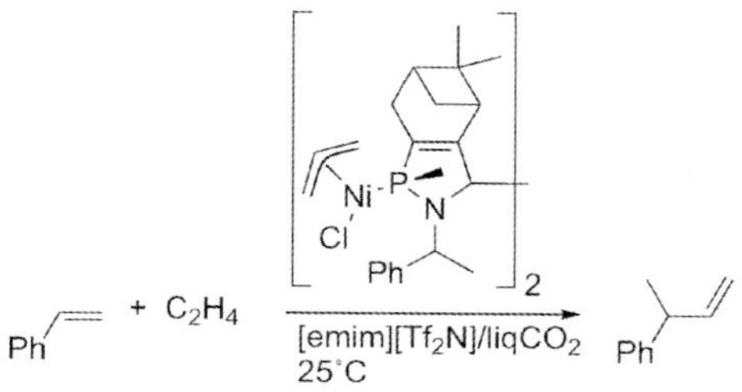

Scheme 4.15 Hydrovinylation of styrene in an IL–CO_2 biphasic medium using Wilke's catalyst.

flow conditions, stable performance could be achieved over more than 60 h on-stream because the concentration of substrate never falls to zero.

4.5.5
To Remove a Kinetic Product

A second phase is sometimes used to extract a kinetic product from reaction phase, in order to prevent the product from reacting further. Although this has not yet been the motivation for the use of a liquid–SCF biphasic medium, it has prompted the development of liquid–liquid and solid–SCF biphasic reactions. For example, Eckert and Liotta's group [239] reported the monoesterification of terephthalic acid, where extraction of the kinetic product monoester prevents esterification at the second acid group (Figure 4.41).

4.5.6
To Control the Concentration of Reagent or Product in the Reacting Phase

In some systems, it is necessary to decrease the concentration of a reagent in the reacting phase, in order to protect a sensitive catalyst (such as a bacterium [240]) from

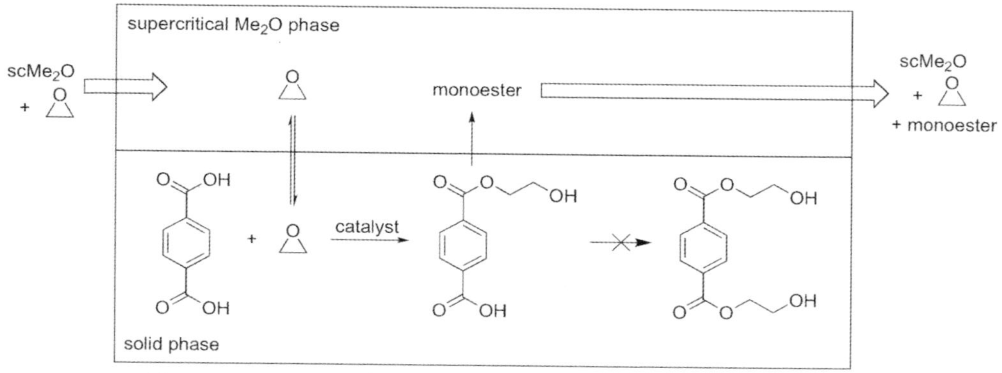

Figure 4.41 Extraction of a kinetic product to avoid further conversion to the thermodynamic product.

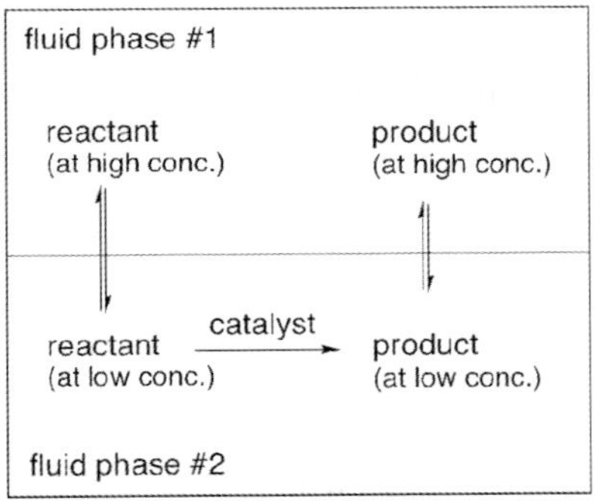

Figure 4.42 A partitioning biphasic reactor. Shown is the case in which both the reactant and the product preferentially partition into the catalyst-free phase. In some situations, only the reactant or only the product partitions in this manner.

being poisoned or inhibited or to avoid intermolecular reagent–reagent reactions (such as dimerization). Biphasic reactions performed for this reason are called partitioning biphasic reactions (Figure 4.42). The reactant partitions preferentially in the fluid that does not contain the catalyst. Nevertheless, the concentration of the reactant in the catalyst-bearing phase is sufficient to allow the reaction to proceed. As soon as some of the reactant has been consumed by the reaction, more reactant partitions into the catalyst-bearing phase from the other phase. The reaction thus continues without the concentration of reactant ever reaching a high level in the catalyst-bearing phase. The same technique can be used to prevent inhibition of the catalyst by the product or a byproduct.

SCFs have not, to our knowledge, been used in partitioning biphasic reactions to prevent a reagent from damaging a catalyst, although other advanced fluids such as ILs [241] have been used.

The removal of catalyst-inhibiting byproducts is an important role of the supercritical fluid in the butene hydration reaction (Figure 4.39), where the extracting ability of the scC_4H_8 prevents the build-up of polymeric residues in the catalyst-containing aqueous phase.

4.5.7
To Permit Emulsion Polymerization

In an emulsion polymerization, the reaction starts with a surfactant-stabilized emulsion of monomer (or monomer plus hydrophobic solvent such as scCO_2) in a continuous phase of water (Figure 4.43a). A water-soluble initiator begins the reaction in the aqueous phase, causing the polymerization of low concentrations of monomer molecules that have partitioned into the aqueous phase. The growing polymer, which is insoluble in the water, forms particles contained within a sphere of surfactant molecules. Thus the aqueous phase remains free of dissolved polymer, remains non-viscous, and is able to conduct heat away from the forming polymer.

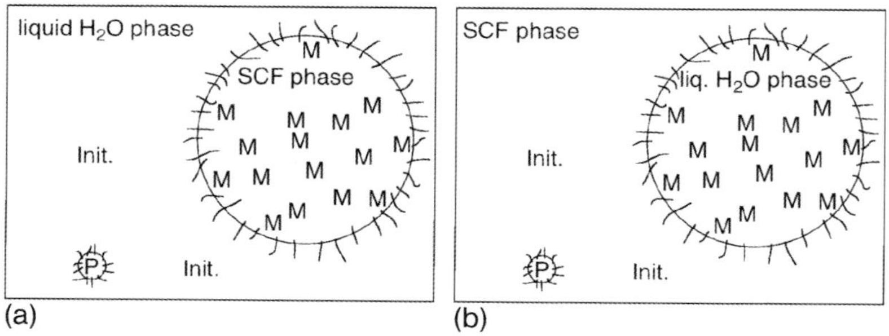

Figure 4.43 (a) An emulsion polymerization in an SCF-in-water emulsion, showing surfactant-stabilized droplets of SCF containing a hydrophobic monomer (M) and newly formed surfactant-stabilized particles of polymer (P) floating in liquid water. A hydrophilic initiator (Init.) is dissolved in the water. (b) An inverse emulsion polymerization. Surfactant-stabilized droplets of water containing a hydrophilic monomer (M) and newly formed surfactant-stabilized particles of polymer (P) are suspended in the supercritical fluid. A hydrophobic initiator is dissolved in the SCF.

In contrast, in a single-phase polymerization, where the monomer is polymerized in a solvent, the viscosity rises, heat transfer is inhibited, and the polymer does not form uniform particles.

In an inverse emulsion polymerization, the roles of the two phases are reversed (Figure 4.43b). An emulsion of water in a hydrophobic solvent (such as $scCO_2$) is used. Surfactant-stabilized micelles of water containing a hydrophilic monomer and newly formed surfactant-stabilized particles of polymer are suspended in the supercritical fluid. A hydrophobic initiator is dissolved in the SCF.

In 1990, Beckman's group [242, 243] reported the inverse emulsion polymerization of acrylamide in a surfactant-stabilized inverse emulsion of water in a supercritical ethane–propane mixture; 4 years later they reported the same process in a water–$scCO_2$ inverse emulsion [244]. These polymerizations used a CO_2-soluble initiator, azobisisobutyronitrile (AIBN), to polymerize a water-soluble monomer, acrylamide.

Polymerizations can also take place in normal (as opposed to inverse) emulsions of $scCO_2$ in water. Romack *et al.* [245] used a water soluble initiator, ammonium persulfate, to perform the emulsion polymerization of a CO_2-soluble monomer, tetrafluoroethylene. With perfluorooctanoate salts as surfactants, they obtained good yields of monodisperse particles. Without surfactants, the yields were poor and the particles were polydisperse and agglomerated.

The physical chemistry of water–SCF and SCF–water emulsions is described in Section 4.7.

4.5.8
To Create Templated Materials

CO_2-in-water emulsions can serve as media and templates for polymerizations to create porous materials [246–248]. In the templating method, a polymerization in the continuous phase of the emulsion preserves the structure of that phase as a solid polymer; after the dispersed phase and any liquid component of the continuous phase have been removed, a highly porous solid is obtained. The porosity is greatest if the volume fraction of the dispersed phase of the emulsion is high (74–90%). Hence such emulsions are called high internal phase emulsions (HIPEs). CO_2 is advantageous because, being a gas, it is far more easily removed from the porous solid product than is a conventional organic solvent [246]. For example, using di- and tri-block copolymers of vinyl acetate and PEG, Cooper's group prepared porous polyacrylamide having a pore volume of 8.7 $cm^3\,g^{-1}$, a density of 0.06 $g\,cm^{-3}$ and a mean pore diameter of 11 μm. The emulsion from which this material was prepared was 90 vol.% CO_2 in water; such emulsions were not possible with PFPE surfactants [248]. Similar porous materials consisting of methacrylated dextran have been prepared by Barbetta's group [249]. Including $Ca(OH)_2$ in the aqueous phase during the polymerization of acrylamide creates a porous polyacrylamide structure with embedded $CaCO_3$ crystals [250].

4.6 Biphasic Liquid–SCF Systems

4.6.1 Solvent Selection

Industrial applications of chemistry in biphasic liquid–SCF media have always, so far, used water as the liquid phase and a supercritical reagent as the SCF phase. However, in academic research laboratories, the SCF phase is typically $scCO_2$. Although it is difficult to make generalizations about supercritical reagents, because they vary widely, one can make generalizations about the liquids that are usually coupled with $scCO_2$. Their polarities increase in the order polymer < ionic liquid < water, although there is some overlap in the ranges of polymers and ionic liquids (Figure 4.44). The phase behavior of example liquids with $scCO_2$ at 40 °C and 15 MPa is summarized in Table 4.4 and described in more detail in the following sections.

4.6.2 Aqueous–SCF Biphasic Systems

Water has low solubility in most moderate-temperature SCFs, so that the product (if it resides in the SCF phase) is not significantly contaminated by water. At 50 °C and 15 MPa, the solubility of water is 0.62 mol% in $scCO_2$ [258] and 0.36 mol% in scC_2H_4 (interpolated from literature data [259]). Water dissolves more in scC_3H_6 because of the higher temperature: 0.90 mol% at 104 °C, 14.6 MPa [260]. The literature on the phase behavior of water–SCF biphasic mixtures is listed in Table 4.5. More information can be obtained from reviews of the phase behavior of CO_2–water [261], propane–water [262, 263], ethane–water [264], and ethene–water [265].

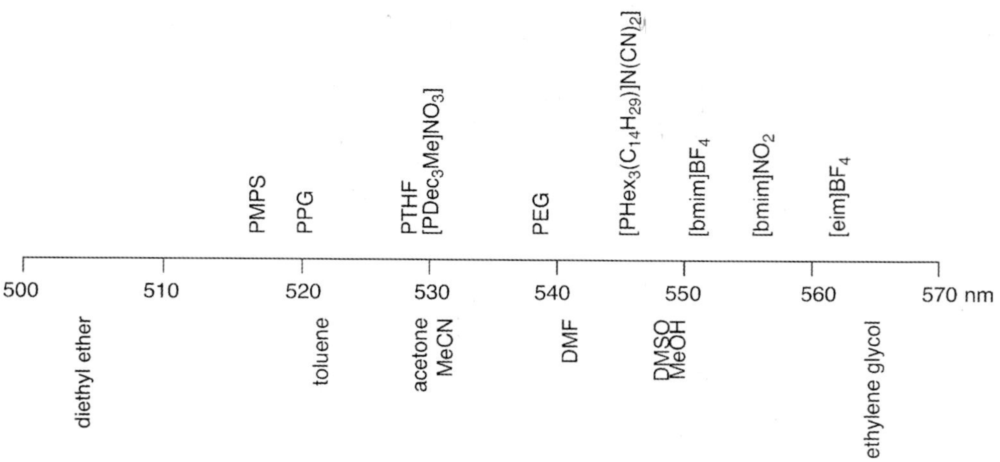

Figure 4.44 A comparison of the polarities of ionic liquids and liquid polymers [86, 254–256] (above the scale) with conventional solvents (shown below the scale). Polarities are indicated by the wavelength of maximum absorbance for the solvatochromic indicator Nile Red; polarity increases to the right. Water is at 593 nm and liquid CO_2 is at 480 nm [257]. Dec = decyl; Hex = hexyl; eim = *N*-ethylimidazolium; PMPS = polymethylphenylsiloxane; PPG = poly (propylene glycol); PTHF = polytetrahydrofuran; PEG = poly(ethylene glycol).

Supercritical fluids also dissolve in the aqueous phase, which is desirable if the SCF is a reagent, because the reagent must come into contact with the catalyst and/or reagents dissolved in the aqueous phase. The solubilities of supercritical ethane [264], ethene [265], propene [260], and CO_2 [31, 261] in water have been reviewed. The solubility of CO_2 in water is much greater than that of organic SCFs in water (Figure 4.45). The dissolution of CO_2 in water causes a significant pH drop, as discussed in Section 4.3.1.

Partition coefficients determine whether the reagents, products, and catalyst will reside more in one phase or the other. The partition coefficient between $scCO_2$ and

Table 4.4 Comparison of three types of non-volatile liquids that can be paired with $scCO_2$ in a liquid–SCF biphasic medium. [bmim][BF$_4$] and PEG are taken as prototypical examples of ionic liquids and liquid polymers, respectively. Conditions: 40 °C and 15 MPa of CO_2.

Parameter	Water	Ionic liquid ([bmim][BF$_4$])	Liquid polymer (PEG-600)
Solubility of CO_2 in the liquid (wt%)	5.6 [31]	19 [251]	51 [252]
Volumetric expansion of liquid (%)	Small	22 [28]	46^a [30]
Solubility of liquid in CO_2 (wt%)	0.2 [253]	Not detectable	0.12 [252]

^{a}For PEG-400. Calculated by extrapolation from only two data points and therefore considered very approximate.

Table 4.5 Literature on the phase behavior of water–SCF mixtures.

SCF	*T* (°C)	*p* (MPa)	Ref.
CO_2	41–80	8.5–16.1	[266]
	−1 to 100	0.1–100	[31]
	15–40	0–24	[253]
	20–102	2.4–16.7	[13]
	25–75	0.1–70	[258, 267]
	50–80	4–14.1	[268]
	110–350	10–150	[269]
	225–275	114–311	[270]
	50–350	20–350	[271]
N_2O	20–178	4.1	[272]
	25–100	2.4–5.0	[273]
	10–140	0.1–19.6	[274]
CHF_3	25–75	0.1–2.1	[275]
CF_3CFH_2	110	6.9–34.5	[276]
Me_2O	50–220	Up to 50	[277, 278]
C_2H_4	15–25	2.8–11.7	[279]
	35–102	0.5–53.1	[280]
	38–134	3.4–34.5	[259]
	166–300	0.1–94.5	[281]
C_2H_6	1–70	0.4–4.9	[282]
	200–400	20–350	[283]
C_3H_6	12–149	0.5–19.2	[284]
	38–138	0.2–32.4	[260]

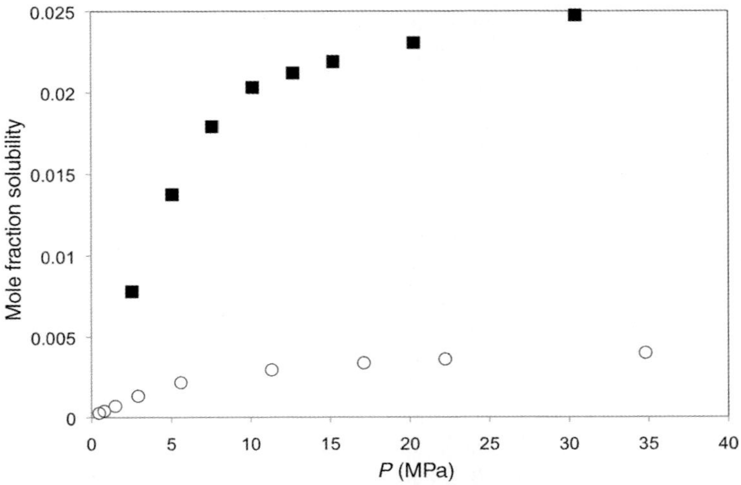

Figure 4.45 Solubility of CO_2 (■) at 50 °C [267] and ethene (○) at 55 °C [280] in water.

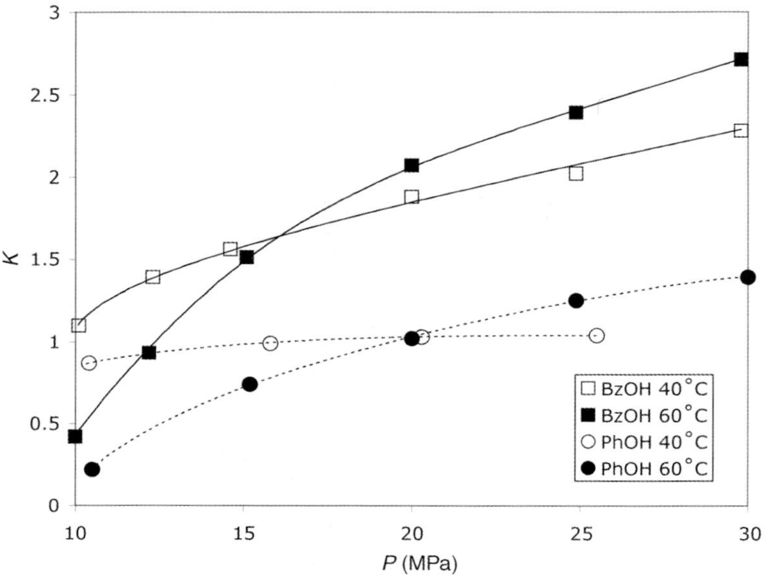

Figure 4.46 The $scCO_2$–water partition coefficients for benzyl alcohol and phenol as a function of pressure and temperature. The crossover pressures are 16 and 20 MPa for benzyl alcohol and phenol, respectively. Data from Brudi *et al.* [287].

water, defined as $[\text{solute}]_{CO_2}/[\text{solute}]_{aq}$, increases with pressure and hydrophobicity of the solute [285, 286]. Acidic conditions decrease the partition coefficient of basic species and basic conditions decrease the coefficient of acidic species [286]. The temperature effect depends on the pressure; below a crossover pressure (specific to the solute), the coefficient decreases with increasing temperature, whereas above the crossover pressure the reverse occurs (Figure 4.46) [287]. At the crossover pressure, the partition coefficient is independent of temperature. The coefficients vary over several orders of magnitude, as illustrated in the following series (data at 15 MPa, 40 °C) [287]:

caffeine $<$	benzoic acid $\approx$	benzyl alcohol $<<$	cyclohexanone $<$	2-hexanone
0.13	1.4	1.6	46	133

Leaching of metal-containing catalysts into the supercritical phase is rarely a problem because there is extensive literature on the design of water-soluble homogeneous catalysts [288, 289] and because the same structural features that ensure water solubility also ensure insolubility in $scCO_2$.

Advantages of using water as the liquid phase are that water is essentially insoluble in most supercritical fluids below 100 °C and that trace water in the product is rarely a significant problem. Catalysts can readily be designed to be soluble in water and insoluble in the supercritical phase. Aqueous–supercritical fluid biphasic reactions have been used in industry for many years, the best example being the Idemitsu Petrochemical direct hydration of butene. As mentioned in Section 4.5.1, there are other industrial examples of liquid water–SCF biphasic reactions, although none of them take advantage of the extracting ability of the supercritical phase; the product and the catalyst both end up in the liquid phase, which negates the separation advantage of the biphasic system.

Although the industrial examples have exclusively used an SCF that is one of the reagents, academic research in this area has emphasized the use of an inert SCF. Much of the early work in this area was of emulsion polymerizations in water-in-SCF emulsions. Biphasic aqueous–inert SCF media for reactions with transition metal catalysts have been reported by many groups. Combining CO_2 with water has been a popular choice, because both solvents are benign and inexpensive, and because catalysts and promoters can be easily chosen or designed in order to be soluble only in the aqueous phase. Arai's group [170], for example, reported that the use of a water–$scCO_2$ biphasic medium for the hydrogenation of cinnamaldehyde gave better conversion and selectivity that the use of a water–toluene mixture (Section 4.4.3).

The acidity of water–$scCO_2$ mixtures can be problematic for some reactions [5, 96, 232] but not for others. Rh colloid-catalyzed hydrogenation of arenes (Scheme 4.16) proceeds very selectively in a water–scC_2H_6 biphasic mixture but not at all in a water–$scCO_2$ mixture, presumably because of the pH drop [96]. In contrast, other reactions benefit from the increased acidity (Section 4.3.1).

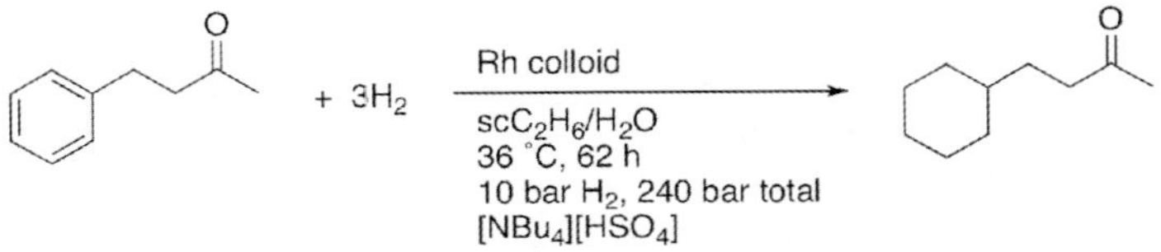

Scheme 4.16 Colloid-catalyzed reduction of an arene in an scC_2H_6–H_2O biphasic medium.

4.6.3 Ionic Liquid–SCF Biphasic Systems

ILs are salts that have a melting point below 100 °C. Potentially millions of ILs are possible from the combinations of available anions and cations (Figure 4.47). With such a large range, there are few generalizations that one can offer about ILs. Although most are non-volatile, some such as dimethylammonium *N,N*-dimethylcarbamate (DIMCARB) [290–293] and the switchable ILs [95, 115, 120] are volatile. Although most are very polar [254, 294] due to the charge separation, some tetraalkylphosphonium ILs have more moderate polarity [256]. Some have significant toxicity [295] and ecotoxicity [296] whereas others are fairly benign [297]. The major environmental concerns with ILs are the energy and materials consumption and volatile solvent emissions during their preparation [298].

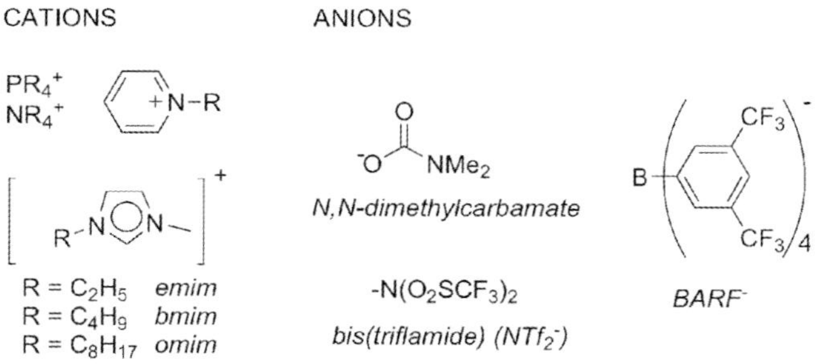

Figure 4.47 Some of the cations and anions used as components of ionic liquids.

The main reason for the green reputation of ILs is their negligible vapor pressure. This reputation is only reasonable if no volatile organic compound (VOC) solvents are used in the original synthesis of the IL and during the separation of the product whenever the IL is used as a reaction medium. If we optimistically assume that future syntheses of ILs will not involve VOC solvents, then it only remains to ensure that VOCs are not used in post-reaction separations either. There are five approaches to separating the product from the IL:

1. If the product is volatile, it can be distilled from the IL.
2. Decant a liquid product if it is immiscible with the ionic liquid. This can be enhanced by the CO_2 expansion of the liquid phase; expansion of a mixture of an IL and an organic liquid causes the two liquids to split into two phases [144, 148, 299, 300].
3. Collect a solid product by filtration if it is insoluble in the ionic liquid. Again, this can be enhanced by the CO_2 expansion of the liquid phase; expansion of a solution of a solid in an IL can cause the precipitation of the solid [141].
4. Have the reaction consume the IL. If the IL is a reactant in the process, then complete conversion results in disappearance of the IL [292, 301].
5. Extract the product from the ionic liquid using a solvent that is not a VOC.

The use of $scCO_2$ for this purpose was made possible by the discovery by Brennecke and Beckman's groups that ILs are completely insoluble in $scCO_2$ [302]. This insolubility is so unique that no other biphasic system can offer the same perfection; the product phase contains *none* of the solvent from the catalyst-bearing phase. This last option is the subject of the remainder of this section.

Whether or not ILs have some slight solubility in $scCO_2$ has been discussed greatly in the years since that initial report. Brennecke's group [303] found no solubility of imidazolium ILs in $scCO_2$. Han's group [304, 305] found that large amounts of organic cosolvents such as 15 mol% ethanol gave [bmim][PF_6] solubility on the order of 0.01 mol% in $scCO_2$ (15 MPa, 40 °C). Tetraalkylphosphonium ILs have significant solubility even in the absence of cosolvents [306]. Imidazolium-type ILs are soluble in $scCHF_3$ without cosolvent [307, 308]. The phase behavior of IL–SCF mixtures is described in a recent review [309].

As mentioned in Section 4.2, the solubility of CO_2 in ILs is relatively low, on a mass percent basis, but high on a mole fraction basis (although not as high as is found in

Table 4.6 Literature on the phase behavior of imidazolium IL–$scCO_2$ mixtures.

Anion	emim	bmim	hmim	omim	Rmmim[a]
BF_4^-		[28, 251, 312]	[313]	[303, 314]	[312]
PF_6^-	[315, 316]	[28, 48, 302–304, 310, 312, 316–318]	[316, 319]	[303]	[312]
NTf_2^-	[312]	[28]	[28, 320]	[28]	[28, 312]
OTf^-		[28]			
CTf_3^-		[28]			
$N(CN)_2^-$		[28]			
RSO_4^-[b]	[303]	[318]			
NO_3^-		[28, 303]			

[a] Rmmim = 3-alkyl-1,2-dimethylimidazolium, where alkyl is ethyl, butyl, or hexyl.
[b] $RSO_4 = MeSO_4$ or $EtSO_4$.

other Class III liquids such as liquid polymers). The solubility increases with decreasing temperature, increasing pressure, increasing fluorophilicity of the anion [$NO_3 < N(CN)_2 < BF_4/PF_6 < OTf < NTf_2 < CTf_3$] [28], and, less importantly, increasing length of the alkyl chains on the cation [309]. As the CO_2 pressure is increased, CO_2 solubility increases up to a "maximum", beyond which further increases in solubility can only be obtained by dramatic rises in pressure [309]. This "maximum" is believed to correspond to the amount of CO_2 that can be placed within the IL liquid structure without significant rearrangement of the ions [310]. Literature on the phase behavior of imidazolium IL–$scCO_2$ mixtures is listed in Table 4.6. Literature on the phase behavior of other ionic liquids with CO_2 is much more limited, and includes [BuPy][BF_4] [303], [$PHex_3(C_{14}H_{29})$]Cl [306], [$PHex_3(C_{14}H_{29})$][$MeSO_3$] [311], and [$PHex_3(C_{14}H_{29})$][$C_{12}H_{25}C_6H_4SO_3$] [311]. IL–$scCHF_3$ data are available for [emim] PF_6 [308] and [bmim]PF_6 [307].

The amount of CO_2 required in order to extract 95% of a product from an IL depends greatly on the nature of the product. Not surprisingly, compounds that have greater miscibility with CO_2 are more readily extracted from ILs with CO_2. For example, 1 mol of cyclohexane is extracted from [bmim][PF_6] by only 1800 mol of CO_2, whereas acetophenone requires 20 000 mol [321, 322]. Ideally, the product would not require too many equivalents of CO_2 to be extracted, whereas other reaction mixture components such as homogeneous catalysts would not be extracted at all. Because many homogeneous catalysts are non-volatile and contain aromatic ligands such as triphenylphosphine, extraction of the catalyst by CO_2 is not a concern. However, unintended extraction of dissociated ligand is unfortunately possible. This can be prevented by modification of the ligand so that it bears a cationic [323] or anionic group [324–326]. If the latter strategy is used, then the counterion should preferably be organic, so that solubility in the IL is maximized [325, 326].

Catalysis using ILs as the catalyst-bearing phase started in the 1960s and 1970s. Parshall in 1972 [327] performed homogeneously catalyzed hydrogenations in molten [NEt_4][$SnCl_3$] ($T_m = 78\,°C$) followed by decantation of the liquid product. Homogeneous catalysis in an IL–SCF biphasic mixture was first performed in

1994 [301], although the IL in that case was used as a reagent rather than for its ability to act as a solvent. The first example of IL–SCF biphasic catalysis to show the separation advantage of the combination of ILs and SCFs was in 2001, with the hydrogenation of alkenes and of CO_2 catalyzed by Rh and Ru complexes in a [bmim][BF_4]–scCO_2 [328]. The combination of ILs and SCFs for catalysis has been reviewed previously [329].

Continuous flow systems are a particularly elegant method for taking advantage of the insolubility of ILs in scCO_2. Industrial water–SCF systems such as the Idemitsu Petrochemical process (Figures 4.39 and 4.40) showed that continuous flow methods could be possible with homogeneous catalysts in a non-volatile phase and with an SCF bringing in the reagents and taking out the products. IL–SCF versions of this concept were reported in 2001 [238, 325, 330, 331]. In the first study [325, 330, 331], which concerned hydroformylation, 1-octene was introduced into the reaction chamber as a solution in scCO_2 along with CO and H_2. The reaction took place in [omim][NTf_2], which contained a rhodium catalyst having a phosphine optimized for solubility in an IL. The product was removed from the system with the scCO_2. A second study [238] described a continuous flow system for the hydrovinylation reaction mentioned in Section 4.4.4. The typical apparatus for IL–scCO_2 experiments is shown in Figure 4.48, although variations on this theme have been reported.

Supported ionic liquids, meaning ILs adsorbed on solid materials such as silica, allow IL solutions to be handled as if they were solid particles. A homogeneous catalyst dissolved in a supported ionic liquid phase (SILP) should have the same reactivity and selectivity as a homogeneous catalyst in a traditional IL. If SILP catalyst particles are placed in a liquid solvent or liquid substrate, then biphasic catalysis can occur, but with significant advantages including (a) a greater interfacial area between the IL and the liquid solvent, (b) a decreased volume of IL (and therefore less expense), and (c) simplified handling of the catalyst. SILP was invented by Mehnert *et al.* [326] and was first applied to batch IL–scCO_2 biphasic catalysis by He's group [332] (using silica-supported molten [Bu_4N]Br as both IL and catalyst for the

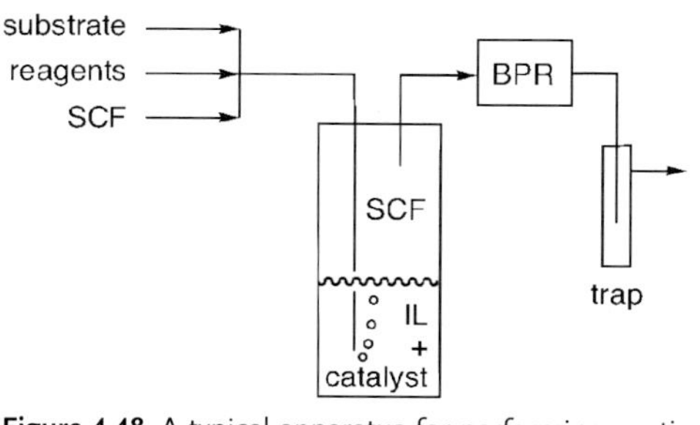

Figure 4.48 A typical apparatus for performing continuous flow experiments with an SCF–IL biphasic solvent system. The SCF is bubbled through the IL to release the substrate and reagents into the IL and to extract out product. After the SCF has been reduced in pressure by the back-pressure regulator (BPR), the product is trapped and the formerly supercritical gas is vented.

cycloaddition of propylene oxide and CO_2) and then Pagliaro's group [333] (using RuO_4^- in silica-supported imidazolium IL as a catalyst for the O_2 oxidation of secondary alcohols to ketones) in 2006. An adaptation of the method for continuous flow catalysis was reported by Cole-Hamilton and co-workers in the following years [184, 185]; 1-octene was hydroformylated over silica gel-supported [omim][NTf_2] containing Rh(acac)$(CO)_2$ and [pmim][$Ph_2PC_6H_4$-*m*-SO_3] (where omim = 3-octyl-1-methylimidazolium and pmim = 3-propyl-1-methylimidazolium). For a detailed discussion of the phase behavior involved, see Section 4.4.4.

4.6.4
Polymer–SCF Biphasic Systems

Many advantages are inherent in a liquid polymer–CO_2 solvent pair. CO_2 and poly(ethylene glycol) (PEG), the most commonly used liquid polymer solvent, are nontoxic, being approved in the USA as food or beverage additives [334]. Liquid polymers are far less expensive than ILs, their main competitor in the area of non-volatile liquids. PEG is available in bulk at approximately US\$1 kg^{-1}. Polymers are tunable over a very wide polarity range. Compared with water, liquid polymers should be less susceptible to acidification by CO_2.

The polymers that will be discussed (Scheme 4.17) are poly(ethylene glycol) (PEG), poly(ethylene glycol) 4-nonylphenyl monoether (PEG-NPE), poly(propylene glycol) (PPG), polytetrahydrofuran)(PTHF), polydimethylsiloxane (PDMS) and polymethylphenylsiloxane (PMPS). Numbers given after an acronym give the nominal average molar mass of the polymer. Variations upon these polymers are possible, including (a) versions with ether or ester groups at the termini, (b) block copolymers, and (c) versions with modified side-chains. These polymers are either low-melting solids or viscous liquids. Around room temperature, PEG and PTHF are solids if the average molar mass is above approximately 700. Commercially available PPG

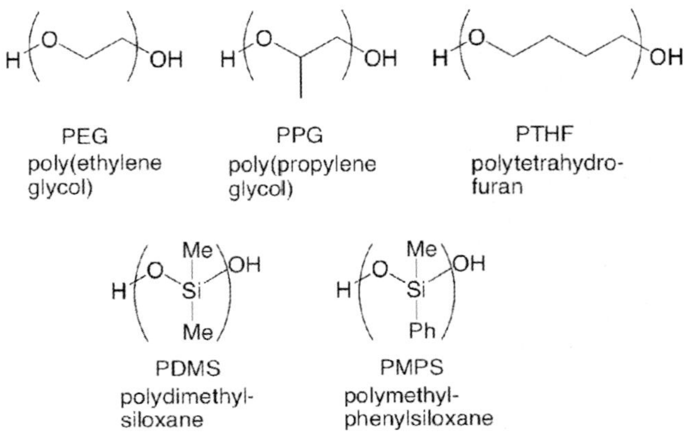

Scheme 4.17 Liquid polymers that can be used in combination with $scCO_2$ for biphasic reactions [86].

fractions (up to 3500 g mol^{-1}) are all liquids. PDMS is a liquid regardless of its molar mass, with pour points typically between −60 and −40 °C [335]. PMPS is a liquid unless it is prepared with carefully controlled tacticity [336].

Polarity varies widely from one polymer to the next, but liquid polymers are, in general, less polar than ionic liquids (Figure 4.44). PDMS has a dielectric constant of only 2.8 [335], meaning that it is close to the polarity of toluene. Whereas data are unavailable for the liquid, solid PEG is known to be far more polar than DMSO with a Kamlet–Taft π^* value of 0.86 and an $E_T(30)$ value from 45.7 [337] to 46.6 kcal mol^{-1}, increasing with molar mass [338]. The $E_T(30)$ value implies a polarity comparable to that of MeCN or DMSO. PPG is significantly less polar [338], with an $E_T(30)$ of up to 38.5 kcal mol^{-1}, matching the polarity of butyl acetate.

PEG and other polyethers are inherently coordinating, whereas siloxanes are not. PEG can bind metal centers in a multidentate structure similar to a crown ether [339]. The coordinating ability can significantly affect reaction performance. The presence of dimethoxylated PEG has been shown to enhance the enantioselectivity and TON of the alkylation of benzaldehydes catalyzed by chiral amino alcohols [340].

Liquid polymers are significantly better than ILs at dissolving permanent gases. For example, the solubility of N_2 in PEG-1500 is 0.4 wt% at 18.8 MPa and 50 °C [341], whereas it is undetectable in [bmim]PF_6 [342]. At 2.2 MPa, the solubility of O_2 is 0.4 wt% in PDMS-77000 (30 °C) [343] and 0.03 wt% in [bmim]PF_6 (25 °C) [342].

Carbon dioxide is very soluble in some of these polymers. For example, in PDMS-308000 [344] and PEG-1500 [345, 346], the CO_2 content of the condensed phase reaches 40 and 29 wt%, respectively, at 26 MPa and 50 °C. The nature of the end groups on PEG has little effect on the phase behavior for longer PEG chains such as PEG-1500 [346]. The solubility of CO_2 in polymers decreases with increasing temperature. Key references for phase behavior data are listed in Table 4.7. The topic of polymer–SCF phase behavior has been reviewed [347].

Volumetric expansion caused by dissolution of CO_2 in liquid polymers is accompanied by significant changes in the physical properties of the condensed phase. The melting point of PEG is decreased by 10–20 °C when CO_2 dissolves in the polymer [47, 346]. Beyond a certain pressure, typically 7–10 MPa, no further decrease in melting point is observed. The viscosity of PEG-400 is lowered by 10-fold upon addition of 25 MPa of CO_2 at 40 °C [348]. In contrast, a similar viscosity drop is observed in PPG-2700 at 35 °C when it is exposed to a CO_2 pressure of only 4 MPa [349]. The viscosity of PDMS-380000 at 50 °C decreases by 55–58% at 9.1 MPa CO_2 [350]. The interfacial tension in a PEG-NPE-2500–scCO_2 biphasic system at 50 °C decreases from just above 17 dyn cm^{-1} to only 9 dyn cm^{-1} when the scCO_2 pressure is increased from 9 to 20 MPa [351]. The diffusion coefficient of CO_2 in PEG-12000 at 62 °C decreases by about half when the CO_2 pressure is raised from 5 to 29 MPa [47]. The polarities of PEG and PPG decrease upon expansion with CO_2 [86], although the decrease is not as large as with traditional polar solvents such as DMSO.

Solubility of the polymer in the SCF is to be avoided, but the molar mass at which the solubility becomes negligible depends strongly on the choice of polymer. The solubility of PEG in scCO_2 falls rapidly as the molar mass increases. The solubilities of PEG-400, PEG-600 and PEG-1000 are 2.1, 0.2 and 0.08 wt%, respectively, under the

Table 4.7 Literature on the phase behavior of liquid polymer–SCF mixtures.

SCF	Polymer	Nominal M_n	Ref.
CO_2	PEG	≤600	[30, 252, 352–354]
	PEG	1000	[252]
	PEG	1500	[341, 345, 346]
	PEG	≥4000	[47, 341, 345, 346]
	PEG 4-nonylphenyl monoether	2500	[355]
	PEG monomethyl ether	≤1000	[354]
	PEG dimethyl ether	500	[354]
	PEG didodecyl ether	350	[356]
	PPG	≤1000	[354, 357]
	PPG	≥2000	[30, 354]
	PPG monoalkyl ethers	1000, 1200	[354, 358]
	PPG dimethyl ether	3500	[358]
	PDMS	–	[344, 359]
C_2H_4	PEG didodecyl ether	350	[356]
C_2H_6	PPG	420, 780, 2000	[360]
C_3H_8	PEG	200, 1500, 4000, 8000	[341]

same conditions (26–27 MPa and 50 °C) [252]. PPG is more soluble than PEG in $scCO_2$, presumably due to the reduced crystallinity of the former. Paradoxically, PTHF is far less soluble than both PEG and PPG [354]. The monoalkyl ether and especially dialkyl ether derivatives of PEG and PPG are far more soluble than the dihydroxy polymers in $scCO_2$ [354]. The average molar mass of the solubilized fraction of a polymer is lower than that of the parent polymer [355, 365]. Lighter fractions of PDMS are fairly soluble, but Garg *et al.* [344]. reported that PDMS-308000 is essentially insoluble. Although there are few data for other liquid polymers, they

Table 4.8 Literature on the physical properties of CO_2-expanded liquid polymers.

Property	Polymer	Reference
Compressibility	PDMS	[344]
Density	PEG	[348]
Melting temperature	PEG	[47, 345, 346, 361]
Polarity	PEG, PPG	[86]
Viscosity	PDMS	[350, 362, 363]
	PPG	[349]
	PEG	[348]
	PDMS	[364]
Volumetric expansion	PEG, PPG	[30]
	PDMS	[344]
Volumetric expansion, viscosity, interfacial tension	PEG-2500-NPE	[351]

are expected to show similar behavior, with the higher molar mass fractions having very low solubility in $scCO_2$. Addition of cosolvents such as ethanol to the $scCO_2$ increases the solubility of PEG in the SCF phase [366].

Partitioning of solutes between the $scCO_2$ phase and the polymer phase has been studied for cross-linked PDMS and polycyanopropylmethylsiloxane (PCPMS). Eckert's group [367, 368] showed that naphthalene, acridine, and 2-naphthol partition preferentially into the polymer phase, but the partition constant decreases from 10^2–10^3 below 7.5 MPa to only 1–5 at higher pressures. One would expect, therefore, that heavy polyaromatic compounds will be difficult to extract from liquid polymers with $scCO_2$. Extraction of lighter compounds is relatively facile. Ethylbenzene (340 mg) was extracted from 10 ml of PEG-900 by $scCO_2$ (15.5 MPa, 60 °C, 2 ml min^{-1}, 4 h), giving 59% mass recovery with <0.5 wt% non-volatile material (water and PEG) in the extracted material. Scaling up the experiment to 7.8 g of ethylbenzene and increasing the molar mass of the PEG to 1500 greatly increased the percent mass recovery of ethylbenzene and decreased the amount of non-volatile material co-extracted [172].

If the ligands on a catalyst in the liquid polymer have multiple aromatic rings, then leaching of the ligand or the catalyst from the polymer phase is less likely to be a problem, although leaching of PPh_3 from a PEG solution of $RhCl(PPh_3)_3$ into $scCO_2$ has been observed [172]. Tagging ligands with PEG or charged groups may be necessary with some systems to prevent leaching.

The first reaction performed in a liquid polymer–SCF biphasic medium was the hydrogenation of styrene in PEG-900–$scCO_2$ (Scheme 4.18) [172]. The product was isolated by extraction with $scCO_2$ and the catalyst solution in PEG was reused, by addition of fresh styrene and H_2, for five cycles. Conversion was >99% in each cycle. Product recovery was 79% in the fifth cycle but lower in earlier cycles. The Rh content of the extracted material was found to be <1 ppm.

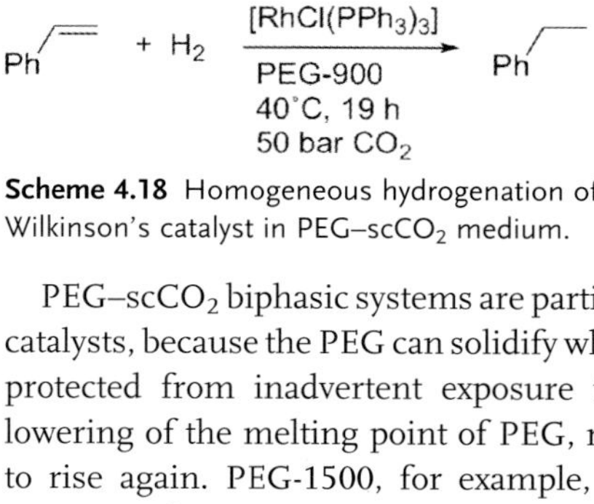

Scheme 4.18 Homogeneous hydrogenation of styrene with Wilkinson's catalyst in PEG–$scCO_2$ medium.

PEG–$scCO_2$ biphasic systems are particularly appropriate for use with air-sensitive catalysts, because the PEG can solidify when the CO_2 is released, so that the catalyst is protected from inadvertent exposure to air. Because CO_2 causes a significant lowering of the melting point of PEG, releasing the CO_2 causes the melting point to rise again. PEG-1500, for example, normally melts around 43 °C but in the presence of CO_2 it melts at a significantly lower temperature [47, 369]. If the reaction is performed in the CO_2-melted PEG and then the CO_2 is released, the catalyst becomes entrapped in a solid mass of PEG and is therefore protected from exposure to air until the next time the catalyst is to be used [86].

Reetz and Wiesenhöfer [370] showed that the $scCO_2$–PEG-1500 combination can be used in a continuous flow manner for the lipase-catalyzed esterification of 2-phenylethanol, with the catalyst having roughly constant activity over almost 25 h (Scheme 4.19). A kinetic resolution, performed in a batch process in $scCO_2$–PEG-1500 with subsequent $scCO_2$ extraction, gave excellent enantioselectivity for both the esterification product and the remaining alcohol (Scheme 4.19).

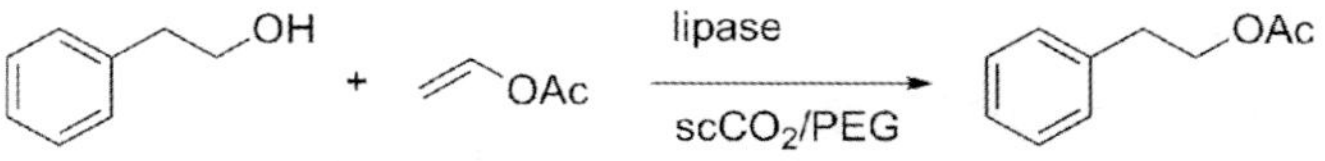

Scheme 4.19 Enzymatic esterification of 2-phenylethanol.

Other reactions performed in liquid polymer–SCF media include asymmetric catalysis [86], CO_2 hydrogenation [86], CO_2 insertion into epoxides [371], oxidation of styrene [372] and alcohols [373, 374], and Pd nanoparticle-catalyzed oxidation of alcohols [375].

4.6.5
Liquid Product–SCF Biphasic Systems

In the traditional hydroformylation (Oxo) process, the cobalt carbonyl catalyst, dissolved in high-boiling liquids or water, is reacted with a supercritical mixture of CO, H_2, and propene (20–35 MPa and 110–180 °C). The effluent from the vessel is butanal product containing Co catalyst.

Cole-Hamilton's group [331, 376, 377] described the use of a continuous flow system in which the hydroformylation catalyst is dissolved in liquid product aldehyde, $scCO_2$ carries alkene, CO, and H_2 into the system, and $scCO_2$ carries aldehyde product and unreacted gases out of the system. The catalyst was designed to be soluble in aldehyde but insoluble in $scCO_2$ (Scheme 4.20 and Figure 4.49). Through careful optimization of fluid density and concentrations (residence time), they were able to perform the hydroformylation of 1-octene continuously in a CO_2-expanded steady-state mixture of octene and nonanals (see also Section 4.5.5). This reactant ELP is the only liquid phase, so that this elegant continuous flow process avoids the need for a second solvent such as an IL or liquid polymer.

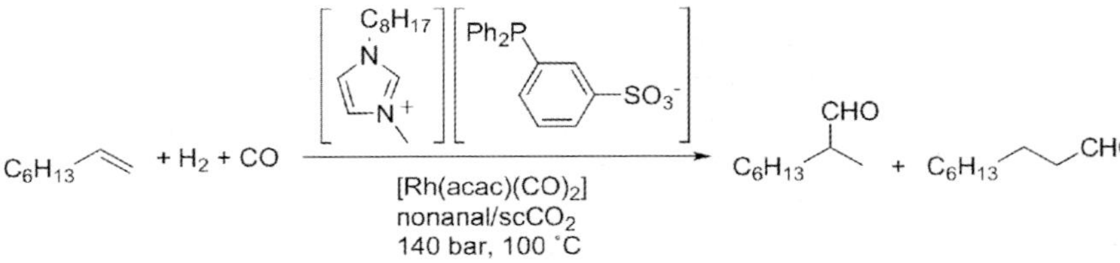

Scheme 4.20 Catalytic system for continuous flow hydroformylation in expanded product–$scCO_2$ phase [377].

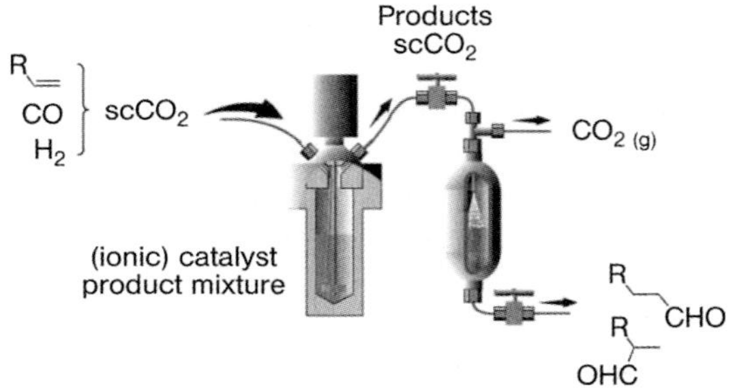

Figure 4.49 Concept of the "solventless" continuous flow hydroformylation of 1-octene with $scCO_2$ as the mobile phase and an ionic Rh complex dissolved in a mixed ELP as stationary phase (Reprinted with permission from Royal Society of Chemistry [377]).

4.7 Biphasic Reactions in Emulsions

4.7.1 Water-in-SCF Inverse Emulsions

Whereas emulsions and inverse emulsions of water and organic solvents have been known for a very long time, the study of water–SCF emulsions is relatively recent, starting with the 1987 report by Gale *et al.* [378] of water-in-scC_2H_6 inverse emulsions stabilized by AOT surfactant (Figure 4.50). They correctly theorized that water–SCF emulsions could have many applications in extraction, separation, and reactions, and that the high tunability of the SCF would allow researchers to break and then reform the emulsion at will, by minor changes in pressure. They showed the latter point by charting the minimum ethane density at which an emulsion could be maintained. Ever since, the primary focus in water–SCF emulsions has been on inverse emulsions (i.e. water-in-SCF emulsions, with the SCF as the continuous phase) rather than on normal emulsions (i.e. SCF-in-water emulsions, with water as the continuous phase). The areas of emulsions [379–381] and inverse emulsion polymerization [382] in water–SCF mixtures have been reviewed.

Inverse emulsion stability is strongly dependent on the SCF density; at higher temperatures it is necessary to use higher pressures, or the density will be insufficient,

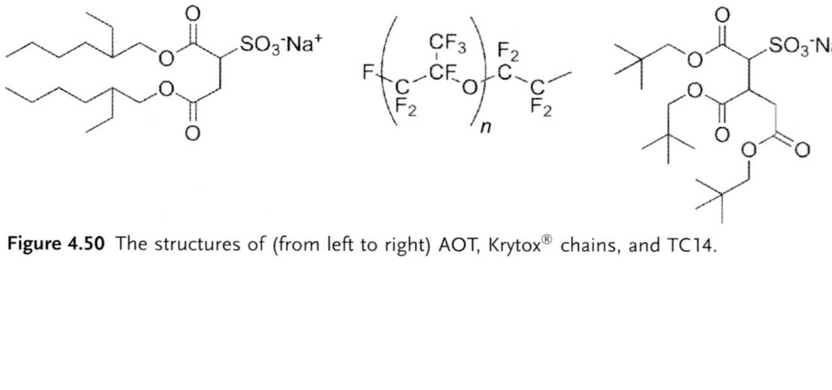

Figure 4.50 The structures of (from left to right) AOT, Krytox® chains, and TC14.

the emulsion will break, and a separate water layer will form. The density below which flocculation of the emulsion occurs is called the critical flocculation density (CFD) [379, 383]. Below the CFD, the surfactant tails attract each other, repel the solvent, and the droplets are able to merge. Above the CFD, the surfactant tails attract the solvent rather than each other, the droplets do not merge, and the emulsion is stable. The CFD is essentially the same as the critical solution density for the chain dissolved in the SCF.

Because of the link between inverse emulsion stability and the solubility of the chain in the SCF, the chain must be designed for optimum solubility in the SCF. Although AOT was satisfactory for water-in-alkane inverse emulsions, water-in-scCO_2 inverse emulsions require surfactants that contain more CO_2-philic chains such as perfluoroalkyl, perfluoropolyether (PFPE), or polydimethylsiloxane chains [379, 384]. Commonly chosen chains for this purpose are the PFPE chains derived from the DuPont product Krytox® [385]. Poly(vinyl acetate) chains are also usable [386, 387]. Eastoe's group has recommended non-fluorinated surfactants containing neopentyl ester groups such as those in TC14 (Figure 4.50) [388].

4.7.2 SCF-in-Water Emulsions

Typically, the solubility of the surfactant determines the continuous phase. Thus, surfactants that are more CO_2 soluble than water soluble will encourage the formation of water-in-CO_2 emulsions (W/C emulsions) whereas surfactants having greater water solubility will promote formation of CO_2-in-water emulsions (C/W emulsions). Thus, the highly CO_2-philic and expensive fluorous and siloxane tails that are optimum for W/C emulsions are unnecessary and typically not as effective for C/W emulsions; oligomers of ethylene oxide are fairly satisfactory [389]. Lower CO_2 densities tend to favor C/W emulsions over W/C emulsions [390]. C/W emulsions with a very high CO_2 loading can be stable; Tan *et al.* [248] used a triblock copolymer of vinyl acetate and PEG (Figure 4.51) as a surfactant to create C/W emulsions stable for 2 days at 97% CO_2 content by volume.

4.7.3 Ionic Liquid-in-SCF Emulsions

IL-in-scCO_2 emulsions (I/C emulsions) have not been studied nearly as much as W/C emulsions, and we are unaware of any work on CO_2-in-IL emulsions (C/I emulsions).

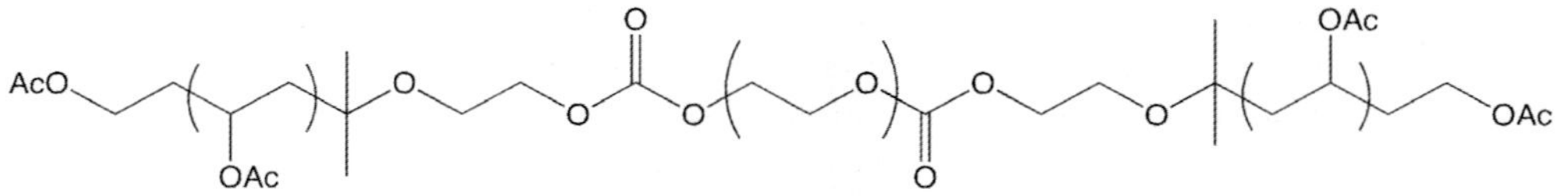

Figure 4.51 Triblock copolymer of vinyl acetate and PEG used as a surfactant for very high CO_2 content C/W emulsions [248].

Liu *et al.* [391] prepared emulsions of 1,1,3,3-tetramethylguanidinium acetate, lactate, and trifluoroacetate in $scCO_2$ using *N*-ethylperfluorooctylsulfonamide ($EtHNSO_2C_8F_{17}$) as the surfactant, but amount of surfactant required was greater than or equal to the amount of IL by mass. Inorganic salts such as $CoCl_2$ that were insoluble in $scCO_2$ or in solutions of $EtHNSO_2C_8F_{17}$ in CO_2 were soluble in the I/C emulsions. Poly(ethylene oxide)-b-poly(1,1,2,2-tetrahydroperfluorodecyl acrylate) can also be used as a surfactant to stabilize I/C emulsions [392].

4.7.4 Applications of Emulsions

If the rate of a biphasic reaction is poor, it can be increased by the use of an emulsion, because emulsions have greater surface areas and consequently greater mass transfer rates. Jacobsen *et al.* showed that adding surfactants to a biphasic reaction (Scheme 4.21) greatly increased the rate [393]. The post-reaction separation of catalyst from product was not impeded because lowering the CO_2 pressure triggered the breaking of the emulsion.

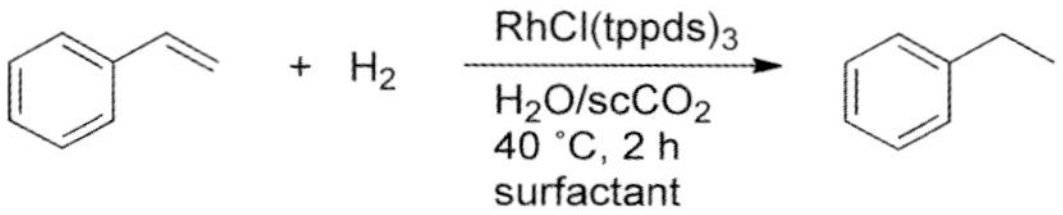

Scheme 4.21 Hydrogenation of styrene in an H_2O–$scCO_2$ emulsion.

Catalysts can act, in some cases, as surfactants. Zhu *et al.* [394] used a cobalt complex as both catalyst and surfactant for the oxidation of toluene (Scheme 4.22).

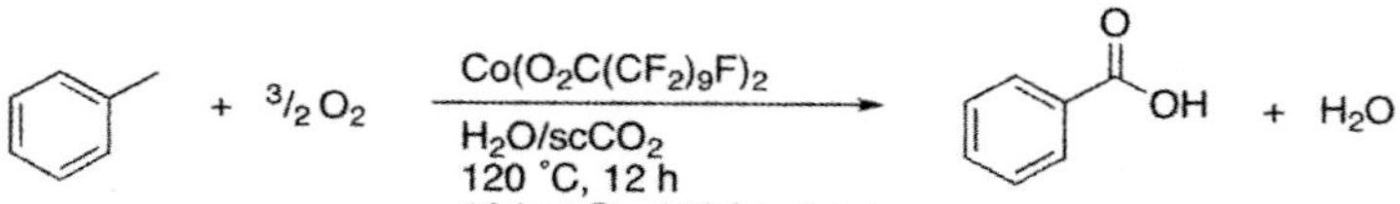

Scheme 4.22 Oxidation using a Co complex as both catalyst and surfactant.

Timko *et al.* [395] showed that the increase in rate could be achieved without the use of a surfactant. They showed that an ultrasound-generated emulsion of CO_2 and water had sufficient stability to increase the rate of a hydrolysis reaction 100-fold.

Microemulsions differ from emulsions in terms of droplet size (2–20 nm versus 0.05–10 μm), appearance (transparent versus opaque), and stability (stable versus unstable) [396]. Microemulsions are thermodynamically stable, whereas emulsions (sometimes called "macroemulsions") are kinetically stable. Reviews on supercritical microemulsions appeared in the late 1990s [397, 398]. Despite the greater stability and surface area of microemulsions, some applications may be better performed with emulsions because of the greater volume of water that can be "solubilized" [399]. For example, the halide exchange reaction between $PhCH_2Cl$ and KBr proceeds more rapidly in a water-in-$scCO_2$ emulsion than in a microemulsion.

The subject of chemical synthesis and reactions in supercritical emulsions has been reviewed several times, and the reader is referred to the relevant papers for more details [379, 380, 382, 399, 400].

References

1 Akien, G.R. and Poliakoff, M. (2009) *Green Chemistry*, **11**, 1083.
2 Jessop, P.G. and Subramaniam, B. (2007) *Chemical Reviews*, **107**, 2666.
3 Hallett, J.P., Kitchens, C.L., Hernandez, R., Liotta, C.L. and Eckert, C.A. (2006) *Accounts of Chemical Research*, **39**, 531.
4 Jessop, P.G. and Leitner, W. (1999) *Chemical Synthesis Using Supercritical Fluids*, Wiley-VCH Verlag GmbH, Weinheim.
5 Leitner, W. (2002) *Accounts of Chemical Research*, **35**, 746.
6 Scurto, A.M., Newton, E., Weikel, R.R., Draucker, L., Hallett, J., Liotta, C.L., Leitner, W. and Eckert, C.A. (2008) *Industrial & Engineering Chemistry Research*, **47**, 493.
7 Phillips, S. and Olesik, S.V. (2002) *Analytical Chemistry*, **74**, 799.
8 Cornils, B., Herrmann, W.A., Vogt, D., Horvath, I., Olivier-Bourbigon, H., Leitner, W. and Mecking, S. (2005) *Multiphase Homogeneous Catalysis*, Wiley-VCH Verlag GmbH, Weinheim.
9 Rajagopalan, B., Wei, M., Musie, G.T., Subramaniam, B. and Busch, D.H. (2003) *Industrial & Engineering Chemistry Research*, **42**, 6505.
10 Andreatta, A.E., Florusse, L.J., Bottini, S.B. and Peters, C.J. (2007) *Journal of Supercritical Fluids*, **42**, 60.
11 Brennecke, J.F. and Eckert, C.A. (1989) *AICHE Journal*, **35**, 1409.
12 Sander, W. (1911) University of Göttingen, Dissertation, Physical Chemistry Institute.
13 Sander, W. (1911–1912) *Z. Phys. Chem., Stoechiom. Verwandtschaftsl.*, **78**, 513.
14 Christov, M. and Dohrn, R. (2002) *Fluid Phase Equilibria*, **202**, 153 and references therein.
15 Leu, A.D., Chung, S.Y.K. and Robinson, D.B. (1991) *Journal of Chemical Thermodynamics*, **23**, 979.
16 Mueller, S.G., Werber, J.R., Al-Dahhan, M.H. and Dudukovic, M.P. (2007) *Industrial & Engineering Chemistry Research*, **46**, 4330.
17 Elvassore, N., Bertucco, A. and Di Noto, V. (2002) *Journal of Chemical and Engineering Data*, **47**, 223.
18 Zevnik, L. and Levec, J. (2007) *Journal of Supercritical Fluids*, **41**, 158.
19 Hutchenson K.W., Scurto A.M. and Subramaniam B. (eds) (2009) *Gas-expanded Liquids and Near-critical Media*, American Chemical Society, Oxford University Press.
20 Kordikowski, A., Schenk, A.P., Van Nielen, R.M. and Peters, C.J. (1995) *Journal of Supercritical Fluids*, **8**, 205.
21 Wandeler, R., Kunzle, N., Schneider, M.S., Mallat, T. and Baiker, A. (2001) *Journal of Catalysis*, **200**, 377.
22 Lee, H.-J., Ghanta, M., Busch, D.H. and Subramaniam, B. (2010) *Chemical Engineering Science*, **65**, 128.
23 van den Hark, S. and Harrod, M. (2001) *Applied Catalysis A-General*, **210**, 207.
24 Thomas, C.A., Bonilla, R.J., Huang, Y. and Jessop, P.G. (2001) *Canadian Journal of Chemistry*, **79**, 719.
25 Christou, G., Young, C.L. and Svejda, P. (1991) *Fluid Phase Equilibria*, **67**, 45.
26 Villard, P. (1898) *Chemistry News*, **78**, 297.
27 Krukonis, V.J., McHugh, M.A. and Seckner, A.J. (1984) *The Journal of Physical Chemistry*, **88**, 2687.
28 Aki, S.N.V.K., Mellein, B.R., Saurer, E.M. and Brennecke, J.F. (2004) *The Journal of Physical Chemistry B*, **108**, 20355.

29 Simon, R. and Graue, D.J. (1965) *Journal of Petroleum Technology*, **17**, 102.

30 Guadagno, T. and Kazarian, S.G. (2004) *The Journal of Physical Chemistry. B*, **108**, 13995.

31 Diamond, L.W. and Akinfiev, N.N. (2003) *Fluid Phase Equilibria*, **208**, 265.

32 Kumelan, J., Tuma, D. and Maurer, G. (2009) *Fluid Phase Equilibria*, **275**, 132.

33 Francis, A.W. (1954) *The Journal of Physical Chemistry*, **58**, 1099.

34 Lazzaroni, M.J., Bush, D., Brown, J.S. and Eckert, C.A. (2005) *Journal of Chemical and Engineering Data*, **50**, 60.

35 Zuniga-Moreno, A., Galicia-Luna, L.A. and Sandler, S.I. (2007) *Journal of Chemical and Engineering Data*, **52**, 1960.

36 Kho, Y.W., Conrad, D.C. and Knutson, B.L. (2003) *Fluid Phase Equilibria*, **206**, 179.

37 Sih, R., Dehghani, F. and Foster, N.R. (2007) *Journal of Supercritical Fluids*, **41**, 148.

38 Sih, R., Armenti, M., Mammucari, R., Dehghani, F. and Foster, N.R. (2008) *Journal of Supercritical Fluids*, **43**, 460.

39 Sih, R. and Foster, N.R. (2008) *Journal of Supercritical Fluids*, **47**, 233.

40 Poling, B.E., Prausnitz, J.M. and O'Connell, J.P. (2000) *The Properties of Gases and Liquids*, 5th edn, McGraw-Hill Professional, New York.

41 van der Gulik, P.S. (1997) *Physica A (Amsterdam)*, **238**, 81.

42 Suckert, J.J. (1895) US Patent 541 462.

43 Doherty, H.L. (1926) *Oil & Gas Journal*, **24**, 90.

44 Dow, D.B. and Calkin, L.P. (1926) Solubility and Effects of Natural Gas and Air in Crude Oils, Reports of Investigations, National Bureau of Mines, Washington, DC.

45 Beecher, C.E. and Parkhurst, I.P. (1926) *Trans. Am. Inst. Min. Met. Eng.*, No. 1608-G.

46 Nalawade, S.P., Picchioni, F. and Janssen, L.P.B.M. (2006) *Progress in Polymer Science*, **31**, 19.

47 Kukova, E., Petermann, M. and Weidner, E. (2004) *Chemie Ingenieur Technik*, **76**, 280.

48 Liu, Z., Wu, W., Han, B., Dong, Z., Zhao, G., Wang, J., Jiang, T. and Yang, G. (2003) *Chemistry - A European Journal*, **9**, 3897.

49 Lu, J., Liotta, C.L. and Eckert, C.A. (2003) *The Journal of Physical Chemistry A*, **107**, 3995.

50 Prins, A. (1915) *Proc. K. Ned. Akad. Wet.*, **17**, 1095.

51 Cleve, E., Bach, E. and Schollmeyer, E. (1998) *Angewandte Makromolekulare Chemie*, **256**, 39.

52 Shieh, Y.-T. and Yang, H.-S. (2005) *Journal of Supercritical Fluids*, **33**, 183.

53 Wang, W.-C.V., Kramer, E.J. and Sachse, W.H. (1982) *Journal of Polymer Science Part B-Polymer Physics*, **20**, 1371.

54 Kazarian, S.G., Sakellarios, N. and Gordon, C.M. (2002) *Chemical Communications*, 1314.

55 Scurto, A.M. and Leitner, W. (2006) *Chemical Communications*, 3681.

56 Hammam, H. and Sivik, B. (1993) *Journal of Supercritical Fluids*, **6**, 223.

57 Wilken, M., Fischer, K. and Gmehling, J. (2001) *Chemie Ingenieur Technik*, **73**, 1300.

58 Diepen, G.A.M. and Scheffer, F.E.C. (1948) *Journal of the American Chemical Society*, **70**, 4081.

59 Chun, B.-Y. and Wilkinson, G.T. (1995) *Industrial & Engineering Chemistry Research*, **34**, 4371.

60 Dittmar, D., Fredenhagen, A., Oei, S.B. and Eggers, R. (2003) *Chemical Engineering Science*, **58**, 1223.

61 Hebach, A., Oberhof, A., Dahmen, N., Koegel, A., Ederer, H. and Dinjus, E. (2002) *Journal of Chemical and Engineering Data*, **47**, 1540.

62 Hsu, J.J.C., Nagarajan, N. and Robinson, R.L. Jr. (1985) *Journal of Chemical and Engineering Data*, **30**, 485.

63 Jaeger, P. and Eggers, R. (2009) *Chemical Engineering and Processing*, **48**, 1173.

64 Einstein, A. (1905) *Annals of Physical*, **17**, 549.

65 Wilke, C.R. and Chang, P. (1955) *AICHE Journal*, **1**, 264.

66 Sassiat, P.R., Mourier, P., Caude, M.H. and Rosset, R.H. (1987) *Analytical Chemistry*, **59**, 1164.

67 Maxey, N.B. (2006) *PhD thesis*, Georgia Institute of Technology.

68 Lin, I.-H. and Tan, C.-S. (2008) *Journal of Supercritical Fluids*, **46**, 112.

69 Morgan, D., Ferguson, L. and Scovazzo, P. (2005) *Industrial & Engineering Chemistry Research*, **44**, 4815.

70 Skrzypczak, A. and Neta, P. (2003) *The Journal of Physical Chemistry A*, **107**, 7800.

71 Eckert, C.A. and Chandler, K. (1998) *Journal of Supercritical Fluids*, **13**, 187.

72 Ochel, H. and Schneider, G.M. (1994) *Ber. Bunsen-Ges. Phys. Chem.*, **98**, 615.

73 Span, R. and Wagner, W. (2003) *International Journal of Thermophysics*, **24**, 111.

74 Lu, J., Boughner, E.C., Liotta, C.L. and Eckert, C.A. (2002) *Fluid Phase Equilibria*, **198**, 37.

75 Weingaertner, H. and Franck, E.U. (2005) *Angewandte Chemie-International Edition*, **44**, 2672.

76 Kauffman, J.F. (2001) *The Journal of Physical Chemistry A*, **105**, 3433.

77 Reichardt, C. (2003) *Solvents and Solvent Effects in Organic Chemistry*, 3rd edn, Wiley-VCH Verlag GmbH, Weinheim.

78 Taft, R.W. and Kamlet, M.J. (1976) *Journal of the American Chemical Society*, **98**, 2886.

79 Kamlet, M.J. and Taft, R.W. (1976) *Journal of the American Chemical Society*, **98**, 377.

80 Kamlet, M.J., Abboud, J.L. and Taft, R.W. (1977) *Journal of the American Chemical Society*, **99**, 6027.

81 O'Neill, M.L., Kruus, P. and Burk, R.C. (1993) *Canadian Journal of Chemistry*, **71**, 1834.

82 Schneider, G.M. (1998) *Journal of Supercritical Fluids*, **13**, 5.

83 Abbott, A.P., Hope, E.G., Minstry, R. and Stuart, A.M. (2009) *Green Chemistry*, **11**, 1530.

84 Fredlake, C.P., Muldoon, M.J., Aki, S.N.V.K., Welton, T. and Brennecke, J.F. (2004) *Physical Chemistry Chemical Physics*, **6**, 3280.

85 Baker, S.N., Baker, G.A., Kane, M.A. and Bright, F.V. (2001) *The Journal of Physical Chemistry B*, **105**, 9663.

86 Heldebrant, D.J., Witt, H., Walsh, S., Ellis, T., Rauscher, J. and Jessop, P.G. (2006) *Green Chemistry*, **8**, 807.

87 Jin, H., Subramaniam, B., Ghosh, A. and Tunge, J. (2006) *AICHE Journal*, **52**, 2575.

88 Lopez-Castillo, Z.K., Aki, S.N.V.K., Stadtherr, M.A. and Brennecke, J.F. (2006) *Industrial & Engineering Chemistry Research*, **45**, 5351.

89 Lopez-Castillo, Z.K., Aki, S.N.V.K., Stadtherr, M.A. and Brennecke, J.F. (2008) *Industrial & Engineering Chemistry Research*, **47**, 570.

90 Zevnik, L. and Levec, J. (2007) *Journal of Supercritical Fluids*, **41**, 335.

91 Hert, D.G., Anderson, J.L., Aki, S.N.V.K. and Brennecke, J.F. (2005) *Chemical Communications*, 2603.

92 Höfener, T. (2008) *PhD thesis*, RWTH Aachen University.

93 Duckett, S.B. and Sleigh, C.J. (1999) *Prog. Nucl. Magn. Reson. Spectrosc.*, **34**, 17.

94 Fogg, P.G.T. and Gerrard, W. (1991) *Solubility of Gases in Liquids: a Critical Evaluation of Gas/Liquid Systems in Theory and Practice*, John Wiley & Sons Ltd, Chichester.

95 Jessop, P.G., Heldebrant, D.J., Xiaowang, L., Eckert, C.A. and Liotta, C.L. (2005) *Nature*, **436**, 1102.

96 Bonilla, R.J., James, B.R. and Jessop, P.G. (2000) *Chemical Communications*, 941.

97 Roosen, C., Ansorge-Schumacher, M., Mang, T., Leitner, W. and Greiner, L. (2007) *Green Chemistry*, 455.

98 West, K.N., Wheeler, C., McCarney, J.P., Griffith, K.N., Bush, D., Liotta, C.L. and Eckert, C.A. (2001) *The Journal of Physical Chemistry A*, **105**, 3947.

99 Chamblee, T.S., Weikel, R.R., Nolen, S.A., Liotta, C.L. and Eckert, C.A. (2004) *Green Chemistry*, **6**, 382.

100 Song, J., Hou, M., Jiang, T., Han, B., Li, X., Liu, G. and Yang, G. (2007) *The Journal of Physical Chemistry A*, **111**, 12007.

101 Blanchard, L.A. and Brennecke, J.F. (2001) *Green Chemistry*, **3**, 17.

102 Liu, S., Wang, Y., Jiang, J. and Jin, Z. (2009) *Green Chemistry*, **11**, 1397.

103 Gao, G., Tao, Y. and Jiang, J. (2008) *Green Chemistry*, **10**, 439.

104 Ganchegui, B. and Leitner, W. (2007) *Green Chemistry*, **9**, 26.

105 Dijkstra, Z.J., Doornbos, A.R., Weyten, H., Ernsting, J.M., Elsevier, C.J. and Keurentjes, J.T.F. (2007) *Journal of Supercritical Fluids*, **41**, 109.

106 Yu, K.M.K., Curcic, I., Gabriel, J. and Tsang, S.C.E. (2008) *ChemSusChem*, **1**, 893.

107 Thayer, A. (2009) *Chemical & Engineering News*, **87**, 18.

108 Wittmann, K., Wisniewski, W., Mynott, R., Leitner, W., Kranemann, C.L., Rische, T., Eilbracht, P., Kluwer, S., Ernsting, J.M. and Elsevier, C.L. (2001) *Chemistry - A European Journal*, **7**, 4584.

109 Fu, G.C., Nguyen, S.T. and Grubbs, R.H. (1993) *Journal of the American Chemical Society*, **115**, 9856.

110 Fürstner, A., Koch, D., Langemann, K., Leitner, W. and Six, C. (1997) *Angewandte Chemie (International Edition in English)*, **36**, 2466.

111 Xie, X., Liotta, C.L. and Eckert, C.A. (2004) *Industrial & Engineering Chemistry Research*, **43**, 7907.

112 Selva, M., Tundo, P., Perosa, A. and Dall'Acqua, F. (2005) *The Journal of Organic Chemistry*, **70**, 2771.

113 Dunetz, J.R., Ciccolini, R.P., Fröling, M., Paap, S.M., Allen, A.J., Holmes, A.B., Tester, J.W. and Danheiser, R.L. (2005) *Chemical Communications*, 4465.

114 Dunetz, J.R., Ciccolini, R.P., Tester, J.W. and Danheiser, R.L. (2007) US Patent US – 2007032385.

115 Phan, L., Li, X., Heldebrant, D.J., Wang, R., Chiu, D., John, E., Huttenhower, H., Pollet, P., Eckert, C.A., Liotta, C.L. and Jessop, P.G. (2008) *Industrial & Engineering Chemistry Research*, **47**, 539.

116 Yamada, T., Lukac, P.J., Yu, T. and Weiss, R.G. (2007) *Chemistry of Materials*, **19**, 4761.

117 Yu, T., Yamada, T., Gaviola, G.C. and Weiss, R.G. (2008) *Chemistry of Materials*, **20**, 5337.

118 Heldebrant, D.J., Yonker, C.R., Jessop, P.G. and Phan, L. (2008) *Energy & Environmental Science*, **1**, 487.

119 Hart, R., Pollet, P., Hahne, D.J., John, E., Llopis-Mestre, V., Blasucci, V., Huttenhower, H., Leitner, W., Eckert, C.A. and Liotta, C.L. (2009) *Tetrahedron*, submitted.

120 Phan, L., Andreatta, J.R., Horvey, L.K., Edie, C.F., Luco, A.-L., Mirchandi, A., Darensbourg, D.J. and Jessop, P.G. (2008) *The Journal of Organic Chemistry*, **73**, 127.

121 Blasucci, V., Dilek, C., Huttenhower, H., John, E., Llopis-Mestre, V., Pollet, P., Eckert, C.A. and Liotta, C.L. (2009) *Chemical Communications*, 116.

122 Li, W., Zhang, Z., Han, B., Hu, S., Song, J., Xie, Y. and Zhou, X. (2008) *Green Chemistry*, **10**, 1142.

123 Bara, J.E., Camper, D.E., Gin, D.L. and Noble, R.D. (2009) *Accounts of Chemical Research*, doi: 10.1021/ar9001747.

124 Vinci, D., Donaldson, M., Hallett, J.P., John, E.A., Pollet, P., Thomas, C.A., Grilly, J.D., Jessop, P.G., Liotta, C.L. and Eckert, C.A. (2007) *Chemical Communications*, 1427.

125 Phan, L. and Jessop, P.G. (2009) *Green Chemistry*, **11**, 307.

126 Liu, Y., Jessop, P.G., Cunningham, M., Eckert, C.A. and Liotta, C.L. (2006) *Science*, **313**, 958.

127 Desset, S.L. and Cole-Hamilton, D.J. (2009) *Angewandte Chemie-International Edition*, **48**, 1472.

128 Rajagopalan, B. (2007) Dissertation, University of Kansas.

129 Jin, H. and Subramaniam, B. (2003) *Chemical Engineering Science*, **58**, 1897.

130 Mitchell, J.K. (1838) *Journal of the Franklin Institute-Engineering and Applied Mathematics*, **22**, 289.

131 Naamlooze Vennootschap de Bataafsche Petroleum Maatschappij (1936) French Patent 804 347.

132 Pilat, S. and Godlewicz, M. (1937) Canadian Patent 363 364.

133 Pilat, S. and Godlewicz, M. (1940) US Patent 2 188 013.

134 Gallagher, P.S., Coffey, M.P., Krukonis, V.J. and Klasutis, N. (1989) in *Supercritical Fluid Science and Technology* (eds K.P. Johnston and J.M.L. Penninger), American Chemical Society, Washington, DC, p. 334.

135 Kikic, I., Lora, M. and Bertucco, A. (1997) *Industrial & Engineering Chemistry Research*, **36**, 5507.

136 de la Fuente Badilla, J.C., Peters, C. and de Swaan Arons, J. (2000) *Journal of Supercritical Fluids*, **17**, 13.

137 Striolo, A., Elvassore, N., Parton, T. and Bertucco, A. (2003) *AICHE Journal*, **49**, 2671.

138 Eckert, C., Liotta, C., Ragauskas, A., Hallett, J., Kitchens, C., Hill, E. and Draucker, L. (2007) *Green Chemistry*, **9**, 545.

139 Adrian, T., Freitag, J. and Maurer, G. (2000) *Biotechnology & Bioengineering*, **69**, 559.

140 Wei, M., Musie, G.T., Busch, D.H. and Subramaniam, B. (2002) *Journal of the American Chemical Society*, **124**, 2513.

141 Kroon, M.C., Spronsen, J.v., Peters, C.J., Sheldon, R.A. and Witkamp, G.-J. (2006) *Green Chemistry*, **8**, 246.

142 Baker, L.C.W. and Anderson, T.F. (1957) *Journal of the American Chemical Society*, **79**, 2071.

143 Adrian, T., Wendland, M., Hasse, H. and Maurer, G. (1998) *Journal of Supercritical Fluids*, **12**, 185.

144 Mellein, B.R. and Brennecke, J.F. (2007) *The Journal of Physical Chemistry. B*, **111**, 4837.

145 Lu, J., Lazzaroni, M.J., Hallett, J.P., Bommarius, A.S., Liotta, C.L. and Eckert, C.A. (2004) *Industrial & Engineering Chemistry Research*, **43**, 1586.

146 Hallett, J.P., Ford, J.W., Jones, R.S., Pollet, P., Thomas, C.A., Liotta, C.L. and Eckert, C.A. (2008) *Industrial & Engineering Chemistry Research*, **47**, 2585.

147 Abbott, A.P., Hope, E.G., Mistry, R. and Stuart, A.M. (2009) *Green Chemistry*, **11**, 1536.

148 Scurto, A.M., Aki, S. and Brennecke, J.F. (2002) *Journal of the American Chemical Society*, **124**, 10276.

149 Najdanovic-Visak, V., Serbanovic, A., Esperança, J.M.S.S., Guedes, H.J.R., Rebelo, L.P.N. and da Ponte, M.N. (2003) *Chem. Phys. Chem.*, **4**, 520.

150 Scurto, A.M., Aki, S.N.V.K. and Brennecke, J.F. (2003) *Chemical Communications*, 572.

151 Zhang, Z., Wu, W., Han, B., Jiang, T., Wang, B. and Liu, Z. (2005) *The Journal of Physical Chemistry B*, **109**, 16176.

152 West, K.N., Hallett, J.P., Jones, R.S., Bush, D., Liotta, C.L. and Eckert, C.A. (2004) *Industrial & Engineering Chemistry Research*, **43**, 4827.

153 Horvath, I.T., Kiss, G., Cook, R.A., Bond, J.E., Stevens, P.A., Rábai, J. and Mozeleski, E.J. (1998) *Journal of the American Chemical Society*, **120**, 3133.

154 Jessop, P.G., Olmstead, M.M., Ablan, C.D., Grabenauer, M., Sheppard, D., Eckert, C.A. and Liotta, C.L. (2002) *Inorganic Chemistry*, **41**, 3463.

155 Ablan, C.D., Hallett, J.P., West, K.N., Jones, R.S., Eckert, C.A., Liotta, C.L. and Jessop, P.G. (2003) *Chemical Communications*, 2972.

156 Devetta, L., Giovanzana, A., Canu, P., Bertucco, A. and Minder, B.J. (1999) *Catalysis Today*, **48**, 337.

157 Bertucco, A., Canu, P., Devetta, L. and Zwahlen, A.G. (1997) *Industrial & Engineering Chemistry Research*, **36**, 2626.

158 Chouchi, D., Gourgouillon, D., Courel, M., Vital, J. and da Ponte, M.N. (2001) *Industrial & Engineering Chemistry Research*, **40**, 2551.

159 Chan, J.C. and Tan, C.S. (2006) *Energy Fuels*, **20**, 771.

160 Chen, Y.-C. and Tan, C.-S. (2007) *Journal of Supercritical Fluids*, **41**, 272.

161 Chatterjee, M., Ikushima, Y. and Zhao, F.Y. (2003) *New Journal of Chemistry*, **27**, 510.

162 Licence, P., Ke, J., Sokolova, M., Ross, S.K. and Poliakoff, M. (2003) *Green Chemistry*, 5, 99.

163 Xu, D., Carbonell, R.G., Roberts, G.W. and Kiserow, D.J. (2005) *Journal of Supercritical Fluids*, **34**, 1.

164 Xu, D., Carbonell, R.G., Kiserow, D.J. and Roberts, G.W. (2005) *Industrial & Engineering Chemistry Research*, **44**, 6164.

165 Bogel-Lukasik, E., Fonseca, I., Bogel-Lukasik, R., Tarasenko, Y.A., Nunes da Ponte, M., Paiva, A. and Brunner, G. (2007) *Green Chemistry*, **9**, 427.

166 Bogel-Lukasik, E., Bogel-Lukasik, R., Kriaa, K., Fonseca, I., Tarasenko, Y. and Nunes da Ponte, M. (2008) *Journal of Supercritical Fluids*, **45**, 225.

167 Wandeler, R., Kunzle, N., Schneider, M.S., Mallat, T. and Baiker, A. (2001) *Chemical Communications*, 673.

168 Zhao, F., Fujita, S.-i., Sun, J., Ikushima, Y. and Arai, M. (2004) *Chemical Communications*, 2326.

169 Fujita, S.-I., Akihara, S., Zhao, F., Liu, R., Hasegawa, M. and Arai, M. (2005) *Journal of Catalysis*, **236**, 101.

170 Bhanage, B.M., Ikushima, Y., Shirai, M. and Arai, M. (1999) *Chemical Communications*, 1277.

171 Jessop, P.G., DeHaai, S., Wynne, D.C. and Nakawatase, D. (2000) *Chemical Communications*, 693.

172 Heldebrant, D.J. and Jessop, P.G. (2003) *Journal of the American Chemical Society*, **125**, 5600.

173 Cimpeanu, V., Kocevar, M., Parvulescu, V.I. and Leitner, W. (2009) *Angewandte Chemie-International Edition*, **48**, 1085.

174 Combes, G., Coen, E., Dehghani, F. and Foster, N. (2005) *Journal of Supercritical Fluids*, **36**, 127.

175 Jessop, P.G., Stanley, R., Brown, R.A., Eckert, C.A., Liotta, C.L., Ngo, T.T. and Pollet, P. (2003) *Green Chemistry*, **5**, 123.

176 Solinas, M., Pfaltz, A., Cozzi, P.G. and Leitner, W. (2004) *Journal of the American Chemical Society*, **126**, 16142.

177 Jin, H. and Subramaniam, B. (2004) *Chemical Engineering Science*, **59**, 4887.

178 Fang, J., Jin, H., Ruddy, T., Pennybaker, K., Fahey, D. and Subramaniam, B. (2007) *Industrial & Engineering Chemistry Research*, **46**, 8687.

179 Kainz, S. and Leitner, W. (1998) *Catalysis Letters*, **55**, 223.

180 Francio, G., Wittmann, K. and Leitner, W. (2001) *Journal of Organometallic Chemistry*, **621**, 130.

181 Ke, J., Han, B.X., George, M.W., Yan, H.K. and Poliakoff, M. (2001) *Journal of the American Chemical Society*, **123**, 3661.

182 Hou, Z., Han, B., Liu, Z., Jiang, T. and Yang, G. (2002) *Green Chemistry*, **4**, 467.

183 Ke, J., George, M.W., Poliakoff, M., Han, B.X. and Yan, H.K. (2002) *The Journal of Physical Chemistry B*, **106**, 4496.

184 Hintermair, U., Gong, Z., Serbanovic, A., Muldoon, M.J., Santini, C.C. and Cole-Hamilton, D.J. (2010) *Dalton Transactions*, Submitted.

185 Hintermair, U., Gong, X., Muldoon, M.J., Santini, C.C. and Cole-Hamilton, D.J. (2009) *Green Chemistry*, submitted.

186 Hemminger, O., Marteel, A., Mason, M.R., Davies, J.A., Tadd, A.R. and Abraham, M.A. (2002) *Green Chemistry*, **4**, 507.

187 Stobrawe, A., Makarczyk, P., Maillet, C., Muller, J.-L. and Leitner, W. (2008) *Angewandte Chemie-International Edition*, **47**, 6674.

188 del Moral, D., Banet Osuna, A.M., Cordoba, A., Moreto, J.M., Veciana, J., Ricart, S. and Ventosa, N. (2009) *Chemical Communications*, 4723.

189 Seki, T. and Baiker, A. (2009) *Chemical Reviews*, **109**, 2409.

190 Wu, X.-W., Oshima, Y. and Koda, S. (1997) *Chemistry Letters*, 1045.

191 Loeker, F. and Leitner, W. (2000) *Chemistry - A European Journal*, **6**, 2011.

192 Srinivas, P. and Mukhopadhyay, M. (1997) *Industrial & Engineering Chemistry Research*, **33**, 3118.

193 Yoo, J.S., Jhung, S.H., Lee, K.H. and Park, Y.S. (2002) *Applied Catalysis A-General*, **223**, 239.

194 Burri, D.R., Jun, K.-W., Yoo, J.S., Lee, C.W. and Park, S.-E. (2002) *Catalysis Letters*, **81**, 169.

195 Baek, S.-C., Roh, H.-S., Chavan, S.A., Choi, M.-H., Jun, K.-W., Park, S.-E., Yoo, J.S. and Kim, K.-J. (2003) *Applied Catalysis A-General*, **244**, 19.

196 Choi, M.-H., Baek, S.-C., Chavan, S.A., Roh, H.-S., Jun, K.-W., Park, S.-E. and Yoo, J.S. (2003) *Applied Catalysis A-General*, **247**, 303.

197 Wei, M., Musie, G.T., Busch, D.H. and Subramaniam, B. (2004) *Green Chemistry*, **6**, 387.

198 Theyssen, N., Hou, Z. and Leitner, W. (2006) *Chemistry - A European Journal*, **12**, 3401.

199 Maayan, G., Ganchegui, B., Leitner, W. and Neumann, R. (2006) *Chemical Communications*, 2230.

200 Bolm, C., Palazzi, C., Francio, G. and Leitner, W. (2002) *Chemical Communications*, 1588.

201 Caravati, M., Meier, D.M., Grunwaldt, J.-D. and Baiker, A. (2006) *Journal of Catalysis*, **240**, 126.

202 Jenzer, G., Schneider, M.S., Wandeler, R., Mallat, T. and Baiker, A. (2001) *Journal of Catalysis*, **199**, 141.

203 Kimmerle, B., Grunwaldt, J.-D. and Baiker, A. (2007) *Topics in Catalysis*, **44**, 285.

204 Beckman, E.J. (2003) *Green Chemistry*, **5**, 332.

205 Hâncu, D., Green, H. and Beckman, E.J. (2002) *Industrial & Engineering Chemistry Research*, V41, 4466.

206 Karmee, S.K., Roosen, C., Kohlmann, C., Lütz, S., Greiner, L. and Leitner, W. (2009) *Green Chemistry*, **11**, 1052.

207 Nolen, S.A., Lu, J., Brown, J.S., Pollet, P., Eason, B.C., Griffith, K.N., Glaser, R., Bush, D., Lamb, D.R., Liotta, C.L., Eckert, C.A., Thiele, G.F. and Bartels, K.A. (2002) *Industrial & Engineering Chemistry Research*, **41**, 316.

208 Yao, H. and Richardson, D.E. (2000) *Journal of the American Chemical Society*, **122**, 3220.

209 Wang, X., Venkataramanan, N.S., Kawanami, H. and Ikushima, Y. (2007) *Green Chemistry*, **9**, 1352.

210 Ford, J.W., Lu, J., Liotta, C.L. and Eckert, C.A. (2008) *Industrial & Engineering Chemistry Research*, **47**, 632.

211 Ford, J.W., Janakat, M.E., Lu, J., Liotta, C.L. and Eckert, C.A. (2008) *The Journal of Organic Chemistry*, **73**, 3364.

212 Song, J., Hou, M., Liu, G., Zhang, J., Han, B. and Yang, G. (2009) *The Journal of Physical Chemistry B*, **113**, 2810.

213 Licence, P., Gray, W.K., Sokolova, M. and Poliakoff, M. (2005) *Journal of the American Chemical Society*, **127**, 293.

214 Buchmueller, K., Dahmen, N., Dinjus, E., Neumann, D., Powietzka, B. and Pitter, S. (2003) *Green Chemistry*, **5**, 218.

215 Chanfreau, S., Cognet, P., Camy, S. and Condoret, J.S. (2008) *Journal of Supercritical Fluids*, **46**, 156.

216 Zwolak, G., Jayasinghe, N.S. and Lucien, F.P. (2006) *Journal of Supercritical Fluids*, **38**, 420.

217 Forster, D.J., Heuts, J.P.A., Lucien, F.P. and Davis, T.P. (1999) *Macromolecules*, **32**, 5514.

218 Rosen, C., Müller, P., Greiner, L. (2008) *Applied Microbiologic Chemistry*, **81**, 607.

219 Lubary, M., ter Horst, J.H., Hofland, G.W. and Jansens, P.J. (2009) *Journal of Agricultural and Food Chemistry*, **57**, 116.

220 Bestian, H., Friedrich, H.-J. and Horn, O. (1972) in *Chemische Technologie, Band 4, Organische Technologie II* (eds K. Winnacker and L. Küchler), Carl Hanser Verlag, Munich, p. 1.

221 Asinger, F. (1968) *Mono-olefins: Chemistry and Technology*, English edn, Pergamon Press, Oxford.

222 Logsdon, J.E. and Loke, R.A. (2002) in *Kirk–Othmer Encyclopedia of Chemical Technology*, John Wiley & Sons Inc., New York.

223 Neier, W. and Woellner, J. (1972) *Hydrocarbon Processing*, **51**, 113.

224 Izumi, Y., Kawasaki, Y. and Tani, M. (1973) US Patent 3 758 615.

225 Onoue, Y. and Izumi, Y. (1974) *Chemical Economy and Engineering Review*, **6**, 47.

226 Wagner, F.S., Jr. (2002) Acetic Acid in *Kirk–Othmer Encyclopedia of Chemical Technology*, John Wiley & Sons Inc., New York.

227 Hohenschutz, H., von Kutepow, N. and Himmele, W. (1966) *Hydrocarbon Processing*, **45**, 141.

228 Falbe, J. (1970) *Carbon Monoxide in Organic Synthesis*, Springer-Verlag, New York.

229 Paulik, F.E. and Roth, J.F. (1968) *Chemical Communications*, 1578a.

230 Jones, J.H. (2000) *Platinum Metals Review*, **44**, 94.

231 Hâncu, D. and Beckman, E.J. (2001) *Green Chemistry*, **3**, 80.

232 McCarthy, M., Stemmer, H. and Leitner, W. (2002) *Green Chemistry*, **4**, 501.

233 Ihmels, E.C., Fischer, K. and Gmehling, J. (2005) *Fluid Phase Equilibria*, **228–229** 155.

234 Muto, T. (2003) *Shokubai*, **45**, 591.

235 Yamada, T. and Muto, T. (1991) *Sekiyu Gakkaishi*, **34**, 201.

236 Yamada, T., Muto, T. and Yamaguchi, K. (1987) presented at *AIChE 1987 Summer National Meeting*.

237 Mahajani, S.M., Shanna, M.M. and Sridhar, T. (2002) *Chemical Engineering Science*, **57**, 4877.

238 Bösmann, A., Franciò, G., Janssen, E., Solinas, M., Leitner, W. and Wasserscheid, P. (2001) *Angewandte Chemie-International Edition*, **40**, 2697.

239 Brown, J.S., Lesutis, H.P., Lamb, D.R., Bush, D., Chandler, K., West, B.L., Liotta, C.L., Eckert, C.A., Schiraldi, D. and Hurley, J.S. (1999) *Industrial & Engineering Chemistry Research*, **38**, 3622.

240 Daugulis, A.J. (2001) *Trends in Biotechnology*, **19**, 457.

241 Baumann, M.D., Daugulis, A.J. and Jessop, P.G. (2005) *Applied Microbiology and Biotechnology*, **67**, 131.

242 Beckman, E.J. and Smith, R.D. (1990) *The Journal of Physical Chemistry*, **94**, 345.

243 Beckman, E.J. and Smith, R.D. (1990) *Journal of Supercritical Fluids*, **3**, 205.

244 Adamsky, F.A. and Beckman, E.J. (1994) *Macromolecules*, **27**, 312.

245 Romack, T.J., Combes, J.R. and Desimone, J.M. (1995) *Macromolecules*, **28**, 1724.

246 Butler, R., Davies, C.M., Hopkinson, I. and Cooper, A.I. (2002) *Polym. Prepr.*, **43**, 744.

247 Lee, J.-Y., Tan, B. and Cooper, A.I. (2007) *Macromolecules*, **40**, 1955.

248 Tan, B., Lee, J.-Y. and Cooper, A.I. (2007) *Macromolecules*, **40**, 1945.

249 Palocci, C., Barbetta, A., La Grotta, A. and Dentini, M. (2007) *Langmuir*, **23**, 8243.

250 Bing, Z., Lee, J.Y., Choi, S.W. and Kim, J.H. (2007) *European Polymer Journal*, **43**, 4814.

251 Kroon, M.C., Shariati, A., Costantini, M., Spronsen, J.v., Witkamp, G.-J., Sheldon, R.A. and Peters, C.J. (2005) *Journal of Chemical and Engineering Data*, **50**, 173.

252 Daneshvar, M., Kim, S. and Gulari, E. (1990) *The Journal of Physical Chemistry*, **94**, 2124.

253 King, M.B., Mubarak, A., Kim, J.D. and Bott, T.R. (1992) *Journal of Supercritical Fluids*, **5**, 296.

254 Carmichael, A.J. and Seddon, K.R. (2000) *Journal of Physical Organic Chemistry*, **13**, 591.

255 Ogihara, W., Aoyama, T. and Ohno, H. (2004) *Chemistry Letters*, **33**, 1414.

256 Chen, H., Kwait, D.C., Gönen, Z.S., Weslowski, B.T., Abdallah, D.J. and Weiss, R.G. (2002) *Chemistry of Materials*, **14**, 4063.

257 Deye, J.F., Berger, T.A. and Anderson, A.G. (1990) *Analytical Chemistry*, **62**, 615.

258 Wiebe, R. and Gaddy, V.L. (1941) *Journal of the American Chemical Society*, **63**, 475.

259 Anthony, R.G. and McKetta, J.J. (1967) *Journal of Chemical and Engineering Data*, **12**, 17.

260 Li, C.C. and McKetta, J.J. (1963) *Journal of Chemical and Engineering Data*, **8**, 271.

261 Scharlin, P. (1996) *Carbon Dioxide in Water and Aqueous Electrolyte Solutions*,

IUPAC Solubility Data Series No. 62, Oxford University Press, Oxford.

262 Harmens, A. and Sloan, E.D. (1990) *The Canadian Journal of Chemical Engineering*, **68**, 151.

263 Hayduk, W. (1986) *Propane, Butane and 2-Methylpropane*, IUPAC Solubility Data Series No. 24, Oxford University Press, Oxford.

264 Hayduk, W. (1982) *Ethane*, IUPAC Solubility Data Series No. 9, Pergamon Press, Oxford.

265 Hayduk, W. (1994) *Ethene*, IUPAC Solubility Data Series No. 57, Oxford University Press, Oxford.

266 D'Souza, R., Patrick, J.R. and Teja, A.S. (1988) *Canadian Journal of Chemistry Eng.*, **66**, 319.

267 Wiebe, R. and Gaddy, V.L. (1939) *Journal of the American Chemical Society*, **61**, 315.

268 Bamberger, A., Sieder, G. and Maurer, G. (2000) *Journal of Supercritical Fluids*, **17**, 97.

269 Takenouchi, S. and Kennedy, G.C. (1964) *The American Journal of the Medical Sciences*, **262**, 1055.

270 Mather, A.E. and Franck, E.U. (1992) *The Journal of Physical Chemistry*, **96**, 6.

271 Todheide, K. and Franck, E.U. (1963) *Zeitschrift fur Physiologische Chemie*, **37**, 387.

272 Schröder, W. (1973) *Chemie Ingenieur Technik*, **45**, 603.

273 King, A.D., Jr. and Coan, C.R. (1971) *Journal of the American Chemical Society*, **93**, 1857.

274 Jou, F.Y., Carroll, J.J., Mather, A.E. and Otto, F.D. (1992) *Zeitschrift fur Physiologische Chemie*, **117**, 225.

275 Parmelee, H.M. (1953) *Refrigeration Engineering*, **61**, 1341.

276 Jackson, K., Bowman, L.E. and Fulton, J.L. (1995) *Analytical Chemistry*, **67**, 2368.

277 Pozo, M.E. and Streett, W.B. (1983) *Fluid Phase Equilibria*, **14**, 219.

278 Pozo, M.E. and Streett, W.B. (1984) *Journal of Chemical and Engineering Data*, **29**, 324.

279 Diepen, G.A.M. and Scheffer, F.E.C. (1950) *Recueil des Travaux Chimiques des Pays-Bas Belgica*, **69**, 604.

280 Bradbury, E.J., McNulty, D., Savage, R.L. and McSweeney, E.E. (1952) *Industrial & Engineering Chemistry*, **44**, 211.

281 Sanchez, M. and Lentz, H. (1973) *High Temperatures-High Pressures*, **5**, 689.

282 Mohammadi, A.H., Chapoy, A., Tohidi, B. and Richon, D. (2004) *Industrial & Engineering Chemistry Research*, **43**, 5418.

283 Danneil, A., Todheide, K. and Franck, E.U. (1967) *Chemie Ingenieur Technik*, **39**, 816.

284 Kobayashi, R. and Katz, D.L. (1953) *Journal of Industrial and Engineering Chemistry*, **45**.

285 Wagner, K.-D., Brudi, K., Dahmen, N. and Schmieder, H. (1999) *Journal of Supercritical Fluids*, **15**, 109.

286 Curren, M.S.S. and Burk, R.C. (2000) *Journal of Chemical and Engineering Data*, **45**, 746.

287 Brudi, K., Dahmen, N. and Schmieder, H. (1996) *Journal of Supercritical Fluids*, **9**, 146.

288 Cornils, B. and Herrmann, W.A. (2004) *Aqueous Phase Organometallic Catalysis*, 2nd edn, Wiley-VCH Verlag GmbH, Weinheim.

289 Pinault, N. and Bruce, D.W. (2003) *Coordination Chemical Reviews*, **241**, 1.

290 Bhatt, A.I., Bond, A.M., MacFarlane, D.R., Zhang, J., Scott, J.L., Strauss, C.R., Iotov, P.I. and Kalcheva, S.V. (2006) *Green Chemistry*, **8**, 161.

291 Hess, U., Dunkel, S. and Muller, B. (1993) *Pharmazie*, **48**, 591.

292 Jessop, P.G., Hsiao, Y., Ikariya, T. and Noyori, R. (1996) *Journal of the American Chemical Society*, **118**, 344.

293 Schroth, W., Schaedler, H.D. and Joerg, A. (1989) *Zeitschrift Fur Chemie*, **29**, 56.

294 Reichardt, C. (2005) *Green Chemistry*, **7**, 339.

295 Jastorff, B., Störmann, R., Ranke, J., Mölter, K., Stock, F., Oberheitmann, B., Hoffmann, W., Hoffmann, J., Nüchter,

M., Ondruschka, B. and Filser, J. (2003) *Green Chemistry*, **5**, 136.

296 Bernot, R.J., Brueseke, M.A., Evans-White, M.A. and Lamberti, G.A. (2005) *Environmental Toxicology and Chemistry/Setac*, **24**, 87.

297 Stolte, S., Matzke, M., Arning, J., Böschen, A., Pitner, W.-R., Welz-Biermann, U., Jastorff, B. and Ranke, J. (2007) *Green Chemistry*, **9**, 1170.

298 Zhang, Y., Bakshi, B.R. and Demessie, E.S. (2008) *Environmental Science & Technology*, **42**, 1724.

299 Zhang, Z., Wu, W., Liu, Z., Han, B., Gao, H. and Jiang, T. (2004) *Physical Chemistry Chemical Physics*, **6**, 2352.

300 Aki, S.N.V.K., Scurto, A.M. and Brennecke, J.F. (2006) *Industrial & Engineering Chemistry Research*, **45**, 5574.

301 Jessop, P.G., Hsiao, Y., Ikariya, T. and Noyori, R. (1994) *Journal of the American Chemical Society*, **116**, 8851.

302 Blanchard, L.A., Hancu, D., Beckman, E.J. and Brennecke, J.F. (1999) *Nature*, **399**, 28.

303 Blanchard, L.A., Gu, Z. and Brennecke, J.F. (2001) *The Journal of Physical Chemistry B*, **105**, 2437.

304 Wu, W., Zhang, J., Han, B., Chen, J., Liu, Z., Jiang, T., He, J. and Li, W. (2003) *Chemical Communications*, 1412.

305 Wu, W., Li, W., Han, B., Jiang, T., Shen, D., Zhang, Z., Sun, D. and Wang, B. (2004) *Journal of Chemical and Engineering Data*, **49**, 1597.

306 Hutchings, J.W., Fuller, K.L., Heitz, M.P. and Hoffmann, M.M. (2005) *Green Chemistry*, **7**, 475.

307 Shariati, A., Gutkowski, K. and Peters, C.J. (2005) *AICHE Journal*, **51**, 1532.

308 Shariati, A. and Peters, C.J. (2003) *Journal of Supercritical Fluids*, **25**, 109.

309 Shariati, A., Raeissi, S. and Peters, C.J. (2007) in *Developments and Applications in Solubility* (ed. T.M. Letcher), Royal Society of Chemistry, Cambridge, p. 131.

310 Huang, X., Margulis, C.J., Li, Y. and Berne, B.J. (2005) *Journal of the American Chemical Society*, **127**, 17842.

311 Zhang, S., Chen, Y., Ren, R.X.-F., Zhang, Y., Zhang, J. and Zhang, X. (2005) *Journal of Chemical and Engineering Data*, **50**, 230.

312 Cadena, C., Anthony, J.L., Shah, J.K., Morrow, T.I., Brennecke, J.F. and Maginn, E.J. (2004) *Journal of the American Chemical Society*, **126**, 5300.

313 Constantini, M., Toussaint, V.A., Shariati, A., Peters, C.J. and Kikic, I. (2005) *Journal of Chemical and Engineering Data*, **50**, 52.

314 Gutkowski, K.I., Shariati, A. and Peters, C.J. (2006) *Journal of Supercritical Fluids*, **39**, 187.

315 Shariati, A. and Peters, C.J. (2004) *Journal of Supercritical Fluids*, **29**, 43.

316 Shariati, A. and Peters, C.J. (2005) *Journal of Supercritical Fluids*, **34**, 171.

317 Kamps, A.P.-S., Tuma, D., Xia, J. and Maurer, G. (2003) *Journal of Chemical and Engineering Data*, **48**, 746.

318 Kumelan, J., Pérez-Salado Kamps, Á., Tuma, D. and Maurer, G. (2006) *Journal of Chemical and Engineering Data*, **51**, 1802.

319 Shariati, A. and Peters, C.J. (2004) *Journal of Supercritical Fluids*, **30**, 139.

320 Kumelan, J., Pérez-Salado Kamps, Á., Tuma, D. and Maurer, G. (2006) *Journal of Chemical Thermodynamics*, **38**, 1396.

321 Blanchard, L.A. and Brennecke, J.F. (2001) *Industrial & Engineering Chemistry Research*, **40**, 287.

322 Blanchard, L.A. and Brennecke, J.F. (2001) *Industrial & Engineering Chemistry Research*, **40**, 2550.

323 Brasse, C.C., Englert, U., Salzer, A., Waffenschmidt, H. and Wasserscheid, P. (2000) *Organometallics*, **19**, 3818.

324 Chauvin, Y., Mussmann, L. and Olivier, H. (1995) *Angewandte Chemie (International Edition in English)*, **34**, 2698.

325 Sellin, M.F., Webb, P.B. and Cole-Hamilton, D.J. (2001) *Chemical Communications*, 781.

326 Mehnert, C.P., Cook, R.A., Dispenziere, N.C. and Afeworki, M. (2002) *Journal of the American Chemical Society*, **124**, 12932.

327 Parshall, G.W. (1972) *Journal of the American Chemical Society*, **94**, 8716.

328 Liu, F.C., Abrams, M.B., Baker, R.T. and Tumas, W. (2001) *Chemical Communications*, 433.

329 Keskin, S., Kayrak-Talay, D., Akman, U. and Hortaçsu, Ö. (2007) *Journal of Supercritical Fluids*, **43**, 150.

330 Webb, P.B., Sellin, M.F., Kunene, T.E., Williamson, S., Slawin, A.M.Z. and Cole-Hamilton, D.J. (2003) *Journal of the American Chemical Society*, **125**, 15577.

331 Webb, P.B., Kunene, T.E. and Cole-Hamilton, D.J. (2005) *Green Chemistry*, **7**, 373.

332 Wang, J.-Q., Kong, D.-L., Chen, J.-Y., Cai, F. and He, L.-N. (2006) *Journal of Molecular Catalysis A: Chemical*, **249**, 143.

333 Ciriminna, R., Hesemann, P., Moreau, J.J.E., Carraro, M., Campestrini, S. and Pagliaro, M. (2006) *Chemistry - A European Journal*, **12**, 5220.

334 *Code of Federal Regulations*, Title 21, Vol. 3, CITE 21CFR172.820, FDA, Washington, DC. 2001.

335 Noll, W. (1968) *Chemistry and Technology of the Silicones*, Academic Press, Orlando, FL.

336 Momper, B., Wagner, T., Maschke, U., Ballauff, M. and Fischer, E.W. (1990) *Polymer Communications*, **31**, 186.

337 Spange, S., Vilsmeier, E., Fischer, K., Reuter, A., Prause, S., Zimmermann, Y. and Schmidt, C. (2000) *Macromolecular Rapid Communications*, **21**, 643.

338 Ohno, H. and Kawanabe, H. (1996) *Polymers for Advanced Technologies*, **7**, 754.

339 Bailey, F.E. and Koleske, J.V. (1976) *Poly (Ethylene Oxide)*, Academic Press, New York.

340 Rudolph, J., Hermanns, N. and Bolm, C. (2004) *The Journal of Organic Chemistry*, **69**, 3997.

341 Wiesmet, V., Weidner, E., Behme, S., Sadowski, G. and Arlt, W. (2000) *Journal of Supercritical Fluids*, **17**, 1.

342 Anthony, J.L., Maginn, E.J. and Brennecke, J.F. (2002) *The Journal of Physical Chemistry B*, **106**, 7315.

343 Cannon, P., St. Pierre, L.E. and Miller, A.A. (1960) *Journal of Chemical and Engineering Data*, **5**, 236.

344 Garg, A., Gulari, E. and Manke, C.W. (1994) *Macromolecules*, **27**, 5643.

345 Weidner, E., Wiesmet, V., Knez, Z. and Skerget, M. (1997) *Journal of Supercritical Fluids*, **10**, 139.

346 Weidner, E., Wiesmet, V. and Maurer, G. (eds) (2004) in *Thermodynamic Properties of Complex Fluid Mixtures*, Wiley-VCH Verlag GmbG, Weinheim, p. 511.

347 Kirby C.F. and McHugh, M.A. (1999) *Chemical Reviews*, **99**, 565.

348 Gourgouillon, D., Avelino, H., Fareleira, J. and da Ponte, M.N. (1998) *Journal of Supercritical Fluids*, **13**, 177.

349 Flichy, N.M.B., Lawrence, C.J. and Kazarian, S.G. (2003) *Industrial & Engineering Chemistry Research*, **42**, 6310.

350 Gerhardt, L.J., Manke, C.W. and Gulari, E. (1997) *Journal of Polymer Science Part B-Polymer Physics*, **35**, 523.

351 Dimitrov, K., Boyadzhiev, L. and Tufeu, R. (1999) *Macromolecular Chemistry and Physics*, **200**, 1626.

352 Lopes, J.A., Gourgouillon, D., Pereira, P.J., Ramos, A.M. and da Ponte, M.N. (2000) *Journal of Supercritical Fluids*, **16**, 261.

353 Gourgouillon, D. and da Ponte, M.N. (1999) *Physical Chemistry Chemical Physics*, **1**, 5369.

354 Drohmann, C. and Beckman, E.J. (2002) *Journal of Supercritical Fluids*, **22**, 103.

355 Dimitrov, K., Boyadzhiev, L., Tufeu, R., Cansell, F. and Barth, D. (1998) *Journal of Supercritical Fluids*, **14**, 41.

356 Yee, G.G., Fulton, J.L. and Smith, R.D. (1992) *Langmuir*, **8**, 377.

357 Parks, K.L. and Beckman, E.J. (1996) *Polymer Engineering and Science*, **36**, 2404.

358 Kilic, S., Michalik, S., Wang, Y., Johnson, J.K., Enick, R.M. and Beckman, E.J. (2007) *Macromolecules*, **40**, 1332.

359 Li, J.T., Zhang, M. and Kiran, E. (1999) *Industrial & Engineering Chemistry Research*, **38**, 4486.

360 Martin, T.M., Gupta, R.B. and Roberts, C.B. (2000) *Industrial & Engineering Chemistry Research*, **39**, 185.

361 Madsen, L.A. (2006) *Macromolecules*, **39**, 1483.

362 Bae, Y.C. and Gulari, E. (1997) *Journal of Applied Behavioral Science*, **63**, 459.

363 Gerhardt, L.J., Garg, A., Manke, C.W. and Gulari, E. (1998) *Journal of Polymer Science Part B-Polymer Physics*, **36**, 1911.

364 Royer, J.R., Gay, Y.J., Adam, M., DeSimone, J.M. and Khan, S.A. (2002) *Polymer*, **43**, 2375.

365 Daneshvar, M. and Gulari, E. (1992) *Journal of Supercritical Fluids*, **5**, 143.

366 Mishima, K., Matsuyama, K. and Nagatani, H. (1999) *Fluid Phase Equilibria*, **161**, 315.

367 Brantley, N.H., Bush, D., Kazarian, S.G. and Eckert, C.A. (1999) *The Journal of Physical Chemistry B*, **103**, 10007.

368 Kazarian, S.G., Vincent, M.F., West, B.L. and Eckert, C.A. (1998) *Journal of Supercritical Fluids*, **13**, 107.

369 Weidner, E., Steiner, R. and Knez, Z. (1996) in *High Pressure Chemical Engineering* (eds P.R. von Rohr and C. Trepp), Elsevier, Amsterdam, p. 223.

370 Reetz, M.T. and Wiesenhöfer, W. (2004) *Chemical Communications*, 2750.

371 Du, Y., Wang, J.-Q., Chen, J.-Y., Cai, F., Tian, J.-S., Kong, D.-L. and He, L.-N. (2006) *Tetrahedron Letters*, **47**, 1271.

372 Wang, J.-Q., Fei Cai, E.W. and He, L.-N. (2007) *Green Chemistry*, 882.

373 He, J., Wu, T., Jiang, T., Zhou, X., Hu, B. and Han, B. (2008) *Catalysis Communications*, **9**, 2239.

374 Dapurkar, S.E., Kawanami, H., Suzuki, T.M., Yokoyama, T. and Ikushima, Y. (2008) *Chemistry Letters*, **37**, 150.

375 Hou, Z., Theyssen, N., Brinkmann, A. and Leitner, W. (2005) *Angewandte Chemie-International Edition*, **44**, 1346.

376 Webb, P.B. and Cole-Hamilton, D.J. (2004) *Chemical Communications*, 612.

377 Frisch, A.C., Webb, P.B., Zhao, G., Muldoon, M.J., Pogorzelec, P.J. and Cole-Hamilton, D.J. (2007) *Dalton Transactions*, 5531.

378 Gale, R.W., Fulton, J.L. and Smith, R.D. (1987) *Journal of the American Chemical Society*, **109**, 920.

379 Johnston, K.P., Jacobsen, G.B., Lee, C.T., Meredith, C., Da Rocha, S.R.P., Yates, M.Z., DeGrazia, J. and Randolph, T.W. (1999) in *Chemical Synthesis Using Supercritical Fluids* (eds P.G. Jessop and W. Leitner), Wiley-VCH Verlag GmbH, Weinheim, p. 127.

380 Goetheer, E.L.V., Vorstman, M.A.G. and Keurentjes, J.T.F. (1999) *Chemical Engineering Science*, **54**, 1589.

381 Johnston, K.P., Da Rocha, S.R.P., Lee, C.T., Li, G. and Yates, M.Z. (2004) in *Supercritical Fluid Technology for Drug Product Development* (eds York P., Kompella U.B. and Shekunov B.Y.), Marcel Dekker, New York, p. 213.

382 Beckman, E. (2005) in *Supercritical Carbon Dioxide in Polymer Reaction Engineering* (eds M.F. Kemmere and T. Meyer), Wiley-VCH Verlag GmbH, Weinheim, p. 139.

383 Meredith, J.C., Sanchez, I.C., Johnston, K.P. and Pablo, J.J.d. (1998) *Journal of Chemical Physics*, **109**, 6424.

384 da Rocha, S.R.P., Dickson, J., Cho, D., Rossky, P.J. and Johnston, K.P. (2003) *Langmuir*, **19**, 3114.

385 Yazdi, A.V., Lepilleur, C., Singley, E.J., Liu, W., Adamsky, F.A., Enick, R.M. and Beckman, E.J. (1996) *Fluid Phase Equilibria*, **117**, 297.

386 Fan, X., Potluri, V.K., McLeod, M.C., Wang, Y., Liu, J., Enick, R.M., Hamilton, A.D., Roberts, C.B., Johnson, J.K. and Beckman, E.J. (2005) *Journal of the American Chemical Society*, **127**, 11754.

387 Tan, B., Lee, J.-Y. and Cooper, A.I. (2006) *Macromolecules*, **39**, 7471.

388 Hollamby, M.J., Trickett, K., Mohamed, A., Cunmmings, S., Tabor, R.F., Myakonkaya, O., Gold, S., Rogers, S., Heenan, R.K. and Eastoe, J. (2009) *Angewandte Chemie-International Edition*, **48**, 4993.

389 da Rocha, S.R.P., Psathas, P.A., Klein, E. and Johnston, K.P. (2001) *Journal of Colloid and Interface Science*, **239**, 241.

390 Lee, C.T., Jr. Psathas, P.A., Johnston, K.P., deGrazia, J. and Randolph, T.W. (1999) *Langmuir*, **15**, 6781.

391 Liu, J., Cheng, S., Zhang, J., Feng, X., Fu, X. and Han, B. (2007) *Angewandte Chemie-International Edition*, **46**, 3313.

392 Lacroix-Desmazes, P., Hesemann, P., Boutevin, B. and Moreau, J.J.E. (2005) *Polym. Prepr.*, **46**, 655.

393 Jacobson, G.B., Lee, C.T., Johnston, K.P. and Tumas, W. (1999) *Journal of the American Chemical Society*, **121**, 11902.

394 Zhu, J., Robertson, A. and Tsang, S.C. (2002) *Chemical Communications*, 2044.

395 Timko, M.T., Diffendal, J.M., Tester, J.W., Smith, K.A., Peters, W.A., Danheiser, R.L. and Steinfeld, J.I. (2003) *The Journal of Physical Chemistry A*, **107**, 5503.

396 Klier, J. (2005) *Microemulsions* in *Kirk–Othmer Encyclopedia of Chemical Technology*, Vol. 16, John Wiley & Sons Inc., p. 419.

397 Bartscherer, K.A., Renon, H. and Minier, M. (1995) *Fluid Phase Equilibria*, **107**, 93.

398 Jackson, K. and Fulton, J.L. (1998) in *Supercritical Fluid Cleaning* (eds J. McHardy and S.P. Sawan), Noyes, Westwood, NJ, p. 87.

399 Jacobson, G.B., Lee, C.T., Jr. da Rocha, S.R.P. and Johnston, K.P. (1999) *The Journal of Organic Chemistry*, **64**, 1207.

400 Johnston, K.P. and da Rocha, S.R.P. (2009) *Journal of Supercritical Fluids*, **47**, 523.

5
Synthetic Organic Chemistry in Supercritical Fluids

Christopher M. Rayner, Paul M. Rose, and Douglas C. Barnes

5.1
Introduction

Research in the field of synthetic transformations in supercritical fluids (SCFs), particularly supercritical carbon dioxide ($scCO_2$), has expanded enormously over the past decade. As more research groups have become involved, the range of transformations that can be carried out has greatly increased, and the scope and limitations are now much better understood. There are numerous previous reviews which have focused on describing the different classes of reactions that can be carried out in SCFs, most notably CO_2. Many of these are extensive, and provide an excellent introduction to the area. Topics covered include homogeneous catalysis [1], heterogeneous catalysis [2], and SCFs as solvent replacements in chemical synthesis [3, 4] and in synthetic organic chemistry [5, 6]. There are also two books published in the area, one of which describes many aspects of chemical synthesis in SCFs [7], while the other concentrates on more physical aspects [8].

The purpose of this chapter is not to give an exhaustive account of all the different kinds of reactions that can be carried out in SCFs, but more to concentrate on those that are most likely to be of interest to synthetic organic chemists. This includes some of the more subtle, but vitally important aspects of reactions in SCFs, which provide real advantages compared with reactions in conventional solvents.

There are two main motivations for studying reactions in SCFs. One of the main ones, which is of ever increasing importance, is the environmental benefit of the two most commonly used SCFs, namely water and CO_2. Such virtues have been described in detail in many previous reviews and articles, and are particularly relevant to large-scale synthetic processes and purification methods (e.g. chromatography) [9]. The other major motivation for investigating reactions in SCFs is their versatility, particularly the potential for varying the solvent properties by varying temperature, pressure, and density of the SCF. In contrast to conventional solvents, where "quantum" leaps in solvent properties are achieved by switching from one solvent to another, the gradual change in properties over a wide range of temperatures and pressures for SCFs makes it possible to have much greater control over the influence

Handbook of Green Chemistry, Volume 4: Supercritical Solvents. Edited by Walter Leitner and Philip G. Jessop

ISBN: 978-3-527-32590-0

a solvent may have on a reaction, particularly in terms of rate and selectivity. Such control can be a very powerful tool, especially for complex synthetic reactions, where numerous products can potentially be formed, but by fine tuning the SCF solvent, a single pure product can be obtained.

Hence the focus of this review is on organic chemistry, and the influence that solvent tuning and pressure effects can have on the outcome of a reaction. One problem in trying to write a review such as this is that almost every reaction in a SCF will be affected to some extent by variations of solvent. This could be something as simple as the rate of a reaction, or a more subtle complex effect, such as a variation of enantioselectivity in an asymmetric process. In some cases, the reason for the effect may be well established; in others there may simply be a tentative suggestion for the origin of the effect, with no direct evidence. We have focused on literature published from 1998, although particularly pertinent examples earlier than this are included for completeness. It is assumed that the reader is familiar with SCFs and their properties, particularly with regard to temperature, pressure, and density variations. If this is not the case, then they are encouraged to consult previous chapters in this book, or some of the more extensive excellent reviews in the area [1–3, 5–8].

5.2
Hydrogenation in Supercritical Fluids

The ability of SCFs to bring together catalysts, reagents, and high concentrations of light gases such as H_2 and CO makes them ideal for transformations such as hydrogenation, hydroformylation, and oxidation. Often these reactions proceed readily, although they can be more complex than envisaged, particularly with regard to phase behavior and its effects. The main reason for this is that the added gas (H_2, CO, or O_2), if present at reasonable concentrations, can have a significant effect on the phase behavior of the reaction mixture, and this is likely to change as the reaction proceeds, particularly in a batch (rather than continuous) process.

The hydrogenation of propene in $scCO_2$ illustrates the point. With an initial mole fraction of 0.40, the critical pressure for the reaction mixture decreased by $\sim$8 MPa from start to completion of the reaction, whereas the critical temperature increased by 7 K. Even for such a simple, well-characterized system, the behavior is complex, but the observations highlight the potential impact that changes of composition of a reaction mixture can have on phase behavior [10]. As phase behavior can have a significant impact on reaction outcomes (e.g. reagents no longer being in contact with a catalyst due to phase separation), such effects have to be considered when attempting to rationalize observations, and although this kind of reaction is particularly susceptible to such effects, they are likely to be operating to a certain extent in a wide range of processes where composition changes as the reaction proceeds. Such effects also occur in other reactions, such as hydroformylation [11], and some which do not involve light gases, such as enzymatic processes [12] and epoxidation reactions [10, 13]. Computational approaches to the determination of critical points of mixtures have also been published [14]. Note that in some cases, alternative sources of hydrogen can be used, such as formate [15].

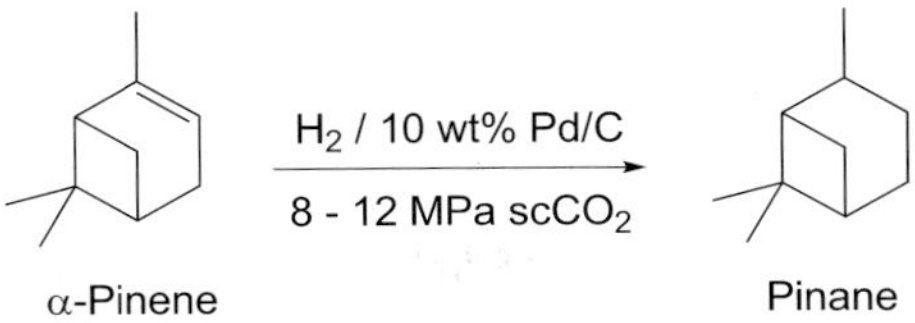

Scheme 5.1 Hydrogenation of pinene in $scCO_2$.

Although much is made of trying to achieve a homogeneous reaction mixture in $scCO_2$, this is not always beneficial. For example, in the hydrogenation of α-pinene using a simple Pd/C heterogeneous catalyst, much faster reaction rates were observed when a two-phase system was present, rather than a single homogeneous phase (Scheme 5.1). This was rationalized on the basis of the formation of an expanded liquid phase, where CO_2 (and, more importantly, hydrogen) dissolve in relatively large quantities in the liquid pinene phase in contact with the catalyst [16]. Similar effects have also been rationalized by suggesting that adsorption of pinene on the metal surface was the rate determining step. Although initially a single-phase reaction (other than the heterogeneous catalyst), at intermediate pressures a phase change occurred (from one phase to two phases) as the reaction proceeded, and this appeared to enhance the efficiency of the reaction [17].

Citral represents an interesting substrate for studies of selective hydrogenation. CO_2 pressure has a significant impact on the product distribution. With a Ru–triphenylphosphine catalyst, C=O reduction to geraniol and nerol was enhanced from 27 to 75% on increasing the CO_2 pressure from 6 to 16 MPa, with selectivity for subsequent reduction to citronellol (the predominant product at 6 MPa) decreasing from 70 to 20% (Scheme 5.2). Optimum conversion was obtained at 12 MPa,

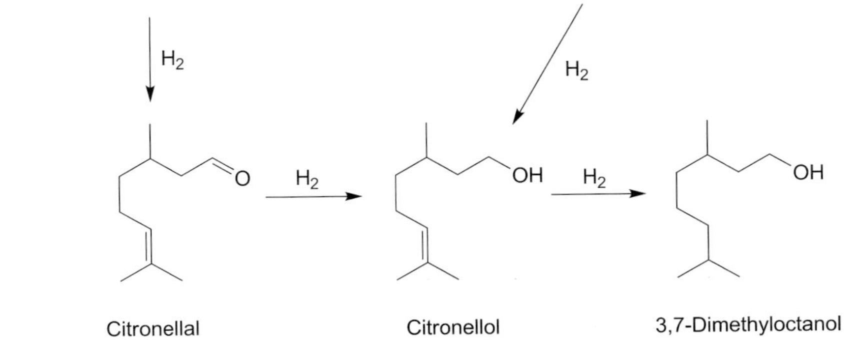

Scheme 5.2 Possible products from the hydrogenation of citral.

which through visual observation was the lowest pressure at which a single homogeneous phase existed [18]. Similar observations have been reported with other catalyst systems, with CO_2 giving greatly enhanced selectivity compared with conventional solvents [19].

A Pd/Al_2O_3-based catalyst gave high selectivity for C=C bond reduction. Reaction rates in $scCO_2$ were strongly dependent on phase behavior, but were up to two orders of magnitude faster than in conventional solvents, and could also be carried out under continuous flow conditions [20]. Using micelle-hosted Ru or Pd nanoparticles, selective reduction of the α,β-unsaturated C=C bond was observed, and this was particularly favorable at higher pressures when a homogeneous reaction was achieved [21]. At lower pressures, the fully reduced product, 3,7-dimethyloctanol, was the major product. Using Ni(II)-based catalysts, high selectivity for C=O reduction is observed, whereas for Ni(0), selective reduction of the α,β-unsaturated C=C bond is observed, and this selectivity was shown for a range of related substrates [22].

Hydrogenation of limonene also has potential for interesting selectivity effects, containing two rather different C=C bonds (Scheme 5.3). Using a heterogeneous Pt/C catalyst, at 16 MPa total pressure (4 MPa H_2) a single supercritical phase was formed, and the more accessible exocyclic double bond was preferentially hydrogenated to give 1-menthene. This intermediate rapidly disappeared in the first 5 min of reaction by undergoing rapid hydrogenation of the remaining double bond to produce a mixture of the *cis*- and *trans*-menthane isomers. At lower pressure (12.5 MPa including 4 MPa H_2), a biphasic vapor–liquid system was visible, and gave very different results. The initial hydrogenation of limonene was significantly slower than under homogeneous conditions, but after 20 min, the main product was the 1-menthene (70%), along with 20% of menthane (~1 : 1 mixture of isomers) and a small amount of unreacted limonene (10%). Analysis of phase behavior and likely hydrogen concentrations suggested that these results could be accounted for by a

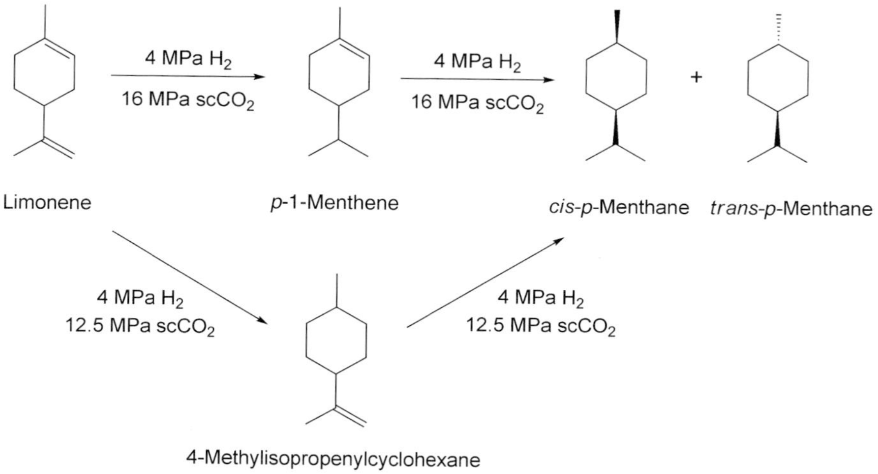

Scheme 5.3 Hydrogenation of limonene in $scCO_2$.

~1 : 1 molar ratio of hydrogen to substrate in the liquid phase (which was circulated predominantly in contact with the catalyst) of the two-phase system [23].

Another classic substrate for hydrogenation studies is isophorone. Selectivity for hydrogenation of the alkene is desirable as this product is used commercially as a solvent. Recent studies have shown that the reaction selectivity for isophorone hydrogenation was highly catalyst dependent (Scheme 5.4). Use of charcoal-supported rhodium or platinum catalysts selectively reduced the C=C bond, but subsequent reduction of the carbonyl group was also apparent. Supported palladium catalysts were less active, but cleanly reduced the C=C bond to give dihydroisophorone with high selectivity, and conversions were improved at elevated hydrogen and CO_2 pressures [24].

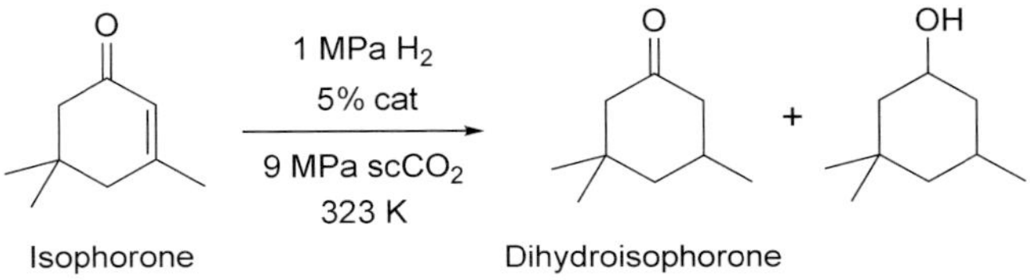

Scheme 5.4 Hydrogenation of isophorone in $scCO_2$.

Hydrogenation of isophorone has also been achieved in a continuous reactor using a supported 5%Pd/Deloxan catalyst as long as the reaction temperature was below 200 °C. A range of other substrates (aromatics, carbonyl compounds, ethers, nitroaromatics, imines, oximes, alkenes, and alkynes) were also hydrogenated under similar conditions with variable levels of selectivity, although for nitrogen-based functionality, supercritical propane was the solvent of choice to prevent problematic carbamate formation between the product amines and CO_2 [25].

The hydrogenation of ketoisophorone (4-oxoisophorone) has also been studied in $scCO_2$ alongside cosolvent combinations and correlation with phase behavior (Scheme 5.5). Reaction rates were found to be similar to those in conventional solvents, but different selectivities were observed. A wide range of CO_2–cosolvent mixtures were investigated but did not show any marked advantages. The main product was usually formed by hydrogenation of the C=C bond in $scCO_2$ and selectivity was highest at low CO_2 pressures (8.0 MPa) where the reaction was biphasic (along with the heterogeneous catalyst), although conversion was lower. In solvents such as MeOH, the dominant product was obtained by the hydrogenation

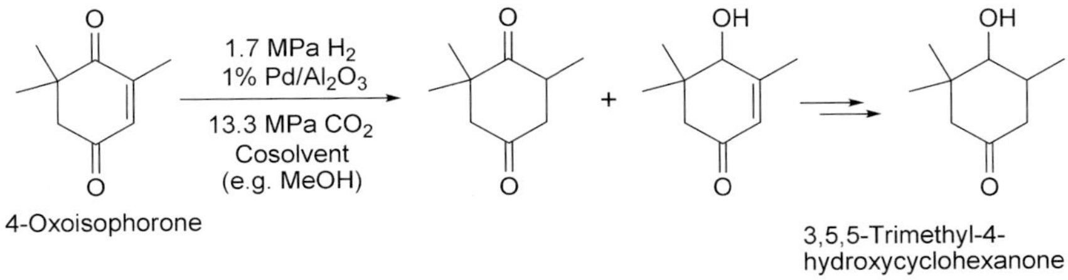

Scheme 5.5 Hydrogenation of 4-oxoisophorone in $scCO_2$.

of the more sterically encumbered C=O group alongside substantial C=C hydrogenation. Interestingly, catalyst deactivation was found to be significantly lower in $scCO_2$ compared with conventional solvents [26].

Hydrogenation of maleic anhydride over Pd/Al_2O_3 catalyst can give either γ-butyrolactone or succinic anhydride as the major products (Scheme 5.6). At relatively high temperature (473 K), high selectivity (>80%) for γ-butyrolactone was observed provided that reasonably high pressures (12 MPa CO_2 + 2.1 MPa H_2) were used. At lower temperatures and/or pressures, substantial amounts of succinic anhydride were formed, and indeed this was often the major product (e.g. 95:5 at 423 K) [27].

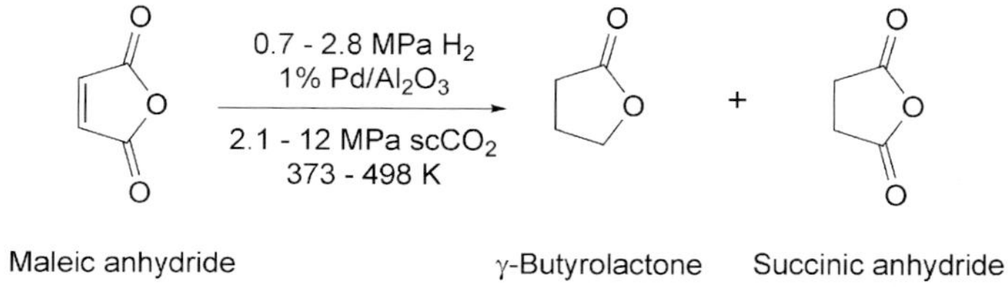

Scheme 5.6 Hydrogenation of maleic anhydride in $scCO_2$.

Numerous studies have been published on the hydrogenation of cinnamaldehyde. It is of particular interest because of the variety of products that can be formed, depending on whether the C=C or C=O bonds (or both) are hydrogenated (Scheme 5.7). With heterogeneous catalysts, the selectivity of the reaction was very dependent on the nature of the catalyst, but in some cases showed a significant enhancement in C=O reduction around 10 MPa, which was also where optimum conversion was observed [28, 29]. It may be no coincidence that optimum selectivities in other totally unrelated reactions are also often observed around this pressure (see below), as it is often where reactions reach homogeneity (CO_2 at lower pressures not being sufficiently strong to solubilize all reagents). Importantly, the selectivity observed was much higher than could be achieved in conventional solvents, although with some catalysts only small variations in selectivity were observed [30].

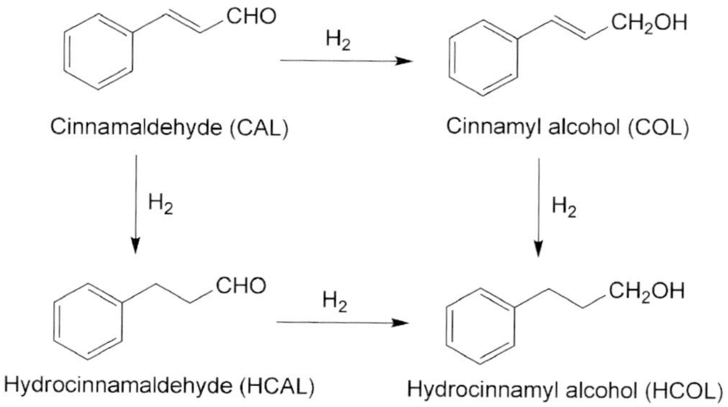

Scheme 5.7 Products from hydrogenation of cinnamaldehyde.

Using a Ru–Pt bimetallic catalyst, hydrogenation of cinnamaldehyde to cinnamyl alcohol was completely selective, which was reportedly the best selectivity ever achieved. Optimum selectivity was around 7.5 MPa, with significant amounts of alkene hydrogenation being observed at extremes of the pressures investigated (6 or 17 MPa) [29].

Interestingly, high levels of selectivity were reported in CO_2-expanded cinnamaldehyde with a homogeneous catalyst, in which the reaction mixture comprised a CO_2-rich gas phase and a relatively large cinnamaldehyde liquid phase expanded by dissolution of CO_2 (and H_2) (Scheme 5.8). Increased pressure up to 16 MPa gave enhanced conversion consistent with an increase in the H_2 concentration in the cinnamaldehyde phase, whereas enhanced selectivity was attributed to an interaction between CO_2 and cinnamaldehyde (observed spectroscopically) which increased the polarity of the C=O bond and its susceptibility to reduction [31]. With alternative homogeneous catalyst systems, poorer selectivity was observed [32]. With Pd/C catalyst, good selectivity for C=C reduction was observed for cinnamaldehyde and crotonaldehyde, although there was no significant variation with pressure [33].

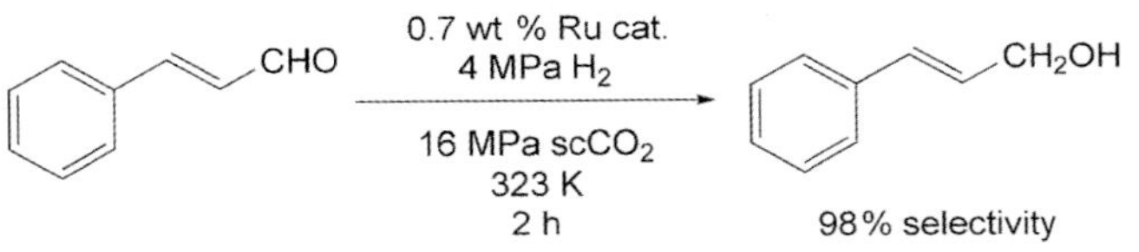

Scheme 5.8 Selective hydrogenation of cinnamaldehyde in $scCO_2$.

Furthermore, interesting selectivities were observed in the hydrogenation of levulinic acid to γ-valerolactone (GVL) (Scheme 5.9). Complete conversion to GVL was observed at 473 K using 3 equiv. of H_2 (10 MPa CO_2) in a continuous reactor system. After reduction of the ketone group, the intermediate hydroxy acid underwent rapid ring closure with loss of water to form the product. If the reaction was carried out under aqueous conditions, the required product (which is usually water soluble) could be separated by incorporating a simple liquid separator between the reactor and the back-pressure regulator. Under CO_2 pressure, the GVL formed a distinct gas-expanded liquid phase, which separated from the water layer and could be physically separated using a run-off [34].

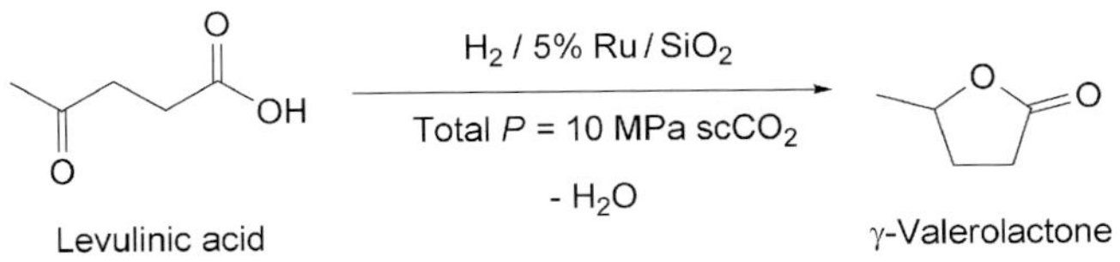

Scheme 5.9 Hydrogenation of levulinic acid in $scCO_2$.

In the hydrogenation of benzyl 4-methoxycinnamate in $scCO_2$, there are two likely processes, hydrogenation of the alkene and/or hydrogenolysis of the benzyl group (Scheme 5.10). The selectivity of the reaction towards C=C hydrogenation was much more sensitive to catalyst structure than pressure; however, optimal conversion was

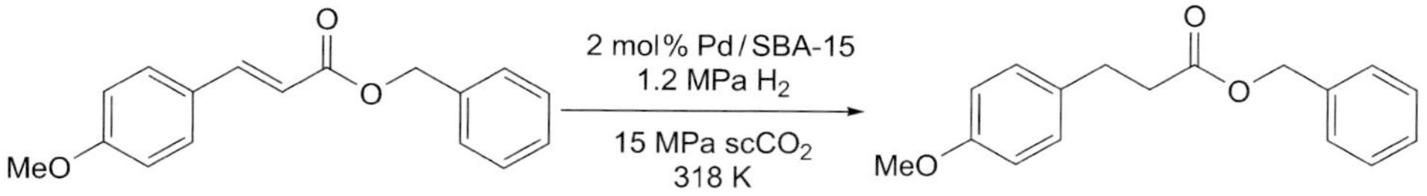

Scheme 5.10 Hydrogenation of benzyl 4-methoxycinnamate.

observed at 15 MPa (100%) with significantly lower conversions obtained at lower (33% at 8 MPa, 66% at 120 MPa) and higher (55% at 20 MPa) pressures [35]. In this case, the reduced rate at higher pressure was consistent with the dilution effect of higher CO_2 concentrations when the H_2 concentration was kept constant (1.2 MPa).

The selective hydrogenation of 2-butyne-1,4-diol was shown to be promoted by the stainless-steel reactor wall using either $scCO_2$ or conventional solvents (Scheme 5.11). At low pressure, the butenediol was the major product, but conversion to the butanediol was dominant at higher pressures. A study of the solubility of the various solutes showed that the reaction was biphasic when carried out in $scCO_2$, and hydrogenation of the alkyne to the alkene was faster in ethanol than in CO_2, but alkene to alkane reduction was fastest in CO_2. Clearly, a range of factors were important in determining rates of reaction, including H_2 concentration (or solubility in ethanol), and the relative solubility of substrates (alkyne, alkene, alkane), as their location (in the liquid or supercritical phase) determined the surface area of catalyst (the reactor wall) to which they were exposed [33, 36].

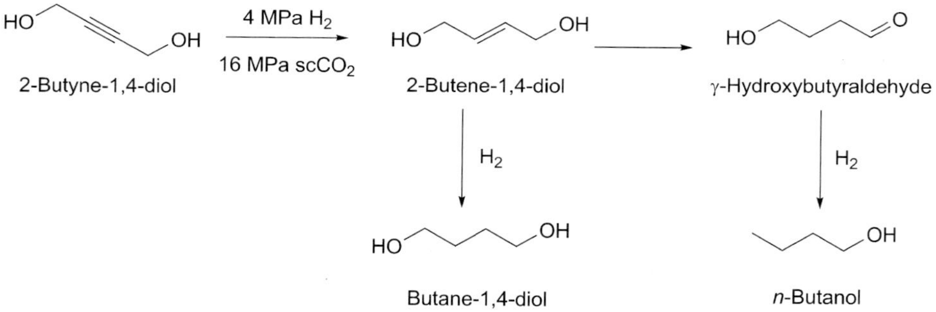

Scheme 5.11 Selective hydrogenation of 2-butyne-1,4-diol in $scCO_2$.

Selectivity in the semihydrogenation of propargylic alcohols has also been observed using an amorphous $Pd_{81}Si_{19}$ catalyst (Scheme 5.12). Selectivity to the alkene, isophytol, was essentially 100% at low conversions, but reduced to 77% at around 70% conversion due to complete hydrogenation to the alkane. As might be expected, conversion increased with hydrogen concentration, but selectivity decreased [37].

A range of aromatic substrates have been hydrogenated in $scCO_2$. The hydrogenation of naphthalene can occur in two stages, either to the partially hydrogenated tetralin or to fully hydrogenated decalin (Scheme 5.13). Although conversion remained relatively unaffected, increasing the CO_2 pressure from 10 to 22 MPa (at a constant H_2 pressure of 6 MPa) resulted in a decrease in the selectivity to decalin from ~60 to ~42%. The authors suggested that desorption of the intermediate

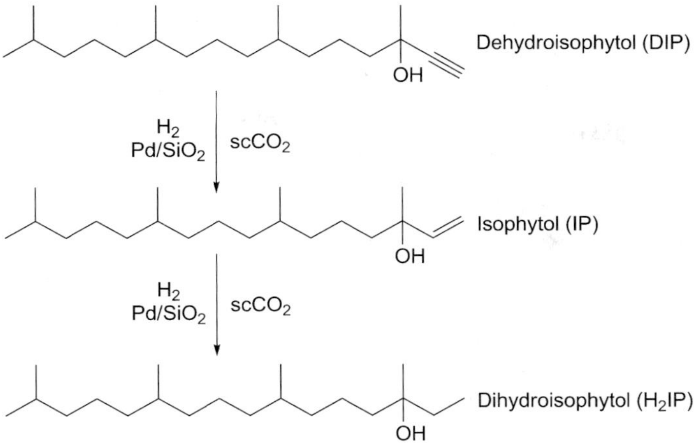

Scheme 5.12 Selective hydrogenation of dehydroisophytol.

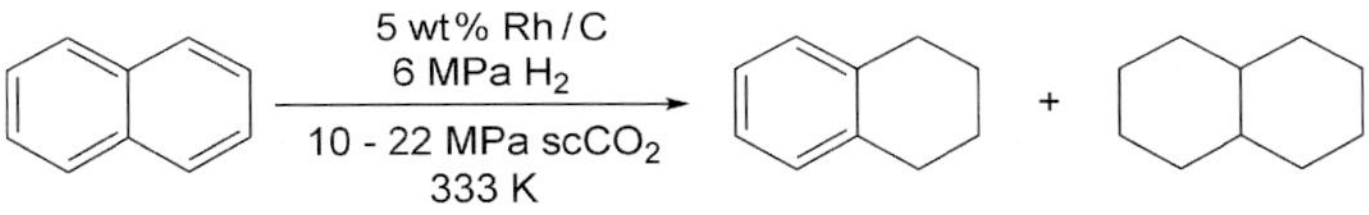

Scheme 5.13 Hydrogenation of naphthalene in $scCO_2$.

tetralin from the catalyst surface increased with CO_2 pressure due to its enhanced solubility, thus reducing the degree of further hydrogenation to decalin [38]. Interestingly, when the reaction was carried out in *n*-heptane, much lower levels of decalin formation were observed. This was consistent with higher levels of surface hydrogen being available in $scCO_2$ compared with heptane, with saturation apparently achieved at 6 MPa (45% conversion), compared with 9 MPa for heptane (24% conversion).

Selectivity in the hydrogenation of 4-*tert*-butylphenol has also been studied (Scheme 5.14). At low pressures, conversion was limited due to poor substrate solubility; significant solubility was observed around 10 MPa, but only near complete solubility around 25 MPa. Interestingly, optimum selectivity for *cis*-4-*tert*-butylcyclohexanol was around 10 MPa, and again the results were rationalized in terms of desorption of partially hydrogenated intermediates at higher pressures allowing reorientation of substrate, which opened up a route to alternative product formation [39]. The hydrogenation of phenol [40] itself and naphthols [41] has also been reported.

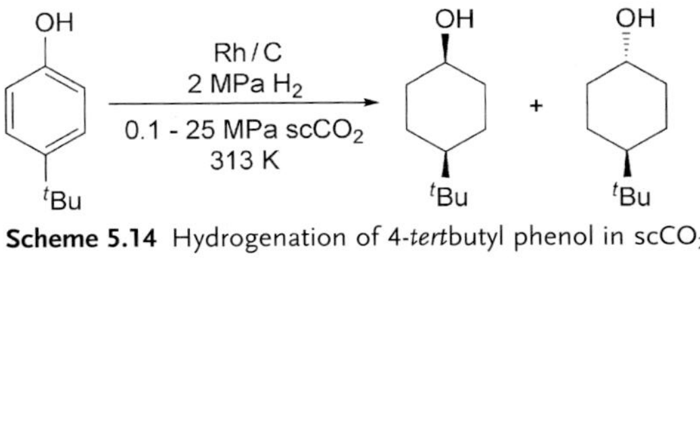

Scheme 5.14 Hydrogenation of 4-*tert*butyl phenol in $scCO_2$.

Hydrogenation of 1- and 2-phenylethanols using a variety of heterogeneous catalysts has also been reported. A charcoal-supported ruthenium catalyst showed enhanced selectivity in favor of benzene ring hydrogenation, which was also favored at increased H_2 pressures [42].

Numerous studies on the hydrogenation of other aromatic substrates have been published, including biphenyl [43], benzoic acid [44], and polystyrene [45]. In almost all cases, reaction rates increased with increase in CO_2 pressure, and were significantly better than reactions in conventional solvents.

Hydrogenation of aromatic nitro compounds has also been reported, and showed interesting solubility effects. The solubility of substrates generally increased as the CO_2 pressure was increased, but decreased with increasing H_2 pressure, although the changes in selectivity of the reaction were relatively small. Reactions in $scCO_2$ invariably gave higher selectivities than those in ethanol, although the effects were often small [46].

The hydrogenation of halonitroaromatics is an interesting example of how selectivity can be enhanced in $scCO_2$ over conventional solvents (Scheme 5.15). For example, in the reduction of 2-chloronitrobenzene using a Pt/C catalyst, the rate of nitro group hydrogenation was markedly enhanced in $scCO_2$ compared with a neat reaction, and the competing dehalogenation reaction was significantly suppressed [47].

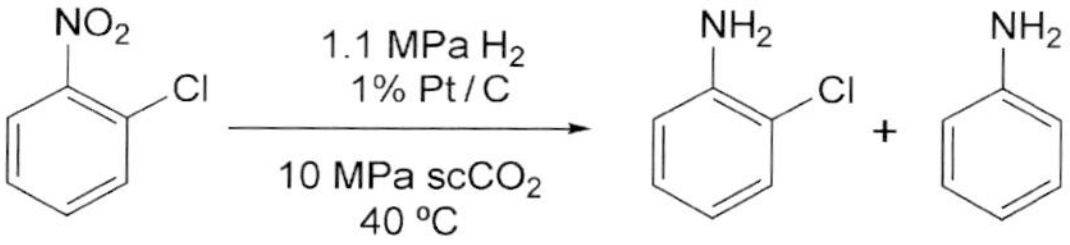

Scheme 5.15 Selective hydrogenation of 2-chloronitrobenzene in $scCO_2$.

Although beyond the scope of this chapter, it is worth mentioning that the hydrogenation of various vegetable oils has been investigated in a range of supercritical solvents, such as CO_2 [48], dimethyl ether [49], and propane [50].

5.2.1
Asymmetric Hydrogenation and Related Reactions

Much of the commercial success of the use of high-pressure CO_2 for reaction chemistry relies on the efficient and selective preparation of high-value products. Enantioselective reactions can potentially lead to such products, and have been the subject of a significant number of studies.

A variety of fluorinated phosphine ligands have been prepared which have significantly increased solubility in $scCO_2$ compared with common phosphines such as PPh_3. However, although solubility may be improved, this does not automatically lead to better conversions and selectivities. In a study of the hydrogenation of dimethyl itaconate, it was shown that the highest rates and selectivities were actually observed in MeOH rather than $scCO_2$, even using extensively fluorinated phosphines or related ligands (Scheme 5.16). The origin of this was attributed to the sensitivity of the reaction to solvent polarity, and the substantially lower polarity of CO_2 compared with MeOH [51].

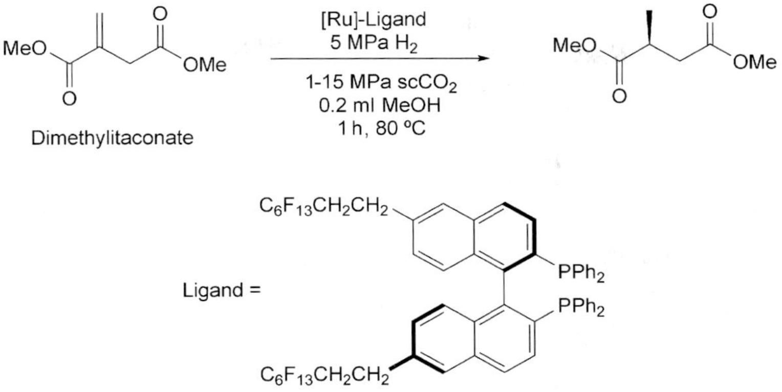

Scheme 5.16 Asymmetric hydrogenation of dimethyl itaconate.

Other related fluorinated phosphines have been used in the hydrogenation of 2-acetamidomethyl acrylate and dimethyl itaconate (Scheme 5.17). The acetamidomethyl acrylate system was biphasic, containing a liquid and a gas phase, whereas for the dimethyl itaconate this was initially opaque but rapidly became transparent and homogeneous within a few minutes. High enantioselectivities were observed, although they were slightly lower than those observed in dichloromethane [52]. Mechanistic aspects of such reactions in $scCO_2$ have been discussed [53].

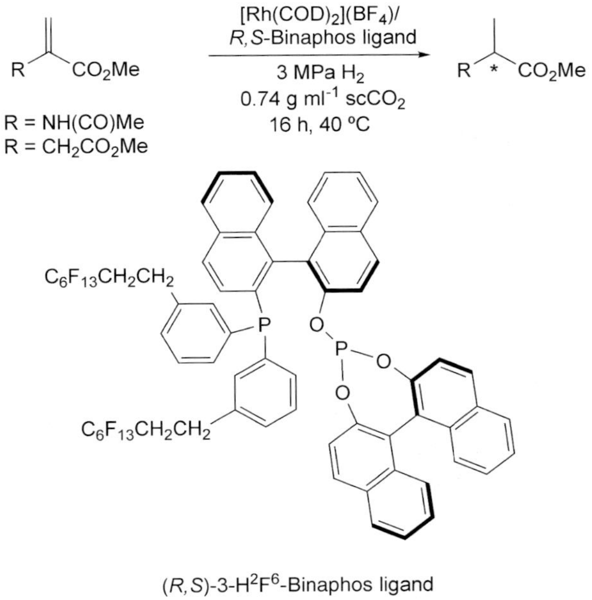

Scheme 5.17 Asymmetric hydrogenation using a fluorinated Binaphos ligand in $scCO_2$.

Asymmetric hydrogenations have also been carried out in continuous processes using solid-supported metal complexes (Scheme 5.18). Hydrogenation of dimethyl

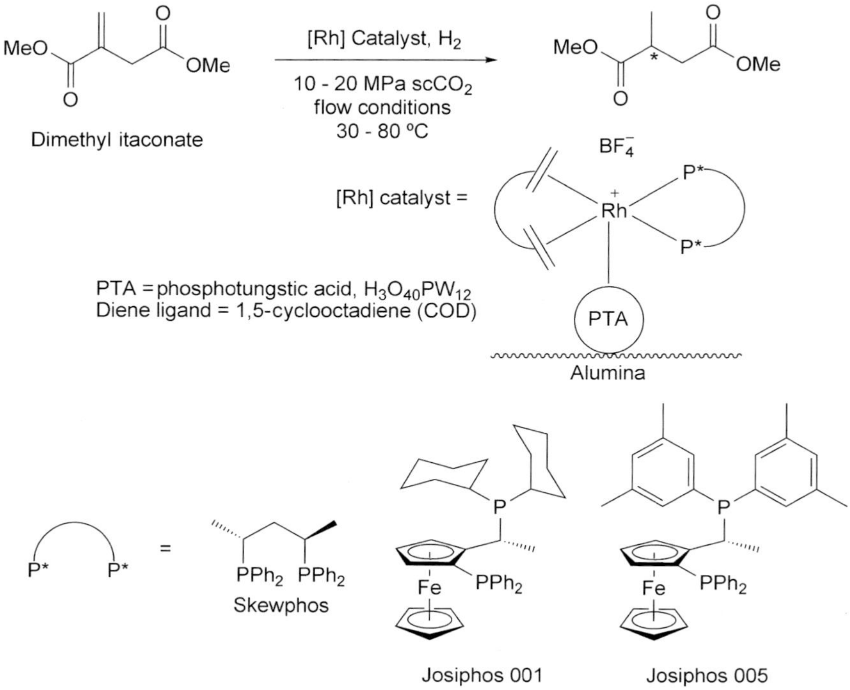

Scheme 5.18 Asymmetric hydrogenation using solid-supported metal complexes.

itaconate using an Rh(Skewphos) complex bound to an alumina–phosphotungstic acid composite gave moderate enantioselectivities (up to 63%), which were slightly lower than those obtained in conventional solvents (65–70% in EtOH) [54]. A wide range of alternative ligands were subsequently screened, with up to 83% being achievable with the commercially available Josiphos ligand [55].

Hydrogenations, including asymmetric variants, have also been demonstrated in supercritical 1,1,1,2-tetrafluoroethane (HFC134a). Although this solvent may not have the more obvious benefits of CO_2, it does not require fluorinated ligands to give high conversions and enantioselectivities. For example, using a rhodium–monophos catalyst system gave results comparable to those obtained in conventional solvents (Scheme 5.19). Interestingly, enantioselectivity showed no significant variation with pressure at constant mole fraction, consistent with previous reports showing that solvent polarity (which varies widely with pressure for HFC134a) had no significant influence on the outcome of the reaction [56].

Asymmetric hydrogenation of imines can also be efficient in $scCO_2$ using cationic iridium catalysts, although there are limitations. Use of chiral phosphinodihydrooxazole ligands modified with perfluoroalkyl groups with a CO_2-philic tetrakis-3,5-bis(trifluoromethyl)phenylborate (BARF) anion, gave essentially a quantitative yield of the amine with up to 81% enantiomeric excess (*ee*) for reduction of *N*-(1-phenylethylidene)aniline, which was comparable to results obtained in dichloromethane

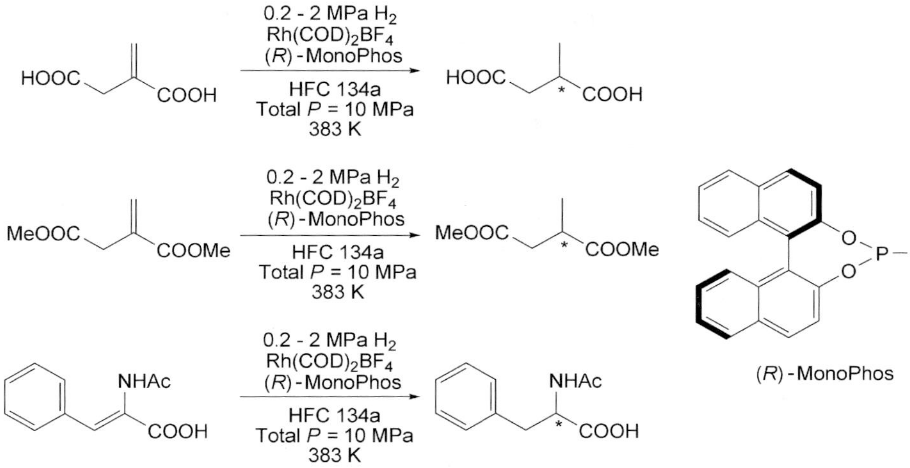

Scheme 5.19 Asymmetric hydrogenation using monophos complexes.

(Scheme 5.20). However, with the corresponding benzylamine-derived substrate, *N*-(1-phenylethylidene)benzylamine, only low conversions (<30%) were obtained, which was rationalized by invoking the formation of highly coordinating carbamate salt by reaction of the amine product with CO_2, which interfered with the catalytic cycle (other explanations may also be possible) [57]. Aromatic amines are known to have a very low tendency to form such carbamate salts, which may account for the difference in reactivity. Although carbamate formation appeared to cause problems in this reaction, there are notable occasions where this can be exploited by using CO_2 as an "*in situ*" protecting group for amines (see below).

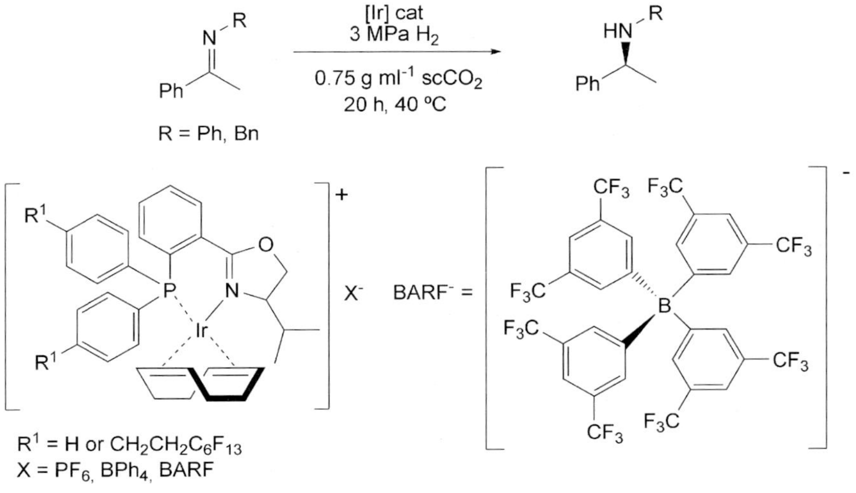

Scheme 5.20 Asymmetric hydrogenation of imines in $scCO_2$.

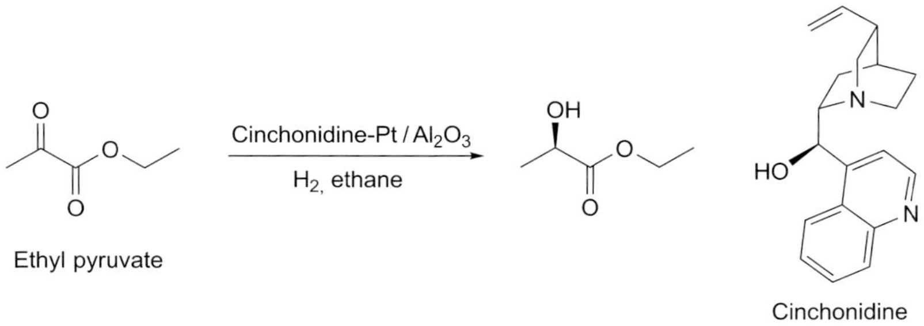

Scheme 5.21 Asymmetric hydrogenation of reactive ketones in $scCO_2$.

Asymmetric hydrogenation of reactive ketones has also been reported using a cinchonidine-modified Pt/Al_2O_3 catalyst (Scheme 5.21). Reduction of ethyl pyruvate in a continuous flow reactor gave only low conversions (<10%) and modest *ee* (52%) using $scCO_2$ as solvent. However, if this was changed to supercritical ethane ($T_c = 32.2\,°C$, $P_c = 4.88\,MPa$) then conversions rose to ~65% and *ee* to ~72%. Conversions were greatest at higher pressures (11.4 MPa) when a single-phase system was present, which also gave the highest enantioselectivities with an H_2 to ethyl pyruvate feed ratio of 2:1 [58]. A subsequent detailed study of the phase behavior of this reaction permitted rationalization of the changes in conversion and enantioselectivity in more detail [59]. The poor performance of the catalyst in CO_2 may be attributed to poisoning of the platinum catalyst by CO formed *in situ* by a water gas shift reaction between CO_2 and H_2 (to give CO and H_2O).

5.3 Hydroformylation and Related Reactions in Supercritical Fluids

Hydroformylation is a reaction of considerable industrial importance, and it has been underutilized from the viewpoint of synthetic organic chemistry. The high solubility of H_2 and CO in $scCO_2$ (along with an appropriate catalyst), resulted in extensive research on hydroformylation in supercritical fluids. There is a wide range of opportunities for the investigation of selectivity effects, in terms of linear to branched product formation, and also enantioselectivity where appropriate.

Rhodium complexes modified by simple trialkylphosphines can be used to carry out homogeneous hydroformylation in $scCO_2$. For example, the catalyst derived from PEt_3 is more active and slightly more selective for linear products in $scCO_2$ than in toluene (Scheme 5.22). Other ligands, which had lower solubility in $scCO_2$, were less effective [60]. $Ru_3(CO)_{12}$ has also been found to be sufficiently soluble in $scCO_2$ to be an active catalyst for the hydroformylation of ethylene, although it required temperatures in excess of 70 °C, and showed an induction period of a few hours before reaction rate increased exponentially [61].

The effect of process parameters on the hydroformylation of 1-hexene using a heterogeneous Rh catalyst have been determined. Aldehyde yield and regioselectivity were both dependent on catalyst support structure, solvent pressure, and

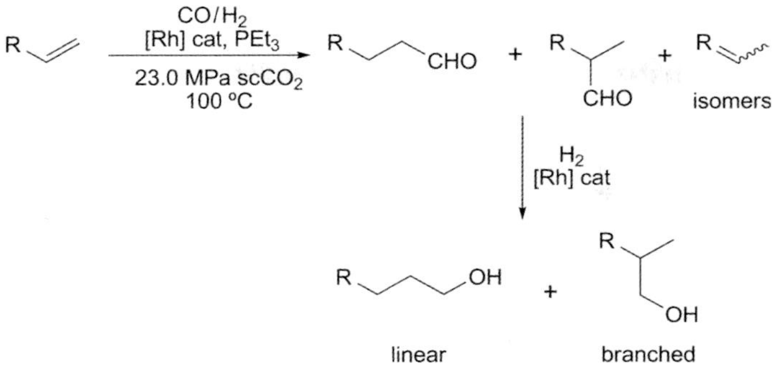

Scheme 5.22 Hydroformylation in $scCO_2$.

temperature [62]. Similarly, in the hydroformylation of styrene, the reaction rate and selectivity were both affected by changes in pressure at constant temperature, and this could be explained in terms of transition state theory [63].

The hydroformylation of acrylates is usually a particularly slow reaction. Interestingly, if the reaction is carried out in $scCO_2$ using an Rh catalyst with a fluorous polymeric phosphine ligand, the reaction is rapid and chemoselective, and indeed acrylate hydroformylation occurs in preference to alkene hydroformation (Scheme 5.23) [64]. The faster rates have been rationalized by interactions between CO_2 and the acrylate carbonyl group preventing formation of a thermodynamically stable five- or six-membered ring between the same carbonyl group and the rhodium center.

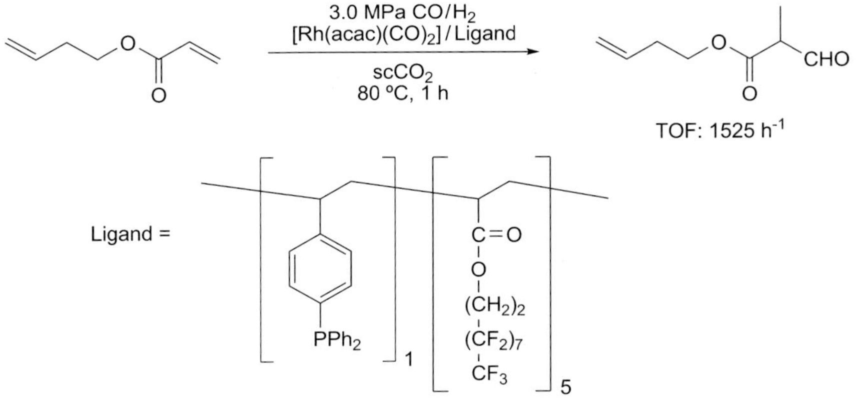

Scheme 5.23 Hydroformylation of acrylates in $scCO_2$.

Asymmetric hydroformylation has also been reported in $scCO_2$. Using perfluorinated Binaphos derivatives, a variety of substrates could be hydroformylated in $scCO_2$ with rates and enantioselectivities comparable to those obtained in benzene (Scheme 5.24) [52].

The interaction of amines with CO_2 to form carbamic acid derivatives can be a problem in some organometallic reactions. However, in other cases, it can act as an

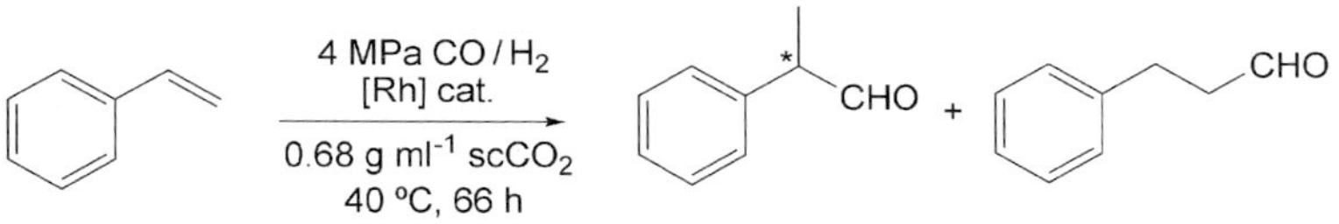

Scheme 5.24 Asymmetric hydroformylation of styrene in $scCO_2$.

in situ protecting group for amine functionality and prevent possible deleterious interactions between an amine and a metal center. This can lead to more efficient reactions, or alternative pathways. An example is in the rhodium-catalyzed hydroaminomethylation of alkenes, which occurs under conditions very similar to those for hydroformylation, and leads to the formation of amines. These result from olefin insertion into the Rh–H bond, CO insertion to give a rhodium acyl species, and hydrogenation of the rhodium acyl species to give an aldehyde, which then reacts with an amine to give an imine, which is hydrogenated to the amine (Scheme 5.25). However, in the case of ethylmethallylic amines, the major product is a cyclic amide, which is formed as a competing cyclization reaction (insertion of N–H into the adjacent acylrhodium moiety) is faster than the hydrogenation. If, however, the reaction is carried out in $scCO_2$, the cyclization reaction is suppressed by the formation of a carbamate species on the nitrogen, allowing selective hydrogenation of the acylrhodium species to the aldehyde, which eventually leads to amine formation. The intermediacy of the carbamate was demonstrated by high-pressure ^{1}H NMR spectroscopy [65].

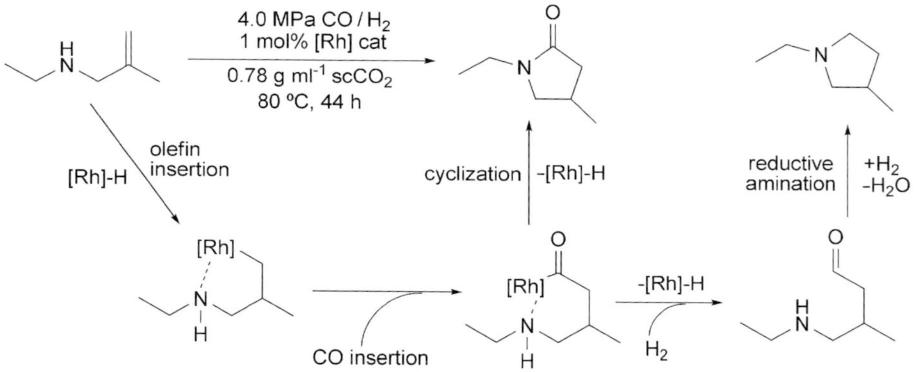

Scheme 5.25 Rhodium-catalyzed hydroaminomethylation of alkenes in $scCO_2$.

In related chemistry, the fixation of carbon dioxide into oxazolones and oxazolidinones by the copper(I)-catalyzed addition of propargylic alcohols to amines in $scCO_2$ was optimized by adjustment of pressure and temperature [66]. The reagents selected for optimization investigations were 2-methyl-3-butyn-2-ol and *n*-butylamine to afford the corresponding 4-methyloxazol-2-one product (Scheme 5.26). It was expected that higher pressures of CO_2 may favor enhanced product yield, but in fact optimum yield was observed at the lower pressure (89%, 8 MPa, 60 °C). These conditions were then successfully transferred to a range of substrates to confirm smooth reaction of primary and secondary propargylic alcohols with primary amines.

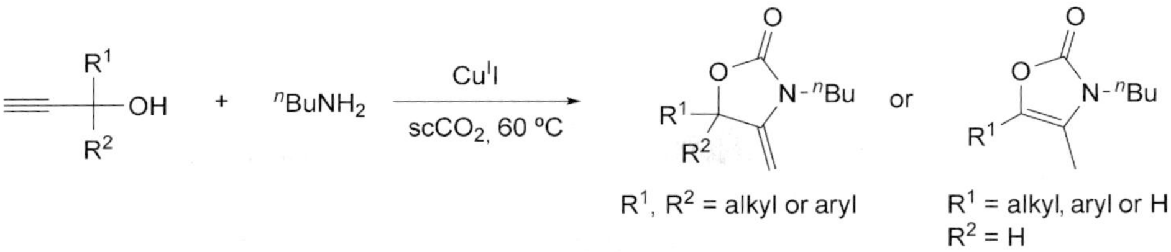

Scheme 5.26 Addition of propargylic alcohols and amines to CO_2.

5.4 Oxidation Reactions in Supercritical Fluids

Dioxygen and air are seen as ideal oxidants as they are inexpensive, readily available, and produce water as the only by-product [67]. The high miscibility of oxygen with $scCO_2$ potentially offers rate benefits, as observed for hydrogenation and hydroformylation. The stability of carbon dioxide towards oxidation, and avoidance of combustible solvents, suggest that $scCO_2$ may be a particularly useful reaction medium for such reactions of organic substrates.

A range of very challenging oxidations have been investigated in $scCO_2$. Examples include aerobic oxidation of cyclohexane in the presence of an iron–porphyrin catalyst and acetaldehyde [68] and halogenated porphyrin systems for the aerobic oxidation of cyclohexene [69].

Catalytic partial oxidation of alcohols is an environmentally friendly process, but is rather slow in organic solvents, and because of safety concerns the process is unattractive even on a laboratory scale. To overcome these restrictions, the organic solvent may be replaced by $scCO_2$ [2, 7].

A range of catalyst systems have been developed to effect this transformation. For example, poly(ethylene glycol) (PEG)-stabilized Pd nanoparticles catalyze the biphasic aerobic oxidation of a wide variety of alcohols with high selectivity and efficiency (Scheme 5.27). They can also be reused or alternatively used in a continuous reaction system [70].

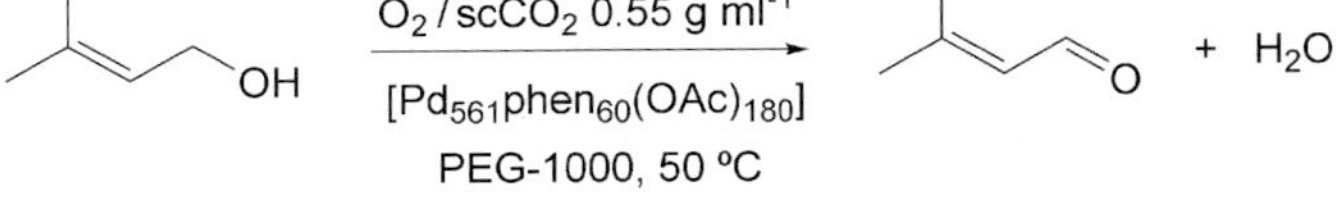

Scheme 5.27 Oxidation of alcohols using PEG-stabilized Pd nanoparticles.

For the oxidation of weakly polar, water-insoluble alcohols, $scCO_2$ is particularly useful, due to its low polarity. Oxidation of alcohols to carbonyl compounds was shown to proceed with high selectivity at a high rate [71]. Experiments were performed in a continuous fixed-bed reactor over a promoted noble metal catalyst (4% Pd/1% Pt/5% Bi/C). Transformation of primary alcohols to aldehydes under similar conditions was fast but non-selective. However, benzyl alcohol and related substrates could be oxidized to the corresponding aldehydes with >99% selectivity. Similar results have also been reported using O_2 with a Pt/C or Pd/C catalyst, with catalytic activity increasing with more hydrophobic carbon surfaces as supports; [72] and utilizing novel bimetallic Ru–Ni complexes which act as homogeneous catalysts [73].

A common method for the utilization of O_2 in transition metal-catalyzed epoxidations is the use of an aldehyde as a sacrificial oxygen transfer agent. Such reactions have been shown to occur efficiently in $scCO_2$ [74]. High reaction rates and selectivities for the epoxidation of internal double bonds were achieved without the addition of a metal catalyst. Quantitative conversion of *cis*-cyclooctene, with almost complete selectivity towards epoxycyclooctane, was reported with the use of 2-propionaldehyde (Scheme 5.28). Under similar conditions, cyclohexene and *trans*-3-hexene were quantitatively converted to epoxycyclohexane (91% selectivity) and *trans*-3-epoxyhexane (>98% selectivity), respectively.

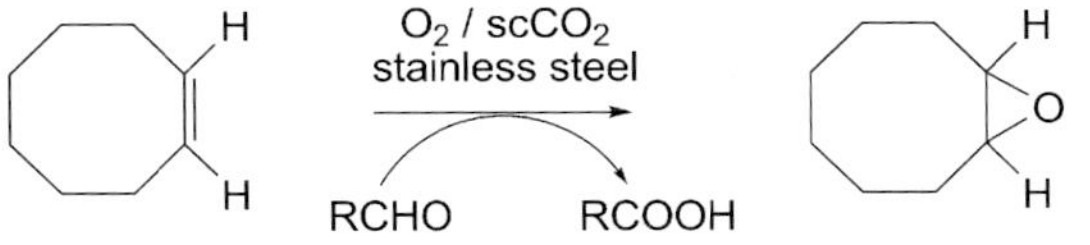

Scheme 5.28 Epoxidation of alkenes using a sacrificial aldehyde in $scCO_2$.

The most intriguing aspect of this work was that the reaction was believed to be catalyzed by the stainless-steel reactor walls, which may facilitate the initial formation of acylperoxy radicals, allowing the oxidation to occur via a non-catalytic radical pathway.

Alkene epoxidation has been reported in $scCO_2$, using *tert*-butyl hydroperoxide (TBHP) and catalytic $Mo(CO)_6$ [75]. Using aqueous TBHP solution (70 wt%), *trans*-diols were formed, whereas if anhydrous solutions of TBHP in decane were used, then the epoxide was formed. *cis*-Alkenes gave much faster reaction rates than *trans*-alkenes, with cyclooctene giving quantitative conversion to the epoxide. No reaction was observed with *trans*-2-heptene or *trans*-stilbene.

The diastereoselective epoxidation of olefins in $scCO_2$ has also been reported. By combining the Jacobsen-type salen ligand with a vanadyl system, a salen catalyst was developed, utilizing stoichiometric TBHP [76]. A range of allylic alcohols were epoxidized using this catalytic system, giving good yields (50–100%) after 24 h at 40 °C. Good diastereoselectivity for the *erythro* product was observed for alkenes with secondary alcohols (Scheme 5.29).

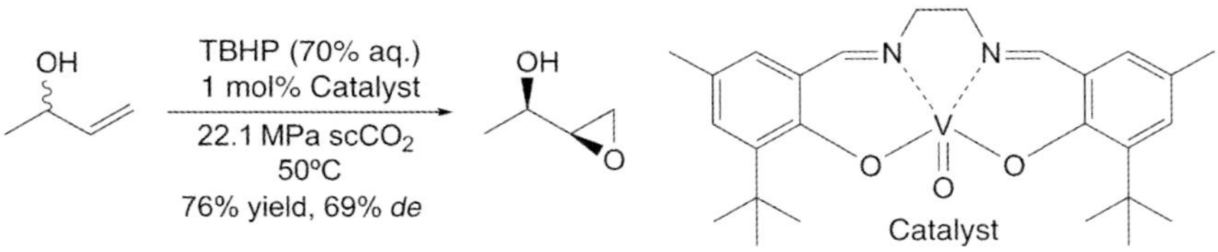

Scheme 5.29 Diastereoselective epoxidation in $scCO_2$.

Liquid CO_2 is also a potentially attractive medium for epoxidation [77]. Reaction of allylic alcohols with anhydrous TBHP and catalytic oxovanadium(V) tri(isopropoxide) was efficient in liquid CO_2 at 25 °C. The reaction rate was found to be three times

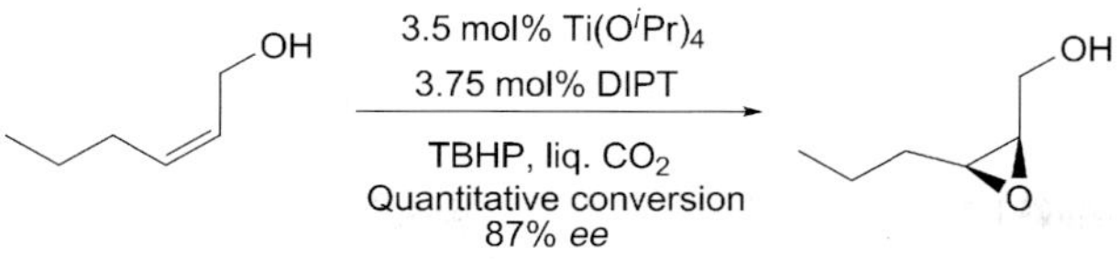

Scheme 5.30 Asymmetric Sharpless epoxidation in liquid CO_2.

faster than that in hexane solution. Using titanium(IV) tetra(isopropoxide) and a chiral diisopropyl L-tartrate (DIPT) ligand, the asymmetric Sharpless epoxidation of allylic alcohols was also shown to be effective if carried out at 0 °C in liquid CO_2 (Scheme 5.30).

The selective oxidation of sulfides to sulfoxides can be effected using TBHP as the oxidant and Amberlyst 15 as a heterogeneous acid catalyst [78]. Using this method, diastereoselective sulfoxidation of cysteine derivatives in $scCO_2$ was achieved after pressure optimization [79] giving the *anti* diastereomer as the sole product (Scheme 5.31). Such high selectivity is remarkable as the corresponding oxidation in conventional solvents (toluene and dichloromethane) showed no diastereoselectivity.

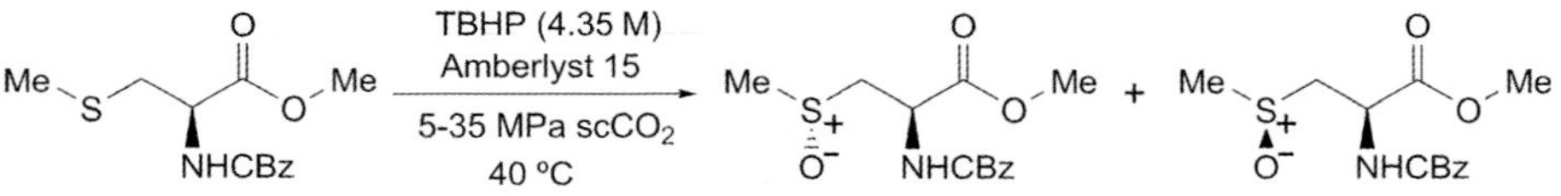

Scheme 5.31 Diastereoselective sulfoxidation in $scCO_2$.

Other epoxidation reactions utilize hydrogen peroxide as the oxidant, as it is cheap and readily available. Unfortunately, it has limited use without activation by additional catalysts or conversion to a more reactive peroxy acid. An interesting approach exploits the *in situ* formation of peroxycarbonic acid from H_2O_2 and CO_2 [80]. The epoxidations of cyclohexene (Scheme 5.32) and sodium 3-cyclohexen-1-carboxylate were investigated in a biphasic system composed of a $scCO_2$–olefin phase and an aqueous H_2O_2 phase, to yield the corresponding epoxides and diols. Epoxide yield

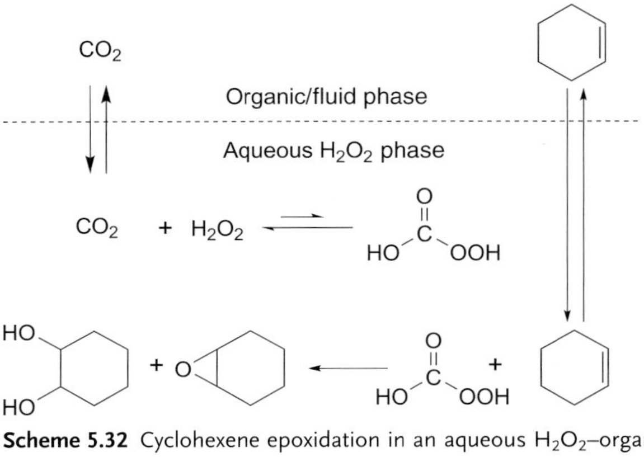

Scheme 5.32 Cyclohexene epoxidation in an aqueous H_2O_2–organic fluid phase.

was improved by the addition of dimethylformamide (DMF) to increase the aqueous solubility of the olefin, suggesting that the epoxidation occurred in the aqueous phase. The highest yield was achieved in the epoxidation of sodium 3-cyclohexen-1-carboxylate (89%) and the process may therefore be best exploited using water-soluble alkenes.

Hydrogen peroxide has also been used as the primary oxidant for olefin epoxidations in $scCO_2$ with the use of manganese 5,10,15,20-tetrakis(2′,6′-dichlorophenyl) porphyrinate as catalyst and hexafluoroacetone hydrate (HFAH) as cocatalyst in the presence of 4-*tert*-butylpyridine [81].

Using $scCO_2$ under continuous flow conditions, successful Baeyer–Villiger oxidation of ketones was shown to be dependent upon pressure and flow. A range of cyclic ketones were oxidized to the corresponding lactones by predissolving in CO_2, then flowing over hydrated silica-supported potassium peroxomonosulfate (h-$SiO_2{\cdot}KHSO_5$). Using this technique, no hydrolysis of the reaction products was observed. In one example, conversion of cyclohexanone to caprolactone rose from 65 to 91% as the pressure was increased from 15 to 30 MPa, whilst the lowest flow rate investigated, across the range 0.13–0.33 ml min^{-1}, afforded the highest conversion (>90%) (Scheme 5.33) [82].

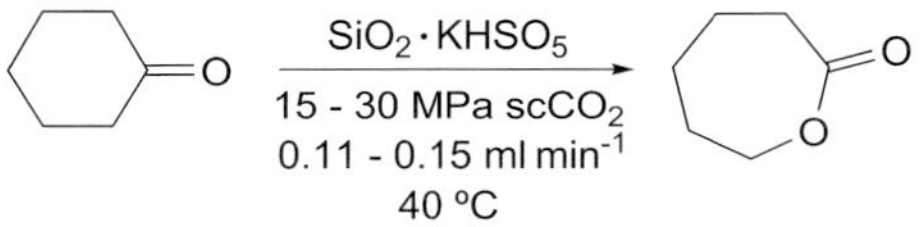

Scheme 5.33 Baeyer–Villiger oxidation of ketones in $scCO_2$.

5.5 Palladium-mediated Coupling Reactions in Supercritical Fluids

Palladium-catalyzed coupling reactions are among the most powerful methods of C–C bond formation. Catalyst systems successful under more conventional conditions usually give poor yields in $scCO_2$, and considerable efforts have been made to develop more effective palladium sources, ligands, and other additives. A common approach is the utilization of perfluorinated ligands, which have substantially higher solubility in $scCO_2$ than their non-fluorinated analogues. For example, the perfluoro-tagged palladium complexes **1** and **2** were shown to be highly effective catalysts for Stille coupling reactions in $scCO_2$ [83]. Lipophilic tetrabutylammonium chloride replaced the more usual lithium chloride, which was insoluble in $scCO_2$. The couplings were compared with those mediated by $Pd(PPh_3)_2Cl_2$, and in both cases better yields were achieved with the fluorous ligand (95%) than the non-fluorinated complex (87%) [84]. Interestingly, the acceptable coupling yield using $Pd(PPh_3)_2Cl_2$ catalyst suggests that it (or at least an active catalyst species) had a reasonable activity under the reaction conditions.

In related research, the solubilities of a range of fluorinated phosphines have been determined, and the influence of solubility on the yield of Pd-mediated couplings has

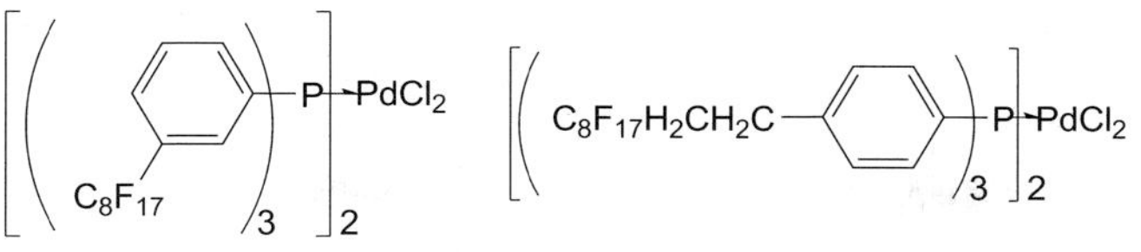

1 2

been investigated [85]. Interestingly, the optimum ligand depended on the $scCO_2$ pressure, with bis(pentafluorophenyl)phenylphosphine yielding the best results at 12 MPa (homogeneous reaction), whereas the optimum pressure for reactions with tris(*p*-trifluoromethylphenyl)phosphine was 8 MPa (biphasic). In addition to fluorous tags, it has also been shown that siloxanes can be used in a similar manner, and also give good results for a variety of Pd-catalyzed coupling reactions [86].

Partially fluorinated, dendrimer-encapsulated (DEC) palladium nanoparticles have been shown to be effective catalysts for Heck couplings in $scCO_2$ of methyl acrylate to iodobenzene (Scheme 5.34) [87]. The metal catalysts were a composite of functionalized polypropylenimine (PPI) dendrimers having perfluoro-2,5,8,11-tetramethyl-3,6,9,12-tetraoxapentadecanoyl perfluoropolyether chains covalently attached to their periphery. It is the perfluoro-functionalized portion of the composite that is responsible for making them soluble in $scCO_2$. Interestingly, this reaction shows opposite regioselectivity in the Heck coupling reaction, giving exclusively methyl 2-phenylacrylate rather than the more conventional cinnamate product [88].

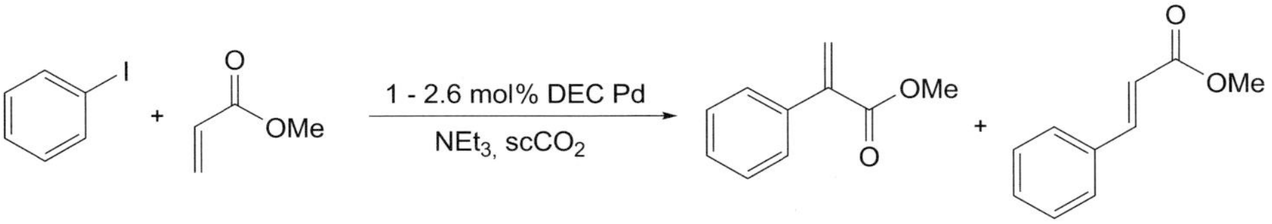

Scheme 5.34 Heck reaction using dendrimer-encapsulated Pd nanoparticles in $scCO_2$.

Although using ligands with specific structural features to enhance their solubility in $scCO_2$ usually works well, they can be very expensive, and there can also be potential environmental and practical issues associated with their use. If palladium-mediated coupling reactions are to be widely adopted, it would be beneficial if relatively simple, commercially available reagents could be used. With this in mind, it has been shown that Stille, Heck, and Suzuki couplings can be carried out efficiently using commercially available ligands (e.g. PBu_3, PCy_3) and fluorinated palladium sources in low catalyst loadings at moderate temperatures [89]. Palladium trifluoroacetate and hexafluoroacetylacetanoate gave the best yields, superior to previously reported results. Palladium-catalyzed biaryl couplings of iodobenzenes were also affected under similar conditions in the absence of an alkene [90]. It was shown that electron-donating groups (e.g. MeO) promoted the couplings whereas strong electron withdrawing groups (e.g. NO_2, CO_2Me) limited yields due to reductive deiodination.

It was also demonstrated that both Heck and Suzuki reactions could be carried out on polymer-tethered substrates in $scCO_2$ using palladium acetate with tri-*tert*-

butylphosphine and various bases [91]. The reactions proceeded in good to excellent yields on both polymer-tethered substrates and untethered analogs.

The use of solid-supported reagents has also been investigated for Heck and Suzuki couplings in $scCO_2$, as illustrated by the utilization of polystyrene-supported amines as the base component and resin-supported phosphine–palladium catalysts [92]. The two solid-supported agents were not used in combination; rather, one reaction set demonstrated the potential of the polystyrene-supported amine in developing good to excellent yields in the coupling of butyl acrylate to aryl iodides, whereas the other sought to optimize which base should be used in the coupling of butyl acrylate to iodobenzene, in some cases achieving quantitative yields. Further work using immobilized palladium catalysts in Suzuki coupling reactions has also been reported [93], including application in continuous flow Suzuki couplings. For example, passing a stream of $scCO_2$ containing *p*-tolylboronic acid, iodobenzene, tetrabutylammonium methoxide, and some methanol cosolvent through a polyurea-encapsulated palladium source [94] as a fixed bed gave conversions of up to 81% after optimization [95].

A key driver for the implementation of $scCO_2$ technology is if it allows control of reactions which is otherwise not possible. A good example of this is the intramolecular Heck coupling of alkenyl aryliodides [6, 96]. In such reactions, the endocyclic alkene is usually the predominant product resulting from cyclization and subsequent double bond isomerization. However, if the reaction was carried out in $scCO_2$, the double bond isomerization was prevented and the exocyclic alkene was the dominant product (Scheme 5.35).

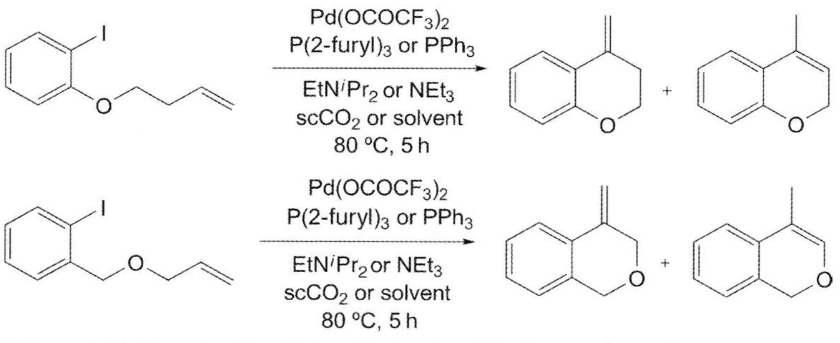

Scheme 5.35 Control of double bond migration following Heck coupling in $scCO_2$.

Pd-mediated coupling reactions have also been shown to allow aromatic amination in $scCO_2$ (Scheme 5.36) [97]. Carbamic acid formation was in part avoided by use of *N*-silylamine coupling partners as surrogates for the free amines, along with restricting the couplings to aniline derivatives, which are usually not nucleophilic enough to allow carbamate formation. Silylated non-aromatic amines led to silylcarbamate formation when exposed to CO_2.

Carbamate and/or carbamic acid formation can also be exploited in Pd-catalyzed reactions. For example, the classic coupling of iodobenzene with methyl acrylate proceeded very efficiently using a $Pd(OCOCF_3)_2$–tris(2-furyl)phosphine (TFP)

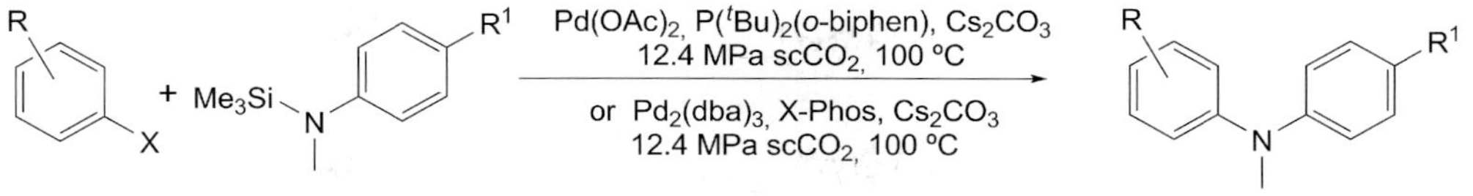

Scheme 5.36 Pd-catalyzed aromatic amination in $scCO_2$.

catalyst system in both toluene and $scCO_2$. However with 4-iodoaniline, the reaction was significantly slower in toluene, and only reached 67% conversion after 22 h with extensive Pd decomposition. When the equivalent reaction was carried out in $scCO_2$, it was complete within 14 h, with no sign of Pd decomposition (Scheme 5.37). In this case, it was unlikely that a formal carbamate derivative was formed, and a Lewis acid–Lewis base complex between the CO_2 and aniline is more likely. This would have two possible effects. First, it dramatically reduced the possibility of the aniline nitrogen coordinating to Pd and reducing its efficiency. Second, if the aniline lone pair was coordinated to CO_2, then its donation of electron density into the benzene ring was reduced. As it is known that electron-rich iodoarenes react significantly more slowly in Heck reactions than electron-deficient iodoarenes, then again this effect should promote the coupling reaction [6].

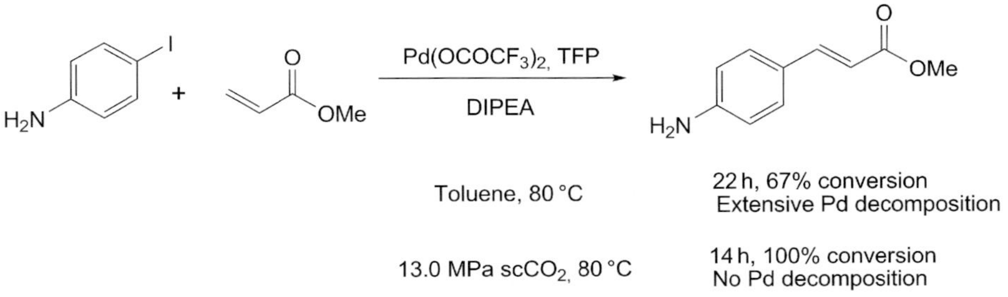

Scheme 5.37 Pd-catalyzed Heck coupling of iodoaniline in $scCO_2$.

Similar observations were made for the coupling of aryl bromides, although in this case 1,1′-bis(diphenylphosphino)ferrocene (dppf) was used as ligand rather than TFP (Scheme 5.38). Coupling of 4-aminostyrene to bromobenzene occurred efficiently in $scCO_2$, but gave negligible yields in toluene or in the absence of solvent. Similarly, the Suzuki reaction between benzeneboronic acid and 4-bromodibenzylamine worked well in $scCO_2$, but again gave negligible yields in toluene, and only a low yield in the absence of solvent [6].

Biphasic $scCO_2-H_2O$ systems have been employed for Heck coupling reactions, utilizing water-soluble catalysts [98]. The coupling of iodobenzene and butyl acrylate was investigated using $Pd(OAc)_2$ and triphenylphosphine trisulfonate sodium salt (TPPTSS) (Scheme 5.39). Without cosolvent, the catalyst remained insoluble and yields were very poor (<5% at 8 MPa). Adding water to the reaction increased the rate and conversion (up to 18%), whereas use of a more CO_2-philic solvent such as ethylene glycol offered a slight improvement on conversion (up to 29%).

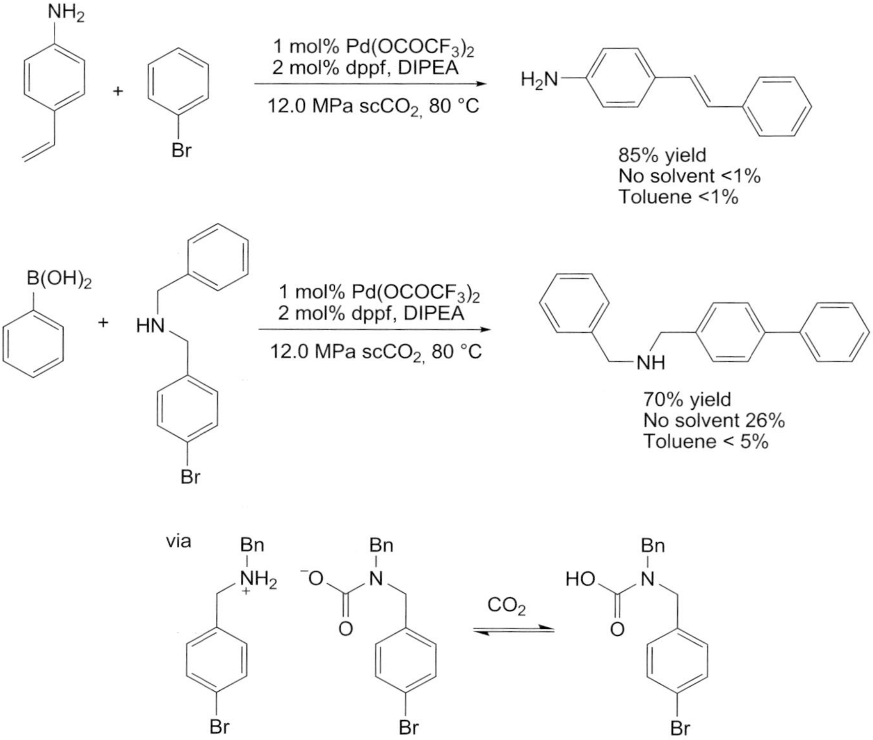

Scheme 5.38 Carbamate protection of amines during Pd-mediated coupling reactions in $scCO_2$.

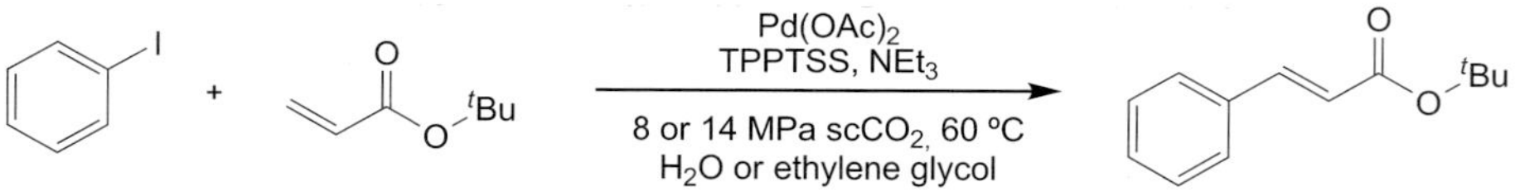

Scheme 5.39 Biphasic Pd-catalyzed coupling in $scCO_2$ and water.

Use of a biphasic system in the Heck coupling of ethylene and a variety of aromatic compounds (Scheme 5.40) afforded an interesting and advantageous method of reducing side reactions [99]. The majority of the reaction occurred in the lower liquid phase (mainly aryl halide, catalyst, and organic base), whereas the $scCO_2$ phase removed the initial styrene product from the reactive liquid phase, preventing further Heck coupling. This occurred due to the relatively high solubility of styrenes in $scCO_2$ compared with the aryl halide precursor.

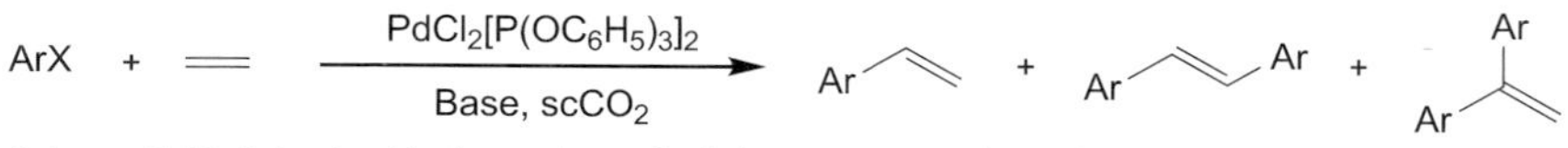

Scheme 5.40 Selective Heck coupling of ethylene in $scCO_2$ through phase separation.

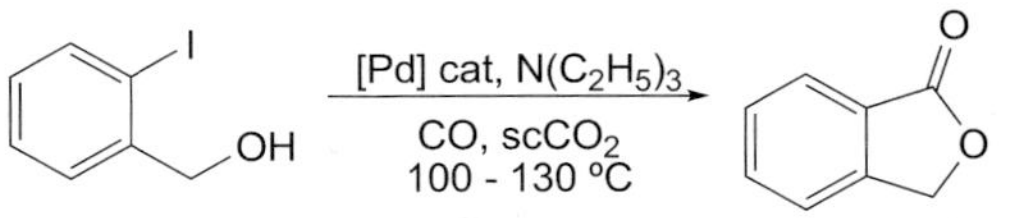

Scheme 5.41 Pd-catalyzed carbonylation of 2-iodobenzyl alcohol in $scCO_2$.

The palladium-catalyzed carbonylation of 2-iodobenzyl alcohol (Scheme 5.41) was reported using a supercritical mixture of CO (1 MPa) and CO_2 (20 MPa) [100]. The bisphosphite complex $PdCl_2[P(OEt)_3]_2$ was used as catalyst with triethylamine as base to give phthalide product.

The Pd-catalyzed carbonylation of amines has also been studied in $scCO_2$, where it was found that the concentration of O_2 in the reaction media had an effect on the chemoselectivity of the reaction [101]. Carbonylation of *n*-butylamine using $PdCl_2$ catalyst (Scheme 5.42) gave 1,1′-oxalyldibutylamine or *N*-*n*-butyl carbamate. When using conventional solvents, it was necessary to add O_2 to prevent unwanted formation of the carbamate, whereas in the supercritical system, the reaction proceeded in 98% yield with a high oxalate to carbamate ratio of 96:4. When O_2 was added to the system, the carbamate product was obtained almost exclusively.

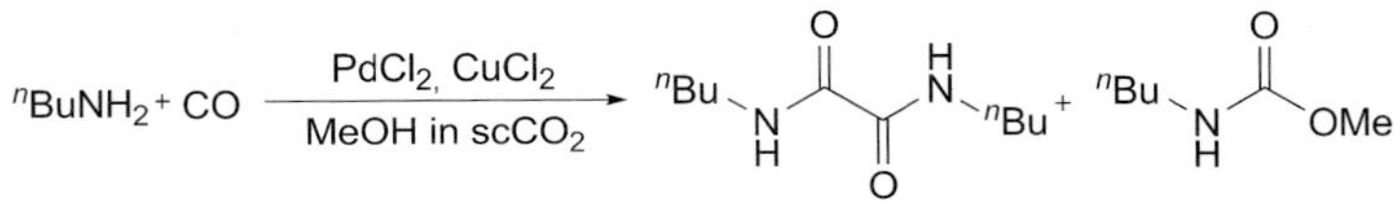

Scheme 5.42 Pd-catalyzed carbonylation of amines in $scCO_2$.

The prototypical Wacker reaction is a palladium-catalyzed oxidation of ethylene to acetaldehyde. It can also be applied to higher terminal alkenes, resulting in the corresponding methyl ketones. It has been reported in $scCO_2$, employing methanol as a cosolvent to aid dissolution of the metal catalysts [102]. Selectivity towards methyl ketone formation (Scheme 5.43) was as high as 90.1%, with 63.5% conversion after 12 h. The reaction was reported to be homogeneous, even though the metal catalysts are only partially soluble in MeOH–$scCO_2$. The monophasic nature was attributed to organopalladium species formed during the catalytic cycle, which were believed to be soluble.

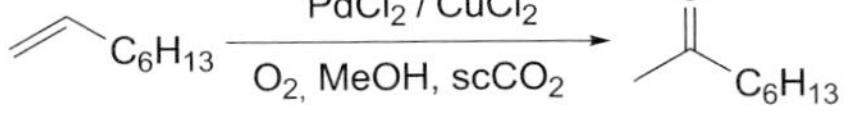

Scheme 5.43 Wacker oxidation of alkenes in $scCO_2$.

A related palladium(II)-catalyzed oxidation of acrylic esters to acetals in $scCO_2$ has been reported [103]. Methyl acrylate was oxidized in the presence of dioxygen and excess methanol, catalyzed by $PdCl_2$ with CuCl or $CuCl_2$ as cocatalyst, to generate the dimethylacetal as the major product (Scheme 5.44).

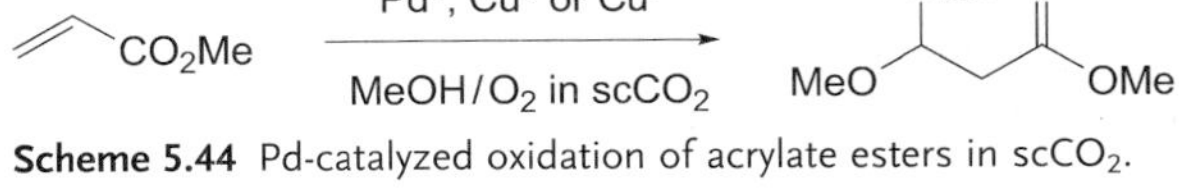

Scheme 5.44 Pd-catalyzed oxidation of acrylate esters in $scCO_2$.

5.6
Miscellaneous Catalytic Reactions in Supercritical Fluids

5.6.1
Metal-catalyzed Processes

A number of miscellaneous catalytic synthetic processes utilizing $scCO_2$ and other SCFs as reaction media have been reported. Although these topics have not attracted as much attention as those previously discussed, several important synthetic transformations have been reported, and a number of interesting observations have been made.

The coupling of two alkynes with CO_2 in the presence of nickel gives 2-pyrones. In $scCO_2$, a catalyst generated from $Ni(cod)_2$ and 1,4-diphenylphosphinobutane $[Ph_2P(CH_2)_4PPh_2]$ combined 3-hexyne and CO_2 to form tetraethyl-2-pyrone (Scheme 5.45) [7, 104]. In related chemistry, the cyclotrimerization of alkynes using $CpCo(CO)_2$ in $scCO_2$ has also been reported [60].

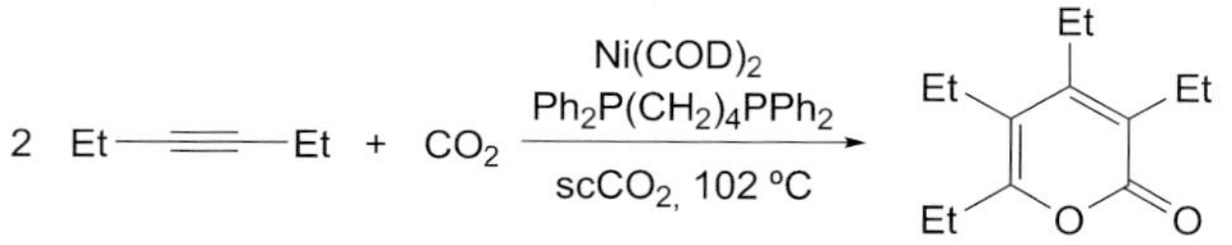

Scheme 5.45 Ni-mediated pyrone formation in $scCO_2$.

The co-cyclization of an alkyne with an alkene and carbon monoxide leading to cyclopentanones is known as the Pauson–Khand reaction, and this has been carried out in $scCO_2$ using dicobalt octacarbonyl as catalyst [105]. The intramolecular reaction of enyne **3** was optimized to give an 85% yield of **4** using a 2.5% catalyst loading, a CO pressure of 15 bar, a total pressure of 18 MPa and a temperature of 69 °C (Scheme 5.46). The reaction was also successful for a number of substituted enynes. An intermolecular reaction was also possible with phenylacetylene being coupled to norbornadiene (excess) to give the product in 87% yield (Scheme 5.47).

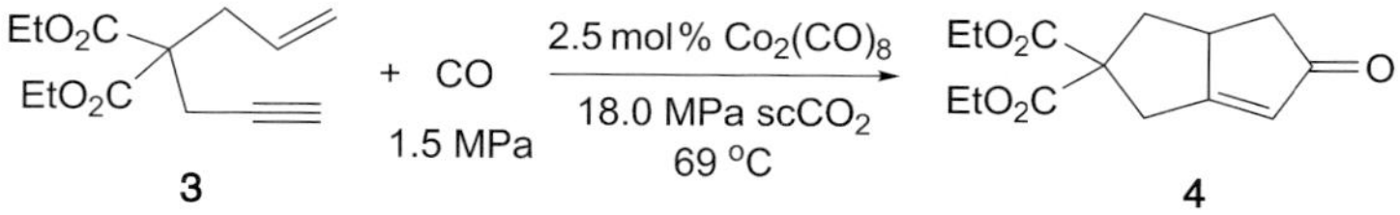

3 **4**

Scheme 5.46 Intramolecular Pauson–Khand reaction in $scCO_2$.

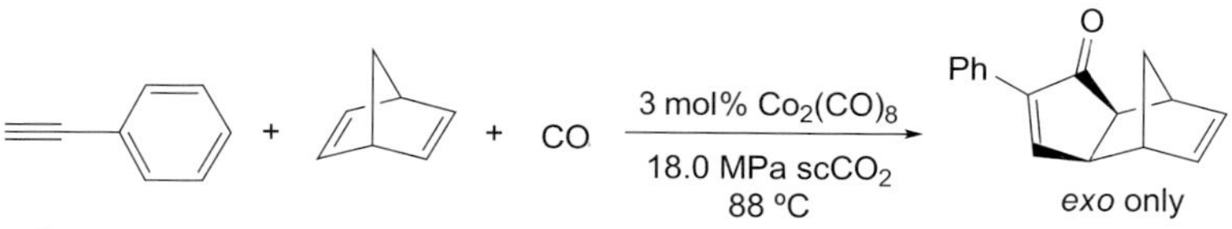

Scheme 5.47 Intermolecular Pauson–Khand reaction in $scCO_2$.

Alkene metathesis reactions in compressed CO_2 media show remarkable effects [106]. Ring-closing metathesis (RCM) of **6** was extremely sensitive to CO_2

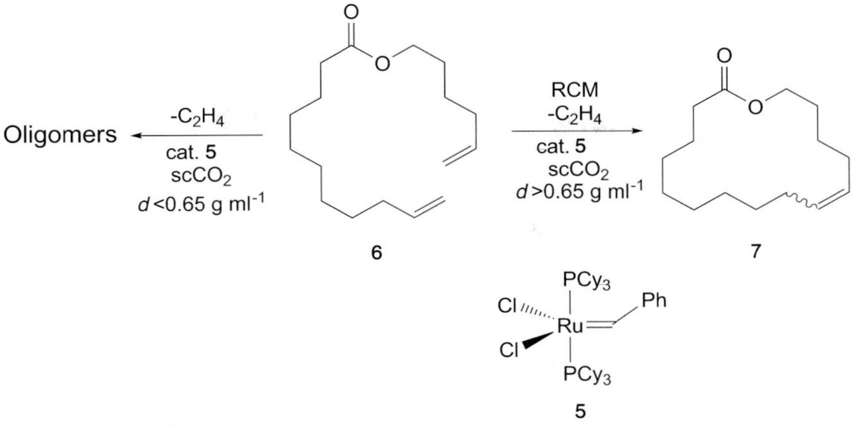

Scheme 5.48 Ring-closing metathesis in $scCO_2$.

density (pressure), with the 16-membered ring **7** being formed in excellent yield (>90%) at densities >0.65 g ml^{-1}, whereas mainly oligomers (70% with <10% **7**) were produced at low densities (Scheme 5.48). This density dependence was explained as a dilution effect, as at higher densities there are in effect more solvent molecules and the more dilute conditions favor the intramolecular reaction and *vice versa*. A number of other cyclizations were performed in good yield. An interesting observation was that catalyst **5**, which is normally deactivated in the presence of basic N–H groups, was active under such conditions in CO_2 solution. This was attributed to the *in situ* protection of the amine by reversible formation of the corresponding carbamic acid.

Catalytic asymmetric cyclopropanation reactions in $scCHF_3$ have been shown to be dependent on fluid pressure [107]. The cyclopropanation reaction between styrene and methyl phenyldiazoacetate, catalyzed by dirhodium species **8**, gave diastereomeric products **9** and **10** (Scheme 5.49). In conventional solvent, the *ee* of the major product **9** increased as the solvent polarity decreased (61% in CH_2Cl_2, 85% in pentane). However, in $scCHF_3$ the dielectric constant of the fluid changes significantly with pressure, particularly around the critical point. This was shown

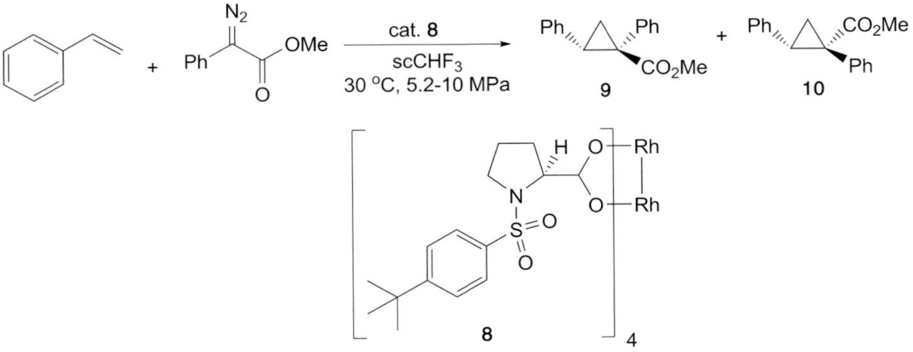

Scheme 5.49 Catalytic asymmetric cyclopropanation in $scCHF_3$.

experimentally as the selectivity of this reaction changes from 40% *ee* at 10 MPa to 77% *ee* at 5.2 MPa, in $scCHF_3$ at 30 °C. This effect was shown to be due to polarity, as the reaction was repeated in $scCO_2$ over a similar pressure range. Although selectivity was excellent (84% *ee* is equivalent to the best reported in conventional solvent), no dramatic pressure dependence was observed, which was expected as the dielectric constant of $scCO_2$ is not affected by pressure.

In more recent asymmetric cyclopropanations in $scCO_2$ using bisoxazoline (Box)–Cu(II) complexes as catalysts, it was shown that a substantial variation of selectivity with pressure is possible, and the selectivity also is sensitive to the structure of the ligand [6]. The results (enantioselectivity for formation of *trans*-cyclopropane) of the reaction between ethyl diazoacetate and styrene are summarized in Figure 5.1 and Table 5.1.

As expected, the *tert*-leucinol-derived ligands **11** and **12** gave high enantioselectivities. It is interesting, however, that only ligand **11** gave the highest enantioselectivity in conventional solvents (chloroform, hexane, toluene, or neat), whereas for all the others, higher enantioselectivity could be observed in $scCO_2$ after optimization (Table 5.1). In the $scCO_2$ reactions, the optimum selectivities varied considerably, and were not necessarily at the one- to two-phase boundary as is common in many of the previous reactions. The majority of these reactions were single phase, with only the 75 bar reactions showing an additional liquid layer at the bottom of the reactor. It is

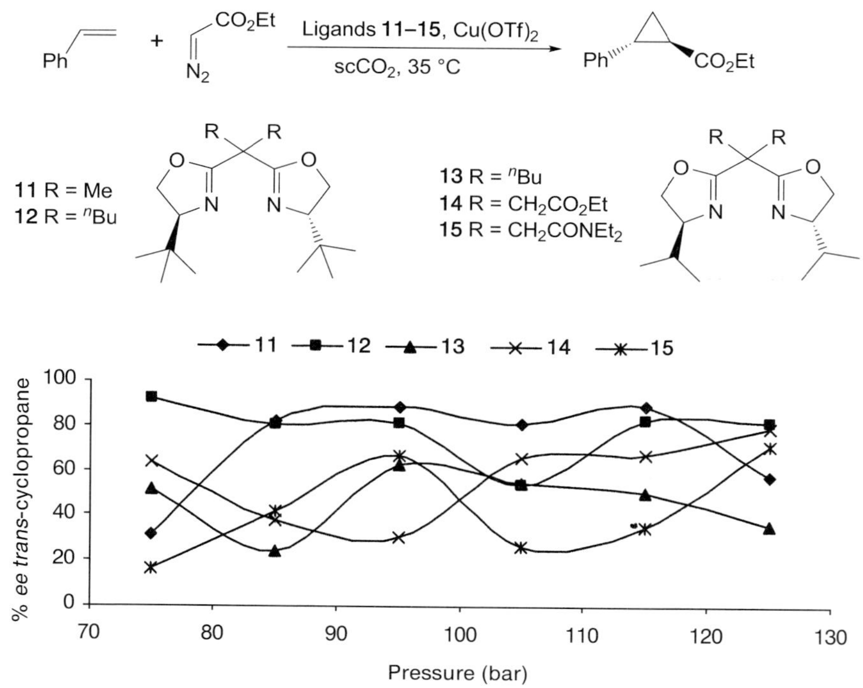

Figure 5.1 Variation of the enantioselectivity of cyclopropanations with $scCO_2$ pressure and Box ligand.

Table 5.1 Optimum enantioselectivity of cyclopropanations in $scCO_2$ and conventional solvents

Ligand	Optimum *ee* (%)	
	Conventional solvent	$scCO_2$ (pressure)
11	96 (hexane)	90 (115 bar)
12	90 ($CHCl_3$)	92 (75 bar)
13	57 (hexane)	63 (95 bar)
14	70 (hexane)	80 (125 bar)
15	62 (toluene)	72 (125 bar)

clear that the substituents on the methylene carbon of the Box ligand have a significant effect on the enantioselectivity (cf. ligands **13** and **14**), and that the pressure for optimum selectivity depends very much on the ligand, with a complex range of effects operating.

In other asymmetric processes, the catalytic enantioselective hydrovinylation of styrenes in $scCO_2$, using a catalytic system of η^3-allyl-Ni(**16**)(BARF), where **16** is a dimeric azaphospholene ligand and the BARF anion has been described previously (Scheme 5.20) [108], gave quantitative conversion, high regioselectivity (>70% and higher than in CH_2Cl_2 at the higher temperatures) and excellent enantioselectivity (83.2–91.6%) (Scheme 5.50).

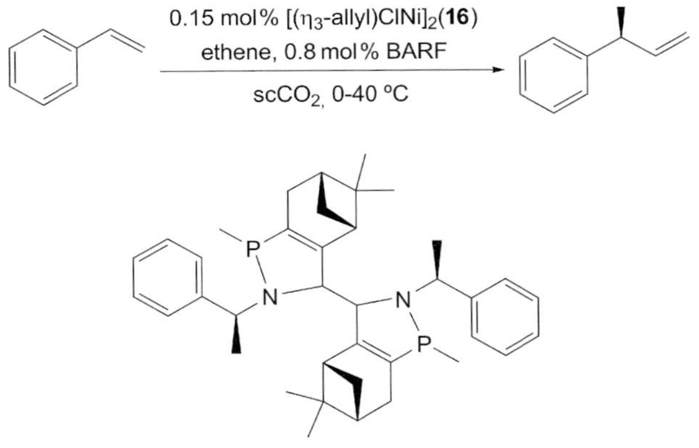

16

Scheme 5.50 Enantioselective vinylation of styrene in $scCO_2$.

The rhodium-catalyzed hydroboration of styrenes with catecholborane (HBcat), rhodium catalyst precursor **17**, and phosphorus ligand **18** gave a quantitative conversion in $scCO_2$ (Scheme 5.51). Reactivities with other ligands were also reported but gave lower yields and/or selectivities for reaction in $scCO_2$ [109]. The catalytic reaction was homogeneous and exhibited higher rates and regioselectivity compared with the equivalent reaction performed in liquid perfluoromethylcyclohexane or THF solvents.

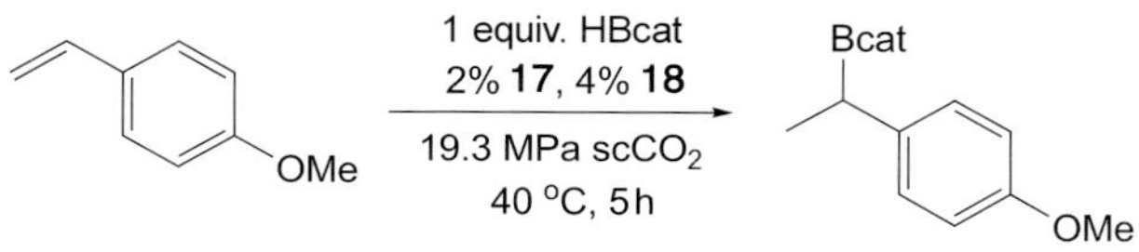

(hfacac)Rh(cyclooctene)$_2$ — **17**

$Cy_2P(CH_2CH_2C_6F_{13})$ — **18**

Scheme 5.51 Asymmetric hydroboration in $scCO_2$.

5.6.2 Base-catalyzed Processes

The Baylis–Hillman reaction has been carried out in $scCO_2$ and gives better conversions and reaction rates than comparable solution-phase reactions (Scheme 5.52) [110]. Interestingly, if the reaction is carried out in the presence of an alcohol, the major product is an ether resulting from a novel three-component coupling reaction, which occurs only in the presence of CO_2.

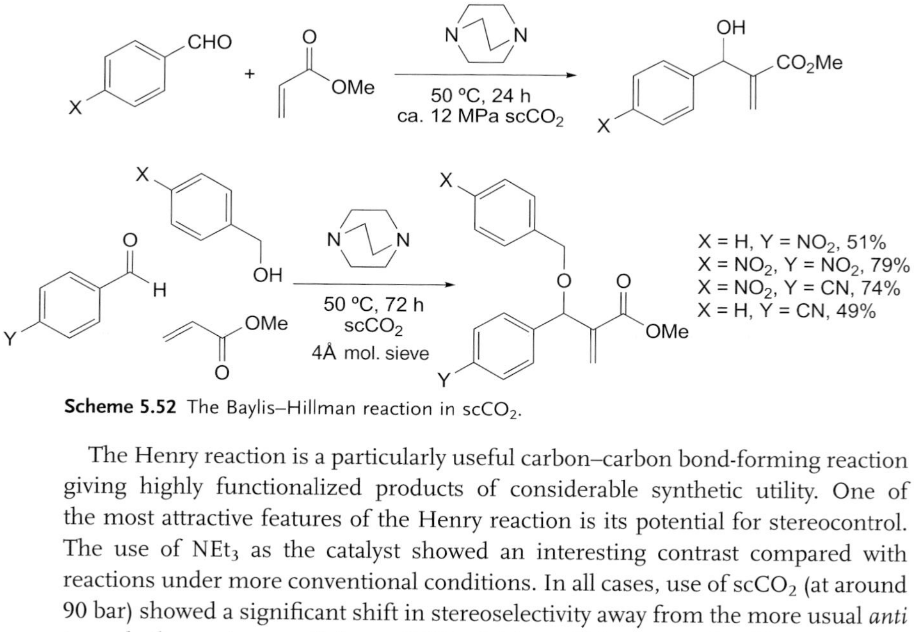

Scheme 5.52 The Baylis–Hillman reaction in $scCO_2$.

The Henry reaction is a particularly useful carbon–carbon bond-forming reaction giving highly functionalized products of considerable synthetic utility. One of the most attractive features of the Henry reaction is its potential for stereocontrol. The use of NEt_3 as the catalyst showed an interesting contrast compared with reactions under more conventional conditions. In all cases, use of $scCO_2$ (at around 90 bar) showed a significant shift in stereoselectivity away from the more usual *anti* towards the *syn* isomer (Scheme 5.53), which could be manipulated by working at

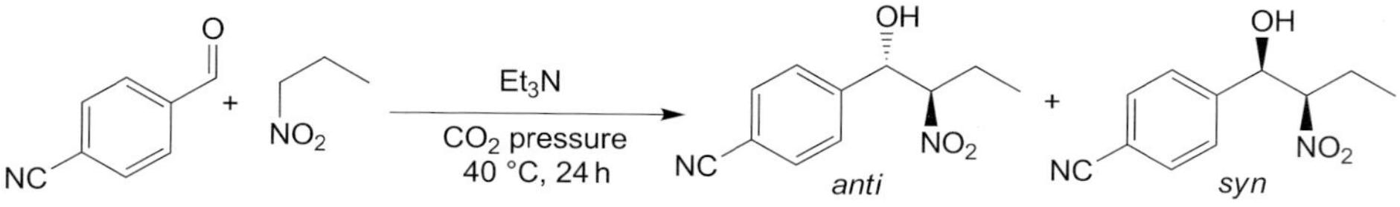

Scheme 5.53 The Henry reaction in $scCO_2$.

different pressures. At low CO_2 pressures, rapid equilibration of the kinetic product mixture to the thermodynamically more stable *anti* isomer was dominant, whereas at higher pressures, the reaction was significantly slower, particularly under supercritical conditions, and kinetic control of the reaction was dominant with the product distribution tending towards ~10% in favor of the *syn* isomer [111].

The use of $scCO_2$ as both the reaction medium and carbonyl source for the production of carbamates from amines is an attractive process (Scheme 5.54) [112]. Using potassium carbonate in conjunction with an ammonium salt catalyst, a number of primary and secondary aliphatic and also aromatic amines reacted well with butyl chloride to give the corresponding carbamate in yields in the range 72–90%. Use of potassium phosphate in place of potassium carbonate demonstrated that the carbonyl source was the CO_2 and not the carbonate. This novel methodology is an attractive catalytic one-pot alternative to the use of phosgene in urethane synthesis.

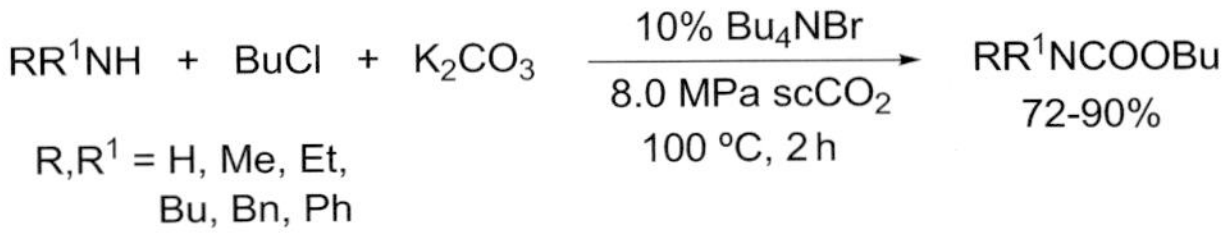

Scheme 5.54 Carbamate synthesis in $scCO_2$.

In a related process, vinylcarbamates have been synthesized using carbon dioxide, secondary amines, and terminal alkynes [113]. A comparison of conversion and selectivity for $[RuCl_2(C_5H_5N)_4]$- and $[RuCl_2(\eta^6\text{-}C_6H_6)(PMe_3)]$-catalyzed vinylcarbamate synthesis in $scCO_2$ and toluene was studied (Scheme 5.55). In general, conversions were significantly higher in $scCO_2$ than in toluene, with the latter catalyst displaying the highest activity.

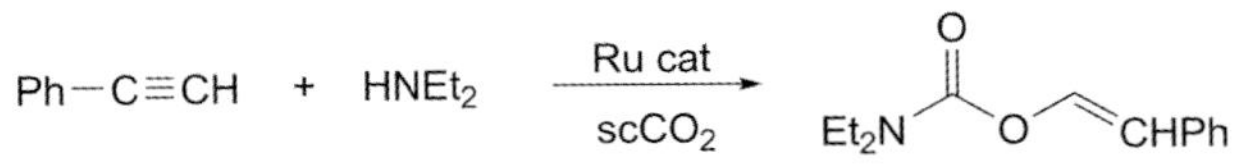

Scheme 5.55 Vinylcarbamate synthesis in $scCO_2$.

There have also been extensive investigations on the formation of carbonates using CO_2 as a feedstock. For example, synthesis of dimethyl carbonate can be achieved using methanol, methyl iodide, a base, and CO_2 [114]. Alternative approaches utilize just methanol and CO_2, with an appropriate catalyst [e.g. $Bu_2Sn(OMe)_2$] [115], or from an epoxide and CO_2, although only a limited range of substrates (e.g. propylene oxide, ethylene oxide, styrene oxide) have been investigated [116]. A related process with aziridines to give 2-oxazolidinones has also been reported [117].

5.6.3
Acid-Catalyzed Processes

A range of Brønsted- and Lewis acid-catalyzed processes have been investigated in SCFs. Using a fixed-bed heterogeneous polysiloxane-supported acid catalyst, continuous Friedel–Crafts alkylation of aromatics has been demonstrated [118].

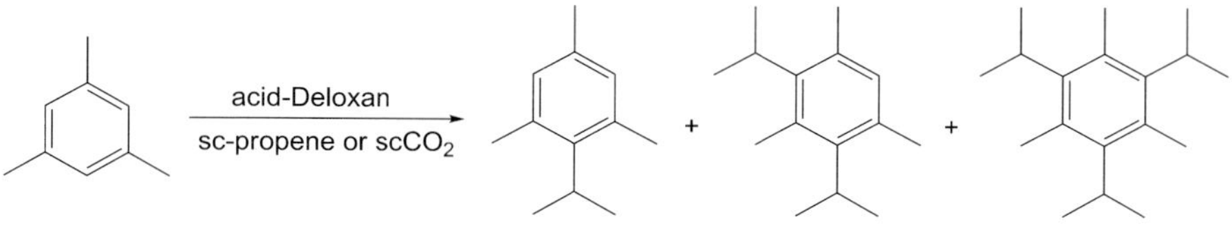

19 20 21

Scheme 5.56 Friedel–Crafts alkylation of mesitylene in SCFs.

Mesitylene was alkylated in supercritical propene ($T_c = 91.9\,°C$, $P_c = 46.0$ bar), which acted both as solvent and alkylating agent (Scheme 5.56). Temperatures required were high (160 °C) however because of the short residence time, and although the monoalkylated species **19** was the major product (25% yield, 6% **20**), selectivity was poor. Selectivity was much improved, however, if $scCO_2$ was used as the reaction medium with 2-propanol as the alkylating agent. At a molar ratio mesitylene to 2-propanol of 2 : 1, a pressure of 20 MPa, a catalyst temperature of 250 °C, and a flow rate of 0.60 g min^{-1}, **19** was produced as the only product with a conversion of 42%, at a rate of 0.13 g min^{-1}. Alkylation of anisole also gave similar results, with increased selectivity achieved in the $scCO_2$–2-propanol system [119]. Use of *m*-cresol as starting material provides a selective route for the synthesis of thymol [120].

Using the same continuous flow reactor apparatus, the acid-catalyzed dehydration of alcohols in $scCO_2$ was also investigated [121]. The dehydration of 1,4-butanediol gave tetrahydrofuran with a quantitative conversion at a flow rate of 0.5 ml min^{-1} through a 10 ml reactor (Scheme 5.57). The dehydration of a number of other diols, to give various cyclic ethers and acetals in more moderate yields, was also reported.

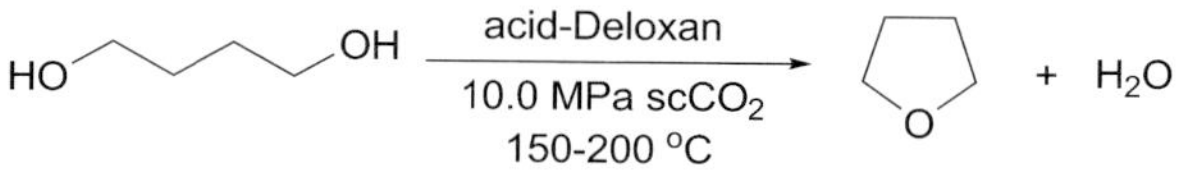

Scheme 5.57 Acid-catalyzed etherification in $scCO_2$.

A variety of Lewis acid-catalyzed alkylation reactions have been reported in $scCO_2$ which are aided by the addition of PEG derivatives. This includes the Mannich and aldol reactions of silyl enolates with aldehydes and imines in $scCO_2$ (Scheme 5.58) [122, 123]. PEGs act as surfactants and form colloidal dispersions in $scCO_2$, manifested as emulsions, and can be shown to accelerate reactions. In the case of scandium-catalyzed aldol reactions of silyl enolates with aldehydes, poly (ethylene glycol) dimethyl ether [$PEG(OMe)_2$, average $M_w = 500$] was found to be more effective than PEG itself [122]. An alternative additive, 1-dodecyloxy-4-heptadecafluorooctylbenzene (**22**), was subsequently reported to work as an efficient surfactant to accelerate aldol, Mannich, and Friedel–Crafts alkylation of indoles in $scCO_2$ (Scheme 5.59) [124]. This additive was designed to incorporate a CO_2-philic unit (perfluoroalkyl chain), and a lipophilic unit (alkyl chain) in the same molecule.

Water under an atmosphere of pressurized CO_2 is mildly acidic due to the formation of carbonic acid (Scheme 5.60). Simple pressure release at the end of the reaction raises the pH back to levels which require minimal neutralization.

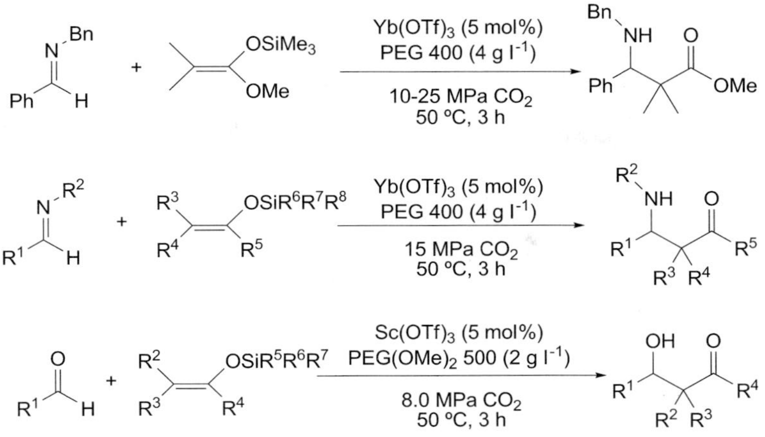

Scheme 5.58 Lewis acid-catalyzed alkylation reactions in $scCO_2$.

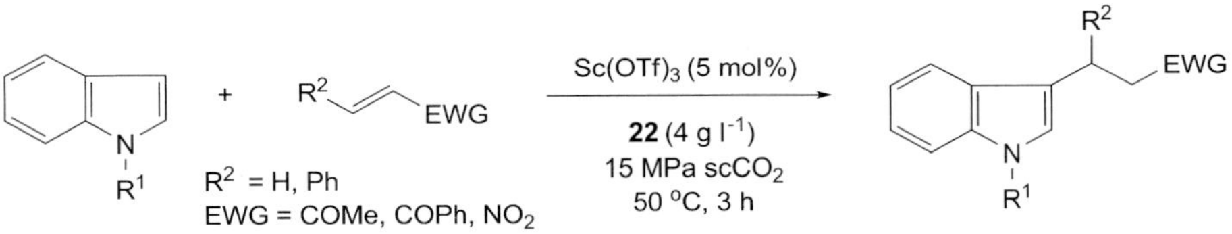

Scheme 5.59 Friedel–Crafts alkylation of indole derivatives in $scCO_2$.

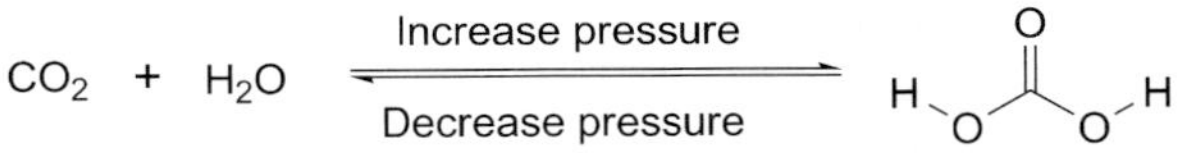

Scheme 5.60 Carbonic acid formation from water and CO_2.

The acidic environment of water under CO_2 has been shown to catalyze a range of hydrolysis reactions. Representative examples are shown in Scheme 5.61. Thus, mixing a ketal with water, pressurizing with 2 MPa CO_2, and heating to 65 °C gave complete hydrolysis to cyclohexanone and ethylene glycol within 4 h. This was also successful with more complex ketals such as those derived from D-mannitol, acetals of unsaturated aldehydes, epoxides also undergo hydrolysis to diols, and *N*-Boc deprotection could also be effected on specific substrates. In all cases, little or no reaction was observed under comparable conditions in the absence of CO_2. Note that these reactions were best at subcritical pressures, which prevented two phases from being formed. However, increasing to supercritical pressure and temperatures post-reaction allowed selective extraction of the more lipophilic products into the $scCO_2$ phase.

5.7 Cycloaddition Reactions in Supercritical Fluids

The Diels–Alder reaction has been the subject of numerous studies in $scCO_2$ media, from both synthetic and mechanistic points of view. In almost all cases, the most

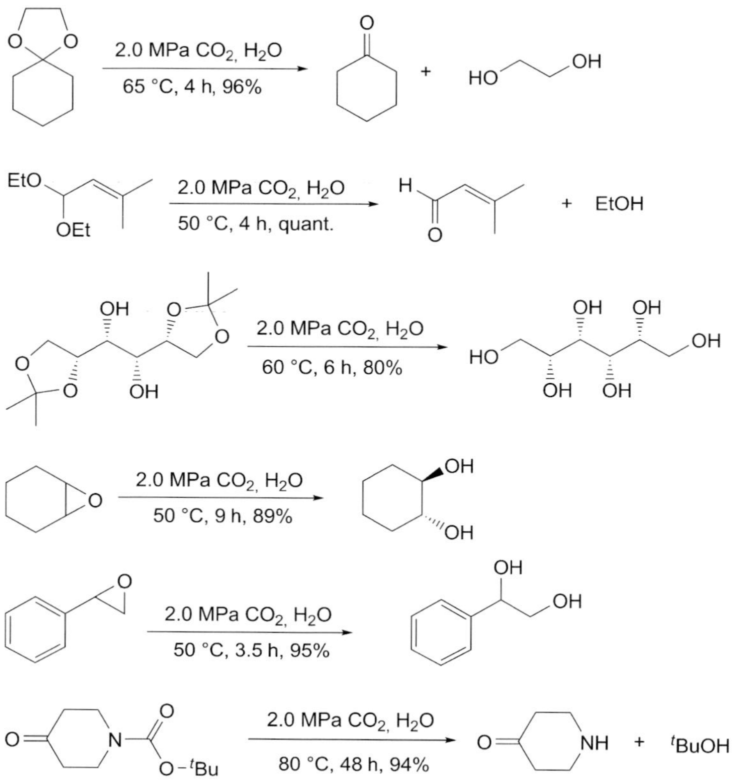

Scheme 5.61 Carbonic acid-catalyzed hydrolysis of acid-sensitive functional groups.

significant differences (usually enhanced rates and/or selectivities) were observed around the critical point of the reaction in CO_2 [125]. Some of the most important results have been observed in the presence of Lewis acid catalysts such as scandium trifluoromethanesulfonate, and again, optimum selectivity (in this case the *endo*:*exo* ratio) was usually observed around the critical point of the reaction mixture (Scheme 5.62) [126].

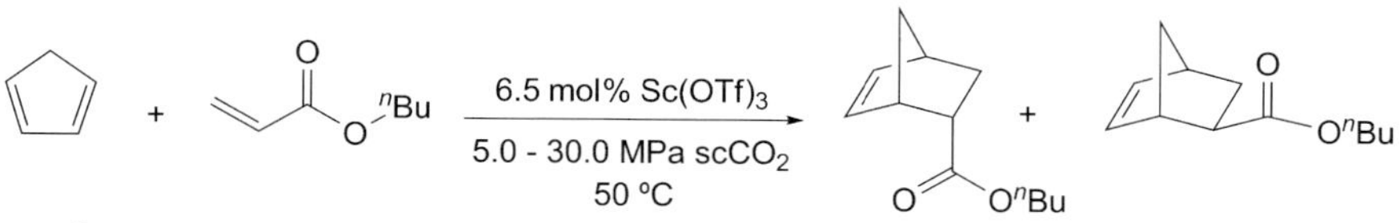

Scheme 5.62 Diels–Alder reaction in $scCO_2$ catalyzed by $Sc(OTf)_3$.

Other scandium perfluoroalkanesulfonates are also efficient and have enhanced solubility in $scCO_2$ [123]. The optimum catalyst, $Sc(OSO_2C_8F_{17})_3$, was used to catalyze an aza-Diels–Alder reaction. Danishefsky's diene **23** reacted with imine **24** to give the aza-Diels–Alder adduct **25** in 99% yield (Scheme 5.63). Silica gel has also been reported to be an efficient catalyst for Diels–Alder reactions in $scCO_2$ [127].

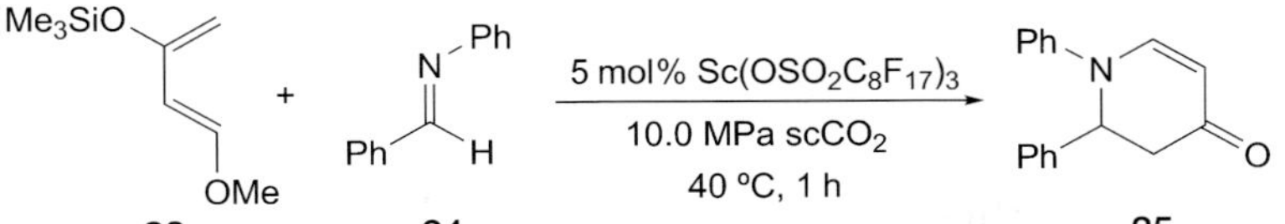

23 24 25

Scheme 5.63 Aza-Diels–Alder reactions in $scCO_2$.

Diels–Alder reactions involving chiral auxiliaries have also been reported [128]. As part of a study of solvent effects on stereoselectivity, the reaction between cyclopentadiene and the dienophile **26** was performed in $scCO_2$ (Scheme 5.64). In conventional solvents, the diastereomeric excess (*de*) generally increased with polarity (58% *de* in CCl_4, 92% *de* in water), whereas in $scCO_2$, the best selectivity was observed around the critical point (65% conversion, 93% *de* at 33 °C, 7.4 MPa) although a better overall result was obtained at slightly higher temperature and pressure (100% conversion, 92% *de* at 43 °C, 7.8 MPa).

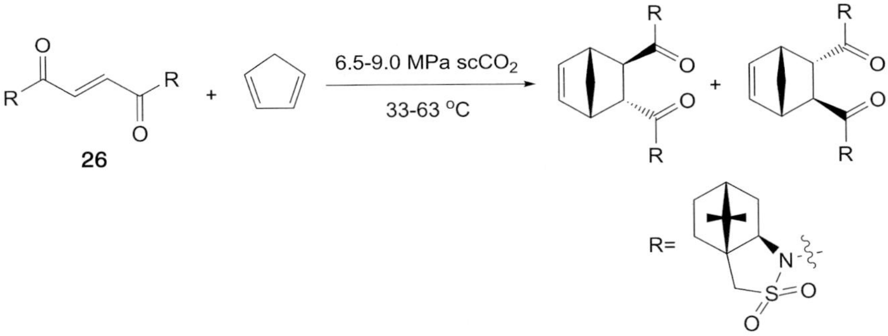

26

Scheme 5.64 Auxiliary controlled Diels–Alder reactions in $scCO_2$.

Other auxiliary controlled Diels–Alder reactions have also been reported in $scCO_2$ using rare earth metal catalysts [129, 130]. A selection of rare earth triflates were shown to catalyze the reaction of cyclopentadiene with a chiral dienophile (Scheme 5.65). The reaction proceeded rapidly in $scCO_2$, to give the *endo* adduct with higher diastereoselectivity than obtained in dichloromethane [130].

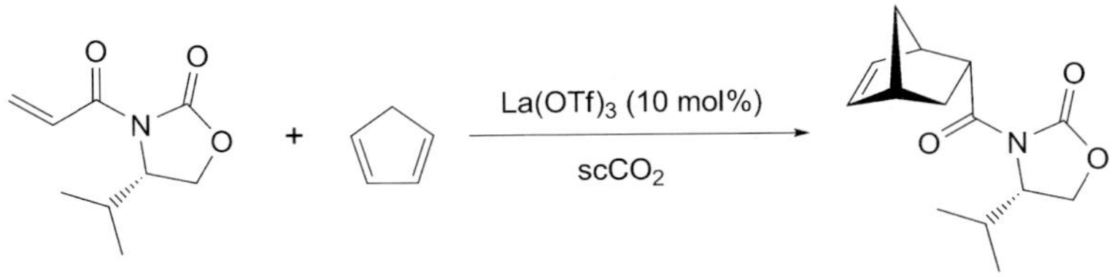

Scheme 5.65 La(OTf)$_3$-catalyzed asymmetric Diels–Alder reaction in $scCO_2$.

More recently, a number of interesting investigations into 1,3-dipolar [3 + 2] cycloadditions in $scCO_2$ have emerged. One pertinent example describes the addition of mesitonitrile oxide (MesCNO) to various dipolarophiles [131]. Quantitative

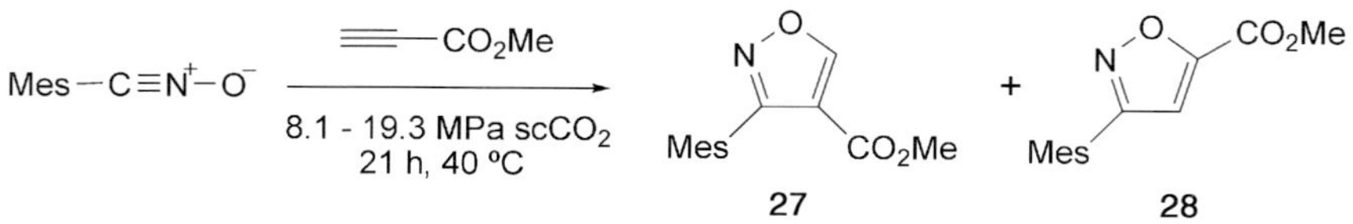

Scheme 5.66 1,3-Dipolar cycloaddition in $scCO_2$.

conversion was obtained for a broad range of dipolarophiles including electron-deficient, electron-rich, hindered, and strained examples. Effects of CO_2 pressure on the regioselectivity of the cycloaddition of MesCNO to propiolate were examined, with 4-methoxycarbonylisoxazole being the principal regioisomeric product (Scheme 5.66). The solvent density was varied (0.294–0.833 g ml^{-1} at 40 °C) by adjustment of the pressure (8.1–19.3 MPa), and the product ratio **27**:**28** was shown to be tuned between 2.5 : 1 and 3.8 : 1, reaching its maximum at a solvent density of 0.611 g ml^{-1} (9.8 MPa).

In the 1,3-dipolar cycloaddition of methyl propiolate to 3-phenylsydnone, two regioisomers [3-carbomethoxy-1-phenylpyrazole (**29**) and 4-carbomethoxy-1-phenylpyrazole (**30**), Scheme 5.67] are produced [132]. Experiments in a conventional solvent (toluene) displayed a mean selectivity of 3.62 in favor of isomer **29**. When CO_2 was employed as reaction medium (7.6–30.4 MPa), mean selectivity was tuned between 3.14 and 6.56, with temperature optimization also playing an important role. The optimum selectivity was achieved at the highest pressure and lowest temperature (30.4 MPa, 353 K); unfortunately, this increase in regioselectivity was accompanied by a drastic reduction in yield of 50% compared with the lower pressure, higher temperature (7.6 MPa, 423 K) case. Carbon dioxide was also a product of this particular reaction, and it was suggested by the authors that increased pressure of the solvent CO_2 could be expected to affect the yield.

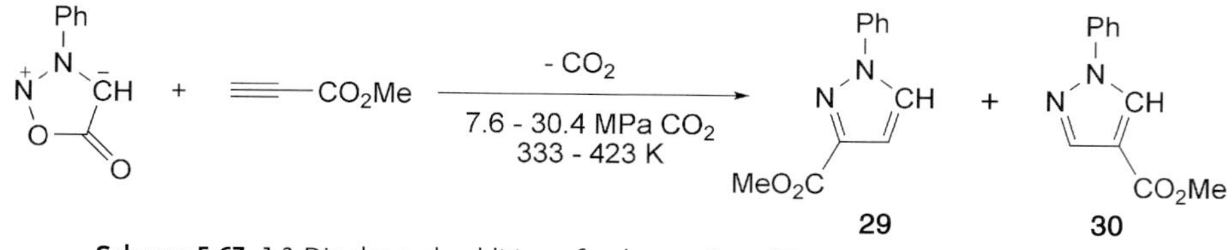

Scheme 5.67 1,3-Dipolar cycloaddition of sydnones in $scCO_2$.

5.8 Photochemical Reactions in Supercritical Fluids

Stereoselective photolytic denitrogenation of labeled *exo*-[$^2H_2$]diazabicyclo[2.2.1] heptene (**31**) has been shown to display pressure sensitivity in supercritical CO_2 and C_2H_6 (Scheme 5.68) [133]. Research on the gas-phase thermolysis [134] and liquid-phase photolysis [135] of the deuterated DBH derivative showed that the inverted *exo*-deuterated housane bicyclo[2.2.1]pentane (**32**) was formed as the major product. Of mechanistic interest is whether the diazenyl (**34**) or the denitrogenated

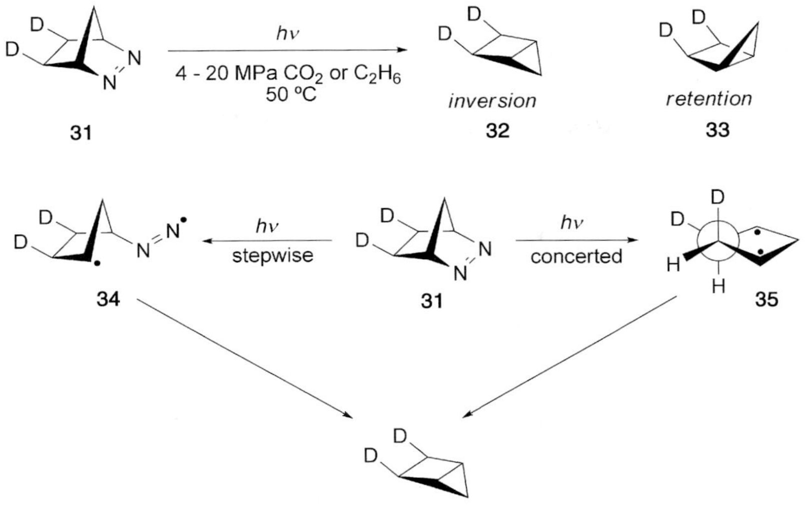

Scheme 5.68 Photochemical denitrogenation of isotopically labeled diazabicyclo[2.2.1]heptene.

cyclopentane-1,3-diyl (**35**) diradical mediates formation of the diastereomeric housanes **32** and **33**, via stepwise or concerted pathways, respectively, as both routes would favor inversion over retention in this example.

Consistent with previous research, the authors described a dependence of viscosity, and therefore friction of the medium, on the potential pathways, supporting the case for the singlet diazenyl diradical intermediate **34** as $k_{inv} > k_{ret}$. The use of supercritical CO_2 and C_2H_6 offered the ability to investigate the effect of viscosity on the $k_{inv/ret}$ ratio by adjustment of the pressure. Indeed, preference for the inversion process was favored at lower pressures in both solvents. A significant increase in viscosity accompanied an increase in pressure from 4 to 20 MPa (fourfold in CO_2, fivefold in C_2H_6), leading to a 2.3-fold decrease in the k_{inv}/k_{ret} ratio.

Photoaddition of alkyl alcohols to 1,1-diphenyl-1-alkenes offers potential for stereocontrol by modification of the supercritical reaction conditions. Enantiodifferentiating anti-Markovnikov addition was reported as displaying a substantial pressure dependence at different temperatures close to the critical density (Scheme 5.69) [136]. At 31 °C, the *ee* varied inconsistently (22–31%) across the pressure range (7.4–18.6 MPa), as might be expected so close to the critical temperature. Above 35 °C, *ee* correlated with density, as a function of pressure, with a sharp increase at pressures giving rise to the unit reduced density ($d_r = 1$) at the temperatures investigated (8.1, 9.9 and 14.2 MPa at 35, 45 and 70 °C, respectively). It should be noted that presence of methanol gave rise to a marked increase in the polarity of the $scCO_2$ system, which was conducive to this polar photoaddition, where the importance of solvent polarity control around the exciplex is paramount for enantiocontrol. When the study was extended to include higher chain alkyl alcohols [137], the steric bulk of the alcohol became a factor. Longer

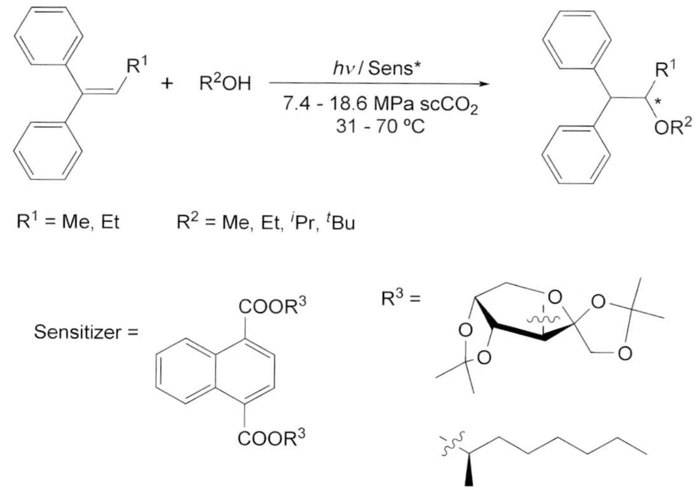

Scheme 5.69 Asymmetric photochemical addition of alcohols to 1,1-diphenylalkenes.

periods of irradiation were necessary, up to 24 h. However, for ^{t}BuOH, the formation of unwanted side products such as benzophenone occurred at the expense of the required adduct.

The photochemical addition of carbon dioxide to anthracene was conducted in acetonitrile, DMF, and supercritical CO_2 [138]. The reaction proceeded through a photoinduced electron transfer mechanism. When the reaction was performed in conventional non-polar aprotic solvents, with a continuous flow of gaseous CO_2, low amounts of the dihydrocarboxylic acid **36** were noted, in agreement with existing results in the literature (trace amounts and 11% in acetonitrile and DMF, respectively) [139]. In these reaction systems, the concentration of CO_2 was found to be a limiting factor, and the solvent (DMA or MeCN) was also participating as the necessary hydrogen donor. When the system was transferred to $scCO_2$, addition of a hydrogen donor was necessary (cyclohexane) and DMA was selected as a cosolvent additive to ensure homogeneity (Scheme 5.70). Moderate conditions were employed for the initial trials (35 °C, 11.7 MPa, 5 h of photoirradiation), resulting in significantly improved results (34.5% dihydrocarboxylic acid) over traditional media. The influence of pressure was explored (8.2–24.1 MPa), and it was found that the yields were indeed further improved as pressure was raised (47%, 24.1 MPa). Additional

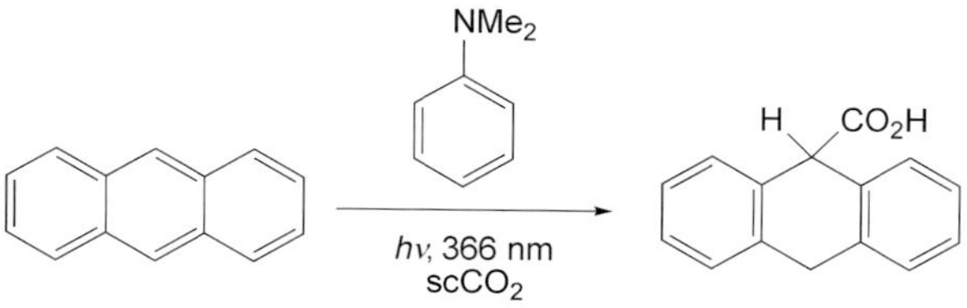

36

Scheme 5.70 Photochemical addition of CO_2 to anthracene.

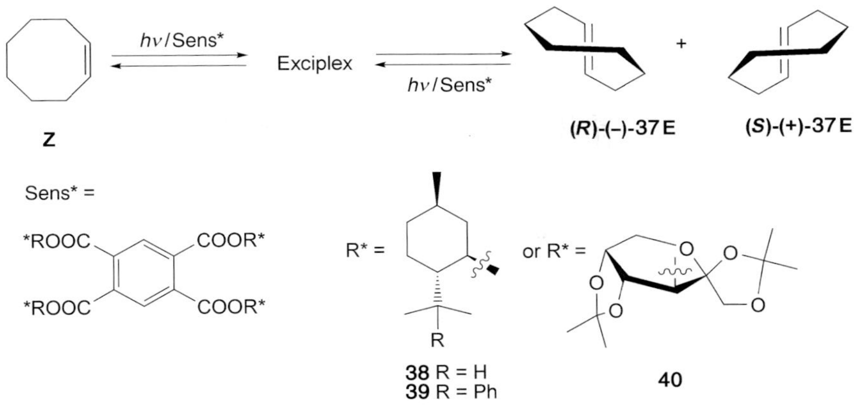

Scheme 5.71 Enantiodifferentiating photoisomerization in $scCO_2$.

yield optimization was made possible by the use of an alternative hydrogen donor (2-propanol) and the reaction was even found to proceed to complete consumption of anthracene after only 2 h of photoirradiation and a relatively moderate pressure (13.8 MPa, 57%). The authors' interpretation of the particular effect of pressure tuning of the reaction system was that an increased concentration of CO_2 is favorable for product formation, rather than a true thermodynamic effect on the rate constants for the electron transfer reaction.

Enantiodifferentiating photoisomerization of (*Z*)-cyclooctene has been shown to proceed in $scCO_2$, with an interesting pressure dependence upon the *ee* [140] The photoisomerization was sensitized by enantiopure benzenetetracarboxylates (Scheme 5.71) at mild temperature (45 °C) and was complete after only 1 h of irradiation. A non-linear relationship between pressure and *ee* value was reported over the range 7.8–23 MPa, with a notable maximum highlighted at 9.8 MPa (0.468 g ml^{-1}, 45% *ee*) when the terpenoid ester **38** was employed. The results, displayed in terms of relative rate constants (k_S/k_R), show a notable enhancement of *ee* in the low-pressure region (8–11 MPa) of the investigation, where solvent clustering was believed to be beneficial.

Carbon dioxide has been demonstrated to be a useful potential replacement for benzene as solvent in the photoinduced addition of 1,4-benzoquinone to butyraldehyde in the presence of benzophenone (a sensitizer) (Scheme 5.72) [141]. This is an interesting development with regard to the previously published general photochemical process toward the synthesis of acylhydroquinones [142], which is an alternative to the classic Friedel–Crafts acylation reaction, avoiding the use of acid

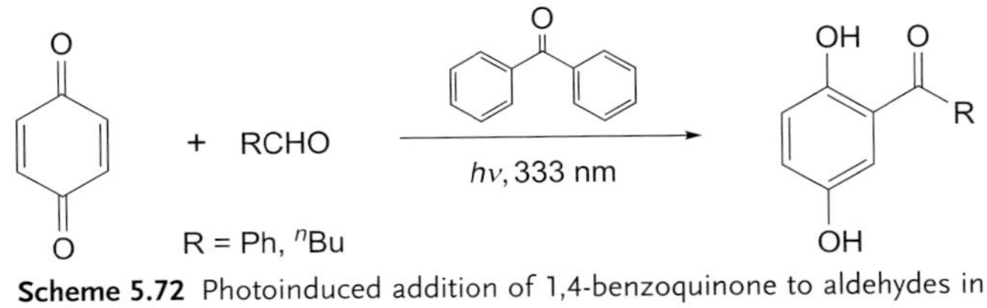

Scheme 5.72 Photoinduced addition of 1,4-benzoquinone to aldehydes in $scCO_2$.

chlorides and strong Lewis acids such as $AlCl_3$. At 50 °C the product yield was shown to increase with increasing pressure, reaching a maximum of 65% at approximately 56 MPa. Solubility of the reagents was highlighted as being essential to promote higher yields. To this end, addition of an alcoholic cosolvent (5% ^{t}BuOH) proved beneficial. Without cosolvent modifier, a yield of 60% was obtained after 48 h at 40 MPa pressure, whereas at a similar pressure with addition of polar modifier the yield increased to 81%. This also represented an increase over the yield in modified benzene (5% ^{t}BuOH, 74%).

5.9 Radical Reactions in Supercritical Fluids

As SCFs are low-density fluids with low viscosity resulting in high rates of diffusion, they offer potential benefits for radical processes. A significant proportion of the radical reactions studied have been in polymerization-type applications [7, 143]. Although radical reactions were some of the first synthetic processes to be considered in SCFs, relatively few literature examples exist.

Many free radical halogenation reactions have historically been carried out in CCl_4, but $scCO_2$ is an attractive alternative. The reaction of molecular bromine with toluene and ethylbenzene forms the corresponding benzylic bromides in good yields (Scheme 5.73), and Ziegler bromination using *N*-bromosuccinimide (NBS) was also successful [144].

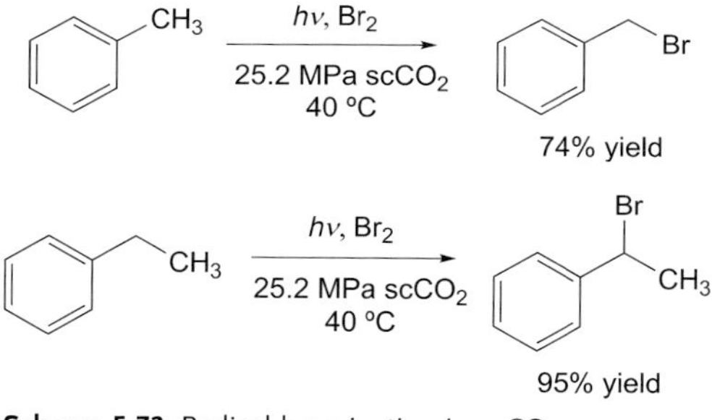

Scheme 5.73 Radical bromination in $scCO_2$.

$ScCO_2$ is also a practical medium for free-radical carbonylation of organic halides to ketones or aldehydes (Scheme 5.74) [145]. Using a silane-mediated carbonylation of an alkyl halide, alkene, and CO using azobisisobutyronitrile (AIBN) initiator gave yields

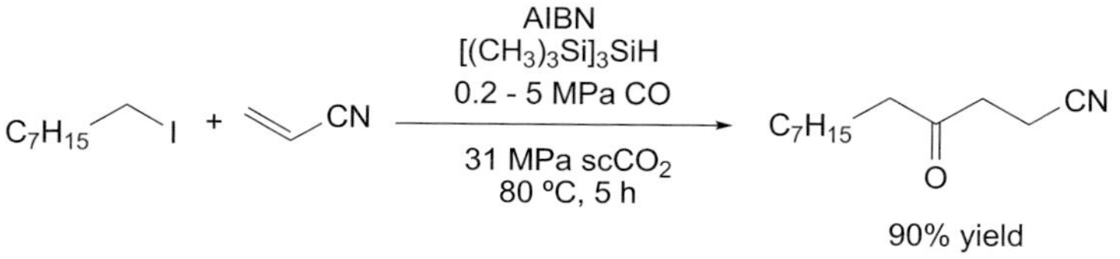

Scheme 5.74 Free-radical carbonylation in $scCO_2$.

comparable to those obtained in benzene [146]. Related intramolecular reactions also proceeded efficiently, and showed interesting pressure-dependent selectivity.

The use of tin hydride reagents in $scCO_2$ has also been reported [147]. Both tributyltin hydride (**41**) and *tris*(perfluorohexylethyl)tin hydride (**42**) were investigated, **42** being miscible under the reaction conditions (90 °C and 27.2 MPa) whereas **41** was insoluble. Bromoadamantane was reduced by **42** (initiated by AIBN) under $scCO_2$ conditions to give adamantane in 90% yield after 3 h (Scheme 5.75). The work-up for this reaction was particularly clean by partitioning between benzene and perfluorohexane. Surprisingly, despite its insolubility, **41** also facilitated reduction, with adamantane being isolated in 88% yield. Reaction of steroidal bromides, iodides, and selenides also gave the corresponding reduced products in high yields (85–98%).

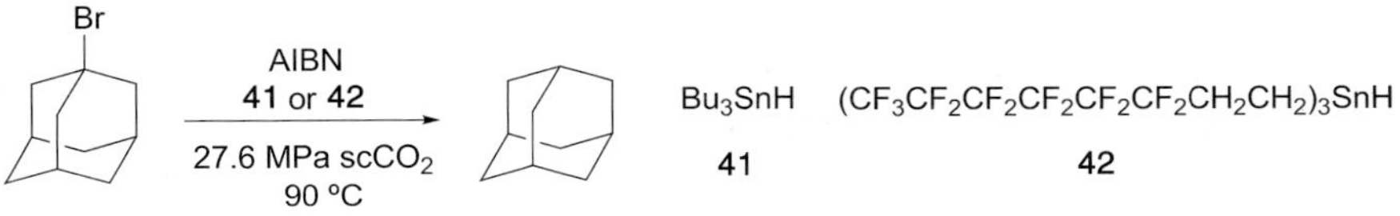

Scheme 5.75 Tin hydride reduction in $scCO_2$.

Several radical cyclization reactions were also studied. Reduction of 1,1-diphenyl-6-bromo-1-hexene (**43**) with **42** gave the 5-*exo* product **44** in 87% yield (Scheme 5.76). Similarly, reduction of aryl iodide **45** with **42** gave quantitative conversion to cyclized product **46** (Scheme 5.76). Interestingly, no reaction was observed with **41** in either case.

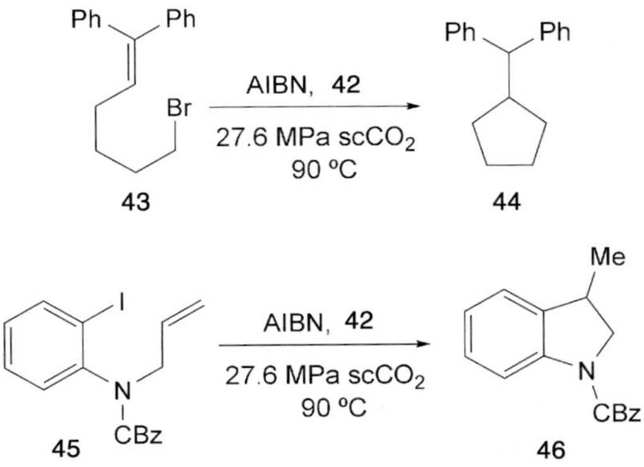

Scheme 5.76 Radical cyclization reactions in $scCO_2$.

5.10 Biotransformations in Supercritical Fluids

The prospect of using enzymes as heterogeneous catalysts in SCF media has created a significant interest since the mid-1980s. The low viscosity and high diffusion rates offer the potential to increase the rate of mass transfer-controlled reactions, and because enzymes are not usually soluble in SCFs, dispersion of the free enzyme potentially allows simple separation without the need for immobilization [148]. A number of factors and questions have been covered in the literature, such as the stability of enzymes in SCFs, the role of water in an essentially non-aqueous environment, temperature and pressure effects, and the potential enhancement due to high diffusivity. Further information, particularly regarding experimental procedures and theoretical kinetics, can be obtained from alternative recent reviews [7, 148].

A number of biocatalytic reactions have been investigated in $scCHF_3$ and $scCO_2$, with examples exhibiting degrees of stereocontrol. The substantial polarity change with pressure (or temperature) of $scCHF_3$ afforded an opportunity for tunable selectivity [149] and activity [150]. Although there are fewer examples of such tunable product control in $scCO_2$, interesting examples of a range of biotransformations have been reported.

In the Novozym-catalyzed enantioselective acetylation of racemic 1-(*p*-chlorophenyl)-2,2,2-trifluoroethanol (Scheme 5.77), the enantiomeric selectivity (*E*) increased continuously from 10 to 50 as the pressure was decreased from 19 to 8 MPa at 55 °C, with the *S*-enantiomer reacting preferentially [151]. Analogous studies at reaction temperatures of 31, 40, and 60 °C highlighted a rapid change in *E* at the lower temperatures within a small range of pressures around 10 MPa. At higher reaction temperatures (55 and 60 °C), changes in *E* values were only gradual. It was claimed that both temperature and density affected the stereochemical outcome. Studies of the enantioselective esterification of (*R*)-1-phenylethanol in fluoroform showed that control of the dielectric constant, by variation of pressure and temperature of the solvent, directly affected stereocontrol [150].

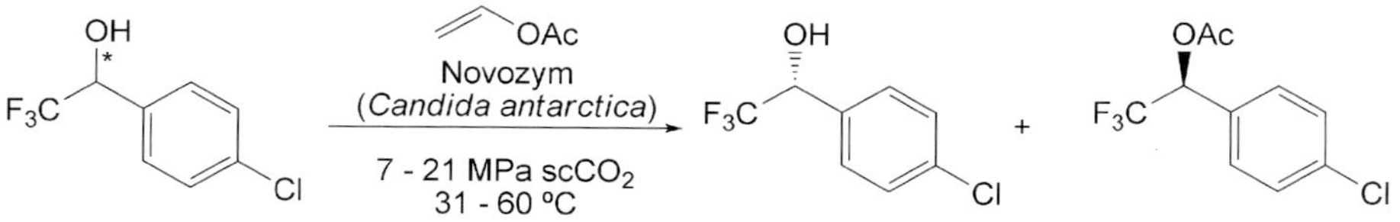

Scheme 5.77 Enzymatic resolution of 1-(*p*-chlorophenyl)-2,2,2-trifluoroethanol in $scCO_2$.

The enzyme alkaline phosphatase EC 3.1.3.1 was found to be active in $scCO_2$ for the hydrolytic formation of *p*-nitrophenol from disodium *p*-nitrophenylphosphate at 35 °C and 10 MPa, with a water concentration of 0.1 vol.% [152]. More synthetically relevant examples have also been reported. The kinetic resolution of racemic 3-(4-methoxyphenyl)glycidic acid methyl ester (**47**) by immobilized *Mucor miehei* lipase (Lipozyme IM 20) in $scCO_2$ (Scheme 5.78) [153] gave **48** with 87% *ee* at a conversion of 53% after 5 h. A range of substrate and water concentrations (0.15–0.5 vol.%) were investigated, and were found to have no effect on the stereoselectivity of the reaction. The reaction rate was found to be considerably faster than that in a toluene–water mixture, where a 75% *ee* was observed after 21 h, corresponding to a five fold increase in rate in $scCO_2$. Other interesting substrates that have been resolved by enzymatic hydrolysis in $scCO_2$ include naproxen methyl ester [154], 2-benzyl 1,3-propanediacetate [155], benzoylbenzoin [156], and *P*-chiral hydroxymethanephosphinates [157].

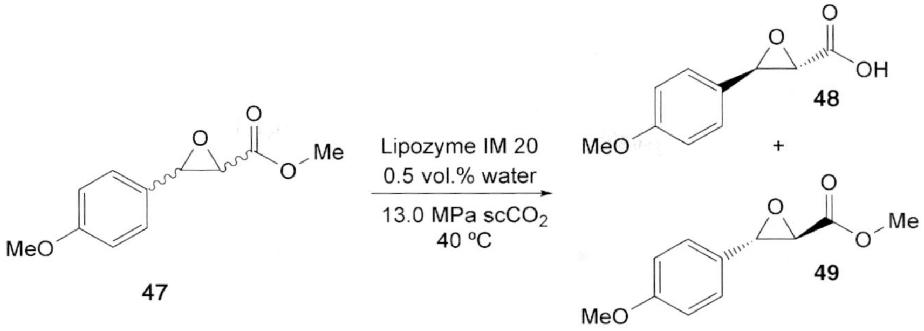

Scheme 5.78 Kinetic resolution of epoxy esters in $scCO_2$.

The transesterification of *N*-acetyl-L-phenylalanine chloroethyl ester (**50**) with ethanol catalyzed by *subtilisin* Carlsberg (Scheme 5.79) [158], gave quantitative conversion after 45 min with 2.5 vol.% ethanol in $scCO_2$. The enzyme was recovered and was used in repeated transformations with no appreciable loss of activity after three cycles. Enzyme stability was further confirmed by the increase in rate observed with an increase in temperature to 80 °C. Comparison of the rate with that obtained in conventional organic solvents showed that the reaction was significantly faster in $scCO_2$.

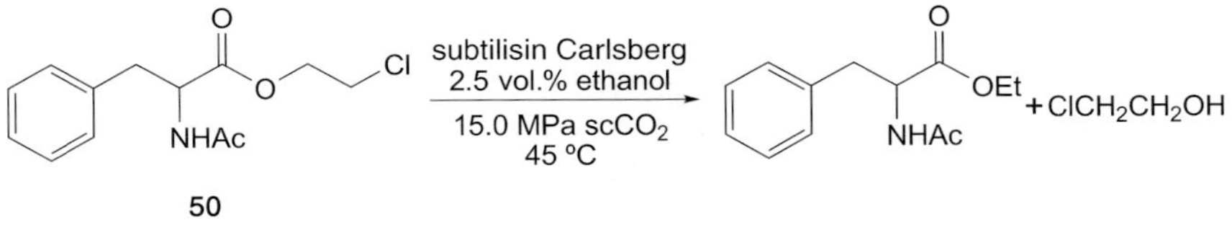

Scheme 5.79 Enzyme-mediated transesterification in $scCO_2$.

Lipid-coated β-D-galactosidase (prepared from *Bacillus circulans*) [159] was found to be soluble in $scCO_2$ over a wide range of temperatures and pressures. In the transacetalization of 1-*O*-*p*-nitrophenyl-β-D-galactopyranoside (**51**) with 5-phenyl-1-pentanol, the reaction was 25-fold faster in $scCO_2$ (40 °C, 15 MPa) than in diisopropyl ether, producing **52** in 72% yield after 3 h (Scheme 5.80). Temperature and pressure were shown to have a drastic effect around the critical point. On replacement of the lipid-coated enzyme (LCE) with the native β-D-galactosidase, no reaction was observed.

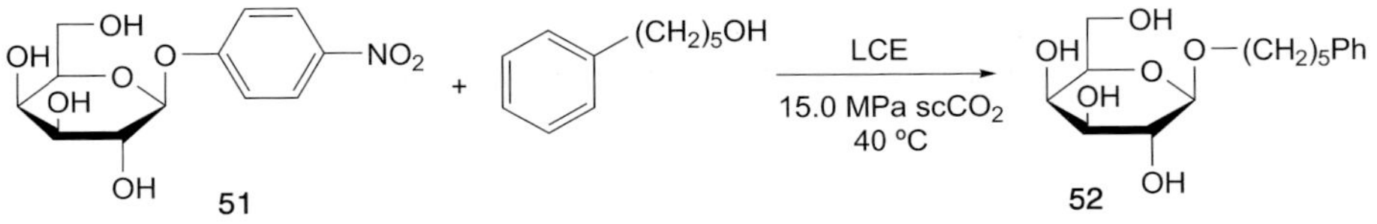

Scheme 5.80 Enzyme-mediated transacetalization in $scCO_2$.

Using an alternative LCE prepared from *Rhizopus delemar*, the enzymatic esterification of lauric acid (**53**) with the glyceride **54** at 40 °C, proceeded 5–10-fold faster in $scCO_2$ (20 MPa) than in benzene at atmospheric pressure (Scheme 5.81) [160]. In

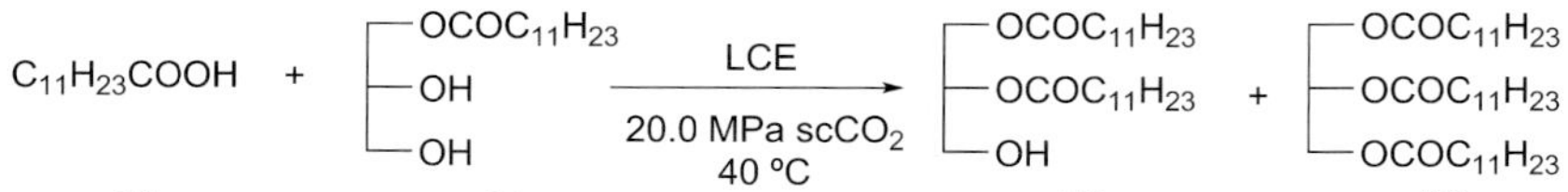

53 54 55 56

Scheme 5.81 Enzyme-mediated esterification in $scCO_2$.

$scCO_2$, di- and triglycerides **55** and **56** were produced in 90% yield after 3 h. The free lipase was found to be inactive under SCF conditions, whereas the LCE was found to maintain stability for over 80 h. If the temperature was reduced to below 31 °C at constant pressure (20 MPa), the esterification almost ceased. The solution under these conditions was then liquid phase and the LCE had a much reduced solubility. Reactivity increased with temperature until 45 °C, where it started to fall. This was attributed to denaturation at elevated temperatures. The rates also increased on increasing the pressure from 7.5 to 15 MPa (at a constant temperature of 40 °C), before falling with further increase in pressure. By modifying the temperature and pressure, it was observed that the enzyme activity could be reversibly controlled for at least five cycles.

The synthesis of terpene esters from primary terpene alcohols and acyl donors can be catalyzed by *Candida cylindracea* lipase (CCL) in $scCO_2$ [161]. The rate of esterification of *n*-valeric acid with citronellol in $scCO_2$ (Scheme 5.82) at constant temperature (35 °C) showed a dramatic pressure dependence around the critical pressure. It was proposed, therefore, that the mechanism of this esterification in $scCO_2$ involved CO_2 interacting with and activating the enzyme in a Lewis acidic manner.

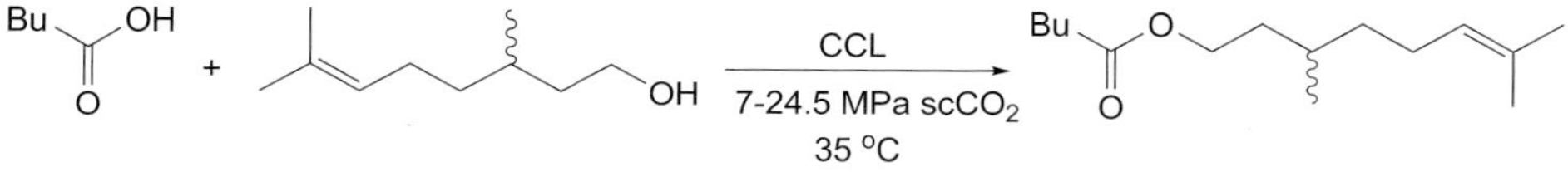

Scheme 5.82 Esterification of citronellol in $scCO_2$.

In the CCL-catalyzed esterification of oleic acid with citronellol in $scCO_2$ forming 3,7-dimethyl-6-octenyl ester (Scheme 5.83), the stereoselectivity exhibited a similar pressure dependence. Around the critical pressure, moderate selectivity was observed (26% *ee* at 7.6 MPa), increasing drastically to >98% *ee* at 8.4 MPa, before falling to 4% *ee* at 10.2 MPa. The weight of the CCL was found to change significantly over the pressure range, showing a large increase in the range 7.7–8.7 MPa, suggesting that a large number of CO_2 molecules were becoming reversibly absorbed by the enzyme. Fourier transform infrared (FT-IR) experiments also suggested a significant change

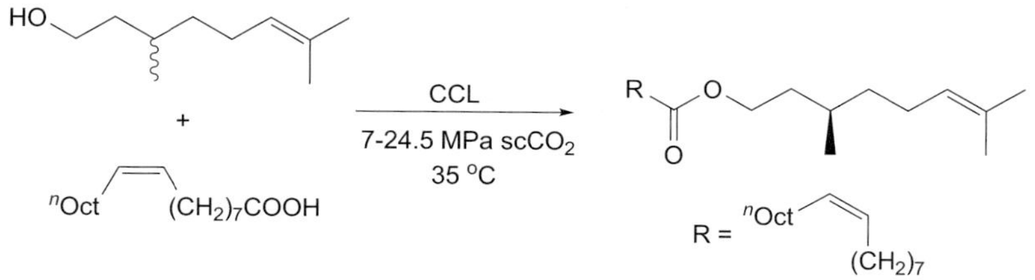

Scheme 5.83 Enantioselective esterification of oleic acid with citronellol in $scCO_2$.

in structure over this pressure range, and the authors suggested that this corresponded to a change in conformation whereby the α-helix "lid" covering the active site swung wide open and allowed easier access for the substrate.

The enzymatic kinetic resolution of 1-phenylethanol with vinyl acetate in $scCO_2$ (Scheme 5.84) [162] using Novozym 435 (EC 3.1.1.3 from *Candida antarctica* B) was found to be excellent and independent of temperature, with (*R*)-1-phenylethylacetate produced in >99% *ee* achieved at 50% conversion. A related resolution of 1-(*p*-chlorophenyl)-2,2,2-trifluoroethanol under very similar conditions has also been reported [163]. The reaction can also be carried out as a continuous process in $scCO_2$ using cross-linked enzyme aggregates (CLEAs) [164], or using immobilized lipase in a membrane reactor [165].

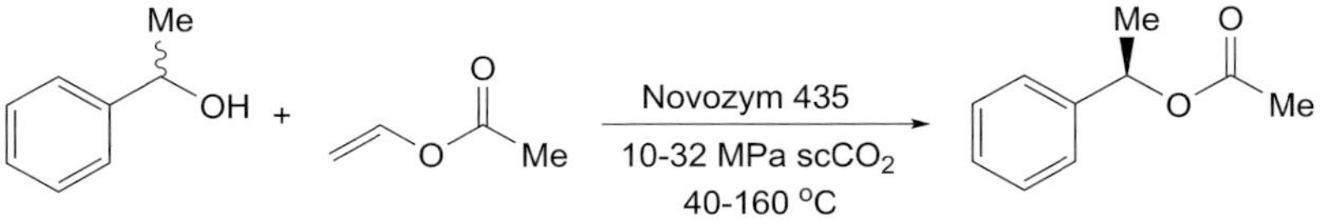

Scheme 5.84 Kinetic resolution of 1-phenylethanol in $scCO_2$.

The enantioselective acetylation of a number of racemic alcohols using an immobilized lipase (*Pseudomonas* sp. from Amano P) in $scCO_2$ has been reported [166]. The enzyme was found to be stable over a large pressure range (15–25 MPa) and the rate of reaction for the esterification of a range of alcohols was found to be significantly faster in $scCO_2$ than in conventional solvents. The reaction of racemic *trans*-3-penten-2-ol reached 50% conversion after 250 min at 40 °C, compared with the rate in toluene (the best of the organic solvents), where <20% conversion was achieved in this time (Scheme 5.85). The enantioselectivity was also higher in $scCO_2$, with the product being produced with 89% *ee*, at 53% conversion, compared with 65% *ee* at 19% conversion in toluene.

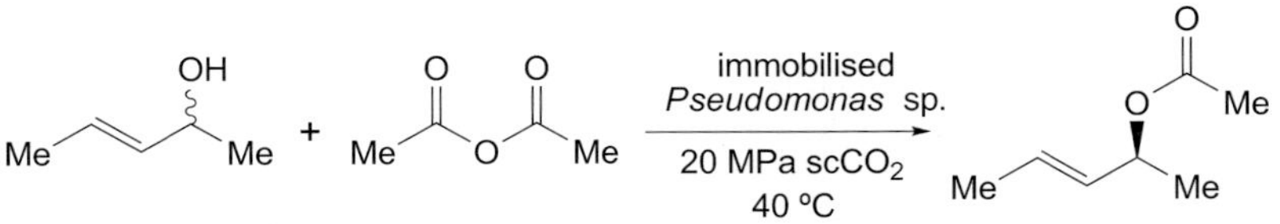

Scheme 5.85 Enantioselective acylation of alcohols in $scCO_2$.

In an unusual enzyme-catalyzed process, pyrrole was successfully converted to pyrrole-2-carboxylate in $scCO_2$, using cells of *Bacillus megaterium* PYR 2910 to catalyze the fixation of CO_2 [167]. The reaction was conducted (Scheme 5.86) by adding CO_2 (10 MPa) to the system at 40 °C to give a moderate yield after 1 h (54%).

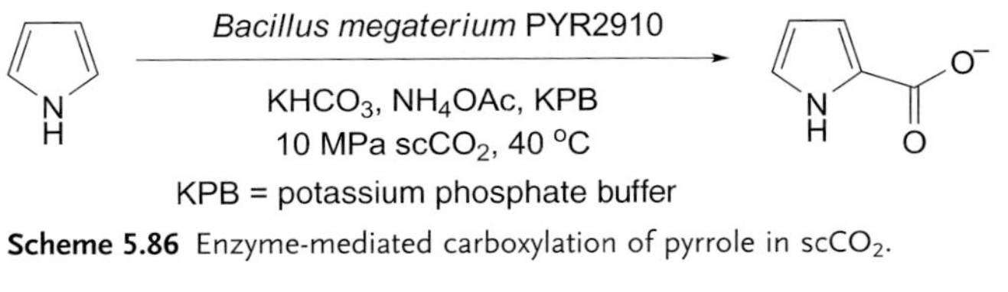

Scheme 5.86 Enzyme-mediated carboxylation of pyrrole in $scCO_2$.

This yield is much higher than for the reaction at atmospheric pressure (7%) under otherwise identical conditions. The maximum yield (77–79%) was reached at pressures below the critical pressure for CO_2 (4–7 MPa).

In the lipase-catalyzed synthesis of fatty acid esters, optimized conversion of oleic acid into lauryl oleate (79% at 10 MPa, 323 K) has been reported to display notable pressure dependence [168]. It was postulated that pressure may directly affect the enzymic structure and function, and therefore activity, in a CO_2-expanded liquid phase. Similarly, the lipase (Lipozyme RM IM)-catalyzed esterification of free fatty acid (FFA) with *n*-octanol to *n*-octyl oleate was successfully optimized through investigations into varied $scCO_2$ parameters, and indeed other process conditions [169]. Through visual interpretation of the reaction mixture, it was elucidated that a CO_2-expanded phase predominated at pressures below 20 MPa, ultimately proving beneficial to the reaction progress. The reaction took place in this "expanded" liquid (subcritical) phase in contact with the heterogeneous catalyst. Beyond 28 MPa, supercriticality was achieved – such conditions were found to be detrimental to the reaction progress. The influence of pressure in the range 8–30 MPa was also investigated. A maximum conversion of FFA (88%) was achieved (10 MPa, 323.15 K), where it was shown that CO_2 was highly soluble in the aqueous phase. In this case, density tuning of CO_2 was not the only effect of pressure modification. As the pressure was increased above 10 MPa, the yield was found to decrease due to a decrease in the partition rate of the substrates in the liquid phase and the increase in reaction volume. In addition, the reaction kinetics were found to be enhanced in the optimized system compared with analogous neat systems, due to superior mass transfer. Furthermore, lower enzyme concentrations were made possible (5.45% w/w substrates), adding further weight to the appeal of this system.

5.11 Conclusion

It is clear that SCFs are fascinating media in which to conduct synthetic organic chemistry, and this chapter is testament to the wide variety of synthetic processes that can be carried out under such conditions. The unique properties of SCFs allow for control of reaction processes often not available when using conventional solvents, including aspects such as reaction selectivity, control of reaction rates, and product isolation, and in many cases can also give useful mechanistic information that can be valuable in optimizing synthetic processes. Although examples of commercial processes are limited to date, it is clear that SCFs have real potential for the future and their importance will continue to grow.

Acknowledgments

We wish to thank Eli Lilly Ltd (Windlesham, UK) and the University of Leeds for a PhD studentship (D.C.B.) and DEFRA for postdoctoral funding (P.M.R.).

References

1 (a) Jessop, P.G., Ikariya, T. and Noyori, R. (1999) *Chemical Reviews*, **99**, 475–493; (b) Leitner, W. (2002) *Accounts of Chemical Research*, **35**, 746–756; (c) Stewart, I.H. and Derouane, E.G. (1999) *Current Topics in Catalysis*, 17–38.

2 Baiker, A. (1999) *Chemical Reviews*, **99**, 453–473.

3 (a) Tester, J.W., Danheiser, R.L., Weinstein, R.D., Renslo, A., Taylor, J.D. and Steinfeld, I.J. (2000) *ACS Symposium Series*, **767**, 270–291; (b) Clarke, D., Mohammed, A.A., Anthony, A.C., Parratt, A., Rose, P., Schwinn, D., Bannwarth, W. and Rayner, C.M. (2004) *Current Topics in Medicinal Chemistry*, **4**, 729–771.

4 Mikami, K. (ed.) (2005) *Green Reaction Media in Organic Synthesis*, Blackwell, Oxford.

5 (a) Oakes, R.S., Clifford, A.A. and Rayner, C.M. (2001) *Journal of the Chemical Society, Perkin Transactions 1*, 917–941; (b) Beckman, E.J. (2004) *Journal of Supercritical Fluids*, **28**, 121–191.

6 Rayner, C.M. (2007) *Organic Process Research and Development*, **11**, 121–132.

7 Jessop, P.G. and Leitner, W. (1999) *Chemical Synthesis using Supercritical Fluids*, Wiley-VCH Verlag GmbH, Weinheim.

8 Clifford, A.A. (1998) *Fundamentals of Supercritical Fluids*, Oxford University Press, Oxford.

9 (a) Searle, P.A., Glass, K.A. and Hochlowski, J.E. (2004) *Journal of Combinatorial Chemistry*, **6**, 175–180; (b) Welch, C.J., Leonard, W.R., Jr. DaSilva, J.O., Biba, M., Albaneze-Walker, J., Henderson, D.W., Laing, B. and Mathre, D.J. (2005) *LC-GC Europe*, **18**, 264–272; (c) van Ginneken, L. and Weyten, H. (2003) *Carbon Dioxide Recovery and Utilization*, 137–148; (d) Nicoud, R.-M., Clavier, J.-Y. and Perrut, M. (1999) *Chromatography: Principles and Practice*, **2**, 397–433.

10 Ke, J., George, M.W., Poliakoff, M., Han, B. and Yan, H. (2002) *Journal of Physical Chemistry B*, **106**, 4496–4502.

11 Ke, J., Han, B., George, M.W., Yan, H. and Poliakoff, M. (2001) *Journal of the American Chemical Society*, **123**, 3661–3670.

12 (a) Chrisochoou, A., Schaber, K. and Bolz, U. (1995) *Fluid Phase Equilibria*, **108**, 1–14; Chrisochoou, A.A., Schaber, K. and Stephan, K. (1997) *Journal of Chemical and Engineering Data*, **42**, 551–557; (b) Chrisochoou, A.A., Schaber, K. and Stephan, K. (1997) *Journal of Chemical and Engineering Data*, **42**, 558–561.

13 Stradi, B.A., Stadtherr, M.A. and Brennecke, J.F. (2001) *Journal of Supercritical Fluids*, **20**, 1–13; Stradi, B.A., Kohn, J.P., Stadtherr, M.A. and Brennecke, J.F. (1998) *Journal of Supercritical Fluids*, **12**, 109–122.

14 Stradi, B.A., Brennecke, J.F., Kohn, J.P. and Stadtherr, M.A. (2001) *AIChE Journal*, **47**, 212–221.

15 Hyde, J.R., Walsh, B., Singh, J. and Poliakoff, M. (2005) *Green Chemistry*, **7**, 357–361.

16 Chouchi, D., Gourgouillon, D., Courel, M., Vital, J. and Nunes da Ponte, M. (2001) *Industrial and Engineering Chemistry Research*, **40**, 2551–2554.

17 Milewska, A., Osuna, A.M.B., Fonseca, I.M. and Nunes da Ponte, M. (2005) *Green Chemistry*, **7**, 726–732.

18 Liu, R., Zhao, F., Fujita, S.-I. and Arai, M. (2007) *Applied Catalysis, A: General*, **316**, 127–133.

19 Chatterjee, M., Zhao, F.Y. and Ikushima, Y. (2004) *Advanced Synthesis and Catalysis*, **346**, 459–466.

20 Burgener, M., Furrer, R., Mallat, T. and Baiker, A. (2004) *Applied Catalysis, A: General*, **268**, 1–8.

21 Meric, P., Yu, K.M.K., Kong, A.T.S. and Tsang, S.C. (2006) *Journal of Catalysis*, **237**, 330–336.

22 Chatterjee, M., Chatterjee, A., Raveendran, P. and Ikushima, Y. (2006) *Green Chemistry*, **8**, 445–449.

23 Bogel-Lukasik, E., Fonseca, I., Bogel-Lukasik, R., Tarasenko, Y.A., Nunes da Ponte, M., Paiva, A. and Brunner, G. (2007) *Green Chemistry*, **9**, 427–430.

24 Sato, T., Rode, C.V., Sato, O. and Shirai, M. (2004) *Applied Catalysis, B: Environmental*, **49**, 181–185.

25 Hitzler, M.G., Smail, F.R., Ross, S.K. and Poliakoff, M. (1998) *Organic Process Research and Development*, **2**, 137–146.

26 Pillai, U.R. and Sahle-Demessie, E. (2003) *Industrial and Engineering Chemistry Research*, **42**, 6688–6696.

27 Pillai, U.R., Sahle-Demessie, E. and Young, D. (2003) *Applied Catalysis, B: Environmental*, **43**, 131–138.

28 (a) Chatterjee, M., Zhao, F.Y. and Ikushima, Y. (2004) *Applied Catalysis, A: General*, **262**, 93–100; (b) Chatterjee, M., Ikushima, Y. and Zhao, F.Y. (2002) *Catalysis Letters*, **82**, 141–144; (c) Xi, C., Wang, H., Liu, R., Cai, S. and Zhao, F. (2007) *Catalysis Communications*, **9**, 140–145.

29 Chatterjee, M., Ikushima, Y. and Zhao, F. (2003) *New Journal of Chemistry*, **27**, 510–513.

30 Zhao, F., Ikushima, Y., Shirai, M., Ebina, T. and Arai, M. (2002) *Journal of Molecular Catalysis A: Chemical*, **180**, 259–265.

31 Zhao, F., Fujita, S.-i., Sun, J., Ikushima, Y. and Arai, M. (2004) *Chemical Communications*, 2326–2327.

32 Zhao, F., Ikushima, Y., Chatterjee, M., Sato, O. and Arai, M. (2003) *Journal of Supercritical Fluids*, **27**, 65–72.

33 Zhao, F., Ikushima, Y. and Arai, M. (2003) *Green Chemistry*, **5**, 656–658.

34 Bourne, R.A., Stevens, J.G., Ke, J. and Poliakoff, M. (2007) *Chemical Communications*, 4632–4634.

35 Lee, S.-S., Park, B.-K., Byeon, S.-H., Chang, F. and Kim, H. (2006) *Chemistry of Materials*, **18**, 5631–5633.

36 Zhao, F., Ikushima, Y. and Arai, M. (2004) *Catalysis Today*, **93–95**, 439–443.

37 Tschan, R., Wandeler, R., Schneider, M.S., Burgener, M., Schubert, M.M. and Baiker, A. (2002) *Applied Catalysis, A: General*, **223**, 173–185.

38 (a) Shirai, M., Rode, C.V., Mine, E., Sasaki, A., Sato, O. and Hiyoshi, N. (2006) *Catalysis Today*, **115**, 248–253; (b) Hiyoshi, N., Mine, E., Rode, C.V., Sato, O. and Shirai, M. (2006) *Applied Catalysis, A: General*, **310**, 194–198; (c) Hiyoshi, N., Mine, E., Rode, C.V., Sato, O. and Shirai, M. (2006) *Chemistry Letters*, **35**, 188–189; (d) Hiyoshi, N., Inoue, T., Rode, C.V., Sato, O. and Shirai, M. (2006) *Catalysis Letters*, **106**, 133–138.

39 (a) Hiyoshi, N., Mine, E., Rode, C.V., Sato, O., Ebina, T. and Shirai, M. (2006) *Chemistry Letters*, **35**, 1060–1061; (b) Hiyoshi, N., Rode, C.V., Sato, O., Tetsuka, H. and Shirai, M. (2007) *Journal of Catalysis*, **252**, 57–68.

40 Rode, C.V., Joshi, U.D., Sato, O. and Shirai, M. (2003) *Chemical Communications*, 1960–1961.

41 Mine, E., Haryu, E., Arai, K., Sato, T., Sato, O., Sasaki, A., Rode, C.V. and Shirai, M. (2005) *Chemistry Letters*, **34**, 782–783.

42 Sato, T., Sato, O., Arai, K., Mine, E., Hiyoshi, N., Rode, C.V. and Shirai, M. (2006) *Journal of Supercritical Fluids*, **37**, 87–93.

43 (a) Hiyoshi, N., Rode, C.V., Sato, O. and Shirai, M. (2005) *Applied Catalysis, A: General*, **288**, 43–47; (b) Hiyoshi, N., Rode, C.V., Sato, O. and Shirai, M. (2004) *Journal of the Japan Petroleum Institute*, **47**, 410–411.

44 Wang, H. and Zhao, F. (2007) *International Journal of Molecular Sciences*, **8**, 628–634.

45 Xu, D., Carbonell, R.G., Roberts, G.W. and Kiserow, D.J. (2005) *Journal of Supercritical Fluids*, **34**, 1–9.

46 (a) Zhao, F., Fujita, S.-i., Sun, J., Ikushima, Y. and Arai, M. (2004) *Catalysis Today*, **98**, 523–528; (b) Zhao, F., Zhang, R., Chatterjee, M., Ikushima, Y. and Arai, M. (2004) *Advanced Synthesis and Catalysis*, **346**, 661–668; (c) Zhao, F.,

Ikushima, Y. and Arai, M. (2004) *Journal of Catalysis*, **224**, 479–483.

47 Ichikawa, S., Tada, M., Iwasawa, Y. and Ikariya, T. (2005) *Chemical Communications*, 924–926.

48 King, J.W., Holliday, R.L., List, G.R. and Snyder, J.M. (2001) *Journal of the American Oil Chemists' Society*, **78**, 107–113.

49 Santana, A., Larrayoz, M.A., Ramirez, E., Nistal, J. and Recasens, F. (2007) *Journal of Supercritical Fluids*, **41**, 391–403.

50 (a) van den Hark, S. and Haerroed, M. (2001) *Industrial and Engineering Chemistry Research*, **40**, 5052–5057; (b) Rovetto, L.J., Bottini, S.B., Brignole, E.A. and Peters, C.J. (2003) *Journal of Supercritical Fluids*, **25**, 165–176.

51 Hu, Y., Birdsall, D.J., Stuart, A.M., Hope, E.G. and Xiao, J. (2004) *Journal of Molecular Catalysis A: Chemical*, **219**, 57–60.

52 Francio, G., Wittmann, K. and Leitner, W. (2001) *Journal of Organometallic Chemistry*, **621**, 130–142.

53 Lange, S., Brinkmann, A., Trautner, P., Woelk, K., Bargon, J. and Leitner, W. (2000) *Chirality*, **12**, 450–457.

54 Stephenson, P., Licence, P., Ross, S.K. and Poliakoff, M. (2004) *Green Chemistry*, **6**, 521–523.

55 Stephenson, P., Kondor, B., Licence, P., Scovell, K., Ross, S.K. and Poliakoff, M. (2006) *Advanced Synthesis and Catalysis*, **348**, 1605–1610.

56 Abbott, A.P., Eltringham, W., Hope, E.G. and Nicola, M. (2005) *Green Chemistry*, **7**, 721–725.

57 Kainz, S., Brinkmann, A., Leitner, W. and Pfaltz, A. (1999) *Journal of the American Chemical Society*, **121**, 6421–6429.

58 Wandeler, R., Kunzle, N., Schneider, M.S., Mallat, T. and Baiker, A. (2001) *Chemical Communications*, 673–674.

59 Wandeler, R., Kunzle, N., Schneider, M.S., Mallat, T. and Baiker, A. (2001) *Journal of Catalysis*, **200**, 377–388.

60 Sellin, M.F., Bach, I., Webster, J.M., Montilla, F., Rosa, V., Aviles, T., Poliakoff, M. and Cole-Hamilton, D.J. (2002) *Journal of the Chemical Society, Dalton Transactions*, 4569–4576.

61 Erkey, C., Lozano Diz, E., Suss-Fink, G. and Dong, X. (2002) *Catalysis Communications*, **3**, 213–219.

62 Tadd, A.R., Marteel, A., Mason, M.R., Davies, J.A. and Abraham, M.A. (2003) *Journal of Supercritical Fluids*, **25**, 183–196.

63 Lin, B. and Akgerman, A. (2001) *Industrial and Engineering Chemistry Research*, **40**, 1113–1118.

64 Hu, Y., Chen, W., Osuna, A.M.B., Iggo, J.A. and Xiao, J. (2002) *Chemical Communications*, 788–789.

65 Wittmann, K., Wisniewski, W., Mynott, R., Leitner, W., Kranemann, C.L., Rische, T., Eilbracht, P., Kluwer, S., Ernsting, J.M. and Elsevier, C.J. (2001) *Chemistry – A European Journal*, **7**, 4584–4589.

66 Jiang, H., Zhao, J. and Wang, A. (2008) *Synthesis*, 763–769.

67 Jorgensen, K.A. (1989) *Chemical Reviews*, **89**, 431–458.

68 Wu, X.-W., Oshima, Y. and Koda, S. (1997) *Chemistry Letters*, 1045–1046.

69 Birnbaum, E.R., Lacheur, R.M.L., Horton, A.C. and Tumas, W. (1999) *Journal of Molecluar Catalysis A: Chemical*, **139**, 11–24.

70 Hou, Z., Theyssen, N., Brinkmann, A. and Leitner, W. (2005) *Angewandte Chemie International Edition*, **44**, 1346–1349.

71 Jenzer, G., Sueur, D., Mallat, T. and Baiker, A. (2000) *Chemical Communications*, 2247–2248.

72 Steele, A.M., Zhu, J. and Tsang, S.C. (2001) *Catalysis Letters*, **73**, 9–13.

73 Kuiper, J.L., Shapley, P.A. and Rayner, C.M. (2004) *Organometallics*, **23**, 3814–3818.

74 Loeker, F. and Leitner, W. (2000) *Chemistry – A European Journal*, **6**, 2011–2015.

75 Haas, G.R. and Kolis, J.W. (1998) *Organometallics*, **17**, 4454–4460.

76 Haas, G.R. and Kolis, J.W. (1998) *Tetrahedron Letters*, **39**, 5923–5926.

77 Pesiri, D.R., Morita, D.K., Glaze, W. and Tumas, W. (1998) *Journal of the Chemical*

Society, Chemical Communications, 1015–1016.
78 Oakes, R. S. (2000) PhD thesis, University of Leeds.
79 Oakes, R.S., Clifford, A.A., Bartle, K.D., Thornton-Pett, M. and Rayner, C.M. (1999) *Journal of the Chemical Society, Chemical Communications*, 247–248.
80 Nolen, S.A., Lu, J., Brown, J.S., Pollet, P., Eason, B.C., Griffith, K.N., Glaser, R., Bush, D., Lamb, D.R., Liotta, C.L., Eckert, C.A., Thiele, G.F. and Bartels, K.A. (2002) *Industrial and Engineering Chemistry Research*, **41**, 316–323.
81 Campestrini, S. and Tonellato, U. (2001) *Advanced Synthetic Catalysis*, **343**, 819–825.
82 Gonzalez-Nunez, M.E., Mello, R., Olmos, A. and Asensio, G. (2006) *Journal of Organic Chemistry*, **71**, 6432–6436.
83 Osswald, T., Schneider, S., Wang, S. and Bannwarth, W. (2001) *Tetrahedron Letters*, **42**, 2965–2967.
84 Schneider, S. and Bannwarth, W. (2000) *Angewandte Chemie International Edition*, **39**, 4142–4145.
85 Bhanage, B.M., Fujita, S.-i. and Arai, M. (2003) *Journal of Organometallic Chemistry*, **687**, 211–218.
86 (a) Saffarzadeh-Matin, S., Kerton, F.M., Lynam, J.M. and Rayner, C.M. (2006) *Green Chemistry*, **8**, 965–971; (b) Saffarzadeh-Matin, S., Chuck, C.J., Kerton, F.M. and Rayner, C.M. (2004) *Organometallics*, **23**, 5176–5181.
87 Yeung, L.K., Lee, C.T., Jr. Johnston, K.P. and Crooks, R.M. (2001) *Chemical Communications*, 2290–2291.
88 Yeung, L.K. and Crooks, R.M. (2001) *Nano Letters*, **1**, 14–17.
89 Shezad, N., Oakes, R.S., Clifford, A.A. and Rayner, C.M. (1999) *Tetrahedron Letters*, **40**, 2221–2224.
90 Shezad, N., Clifford, A.A. and Rayner, C.M. (2002) *Green Chemistry*, **4**, 64–67.
91 Early, T.R., Gordon, R.S., Carroll, M.A., Holmes, A.B., Shute, R.E. and McConvey, I.F. (2001) *Chemical Communications*, 1966–1967.
92 Gordon, R.S. and Holmes, A.B. (2002) *Chemical Communications*, 640–641.
93 (a) Ramarao, C., Ley, S.V., Smith, S.C., Shirley, I.M. and DeAlmeida, N. (2002) *Chemical Communications*, 1132–1133; (a) Ley, S.V., Ramarao, C., Gordon, R.S., Holmes, A.B., Morrison, A.J., McConvey, I.F., Shirley, I.M., Smith, S.C. and Smith, M.D. (2002) *Chemical Communications*, 1134–1135.
94 Lee, C.K.Y., Holmes, A.B., Ley, S.V., McConvey, I.F., Al-Duri, B., Leeke, G.A., Santos, R.C.D. and Seville, J.P.K. (2005) *Chemical Communications*, 2175–2177.
95 Leeke, G.A., Santos, R.C.D., Al-Duri, B., Seville, J.P.K., Smith, C.J., Lee, C.K.Y., Holmes, A.B. and McConvey, I.F. (2007) *Organic Process Research and Development*, **11**, 144–148.
96 Shezad, N., Clifford, A.A. and Rayner, C.M. (2001) *Tetrahedron Letters*, **42**, 323–325.
97 (a) Smith, C.J., Tsang, M.W.S., Holmes, A.B., Danheiser, R.L. and Tester, J.W. (2005) *Organic and Biomolecular Chemistry*, **3**, 3767–3781; (b) Smith, C.J., Early, T.R., Holmes, A.B. and Shute, R.E. (2004) *Chemical Communications*, 1976–1977.
98 Bhanage, B.M., Ikushima, Y., Shirai, M. and Arai, M. (1999) *Tetrahedron Letters*, **40**, 6427–6430.
99 Kayaki, Y., Noguchi, Y. and Ikariya, T. (2000) *Chemical Communications*, 2245–2246.
100 (a) Ikariya, T., Kayaki, Y., Kishimoto, Y. and Noguchi, Y. (2000) *Progress in Nuclear Energy*, **37**, 429–434; (b) Kayaki, Y., Noguchi, Y., Iwasa, S., Ikariya, T. and Noyori, R. (1999) *Chemical Communications*, 1235–1236.
101 Li, J., Jiang, H. and Chen, M. (2001) *Green Chemistry*, **3**, 137–139.
102 Jiang, H., Jia, L. and Li, J. (2000) *Green Chemistry*, **2**, 161–164.
103 Jia, L., Jiang, H. and Li, J. (1999) *Chemical Communications*, 985–986.
104 Reetz, M.T., Konen, W. and Strack, T. (1993) *Chimia*, **47**, 493.

105 Jeong, N., Hwang, S.H., Lee, Y.W. and Lim, J.S. (1997) *Journal of the American Chemical Society*, **119**, 10549–10550.

106 Furstner, A., Koch, D., Langemann, K., Leitner, W. and Six, C. (1997) *Angewandte Chemie (International Edition in English)*, **36**, 2466–2469.

107 Wynne, D.C., Olmstead, M.M. and Jessop, P.G. (2000) *Journal of the American Chemical Society*, **122**, 7638–7647.

108 Wegner, A. and Leitner, W. (1999) *Chemical Communications*, 1583–1584.

109 Carter, C.A.G., Baker, R.T., Nolan, S.P. and Tumas, W. (2000) *Chemical Communications*, 347–348.

110 Rose, P.M., Clifford, A.A. and Rayner, C.M. (2002) *Chemical Communications*, 968–969.

111 Parratt, A.J., Adams, D.J., Clifford, A.A. and Rayner, C.M. (2004) *Chemical Communications*, 2720–2721.

112 (a) Yoshida, M., Hara, N. and Okuyama, S. (2000) *Chemical Communications*, 151–152; (b) Selva, M., Tundo, P. and Perosa, A. (2002) *Tetrahedron Letters*, **43**, 1217–1219; (c) Selva, M., Tundo, P., Perosa, A. and Dall'Acqua, F. (2005) *Journal of Organic Chemistry*, **70**, 2771–2777.

113 Rohr, M., Geyer, C., Wandeler, R., Schneider, M., Murphy, A. and Baiker, A. (2001) *Green Chemistry*, **3**, 123–125.

114 Fujita, S.-i., Bhanage, B.M., Arai, M. and Ikushima, Y. (2001) *Green Chemistry*, **3**, 87–91.

115 (a) Sakakura, T., Saito, Y., Okano, M., Choi, J.-C. and Sako, T. (1998) *Journal of Organic Chemistry*, **63**, 7095–7096; (b) Kohno, K., Choi, J.-C., Ohshima, Y., Yili, A., Yasuda, H. and Sakakura, T. (2008) *Journal of Organometallic Chemistry*, **693**, 1389–1392.

116 (a) Bhanage, B.M., Fujita, S.-i., Ikushima, Y., Torii, K. and Arai, M. (2003) *Green Chemistry*, **5**, 71–75; (b) Yasuda, H., He, L.-N. and Sakakura, T. (2002) *Journal of Catalysis*, **209**, 547–550; (c) Kawanami, H. and Ikushima, Y. (2000) *Chemical Communications*, 2089–2090; (d) Sako, T., Fukai, T., Sahashi, R., Sone, M. and Matsuno, M. (2002) *Industrial and Engineering Chemistry Research*, **41**, 5353–5358; (e) Du, Y., Kong, D.-L., Wang, H.-Y., Cai, F., Tian, J.-S., Wang, J.-Q. and He, L.-N. (2005) *Journal of Molecular Catalysis A: Chemical*, **241**, 233–237; (f) Yasuda, H., He, L.-N., Takahashi, T. and Sakakura, T. (2006) *Applied Catalysis, A: General*, **298**, 177–180; (g) Wang, J.-Q., Kong, D.-L., Chen, J.-Y., Cai, F. and He, L.-N. (2006) *Journal of Molecular Catalysis A: Chemical*, **249**, 143–148; (h) Jutz, F., Grunwaldt, J.-D. and Baiker, A. (2008) *Journal of Molecular Catalysis A: Chemical*, **279**, 94–103.

117 Kawanami, H. and Ikushima, Y. (2002) *Tetrahedron Letters*, **43**, 3841–3844.

118 Hitzler, M., Smail, F.R., Ross, S.K. and Poliakoff, M. (1998) *Journal of the Chemical Society, Chemical Communications*, 359–360.

119 Amandi, R., Licence, P., Ross, S.K., Aaltonen, O. and Poliakoff, M. (2005) *Organic Process Research and Development*, **9**, 451–456.

120 Amandi, R., Hyde, J.R., Ross, S.K., Lotz, T.J. and Poliakoff, M. (2005) *Green Chemistry*, **7**, 288–293.

121 Gray, W.K., Smail, F.R., Hitzler, M.G., Ross, S.K. and Poliakoff, M. (1999) *Journal of the American Chemical Society*, **121**, 10711–10718.

122 Komoto, I. and Kobayashi, S. (2001) *Chemical Communications.*, 1842–1843.

123 Matsuo, J., Tsuchiya, T., Odashima, K. and Kobayashi, S. (2000) *Chemistry Letters*, 178–179.

124 Komoto, I. and Kobayashi, S. (2002) *Organic Letters*, **4**, 1115–1118.

125 (a) Ikushima, Y., Saito, N., Sato, O. and Arai, M. (1994) *Bulletin of the Chemical Society of Japan*, **67**, 1734–1736; (b) Ikushima, Y., Saito, N. and Arai, M. (1991) *Bulletin of the Chemical Society of Japan*, **64**, 282–284; (c) Isaacs, N.S. and Keating, N. (1992) *Journal of the Chemical Society, Chemical Communications*, 876–877; (d) Weinstein, R.D., Renslo,

A.R., Danheiser, R.L., Harris, J.G. and Tester, J.W. (1996) *The Journal of Physical Chemistry*, **100**, 12337–12341; (e) Renslo, A.R., Weinstein, R.D., Tester, J.W. and Danheiser, R.L. (1997) *Journal of Organic Chemistry*, **62**, 4530–4533; (f) Clifford, A.A., Pople, K., Gaskill, W.J., Bartle, K.D. and Rayner, C.M. (1998) *Jornal of the Chemical Society, Faraday Transactions*, **94**, 1451–1456;
(g) Reaves, J.T. and Roberts, C.B. (1999) *Chemical Engineering Communications*, **171**, 117–134; (h) Clifford, A.A., Pople, K., Gaskill, W.J., Bartle, K.D. and Rayner, C.M. (1997) *Chemical Communications*, 595–596.

126 Oakes, R.S., Happenstall, T.J., Shezad, N., Clifford, A.A. and Rayner, C.M. (1999) *Chemical Communications*, 1459–1460.

127 Weinstein, R.D., Renslo, A.R., Danheiser, R.L. and Tester, J.W. (1999) *Journal of Physical Chemistry B*, **103**, 2878–2887.

128 Chapuis, C., Kucharska, A., Rzepecki, P. and Jurczak, J. (1998) *Helvetica Chimica Acta*, **81**, 2314–2325.

129 Fukuzawa, S.-I., Matsuzawa, H. and Metoki, K. (2001) *Synlett*, 709–711.

130 Fukuzawa, S.-I., Metoki, K., Komuro, Y. and Funazukuri, T. (2002) *Synlett*, 134–136.

131 Lee, C.K.Y., Holmes, A.B., Al-Duri, B., Leeke, G.A., Santos, R.C.D. and Seville, J.P.K. (2004) *Chemical Communications*, 2622–2623.

132 Totoe, H., McGowin, A.E. and Turnbull, K. (2000) *Journal of Supercritical Fluids*, **18**, 131–140.

133 Adam, W., Diedering, M. and Trofimov, A.V. (2001) *Chemical Physics Letters*, **350**, 453–458.

134 Roth, W.R. and Martin, M. (1967) *Tetrahedron Letters*, 4695–4698.

135 Allred, E.L. and Smith, R.L. (1967) *Journal of the American Chemical Society*, **89**, 7133–7134.

136 (a) Nishiyama, Y., Kaneda, M., Saito, R., Mori, T., Wada, T. and Inoue, Y. (2004) *Journal of the American Chemical Society*, **126**, 6568–6569; (b) Nishiyama, Y., Wada, T., Mori, T. and Inoue, Y. (2007) *Chemistry Letters*, **36**, 1488–1489.

137 Nishiyama, Y., Kaneda, M., Asaoka, S., Saito, R., Mori, T., Wada, T. and Inoue, Y. (2007) *Journal of Physical Chemistry A*, **111**, 13432–13440.

138 Chateauneuf, J.E., Zhang, J., Foote, J., Brink, J. and Perkovic, M.W. (2002) *Advances in Environmental Research*, **6**, 487–493.

139 Tazuke, S. and Ozawa, H. (1975) *Journal of the Chemical Society, Chemical Communications*, 237–238.

140 Saito, R., Kaneda, M., Wada, T., Katoh, A. and Inoue, Y. (2002) *Chemistry Letters*, 860–861.

141 Pacut, R., Grimm, M.L., Kraus, G.A. and Tanko, J.M. (2001) *Tetrahedron Letters*, **42**, 1415–1418.

142 Kraus, G.A. and Kirihara, M. (1992) *Journal of Organic Chemistry*, **57**, 3256–3257.

143 Kendall, J.L., Canelas, D.A., Young, J.L. and DeSimone, J.M. (1999) *Chemical Reviews*, **99**, 543–563.

144 Tanko, J.M. and Blackert, J.F. (1994) *Science*, **263**, 203–205.

145 Ryu, I. and Sonada, N. (1996) *Angewewandte Chemie (International Edition in English)*, **35**, 1051–1066.

146 Ikariya, T., Kayaki, Y., Kishimoto, Y. and Noguchi, Y. (2000) *Progress in Nuclear Energy*, **37**, 429–434.

147 Hadida, S., Super, M.S., Beckman, E.J. and Curran, D.P. (1997) *Journal of the American Chemical Society*, **119**, 7406–7407.

148 Mesiano, A.J., Beckman, E.J. and Russell, A.J. (1999) *Chemical Reviews*, **99**, 623–633.

149 Matsuda, T., Watanabe, K., Harada, T. and Nakamura, K. (2004) *Catalysis Today*, **96**, 103–111.

150 Mori, T., Funasaki, M., Kobayashi, A. and Okahata, Y. (2001) *Chemical Communications*, 1832–1833.

151 Matsuda, T., Kanamaru, R., Watanabe, K., Kamitanaka, T., Harada, T. and Nakamura, K. (2003) *Tetrahedron: Asymmetry*, **14**, 2087–2091.

152 Randolph, T.W., Blanch, H.W., Prausnitz, J.M. and Wilke, C.R. (1985) *Biotechnology Letters*, **7**, 325–328.

153 Rantakyla, M., Alkio, M. and Aaltonen, O. (1996) *Biotechnology Letters*, **18**, 1089–1094.

154 Salgin, U., Salgin, S. and Takac, S. (2007) *Journal of Supercritical Fluids*, **43**, 310–316.

155 Mase, N., Sako, T., Horikawa, Y. and Takabe, K. (2003) *Tetrahedron Letters*, **44**, 5175–5178.

156 Celebi, N., Yildiz, N., Demir, A.S. and Calimli, A. (2007) *Journal of Supercritical Fluids*, **41**, 386–390.

157 Albrycht, M., Kielbasinski, P., Drabowicz, J., Mikolajczyk, M., Matsuda, T., Harada, T. and Nakamura, K. (2005) *Tetrahedron: Asymmetry*, **16**, 2015–2018.

158 Pasta, P., Mazzola, G., Carrea, G. and Riva, S. (1989) *Biotechnology Letters*, **11**, 643–648.

159 Mori, T. and Okahata, Y. (1998) *Journal of the Chemical Society, Chemical Communications*, 2215–2216.

160 Mori, T., Kobayashi, A. and Okahata, Y. (1998) *Chemistry Letters*, 921–922.

161 Ikushima, Y., Saito, N., Hatakeda, K. and Sato, O. (1996) *Chemical Engineering Science*, **51**, 2817–2822.

162 Overmeyer, A., Schrader-Lippelt, S., Kasche, V. and Brunner, G. (1999) *Biotechnology Letters*, **21**, 65–69.

163 Matsuda, T., Kanamaru, R., Watanabe, K., Harada, T. and Nakamura, K. (2001) *Tetrahedron Letters*, **42**, 8319–8321.

164 Hobbs, H.R., Kondor, B., Stephenson, P., Sheldon, R.A., Thomas, N.R. and Poliakoff, M. (2006) *Green Chemistry*, **8**, 816–821.

165 Lozano, P., Villora, G., Gomez, D., Gayo, A.B., Sanchez-Conesa, J.A., Rubio, M. and Iborra, J.L. (2004) *Journal of Supercritical Fluids*, **29**, 121–128.

166 Cernia, E., Palocci, C., Gasparrini, F., Misiti, D. and Fagnano, N. (1994) *Journal of Molecular Catalysis*, **89**, L11–L18.

167 Matsuda, T., Ohashi, Y., Harada, T., Yanagihara, R., Nagasawa, T. and Nakamura, K. (2001) *Chemical Communications*, 2194–2195.

168 Knez, Z., Laudani, C.G., Habulin, M. and Reverchon, E. (2007) *Biotechnology and Bioengineering*, **97**, 1366–1375.

169 Laudani, C.G., Habulin, M., Knez, Z., Della Porta, G. and Reverchon, E. (2007) *Journal of Supercritical Fluids*, **41**, 92–101.

6
Heterogeneous Catalysis

Roger Gläser

6.1
Introduction and Scope

Catalysis is one of the 12 principles of green chemistry [1, 2]. It stands for the efficient use of raw materials for a given chemical conversion, that is, the use of a catalyst instead of a stoichiometrically consumed reagent. The utilization of supercritical fluids such as carbon dioxide or water as "safer solvents" for chemical reactions corresponds to another principle of green chemistry. The combination of both is, therefore, an attractive way to design chemical conversions according to the basic ideas of green chemistry. Indeed, supercritical fluids not only offer environmental benefits, but also have proven to offer substantial additional advantages in numerous instances of catalytic conversions. This chapter focuses on the utilization of supercritical fluids (SCFs) and the closely related gas-expanded liquids (GXLs) in heterogeneous catalysis.

In most cases, especially in large-scale industrial applications, heterogeneous catalysis involves a gas phase of the reactants and products and a solid catalyst. In that sense, this chapter is limited to conversions over solid catalysts in SCF and GXL reaction media. Other cases that are closely related to or may, in a rigorous sense, be considered a part of heterogeneous catalysis such as catalysis in bi- or multi-phase systems, phase-transfer catalysis, enzymatic catalysis, or electrocatalysis will not be treated here, but are covered in other chapters of this Handbook. Also, the application of large metal clusters that could also be viewed as solid catalysts is not included in this chapter.

While several examples exist where solid catalysts are employed in a liquid reaction medium, a rather small number of processes occurs under supercritical conditions. Two prominent examples are the ammonia synthesis via the Haber–Bosch process [3] and the early methanol synthesis over zinc–chromium oxide catalysts [4]. Under the conditions of these processes, however, the reaction mixture is far remote from the critical point and should rather be considered a compressed gas. The peculiar properties of supercritical fluids are, however, most apparent in the vicinity of the critical point, that is, at $(1.05–1.20)T_c$ and $(0.9–2.0)p_c$, the so-called "near-critical

Handbook of Green Chemistry, Volume 4: Supercritical Solvents. Edited by Walter Leitner and Philip G. Jessop

ISBN: 978-3-527-32590-0

region." In this chapter, we will therefore mostly refer to this region when using the term "supercritical fluids" (SCFs). For the fundamental properties and the phase behavior of SCFs and also those of GXLs, the reader is referred to Chapters 3 and 4.

The major focus of the present chapter is to describe the peculiarities, advantages and limitations of utilizing SCFs and the closely related GXLs in heterogeneous catalysis over solid materials. It is organized into two main sections. The first section treats the specific advantages of SCFs and GXLs for chemical conversions over solid catalysts. In the subsequent section, selected examples are discussed that highlight these specific advantages for the two kinds of solvent systems. Within the scope of this chapter, a thorough coverage of the extensive amount of literature on heterogeneously catalyzed conversions in SCFs is neither suitable nor attempted. Whenever appropriate, the reader will be referred to the original literature or to recommended review articles. In particular, reference is made to the other chapters of this handbook that include important aspects of heterogeneous catalysis, that is, those on "Synthetic Organic Chemistry in SCFs" (Chapter 5) and "Chemistry in Supercritical Water" (Chapter 13). Finally, it should be mentioned that supercritical fluids can be successfully applied in the preparation (Chapter 9) and the recycling [5–7] of solid catalysts. These and related applications will also not be further treated here.

6.2
General Aspects of Heterogeneous Catalysis in SCFs and GXLs

In the typical case of heterogeneous catalysis on the surface of a porous solid, a sequence of steps is involved as depicted schematically in Figure 6.1 [8]. Before the chemical reaction occurs, the reactants have to diffuse through a boundary layer to the outer surface of the catalyst particle (film diffusion), then through the pores of the catalyst particle (pore diffusion), and finally have to be adsorbed on the catalytically active site. Likewise, the products have to desorb from the active site and diffuse

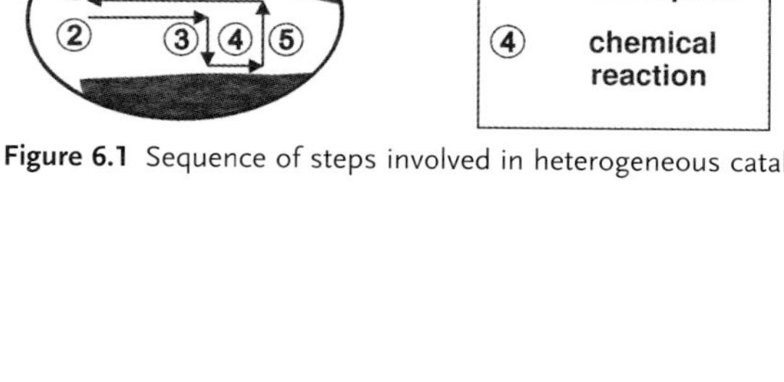

Figure 6.1 Sequence of steps involved in heterogeneous catalysis over porous solids (cf. [8]).

though the pore and the boundary layer, eventually to reach the bulk fluid phase. The respective mass transfer steps may significantly influence or even totally govern the overall reaction rate. If the reaction is strongly exothermic or endothermic, effects of the heat transfer from the bulk to or away from the catalyst surface also have to be taken into account.

The strong influence of mass and heat transfer effects is among the main differences between homogeneous and heterogeneous catalysis. Consequently, the effect of the reaction medium or a solvent is often very different. Whereas in homogeneous catalysis the solvent may directly interact with the catalyst and also with the intermediates and transition states of a catalytic reaction, the solvent of a heterogeneously catalyzed conversion may, in principle, affect all the steps in the sequence discussed above (cf. Figure 6.1). Depending on the nature of the catalyst and, in particular, its surface hydrophobicity/hydrophilicity [9] and the size distribution and geometry of its pore system, the impact of the reaction medium on a heterogeneously catalyzed conversion may vary widely, from very strong to almost negligible. Whether or not a heterogeneously catalyzed conversion may benefit from the application of an SCF or a GXL as the reaction medium is strongly determined by, in addition to the reaction system itself, the catalyst and its particular properties. In the following, the most important advantages that can be taken by utilizing SCFs or GXLs in heterogeneous catalysis on porous solids will be highlighted.

6.2.1 Utilization of SCFs in Heterogeneous Catalysis

6.2.1.1 General Considerations

With reference to Chapter 3, the most important properties of SCFs from the viewpoint of heterogeneous catalysis will be briefly reviewed in the following. On the basis of these properties, the most important advantages of using SCFs as reaction media for heterogeneously catalyzed conversions will be discussed subsequently.

Similarly to the application in homogeneous catalysis, the interest in SCFs as solvents and reaction media in heterogeneous catalysis is largely based on the fact that their behavior and properties can be continuously changed from liquid- to gas-like without the occurrence of phase transitions. As mentioned above, this tunability of the physico-chemical properties of SCFs with only small changes in pressure or temperature is most pronounced in the vicinity of the critical point, that is, at $(1.05–1.20)T_c$ and $(0.9–2.0)p_c$, the so-called "near-critical region." In this range of conditions, the properties of SCFs can be varied without changing the molecular functionality of these solvents. It should be noted, however, that the attraction of SCFs generally does not evolve from their fundamentally different properties from those of conventional liquids and gases. It rather lies in the unique combination of typical liquid- and gas-like properties and the strong dependence of the properties on pressure and temperature in the vicinity of the critical point.

As the reaction mixtures in heterogeneous catalysis are multi-component systems by definition, the particular phase behavior of SCFs is another aspect of utmost importance for conversions over solid catalysts. SCFs can not only solubilize higher molecular weight solutes in significant concentrations, but are also mostly miscible with permanent gases such as oxygen, hydrogen, and carbon monoxide over a wide range of compositions (see Chapter 3). Hence, chemical conversions may be carried out in a single, supercritical phase whereas, under conventional conditions, a multiphase reaction system would be involved. Since mass transfer processes to or away from the (outer or inner) catalyst surface are often rate-limiting steps, this is a particular advantage for heterogeneously catalyzed conversions. This advantage is particularly evident in cases where, conventionally, a gas is to be transported into a liquid reaction phase, such as in hydrogenations, oxidations, hydroformylations, or Fischer–Tropsch synthesis (see Section 6.3.1)

Moreover, mass transfer restrictions on the reaction rate may be alleviated in supercritical media due to their lower viscosity and, thus, higher diffusivity than in conventional liquids. Additionally, the almost negligible surface tension of SCFs often allows good wettability of the catalyst surface and facile penetration of the catalyst pores. Also, the heat transfer properties of SCFs are favorable for chemical reactions. For instance, the heat capacity of CO_2 passes through a maximum in the near-critical region. This allows the adiabatic temperature rise in strongly exothermic reactions such as the combustion of hydrogen to be significantly reduced [10].

In order to fully utilize the favorable solvent properties of SCFs, the phase behavior at higher pressures and at temperatures relevant for the often complex multicomponent mixtures involved in a given catalytic application has to be known with sufficient accuracy and reliability. A treatment of the phase behavior of binary systems based on the classification by Scott and van Konyenburg [11] can be found in a recent review [12]. This review also covers the experimental observation of high-pressure phase equilibria and their relation to catalytic reactions.

It should also be mentioned that the phase behavior within the pores of solids may be significantly different from the bulk phase behavior. This difference is especially important for mesopores (pore diameter 2–50 nm [13]) where, in addition to a dense phase adsorbed on the pore walls, other phases such as liquids or solids and phase transitions between them can occur within the pores. This difference is a result of the confinement of the phases within the pores. In contrast to mesopores, micropores (pore diameters below 2 nm [13]) are, for most substances, too small for another phase except a dense adsorbed phase to exist. In macropores (pore diameters above 50 nm [13]), the influence of the walls on the phase behavior within the pores is almost negligible. The deviation of conditions for phase equilibria or transitions under confinement in mesopores is a general phenomenon and not limited to particular phases. For instance, both the liquid–gas and the solid–liquid coexistence curves and, consequently, the critical and the triple point are shifted towards lower temperatures and pressures compared with the bulk with decreasing pore diameter (Figure 6.2) [14].

The chemical nature of the SCFs used as reaction media in heterogeneous catalysis is not greatly different from those applied for other chemical conversion, for example, in homogeneous catalysis. Likewise, in addition to chemical compatibility with the

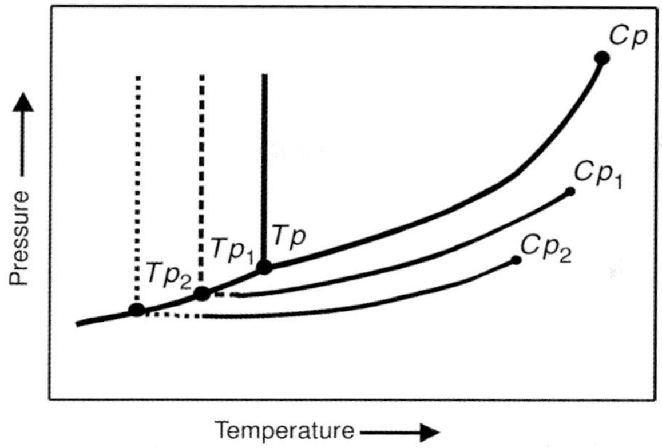

Figure 6.2 Schematic pressure–temperature diagram with critical point (*Cp*) and triple point (*Tp*) for a bulk fluid (full lines, *Cp*, *Tp*) and for a fluid confined within pores of decreasing diameter [$dp(Cp_1, Tp_1) > dp(Cp_2, Tp_2)$]. After [14].

reaction system including the catalyst, the most important criteria for their selection are environmental impact, safety considerations such as flammability or toxicity, and the specific conditions for reaching a single-phase, supercritical state. Thus, carbon dioxide, water, and lower hydrocarbons such as ethane and propane are by far the most frequently studied and used supercritical solvents in heterogeneous catalysis. The non-toxic and non-flammable character, the chemical inertness under most synthesis conditions, the low cost, abundant availability, and low environmental impact often make CO_2 the first choice as a solvent or reaction medium [15–17].

As mentioned in Chapter 3, some physical properties show an anomalous behavior at the critical point. Among those is the well-known phenomenon of the critical opalescence as a result of the high compressibility, strong density fluctuations, and spatial inhomogeneities [18]. Another anomalous phenomenon near the critical point is the much higher density of solvent molecules around a solute molecule than the average density in the bulk. This "local density enhancement" (also referred to as "clustering" [19] or "molecular charisma" [20]) should also be expected to occur at solid surfaces. However, corresponding effects on the rate or the selectivity of heterogeneously catalyzed reactions near the critical point have, so far, not been unambiguously proven (cf. Section 6.3.1.5).

6.2.1.2 **Rate Enhancement**

The unique properties of SCFs often result in higher reaction rates than under conventional gas- and/or liquid-phase conditions. These rate enhancements may be a result of the favorable phase behavior, that is, higher reactant concentrations, of improved heat and mass transfer within the supercritical reaction phase or within the pore-confined phase, or of eliminated interphase mass transfer resistances. In addition, the reaction rate may be influenced by the kinetic pressure effect: According to transition-state theory, the mole fraction-based rate constant k_x is largely determined by the activation volume $\Delta v^{\neq}$, that is, the difference in the partial molar volumes of the transition state and the reactants (Equation 6.1).

$$\left(\frac{\partial \ln k_x}{\partial p}\right)_T = -\frac{\Delta v^{\neq}}{RT} \tag{6.1}$$

$$\left(\frac{\partial \ln K_x}{\partial p}\right)_T = -\frac{\Delta v_R}{RT} \tag{6.2}$$

Since the partial molar volumes diverge at the critical point (cf. Chapter 3), the activation volumes and the rate changes may amount to several orders of magnitude for homogeneous reactions [21]. Likewise, the position of the chemical equilibrium as expressed by K_x may be strongly pressure dependent near the critical point (Equation 6.2, where Δv_{R} is the reaction volume, that is, the difference in the partial molar volumes of the products and the reactants). Similarly strong pressure effects on the rate of heterogeneously catalyzed conversions have not been reported, however, possibly due to the lack of pronounced solvation at solid surfaces.

A rather attractive, although less intensively studied, opportunity for utilizing the favorable properties of SCFs emerges from the above-mentioned difference in the critical point for the bulk and the pore-confined phases in mesoporous catalysts, respectively. While a bulk liquid phase with high solubility for higher molecular weight reactants may exist around the catalyst particle, the phase within the catalyst pores may already be present in the supercritical state. Thus, high pore effectiveness factors due to rapid diffusion inside the catalyst pores and, consequently, higher rates may be achieved for reactions in bulk liquid phases. An example of this type of rate enhancement by tuning the diffusivity within the pore-confined phase will be given in Section 6.3.1.5.

Figure 6.3 shows the extent of the diffusivity increase for two controlled pore glasses (CPGs) with different mean pore diameters in the mesopore range (pore diameter 6 or 15 nm) as measured by pulsed-field gradient NMR spectroscopy [22]. Starting from a liquid phase around and inside the porous materials, the diffusivity for both pore glasses follows an Arrhenius-type dependence on increasing temperature. At a distinct temperature below the bulk critical temperature, however, the diffusivity inside the pores of both materials increases significantly with only a small temperature increase. This jump in diffusivity is an indication of a phase transition from liquid to supercritical within the pores of the materials. As expected, the increase in the diffusivities in the CPGs is related to their pore diameter: the jump is more pronounced and its temperature, that is, the pore critical temperature T_{cp}, is closer to that of the bulk critical temperature T_{c} for the material with the larger pores. The diffusivity change can, therefore, be directly assigned to the shift of the phase transition caused by confinement of the phases in the mesopores.

6.2.1.3 **Selectivity Tuning**

The tunability of the solvent properties of SCFs by pressure, temperature or cosolvents offers the opportunity to control the catalytic selectivity. This can emerge from the tuning of mass transfer, for example, through the transport of a reaction

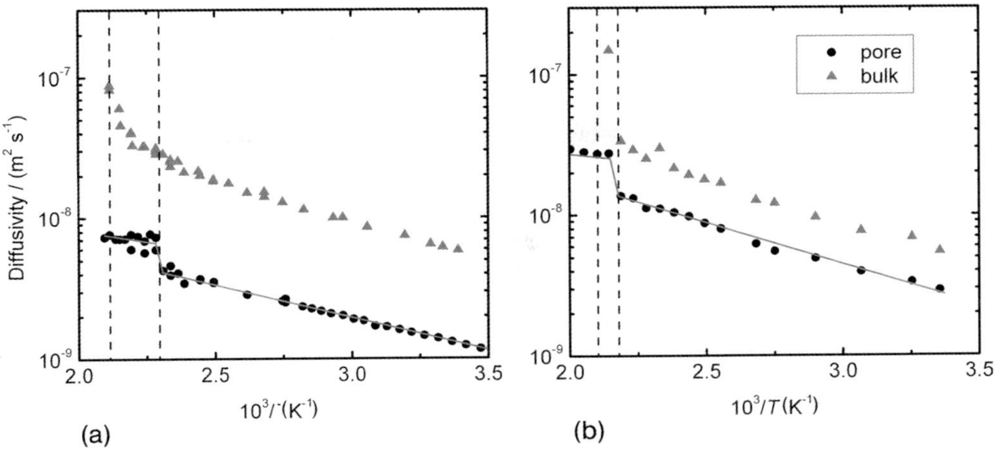

Figure 6.3 Diffusivity of *n*-pentane as a function of inverse temperature (Arrhenius plot) in the bulk and confined within the pores of controlled pore glasses (CPGs) with different pore diameters [(a) $d_P = 6$ nm and (b) $d_P = 15$ nm]. The solid line is calculated by assuming a transition to the supercritical state at the pore critical point T_{cp}. The vertical dashed lines show the positions of the bulk (left line) and the pore (right line) critical points. After [22].

intermediate away from the catalyst surface before it can undergo a consecutive reaction. It can also result from tuning the heat transfer, for example, in conversions where parallel or consecutive reactions with different exothermicity or endothermicity are involved. Selectivity tuning may also be based on the kinetic pressure effect, for example, when two parallel reactions involve opposing activation volumes (Figure 6.4) [23]. Whether or not the properties of supercritical reaction phases may affect or even be applied to control shape selectivity in heterogeneously catalyzed conversions over microporous solids such as zeolites is still a question of considerable debate. This discussion will be referred to in Section 6.3.1.5.

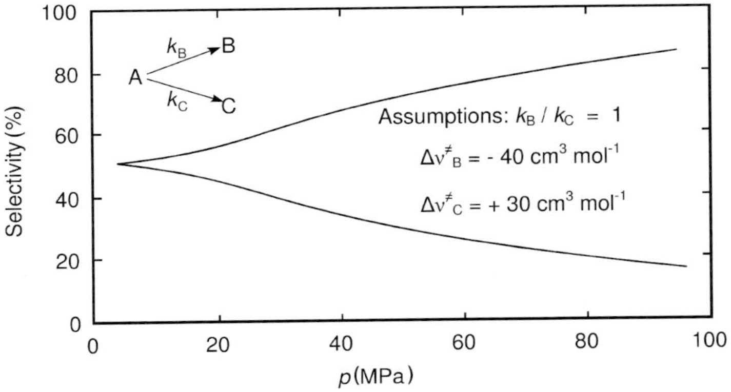

Figure 6.4 Selectivity for two products (B, C) formed in irreversible parallel reactions from the same reactant (A) as a function of pressure. After [23].

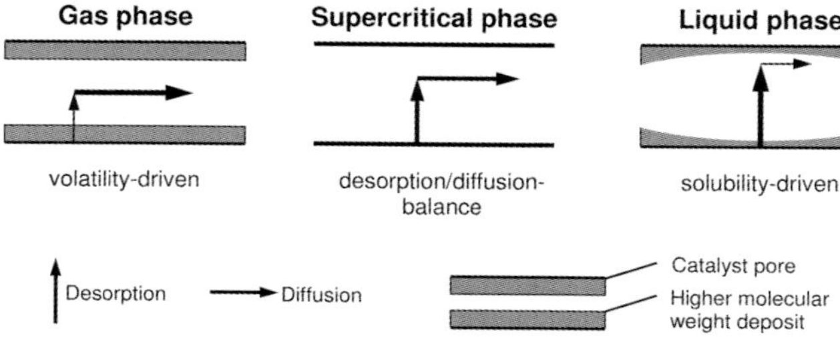

Figure 6.5 Balance of desorption and diffusion for removal of higher molecular weight deposits from a catalyst pore in gaseous, liquid, and supercritical reaction phases. After [25].

6.2.1.4 **Lifetime/Stability Enhancement**

The catalyst lifetime may be significantly enhanced with the aid of supercritical reaction media. If deactivation occurs by deposition of higher molecular weight compounds such as coke on the catalyst surface, the deposits may be efficiently removed during the reaction ("*in situ* extraction") due to the liquid-like solubility and the gas-like transport properties of SCFs. Improved activity maintenance may be particularly achieved for meso- and macroporous catalysts (Figure 6.5) [24]. While the removal of high-boiling deposits in gases is desorption limited (volatility driven) and in liquids diffusion limited (solubility driven), the balance between desorption and diffusion is more favorable in SCFs. Thus, an optimum density exists in the supercritical region at which the solubility and the transport of the deposits out of the catalyst pores are maximized [25].

This is further exemplified in Figure 6.6, which shows results from a model prediction of the temporal effectiveness factor as a function of reduced density

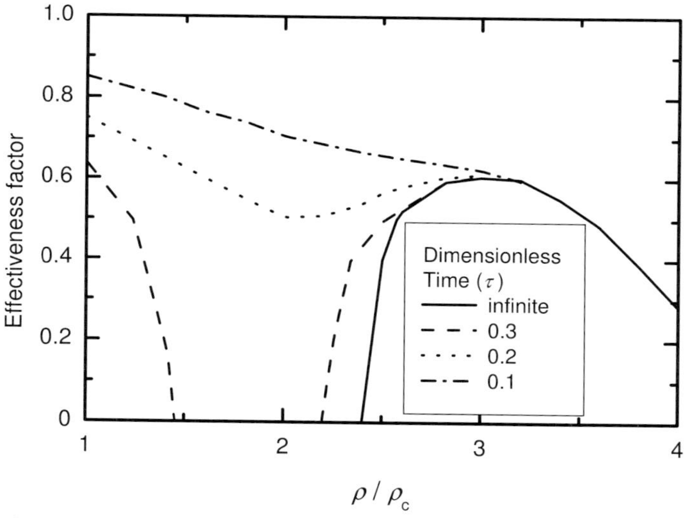

Figure 6.6 Catalyst pore effectiveness factor as a function of reduced density for different dimensionless times. After [25].

for the isomerization of 1-hexene on a porous Pt/γ-Al_2O_3 catalyst [25]. The catalyst is subject to a progressively rapid deactivation with increasing reduced density up to a value of ~2.4. The sharp activity increase above this value is due to the alleviation of pore-mouth restrictions following onset of coke solubilization in the supercritical reaction mixture. The model predicts an optimum at a reduced density of ~3 in the supercritical region at which catalyst activity is maintained at a maximum value. At lower than the optimum density, the reaction rate is limited by coke extraction, whereas above the optimum density the effectiveness factor is nearly that obtained in the absence of deactivation. This implies that the reaction rate (or catalyst activity) is limited by pore diffusion limitations only.

6.2.1.5 Reactor and Process Design

Due to the rate enhancement and the higher densities of the supercritical versus gaseous reaction phases, reactors may be designed with smaller volumes to achieve the same overall performance. This obvious advantage for continuously operated reactors offers significant potential for process intensification by applying SCFs. Despite these advantages of SCFs as media for catalytic reactions, there are some serious drawbacks that have often prevented their industrial application (see Chapter 2). Most of all, it is the high capital costs associated with the proper equipment for safely operating at the higher pressures required to reach and maintain supercritical process conditions.

6.2.2 Utilization of GXLs in Heterogeneous Catalysis

The physico-chemical properties of GXLs were already introduced in Chapter 4. As for SCFs above, only a short review of the fundamental properties of GXLs with relevance to heterogeneous catalysis will be given here. On that basis, the general advantages of utilizing GXLs in chemical conversions over solid catalysts will be outlined briefly.

Generally, the properties of GXLs lie between those of conventional organic liquids and those of SCF phases. With respect to the latter, the liquid phase that occurs when the concentration of a cosolvent exceeds its solubility in the SCF phase may be viewed as a GXL. Hence, GXLs are, by definition, part of a multi-component and multi-phase system. The higher content of organic solvent in GXLs compared with cosolvent-modified SCFs allows a much wider range of polarities, dielectric constants, and hydrogen bonding behavior to be realized in this type of solvent system. Besides the choice of organic solvent, the solvent polarity of GXLs may strongly depend on the pressure and the mole fraction of the gas dissolved in the liquid phase. Consequently, if the presence of an organic solvent can be tolerated or even is required, GXLs may be a particularly interesting alternative to SCFs.

Similarly to SCFs, the properties of GXLs can be tuned to a large extent by pressure and temperature. This tunability for GXLs is based on the varying fraction of the gas dissolved in the expanded liquid phase. In most cases, carbon dioxide is used to create GXLs due to its high solubility in liquid organic solvents (for some solvents over 80 wt% below 10.0 MPa; cf. Chapter 4). The high fraction of CO_2 in both the liquid and the gas phase may shift the flammability or explosion limits away from those of the pure organic solvent. Operation in CO_2-expanded liquids may, therefore, be considered to be intrinsically safe. Another advantage of GXLs over SCFs is that they typically exist at lower, subcritical pressures ($p < 10$ MPa). Therefore, the high pressures needed occasionally for obtaining a single-phase supercritical mixture for a catalytic conversion, often above 20 MPa, can be avoided.

The partly large amount of gas dissolved in the organic liquid phase results in a volume expansion ("swelling") and a viscosity decrease with respect to the pure liquid organic solvent (cf. Chapter 4). As a result, the diffusion within GXLs may be significantly faster than in conventional liquids. For instance, the self-diffusion coefficient of acetone in a liquid mixture with CO_2 may be higher by about a factor of three than in the single-component liquid [26]. Evidently, the diffusivity enhancement is directly related to the CO_2 content in the GXL phase [27].

Also, the solubility of permanent gases in GXLs is largely higher than that in the conventional liquid organic phase. For instance, the solubility of oxygen in CO_2-expanded acetonitrile at 30 °C exceeds that in the absence of CO_2 by two orders of magnitude and is comparable to that of liquid CO_2 [28, 29]. The combination of the higher solubility for permanent gases with the improved mass transfer properties makes GXLs attractive media for heterogeneously catalyzed conversions that are conventionally carried out in multiphase reaction systems. Hence GXLs are particularly attractive as solvents for catalytic hydrogenations, oxidations, and hydroformylations where mass transfer of hydrogen or oxygen into an organic liquid phase often limits conversion and/or selectivity (cf. Section 6.3.2).

Moreover, the higher density of the GXLs compared with SCFs makes them generally more powerful solvents for higher molecular weight compounds. Accordingly, GXLs have the potential to dissolve higher molecular weight deposits from the surface of a solid catalyst that would otherwise lead to its deactivation. A related application that demonstrates this potential is the use of GXLs as solvents for the extraction of high-boiling solutes from solid matrices such as sediment [30, 31].

6.3
Selected Examples of Heterogeneously Catalyzed Conversions in SCFs and GXLs

6.3.1
Conversions in SCFs

Heterogeneous catalysis in SCFs has advanced to become an important and vibrant field of contemporary research. Both the variety of chemical reactions and the spectrum of solid catalysts studied in SCFs are very broad and most diverse. Table 6.1 gives an overview of the most intensively studied types of heterogeneously catalyzed reactions and some pertinent examples. Due to the multitude of heterogeneously catalyzed reactions that have been studied in SCFs, only some selected cases can be treated here.

Table 6.1 Heterogeneously catalyzed conversions carried out in SCFs.

Reaction	Catalyst	Supercritical fluid	T_R (°C)	p_R (MPa)	Ref.
Hydrogenation					
Cyclohexene to cyclohexane	Pd/C	CO_2	70	13.8	[32]
Phenylacetylene to styrene	$Pd_{81}Si_{19}$	CO_2	30–45	3.0–19.0	[33]
Polystyrene	$Ni/SiO_2–Al_2O_3$	CO_2	150	21.0	[34]
Cyclohexene, isophorone, acetophenone, benzaldehyde, nitrobenzene	Pd, Pt, Ru on Deloxane	CO_2	100–400	6.0–14.0	[35]
Fats and oils	Pd, Pt on Deloxane	CO_2	70–170	8.0–16.0	[36]
Fatty acid methyl esters to alcohols	Cu–chromite	Propane	100–190	10.0–18.0	[37, 38]
Maleic anhydride to succinic anhydride or γ-butyrolactone	Pd/Al_2O_3	CO_2	50–175	4.0–15.0	[39]
Ciral to geraniol and nerol	Pt/MCM-41	CO_2	35–80	4.0–21.0	[40]
Cinnamaldehyde to cinnamyl alcohol and hydrocinnamyl alcohol	Pt/C	CO_2	50	10.0–16.0	[41]
Diethylitaconate to dimethyl (*R*)-2-methylsuccinate	(Rh-complex, $H_3PW_{12}O_{40}$) on Al_2O_3	CO_2	30–100	6.0–12.0	[42]
Ethyl pyruvate to ethyl (*R*)-lactate	Cinchonidine-modified Pt/Al_2O_3	Ethane, propane	40–100	7.0–25.0	[43, 44]
Substituted nitrobenzenes to anilines	Pt/C	CO_2	50	100–24.0	[45]
Polyaromatic hydrocarbons (PAHs)	Pd on HDPE, Fe, γ-Al_2O_3	CO_2	90–150	16.0–15.0	[46, 47]
Dehydrogenation					
C_{10}–C_{14}*n*-alkanes to monoalkenes	Pt,Sn/γ-Al_2O_3	Reactant	400–470	0.1–4.4	[48]
1-Phenylethanol to acetophenone	Pd/Al_2O_3	CO_2	80–165	3.0–19.0	[49]

(*Continued*)

Table 6.1 (*Continued*)

Reaction	Catalyst	Supercritical fluid	T_R (°C)	p_R (MPa)	Ref.
Fischer–Tropsch synthesis					
$CO + H_2$ (synthesis gas) to liquid hydrocarbons	Co,La/SiO_2, Ru/γ-Al_2O_3	*n*-hexane	210–240	4.5	[50–52]
	Fe with Cu, K and Si	*n*-hexane	240	3.5–7.0	[53]
	Co/Al_2O_3	*n*-hexane	210–260	3.0–8.0	[54, 55]
	Co/SiO_2	C_5–C_{10} *n*-alkanes	240	4.5	[56]
	Co,Ru/γ-Al_2O_3	*n*-hexane	235–250	4.5–6.5	[57, 58]
	Co/USY, ZSM-5, MCM-22	*n*-hexane	230	4.5	[59]
Hydroformylation					
$CO + H_2$ + propene	Rh/C	CO_2	100–300	17.0	[60]
$CO + H_2$ + 1-hexene	Rh(Fe)/ modified SiO_2	CO_2	100	17.0	[61]
	Rh/Pt complexes on SiO_2, MCM-41, zeolite MCM-20	CO_2	60–100	13.6–28.7	[62–65]
	Rh complex/ polymer	CO_2	70	12.0–24.0	[66]
$CO + H_2$ + 1-octene	Ru complex/ SiO_2	CO_2	80–90	12.0–18.0	[67]
Oxidation					
Methane to methanol and H_2CO	Cu, Ag–Cu, Ag, Ag–Au	H_2O			[68]
Propane to oxygenates	CoO_x/SiO_2	CO_2	375–450	22.0–35.0	[69]
	$(VO)_2P_2O_7$	CO_2	280–350	2.6– 11.3	[70]
Cyclohexane to cyclohexanol/-one	Oxides of Mn, Co, Fe on SiO_2, Al_2O_3, mordenite, AlPO-31, MCM-41	CO_2	80–400	8.0–16.0	[71]
	[Co]APO-5	CO_2	125		[72]
	$Ag_5PMo_{10}V_2O_{40}$	CO_2	180–215	14.0	[73]
Propene + H_2 + O_2 to propene oxide	Pd,Pt/TS-1	CO_2	45	13.1	[74]
Toluene to benzaldehyde	CoO (MoO_3)/ Al_2O_3	CO_2	20–220	8.0	[75]
Methanol to formaldehyde	Fe_2O_3-MoO_3, Fe_2O_3-SiO_2	CO_2	150–230	9.0	[76, 77]

Table 6.1 (*Continued*)

Reaction	Catalyst	Supercritical fluid	T_R (°C)	p_R (MPa)	Ref.
1-, 2-propanol to aldehyde, ketone	Pt/C	CO_2	40	10.0–19.0	[78]
benzyl alcohol to benzaldehyde	Pd/Al_2O_3, Pd/C	CO_2	60–100	6.0–17.0	[79–82]
	Pd, Pt, or Ru on Al_2O_3	CO_2	80–140	12.0–15.0	[83]
	Au/TiO_2	CO_2	80–120	13.3–16.7	[84]
	Sol–gel-entrapped perruthenate	CO_2	5	22.0	[85–90]
Geraniol to citral	Pd/Al_2O_3, Pd/SiO_2	CO_2	80	3.0–19.5	[91]
Ethanol to CO, CO_2, H_2O	Pt/TiO_2	CO_2	50–300	9.0	[92]
Toluene, tetralin to CO, CO_2, H_2O	$Pt/\gamma\text{-}Al_2O_3$	CO_2	300–390	80–10.7	[93]
Alkylation					
Isobutane/olefins	Zeolites MCM-22, MCM-36, MCM-41	Reactant	120–150	3.5–5.0	[94]
Isobutane/1-butene	H-USY, sulfated zirconia	CO_2	50–140	3.5–15.5	[95]
Isobutane/2-butene	Zeolites H-USY, H-Beta	Reactant	140	8.0	[96]
Isobutane/1-, 2-butene	$H_3PW_{12}O_{40}$	Reactant	145	4.0–9.0	[97]
Benzene + ethene	Zeolite H-Beta	CO_2	250–260	5.4–7.4	[98]
Benzene + propene	Zeolite H-Beta	Reactant	150–250	0.1–20.1	[99]
Naphthalene + 2-propanol	Zeolites La,Na-Y, H-mordenite	Reactant	270–320	4.6–6.5	[100]
Biphenyl + methanol	SAPO-11	Methanol	300–450	0.2–15.5	[101]
Anisole + paraformaldehyde	Zeolites H-Beta, H-ZSM-5, H-mordenite	CO_2	60	10.0	[102]
m-Cresol, phenol + methanol	ZrO_2, $Fe\text{–}V/SiO_2$	Methanol	350–500	0.1–15.0	[103]

(*Continued*)

Table 6.1 (*Continued*)

Reaction	Catalyst	Supercritical fluid	T_R (°C)	p_R (MPa)	Ref.
phenol + *tert*-butanol	Zeolites H-Y, H-Beta, RE(OTf)$_3$, (RE = rare earth), $H_3PW_{12}O_{40}/SiO_2$, Al_2O_3, MCM-41	CO_2	90–150	8.5–15.0	[104, 105]
Isomerization					
1-Hexene to *cis-/trans*-2-and 3-hexene	Pt/γ-Al_2O_3	Reactant	235–310	3.7–7.3	[106]
	γ-Al_2O_3	Reactant	220–250	0.5–8.0	[107, 108]
	shell catalyst	Reactant	115–315	0.2–8.1	[111]
n-Butane to isobutane	sulfated zirconia, H-mordenite, $H_3PW_{12}O_{40}/SiO_2$	Reactant	215–300	0.6–13.8	[109, 110]
o-, *m*-, *p*-Xylene to *p*-xylene	Zeolites, SiO_2–Al_2O_3	Reactant	362–371	3.5	[112]

Several reviews have summarized the state-of-the art in this field [12, 16, 17, 24, 113–120]. Exhaustive lists of heterogeneously catalyzed conversions in SCFs are included in papers by Baiker and co-workers [12, 113] and more recent papers by Kruse and Vogel [17, 119, 120]. While one of the earlier studies focused on *in situ* spectroscopy and on monitoring of high-pressure phase behavior [12], the later work reviews the literature based on the nature of the SCFs applied as the reaction media. The similarly recommendable reviews of Subramaniam *et al.* are devoted to the activity enhancement of porous catalysts in SCFs [24] and to the rational design of reactors for catalytic conversions in dense-phase CO_2 [116].Chapters 5 and 13, devoted to “Synthetic Organic Chemistry in SCFs” and to “Chemistry in Near- and Supercritical Water”, respectively, also include several applications of heterogeneous catalysis. The importance of heterogeneous catalysis for the potential of supercritical fluid applications in industry are highlighted in Chapter 2. The examples of conversions discussed in the following sections were predominantly selected to illustrate the potential and limitations of the utilization of SCFs in catalysis by solids.

6.3.1.1 **Hydrogenations**

Heterogeneously catalyzed hydrogenations are among the most extensively studied conversions in SCFs. For hydrogenations, the advantages of operating in an environmentally benign supercritical, single-phase reaction system with high hydrogen solubility, no interphase transport restrictions and an efficient transfer of the exothermic heat of reaction away from the catalyst are particularly evident. Conventionally, hydrogenations often involve three-phase reactors where gaseous hydrogen is contacted with a slurry of the catalyst in a liquid containing the dissolved or pure reactant. Since hydrogenations are mostly fast, the reaction rate is frequently limited by the low hydrogen solubility in the liquid and by the hydrogen transfer to the active sites of the catalyst. Obviously, SCFs offer the possibility of avoiding these problems. Moreover, the costly separations to obtain the pure hydrogenation products may be more easily achieved by depressurization from a supercritical solution.

A large variety of unsaturated organic compounds can be hydrogenated over solid catalysts in SCFs. This ranges from unsaturated hydrocarbons, fats and oils (see below) over aromatic and aliphatic aldehydes and ketones, epoxides, and nitriles to aromatic substrates, such as substituted nitroaromatics and polyaromatic hydrocarbons, or even polymers (cf. Table 6.1). Several examples of catalytic hydrogenations have already been treated in Chapter 5. A short compilation of rate data for catalytic hydrogenations in SCFs may be found in [121].

The influence of the high-pressure phase behavior in heterogeneously catalyzed hydrogenations has been the subject of a number of investigations [41, 122]. For instance, the semihydrogenation of phenylacetylene to styrene over an amorphous $Pd_{81}Si_{19}$ catalyst was systematically studied by Tschan *et al.* [33]. In the near-critical region, the conversion at 55 °C increased strongly with pressure. It reached a maximum close to the transition of the reaction mixture to supercritical conditions around 13.0 MPa. This was attributed to the improved mass transfer of hydrogen to the catalyst under single-phase supercritical conditions. A high selectivity for styrene was reached which decreased with pressure from 100% at subcritical to ~80% at supercritical conditions due to overhydrogenation.

In the supercritical single-phase region at high pressure, it is possible to tune reaction conditions such as pressure, temperature, and composition, independently. This advantage was used by Hitzler *et al.* to optimize the selectivity for a desired product within a series of successive hydrogenations [35]. The Pd-catalyzed selective conversion of anisole to methylcyclohexanol or methylcyclohexane in supercritical CO_2 or that of nitrobenzene to aniline in supercritical propane are two examples.

When the hydrogenation of ethyl pyruvate was carried out in supercritical CO_2 instead of ethane, deactivation of the Pt/Al_2O_3 catalyst was observed [43]. This was explained by platinum poisoning by CO formed by the reverse water gas shift (RWGS) reaction between the supercritical CO_2 with hydrogen. However, no deactivation of supported Pd catalysts was reported by Hitzler *et al.* for the hydrogenation of several different unsaturated organic substrates in supercritical CO_2 [35]. A detailed study of Arunajatesan *et al.* showed that the deactivation of a Pd/C catalyst in the hydrogenation of cyclohexene could be avoided when the concentration of peroxides in the feed was reduced from an initial value of 180 ppm to <6 ppm by an alumina trap (Figure 6.7) [32]. Also, no CO was detected in the product mixture. In a recent study, the presence of CO on supported platinum group metals in a mixture of hydrogen and dense CO_2 at 50–90 °C was unambiguously proven by *in situ* attenuated total reflection (ATR) IR spectroscopy [123]. However, the metal surface coverage may

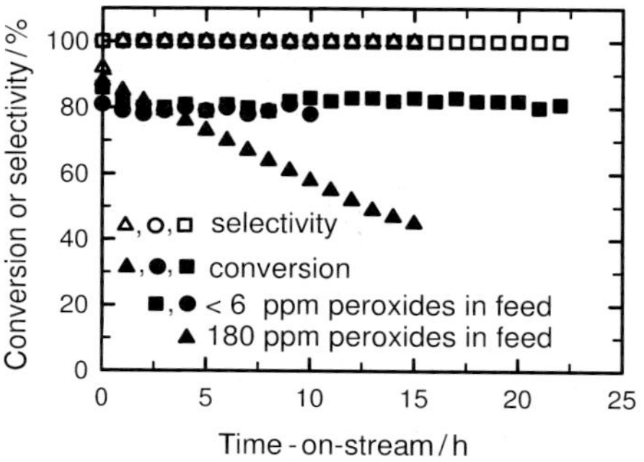

Figure 6.7 Effect of organic peroxide impurities on the conversion and selectivity of the hydrogenation of cyclohexene to cyclohexane over Pd/C in supercritical CO_2 [$T_R = 70\,°C$, $p_R = 13.8$ MPa, weight hourly space velocity ($WHSV$) $= 20\,h^{-1}$; $n_{H_2}/n_{cyclohexene} = 1$]. After [32].

be low and have only a minor effect on the rate of a hydrogenation reaction. Nevertheless, it is also possible that the poisoning of a supported noble metal by CO or consecutive hydrogenation products such as methanol or formates considerably affects the selectivity, rate, or catalyst stability of hydrogenations in supercritical CO_2. This holds especially for structure-sensitive conversions.

Functionalized anthraquinones dissolved in liquid CO_2 were hydrogenated at room temperature over 1 wt% Pd/Al_2O_3 as a first step of a reaction cycle for an environmentally benign, one-pot, and intrinsically safe approach to the production of H_2O_2 from the elements [124]. In an attempt to synthesize H_2O_2 directly from H_2–O_2 mixtures in supercritical CO_2, several supported Au and Pd catalysts were tested [125]. However, at temperatures near the critical point of CO_2, the rapid decomposition of H_2O_2 prohibited the achievement of yields higher than 0.01 wt%.

The importance of sensitive temperature control of a catalytic fixed-bed reactor was shown for the hydrogenation of 4-vinylcyclohex-1-ene over supported Pd catalyst [126]. The exothermic hydrogenation of the reactant may cause inhomogeneous temperature profiles along the catalyst bed (Figure 6.8). In the high-temperature zone, catalyst deactivation and, surprisingly, even the dehydrogenation of the reactant despite the large excess of hydrogen may occur. Careful control of the reaction temperature is, therefore, imperative for an unambiguous interpretation of results obtained in continuous flow fixed-bed reactors.

The hydrogenation of sunflower oil over supported noble metal catalysts is another example that demonstrates the advantage of operating in a single supercritical reaction phase. Although the reaction rate over Pt–K/TiO_2 or Pd/Al_2O_3 is controlled by diffusion of the oil to the catalyst, higher reaction rates in supercritical propane than in conventional biphasic conditions are achieved [127, 128]. The formation of unwanted *trans*-fatty acids can be successfully suppressed over an eggshell Pd/Al_2O_3 catalyst using supercritical dimethyl ether as the reaction medium [129].

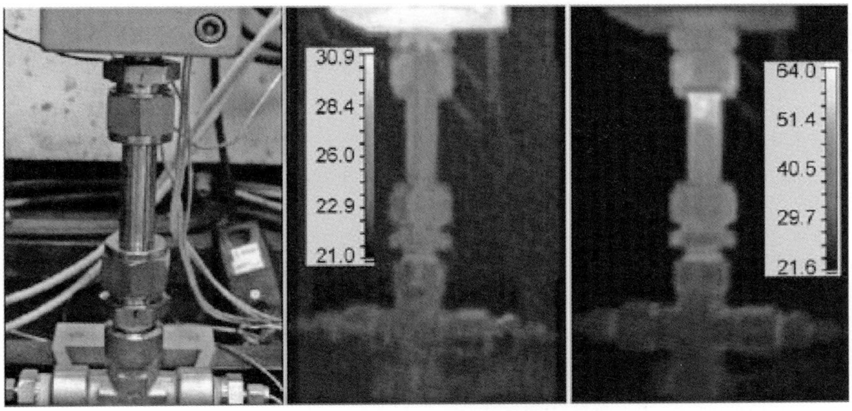

(a) (b) (c)

Figure 6.8 (a) Fixed-bed reactor used for the continuous hydrogenation of 4-vinylcyclohex-1-ene over 2 wt% Pd/ SiO_2–Al_2O_3. Parts (b) and (c) show thermal images of the same reactor with 4-vinylcyclohex-1-ene and carbon dioxide (Part b) and with hydrogen (Part c) flowing and all other conditions kept constant, respectively. After [126].

It should also be mentioned that asymmetric hydrogenations may be carried out successfully in SCFs over solid catalysts (cf. Chapter 5). One of the first examples was the enantioselective hydrogenation of ethyl pyruvate to ethyl (*R*)-lactate over a cinchonidine-modified Pt/Al_2O_3 catalyst in supercritical ethane [43, 44]. For the ethyl pyruvate hydrogenation in supercritical ethane, an increase of the enantiomeric excess (*ee*) was found upon increasing the catalyst/substrate ratio. Interestingly, the same change causes a decrease in the *ee* in liquid solvents where mass transfer of hydrogen becomes rate limiting at higher catalyst/substrate ratios [43].

Due to the obvious advantages of heterogeneously catalyzed hydrogenations in SCFs, several industrial applications were developed. For instance, the hydrogenation of edible oils and fatty acids for hardening (see above) was investigated by Degussa over commercial supported palladium catalysts in a continuous fixed-bed reactor [36]. Compared with reactions in a conventional trickle-bed process, 15 times higher space-time yields and a three times higher catalyst productivity were achieved in supercritical CO_2. A process for the hydrogenation of fatty acid methyl esters (FAMEs) was developed to the pilot-plant scale by Poul Møller Consulting in cooperation with Chalmers University of Technology, Sweden. This process uses a copper chromite catalyst and propane as a supercritical solvent [37, 38, 130].

More recently, a multi-purpose plant was built by Thomas Swan & Co. in collaboration with the University of Nottingham in the UK for the hydrogenation of isophorone to trimethylcyclohexanone over a supported palladium catalyst [16, 131, 132]. Other conversions are also carried out in the same plant, which went on-stream in June 2002 with a production capacity of $\sim$100 kg h^{-1} (1000 t yr^{-1}). With supercritical CO_2 as the reaction medium, the undesired overhydrogenation of isophorone to 3,3,5-trimethylcyclohexanone (TMCH) or trimethylcyclohexane can be largely suppressed [35].

As was shown in the same plant, different products can be obtained starting from the same reactant depending on the reaction pressure and temperature in the supercritical region. For instance, the reduction of acetophenone yields 1-phenylethanol at low pressure and the fully hydrogenated 1-cyclohexylethanol at higher pressures [16]. At variance with supercritical operation, the conversion in liquid acetone is slow, whereas in the gas phase at above 200 °C, it proceeds with poor selectivity for the desired product. Furthermore, the catalyst lifetime in the gas-phase conversion is lower by a factor of 30 with respect to operation at supercritical conditions. Thus, in supercritical CO_2 7.5 kg of isophorone can be hydrogenated with 1 g of catalyst before the onset of a significant catalyst deactivation.

In addition to hydrogenations, dehydrogenations may also benefit from the unique properties of SCFs. For instance, higher conversions by a factor of two with respect to that expected for the thermodynamic equilibrium were reached when a mixture of C_{10}–C_{14} alkanes was dehydrogenated close to its critical point over Pt–Sn/γ-Al_2O_3 [48]. Both a shift of the equilibrium and an enhanced rate of hydrogen transfer from the catalyst into the supercritical reaction mixture are possible explanations for this effect. Improved hydrogen transport away from the catalyst surface was also invoked to explain the improved rate in the palladium-catalyzed dehydrogenation of 1-phenylethanol in dense CO_2 [49].

6.3.1.2 **Fischer–Tropsch Synthesis**

The Fischer–Tropsch synthesis, that is, the conversion of a carbon monoxide–hydrogen mixture (synthesis gas) to hydrocarbons over Ni, Fe, Co or Ru catalysts [133], is predominantly applied for the production of transportation fuels. It has attracted considerable attention in recent years owing to, above all, the possibility of obtaining sulfur- and nitrogen-free (ultra-clean) fuels and of providing the second step for a pathway from coal, natural gas, or biomass-based feedstocks via gasification to liquid fuels. Conventionally, the Fischer–Tropsch synthesis is carried out in the gas phase or in liquid suspension (slurry phase) of the solid catalyst. In the earlier gas-phase process, local overheating of the catalyst due to the strongly exothermic reaction can occur and high yields of the unwanted methane lower the yields of the desired long-chain hydrocarbon products. Moreover, pore blocking by high-boiling products can lead to catalyst deactivation. The slurry-phase process developed later applies a high-boiling hydrocarbon as a liquid reaction medium. Although an excellent temperature distribution is achieved, the conversion rate is often limited by the transport of the synthesis gas into the liquid reaction mixture. It is evident that SCFs offer an attractive alternative to either the gas- or the slurry-phase processes for the heterogeneously catalyzed Fischer–Tropsch synthesis.

With the critical temperature in the range of typical reaction conditions, namely 200–300 °C, the readily available C_5–C_8 *n*-alkanes are mostly used as reaction solvents for Fischer–Tropsch synthesis under supercritical conditions. The advantages of this processing option include more efficient heat removal from the catalyst produced by the highly exothermic reaction and the suppression of catalyst deactivation by improved solubility and accelerated diffusion of higher molecular weight products (wax) that would otherwise block the catalyst pores. Further, through the improved mass transport properties in the supercritical reaction phase, the hydrogenation of the unwanted alkene fraction can be efficiently suppressed. Concentration profiles of hydrogen and carbon monoxide within the catalyst particles are reduced, leading to a "homogeneous" conversion throughout the catalyst.

In early studies of the Fischer–Tropsch synthesis in supercritical *n*-hexane ($T_c = 233.7\,°C$, $p_c = 2.97\,MPa$), Fujimoto and co-workers found a higher activity of a Co,La/SiO_2 catalyst than in a gaseous (nitrogen) or in a liquid (*n*-hexadecane) reaction phase [50–52]. Later, Bochniak and Subramaniam [53] showed that the improved activity under supercritical conditions can be rationalized in terms of an increased catalyst effectiveness factor. Thus, pore-diffusion limitations are alleviated by an efficient extraction of higher molecular weight products (cf. Figures 6.5 and 6.6). The rapid product desorption under supercritical conditions also suppresses hydrogenation of the primary 1-alkenes leading to a higher selectivity than in the gas- or liquid-phase conversion.

In the supercritical phase Fischer–Tropsch synthesis, the product selectivity depends on the nature of the supercritical solvent [53] and also on the reaction conditions, especially temperature and pressure [53, 54, 134]. The carbon number distribution of the products in the supercritical phase is shifted to higher carbon numbers compared with that in the gas phase (Figure 6.9a) [54]. The higher solubility and the more rapid removal of higher molecular weight hydrocarbons from the catalyst surface results in more active sites that are available on the catalyst surface. Also, the chain growth probability for hydrocarbons with a carbon chain length over

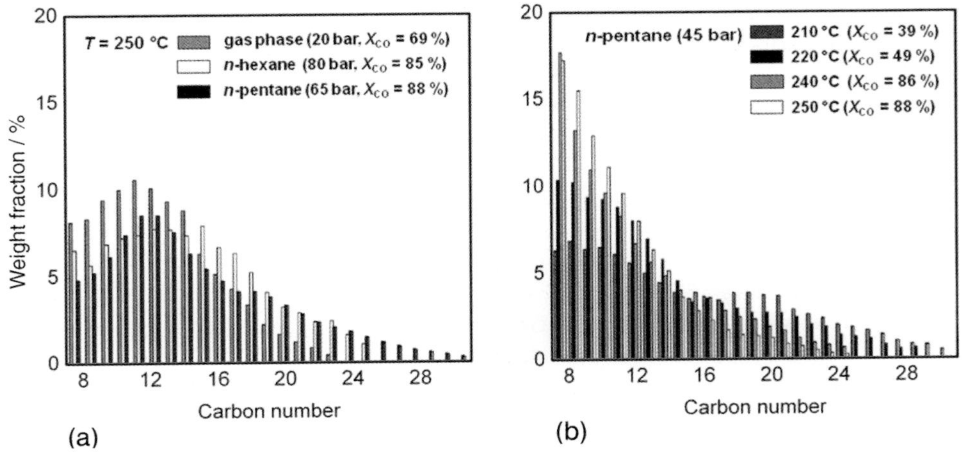

Figure 6.9 Weight fraction of hydrocarbon products as a function of carbon number in the Fischer–Tropsch synthesis ($n_{H_2}/n_{CO} = 2$) over 15 wt% Co/Al_2O_3 in the gas phase and in supercritical *n*-hexane and *n*-pentane (a) and at different temperatures in supercritical pentane (b). After [54].

10 is higher for the supercritical (0.87 and 0.85 for *n*-hexane and *n*-pentane, respectively) than the gas-phase conversion (0.75). Note that for *n*-pentane as the supercritical solvent, the fraction of hydrocarbons with a chain length above 20 is slightly higher than for *n*-hexane, although the overall density of the fluid phases is comparable. This may be due to accelerated diffusion of the higher molecular weight products away from the catalyst surface in the lighter hydrocarbon solvent or due to a pressure effect on the reaction kinetics [54]. A much stronger effect than that of pressure is exerted by the reaction temperature (Figure 6.9b). As expected, the product distribution shifts to lower hydrocarbons with increasing temperature.

Another peculiarity of the Fischer–Tropsch reaction under supercritical conditions is the deviation from the expected Anderson–Schulz–Flory (ASF) distribution of the hydrocarbon products [133]. Thus, the formation rate of C_7–C_{28} hydrocarbons (middle distillate) was found to be independent of the carbon number with a high chain growth probability of >0.96 [135]. In a detailed study, Elbashir and Roberts related the extent of this deviation to the physical properties of the reaction mixture at near-critical conditions: The enhanced solubility of higher molecular weight products results in more vacant sites at the catalyst surface for readsorption and further incorporation of the middle-distillate alkenes into the chain-growth reaction [55].

In order to promote the formation of the desired isoalkanes in the Fischer–Tropsch synthesis in supercritical *n*-hexane, a conventional Co/SiO_2 catalyst was combined with the bifunctional zeolite Pd/Beta [136]. This additional catalyst converts the initially formed Fischer–Tropsch hydrocarbons by hydrocracking and isomerization into isoalkanes. Isoalkane selectivities over 70% without significant hydroisomerization of the *n*-hexane could be achieved. Interestingly, the isoalkane selectivity was only 23% and the product distribution was shifted to lower hydrocarbons when cobalt was directly supported on acidic microporous zeolites such as dealuminated MCM-22 [59]. The acid-catalyzed cracking of the Fischer–Tropsch products to lower hydrocarbons and (undesired) alkenes is responsible for this result. In this particular case, higher yields for the desired isoalkane products were achieved in the liquid rather than in supercritical *n*-hexane as the solvent. This was explained by the higher reaction temperature and the high catalyst lifetime and cracking activity under supercritical conditions. Nevertheless, the zeolite-based catalyst was much more active than the previously studied cobalt catalysts with macroporous SiO_2 or Al_2O_3 as supports.

6.3.1.3 Hydroformylations

Hydroformylation, that is, the conversion of CO and hydrogen with an alkene, yields linear and branched aldehydes that are industrially important as intermediates for a large range of oxygenated products. The conventional industrial process (also referred to as “oxo synthesis”) is typically homogeneously catalyzed by cobalt or rhodium complexes in liquid or biphasic reaction mixtures [137]. As for hydrogenations or Fischer–Tropsch synthesis, supercritical reaction media, mostly CO_2, offer the opportunity to supply all reactants (and products) in a single phase and, thus, alleviate mass transfer restrictions of the gaseous CO–hydrogen mixture into the reaction phase.

The approach to heterogeneously catalyzed hydroformylation in SCFs was, so far, entirely based on the immobilization of promising complex catalysts on solid supports such as SiO_2, Al_2O_3, molecular sieves, dendrimers, or polymers. The stability of the catalyst towards leaching of the active metal component into the supercritical reaction phase is, therefore, of utmost importance. While polymers and related supports can easily be modified for anchoring the active complex, the structural conformation and the shape and size of the catalyst particles may depend strongly on the supercritical fluid and the specific reaction conditions. Consequently, more studies were devoted to inorganic supports. The current state-of-the-art in hydroformylation catalysis in SCFs was summarized in a recent review [138].

As shown earlier by Abraham's group, a rhodium catalyst prepared by impregnation of rhodium complexes onto modified silica was unstable towards leaching of the active metal in to the supercritical reaction phase [61]. Stable rhodium or platinum catalysts for the hydroformylation of 1-hexene in supercritical CO_2 can, however, be obtained by covalent bonding of the complexes to a silica support [62, 67]. Their activity in supercritical CO_2 was lower than that for homogeneous catalysis in liquid solution, but higher than that for heterogeneous gas-phase catalysis [63]. Interestingly, the most active catalyst was a rhodium complex supported on an ordered mesoporous MCM-41-type material [65]. For this catalyst, the yield ratio of linear and branched aldehyde was 8.8 (100 °C, 24.0 MPa). The formation of the undesired branched isomer (and also of the hydrogenation products) can be further reduced with the microporous zeolite MCM-20 as the support (yield ratio of linear to branched aldehyde of 15.8) [65]. However, the slow diffusion within the zeolitic micropores leads to a decrease in catalyst activity.

Recently, the strategy of immobilizing metal catalyst complexes by dissolving them in an ionic liquid supported on porous solids (supported ionic liquid phase [139]) was applied to hydroformylation catalysis of 1-octene in supercritical CO_2 [140]. As opposed to the conversion with gaseous reaction phases, the products could be extracted rapidly and diffusion limitations of the reactant to the rhodium catalysts in the ionic liquid phase were absent in the supercritical solvent.

6.3.1.4 Oxidations

The potential of SCFs to realize the goals of green chemistry is particularly evident for heterogeneously catalyzed oxidations. Carbon dioxide and water are especially attractive reaction media to replace conventional liquid organic solvents. The low environmental impact and the chemical inertness of these media under most process conditions render the operation of reactions in these SCFs "intrinsically safe". Moreover, the efficient dissipation of heat from the catalyst into the bulk fluid is a clear advantage of SCFs for the partly highly exothermic oxidations. As discussed above, for example for hydrogenations (Section 6.3.1.1), the miscibility of all reactants and products with molecular oxygen or air in a single supercritical reaction phase may lead to higher rates than under conventional conditions often involving oxygen transfer into a liquid catalyst suspension. Higher selectivities may be achieved in partial oxidations, if the desired product can be removed from the catalyst surface before it undergoes consecutive reactions to undesired overoxidized products.

Despite the obvious advantages, heterogeneously catalyzed oxidations in SCFs became a focus of research only in recent years. The most intensively studied reactions are partial oxidations of alcohols and, to a lesser extent, of hydrocarbons in supercritical CO_2 (cf. Table 6.1). Catalytic total oxidation in supercritical CO_2 has received less attention. This was investigated more deeply, however, in supercritical water (supercritical water oxidation, SCWO) [118, 141]. Generally, partial and total oxidations in supercritical water were less intensely studied due to the limited stability of several oxidation catalysts such as metals, especially in the presence of oxygen (see Chapter 13). In the majority of the studies on heterogeneous catalysis in SCFs, molecular oxygen is used as the oxidizing agent. The solid-catalyzed gasification of carbon-containing material such as natural gas, coal, or biomass with water which also formally belong to partial oxidation are included in Chapter 13 and will not be treated further here. Instead, some selected examples of solid-catalyzed oxidations of alcohols and hydrocarbons in supercritical CO_2 will be briefly discussed in the following.

A paper by Baiker and co-workers in 2000 [142] marked the beginning of a large and still growing number of studies on the heterogeneously catalyzed alcohol oxidation in SCFs (cf. Table 6.1). As for hydrogenations, the high-pressure phase behavior also plays an important role in partial oxidations. For instance, the effect of pressure on the rate of benzyl alcohol oxidation with molecular oxygen over a Pd/Al_2O_3 catalyst in dense CO_2 is shown in Figure 6.10 [80]. Upon transition from a two- into a single-phase system between 14 and 15 MPa the reaction rate increases by a factor of $\sim$2. Concomitantly, the selectivity for benzaldehyde decreases only slightly (Figure 6.10). In the single supercritical phase at 15 MPa, the turnover frequency is 1585 h^{-1}, which is almost 80 times higher than that for the conversion in liquid toluene as the solvent (20 h^{-1}).

In several studies on noble metal-catalyzed alcohol oxidation in supercritical CO_2, a maximum in the ketone or aldehyde yield depending on the oxygen content in the reaction mixture has been reported [79, 81, 143]. The decreasing product yield at higher oxygen concentration can be explained by catalyst deactivation due to over-

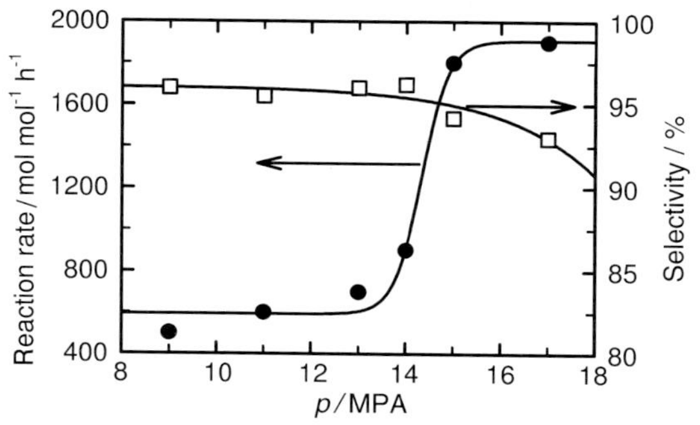

Figure 6.10 Reaction rate and benzaldehyde selectivity in the continuous conversion of benzyl alcohol over 0.5 wt% Pd/Al_2O_3 as a function of pressure ($T = 80\,°C$, $W/F_{alcohol} = 20\,g\,h\,mol^{-1}$; $n_{O_2}/n_{alcohol} = 0.5$). After [80].

oxidation of the noble metal surface. It can occur at temperatures as low as 40 °C [76]. This has been proven for a 0.5 wt % Pd/Al_2O_3 catalyst in the partial oxidation of benzyl alcohol by *in situ* extended X-ray absorption fine structure (EXAFS) spectroscopy [79]. Interestingly, a similar activity loss with increasing oxygen content was not observed for the oxidation of cinnamyl alcohol, where the noble metal surface remained in a reduced state over a much wider range of conditions [144].

In situ EXAFS studies further revealed the presence of a single-phase reaction mixture inside the pores of the 0.5 wt % Pd/Al_2O_3 catalyst [80]. The accelerated mass transfer within this single phase is responsible for the rate increase under supercritical conditions. At variance, higher rates for the oxidation of cinnamyl alcohol [144] and geraniol [91] were observed in the two-phase region of the reaction mixtures. For the geraniol oxidation, this was explained in terms of limited mass transfer by adsorbed water in the catalyst pores under single-phase supercritical conditions.

As water is an inevitable by-product of alcohol oxidation with oxygen, its removal from the catalyst surface supports rapid conversion. Towards that purpose, the surface hydrophobicity of silica gels as supports for the redox active catalyst tetrapropylammonium perruthenate TPA[RuO_4] was increased by modification with methylated and fluorinated alkyl groups [86, 87, 89, 145]. A more than 75-fold higher activity with respect to these catalysts was reported when the perruthenate ion was electrically grafted to a (hydrophobic) surface modified with imidazolium cations ("supported ionic liquids") [146]. This concept can be transferred to the modification of polymers to immobilize perruthenate used as catalyst in benzyl alcohol oxidation [147]. Increased surface hydrophobicity was also the key to higher aldehyde yields in the oxidation of 9-anthracenemethanol over PTFE-modified versus an untreated 5.0 wt % Pt/C catalyst [148] and of benzyl alcohol over Pd_{55} nanoparticles immobilized on silica modified with poly(ethylene glycol) (PEG) moieties [149]. Moreover, the PEG-modified catalysts proved to be considerably more stable than that with the plain and unmodified silica surface.

The high-pressure phase behavior was utilized in an innovative concept for the epoxidation of propene with H_2O_2 over the bifunctional microporous catalyst Pd,Pt/TS-1 [74, 150]. In a single, supercritical CO_2 phase, the H_2O_2 can first be generated from the elements on the noble metal (see Section 6.3.1.1) and then be converted with propene to propene oxide and water on the titanium-substituted silicalite-1 (TS-1). Currently, propene conversions do not reach values above 10%, while the selectivity to propene oxide is higher than 90% [74].

Also, the more challenging, heterogeneously catalyzed oxidation of alkanes by molecular oxygen was attempted in supercritical CO_2. With methane over metal catalysts such as Cu, Ag, and Ag/Au in supercritical water [68] or with propane over CoO_x/SiO_2 [69] or $(VO)_2P_2O_7$ [70], the yields of oxyfuntionalization products did not exceed 20%. Several solid catalysts were evaluated for the conversion of cyclohexane with molecular oxygen in supercritical CO_2 such as supported transition metal oxides [71] and transition metal-substituted microporous aluminophosphates [Mn]APO-5 [151] and [Co]APO-5 [72]. So far, the highest selectivities for cyclohexanol/cyclohexanone and diacids of 28% and 40%, respectively, at 10% conversion at 180 °C and 14.0 MPa in methanol-modified CO_2 were recently reported for the novel polyoxometallate catalyst $Ag_5PMo_{10}V_2O_{40}$ [73]. The proton form of this catalyst is also active for the partial oxidation of activated alkyl aromatics such as anthracene and xanthene in supercritical CO_2 with yields of up to 70% (80 °C, 17.0 MPa) [152].

6.3.1.5 **Alkylations**

In addition to hydrogenations, alkylations are among the most intensively studied conversions on solid catalysts in supercritical reaction phases. Most of these cases belong to the classes of alkene/alkane alkylations or Friedel–Crafts alkylations. Mostly, solid acids, in particular acidic zeolites, are applied as catalysts. The most important benefit of using SCFs as media for these reactions is the alleviation or prevention of catalyst deactivation due to coke formation from the alkenes and/or aromatics frequently involved in alkylation reactions [24]. The extraction of coke precursors by SCFs can be achieved either during the reaction (*in situ*) or in a separate step (*ex situ*).

An example where the deactivation by coking has, until now, prohibited an industrial application of a solid acid catalyst is the alkylation of isobutane by light alkenes such as butenes [153]. This conversion supplies highly branched isoalkanes as high-octane gasoline components. Despite numerous efforts to develop a solid-catalyzed process, the isobutane alkylation is currently carried out using liquid H_2SO_4 or HF as the catalyst. In a 1994 patent, the application of supercritical reaction conditions to extend the lifetime of zeolite catalysts in isobutane/alkene alkylation was reported [94]. However, at the temperatures >135 °C (T_c for isobutane) required to reach the supercritical state, unwanted side reactions, especially cracking and oligomerization, are favored over the desired alkylation.

As shown by Clark and Subramaniam [95], the critical temperature of an isobutane–1-butene mixture (molar ratio 9 : 1) could be decreased to 40 °C by addition of a fivefold molar excess of CO_2. Whereas a USY zeolite deactivates in the conversion of the undiluted feed at 140 °C, it is stable under supercritical conditions at 50 °C in the presence of CO_2 over 2 days on-stream (Figure 6.11). However, the 1-butene conversion was 20% and the alkylate fraction in the product was only 5–10%. Moreover, the product was strongly diluted with CO_2. At higher butene conversion, a USY zeolite is subject to deactivation even under supercritical conditions [154]. Cyclic regeneration of the zeolite catalyst in supercritical isobutane allows most of the coke to be removed and a stable butene conversion of >92% to be obtained for more than 210 h on-stream [155]. Nevertheless, the alkylate composition does not comply with industrial standards.

Several studies have been directed towards the influence of the phase behavior on the alkylation of aromatics with light alkenes. For instance, Shi *et al.* [98] systematically varied the phases present in benzene alkylation with ethene over the acidic zeolite H-Beta by changing pressure and feed composition in the temperature range 240–260 °C. The authors found a maximum for the reaction rate slightly above the critical pressure and a decreasing rate with increasing pressure. An additional, less pronounced rate maximum was observed for a two-phase liquid–gas feed mixture. Similarly, the highest catalyst lifetime in the conversion of

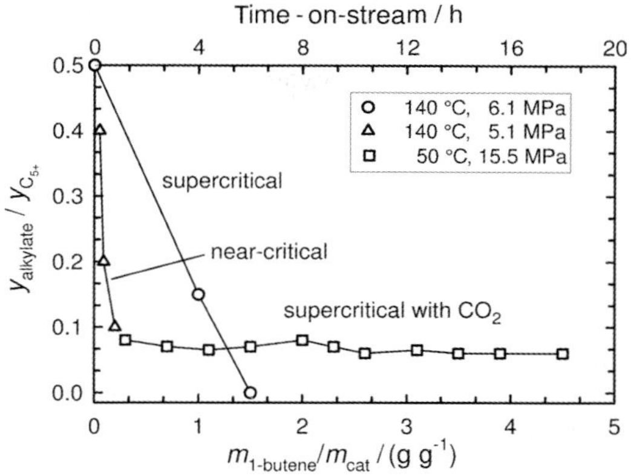

Figure 6.11 Alkylate fraction in the C_{5+} products from the alkylation of isobutane with 1-butene over zeolite H-USY under near-critical (140 °C, 5.1 MPa) and supercritical conditions in the absence (140 °C, 6.1 MPa) and presence of CO_2 (50 °C, 15.5 MPa). After [95].

benzene with propene was found for conditions slightly above the critical point of the reaction mixture [100]. The authors of this study rightly stress that the critical point of the reaction mixtures changes significantly upon progress of the conversion. In this case, the critical point changes from 278.4 °C and 5.34 MPa (for benzene + propene with a propene mole fraction of 0.2) to 316.7 °C and 4.54 MPa (for benzene + cumene). These results again show that applying single-phase, supercritical conditions alone does not necessarily result in high reaction rates or catalyst lifetimes, but that a sensitive tuning is required for rate or stability optimization.

Shape-selectivity effects have been observed for a number of alkylations of aromatics over acidic zeolites or related microporous materials at supercritical reactions conditions [99, 101, 102, 156–158]. The question of whether the physical state of the reaction medium surrounding the catalyst has a direct influence on the processes occurring inside zeolitic micropores is discussed controversially. For instance, the shape-selective conversion of naphthalene with 2-propanol over zeolite H-mordenite in supercritical CO_2 led to a high yield ratio of the more slender 2,6- vs the more bulky 2,7-diisopropylnaphthalene [99]. Since, however, the isomer distribution was almost identical with that for the liquid-phase reaction at the same temperature, a direct influence of the supercritical reaction medium on the shape-selective conversion inside the zeolite pores was rendered unlikely. A recent study on isobutene alkylation supports the absence of a direct influence of the fluid state around the catalyst on the inner pores of microporous zeolites: The higher catalyst stability of zeolite H-Beta vs H-Y in isobutene alkylation with 2-butene was attributed to the removal of coke precursors from the outer surface from the smaller zeolite crystals [96]. The coke formed within the pores of zeolite H-Y with a similarly spacious pore system to zeolite H-Beta, but with larger crystals, could not be removed under supercritical conditions. An influence of supercritical conditions on the processes occurring inside the zeolitic micropores was, however, discussed to explain, for example, shape-selectivity effects on the formation of coke precursors extracted from the acidic zeolite catalysts H-ZSM-5 or H-mordenite during the conversion of supercritical ethylbenzene [156], the threefold higher rate for the hydroxyalkylation of phenol over zeolite H-mordenite in supercritical CO_2 than in liquid toluene [102], or the effect of supercritical reaction conditions on the transalkylation of diisopropylbenzene over the zeolites H-Y, H-Beta and H-mordenite [159].

In the disproportionation of toluene to benzene and xylenes over zeolite H-ZSM-5, Collins *et al.* [160] found a maximum of the *p*-xylene selectivity depending on pressure close to the critical point ($T_c = 318.7\,°C$, $p_c = 4.1\,MPa$) (Figure 6.12). This maximum was essentially absent when the reaction was carried out at a temperature slightly above the critical point, namely at 325 °C ($T_r = 1.01$) instead of 320 °C ($T_r = 1.002$). The conjecture to rationalize this unusual result was that a phenomenon occurring in the immediate vicinity of the critical point was involved: the clustering of solvent molecules around the *p*-xylene formed initially under kinetic control could prevent readsorption, and, as a consequence, isomerization to the thermodynamically favored *meta*-isomer was prevented – hence, a maximum of the selectivity for the *para*-isomer close to the critical point. This example is, however, the only one known so far where the clustering of solvent molecules around a solute in a heterogeneously catalyzed conversion was invoked. Note that, close to the critical point, a pronounced maximum in the excess adsorption for zeolites may occur [161]. Therefore, an influence of this anomaly on the adsorption of the reactants and products or on the shape-selectivity effects on the zeolite also cannot be excluded.

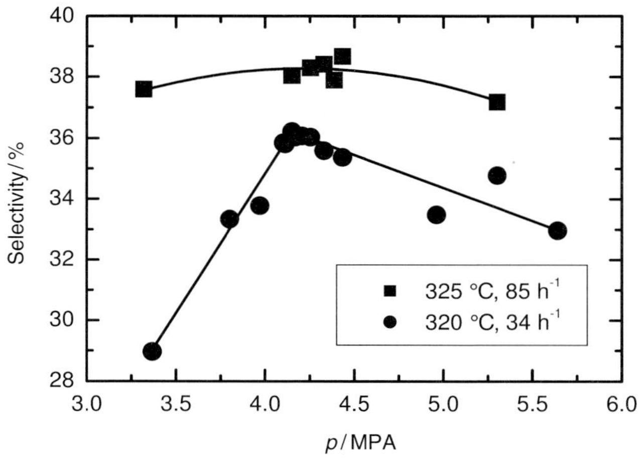

Figure 6.12 Selectivity for *p*-xylene as a function of pressure for different temperatures (and *WHSV*) in the disproportionation of toluene ($T_c = 318.7\,°C$, $p_c = 4.1\,MPa$) over zeolite H-ZSM-5. After [160].

6.3.1.6 **Isomerizations**

The double-bond isomerization of 1-hexene to *cis/trans*-2- and 3-hexene has been used as an example to study the effects of supercritical reaction conditions in conversions over porous catalysts. In particular, the favorable balance between efficient desorption of hexene oligomers that can lead to catalyst coking at low gas-like densities and rapid pore diffusion that can limit the reaction rate at high liquid-like densities was shown for this reaction (see Section 6.2.1.4) [24, 25, 107, 108]. Moreover, the yield ratio of *cis*- to *trans*-2-hexene over a γ-Al_2O_3 shell-type catalyst was found to increase with pressure for temperatures above the critical point, whereas it decreased with pressure for subcritical temperatures (Figure 6.13) [108]. This was attributed to a faster desorption of *cis*-2-hexene as the primary product from the catalyst surface under supercritical than under subcritical, liquid-phase conditions. Hence, the high, liquid-like densities favor the consecutive reaction of the *cis*- to the thermodynamically favored *trans*-isomer.

1-Hexene isomerization is one of the rare cases for which the effective diffusivity of a reactant in the pores of a catalyst under near-critical conditions has been determined experimentally [106]. For the conversion over the mesoporous catalyst 0.6 wt% Pt/γ-Al_2O_3 ($d_{P,average} = 5$ nm), it was calculated from the pore effectiveness factor under isothermal, diffusion-limited conditions and its relation to the Thiele modulus. A maximum for the effective diffusivity was found close to the critical point of 1-hexene ($T_c = 231\,°C$, $p_c = 3.17$ MPa) (Figure 6.14). Note that this maximum is observed clearly below the critical density of the bulk reaction medium. In view of the impact of the pore diameter on the phase behavior inside mesoporous catalysts (Section 6.2.1.2), it is likely that this diffusivity maximum corresponds to conditions under which the pore-confined phase reaches the supercritical state.

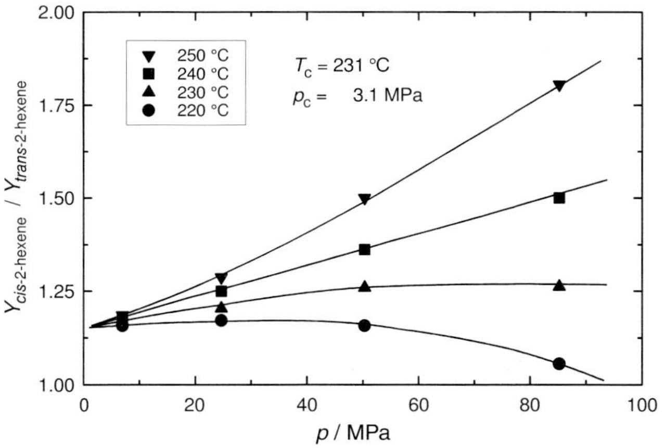

Figure 6.13 Yield ratio of *cis*- and *trans*-2-hexene in the isomerization of 1-hexene over a γ-Al_2O_3 shell-type catalyst at different super- and subcritical temperatures as a function of pressure. After [108].

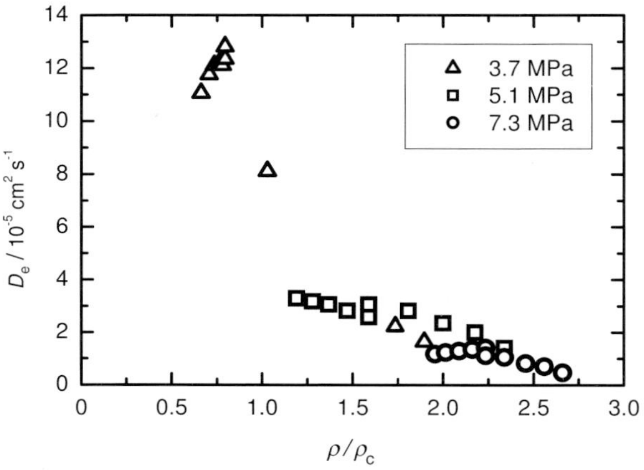

Figure 6.14 Effective diffusivity of 1-hexene as a function of reduced density calculated from the Thiele modulus for the isomerization of 1-hexene over 0.6 wt% Pt/γ-Al_2O_3 at different temperatures (235–305 °C) and pressures. After [106].

Although the effective diffusivity in the pores was much lower than in the bulk, it could be varied over two orders of magnitude by relatively small pressure changes in the near-critical region.

Several studies have focused on the isomerization of *n*- to isobutane under supercritical conditions (cf. Table 6.1). The highest reaction rate on a sulfated zirconia catalyst was found at 215 °C and 4.0 MPa close to the critical point of *n*-butane ($T_c = 152\,°C$, $p_c = 3.8$ MPa). Under these conditions, deactivation due to coking was much lower than in the gas phase and a stable conversion was observed [109, 110]. Although this conversion was lower than in the gas phase, higher production capacities for the isomerization products were achieved as a result of the higher feed density under supercritical conditions.

6.3.1.7 **Miscellaneous**

Numerous other heterogeneously catalyzed conversions have been carried out in SCFs as reaction media. These include hydrocarbon cracking [162, 163], etherifications [164], esterifications [165–168], transesterifications [169, 170], and carbonylations [171, 172], to name just a few. Especially worthy of mention are reactions where the supercritical fluid acts both as a solvent and as a reactant. Examples of this type of conversion are the amination of alcohols with supercritical ammonia over unsupported Co–Fe [173, 174], the synthesis of cyclic carbonates from epoxides and supercritical CO_2 over immobilized Co [175] or Mn complexes [176], also as intermediates to dimethyl carbonate [177, 178], and the synthesis of *N,N*-dimethylformamide from supercritical CO_2, hydrogen, and dimethylamine over a ruthenium-containing silica aerogel [179].

6.3.2 Conversions in GXLs

Based on the obvious advantages of GXLs for catalytic conversions (cf. Section 6.2.2), several studies have been devoted to the investigation of the potential of these solvent systems in reactions over solid catalysts. In many of these cases, the GXL phase occurs as part of a biphasic reaction mixture upon varying the experimental conditions, mostly pressure, in studies of conversions in SCFs. Consequently, a direct comparison of the conversion in GXLs to that in SCFs is possible. Frequently, but not exclusively, the conversion in the GXL phase offers advantages such as higher rates, selectivities, or catalyst stability over that at single-phase supercritical conditions. Nevertheless, the number of reports on the application of GXLs in heterogeneous catalysis is limited, especially when compared with those of investigations using SCFs. To date, examples of conversions in GXLs as reaction media are limited to hydrogenation, hydroformylation, oxidation, and the cycloaddition of CO_2 to epoxides. Some of those will be treated in the following.

The first report of the utilization of GXLs in heterogeneous catalysis was published in 1997 by Bertucco *et al.* [180]. They investigated the continuous selective hydrogenation of a mixture of unsaturated ketones with hydrogen in the presence of CO_2 over a 1 wt% Pd/Al_2O_3 eggshell catalyst at pressures of 12–17.5 MPa and temperatures of 150–200 °C. At a constant reactant to CO_2 ratio, the reaction rate increased with pressure. This was attributed to the higher mole fraction of hydrogen in the liquid reaction phase with increasing expansion by CO_2. However, the dilution of the reactants with increasing amounts of CO_2 relative to the reactant in the feed led to lower reaction rates compared with those at constant pressure.

In another example of a hydrogenation, α-pinene was converted in the presence of CO_2 over a 10 wt% Pd/C catalyst at 50 °C [181]. The reaction occurred at a higher rate when the reaction mixture was present as a CO_2-expanded liquid than when it was present as a single supercritical phase (above ∼9.3 MPa). This was explained by the higher concentration of the reactant in the GXL phase (mole fraction of 10%) than in the supercritical mixture (mole fraction of 1%) and, consequently, a higher reactant concentration at the catalyst surface. Higher rates observed for α-pinene hydrogenation at 50 °C in a GXL than in a supercritical phase were also obtained using a 1 wt% Pt/C catalyst [182]. This study proved that the adsorption of the reactant on the noble metal catalyst is the rate-limiting step and, thus, the faster reaction is due to the higher concentration at the catalyst surface. An improved stability of the Pt catalyst towards poisoning by CO is another advantage of the GXL phase as the reaction medium. For the hydrogenation of cinnamaldehyde over a Pt catalyst supported on a rice husk-based carbon, the higher concentration of the reactant in a GXL was used to explain the higher selectivity to cinnamyl alcohol vs the completely saturated hydrocinnamyl alcohol (88% vs <10% at ∼60% conversion) than in a single-phase SCF, although the catalyst activity in the GXL phase was significantly lower [183].

In further studies on metal-catalyzed hydrogenations, however, two groups showed that the reaction rate can be related to the hydrogen solubility in the GXL

Table 6.2 Volume expansion V/V_0, hydrogen solubility, and rate constant k for the hydrogenation of polystyrene in decahydronaphthalene without and with expansion by CO_2 over 5 wt% $Pd/BaSO_4$ and 65 wt% Ni/Al_2O_3-SiO_2 (weight ratio 3 : 100) at 150 °C ($p_{H_2} = 5.1$ MPa, $p_{H_2} + p_{CO_2} = 20.4$ MPa). After [185].

Gas	V/V_0	H_2 solubility (mol cm^{-3})	k (cm^3 g^{-1} s^{-1})
H_2	1.0	2.4×10^{-4}	6.2×10^{-3}
H_2+CO_2	1.3	4.3×10^{-4}	9.7×10^{-3}

reaction phase. Phiong *et al.* [184] performed a detailed kinetic analysis of the hydrogenation of α-methylstyrene over a carbon-supported Pd catalyst. Xu *et al.* [185] studied the hydrogenation of the aromatic rings in polystyrene with 5 wt% $Pd/BaSO_4$ and 65 wt% Ni/Al_2O_3–SiO_2 as catalysts. Both the hydrogen solubility and the hydrogenation rate were higher in CO_2-expanded decahydronaphthalene than in the pure liquid solvent at constant hydrogen partial pressure in the gas phase (Table 6.2). However, the rate increase was not directly proportional to the hydrogen solubility. In a subsequent study, they investigated the deactivation behavior of the two metal catalysts in the GXL phase conversion [186]. In both cases, poisoning of the catalyst occurred. The Pd-based catalyst was poisoned by CO formed by the RWGS reaction from CO_2 hydrogenation, as also reported for many other hydrogenation reactions in supercritical CO_2. Although much less pronounced, the Ni-based catalyst was subject to poisoning by water as a product of CO_2 methanation in the GXL reaction phase.

Hemminger *et al.* [187] compared the hydroformylation of 1-hexene over a 1,5-cyclooctadienylrhodium complex immobilized on a phosphine-modified silica support in liquid toluene, CO_2-expanded toluene and supercritical CO_2. Interestingly, the conversion rate was the highest in the CO_2-expanded liquid phase. However, a substantially higher amount of double-bond isomerization products was formed in the GXL phase. Thus, the aldehyde yield achieved per rhodium atom was similar in the expanded liquid and in supercritical CO_2. As a further result of the isomerization in the expanded liquid, the selectivity for the desired *n*-aldehyde relative to isoaldehydes was lower in the CO_2-expanded toluene. Supercritical CO_2 was, therefore, the most favorable medium for the heterogeneously catalyzed hydroformylation.

A higher reaction rate than in pure supercritical CO_2 was found for the partial oxidation of benzyl alcohol to benzaldehyde over 0.5 wt% Pd/Al_2O_3 when low amounts of toluene were present as a cosolvent ($n_{toluene}/n_{alcohol} = 1$; cf. Figure 6.10) [188]. In the biphasic GXL phase obtained at higher toluene concentrations, however, the reaction was much slower. In contrast, an expanded liquid phase resulted in higher rates than the single-phase SCF for the cycloaddition of CO_2 to liquid propylene or styrene oxide over immobilized $ZnBr_2(Py_2)$ [189] or Mn(salen) complexes [176]. In the latter case, yields of 95% and turnover frequencies of 255 h^{-1} could be reached. Since CO_2 is also a reactant in these conversions, the CO_2

concentrations in the GXL have to be kept sufficiently high ($n_{\text{styrene oxide}}/n_{CO_2} = 4$) to obtain the optimum rate.

GXLs may also aid in reducing leaching of active catalyst components into a liquid reaction mixture. An improved resistance of an iron porphyrin chloride complex supported on the ordered mesoporous material MCM-41 towards leaching of the metal complex into liquid acetonitrile was reported when the solvent was expanded by CO_2 [190]. Thus, in two subsequent experiments on cyclohexene epoxidation with iodosylbenzene using the fresh and the used catalyst, the iron contents of the reaction mixture were 1.19 and 0.81 ppm for the CO_2-expanded solvent, whereas 2.67 and 2.52 ppm of iron were found after reaction in neat acetonitrile. This was attributed to the lower polarity of the expanded versus the pure organic solvent. The lower polarity of the acetonitrile with increasing CO_2 content was also used to explain the decreasing conversion of cyclohexene oxidation with oxygen over the same catalyst at increasing levels of expansion. Nevertheless, at an expansion degree of $\sim$1.4, the cyclohexene conversion could be roughly doubled with respect to the reaction in pure acetonitrile.

6.4 Outlook

Heterogeneous catalysis in SCFs and GXLs offers tremendous opportunities towards reaching the fundamental principles of green chemistry. In addition to the application of environmentally benign and intrinsically safe solvents such as carbon dioxide or water, the major potential of utilizing these innovative reaction media lies in a more efficient use of solid catalysts. In particular, higher reaction rates, the control of selectivity, faster heat transfer to and away from the catalyst, increased catalyst lifetime, and process intensification are obvious advantages that can emerge from the unique properties of SCFs and GXLs as reaction media for heterogeneous catalysis. Although, generally, the solvent effects in heterogeneous catalysis are less diverse and mostly less pronounced than in homogeneous catalysis, probably due to a lack of defined solvation at solid surfaces, the tunable properties of SCFs and GXLs and their particular phase behavior may be successfully applied to achieve improvements over conversions in conventional gas or liquid phases that are otherwise not obtained.

In order to exploit fully the potential of SCFs and GXLs for applications in heterogeneous catalysis in chemical technology and industry, several challenges still lie ahead. Among them is the need for a deeper understanding of the fundamental processes occurring at the active sites of solid catalysts. Especially the effects of the phase state and the nature of innovative solvents on adsorption and mass and heat transfer in phases confined within the pores of the catalysts are still insufficiently understood. Further progress in spectroscopic methods for the *in situ* characterization of working catalysts and computer-aided approaches to simulation and modeling will play an important role towards an understanding of these processes on a molecular level. These will eventually allow a rational and predictable design of catalysts and catalytic conversions and, thus, open up new pathways to elucidate the potential of innovative solvent systems for industrial applications. Without doubt, SCFs and GXLs will be an integral part of the future toolbox of methodologies to arrive at sustainable and green catalytic processes.

References

1 Anastas, P. and Warner, J.C. (1998) *Green Chemistry – Theory and Practice*, Oxford University Press, New York.

2 Anastas, P., Kirchhoff, M.M. and Williamson, T.C. (2001) *Applied Catalysis A: General*, **221**, 3.

3 Schlögl, R. (2008) in *Handbook of Heterogeneous Catalysis*, 2nd edn, vol. **5** (eds G. Ertl, H. Knözinger, F. Schüth and J. Weitkamp), Wiley-VCH Verlag GmbH, Weinheim, p. 2501.

4 Bøgdil Hansen, J. and Højlund Nielsen, P.E. (2008) in *Handbook of Heterogeneous Catalysis*, 2nd edn, vol. 6 (eds G. Ertl, H. Knözinger, F. Schüth and J. Weitkamp) Wiley-VCH Verlag GmbH, Weinheim, p. 2920.

5 Grumett, P. (2003) *Platinum Metals Review*, **47**, 163.

6 Grumett, P. (2003) *Filtration & Separation*, **40**, 16.

7 Iwao, S., Abd El-Fatah, S., Furukawa, K., Seki, T., Sasaki, M. and Goto, M. (2007) *Journal Of Supercritical Fluids*, **42**, 2004.

8 Dittmeyer, R. and Emig, G. (2008) in *Handbook of Heterogeneous Catalysis*, 2nd edn, vol. 3 (eds G. Ertl, H. Knözinger, F. Schüth and J. Weitkamp), Wiley-VCH Verlag GmbH, Weinheim. p. 1727.

9 Gläser, R. and Weitkamp, J. (2002) in *Handbook of Porous Solids*, vol. 1 (eds F. Schüth, K.S.W. Sing and J. Weitkamp), Wiley-VCH Verlag GmbH, Weinheim, p. 395.

10 Jin, H. and Subramaniam, B. (2003) *Chemical Engineering Science*, **58**, 1897.

11 Scott, R.L. and van Konyenburg, P.H. (1970) *Discussions of the Faraday Society*, **49**, 87.

12 Grunwaldt, J.-D., Wandeler, R. and Baiker, A. (2003) *Catalysis Reviews-Science and Engineering*, **45**, 1.

13 Sing, K.S.W., Everett, D.H., Haul, R.A.W., Moscou, L., Pierotti, R.A., Rouquerol, J. and Siemieniewska, T. (1985) *Pure and Applied Chemistry*, **57**, 603.

14 Thommes, M., Köhn, R. and Fröba, M. (2000) *The Journal of Physical Chemistry. B*, **104**, 7932.

15 Beckmann, E. (2004) *Journal Of Supercritical Fluids*, **28**, 121.

16 Ciriminna, R., Carraro, M.L., Campestrini, S. and Pagliaro, M. (2008) *Advanced Synthesis and Catalysis*, **350**, 221.

17 Kruse, A. and Vogel, H. (2008) *Chemical Engineering & Technology*, **31**, 23.

18 Tucker, S.C. (1999) *Chemical Reviews*, **99**, 353.

19 Brennecke, J.F., Debenedetti, P.G., Eckert, C.A. and Johnston, K.P. (1990) *AICHE Journal*, **36**, 1927.

20 Eckert, C.A. and Knutson, B.L. (1993) *Fluid Phase Equilibria*, **83**, 93.

21 Savage, P.E., Gopalan, S., Mizan, T.I., Martino, C.I. and Brock, E.E. (1995) *AICHE Journal*, **41**, 1723.

22 Dvoyashkin, M., Valiullin, R., Kärger, J., Einicke, W.-D. and Gläser, R. (2007) *Journal of the American Chemical Society*, **129**, 10344.

23 Wolf, H. (1980) PhD thesis, Technical University of Munich.

24 Subramaniam, B. (2001) *Applied Catalysis A: General*, **212**, 199.

25 Baptist-Nguyen, S. and Subramaniam, B. (1992) *AICHE Journal*, **38**, 1027.

26 Dariva, C., Coelho, L.A.F. and Oliveira, J.V. (1999) *Fluid Phase Equilibria*, **158–160**, 1045.

27 Groß, T., Chen, L., Buchhauser, J. and Lüdemann, H.-D. (2001) *Physical Chemistry Chemical Physics*, **3**, 2845.

28 Wei, M., Musie, G.T., Busch, D.H. and Subramaniam, B. (2002) *Journal of the American Chemical Society*, **124**, 2513.

29 Musie, G., Wei, M., Subramaniam, B. and Busch, D.H. (2001) *Coordination Chemistry Reviews*, **219–221**, 789.

30 Reighard, T.S. and Olesik, S.V. (1996) *Analytical Chemistry*, **68**, 3612.

31 Yuan, H. and Olesik, S.V. (1997) *Journal of Chromatography*, **764**, 265.

32 Arunajatesan, V., Subramaniam, B., Hutchenson, K.W. and Herkes, F.E. (2001) *Chemical Engineering Science*, **56**, 1363.

33 Tschan, R., Wandeler, R., Schneider, M.S., Schubert, M.M. and Baiker, A. (2001) *Journal of Catalysis*, **204**, 219.

34 Roberts, G.W., Xu, D., Kiserow, D.J. and Carbonell, R.G. (2004) Patent WO 2004/052937, assigned to North Carolina State University.

35 Hitzler, M.G., Smail, F.R., Ross, S.K. and Poliakoff, M. (1998) *Organic Process Research & Development*, **2**, 137.

36 Tacke, T., Wieland, S. and Panster, P. (1996) in *High Pressure Chemical Engineering*, vol. 12 (eds P.R. von Rohr and C. Trepp), Proc. Technol. Proc., Elsevier, Amsterdam, p. 17.

37 van den Hark, S. and Härröd, M. (2001) *Applied Catalysis A: General*, **210**, 207.

38 Härröd, M. and Möller, P. (1999) US Patent 5 962 711, assigned to Poul Møller Ledelses.

39 Pillai, U.R., Sahle-Demessie, E. and Young, D. (2003) *Applied Catalysis B: Environmental*, **43**, 131.

40 Chatterjee, M., Ikushima, Y., Yokoyama, T. and Sato, M. (2008) *Advanced Synthesis and Catalysis*, **350**, 624.

41 Xi, C., Wang, H., Liu, R., Cai, S. and Zhao, F. (2008) *Catalysis Communications*, **9**, 140.

42 Stephenson, P., Licence, P., Ross, S.K. and Poliakoff, M. (2004) *Green Chemistry*, **6**, 521.

43 Minder, B., Mallat, T., Pickel, K.H., Steiner, K. and Baiker, A. (1995) *Catalysis Letters*, **34**, 1.

44 Wandeler, R., Künzle, N., Schneider, M.S., Mallat, T. and Baiker, A. (2001) *Journal of Catalysis*, **200**, 377.

45 Zhao, F., Fujita, S.-i., Sun, J., Ikushima, Y. and Arai, M. (2004) *Catalysis Today*, **98**, 523.

46 Yuan, T. and Marshall, W.D. (2007) *Journal of Environmental Monitoring*, **9**, 1344.

47 Yuan, T., Fournier, A.R., Proudlock, R. and Marshall, W.D. (2007) *Environmental Science & Technology*, **41**, 1983.

48 Wei, W., Sun, Y. and Zhong, B. 1999 *Chemical Communications*, 2499.

49 Burgener, M., Mallat, T. and Baiker, A. (2005) *Journal of Molecular Catalysis A: Chemical*, **225**, 21.

50 Yokota, K. and Fujimoto, K. (1989) *Fuel*, **68**, 255.

51 Yokota, K., Hanakata, Y. and Fujimoto, K. (1990) *Chemical Engineering Science*, **45**, 2743.

52 Fan, L., Yokota, K. and Fujimoto, K. (1992) *AICHE Journal*, **38**, 1639.

53 Bochniak, D.J. and Subramaniam, B. (1998) *AICHE Journal*, **44**, 1889.

54 Huang, X., Elbashir, N.O. and Roberts, C.B. (2004) *Industrial & Engineering Chemistry Research*, **43**, 6369.

55 Elbashir, N.O. and Roberts, C.B. (2005) *Industrial & Engineering Chemistry Research*, **44**, 505.

56 Linghu, W., Li, X., Asami, K. and Fujimoto, K. (2004) *Fuel Processing Technology*, **85**, 11219.

57 Irankhah, A., Haghtalab, A., Farahani, E.V. and Sadaghianizadeh, K. (2007) *Journal of Natural Gas Chemistry*, **16**, 115.

58 Irankhah, A. and Haghtalab, A. (2007) *Chemical Engineering & Technology*, **31**, 525.

59 Ngamcharussrivichai, C., Liu, X., Li, X., Vitidsant, T. and Fujimoto, K. (2008) *Fuel*, **86**, 50.

60 Dharmidikari, S. and Abraham, M.A. (2000) *Journal Of Supercritical Fluids*, **18**, 1.

61 Snyder, G., Tadd, A. and Abraham, M.A. (2001) *Industrial & Engineering Chemistry Research*, **40**, 5317.

62 Marteel, A.E., Tack, T.T., Bektesevic, S., Davies, J.A., Mason, M.R. and Abraham, M.A. (2003) *Environmental Science & Technology*, **37**, 5424.

63 Tadd, A.R., Marteel, A., Mason, M.R., Davies, J.A. and Abraham, M.A. (2002) *Industrial & Engineering Chemistry Research*, **41**, 4514.

64 Tadd, A.R., Marteel, A., Mason, M.R., Davies, J.A. and Abraham, M.A. (2003) *Journal Of Supercritical Fluids*, **25**, 183.

65 Bektesevic, S., Tack, T., Mason, M.R. and Abraham, M.A. (2005) *Industrial & Engineering Chemistry Research*, **44**, 4973.

66 Fujita, S.-I., Akihara, S., Fujisawa, S. and Arai, M. (2007) *Journal of Molecular Catalysis A: Chemical*, **268**, 244.

67 Meehan, N.J., Sandee, A.J., Reek, J.N.H., Kramer, P.C.J., Leeuwen, P.W.N.M. and Poliakoff, M. 2000 *Chemical Communications*, 1497.

68 Bröll, D., Krämer, A. and Vogel, H. (2003) *Chemical Engineering & Technology*, **26**, 733.

69 Kerler, B., Martin, A., Jans, A. and Baerns, M. (2001) *Applied Catalysis A: General*, **220**, 243.

70 Kerler, B., Martin, A., Pohl, M.-M. and Baerns, M. (2002) *Catalysis Letters*, **78**, 259.

71 Armbruster, U., Martin, A., Smejkal, Q. and Kosslick, H. (2004) *Applied Catalysis A: General*, **265**, 237.

72 Zhang, R., Qin, Z., Dong, M., Wang, G. and Wang, J. (2005) *Catalysis Today*, **110**, 351.

73 Yu, K.M.K., Abutaki, A., Zhou, Y., Yue, B., He, H.Y. and Tsang, S.C. (2007) *Catalysis Letters*, **113**, 115.

74 Danciu, T., Beckmann, E.J., Hancu, D., Cochran, R.N., Grey, R., Hajnik, D.M. and Jewson, J. (2003) *Angewandte Chemie-International Edition*, **42**, 1140.

75 Dooley, K.M. and Knopf, F.C. (1987) *Industrial & Engineering Chemistry Research*, **26**, 1910.

76 Wang, C.-T. and Willey, R.J. (1998) *Journal of Non-Crystalline Solids*, **225**, 173.

77 Wang, C.-T. and Willey, R.J. (1999) *Catalysis Today*, **52**, 83.

78 Gläser, R., Josl, R. and Williardt, J. (2003) *Topics in Catalysis*, **22**, 31.

79 Grunwaldt, J.-D., Caravati, M., Ramin, M. and Baiker, A. (2003) *Catalysis Letters*, **90**, 221.

80 Caravati, M., Grunwaldt, J.-D. and Baiker, A. (2005) *Physical Chemistry Chemical Physics*, **7**, 278.

81 Caravati, M., Grunwaldt, J.-D. and Baiker, A. (2004) *Catalysis Today*, **91–92**, 1.

82 Grunwaldt, J.-D., Caravati, M. and Baiker, A. (2006) *The Journal of Physical Chemistry. B*, **110**, 9916.

83 Caravati, M., Grunwaldt, J.-D. and Baiker, A. (2007) *Catalysis Today*, **126**, 27.

84 Kimmerle, B., Grunwaldt, J.-D. and Baiker, A. (2007) *Topics in Catalysis*, **44**, 285.

85 Ciriminna, R., Campestrini, S. and Pagliaro, M. (2003) *Advanced Synthesis and Catalysis*, **345**, 1261.

86 Ciriminna, R., Campestrini, S. and Pagliaro, M. (2004) *Advanced Synthesis and Catalysis*, **346**, 231.

87 Campestrini, S., Carraro, M., Ciriminna, R., Pagliaro, M. and Tonellato, U. (2005) *Advanced Synthesis and Catalysis*, **347**, 825.

88 Pagliaro, M. and Ciriminna, R. (2005) Patent WO 2005/042155, assigned to Consiglio Nazionale delle Ricerche.

89 Ciriminna, R., Campestrini, S. and Pagliaro, M. (2004) *Advanced Synthesis and Catalysis*, **346**, 231.

90 Ciriminna, R., Campestrini, S. and Pagliaro, M. (2006) *Organic and Biomolecular Chemistry*, **4**, 2637.

91 Burgener, M., Tyszewski, T., Ferri, D., Mallat, T. and Baiker, A. (2006) *Applied Catalysis A: General*, **299**, 66.

92 Zhou, L. and Akgerman, A. (1995) *Industrial & Engineering Chemistry Research*, **34**, 1588.

93 Lee, S.-B. and Hong, I.-K. (2003) *Journal of Industrial and Engineering Chemistry*, **5**, 590.

94 Husain, A. (1994) Patent WO 94/03415, assigned to Mobil Oil Corporation.

95 Clark, M.C. and Subramaniam, B. (1998) *Industrial & Engineering Chemistry Research*, **37**, 1243.

96 Mota Salinas, A.L., Sapaly, G., Ben Taarit, Y., Vedrine, J.C. and Essayem, N. (2008) *Applied Catalysis A: General*, **336**, 61.

97 Gayreaud, P.Y., Stewart, I.H., Derouane-Abd Hamid, S.B., Essayem, N., Derouane, E.G. and Védrine, J.C. (2000) *Catalysis Today*, **63**, 223.

98 Shi, Y.F., Gao, Y. and Yuan, W.-K. (2001) *Industrial & Engineering Chemistry Research*, **40**, 4253.

99 Gläser, R. and Weitkamp, J. (1999) in *Proceedings of the 12th International Zeolite Conference* vol. II (eds M.M.J. Treacy, B.K. Marcus and M.E. Bisher), Materials Research Society, Warrendale, PA. p. 1447.

100 Tian, Z., Qin, Z., Wang, G., Dong, M. and Wang, J. (2008) *Journal Of Supercritical Fluids*, **44**, 325.

101 Horikawa, Y., Bulgarevich, D.S., Uchino, Y., Shichijo, Y. and Sako, T. (2005) *Industrial & Engineering Chemistry Research*, **44**, 2917.

102 Álvaro, M., Das, D., Cano, M. and Garcia, H. (2003) *Journal of Catalysis*, **219**, 464.

103 Oku, T., Arita, Y. and Ikariya, T. (2005) *Advanced Synthesis and Catalysis*, **347**, 1553.

104 Kamalakar, G., Komura, K. and Sugi, Y. (2006) *Applied Catalysis A: General*, **310**, 155.

105 Kamalakar, G., Komura, K. and Sugi, Y. (2006) *Industrial & Engineering Chemistry Research*, **45**, 6118.

106 Arunajatesan, V., Wilson, K.A. and Subramaniam, B. (2003) *Industrial & Engineering Chemistry Research*, **42**, 2639.

107 Tiltscher, H., Wolf, H. and Schelchshorn, J. (1981) *Angewandte Chemie (International Edition in English)*, **20**, 892.

108 Tiltscher, H. and Hofmann, H. (1987) *Chemical Engineering Science*, **42**, 959.

109 Sander, B., Thelen, M. and Kraushaar-Czarnetzki, B. (2001) *Industrial & Engineering Chemistry Research*, **40**, 2767.

110 Funamoto, T., Nakagawa, T. and Segawa, K. (2005) *Applied Catalysis A: General*, **286**, 79.

111 Bogdan, V.I., Klimenko, T.A., Kustov, L.M. and Kazansky, V.B. (2004) *Applied Catalysis A: General*, **267**, 175.

112 Amelse, J.A. and Kutz, N.A (1991) US Patent 5 030 788, assigned to Amoco Corporation.

113 Baiker, A. (1999) *Chemical Reviews*, **99**, 453.

114 Wandeler, R. and Baiker, A. (2000) *Cattech*, **4**, 34.

115 Grunwaldt, J.-D. and Baiker, A. (2005) *Physical Chemistry Chemical Physics*, **7**, 3526.

116 Subramaniam, B., Lyon, C.J. and Arunajatesan, V. (2002) *Applied Catalysis B: Environmental*, **37**, 279.

117 Hyde, J.R., License, P., Carter, D. and Poliakoff, M. (2001) *Applied Catalysis A: General*, **222**, 119.

118 Savage, P.E. (2000) *Catalysis Today*, **62**, 167.

119 Kruse, A. and Vogel, H. (2008) *Chemical Engineering & Technology*, **31**, 1241.

120 Kruse, A. and Vogel, H. (2008) *Chemical Engineering & Technology*, **31**, 1391.

121 Ramírez, E., Zgarni, S., Larrayoz, M.A. and Recasens, F. (2002) *Engineering in Life Sciences*, **2**, 257.

122 Pereda, S., Bottini, S.B. and Brignole, E.A. (2005) *Applied Catalysis A: General*, **281**, 129.

123 Burgener, M., Ferri, D., Grunwaldt, J.-D., Mallat, T. and Baiker, A. (2005) *The Journal of Physical Chemistry. B*, **109**, 16794.

124 Hâncu, D. and Beckmann, E.J. (1999) *Industrial & Engineering Chemistry Research*, **38**, 2833.

125 Landon, P., Collier, P.J., Carley, A.F., Chadwick, D., Papworth, A.J., Burrows, A., Kiely, C.J. and Hutchings, G.J. (2003) *Physical Chemistry Chemical Physics*, **5**, 1917.

126 Hyde, J.R., Walsh, B. and Poliakoff, M. (2005) *Angewandte Chemie-International Edition*, **44**, 7588.

127 Piqueras, C., Bottini, S. and Damiani, D. (2006) *Applied Catalysis A: General*, **313**, 177.

128 Piqueras, C.M., Tonetto, G., Bottini, S. and Damiani, D.E. (2008) *Catalysis Today*, **133–135** 836.

129 Santana, A., Larrayoz, M.A., Ramírez, E., Nistal, J. and Recasens, F. (2007) *Journal Of Supercritical Fluids*, **41**, 391.

130 Härröd, M. and Möller, P. (2001) US Patent 6 265 596, assigned to Poul Møller Ledelses.

131 Poliakoff, M. and License, P. (2003–2004) *Chemical Engineering*, **750/751**, 48.

132 Licence, P., Ke, J., Sokolova, M., Ross, S.K. and Poliakoff, M. (2003) *Green Chemistry*, **5**, 99.

133 Dry, M.E. (2008) in *Handbook of Heterogeneous Catalysis*, 2nd edn, vol. 6 (eds G. Ertl, H. Knözinger, F. Schüth, J. Weitkamp,), Wiley-VCH Verlag GmbH, Weinheim, p. 2965.

134 Jacobs, G., Chaudhari, K., Sparks, D., Zhang, Y., Shi, B., Spicer, R., Das, T.K., Li, J. and Davis, B.H. (2003) *Fuel*, **82**, 1251.

135 Tsubaki, N., Yoshii, K. and Fujimoto, K. (2002) *Journal of Catalysis*, **207**, 371.

136 Li, X., Liu, X., Liu, Z.-W., Asami, K. and Fujimoto, K. (2005) *Catalysis Today*, **106**, 154.

137 Cornils, B. and Kuntz, E.G. (2005) in *Multiphase Homogeneous Catalysis*, vol. 1 (eds B. Cornils, W.A. Hermann, I.T. Horávth, W. Leitner, S. Mecking, H. Olivier-Bourbigou and D. Vogt), Wiley-VCH Verlag GmbH, Weinheim, p. 148.

138 Bektesevic, S., Kleman, A.M., Marteel-Parrish, A.E. and Abraham, M.A. (2006) *Journal Of Supercritical Fluids*, 232.

139 Mehnert, C.P. (2005) *Chemistry - A European Journal*, **11**, 50.

140 Hintermair, U., Zhao, G., Santini, C.C., Muldoon, M.J. and Cole-Hamilton, D.J. (2007) *Chemical Communications*, 1462.

141 Ding, Z.Y., Frisch, M.A., Li, L. and Gloyna, E.F. (1996) *Industrial & Engineering Chemistry Research*, **35**, 3257.

142 Jenzer, G., Sueur, D., Mallat, T. and Baiker, A. 2000 *Chemical Communications*, 2247.

143 Jenzer, G., Schneider, M.S., Wandeler, R., Mallat, T. and Baiker, A. (2001) *Journal of Catalysis*, **199**, 141.

144 Caravati, M., Meier, D.M., Grunwaldt, J.-D. and Baiker, A. (2006) *Journal of Catalysis*, **240**, 126.

145 Fidalgo, A., Ciriminna, R., Ilharco, L.M., Campestrini, S., Carraro, M. and Pagliaro, M. (2008) *Physical Chemistry Chemical Physics*, **10**, 2026.

146 Ciriminna, R., Hesemann, P., Moreau, J.J.E., Carraro, M., Campestrini, S. and Pagliaro, M. (2006) *Chemistry - A European Journal*, **12**, 5220.

147 Xie, Y., Zhang, Z., Hu, S., Song, J., Li, W. and Han, B. (2008) *Green Chemistry*, **10**, 278.

148 Steele, A.M., Zhu, J. and Tsang, S.C. (2001) *Catalysis Letters*, **73**, 9.

149 Hou, Z., Theyssen, N. and Leitner, W. (2007) *Green Chemistry*, **9**, 127.

150 Jenzer, G., Mallat, T., Maciejewski, M., Eigenmann, F. and Baiker, A. (2001) *Applied Catalysis A: General*, **208**, 125.

151 Hou, Z., Han, B., Gao, L. and Yang, G. (2002) *Green Chemistry*, **4**, 426.

152 Maayan, G., Gancheguchi, B., Leitner, W. and Neumann, R. (2006) *Chemical Communications*, 2230.

153 Traa, Y. and Weitkamp, J. (2008) in *Handbook of Heterogeneous Catalysis*, 2nd edn, vol. 6 (eds G. Ertl, H. Knözinger, F. Schüth and J. Weitkamp), Wiley-VCH Verlag GmbH, Weinheim, p. 2830.

154 Thompson, D.N., Ginosar, D.M., Burch, K.C. and Zalewski, D.J. (2005) *Industrial & Engineering Chemistry Research*, **44**, 4534.

155 Ginosar, D.M., Thompson, D.N. and Burch, K.C. (2006) *Industrial & Engineering Chemistry Research*, **45**, 567.

156 Niu, F. and Hofmann, H. (1996) in *High-Pressure Chemical Engineering*, vol. 12 (eds P.R. von Rohr and C. Trepp), Proc. Technol. Proc., Elsevier, Amsterdam, p. 145.

157 Marathe, R.P., Mayadevi, S., Pardhy, S.A., Sabne, S.M. and Sivasanker, S. (2002) *Journal of Molecular Catalysis A: Chemical*, **181**, 201.

158 Gläser, R. (2007) *Chemical Engineering & Technology*, **30**, 557.

159 Sotelo, J.L., Calvo, L., Pérez-Veláquez, A., Capilla, D., Cavani, F. and Bolognini, M. (2006) *Applied Catalysis A: General*, **312**, 194.

160 Collins, N.A., Debenedetti, P.G. and Sundaresan, S. (1988) *AICHE Journal*, **34**, 1211.

161 Gao, W., Buttler, D. and Tomasko, D. (2004) *Langmuir*, **20**, 8083.

162 Süer, M.G., Dardas, Z., Ma, Y.H. and Moser, W.R. (1996) *Journal of Catalysis*, **2**, 320.

163 Dardas, Z., Süer, M.G., Ma, Y.H. and Moser, W.R. (1996) *Journal of Catalysis*, **162**, 327.

164 Gray, W.K., Smail, F.R., Hitzler, M.G., Ross, S.K. and Poliakoff, M. (1999) *Journal of the American Chemical Society*, **121**, 10711.

165 Vieville, C., Mouloungui, Z. and Gaset, A. (1993) *Industrial & Engineering Chemistry Research*, **32**, 2065.

166 Tateno, T. and Sasaki, T. (2001) European Patent EP 1 126 011, assigned to Sumitomo Chemical Company, Ltd.

167 Jackson, M.A., Mbaraka, I.K. and Shanks, B.H. (2006) *Applied Catalysis A: General*, **310**, 48.

168 Sakthivel, A., Komura, K. and Sugi, Y. (2008) *Industrial & Engineering Chemistry Research*, **47**, 2538.

169 Demibras, A. (2005) *Progress in Energy and Combustion Science*, **31**, 466.

170 Wang, L. and Yang, J. (2007) *Fuel*, **86**, 328.

171 Sowden, R.J., Sellin, M.F., De Blasio, N. and Cole-Hamilton, D.J. 1999 *Chemical Communications*, 2511.

172 Cole-Hamilton, D.J. and Sellin, M.F. (2001) Patent WO 01/07388, assigned to Thomas Swan & Co. Ltd.

173 Fischer, A., Mallat, T. and Baiker, A. (1999) *Angewandte Chemie-International Edition*, **38**, 351.

174 Fischer, A., Mallat, T. and Baiker, A. (1999) *Journal of Catalysis*, **182**, 289.

175 Lu, X.-B., Xiu, J.-H., He, R., Jin, K., Luo, L.-M. and Feng, X.-J. (2004) *Applied Catalysis A: General*, **275**, 73.

176 Jutz, F., Gruwaldt, J.-D. and Baiker, A. (2008) *Journal of Molecular Catalysis A: Chemical*, **279**, 94.

177 Bhanage, B.M., Fujita, S.-i., Ikushima, Y. and Arai, M. (2001) *Applied Catalysis A: General*, **219**, 259.

178 Chang, Y., Kiang, T., Han, B., Liu, Z., Wu, W., Gao, L., Li, J., Gao, H., Zhao, G. and Huang, J. (2004) *Applied Catalysis A: General*, **263**, 179.

179 Kröcher, O., Köppel, R.A., Fröba, M. and Baiker, A. (1998) *Journal of Catalysis*, **178**, 284.

180 Bertucco, A., Canu, P., Devetta, L. and Zwahlen, A.G. (1997) *Industrial & Engineering Chemistry Research*, **36**, 2626.

181 Chouchi, D., Gourgouillon, D., Courel, M., Vital, J. and Nunes da Ponte, M. (2001) *Industrial & Engineering Chemistry Research*, **40**, 2551.

182 Milewska, A., Banet Osuna, A.M., Fonseca, I.M. and Nunes da Ponte, M. (2005) *Green Chemistry*, **7**, 726.

183 Xi, C., Wang, H., Liu, R., Cai, S. and Zhao, F. (2008) *Catalysis Communications*, **9**, 140.

184 Phiong, H.-S., Lucien, F.P. and Adesina, A.A. (2003) *Journal Of Supercritical Fluids*, **25**, 155.

185 Xu, D., Carbonell, R.G., Roberts, G.W. and Kieserow, D.J. (2005) *Journal Of Supercritical Fluids*, **34**, 1.

186 Xu, D., Carbonell, R.G., Kieserow, D.J. and Roberts, G.W. (2005) *Industrial & Engineering Chemistry Research*, **44**, 6164.

187 Hemminger, O., Marteel, A., Mason, M.R., Davies, J.A., Tadd, A.R. and Abraham, M.A. (2002) *Green Chemistry*, **4**, 507.

188 Caravati, M., Grunwaldt, J.-D. and Baiker, A. (2006) *Applied Catalysis A: General*, **298**, 50.

189 Ramin, M., Grunwaldt, J.-D. and Baiker, A. (2006) *Applied Catalysis A: General*, **305**, 46.

190 Kerler, B., Robinson, R.E., Borovik, A.S. and Subramaniam, B. (2004) *Applied Catalysis B: Environmental*, **49**, 91.

7
Enzymatic Catalysis

Pedro Lozano, Teresa De Diego, and José L. Iborra

7.1
Enzymes in Non-aqueous Environments

Greenness in chemical processes begins with catalysts. Enzymes which act as catalysts of living systems are usually called biocatalysts and can clearly be regarded as green tools for chemical processes. All enzymes are proteins, polymeric macromolecules based on amino acid units with unique sequences, and show a high level of three-dimensional (3-D) structural organization. Some of them require the presence of additional non-protein components (cofactors and coenzymes) before they can function as catalysts. The catalytic activity of an enzyme depends strongly on this 3-D structure or native conformation, which is maintained by a high number of weak internal interactions (e.g. hydrogen bonds, van der Waals), and also interactions with other molecules, mainly water as the natural solvent of living systems. Enzymes are designed to function in aqueous solutions within a narrow range of environmental conditions (pH, temperature, pressure, etc.), a range which, in fact, represents the limits of life. Outside these conditions, enzymes are usually deactivated because they lose their native conformation through unfolding. Developments in genomics, directed evolution, and our exploitation of natural biodiversity have led to improvements in the activity, stability, and specificity of enzymes, accompanied by a huge increase in the number and variety of their industrial applications [1].

From the functional point of view, the tremendous potential of enzymes as practical catalysts cannot be doubted, since their activity and selectivity (stereo-, chemo-, and regioselectivity) for catalyzed reactions are far-ranging. In this context, the market for enantiopure fine chemicals is continuously growing, making enzymes the most suitable catalysts for green synthetic processes. However, the use of biocatalysts in aqueous media is limited because most chemicals of interest are insoluble in water [2]. Furthermore, water is not an inert compound and usually gives rise to undesired side reactions, in addition to bringing about the degradation of substrates.

There are numerous potential advantages to using enzymes in non-aqueous environments or media of reduced water content (e.g. in the presence of organic

Handbook of Green Chemistry, Volume 4: Supercritical Solvents. Edited by Walter Leitner and Philip G. Jessop

ISBN: 978-3-527-32590-0

solvents and/or additives) [3]. These advantages include the much higher solubility that they permit in the case of hydrophobic substrates, the insolubility of enzymes which makes them easy to reuse, and the avoidance of microbial contamination in reactors. Probably the most interesting advantage of using non-aqueous environments for enzyme catalysis is evident when hydrolytic enzymes (e.g. lipases, esterases, proteases, glycosidases) are applied, because of the ability of these enzymes to catalyze synthetic reactions at the highest level of selectivity towards the target substrate. Among the above, lipases (EC 3.1.1.3) are the most useful enzymes to be applied in non-aqueous environments, because they act at the lipid–water interface and, therefore, they do not need water-soluble substrates. In this context, lipases have been used in synthetic organic chemistry to catalyze various kinds of reactions, such as hydrolysis, esterification, transesterification by acidolysis, transesterification by alcoholysis, inter-esterification, and aminolysis (Scheme 7.1).

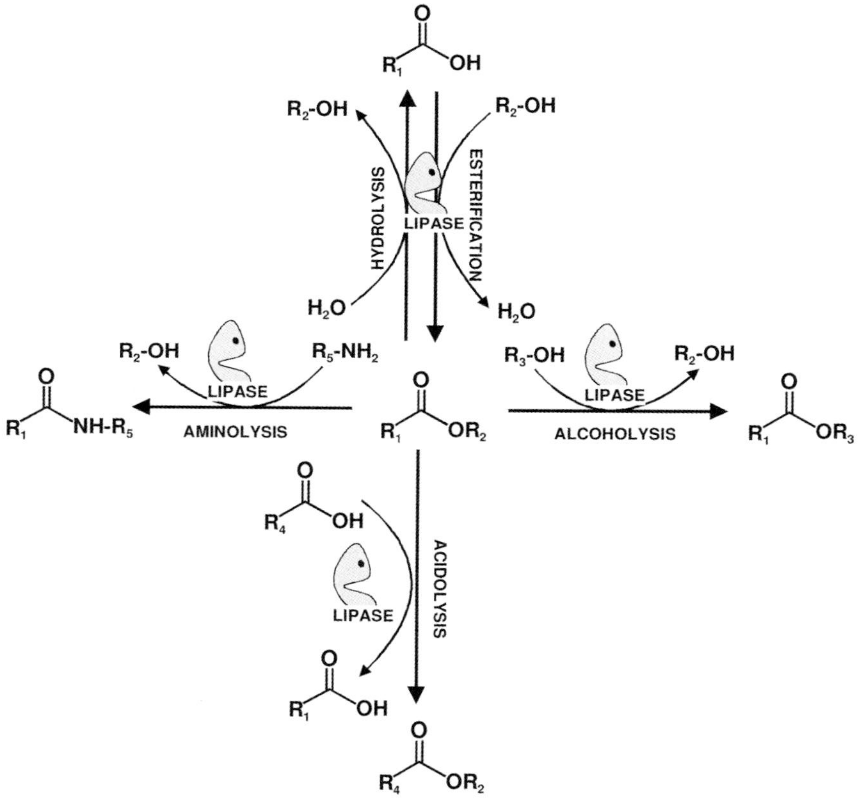

Scheme 7.1 Some common chemical reactions catalyzed by lipases.

Furthermore, the catalytic "promiscuity" of enzymes in non-aqueous environments has been described as being related with the ability of a single active site to catalyze more than one chemical transformation (e.g. lipase B from *Candida*

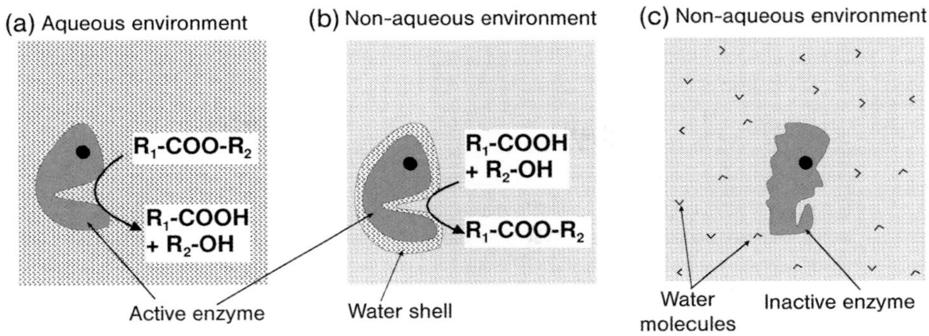

Figure 7.1 Schematic representation about the role of water in different enzyme environments.

antarctica is able to catalyze aldol additions, Michael-type additions, and the formation of Si–O–Si bonds). Other enzymes (e.g. peroxidases, laccases, monooxygenases, dehydrogenases) have also been described as excellent tools for organic synthesis in non-aqueous environments [1, 3].

However, switching from water to a non-aqueous solvent as the reaction medium for enzyme-catalyzed reactions is not always a simple matter, because the native structure of the enzyme can easily be destroyed, resulting in deactivation. Water is the key component of all non-conventional media, because of the importance that enzyme–water interactions have in maintaining of the active conformation of the enzyme (Figure 7.1a). Only a few clusters of water molecules, presumably bound to charged groups on the surface of enzyme molecules, are required for the catalytic function (Figure 7.1b). Thus, hydrophobic solvents typically afford higher enzymatic activity than hydrophilic solvents, because the latter have a tendency to strip some of these essential water molecules from the enzyme molecules (Figure 7.1c). The common hypothesis suggests that when an enzyme is placed in a dry hydrophobic system, it is trapped in the native state, partly due to the low dielectric constant, which greatly intensifies electrostatic forces, enabling it to maintain its catalytic activity. This is one of the most intriguing facets of non-aqueous biocatalysis, the phenomenon of "memory" in anhydrous media [4]. Several factors must be taken into account when choosing a non-aqueous solvent for a given reaction (e.g. the nature of the solvent, the solubility of reagents, enzyme activity, enzyme reusability). However, the use of environmentally benign non-aqueous solvents and efficient biocatalysts for chemical reactions and/or processes is strongly encouraged for developing green chemistry on an industrial scale.

7.2 Supercritical Fluids for Enzyme Catalysis

Solvents are key elements in chemical processes, where they act as media for mass transport, reaction, and product separation. They are responsible for much of the environmental performance of processes in the chemical industry and have a great impact on cost, safety, and health [5]. The search for new, environmentally benign non-aqueous solvents or green solvents, which can easily be recovered/recycled and which permit enzymes to operate efficiently in them, is a priority in the development of integral green chemical processes.

A supercritical fluid (SCF) is defined as a state of matter (molecule or element) at a pressure and temperature higher than its critical point, but below the pressure required to condense it into a solid. The exceptional properties of SCFs as solvents for extraction, reaction, fractionation, and analysis have been studied [6], and are described in other chapters. With respect to enzymatic catalysis, it is notable that proteins are insoluble in all SCFs, and also how the gas-like diffusivities and low viscosities of SCFs enhances mass transport rates of reactants to the active site of enzymes [7]. Additionally, the key feature of SCFs for enzyme catalysis is the tunability of these neoteric solvents, because of the sensitivity of fluid density to both pressure and temperature, mainly in the vicinity of the critical point. All density-dependent solvent properties (e.g. dielectric constant, Hildebrand solubility parameter, partition coefficients) may be greatly affected by small changes in pressure and temperature [6].

The number of fluids studies in supercritical conditions for enzyme catalysis is relatively small owing to the inherent nature of proteins, which unfold and deactivate at high temperatures (Table 7.1). Additionally, SCFs are low-polarity solvents, which preferentially dissolve hydrophobic compounds and have been applied in the biotransformation of these kinds of compounds. The most popular SCF in enzyme catalysis is supercritical carbon dioxide ($scCO_2$), which is considered as a green solvent because it is chemically inert, non-toxic, non-flammable, cheap, and readily available, exhibiting relatively low critical parameters that are compatible with biocatalysis [7]. Other SCFs are less attractive because of their flammability (e.g. ethane, propane), high cost (e.g. CHF_3), or poor solvent power (e.g. SF_6), and are therefore used in few research studies. In addition to these features, it has been demonstrated how the intrinsic activity of enzymes is affected by the nature of the SCF, especially with respect to the preservation of the essential water layer around proteins. Thus, Kamat *et al.* [8] studied the lipase-catalyzed alcoholysis of methyl methacrylate in several SCFs (e.g. sulfur hexafluoride, propane, ethane, ethylene, fluoroform, and carbon dioxide) at the same pressure, observing how the reaction rate increased with increasing hydrophobicity of the SCFs: $SF_6 > C_3H_8 > C_2H_6 > C_2H_4 > CHF_3 > CO_2$. Despite these facts, $scCO_2$ is generally considered a hydro-

Table 7.1 Fluids used under near-critical or supercritical conditions for enzyme catalysis and their critical parameters.

Fluid	T_c (°C)	P_c (bar)
Carbon dioxide (CO_2)	31.3	72.9
Ethane (C_2H_6)	32.3	48.8
Propane (C_3H_8)	96.7	42.5
Butane (C_4H_{10})	152.0	37.5
Fluoroform (CHF_3)	26.2	48.5
Sulfur hexafluoride (SF_6)	45.5	37.7

phobic solvent (e.g. 0.31% w/w water content at 344.8 bar and 50 °C [7]), the capability of $scCO_2$ to strip the essential water molecules from the enzyme microenvironment was suggested as being responsible for this low activity. In the same context, the activity of lipase-catalyzed ethyl butyrate synthesis by esterification was higher in near-critical propane than in $scCO_2$, which was also attributed to the water stripping of the enzyme produced by $scCO_2$ [9]. For the case of subtilisin Carlsberg suspended in CO_2, propane, or mixtures of these solvents at 35 °C and 150 bar, it was also observed how the transesterification activity was dependent on enzyme hydration, the maximum level of activity being obtained at 12% enzyme hydration in all the compressed fluids tried. However, the enzymatic activity was much higher in propane than in either CO_2 or mixtures with 50 and 10% CO_2, also suggesting the adverse effect of CO_2 on the catalytic activity of subtilisin [10].

Even though CO_2 is one of the most widely used SCFs in industry, it is involved in two chemical processes that have been related with enzyme deactivation: the formation of carbamates between CO_2 and the ε-amino group of lysine residues on the surface of the enzyme, and lowering of the pH of the aqueous layer around the enzyme. The first process is related to conformational changes in the 3-D structure of the enzyme, which modify its activity [7, 11], and the second property means that CO_2 may dissolve in the essential hydration shell around the enzyme, thereby altering the pH of the microenvironment through the formation of carbonic acid, hence affecting enzyme activity [12]. Hence treatment with $scCO_2$ has been used to deactivate pectinesterase in orange juice, polyphenol oxidase in grape juice, and α-amylase in liquid food model systems [13].

Another important limitation of SCFs for enzyme catalysis results from their hydrophobic character as solvents, which greatly impedes their ability to carry out biotransformation with polar substrates because of their insolubility. To overcome this limitation, complexation of polar substrates (e.g. glycerol or fructose) with phenylboronic acid or prior adsorption on silica gel has been shown to be an efficient way of facilitating the lipase-catalyzed esterification of oleic acid with these polar substrates in $scCO_2$ [14].

7.3 Enzymatic Reactions in Supercritical Fluids

Lipases and esterases are the most widely used enzymes in biotransformations using $scCO_2$ as reaction medium, because of the excellent ability of this fluid to dissolve and transport hydrophobic compounds [15]. The synthesis of aliphatic esters of different alkyl chain lengths (e.g. from isoamyl acetate to oleyl oleate, terpenic esters, etc. [16, 17]) by esterification (Scheme 7.1 and Table 7.2) is the most popular enzymatic process in $scCO_2$. In this respect, the transesterification (e.g. by alcoholysis, acidolysis, or inter-esterification) approach has also been widely applied to the modification and/or valorization of oils and fats [e.g. isolation of polyunsaturated fatty acids (PUFAs), formation of structured triacylglycerols (TAGs) as nutraceuticals; formation of mono-acylglycerols (MAGs)], including the production of biodiesel [fatty acid methyl esters

Table 7.2 Some examples of lipase-catalyzed reactions in $scCO_2$.

Substrates	Product	Reaction conditions	Yield (%)	Ref.
Isoamyl alcohol + acetic anhydride	Isoamyl acetate	40 °C, 100 bar, 2 h	100	[16a]
Lauric acid + butanol	Butyl laurate	30 °C, 300 bar	98	[16b]
Myristic acid + ethanol	Ethyl myristate	59 °C, 125 bar, 6 h	89	[16c]
Palmitic acid + 1-octanol	Octyl palmitate	55 °C, 80 bar, 6 h	74	[16d]
Oleic acid + ethanol	Ethyl oleate	40 °C, 150 bar, 3 h	95	[16e]
Oleic acid + citronellol	Citronellyl oleate	31.1 °C, 84.1 bar	3.6	[17a]
Valeric acid + citronellol	Citronellyl valerate	35 °C, 75.5 bar	98.9	[17b]
Acetic acid + geraniol	Geranyl acetate	40 °C, 100 bar, 10 h	73	[17c]
Propyl acetate + geraniol	Geranyl acetate	40 °C, 140 bar, 72 h	30	[17d]
1-Butanol + vinyl butyrate	Butyl butyrate	50 °C, 90 bar, 3 h	100	[17e]
Corn oil + caprylic acid	Structured lipids	55 °C, 241 bar, 6 h	62.2	[18a]
Corn oil + methanol	FAMEs	50 °C, 241 bar,	98	[18b]
Palm kern oil + ethanol	FAMEs	55 °C, 136 bar, 4 h	26.4	[18c]
		40°C, 73 bar, 4 h	63.2	
rac-1-Phenylethanol + vinyl acetate	(*R*)-1-Phenylethyl acetate	95 °C, 150 bar, 3 h	48 (*ee* >99%)	[19a]
rac-1-Phenylethanol + vinyl acetate	(*R*)-1-Phenylethyl acetate	40 °C, 90 bar, 3 h	48 (*ee* >99%)	[19b]
rac-1-Phenylethanol + vinyl acetate	(*R*)-1-Phenylethyl acetate	50 °C, 200 bar, 6 h	48 (*ee* >99%)	[19c]
rac-Glycidol + butyric acid	(*S*)-Glycidyl butyrate	35 °C, 140 bar, 10 h	30 (*ee* 83%)	[19d]
rac-Ibuprofen + 1-propanol	(*S*)-Propyl ester of ibuprofen	50 °C, 100 bar, 23 h	75 (*ee* 70%)	[19e]
Divinyl adipate + octafluorooctanediol	Fluorinated polyester	50 °C, 200 bar, 24 h	MW 8230	[20a]
Dicaprolactone	Polycaprolactone	70 °C, 80 bar, 6 h	82	[20b]
ε-Caprolactone	Polycaprolactone	35–65 °C, 81–241 bar, 6–72 h	38–98, MW 23 000–37 000	[20c]
Polycaprolactone + water	ε-Caprolactone	40 °C, 180 bar, 6 h	90	[20d]

(FAMEs)] by the methanolysis of triglycerides (Scheme 7.2) [18]. The advantages of $scCO_2$ for dissolving both immiscible substrates (MeOH and vegetable oil) to permit the enzymatic reaction and to facilitate the recovery of biodiesel have been emphasized.

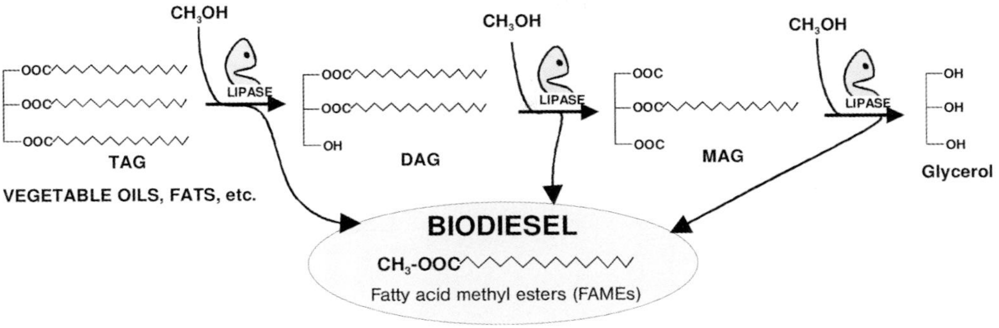

Scheme 7.2 Schematic representation of the lipase-catalyzed synthesis of biodiesel by methanolysis of triacylglycerides.

Currently, the use of lipases for the asymmetric synthesis of esters is one of the most important tools used by organic chemists. The unique properties of $scCO_2$ combined with the catalytic excellence of lipases have allowed the chiral resolution of a large number of racemates (1-phenylethanol, glycidol, ibuprofen, etc.) [19]. Other kinds of reactions on polymeric substrates, where the high diffusivity of $scCO_2$ is the key parameter, such as lipase-catalyzed polymerizations (e.g. polyester synthesis) or depolymerization (e.g. production of ε-caprolactone from polycaprolactone) in $scCO_2$ [20], and also hydrolytic reactions catalyzed by polysaccharide hydrolases (e.g. α-amylase-catalyzed corn starch hydrolysis, cellulase-hydrolyzed cellulose hydrolysis) in $H_2O-scCO_2$ biphasic systems can also be carried out [21].

Several types of enzyme reactors (Figure 7.2), such as stirred-tank, continuous flow, and membrane reactors, have been used with SCFs and with many kinds of enzymatic preparations (e.g. free, immobilized, encapsulated) [7]. The design of SCF bioreactors is another key feature, where mass transfer limitations, environmental conditions (pressure and temperature), and product recovery need to be easily controlled. For example, Marty *et al.* [16e] developed a recycling packed-bed enzyme reactor at the pilot scale for Lipozyme-catalyzed ethyl oleate synthesis by esterification from oleic acid and ethanol in $scCO_2$. The proposed system was coupled with a series of four high-pressure separator vessels, where a pressure cascade was produced by back-pressure valves, allowing continuous recovery of the liquid product at the bottom of each separator, and then recycling unreacted substrates (Figure 7.3). Furthermore, membrane reactors constitute an attempt to integrate catalytic conversion, product separation, and/or concentration and catalyst recovery into a single operation. Thus, enzymatic dynamic membranes, formed by depositing water-soluble polymers (e.g. gelatine, polyethylenimine) on a ceramic porous support, exhibited excellent properties for continuous synthetic processes in $scCO_2$, together with a high operational stability for reuse [17e] (Figure 7.2b).

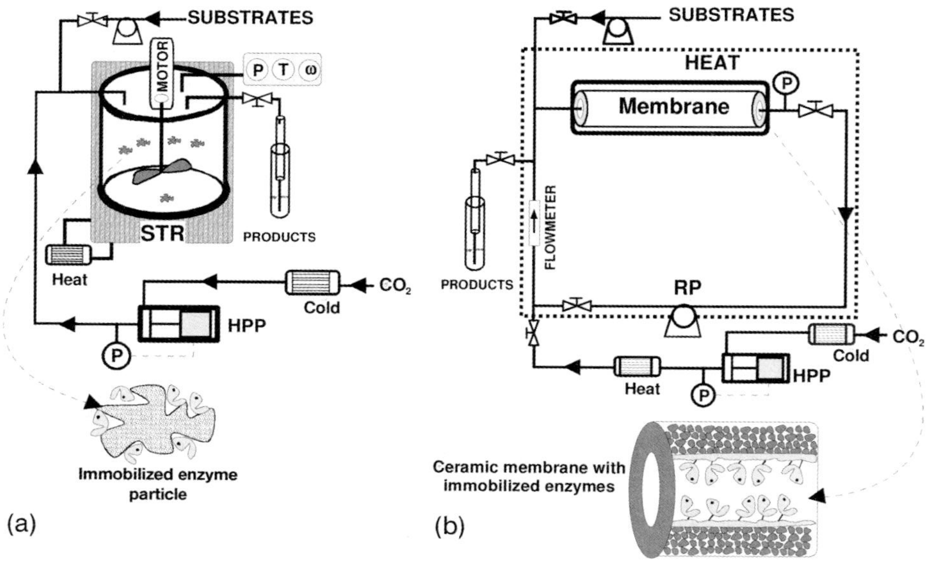

Figure 7.2 High-pressure stirred tank reactor (a) and membrane reactor with recirculation (b) for enzyme-catalyzed transformations in $scCO_2$. HPP, high pressure pump; RP, recirculation pump [17e].

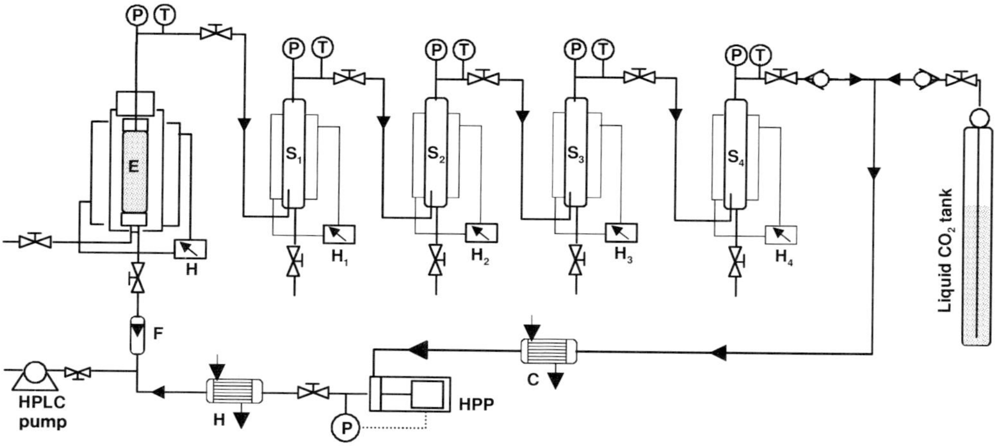

Figure 7.3 Experimental set-up of the recycling packed-bed enzyme reactor at the pilot scale equipped with a high-pressure pump (HPP), enzyme reactor (E), separators (S), heaters (H), pressure control (P), temperature control (T), and flowmeter (F) for Lipozyme-catalyzed ethyl oleate synthesis in $scCO_2$ [16e].

In addition to the many types of reactions described above, carboxylation and asymmetric reduction were reported as two additional kinds of reaction catalyzed by enzymes (other than hydrolases) in $scCO_2$ [22]. However, it should be pointed out that these biotransformations were carried out by whole cells, which were used as biocatalysts. This experimental approach could be considered as similar to the enzyme-catalyzed reactions in reverse micelles, where the cytoplasm of the cell is equivalent to the water pool of the reverse micelle, albeit with a highly complex enzyme content. Whole cells are used in the resting state, offering the advantage of providing any additional coenzyme and cofactors that may be required by enzymes [7b]. For example, the cells of *Bacillus megaterium* were successfully used for the CO_2 fixation reaction on pyrrole in $scCO_2$. The best results were obtained at 76 bar and 40 °C, when a yield 12-fold higher than that at atmospheric pressure was attained (Scheme 7.3a). In the same context, the resting cells of *Geotrichum candidum* immobilized on a water-adsorbing polymer were used for the reduction of various ketones in $scCO_2$. In the case of *o*-fluoroacetophenone (Scheme 7.3b), the reaction was conducted at 100 bar and 40 °C, and an 81% yield [enantiomeric excess (*ee*) >99%] of (*S*)-1-(*o*-fluorophenyl) ethanol enantiomeric product was obtained after 12 h; it was also demonstrated that alcohol dehydrogenase was the enzyme involved in the transformation [22].

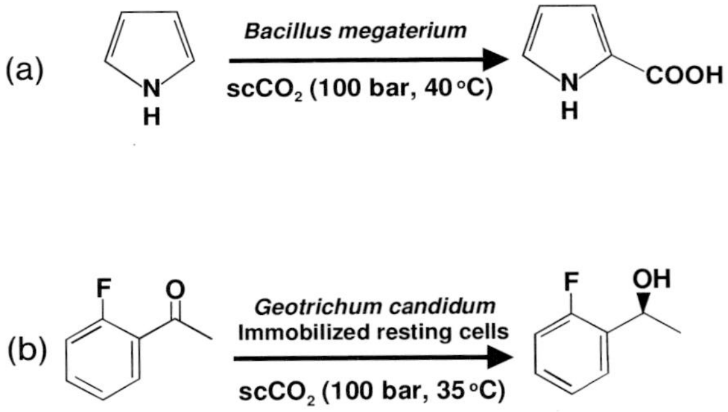

Scheme 7.3 (a) *Bacillus megaterium* cell-catalyzed carboxylation of pyrrole in $scCO_2$; (b) *Geotrichum candidum* cell-catalyzed asymmetric reduction of *o*-fluoroacetophenone in $scCO_2$ [22].

7.4 Reaction Parameters in Supercritical Biocatalysis

Pressure, temperature, and water content are the most important environmental factors affecting enzymatic catalysis in SCFs, particularly their activity, enantioselectivity, and stability. As described in other chapters, SCFs are compressible fluids and so changes in pressure and/or temperature are accompanied by a dramatic change in their density and transport properties, including their partition coefficient, viscosity, diffusivity, and thermal conductivity, all of which indirectly modulate enzyme activity. Pressure and temperature may also directly affect enzyme activity by changing the rate-limiting steps or modulating the selectivity of the enzyme [7, 23].

The effects of changes in pressure and temperature on enzyme-catalyzed reactions in $scCO_2$ have been widely reported. Apart from the specific effect of CO_2 on the proteins described in Section 7.2, pressures above a certain level may also have a negative impact on enzymes due to the direct effect on enzyme conformation. The rapid release of CO_2 dissolved in the bound water of the enzyme during depressurization has been claimed to produce structural change in the enzyme and to cause its inactivation [24]. However, over a limited pressure range of 77–85 bar, the changes in the conformation of lipase resulted in an increase in the rate of esterification reactions [25]. As an example, in the synthesis of butyl butyrate catalyzed by immobilized *Candida antarctica* lipase B (Novozyme 435) in $scCO_2$, it was observed that, within the range 40–60 °C, an increase in temperature improves the enzyme activity at all the pressures applied (80–150 bar). However, at a fixed temperature in the above range, an increase in pressure resulted in a decrease in the synthetic activity of the enzyme, an effect attributed to the increase in the density of $scCO_2$. The best results were obtained at 60 °C and 80 bar [17e]. When lipase-catalyzed hexyl acetate synthesis by transesterification from ethyl acetate and 1-hexanol in near-critical CO_2, ethane and propane was studied at constant pressure, an increase in temperature resulted in an increase in enzyme activity [12b].

In most cases, being close to the critical point of CO_2 has been described as a key condition for improving the efficiency of lipase-catalyzed reactions, even with regard to the preferential enzyme action on one enantiomer in a racemic mixture, or enantioselectivity. For example, in the lipase-catalyzed esterification of *rac*-citronellol with oleic acid in $scCO_2$, an ester product can be obtained with high enantioselectivities (*ee* >99%) simply by manipulating the pressure and temperature around the critical point [16a]. The effect of pressure on enantioselectivity is indeed noteworthy, although the reason for it is not clear. In another example, the effect of pressure (from 80 to 190 bar) on Novozyme 435-catalyzed enantioselective acetylation of *rac*-1-(*p*-chlorophenyl)-2,2,2-trifluoroethanol with vinyl acetate in $scCO_2$ was studied (Scheme 7.4). At 55 °C, the enantioselectivity of the enzyme (*E* value) was gradually decreased from 50 to 10 when the pressure was increased from 80 to 190 bar, regardless of the reaction time, which was related to changes in $scCO_2$ density [26]. Conversely, for the same immobilized lipase-catalyzed continuous kinetic resolution of *rac*-1-phenylethanol, it was demonstrated that changes in pressure did not greatly affect conversion or *E* values [19b]. Similarly, the synthesis of butyl butyrate catalyzed by the same immobilized enzyme was not significantly affected when the pressure of $scCO_2$ was increased to 500 bar [16b]. It may be that the effect of CO_2 pressure on

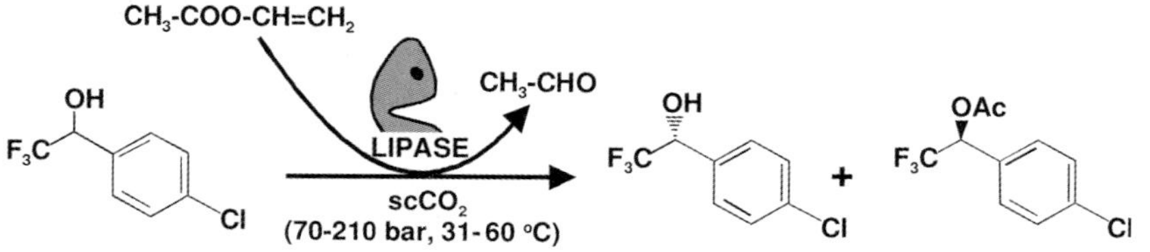

Scheme 7.4 Lipase-catalyzed enantioselective acetylation of 1-(*p*-chlorophenyl)-2,2,2-trifluoroethanol in $scCO_2$ [26].

enzyme activity is closely dependent on the specificity of the enzyme, substrate, and reaction studied.

Temperature influences enzyme activity much more than pressure, not because of the usual increment in reaction rates at higher temperatures, but because of enzyme deactivation processes that take place. The optimal temperature for enzymatic processes in SCFs is related to pressure because both parameters control solvent properties. However, the strong influence that temperature has on deactivation has been related to changes in the hydration level of the enzyme. As happens with enzymes in organic solvents, the water concentration in the supercritical reaction system is a key factor that influences enzyme activity and stability. As mentioned in previous sections, enzymes require a specific amount of bound water molecules to be active, and $scCO_2$ may dissolve up 0.3–0.5% w/w water, depending on the temperature and pressure. On the other hand, if the water content in the supercritical medium is too high or if water is produced in the reaction, the increased humidity may lead to enzyme deactivation. In other words, when immobilized enzymes are used, the support plays a key role in partitioning water molecules between the enzyme microenvironment and the supercritical medium, which affects the enzyme stability. The actual amount of water needed is specific to each SCF–substrate–enzyme system, and must be maintained constant throughout the process [23b, 27]. As an example, the water produced during lipase-catalyzed esterification of fatty acid with 1-butanol in $scCO_2$ was continuously removed by the presence of molecular sieves into the reaction medium [28a]. In the same context, when the effect of water as co-solvent on the transesterification of menhaden oil catalyzed by immobilized 1,3-regiospecific lipase from *Mucor miehei* was studied, the addition of water at up to 4% v/v of the total substrates enhanced the n-PUFA content of triglycerides by up to 46% w/w. However, further addition of water caused the total $n-3$-PUFA content to decrease due to the inhibitory effects of excess of water at the contact of substrates to the enzyme sites [28b]. Similarly, a decrease in the reaction rate was also observed when the solubility of the substrate was reduced due to the increased level of water in $scCO_2$. For example, for the lipase-catalyzed interesterification of myristic acid with trilaurin in $scCO_2$ at 95 bar and 35 °C, it was found that the water content of the CO_2 does not affect the intrinsic activity of the enzyme, whereas a higher water concentration causes a greater degree of unwanted hydrolysis. The selectivity of the reaction for interesterification rather than hydrolysis improves at higher pressures as the strength of the hydrolysis reaction is reduced [28c].

Maintaining water activity at a constant level in enzyme-catalyzed reactions in SCFs is a key factor. The addition of salt hydrates (e.g. $Na_4P_2O_7{\cdot}10H_2O$) to the reaction system is one approach used to control the amount of free water molecules [29a]. A hydrate salt can give off water to the environment based on equilibrium and maintaining the water activity at a constant level. As an example, for the subtilisin-catalyzed transesterification of *N*-acetyl-L-phenylalanine ethyl ester with methanol in supercritical fluoroform, it was found that the maximum reaction rate occurred at the minimum salt concentration. In the same way, by using salt hydrate pairs to control water activity, it was observed how the esterolytic activity of *Fusarium solani pisi* cutinase in supercritical CO_2 and ethylene changed with water activity in a bell-shaped manner [29b].

7.5 Stabilized Enzymes for Supercritical Biocatalysis

The excellent properties of SCFs, especially $scCO_2$, for extracting, dissolving, and transporting chemicals are only tarnished by the denaturative effect that it has on enzymes. In a review published in 1999, Mesiano *et al.* state in the final sentence, " . . .*the advantages of replacing conventional organic solvents with supercritical fluids have not been fully demonstrated yet*" [7a]. Several strategies have been developed to protect enzymes against these adverse effects of $scCO_2$, including immobilization on solid supports, lipid-coated enzymes, the entrapment of enzymes in silica gels, and the use of cross-linking enzyme aggregates [30].

Enzyme immobilization on solid supports is a classical strategy that has been used to improve catalytic rates and, in some cases, to enhance the enzyme stability in non-aqueous media (Figure 7.4a). Solid supports make the biocatalyst more robust under mechanical stress and easier to remove from the reaction and recycle. A large number of solid supports that have been applied to attach enzymes (e.g. Celite, polypropylene beads, α-alumina, silica gel, polystyrene beads) and the resulting immobilized enzyme derivatives showed varying degrees of catalytic activity in SCFs [7]. Several enzymes (e.g. lipases, esterases) are commercially available as immobilized derivatives (e.g. Novozym 435, Lipozyme RM), and are used in many kinds of biotransformations in SCFs, although with different levels of activity and operational stability.

The coating of free enzymes (e.g. lipase from *Rhizopus delemar*) with lipophilic molecules (e.g. didodecyl *N*-D-glucono-L-glutamate; Figure 7.4b) to improve the

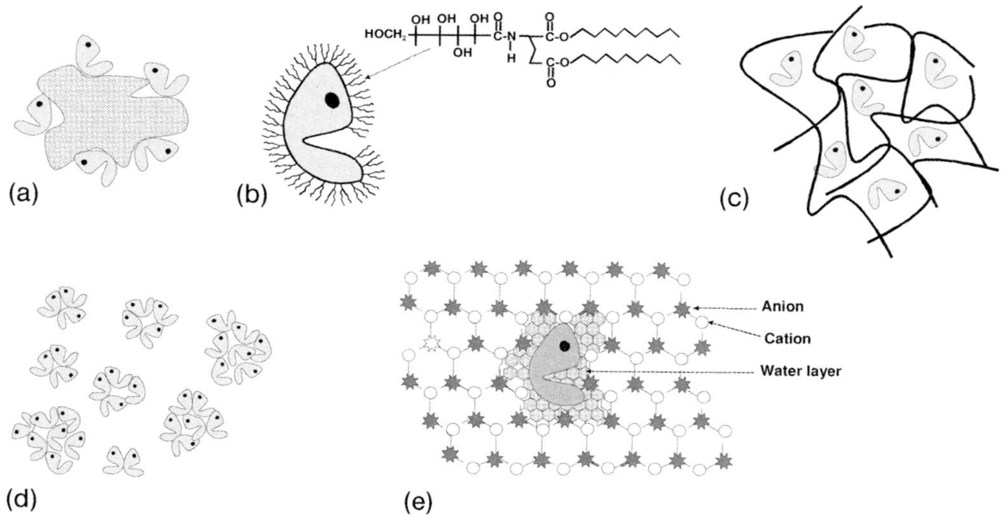

Figure 7.4 Schematic representation of different stabilized enzyme preparations used in SCFs. (a) Immobilized enzyme; (b) enzyme coated with didodecyl *N*-D-glucono-L-glutamate (c) enzymes entrapment into sol–gel matrix; (d) cross-linked enzyme aggregates (CLEAs); (e) enzymes in water-immiscible ILs.

dispersion of enzyme molecules in $scCO_2$ and sc-fluoroform has been successfully applied for the esterification of diglycerides [31a] and the enantioselective acetylation of *rac*-1-phenylethanol [31b], the good results obtained being attributed to the homogeneity of the reaction media.

The entrapment of enzymes (e.g. lipase from *Pseudomonas species*, cutinase from *Fusarium solani pisi*) in silica gel by a sol–gel procedure prior to drying by $scCO_2$ to produce aerogels or by evaporation to produce xerogels has also been tested to develop efficient immobilized enzymes for SCF reaction media (Figure 7.4c). It was reported that the resulting immobilized derivatives were more active in a variety of SCFs than the non-immobilized enzyme, but the sol–gel encapsulation did not prevent the deleterious effect of CO_2 [32]. The encapsulation of lipases into a hydrophilic matrix (e.g. lecitin water-in-oil microemulsion-based organogels formulated with hydroxypropylmethylcellulose and gelatine) was also tested, resulting in active enzyme derivatives for $scCO_2$ reaction media, but with a 50% loss of activity after the third reuse at 35 °C and 110 bar.

On the other hand, enzyme immobilization on a support leads to a "dilution of activity" because of the large portion of non-catalytic mass. The use of carrier-free immobilized enzymes, such as cross-linked enzyme crystals (CLECs), cross-liked dissolved enzymes (CLEs), and cross-linked enzyme aggregates (CLEAs), has also been reported as a way to obtain active and stable biocatalysts for SCF media (Figure 7.4d) [33a]. As an example, the CLEAs of *Candida antarctica* lipase B were able to catalyze the kinetic resolution of *rac*-1-phenylthanol in $scCO_2$ (40 °C, 90 bar) with the same efficiency as the commercial Novozym 435 [33b]. As another example, when CLECs of *Candida antarctica* lipase B were tested for the same reaction in continuous operation in $scCO_2$, a high level of activity and enantioselectivity was observed, but the operational stability was poor because of the dehydration power of the continuous flow of CO_2, being reversed by the activity loss upon the addition of water [33c].

In spite of the advantages obtained with all these stabilization approaches, the best results for enzyme-catalyzed reactions in $scCO_2$ were observed when the biocatalyst was applied in suspension or coated with other green solvents, such as ionic liquids (ILs) [34]. These ILs have emerged as exceptionally interesting non-aqueous reaction media for enzymatic transformations. They are simply salts, and therefore entirely composed of ions, which are liquid below 100 °C or usually close to room temperature. Typical room temperature ILs are based on organic cations, such as 1,3-dialkylimidazolium, *N*-alkylpyridinium, and tetraalkylammonium, paired with a variety of anions that have a strongly delocalized negative charge (e.g. BF_4^-, PF_6^-, bistriflimide), resulting in colorless and easily handled materials of low viscosity with very interesting properties as solvents. Their interest as green solvents resides in their negligible vapor pressure, excellent thermal stability (up to 300 °C in many cases), high ability to dissolve a wide range of organic and inorganic compounds, including gases (e.g. H_2, CO_2) and their non-flammable nature, which can be used to mitigate the problem of volatile organic solvent emissions to the atmosphere. Moreover, their polarity, hydrophilicity/hydrophobicity, and solvent miscibility can be tuned by selecting the appropriate cation and anion [35]. The enzymes (e.g. lipases, proteases, peroxidases, dehydrogenases, glycosidases) display a high level of activity and stereoselectivity in ILs, especially water-immiscible ILs, for synthesizing many different compounds, such as aspartame, aliphatic and aromatic esters, amino acid esters, chiral esters by the (dynamic) kinetic resolution of racemic alcohols, carbohydrate esters, polymers, and terpene esters [36]). Furthermore, free enzyme molecules suspended in these media behave as anchored or immobilized biocatalysts. The IL forms a strong ionic matrix and the added enzyme molecules could be considered as being included rather than dissolved in the media, meaning that ILs should be regarded as liquid enzyme immobilization supports, rather than reaction media, since they enable the enzyme–IL system to be reused in consecutive operation cycles. The excellent stability of free enzymes in water-immiscible ILs for reuse has been widely described, and spectroscopic techniques have demonstrated the ability of these neoteric solvents to maintain the secondary structure and the native conformation of the protein towards the usual unfolding that occurs in non-aqueous environments [37]. The ability of ILs to stabilize enzymes has recently been applied to develop sol–gel immobilized enzyme derivatives. For example, the sol–gel immobilization of lipase from *Candida rugosa* in the presence of ILs protects the enzyme against deactivation by released alcohol during the process and by shrinking the sol–gel process. After incubating this immobilized lipase for 5 days in *n*-hexane at 50 °C, 84% of the initial activity remained, while the residual activity of the immobilized derivative without IL was only 28% [38]. The extremely ordered supramolecular structure of ILs in the liquid phase might be able to act as a mould, maintaining an active three-dimensional structure of the enzyme in aqueous nano-environments (Figure 7.4e), and avoiding the classical thermal unfolding [39]. These observations imply that free enzyme suspended in IL systems may be considered as stable carrier-free-immobilized enzyme derivatives.

7.6
Enzymatic Catalysis in IL–$scCO_2$ Biphasic Systems

Green/sustainable chemistry involves the search to reduce or even eliminate the use of substances in the production of chemical products and reactions which are hazardous to human health and the environment. The low volatility of ILs is the key property that make them green solvents. However, the recovery of solutes dissolved in ILs may represent a breakdown in the greenness of any such process if volatile organic solvents are used to extract them in liquid–liquid biphasic systems. In 1999, Brennecke's group demonstrated the exceptional ability of $scCO_2$ to extract naphthalene from certain ILs based on the 1-butyl-3-methylimidazolium cation because, although $scCO_2$ is highly soluble in the IL phase, the same IL is not measurably soluble in the $scCO_2$ phase [40]. This discovery was crucial for further developments in non-aqueous green processes involving both biotransformation and extraction steps. The combination of both ILs and $scCO_2$ neoteric solvents as reaction media for biocatalysis was first described in 2002, and represented the first operational strategy for the development of integral green processes in non-aqueous environments [41].

The phase behavior of IL–$scCO_2$ systems has been studied for several ILs and under many supercritical conditions, and also in the presence of solutes dissolved in the IL phase. Knowledge of this phase behavior is essential for developing any process because it determines the contact conditions between $scCO_2$ and the solute, and also reduces the viscosity of the IL phase, which enhances the mass transfer rate of the reaction system. Studies of the IL 1-butyl-3-methylimidazolium hexafluorophosphate ([BMim][PF_6])–$scCO_2$ system indicated that it behaves as a biphasic system, in which no measurable amount of [BMim][PF_6] is soluble in the CO_2-rich phase, whereas a large amount of CO_2 is dissolved in the IL-rich phase [42]. When the pressure increases, the solubility of CO_2 into the IL-rich phase increases drastically, reaching solubility values up 0.32 mole fraction at 93 bar and 40 °C, whereas the dependence of CO_2 solubility in [BMim][PF_6] on temperature was low. With respect to the water content of ILs, the increase in CO_2 solubility when ILs are previously dried has been demonstrated. However, the ability of ILs to rehydrate by atmospheric humidity is very high. For example, the estimated water content of [BMim][PF_6] after drying is approximately 0.15% w/w, but this IL is able to absorb several percentage points of water when exposed to the atmosphere [43]. Other IL–$scCO_2$ biphasic systems (e.g. 1-octyl-3-methylimidazolium hexafluorophosphate) showed a similar phase behavior to [BMim][PF_6], the solubility CO_2 in the IL-rich phase being highest for the ILs with fluorinated anions (i.e. [PF_6] and [BF_4]). Additionally, for the case of ILs based on the same anion (e.g. bistriflimide), the solubility of CO_2 increased proportionally with the increase in the alkyl chain length of the cation [44].

In this context, a new concept for continuous biphasic biocatalysis, whereby a homogeneous enzyme solution is immobilized in a liquid phase (working phase), whereas substrates and products reside largely in a supercritical phase (extractive phase), has been proposed as the first approach to integral green bioprocesses in non-aqueous media, that directly provides products (Figure 7.5). The system was tested for two different reactions catalyzed by *Candida antarctica* lipase B (CALB): the synthesis of butyl butyrate from vinyl butyrate and 1-butanol, and the kinetic resolution of *rac*-1-phenylethanol at 150 bar over a range of temperatures (40–100 °C). Under these conditions, the enzyme showed an exceptional level of activity, enantioselectivity (*ee* >99.9%) and operational stability after 11 cycles of 4 h of work [41a]. These results were corroborated in extreme conditions, such as 100 bar and 150 °C [45].

The transport of substrates from the supercritical to the enzyme–IL phase, and then the release of products towards the supercritical phase, are key parameters for controlling the efficiency of the reaction system. In another example where similar ILs based on the same ions were compared, it was demonstrated how the lipase activity in the synthesis of alkyl esters with controlled chain length was improved when the hydrophobicities of ILs with respect to substrates and products were similar, since this favored the mass transfer phenomena between IL and $scCO_2$ phases [46].

A further step towards green biocatalysis in IL–$scCO_2$ biphasic systems was taken by the appropriate selection of reagents, because the selective separation of the synthetic product can be included as an integrated step in the full process [47].

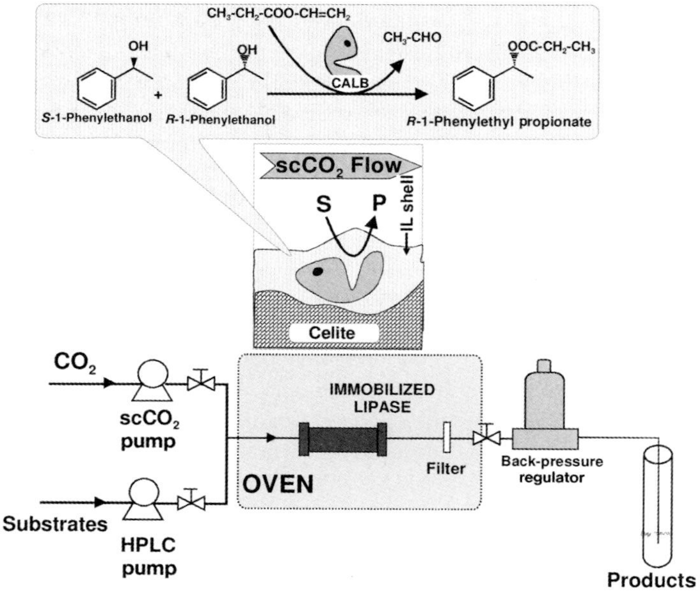

Figure 7.5 Experimental set-up of the continuous green enzyme reactor working in IL–$scCO_2$ biphasic medium [41a].

By using vinyl laurate as acyl donor in the kinetic resolution of *rac*-1-phenylethanol catalyzed by immobilized CALB, the stereoselective synthetic product, (*R*)-1-phenylethyl laurate, can be selectively separated from the non-reacted alcohol with $scCO_2$. This process takes advantage of the fact that the solubility of a compound in $scCO_2$ depends on both the polarity and vapor pressure. Therefore, if the alkyl chain of an ester product is long enough, its low volatility should mean that it is less soluble in $scCO_2$ than the corresponding alcohol. By this strategy, the introduction of an additional separation chamber between the reactor and the back-pressure outlet of the system, accompanied by the selection of an appropriate pressure and temperature, led to the selective separation of the synthetic product from the resulting reaction mixture (66% yield, *ee* >99.9%). As another example, cutinase from *F. solani pisi* immobilized on zeolite NaY was tested in an IL–$scCO_2$ biphasic system for the kinetic resolution of *rac*-2-phenyl-1-propanol. The protective effect of the IL against enzyme deactivation by $scCO_2$ was demonstrated, and also higher activity than observed for the cutinase–IL system. This enhancement in activity was attributed to the CO_2 dissolved in the IL, which would have decreased its viscosity and hence improved the mass transfer of substrates to the enzyme active site [48].

The possibility of developing multicatalytic processes in SCFs was also reported for continuous dynamic kinetic resolution processes in different IL–$scCO_2$ biphasic systems by simultaneously using both immobilized enzyme and an acid catalyst (e.g. silica modified with benzenesulfonic acid, zeolite, etc.) at 40 °C and 10 MPa. Kinetic

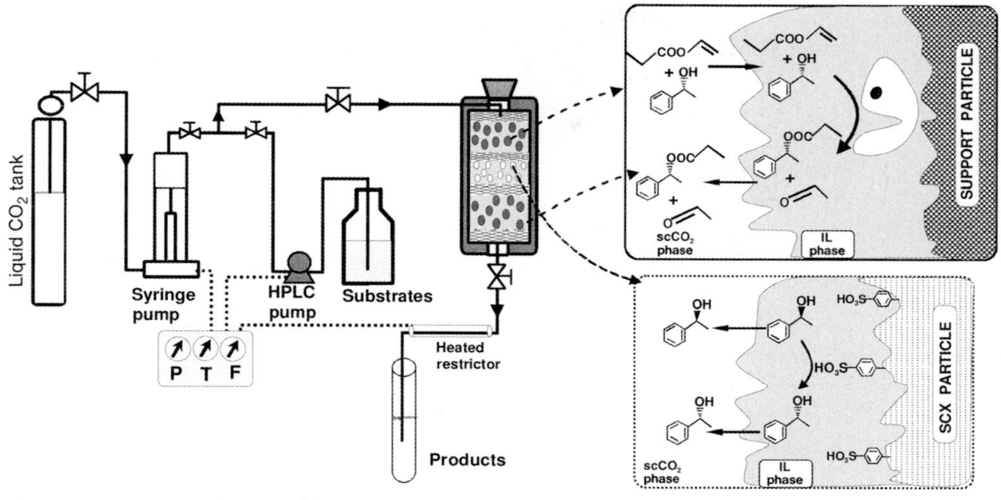

Figure 7.6 Experimental set-up of the continuous chemoenzymatic reactor for dynamic kinetic resolution of *rac*-1-phenylethanol in IL–$scCO_2$ biphasic systems [49].

resolution with enzymes is the most widely used method for separating the two enantiomers of a racemic mixture, although the chemical yield is limited to 50%. To overcome this limitation, the combination of enzymatic kinetic resolution with *in situ* racemization of the undesired enantiomer, using so-called dynamic kinetic resolution (DKR). has been studied (Figure 7.6). Both immobilized lipase and an acid catalyst coated with ILs greatly improved the efficiency of the process, providing a good yield (76%) of (*R*)-1-phenylethyl propionate product and good enantioselectivity (*ee* 91–98%) in continuous operation [49].

Another approach to enzyme catalysis in IL–$scCO_2$ biphasic systems is the development of solid supports to which the IL phase is covalently attached, for which the transfer of some IL properties to its surface has been demonstrated [50a,b]. The adsorption of CALB on this linked IL phase provides excellent immobilized biocatalysts with enhanced activity and increased operational stability for the synthesis of citronellyl butyrate in $scCO_2$, compared with the original strategy based on enzymes coated with ILs [50c].

In the same context of enzyme catalysis in non-aqueous liquid–$scCO_2$ biphasic systems, the discovery that poly(ethylene glycol) (PEG) and $scCO_2$ formed similar biphasic systems to that of ILs–$scCO_2$ has opened up new opportunities for enzymatic processes in SCFs, since PEG has very low solubility in $scCO_2$ whereas CO_2 dissolves readily in PEG [51]. Furthermore, PEG is less expensive than ILs and an accepted additive for foods with a fully evaluated toxicity. By using the lipase-catalyzed kinetic resolution of *rac*-1-phenylethanol as a reaction model, preliminary studies have demonstrated the high suitability of this PEG–$scCO_2$ biphasic medium because enzyme activity and selectivity are maintained after 11 runs at 50 °C and 80 bar.

7.7
Future Trends

Enzymes (and cells) are the best green tools for organic synthesis and their advantages in SCFs are manifold, because they allow rapid reaction rates, simplify product recovery, and permit reuse of the solvent. The weakness of biocatalysts in these unconventional media can probably be overcome by the use of protective agents (e.g. ILs, PEG) to give biocatalytic systems with high levels of activity, enantioselectivity, and, mainly, operational stability, to be used as the core of industrial chemical plants of the near future. Fundamental studies on enzyme catalysis in IL–SCF biphasic systems should be carried out to establish clear criteria for specifically pairing the most appropriate IL–SCF with the corresponding enzyme or bioprocess. Furthermore, the enormous potential of multi-enzymatic and/or multi-chemoenzymatic processes in ILs–SCFs for synthesizing pharmaceutical drugs has only just been realized. Biocatalytic processes based on combinations of neoteric solvents (e.g. ILs with SCFs) could provide a large number of reaction media able to perform a plethora of green catalyzed reactions of industrial interest, and also new processes for product separation. The door leading to the green chemical industry is definitively open.

Acknowledgments

Financial support was provided by grants from the CICYT (CTQ2005-01571-PPQ), SENECA Foundation (02910/PI/05) and BIOCARM (BIO-BMC 06/01-0002).

References

1 (a) Bommarius, A.S. and Riebel, B.R. (2004) *Biocatalysis: Fundamentals and Applications*, Wiley-VCH Verlag GmbH, Weinheim; (b) Reetz, M.T. (2006) *Advances in Catalysis*, **49**, 1–69.

2 (a) Klibanov, A.M. (2001) *Nature*, **409**, 241–246; (b) Patel, R.N. (2001) *Advanced Synthesis and Catalysis*, **343**, 527–546.

3 (a) Hudson, E.P., Eppler, R.K. and Clark, D.S. (2005) *Current Opinion in Biotechnology*, **16**, 637–643; (b) Kazlauskas, R.J. (2005) *Current Opinion in Chemical Biology*, **9**, 195–201; (c) Hult, K. and Berglund, P. (2003) *Current Opinion in Biotechnology*, **14**, 395–400.

4 Fitzpatrick, P.A., Steinmetz, A.C.U., Ringe, D. and Klibanov, A.M. (1993) *Proceedings of the National Academy of Sciences of the United States of America*, **90**, 8653–8657.

5 Capello, C., Fisher, U. and Hungerbühler, K. (2007) *Green Chemistry*, **9**, 927–934.

6 (a) Jessop, P.J. and Leitner, W. (eds) (1999) *Chemical Synthesis Using Supercritical Fluids*, Wiley-VCH Verlag GmbH, Weinheim; (b) for a review on $scCO_2$ properties, see: Beckmann, E.J. (2004) *Journal of Supercritical Fluids*, **28**, 121–191.

7 For reviews on enzyme catalysis in SCFs, see: (a) Hobbs, H.R. and Thomas, N.R. (2007) *Chemical Reviews*, **107**, 2786–2820; (b) Mesiano, A.J., Beckman, E.J. and Russel, A.J. (1999) *Chemical Reviews*, **99**, 623–633.

8 Kamat, S., Barrera, J., Beckman, E.J. and Russell, A.J. (1992) *Biotechnology and Bioengineering*, **40**, 158–166.

9 Habulin, M. and Knez, Z. (2001) *Journal of Chemical Technology and Biotechnology*, **76**, 1260–1266.

10 Borges de Carvalho, I., Correa de Sampaio, T. and Barreiros, S. (1996) *Biotechnology and Bioengineering*, **49**, 399–404.

11 (a) Kamat, S., Critchley, G., Beckman, E.J. and Russell, A.J. (1995) *Biotechnology and Bioengineering*, **46**, 610–620; (b) Striolo, A., Favaro, A., Elvassore, N., Bertucco, A. and Di Notto, V. (2003) *Journal of Supercritical Fluids*, **27**, 283–295.

12 (a) Toews, K.L., Shroll, R.M., Wai, C.M. and Smart, N.G. (1995) *Analytical Chemistry*, **67**, 4040–4043; (b) Almeida, M.C., Ruivo, R., Maia, C., Freire, L., Correa de Sampaio, T. and Barreiros, S. (1998) *Enzyme and Microbial Technology*, **22**, 494–499.

13 (a) Arreola, A.G., Balaban, M.O., Marshal, M.R., Peplow, A.J., Wei, C.I. and Cornell, J.A. (1991) *Journal of Food Science*, **56**, 1030–1033; (b) del Pozo, D., Balaban, M.O. and Talcott, S.T. (2007) *Food Research International*, **40**, 894–899; (c) Yoshimura, T., Furutera, M., Shimoda, M., Ishikawa, H., Miyake, M., Matsumoto, K., Osajima, Y. and Hayakawa, I. (2002) *Journal of Food Science*, **67**, 3227–3231.

14 (a) Castillo, E., Marty, A., Combes, D. and Condoret, J.S. (1994) *Biotechnology Letters*, **16**, 169–174; (b) Stamatis, H., Sereti, V. and Kolisis, F.N. (1998) *Chemical and Biochemical Engineering Quarterly*, **12**, 151–156; (c) Tsitsimpikou, C., Stamatis, H., Sereti, V., Daflos, H. and Kolisis, F.N. (1998) *Journal of Chemical Technology and Biotechnology*, **71**, 309–314.

15 For reviews of application of lipases, see: Hasan, F., Shah, A.A. and Hameed, A. (2006) *Enzyme and Microbial Technology*, **39**, 235–251; (b) Schmid, R.D. and Verger, R. (1998) *Angewandte Chemie (International Edition in English)*, **37**, 1608–1633; (c) Demirbas, A. (2007) *International Journal of Green Energy*, **4**, 15–26; (d) Hayes, D.G. (2004) *Journal of the American Oil Chemists Society*, **81**, 1077–1103; (e) Jackson, M.A., Mbaraka, I.K. and Shanks, B.H. (2006) *Applied Catalysis A: General*, **310**, 48–53.

16 (a) Romero, M.D., Calvo, L., Alba, C., Daneshfar, A. and Ghaziaskar, H.S. (2005) *Journal of Supercritical Fluids*, **33**, 77–84; (b) Steytler, D.C., Moulson, P.S. and Reynolds, J. (1991) *Enzyme and Microbial Technology*, **13**, 221–226; (c) Dumont, T., Barth, D. and Perrut, M. (1993) *Journal of Supercritical Fluids*, **6**, 85–89; (d) Kumar, R., Madras, G. and Modak, J. (2004) *Industrial & Engineering Chemistry Research*, **43**, 7697–7701; (e) Marty, A., Combes, D. and Condoret, J.S. (1994) *Biotechnology and Bioengineering*, **43**, 497–504.

17 (a) Ikushima, Y., Saito, N., Yokoyama, T., Hatakeda, K., Ito, S., Arai, M. and Blanch, H.W. (1993) *Chemistry Letters*, 109–112; (b) Ikushima, Y., Saito, N., Hatakeda, K. and Sato, O. (1996) *Chemical Engineering Science*, **51**, 2817–2822; (c) Peres, C., Da Silva, D.R.G. and Barreiros, S. (2003) *Journal of Agricultural and Food Chemistry*, **51**, 1884–1888; (d) Chulalaksananukul, W., Condoret, J.S. and Combes, D. (1993) *Enzyme and Microbial Technology*, **15**, 691–698; (e) Lozano, P., Villora, G., Gomez, D., Gayo, A.B., Sanchez-Conesa, J.A., Rubio, M. and Iborra, J.L. (2004) *Journal of Supercritical Fluids*, **29**, 121–128.

18 (a) Kim, I.H., Ko, S.N., Lee, S.M., Chung, S.H., Kim, H., Lee, K.T. and Ha, T.Y. (2004) *Journal of the American Oil Chemists Society*, **81**, 537–541; (b) Jackson, M.A. and King, J.W. (1996) *Journal of the American Oil Chemists Society*, **73**, 353–356; (c) Oliveira, D. and Oliveira, J.V. (2001) *Journal of Supercritical Fluids*, **19**, 141–148.

19 (a) Capewell, A., Wendel, V., Bornscheuer, U., Meyer, H.H. and Scheper, T. (1996) *Enzyme and Microbial Technology*, **19**, 181–186; (b) Matsuda, T., Watanabe, K., Harada, T., Nakamura, K., Arita, Y., Misumi, Y., Ichikawa, S. and Ikariya, T. (2004) *Chemical Communications*, 2286–2287; (c) Celia, E., Cernia, E., Palocci, C., Soro, S. and Turchet, T. (2005) *Journal of*

Supercritical Fluids, **33**, 193–199; (d) Martins, J.F., Correa de Sampaio, T., Borges de Carvalho, I. and Barreiros, S. (1994) *Biotechnology and Bioengineering*, **44**, 119–124; (e) Rantakyla, M. and Aaltonen, O. (1994) *Biotechnology Letters*, **16**, 825–830.

20 (a) Mesiano, A.J., Enick, R.M., Beckman, E.J. and Russell, A.J. (2001) *Fluid Phase Equilibria*, **178**, 169–177; (b) Kondo, R., Toshima, K. and Matsumura, S. (2002) *Macromolecular Bioscience*, **2**, 267–271; (c) Loeker, L.F., Duxbury, C.J., Kumar, R., Gao, W., Gross, R.A. and Howdle, S.M. (2004) *Macromolecules*, **37**, 2450–2453; (d) Matsumura, S., Ebata, H., Kondo, R. and Toshima, K. (2001) *Macromolecular Rapid Communications*, **22**, 1325–1329.

21 (a) Lee, H.S., Lee, W.G., Park, S.W., Lee, H. and Chang, H.N. (1993) *Biotechnology Techniques*, **7**, 267–270; (b) Zheng, Y.Z. and Tsao, G.T. (1996) *Biotechnology Letters*, **18**, 451–454.

22 (a) Matsuda, T., Harada, T. and Nakamura, K. (2000) *Chemical Communications*, 1367–1368; (b) Matsuda, T., Ohashi, Y., Harada, T., Yanagihara, R., Nagasawa, T. and Nakamura, K. (2001) *Chemical Communications*, 2914–2915; (c) Matsuda, T., Harada, T. and Nakamura, K. (2004) *Green Chemistry*, **6**, 440–444; (d) Jurcek, O., Wimmerova, M. and Wimmer, Z. (2008) *Coordination Chemistry Reviews*, **252**, 767–781.

23 For reviews on the effects of supercritical parameters in enzyme catalysis, see: (a) Rezaei, K., Temelli, F. and Jenab, E. (2007) *Biotechnology Advances*, **25**, 272–280; (b) Rezaei, K., Jenab, E. and Temelli, F. (2008) *Critical Reviews in Biotechnology*, **27**, 183–195.

24 For examples of enzyme deactivation by depressurization, see: (a) Kasche, V., Schothauer, R. and Brunner, G. (1988) *Biotechnology Letters*, **10**, 569–574; (b) Lozano, P., Avellaneda, A., Pascual, R. and Iborra, J.L. (1996) *Biotechnology Letters*, **18**, 1345–1350; (c) Giessauf, A., Magor, W., Steinberger, D.J. and Marr, R. (1999) *Enzyme and Microbial Technology*, **24**, 577–583.

25 Ikushima, Y., Saito, N., Arai, M. and Blanch, H.W. (1995) *The Journal of Physical Chemistry*, **99**, 8941–8944.

26 (a) Matsuda, T., Harada, T. and Nakamura, K. (2005) *Current Organic Chemistry*, **9**, 299–315; (b) Matsuda, T., Harada, T., Nakamura, K. and Ikariya, T. (2005) *Tetrahedron: Asymmetry*, **16**, 909–915.

27 Knez, Z., Habulin, M. and Primozic, M. (2005) *Biochemical Engineering Journal*, **27**, 120–126.

28 (a) Nagesha, G.K., Manohar, B. and Sankar, K.U. (2004) *Journal of Supercritical Fluids*, **32**, 137–145; (b) Lin, T.J., Chen, S.W. and Chang, A.C. (2006) *Biochemical Engineering Journal*, **29**, 27–34; (c) Miller, D.A., Blanch, H.M. and Prausnitz, J.M. (1991) *Industrial & Engineering Chemistry Research*, **30**, 939–946.

29 (a) Kamat, S.V., Beckman, E.J. and Russell, A.J. (1995) *Critical Reviews in Biotechnology*, **15**, 41–71; (b) Fontes, N., Almeida, M.C., Peres, C., Garcia, S., Grave, J., Aires-Barros, R.M., Soares, C.M., Cabral, J.M.S., Maycock, C.D. and Barreiros, S. (1998) *Industrial & Engineering Chemistry Research*, **37**, 3189–3194.

30 (a) For immobilized enzymes in green solvents see: Lozano, P., de Diego, T. and Iborra, J.L. (2006) *Immobilization of Enzymes and Cells* Methods in Biotechnology Series, vol. 22 (ed. J.M. Guisan), Humana Press, Totowa, NJ. Chapters 22 and 23; (b) for reviews on immobilized enzymes, see: Sheldon, R.A. (2007) *Advanced Synthesis and Catalysis*, **349**, 1289–1307; Mateo, C., Palomo, J.M., Fernandez-Lorente, G., Guisan, J.M. and Fernandez-Lafuente, R. (2007) *Enzyme and Microbial Technology*, **40**, 1451–1463; (c) for nanostructures, see: Kim, J., Grate, J.W. and Wang, P. (2006) *Chemical Engineering Science*, **61**, 1017–1026.

31 (a) Mori, T., Kobayashi, A. and Okahata, Y. (1998) *Chemistry Letters*, 921–923; (b) Mori, T., Funasaki, M., Kobayashi, A. and

Okahata, Y. (2001) *Chemical Communications*, 1832–1833.

32 (a) Novak, Z., Habulin, M., Kremelj, V. and Knez, Z. (2003) *Journal of Supercritical Fluids*, **27**, 169–178; (b) Vidinha, P., Augusto, V., Almeida, M., Fonseca, I., Fidalgo, A., Ilharco, L., Cabral, J.M.S. and Barreiros, S. (2006) *Journal of Biotechnology*, **121**, 23–33.

33 (a) Sheldon, R.A. (2007) *Biochemical Society Transactions*, **35**, 1583–1587; (b) Hobbs, H.R., Kondor, B., Stephenson, P., Sheldon, R.A., Thomas, N.R. and Poliakoff, M. (2006) *Green Chemistry*, **8**, 816–821; (c) Dijkstra, Z.D., Merchant, R. and Keurentjes, J.T.F. (2007) *Journal of Supercritical Fluids*, **41**, 102–108.

34 Lozano, P., De Diego, T. and Iborra, J.L. (2007) *Chemistry today*, **25**, 76–79.

35 For reviews on properties of ILs, see: (a) Dupont, J., de Souza, R.F. and Suarez, P.A.Z. (2002) *Chemical Reviews*, **102**, 3667–3691; (b) Poole, C.F. (2004) *Journal of Chromatography. A*, **1037**, 49–82; (c) Reichardt, C. (2005) *Green Chemistry*, **7**, 339–351.

36 For reviews on biocatalysis in ILs, see: (a) Yang, Z. and Pan, W. (2005) *Enzyme and Microbial Technology*, **37**, 19–28; (b) van Rantwijk, F. and Sheldon, R.A. (2007) *Chemical Reviews*, **107**, 2757–2785; (c) Durand, J., Teuma, E. and Gómez, M. (2007) *Comptes Rendus Chimie*, **10**, 152–177.

37 For examples of enzyme stabilization by ILs, see: (a) Lozano, P., De Diego, T., Carrié, D., Vaultier, M. and Iborra, J.L. (2001) *Biotechnology Letters*, **23**, 1529–1533; (b) Lozano, P., De Diego, T., Guegan, J.P., Vaultier, M. and Iborra, J.L. (2001) *Biotechnology and Bioengineering*, **75**, 563–569; (c) Persson, M. and Bornscheuer, U.T. (2003) *Journal of Molecular Catalysis B: Enzymatic*, **22**, 21–27; (d) Ulbert, O., Belafi-Bako, K., Tonova, K. and Gubicza, L. (2005) *Biocatalysis and Biotransformation*, **23**, 177–183.

38 (a) Lee, S.H., Doam, T.T.N., Ha, S.H. and Koo, Y.M. (2007) *Journal of Molecular Catalysis B: Enzymatic*, **45**, 57–61; (b) Lee, S.H., Doam, T.T.N., Ha, S.H., Chang, W.J. and Koo, Y.M. (2007) *Journal of Molecular Catalysis B: Enzymatic*, **45**, 129–134.

39 For examples of studies on enzyme structure in ILs, see: (a) De Diego, T., Lozano, P., Gmouh, S., Vaultier, M. and Iborra, J.L. (2004) *Biotechnology and Bioengineering*, **88**, 916–924; (b) De Diego, T., Lozano, P., Gmouh, S., Vaultier, M. and Iborra, J.L. (2005) *Biomacromolecules*, **6**, 1457–1464; (c) van Rantwijk, F., Secundo, F. and Sheldon, R.A. (2006) *Green Chemistry*, **8**, 282–286.

40 Blanchard, L.A., Hancu, D., Beckman, E.J. and Brennecke, J.F. (1999) *Nature*, **399**, 28–29.

41 Pioneering works using ILs–scCO_2 Lozano, P., De Diego, T., Carrié, D., Vaultier, M. and Iborra, J.L. (2002) *Chemical Communications*, 692–693; (b) Reetz, M.T., Wiesenhofer, W., Francio, G. and Leitner, W. (2002) *Chemical Communications*, 992–993.

42 Keskin, S., Kayrak-Talay, D., Akman, U. and Hortaçsu, O. (2007) *Journal of Supercritical Fluids*, **43**, 150–180.

43 Blanchard, L.A., Gu, Z.Y. and Brennecke, J.F. (2001) *The Journal of Physical Chemistry. B*, **105**, 2437–2444.

44 Aki, S.N.V.K., Mellein, B.R., Saurer, E.M. and Brennecke, J.F. (2004) *The Journal of Physical Chemistry. B*, **108**, 20355–20365.

45 Lozano, P., De Diego, T., Carrié, D., Vaultier, M. and Iborra, J.L. (2003) *Biotechnology Progress*, **19**, 380–382.

46 Lozano, P., De Diego, T., Gmouh, S., Vaultier, M. and Iborra, J.L. (2004) *Biotechnology Progress*, **20**, 661–669.

47 Reetz, M.T., Wiesenhofer, W., Francio, G. and Leitner, W. (2003) *Advanced Synthesis and Catalysis*, **345**, 1221–1228.

48 Garcia, S., Lourenco, N.M.T., Lousa, D., Sequeira, A.F., Mimoso, P., Cabral, J.M.S., Afonso, C.A.M. and Barreiros, S. (2004) *Green Chemistry*, **6**, 466–470.

49 (a) Lozano, P., De Diego, T., Larnicol, M., Vaultier, M. and Iborra, J.L. (2006) *Biotechnology Letters*, **28**, 1559–1565; (b)

Lozano, P., De Diego, T., Gmouh, S., Vaultier, M. and Iborra, J.L. (2007) *International Journal of Chemical Reactor Engineering*, **5**, A53.

50 (a) Riisager, A., Fehrmann, R., Haumann, M. and Wasserscheid, P. (2006) *Topics in Catalysis*, **40**, 91–102; (b) Burguete, M.I., Galindo, F., Garcia-Verdugo, E., Karbass, N. and Luis, S. (2007) *Chemical Communications*, 3086–3088; (c) Lozano, P., Garcia-Verdugo, E., Piamtongkam, R., Karbass, N., De Diego, T., Burguete, M.I., Luis, S.V. and Iborra, J.L. (2007) *Advanced Synthesis and Catalysis*, **349**, 1077–1084.

51 (a) Heldebrant, D.J. and Jessop, P.G. (2003) *Journal of the American Chemical Society*, **125**, 5600–5601. (b) Reetz, M.T. and Wiesenhofer, W. (2004) *Chemical Communications*, 2750–2751.

8
Polymerization in Supercritical Carbon Dioxide

Uwe Beginn

8.1
General Aspects

8.1.1
Introduction and Scope

It has been stated that the twenty-first century will be the era of polymers and in fact even today synthetic macromolecules are ubiquitous. They are used as materials to replace the "classic" structural materials such as wood, steel, stone, and glass in many applications, and their unique properties constantly generate new opportunities. At the same time, functional polymers used as additives in practically every situation of life and technology became available. In 2006, the yearly worldwide polymer production was about 245 million (metric) tons, with an average growth of ~5% per year between 2001 and 2006 [1]. Our civilization is no longer possible without technology, and technology is no longer possible without polymers. The synthesis, modification, processing, and extraction processes required to manufacture the demanded polymers consume energy, produce waste streams, and contribute to pollution of the environment, although measures are taken by the producers and also legislators to minimize harmful side-effects. However, with its multi-megaton polymer production, we would be well advised to do better. It has been pointed out that performing polymer chemistry in liquid or supercritical carbon dioxide may be one component of the required actions to minimize environmental impacts [2, 3]. Because polymer chemistry in and with carbon dioxide has become a very broad field during the last 10 years [2, 4], the literature overview in this chapter will focus on polymer preparation by means of chain growth reactions. This chapter does not cover step growth polymerizations [5] such as enzymatic polymer synthesis [6, 7], formation of polycarbonates, or nylon, electrochemical, and oxidative polymer synthesis [8]. Review articles on polymer extraction [9], modification [10, 11], and processing [12] in or with carbon dioxide can be found in the cited literature. This chapter focuses on the literature published since 1999, although older references are cited if required for a basic understanding.

Handbook of Green Chemistry, Volume 4: Supercritical Solvents. Edited by Walter Leitner and Philip G. Jessop

ISBN: 978-3-527-32590-0

8.1.2 Supercritical Fluids

Recently, supercritical fluids (SCFs) have attracted interest as reaction solvents or processing agents in macromolecular chemistry. By definition, SCFs are substances at a temperature and pressure higher than their critical values, and which have a density close to or higher than the critical density (Figure 8.1) [9]. In particular supercritical carbon dioxide (sc-CO_2) has been studied extensively for the synthesis and processing of polymers: CO_2 is naturally occurring, abundant, and exists in natural reservoirs. The gas is generated in large quantities as a by-product in ammonia, hydrogen, and ethanol plants and also in electrical power plants that burn fossil fuels. In comparison with other substances (Table 8.1), CO_2 has an easily accessible critical point with a critical temperature (T_c) of 31.1 °C and a critical pressure (P_c) of 73.8 bar [13]. It is an ambient gas, and can be recycled after use. Finally, it is inexpensive and non-flammable. CO_2 can be separated quantitatively from a reaction mixture by simple depressurization, resulting in a dry polymer product. The energy needed for removing 1 m^3 of a liquid solvent from the final product by evaporation at normal pressure (water, 2.3×10^9 kJ m^{-3}; *n*-hexane, 3.7×10^5 kJ m^{-3}) far exceeds that for pressurizing of the same amount of CO_2 from 1 to 250 bar (3.3×10^4 kJ m^{-3}).

Since the ability of a solvent to dissolve solutes depends strongly on its density, changes in temperature or pressure can significantly alter the dissolution properties of the solvent without variation of its composition [9]. The density of CO_2, for example, can be varied from gas-like ($\rho < 0.2$ g cm^{-3}) to liquid-like values ($\rho > 0.6$ g cm^{-3}) as depicted in Figure 8.2a. Furthermore the viscosity of sc CO_2 is much lower than that of liquid solvents and it also varies strongly with changes in pressure and temperature (Figure 8.2b). Because of these features, SCF diffusion coefficients are similar to those of gases. As depicted in Figure 8.3, any small change in temperature or pressure, in particular in the vicinity of the critical point, has a large effect on the diffusivities, a fact that can have important implications for reaction kinetics.

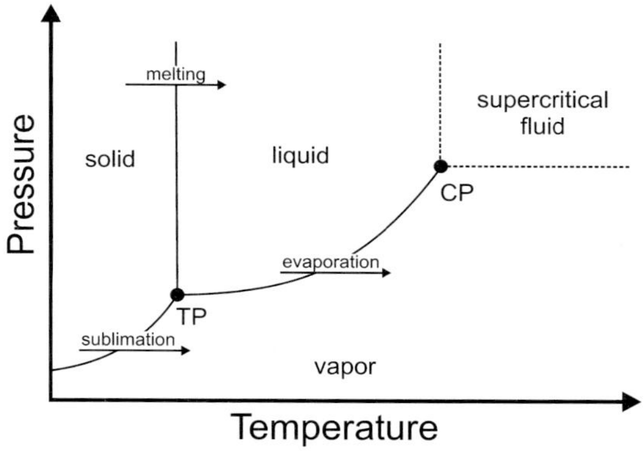

Figure 8.1 Schematic phase diagram of a pure component including its SCF region. TP = triple point, CP = critical point.

Table 8.1 Experimental solubility data for some polymers in CO_2. (Entries in *bold* indicate polymers that are soluble in supercritical carbon dioxide at relative low pressure).

Polymer[a]	Temperature (°C)	Pressure (bar)	$M_w \times 10^{-3}$ (kg mol^{-1})	Ref.
Polydimethylsiloxane	**25–185**	**250–600**	**39–370**	**[42]**
Polyisobutene	50–200	150–750	2–486	[43]
Poly(vinylidene fluoride) (PVDF)	120–220	2000	~200	[44]
PVDF w/acetone	90–220	1700	~200	[44]
PVDF w/DME	100–220	300–1700	~200	[44]
PVDF w/ethanol	100–220	1700	~200	[44]
Poly(1,1-dihydroperfluorooctyl acrylate)	**30–80**	**100–300**	**1200**	**[45, 46]**
Poly(vinyl acetate) (PVAc)	20–160	500–1000	125	[15]
Poly(methyl acrylate)	20–200	1700–2200	31	[15]
Poly(ethyl acrylate)	50–200	1200–3000	119	[15]
Poly(propyl acrylate)	100–180	1200–1500	140	[15]
Poly(butyl acrylate)	80–200	1000–3000	62	[15]
Poly(ethylhexyl acrylate)	150–220	1100–3000	113	[15]
Poly(octadecyl acrylate)	210–260	1000–2600	23	[15]
Poly(butyl methacrylate)	120–230	1100–3000	320	[15]
Poly($PPO_{85mol\%}$-*co*-PPO $carbonate_{15mol\%}$)	**22**	**130–160**	**30**	**[30]**
EMA_{18}	80–280	1500–2800	~100	[15]
EMA_{31}	80–280	1500–2800	~100	[15]
EMA_{41}	80–280	1500–2800	~100	[15]
TFE-HFP_{19}	180–250	1000–3000	210	[15]
TFE-HFP_{48}	170–230	1000–2800	210	[46]
VDF-HFP_{22}	100–230	700–900	191	[15]
VDF-$CTFE_{74}$	120–210	1000–2750	85	[48]
Teflon AF160	50–180	500–1000	400	[15]
Teflon AF240	60–180	500–1000	NA	[49]
Poly(vinyl fluoride) (PVF)	<300	Insol.	120	[12, 50]
Poly(acrylic acid)	<280	Insol.	54	[12]
PEO	140	800–1500	13	[37]
Polyacrylamide	Insol.	Insol.	160	[37]
Polystyrene (PS)	Insol.	Insol.	80	[37]
$TFE_{47.8}$-VAc	80–120	550–1000	53	[51]

[a]Entries that show a polymer with a notation "w/cosolvent" indicate that a cosolvent is added to the solution.

Although CO_2 is a good solvent for many non-polar and some polar compounds of low molecular weight [14], it is a poor solvent for most high molecular weight polymers under mild conditions (<100 °C, <350 bar) [12, 15]. For example, to dissolve poly(vinylidene fluoride) with a molecular weight of 530 kg mol^{-1}, more than 2750 bar and 300 °C are required [16]. The only polymers that show good solubility in CO_2 under mild conditions are amorphous fluoropolymers [12, 17–19], silicones [20], and polyether carbonates [21]. The reason for this peculiar problem will be mentioned in the next section.

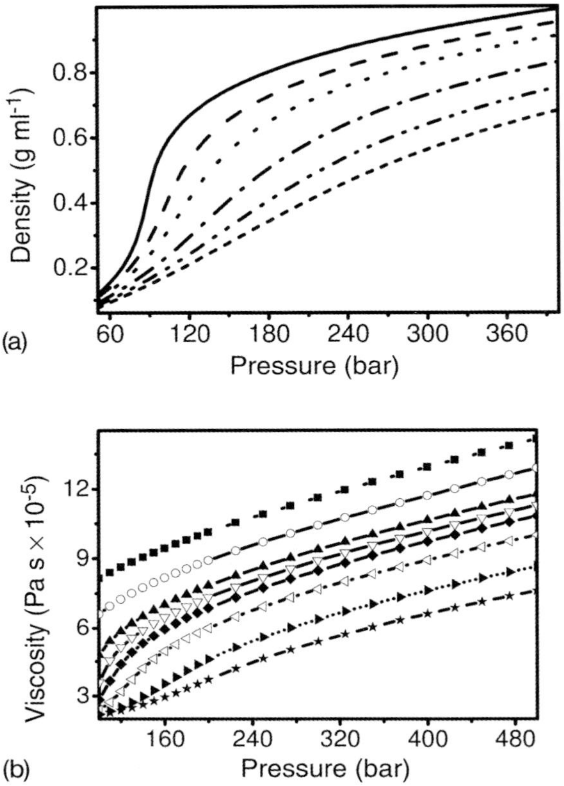

Figure 8.2 Pressure dependence of CO_2 physical properties. (a) Density under isothermal conditions: (—) 50 °C; (- - - -) 50 °C; (· · · ·) 60 °C; (– · – ·) 80 °C; (– · · –) 100 °C; (· · · ·) 120 °C. (b) Viscosity: (■) 20 °C; (○) 30 °C; (▲) 50 °C; (▽) 45 °C; (◆) 50 °C; (◁) 60 °C; (◀) 80 °C; (★) 100 °C. Drawn with data from [11].

8.1.3
Solubility of Macromolecules in $scCO_2$

To form a stable solution of a polymer under thermodynamic equilibrium conditions in an SCF at a given temperature and pressure, the Gibbs free energy of mixing, ΔG_{mix}, must be negative, and of minimum value [22]. The Gibbs free energy of mixing is given by

$$\Delta G_{mix} = \Delta H_{mix} - T\Delta S_{mix} \tag{8.1}$$

where ΔG_{mix} = Gibbs free energy of mixing, ΔH_{mix} = change of mixing enthalpy, ΔS_{mix} = change of mixing entropy, and T = absolute temperature.

Although it is not possible to separate rigorously the impact of enthalpic and entropic contributions to the Gibbs free energy of mixing, the energetic effects will be

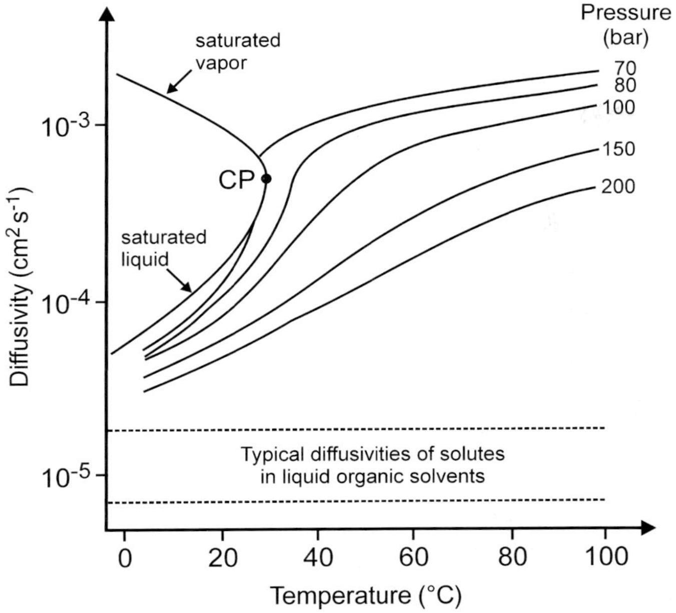

Figure 8.3 Temperature and pressure dependence of low molecular weight solute diffusivity in CO_2. CP = critical point. Redrawn with data from [12].

discussed separately first. For a dense isochoric (no volume changes) SCF solution, ΔH_{mix} equals the change in internal energy of mixing ΔU_{mix}. The dependence of this state function on the mixtures density $\rho(P,T)$, assuming pairwise additivity of interactions, is expressed as follows for an isotropic homogeneous mixture relative to a mixture of ideal gases [23]:

$$\Delta U_{\text{mix}} \approx \frac{2\pi\rho(P,T)}{k_B T} \sum_{i,j} x_i x_j \int \Gamma_{ij}(r,T,\rho) g_{ij}(r,T,\rho) r^2 \mathrm{d}r \tag{8.2}$$

where i,j = index number of components, that is, solvent molecules and polymer segments, x_i = molar fraction of component i, Γ_{ij} = intermolecular pair potential between i and j, $g(r,T,\rho)$ = radial distribution function, r = distance between i and j, $\rho(P,T)$ = solution density, T = absolute temperature, P = pressure, and k_B = Boltzmann constant.

For mixtures of small molecules, an approximate form of the attractive part of the intermolecular potential energy has been derived, namely Equation 8.3, relating $\Gamma_{ij}(r,T,\rho)$ to molecular properties such as the constituents' polarizability α_i, dipole moment μ_i and quadrupole moment Q_i [24]:

$$\Gamma_{ij}(r,T,\rho) \approx -C_1 \frac{\alpha_i \alpha_j}{r^6} - \frac{C_2}{k_B T} \frac{\mu_i \mu_j}{r^6} - \frac{C_3}{k_B T} \frac{\mu_i^2 Q_j^2}{r^8} - \frac{C_4}{k_B T} \frac{\mu_j^2 Q_i^2}{r^8} - \frac{C_5}{k_B T} \frac{Q_i^2 Q_j^2}{r^{10}} + Y \tag{8.3}$$

where C_1–C_5 = constants, Y = a function describing specific complex formation interactions between i and j, and r = distance between i and j.

Non-polar dispersion interactions (the first terms in Equation 8.3) are independent of temperature, and are only governed by the polarizability of the components in solution. The leading terms describing dipolar and quadrupolar interactions in Equation 8.3 are inversely proportional to the temperature, because at elevated temperatures the thermal energy disrupts the configurational alignment of the polar moments of the molecules. Hence it may be possible to dissolve a polar polymer in a non-polar SCF solvent if the temperature is sufficiently high to diminish polar interactions. Note that pressure increases the mixture density, simultaneously reducing the distances r between the components. This is the reason why enthalpic interactions depend predominantly on the solution density and hydrostatic pressure can be applied to polymer–SCF mixtures to obtain a single phase. Specific interactions such as complex formation or hydrogen bonding (Y in Equation 8.3) can also contribute to the attractive pair potential energy. The strength of these "directional" interactions is also temperature sensitive. Equations 8.1–8.3 indicate that the solvent quality of an SCF can be tuned with changes in both pressure and temperature, a degree of flexibility that is not available with liquid solvents [12]. CO_2 has a low polarizability [$\alpha(CO_2) = 2.89 \times 10^{-40}\,C^2\,m^2\,J^{-1}$], which is equal to that of methane [25], so that the first term in Equation 8.3 becomes small. As CO_2 has no static dipole moment, terms 2, 3 and 4 in Equation 8.3 become zero. Thus CO_2 interacts through its quadrupole moment {$Q(CO_2) = 4.5 \times 10^{-31}\,C\,m^{-2}$ [26]}. Normally these interactions do not contribute much to polymer solubility as they are weak, but in some cases they may become important due to the lack of other contributions [14]. Additionally, CO_2 interacts via "complex formation" interactions arising from its Lewis acid–base nature, where the carbon atom of CO_2 acts as an electron acceptor and the polymer chain becomes the electron donor [27, 28]. A known electron donor group in polymers is the carbonyl oxygen of a ketone [27], or a carbonate group, allowing, for example, the copolymer of propene oxide and CO_2 to become CO_2 soluble [29]. The solubility of polymers in scCO_2 is defined not only by the enthalpy but also by the balance between enthalpy and entropy. The mixing entropy change ΔS_{mix} depends on both the combinatorial entropy of mixing and the non-combinatorial contribution associated with the volume change on mixing [30]. The combinatorial entropy $\Delta S_{mix}^{(comb)}$ always promotes the mixing of a polymer with a solvent; however, as stated by Flory over 50 years ago [31], it is inverse proportional to the polymer's degree of polymerization (cf. Equation 8.4). Hence, with increasing molecular weight, the dissolution of macromolecules in any kind of solvent – including scCO_2 – cannot be supported much by the configurational entropy.

$$\frac{\Delta S_{mix}^{(comb)}}{RT} = \frac{\varphi_1}{X_1}\ln(\varphi_1) + \frac{1-\varphi_1}{X_2}\ln(1-\varphi_1) \qquad (8.4)$$

where $\Delta S_{mix}^{(comb)}$ = combinatorial fraction of the mixing entropy, R = gas constant, T = absolute temperature, φ_1 = volume fraction of the solute, and X_1, X_2 = degree of polymerization of the solute and the solvent, respectively.

A main contribution to the entropy of mixing is related to the free volume difference between the polymer and CO_2. Qualitatively, the solvent has to condense around the polymer chain in order to solubilize it [32]. The corresponding decrease in solvent entropy can dominate enthalpic interactions and prevent the formation of homogeneous phases. On the other hand, highly flexible macromolecules may gain entropy by the large variations of the molecular conformation that are available to the solubilized chain molecule. On a qualitative level, the entropy effect can be estimated from the glass transition temperature, T_G, of the polymer [33]. The lower is T_G, the more probable is solubilization of the polymer in CO_2 due to flexibility of the polymeric chain. Concluding from the above equations, it is important to match the physical properties of the polymer with those of CO_2 in order to optimize polymer solubilities by maximizing the solute's polarizability, minimizing its polar contributions (dipole moment) and exploiting complex-formation interactions. The polymer chain should be as flexible as possible and the degree of polymerization should be kept as low as possible [34]. With a given polymer, the experimentalist is free to vary (i) the temperature, (ii) the pressure, and (iii) the polymer concentration. Equation 8.3 demonstrates that temperature is a key factor for adjusting the interaction energies – it can be more pronounced than changing the density through pressure. However, macromolecules should become CO_2 soluble in the limits of high temperature and pressure. It is a typical feature of polymer solutions that the two-phase region, representing undissolved polymer states, becomes extended with lower polymer concentrations.

To dissolve in $scCO_2$, the majority of polymers ($M_n > 10\,kg\,mol^{-1}$) require pressures exceeding 1000 bar at temperatures between 50 and 200 °C. Polyethylene with a molecular weight of 108 $kg\,mol^{-1}$ was found to be completely insoluble at pressures up to 2750 bar and temperatures up to 270 °C [14], most probably because of its high degree of crystallinity. Polystyrene is only soluble up to a degree of polymerization of about 10; longer polystyrenes remain insoluble [9]. It is assumed that the high T_G of 100 °C is responsible as this includes a rather low flexibility of the backbone chain and therefore strong interactions between the monomer units which do not favor dissolution [14]. The solution behavior of polyacrylates was investigated with variation on the length of the side-chain. The shorter the alkyl tail, the better the polymer dissolves in $scCO_2$. Because of the low polarizability of CO_2, high pressures are needed (>100 bar) to dissolve poly(alkyl acrylate)s. Although poly(alkyl methacrylate)s are less polar, they are even less soluble, which is probably due to the higher T_Gs compared with the polyacrylates [9]. In general, prediction of polymer solubility in CO_2 is extremely difficult. It can, however, be distinguished which of the interactions, enthalpic or entropic, is favored, especially when considering polymers derived from the same family. Only the previously mentioned polydimethylsiloxane, poly(propene oxide-*co*-propene oxide carbonate), and amorphous fluoropolymers [poly(1,1-dihydroperfluorooctyl acrylate), hexafluoropropene copolymers] exhibit dissolution pressures well below 300 bar. Polydimethylsiloxane is readily soluble in $scCO_2$ because of its low T_G (−120 °C) and its low polarity [35], while the solubility of the polyether carbonate copolymers was attributed to their low T_G in combination to their electron pair donating capacity [29]. Similar justifications have been given for the solubility of perfluoroalkyl (meth)acrylate, hexafluoropropene, and hexafluoropropene oxide copolymers. In addition to the low T_Gs of the polymers and the weak electron pair donor abilities, the low polymer segment–segment interactions must be mentioned that support solution formation [14]. The incorporation of perfluoro side groups increases the solubility of the polymers in CO_2. An example is statistical copolymers of styrene (29 mol%) with a fluoroacrylate (FOA). Concentrations of 1–5 wt% increased the viscosity of pure CO_2 by a factor of 5–400 [36].

Static and dynamic light scattering experiments with solutions of poly(1,1,2,2-tetrahydroperfluorooctyl methacrylate) (PFOMA) in liquid CO_2 and in scCO_2 revealed that the solvent quality, as expressed by the second virial coefficient, was enhanced with increase in temperature. The virial coefficient and the hydrodynamic coil expansion factor depended on a single interaction parameter, which can be influenced independently by solvent density or temperature [37]. Amphiphilic ureas with fluorinated side-chains were found to be soluble in scCO_2, but can still associate to form gels. Upon removing the CO_2, the materials form free-standing foams [38]. Note that partial fluorination of the polymer increases the solubility but it does not ensure dissolution under moderate conditions [27, 39–41].

From the fact that macromolecules are hardly soluble in CO_2, it should not be inferred that CO_2 does not affect polymers. CO_2 rapidly penetrates the amorphous regions of a polymer, causing plastification, considerable swelling, and morphological changes. Typical equilibrium solubilities of CO_2 in amorphous polymers lie between ~10% (PMMA at 35 °C, 41 bar) and 21% (PDMS at 50 °C, 91 bar) [52], whereas the swelling of semicrystalline polymers is much lower (PVDF, ~1.5% at 80 °C, 40 bar; HDPE, ~1% at 50 °C, 40 bar) [53]. This effect has been exploited for polymer processing, in particular for polymer extraction, crystallization, dying, and chemical modifications [10–12].

8.1.4
Stabilizer Design for Dispersion Polymerizations

Heterogeneous polymerizations [54] appear in cases where in the course of reaction at least one component becomes insoluble [55]. Since most macromolecules are insoluble in CO_2, the polymers will precipitate from solution when formed by polymerization reactions. If no measures are taken to prevent the agglomeration and subsequent precipitation of the polymer, the reaction technique is called "precipitation polymerization" [56]. Note that there is a wide variety of other heterogeneous polymerization techniques such as emulsion polymerization [57], mini-emulsion polymerization [58], microemulsion polymerization [59], dispersion polymerization [55, 60, 61], and suspension (pearl) polymerization [62, 63] that are distinguished by the respective initial state, the reaction mechanism/course of the reaction, and the resulting product properties such as molecular weight distribution and particle shape, size, and size distribution [60, 64]. In connection with free radical polymerization in scCO_2, in addition to homogeneous polymerization reactions, the precipitation and dispersion polymerization reactions represent the most widely studied processes [78].

Electrically uncharged polymer particles will always mutually interact via van der Waals forces. It has been proven for the "symmetrical" case, that is, two particles consisting of the same material separated by a solvent, that these interactions must always be attractive [25]. The attractive potential V_{att} between two particles is proportional to the radius R of the particles and their mutual distance d_{PP}:

$$V_{att} = -A_H \frac{R}{12 d_{pp}} \tag{8.5}$$

The proportionality constant A_H, called the Hamaker constant, of the particle–solvent system depends on the particles material and on the solvent. In non-polar solvents, A_H typically exhibits values between 3.5×10^{-20} and 8×10^{-20} J [65]. From Equation 8.5, it directly follows that particle dispersions are unstable against agglomeration and subsequent precipitation; furthermore, one has to expect more rapid precipitation the larger the particles are and the larger the particles concentration C_P becomes (because $d_{PP} \sim 1/C_P{}^{1/3}$).

Particle agglomeration can only be avoided if a repulsive potential between the particles exists that keeps the particles at bay. Conventional surfactants designed for aqueous systems are insoluble in CO_2 [66], hence new macromolecular stabilizers had to be developed prior to performing successful dispersion polymerizations. Electrostatic repulsion cannot effectively be used in CO_2, because the low polarity of the solvent impedes ion pair dissociation, and hence prevents the formation of sufficiently electrically charged particle surfaces. The alternative stabilization mechanism is the "steric stabilization" caused by the presence of polymer brushes at the particles surface. As indicated by the Flory–Kriegbaum theory, dissolved polymer chains posses an "excluded volume" that cannot simply be entered by other polymer chains [67]. Fixing polymer chains that must be well soluble in the solvent to the particle surface with one end and allowing the second end to move freely in the presence of the other surface-fixed chains creates a solvent-swollen surface layer. This layer contains an "excluded volume" that is forbidden to the solubilized surface chains of other particles. The detailed theory of steric stabilization is a complex matter and beyond the scope of this chapter, in particular since the most complete descriptions cannot be expressed by closed analytical equations. However, elementary principles of stabilizer design can still be learned from the early analytical approaches. In 1979, expressions for the steric repulsion potential V_{rep} were derived, that must outweigh the attractive potential V_{att} (see Equation 8.6) [68].

$$V_{rep} = -2\pi k_B T R \frac{\nu_S^2}{V_1}\left(\frac{1}{2} - \chi\right)\omega^2 S \tag{8.6}$$

where V_{rep} = repulsive potential due to steric stabilization, k_B = Boltzmann constant, T = absolute temperature, R = particle radius, ν_S = volume of a monomer unit of the stabilizer chain, V_1 = molar volume of the solvent, χ = Flory–Huggins interaction parameter between the solvent and the stabilizer polymer, ω = number of stabilizing polymer segments per particle surface area ($\omega = \nu \cdot X_s$, ν = number of stabilizer chains per particle surface area, X_S = stabilizer degree of polymerization), and S = distance dependence of V_{rep}, given by

$$S = 2\left(1 - \frac{d}{2L}\right)^2 \tag{8.7}$$

Equation 8.6 directly indicates that the stabilization will be more effective the more stabilizer chains are present (large ω), and the better the solubility of the stabilizer chains becomes (large χ). It is also of advantage if the volume of the monomer units ν_S of the stabilizer chain exceeds that of the solvent (large ν_S/V_1 in Equation 8.6).

The calculation of the distance dependence S of V_{rep} is complicated as it requires the calculation of the spatial distribution of the stabilizer chain segment distribution. If one assumes low polymer particle concentrations so that the particle–particle collisions are governed by the thermal Brownian movement and not by strong compression forces, the elastic contributions of the polymer brushes can be neglected. In the case of short stabilizing chains, a fairly constant density distribution is obtained and S obeys the simple Equation 8.7, whereas long linear polymer chains cause an approximate distance dependence as expressed by Equation 8.8 [69]:

$$S \approx \frac{9}{\pi F^4} \ln\left(\frac{L}{d}\right)\left\{-\left(1+\frac{F^2}{2}\right)\exp\left(-\frac{3}{F^2}\right)+\frac{F^3}{4}\left(\frac{\pi}{3}\right)^{\frac{1}{2}}\operatorname{erf}\left(\frac{3^{\frac{1}{2}}}{F}\right)+\frac{6}{F^2}\exp\left(-\frac{3}{2F^2}\right)\right.$$
$$\left.\times\left[\frac{F^2}{24}-\frac{F^4}{12}+\left(\frac{1}{16}-\frac{F^2}{12}+\frac{F^4}{12}\right)\left(\frac{\pi}{3}\right)^{\frac{1}{2}} F\exp\left(-\frac{3}{4F^2}\right)\operatorname{erf}\left(\frac{3^{\frac{1}{2}}}{2F}\right)\right]\right\} - B\frac{(L-d)}{f\langle r^2\rangle^{\frac{1}{2}}} \tag{8.8}$$

with

$$F = \frac{\langle r^2\rangle^{\frac{1}{2}}}{d}$$

and $B = 0.7329$ (assuming Gaussian coils), where L = contour length of the stabilizing chain = $X_S b$, b = length of a monomer unit along the chain, d = minimum distance between the two particle surfaces ($d \approx d_{PP} - 2\cdot R$), $L < d < 2L$, f = chain expansion factor due to solubilization, and $\langle r^2\rangle^{\frac{1}{2}}$ = root mean square end-to-end distance of the stabilizer chains.

From Equations 8.7 and 8.8, it follows that the stabilizer becomes more effective with increased degree of polymerization X_S, because the contour length L is direct proportional to X_S. In both equations, larger contour lengths cause higher values of S, and hence increase the repulsive potential ($V_{rep} \sim S$, Equation 8.5). Furthermore, the values of the S-function for long chains [Equation 8.8 exceeds the short-chain function (Equation 8.7]. Note that different equations will be required for non-linear architectures of stabilizers such as branched, hyperbranched, dendritic, or comb-like

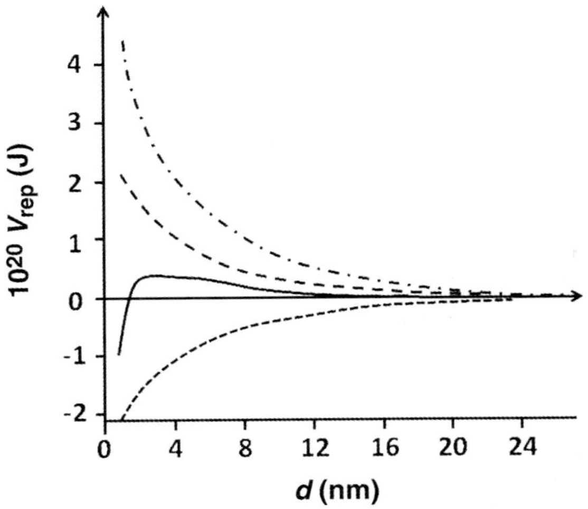

Figure 8.4 Potential between two polyacrylonitrile spheres stabilized with $5.2 \times 10^{-8}\,g\,cm^{-2}$ polystyrene in toluene at T = 155 K (—), 165 K (- - -), and 170 K (- · · · -), compared with the attractive van der Waals potential V_{att}. $R = 100\,nm$, $A = 3 \times 10^{-20}\,J$, $f = 2$, $\chi = 1 - \Theta/T$, $\Theta = 160\,K$, $v_S = 0.91\,cm^3\,g^{-1}$, $V_1 = 107\,cm^3\,mol^{-1}$. Redrawn with data from [69].

macromolecules. Figure 8.4 depicts the resulting potential energy–distance curves with the example of polystyrene ($X_S = 57$)-stabilized polyacrylonitrile particles (R 100 nm) dispersed in toluene [69].

The validity of the theory of steric stabilization in $scCO_2$ has been tested experimentally with dispersions of poly(2-ethylhexyl acrylate) particles in the presence of poly(1*H*,1*H*-perfluooroctyl acrylate) (PFOA), polystyrene-*b*-PFOA, and PFOA-*b*-poly (vinyl acetate) stabilizers [70, 71]. The elementary features of the theory were confirmed, and supported with additional CO_2-specific calculations [72].

The dispersion polymerizations in $scCO_2$ are characterized by the continuous formation of new macromolecules in the homogeneous solution phase subsequent to the formation of stabilized particles. Since these freshly formed polymers adsorb on the already formed polymer particles, the stabilizer polymer must not be permanently fixed to the particles surface, or it will be buried under the continuously arriving macromolecules. Hence the second task in designing a stabilizer is to adjust its anchoring strength to the particle surface. One can guess that the average residence time of a stabilizer on the particle surface should be correlated with the arrival rate of polymer chains from the solution phase. Obviously the right measure must be found between too strong, and too little adherence of the stabilizer to the particle [73]. This point is addressed by the empirically introduced term "anchoring solubility balance" (ASB) [74, 75]. Since all these values are difficult to predict from the present state of knowledge, the stabilizer design remains an empirical task based on educated guesses that can be guided only qualitatively by theory.

8.1.5 Limitations of Polymer Preparation in $scCO_2$

Typical laboratory reaction autoclaves can withstand pressures up to ~300–500 bar, while industrial plants are limited to ~350 bar. From solubility data, it can therefore be fairly concluded that the bulk of polymerizations cannot be performed under homogeneous conditions, since the polymers will not remain soluble under typical reaction conditions ($T < 100\,^\circ C$, $P < 350$ bar) [8]. Since, on the other hand, low molecular weight monomers are readily soluble in CO_2 [76], the reactions must be performed in the form of precipitation, dispersion, suspension, or emulsion polymerizations. The extremely low viscosity of supercritical solutions is favorable for efficient chemical conversion; on the other hand, it favors aggregation, agglomeration, and precipitation of undissolved polymers. Obviously, the selection of a stabilizer is of paramount importance for the success of a heterogeneous polymerization in $scCO_2$. Because of the low dielectric constant of $scCO_2$, electrostatic stabilization as common with aqueous systems cannot be applied, and must be replaced by pure steric stabilization. Since steric stabilization requires macromolecular stabilizers that must be at least partially soluble in the reaction medium, the design and synthesis of suitable stabilizers for CO_2 opened up a new research field [77]. Influencing the viscosity by addition of special viscosity modifiers will be important for solubilizing colloids, but can also help to increase solution densities and thus solubilization power [36, 38]. Although CO_2 is a very stable molecule, not all types of polymer-forming reactions can be carried out in this medium: True carbanionic polymerizations have not yet been reported, because the required strong nucleophiles may add to the carbon dioxide with carboxylate formation [78]. Also, metal-catalyzed polymerization of alkenes using Ziegler–Natta, metallocene, or Philips catalysts have not been reported, since the solvent does not tolerate the highly active catalyst systems. This does not imply, however, that polyolefins cannot be made in $scCO_2$, but one must rely on different catalyst systems [79–81].

At first sight, it seems that polycondensation reactions such as polyester, polyamide, or polycarbonate formation that require the quantitative removal of low molecular weight side products, such as water or CO_2, cannot be performed in a closed-vessel system if no measures are taken to either absorb or to discharge the side products. However, CO_2 swells the formed polymer strongly and extracts the low molecular weight side products very efficiently from the locus of reaction. Hence polycondensation reactions in $scCO_2$ have been shown to be possible [8]. Commercial CO_2 frequently contains ill-defined amounts of helium, carbon monoxide, oxygen, and water. Although the last three can easily be eliminated by adsorbents, helium may remain. However, small percentages of helium reduce the solubility of polymers in the SCF, as proved by supercritical chromatography [82–84] and $scCO_2$ extraction experiments [85]. The importance of the CO_2 purity was demonstrated with the example of dispersion polymerization of methyl methacrylate (MMA) in $scCO_2$ (cf. Section 8.2.1.3). Upon increasing the helium content of the solvent from 0 to 2.4 mol %, the particles diameters obtained increased from 1.64 to 2.66 μm, while the particle size distribution became narrowed [86].

8.2 Polymerization in $scCO_2$

Polymers can be prepared from monomers by chain-growth or step-growth reactions. The first group of mechanisms involves free and controlled radical polymerization, cationic polymerization, anionic polymerization, and metal-catalyzed polymerization reactions. The second group includes polycondensation and polyaddition reactions [87]. The efforts to prepare macromolecules in $scCO_2$ have been reviewed earlier [8, 88–94]. The subsequent text will organize the referred material according to reaction mechanisms and polymer classes.

8.2.1 Radical Polymerization in $scCO_2$

In a chain-growth polymerization reaction, an active species is formed that subsequently grows by sequential addition of monomer molecules. With free radical polymerization (FRP), the active species are electrically neutral radicals that are continuously formed by decomposition of an radical initiator, and are continuously consumed by radical–radical termination reactions (Scheme 8.1) [95].

Initiator decomposition: $$\mathrm{I} \xrightarrow{k_d} 2\,\mathrm{R}^\bullet$$

Initiation: $$\mathrm{R}^\bullet + \mathrm{M} \xrightarrow{k_{ini}} \mathrm{RM}^\bullet$$

Chain growth: $$\mathrm{RM}_i^\bullet + \mathrm{M} \xrightarrow{k_P} \mathrm{RM}_{i+1}^\bullet$$

Transfer: $$\mathrm{RM}_i^\bullet + \mathrm{T} \xrightarrow{k_{tr}} \mathrm{P}_i + \mathrm{T}^\bullet$$

Re-initiation: $$\mathrm{T}^\bullet + \mathrm{M} \xrightarrow{k_{ri}} \mathrm{TM}^\bullet$$

Termination (combination): $$\mathrm{RM}_i^\bullet + \mathrm{RM}_j^\bullet \xrightarrow{k_{tc}} \mathrm{P}_{i+j}$$

(disproportionation): $$\mathrm{RM}_i^\bullet + \mathrm{RM}_j^\bullet \xrightarrow{k_{td}} \mathrm{P}_i + \mathrm{P}_j$$

Scheme 8.1 Reaction steps in a free radical polymerization mechanism. I = initiator, M = monomer, T = transfer agent, $RM^\bullet$ = growing radical, P = unreactive polymer chain.

Under homogeneous conditions, where all constituents such as initiator, monomer, radicals, and polymer are dissolved during the whole reaction time, the reaction can be performed in bulk or in solution. The reaction kinetics of homogeneous solution FRP have been investigated in $scCO_2$, starting with the decomposition kinetics of azobisisobutyronitrile (AIBN). The decomposition rate of the initiator (k_d, Scheme 8.1) was reduced by a factor of 2.5 compared with benzene as solvent, but the initiator efficiency, describing how many growing chain radicals are formed per decomposed initiator molecule, increased by 50% due to the reduced cage effect [96]. A continuous stirred tank reactor was used to determine simultaneously the rate constant of decomposition and the initiator efficiency of diethyl peroxydicarbonate (DEPC). Between 65 and 85 °C, first-order kinetics were confirmed, characterized by an energy of activation $E_{A,d}^{(DEPC)} = 132\,kJ\,mol^{-1}$. The rate constant obeyed the Arrhenius equation $k_D^{(DEPC)} = 6.3 \times 10^{16}\,s^{-1} \times \exp(E_A/RT)$ and the initiator efficiency was $f = 0.6$ [97a].

The light-induced decay and subsequent addition to methyl methacrylate of the photoinitiator trimethylbenzoyldiphenylphosphine oxide (TMDPO) was measured with time-resolved electron spin resonance (ESR) spectroscopy in liquid CO_2 and $scCO_2$ at an excitation wavelength of 308 nm. Measurement of the TREPR line resulted in the rate of initiation $k_{ini} = (5.5 \pm 0.5) \times 10^7\,M^{-1}\,s^{-1}$ in liquid CO_2 and $(6.1 \pm 0.6) \times 10^7\,M^{-1}\,s^{-1}$ in $scCO_2$. These values were close to those obtained in acetonitrile [$(8.1 \pm 1.6) \times 10^7\,M^{-1}\,s^{-1}$] and toluene [$(4.0 \pm 0.4) \times 10^7\,M^{-1}\,s^{-1}$] as the solvent [96]. Since solubility turned out to be an important factor for $scCO_2$-borne reactions, CO_2-soluble azo initiators have been designed and tested. At 215 bar, bis (perfluoroalkylethyl) 4,4′-azobis-4-cyanovalerate decomposes with first-order kinetics and an activation energy of decomposition of $76\,kJ\,mol^{-1}$ between 60 and 80 °C. The decomposition constant at 70 °C was determined as $1.56 \times 10^{-6}\,s^{-1}$ [97b]. The thermal decomposition of diethyl peroxydicarbonate (DEPDC) was investigated by *in situ* attenuated total reflection Fourier transform infrared (ATR-FTIR) spectroscopy in heptane and in $scCO_2$. In both solvents the decomposition followed first-order kinetics with an activation energy of $115\,kJ\,mol^{-1}$ ($T = 40$–74 °C) in heptane and $118\,kJ\,mol^{-1}$ in $scCO_2$ from 40 to 60 °C. When the CO_2 was vented, the unstable intermediate decomposition product monoethyl carbonate was decarboxylated and disintegrated mainly into ethanol [98]. Investigations on the rate of polymerization with the examples of styrene [99], FOA [100], MMA, and butyl acrylate [101] revealed that the rate constant of propagation k_p (Scheme 8.1) remains at the order of magnitude for bulk polymerizations, indicating that CO_2 does not interfere with the polymerization reaction. However, the termination constants k_t (Scheme 8.1) increase in cases where CO_2 is a poor solvent for the generated polymer [102]. Table 8.2 summarizes the ratio of propagation constants measured in $scCO_2$ with that of the bulk polymerization for 11 monomers.

It is a particular advantage of CO_2 that it is inert with respect to radical transfer reactions, that is, k_{tr} (T = solvent) in Scheme 8.1 can be regarded as zero. Hence the free radical polymerization kinetics in the homogeneous solution phase of $scCO_2$ are very similar to those experienced in non-polar liquid solvents [103]. The term "controlled" radical polymerization (CRP) covers radical polymerizations with a strongly suppressed termination step, caused by the presence of equilibrium reactions between "dormant" and "active" species (cf. Scheme 8.2) [104]. Since the equilibrium reactions are designed to lie preferably on the side of the dormant species, the concentration of active radicals is much lower than in conventional free

Table 8.2 Ratio of propagation constants in $scCO_2$ and in bulk ($scCO_2$: 60 vol.% monomer, 40 vol.% CO_2).

Monomer	T (°C)	P_{CO_2} (bar)	$k_P^{CO_2}/k_P^{bulk}$	Ref.
Methyl acrylate	22	1.000	0.60 ± 0.1	[105]
Butyl acrylate	11	200	0.65 ± 0.05	[101]
Dodecyl acrylate	20	200	0.75 ± 0.05	[105]
Methyl methacrylate	30	1.000	0.65 ± 0.05	[101]
Butyl methacrylate	30	200	0.78 ± 0.05	[101]
Glycidyl methacrylate	60	300	0.87 ± 0.05	[105]
Cyclohexyl methacrylate	40	1.000	0.75 ± 0.05	[105]
Isobornyl methacrylate	40	300	0.60 ± 0.05	[105]
3-Hydroxypropyl methacrylate	40	300	1.00 ± 0.05	[106]
Styrene	80	1.000	1.00 ± 0.05	[107]
Vinyl acetate	25	1.000	1.00 ± 0.05	[108]

radical polymerization reactions (FRP, $[R^\bullet] = 10^{-7}$–$10^{-8}\,mol\,l^{-1}$; CRP, $[R^\bullet] < 10^{-9}$–$10^{-12}\,mol\,l^{-1}$) [109]. Since the rate of termination is proportional to the square of the total radical concentration, whereas the propagation rate is proportional to the concentration itself, the termination becomes negligible against polymer chain growth in a certain period of time and monomer conversion.

Dormant / Active: $$RM_jX \;+\; \underset{k_{ass}}{\overset{k_{diss}}{\rightleftharpoons}} \; RM_j^\bullet \; (X^\bullet)$$

Exchange equilibrium: $$RM_i^\bullet + RM_jX \; \overset{k_{exc}}{\rightleftharpoons} \; RM_iX + RM_j^\bullet$$

Scheme 8.2 Equilibrium reactions between dormant and polymerization active species in controlled radical polymerizations.

CRP allows polymers to be prepared with a narrow molecular weight distribution, well-defined end groups, and a certain control of the monomer distribution along the polymer chain in the case that more than one type of monomer is present in the reaction mixture. Hence block, star, graft, and gradient copolymers can be prepared by CRP techniques. However, the advantages are offset by low rates of polymerization and several system-specific disadvantages [110]. For details on the chemistry of the different CRP techniques, such as nitroxide-mediated polymerization (NMP) [111, 112], atom transfer polymerization (ATRP) [113, 114], radical addition fragmentation and transfer (RAFT) [115, 116], and immortal polymerization [117], the interested reader is referred to the cited review articles. The application of CRP techniques will be mentioned with the respective monomers.

8.2.1.1 Side-chain Fluoropolymers

Perfluorinated compounds represent a unique class of matter as they are immiscible with water, do not mix with hydrocarbons [118], form low-energy surfaces [119], and have extreme chemical resistance. These inherent advantages turn into problems upon production, since either environmentally hazardous chlorofluorocarbon (CFC) solvents have to be used in solution polymerizations, or emulsion polymerizations are carried out in water and require perfluorinated surfactants such as perfluorooctylsulfonates (PFOS) or perfluoroalkylcarboxylates [120]. Further problems arise from side reactions of growing fluorinated radicals with solvents that reduce the product quality [121, 122]. Carbon dioxide offers advantages, since it is a non-toxic and non-reactive solvent that can be removed from the final product without an energetic cost [8]. Macromolecules bearing perfluoroalkyl side-chains are of technical importance for the generation of non-sticky surfaces, soil- and water-repelling coatings [122], and textile finishing agents for anti-soil and "easy to clean" applications [123]. Since the polymers become CO_2 soluble below 300 bar upon exceeding a minimum fluoro-side-chain molar fraction (cf. Table 8.1), the substances can be produced under homogeneous solution conditions by FRP in $scCO_2$ [124]. 1,1-Dihydroperfluorooctyl acrylate (FOA) has been polymerized in the presence of AIBN as radical initiator in supercritical CO_2 solution at 60 °C and 207 bar to give high molecular weight PFOA products with high yields that were CO_2 soluble [125]. Styrene monomers with perfluoroalkyl side-chains have also been polymerized in $scCO_2$. The products obtained from these polymerizations were shown to be identical with the products of solution polymerization in 1,2,2-trifluoroethane (Freon-113) [8, 125]. Monomers with fluorinated side groups have been copolymerized with non-fluorinated monomers such as methyl methacrylate (MMA), butyl acrylate (BuA), ethylene, and styrene in $scCO_2$ [88, 96, 126, 127]. FOA was also copolymerized with 2-(dimethylamino)ethyl acrylate (DMAEA) and 4-vinylpyridine to produce CO_2-soluble polymeric amines [128, 129], 3-*O*-methacryloyl-D-glucopyranose [130, 131], and poly(ethylene oxide) monomethacrylate to yield CO_2-philic/hydrophilic macromolecules [132]. Perfluorooctylethyl methacrylate (FOMA) was copolymerized in $scCO_2$ with 2-hydroxyethyl methacrylate, oligoethylene oxide monomethacrylate and dimethylaminoethyl methacrylate (DMAEMA) to obtain CO_2-soluble amphipathic copolymers [133]. Also, the homo- and copolymerization behavior of bis(trifluoromethyl)phenyl acrylate/hexafluoroisopropylidene was investigated [134]. To satisfy the need for CO_2-soluble macromolecular stabilizers, a number of amphipathic polymers were synthesized in conventional solvents and subsequently investigated for CO_2 solubility. Among them are PS-*b*-P[perfluorooctylethyleneoxymethylstyrene), PS-*b*-P[1,1,2,2-tetrahydroperfluorodecyl acrylate] [135], PMMA-*b*-P[FnHmMA] [136], and P(FOA-*co*-PEG-MA) [137]. Controlled radical polymerization reactions have also been performed in $scCO_2$. Perfluoroalkane acrylates and methacrylates such as FOA an FOMA have been polymerized under homogeneous solution conditions by means of ATRP, using a bipyridine ligand that was made CO_2 soluble due to perfluoroalkyl substituents. PFOMA of $M_n = 16\ 800\,g\,mol^{-1}$ and a low polydispersity of M_w/M_n 1.01 was obtained in 83% yield. Note that with CO_2-insoluble ligands, lower yields (<65%) and much broader molecular weight distributions ($M_w/M_n = 5.9$) have been found. The controlled nature of the reaction was demonstrated by well-defined amphiphilic PFOMA-*b*-PMMA diblock copolymer [138]. The particular relevance of these CO_2-soluble homo- and copolymers seems to lie in their potential application as emulsifiers, or stabilizers for emulsion or dispersion polymerization in $scCO_2$ (see below). So far there is no published literature dealing with typical technical applications (e.g. surface treatment, mold release reagents, soil release) of fluorinated side-chain copolymers produced in $scCO_2$.

8.2.1.2 **Fluoroolefin Polymers**

Commercialized perfluorinated bulking materials such as polytetrafluoroethylene (PTFE, Teflon), polychlorotrifluoroethylene (PCTFE), poly(vinylidene difluoride) (PVDF), and poly(vinyl fluoride) (PVF) are prepared in large quantities by aqueous emulsion polymerizations of the respective fluoroolefins [120, 122] (Scheme 8.3). The main disadvantages of this process are (i) the necessary use of environmentally hazardous perfluorinated surfactants [139], (ii) the generation of large waste water streams, (iii) the energy requirement for product drying ($>2.3 \times 10^9$ kJ m^{-3} of evaporated water), and (iv) side reactions due to the high electrophilicity of the growing perfluorinated chain radicals, causing the formation of, for example, acid fluoride or carbonic acid end groups [121].

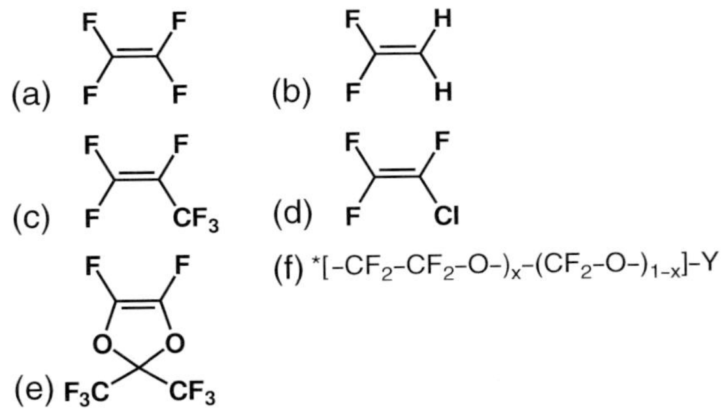

Scheme 8.3 Structural formulas of fluoroolefins (a–e) and perfluoropolyether (PFPE) (f). (a) Tetrafluoroethylene (TFE); (b) 1,1-difluoroethylene = vinylidene difluoride (VDF); (c) hexafluoropropene (HFP); (d) chlorotrifluoroethylene (CTFE); (e) 2,2-bis(trifluoromethyl)-4,5-difluoro-1,3-dioxole.

Polytetrafluoroethylene Polytetrafluoroethylene (PTFE), the perfluorinated analog of polyethylene, as characterized by a T_G of −73 °C [140] and an irreversible melting temperature above 335 °C [141], can be used between −269 and 260 °C. The polymer is tough (187 J m^{-1} at 23 °C) and exhibits a tensile modulus of 340 MPa and an elongation at break of 200–400% at 23 °C. PTFE shows a low friction coefficient of 0.224 W (ASTM D1894 method) and a surface energy of 18.6 mN m^{-1} (advancing contact angle to water: 116°). PTFE is inert to almost any common chemicals, but can be attacked by molten alkali metals and elemental fluorine. Oxygen, alkaline earth metal oxides, and finely dispersed base metals (such as aluminum and magnesium) can react at elevated temperatures. No solvent for this polymer exists [142, 143]. It is produced in the form of a fine powder (average particle diameter 200 nm) that can be sintered, or used as an additive in coating formulations. Note that melt processing is not yet possible because of its high melt viscosity and decomposition at the melting temperature [144]. Because of this high chemical resistance, it is used for inert coatings, chemical processing equipment, and insulation purposes [122, 144, 145].

The monomer, TFE, tends to undergo autopolymerization, which can cause explosions. Upon dilution with CO_2 the mixture becomes stable, but can still be polymerized [146]. The polymerization of TFE in $scCO_2$ has been intensively studied and patented [145]. Due to its high crystallinity and melting temperature, PTFE is essentially insoluble in $scCO_2$. The polymerization proceeds in the form of a dispersion polymerization, where the polymer becomes insoluble. Hence, once formed, the polymer chains will segregate from the homogeneous solution phase and it is of importance if this segregation follows a demixing/crystallization route or if the polymer crystallizes directly. Figure 8.5a depicts a schematic phase diagram containing a region where the binodal (the line of equilibrium between the homogeneous solution and the liquid–liquid two-phase region) intersects the crystallization curve of the polymer. Because of the solubility of the monomer, the reaction will start under homogeneous conditions. Depending on the system composition, the growing polymer chain may enter the two-phase region and liquid–liquid demixing into a polymer-rich gel phase and a polymer-depleted sol phase will occur. In the gel phase, the polymer may either vitrify to form an amorphous precipitate or it can partially crystallize. Note that the composition of sol and gel phases is determined by the shape of the binodal, hence the concentration of the polymer may play a minor role. It is most likely that the binodal is not intersected by the crystallization line (Figure 8.5b). In this case, the polymer will precipitate in the form of semicrystalline particles (path A in Figure 8.5b). By virtue of a rapid rate of polymerization or by fast changes of the reaction conditions, it is possible to overrun crystallization and liquid–liquid demixing may occur in the supersaturated regime of the phase diagram.

As the polymerization proceeds with different reaction kinetics (and mechanisms) in homogeneous solution, in a precipitated amorphous phase or at the surface of crystalline material the evolution of the molecular weight distribution will be complex and the macroscopically observed rate of polymerization must be considered as the sum of the individual reaction rates. The theoretical description of such a precipitation polymerization is a very demanding task, because for each phase (homogenous solution phase, solvent-swollen precipitated amorphous polymer-, and precipitated crystalline polymer surface phase) rate and balance equations must be set up for every species present [147]. This requires at least the knowledge of the distribution of monomer, initiator, radicals, and polymer products over the phases. Moreover, the mutual exchange of species between the phases must be considered, requiring data on the diffusion constants and the size of the contact surfaces between the phases [148]. The growing understanding of these complicated processes was demonstrated by successful modeling calculations that well described the surfactant-free precipitation copolymerization of TFE/vinyl acetate (VAc) in CO_2 [149]. TFE polymerization was found to proceed well at 35 °C with bis(perfluoro-2-*n*-propoxypropionyl) peroxide as the initiator to yield 100% polymer of molecular weight around 150 000–200 000 g mol^{-1} [145, 150]. In particular, the formation of stable end groups should be mentioned. The process was utilized by

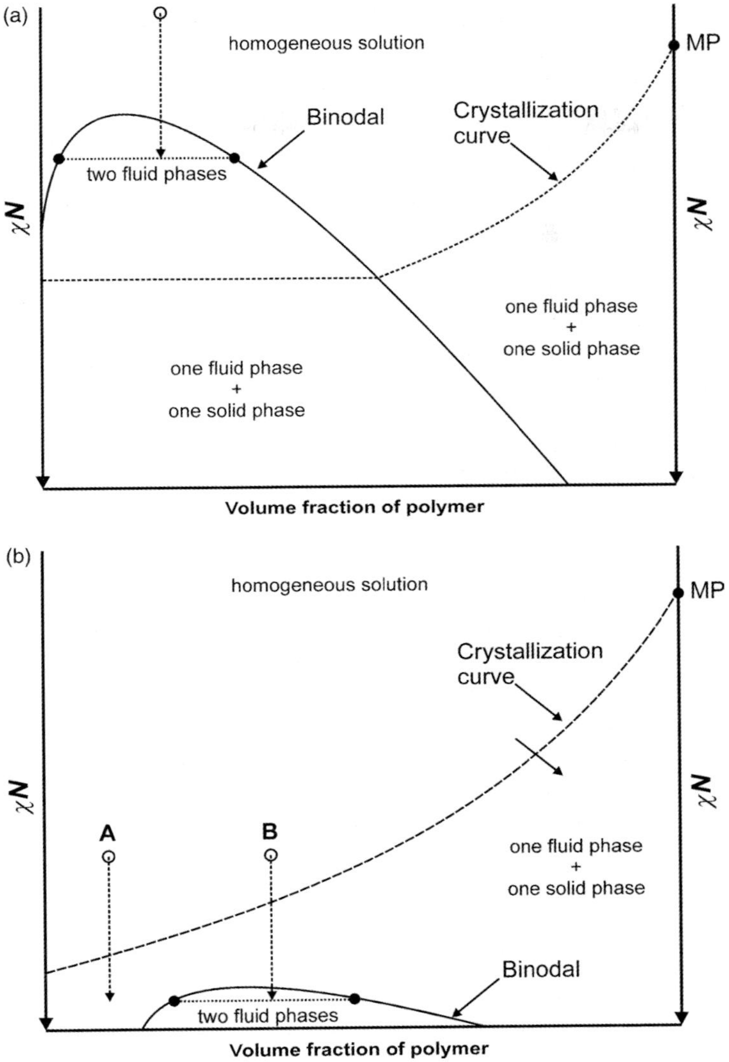

Figure 8.5 Schematic diagram to illustrate the effect of the relative position of the crystallization curve and the liquid–liquid demixing line upon precipitation polymerization. (a) Liquid–liquid demixing of polymer prior to crystallization; (b) crystallization prior to liquid-liquid demixing. MP = melting point of the pure polymer, N = degree of polymerization, χ = polymer–solvent interaction parameter.

DuPont to produce tetrafluoroethylene–hexafluoropropene copolymers (Teflon FEP) by means of an $scCO_2$ process [151, 152]. Recent efforts to polymerize TFE in $scCO_2$ on a pilot-plant scale have been published. The polymerization was initiated by *tert*-amyl perpivalate at 50 °C and 120–130 bar in the presence of perfluoropolyether carbonic acid surfactants. The polymer was reported to be of high crystallinity (80–86%) that was almost unaffected by the surfactant [153].

Poly(Vinylidene Fluoride) Poly(vinylidene fluoride) (PVDF) is a semicrystalline thermoplastic polymer [154], with a degree of crystallinity of 4068% and a density of 1.75–1.78 $g\,cm^{-3}$ [155]. PVDF crystallizes in four different modifications (α–δ) [156]; the most common α form (orthorhombic, *Cm2m*) [156] melts at 183 °C. Technical PVDF exhibits melting ranges between 154 and 180 °C; its maximum continuous use temperature is reported to be as high as 150 °C [155]. With a T_G of −40 °C, PVDF remains tough even at temperatures below 0 °C. In its β-modification (obtained by uniaxial stretching of semicrystalline PVDF [157]), the polymer becomes piezoelectric [158] with piezoelectric coefficients up to 28 $pC\,N^{-1}$ [159], exceeding the values for other polymers by one order of magnitude. Furthermore, the polymer is pyroelectric with a pyroelectric coefficient of 25 $\mu C\,m^{-2}\,K^{-1}$ [159, 160]. Its mechanical strength is superior to those of PTFE and PCTFE. Oriented PVDF films and fibers exhibit Young moduli of 42–58 MPa at 25 °C and 34.5 MPa at 100 °C, limited creep under load (0.02 $ml\,cm^{-1}$ at 13.8 MPa and 25 °C for 104 h), exceptional ability to withstand repeated flexure or fatigue, and good resistance to abrasion (17.6 mg per 1000 cycles, Tabor CS-17, 0.5 kg load) [154, 155]. PVDF is widely used for fabricating pipes and tubing and also in valves and pumps [161], for example in the food and biomedical industries [162, 163], aerospace technology [164, 165], and the nuclear [166] and semiconductor industries [167, 168]. Special coatings for protection against aggressive and corrosive media are based on PVDF [165, 169]. The polymer is also used for column packing, for ultrafiltration membranes [170–174], and porous separator foils are employed for high-energy batteries and lithium ion accumulators [175]. Because of its ability to generate voltage when pressurized or heated, β-PVDF is used in piezoelectric applications [176, 177] and sensors [159, 178, 179]. Commercially, the polymer is produced either in emulsion or suspension polymerization reactions at monomer pressures of 10–200 bar and temperatures from 10 to 130 °C [154]. For emulsion polymerization, fluorinated surfactants have been employed in combination with organic percarbonates and peroxides [180, 181]. The process suffers from the same drawbacks as described for PTFE. Highly regular PVDF with 94–97% alteration of $-CH_2-$ and $-CF_2-$ groups crystallizes well [182] and is soluble only in high-boiling polar solvents such as DMF, DMAC, DMSO, and HMPTA [154]. Solution polymerizations of vinylidene fluoride (VDF) have been reported, initiated by radical initiators or ionizing irradiation such as γ-rays [183, 184], but are not feasible with respect to economy and ecology. Certainly, the manufacture of PVDF might benefit from eliminating the need for additives and minimizing side reactions, reducing waste streams, and minimizing the need for polymer drying. Vinylidene difluoride has been polymerized in the homogeneous phase in $scCO_2$ at 120 °C and 1500 bar in the presence of perfluorohexyl iodide as a transfer agent to keep the molecular weight below the precipitation limit. The molecular weight increased linearly with the reaction time during the initial stage of the polymerization, while the polydispersity M_w/M_n remained below 1.2–1.5, indicating a controlled polymerization reaction [185]. Since pressures beyond

1000 bar exceed technical feasibility, most studies focused on less harsh conditions, where the PVDF formed cannot be dissolved by $scCO_2$. In agreement with the scheme of precipitation polymerizations outlined in the previous section, the homopolymerization of VDF in $scCO_2$ was shown to be of the precipitation polymerization type [186–190], characterized by multimodal molecular weight distributions, low powder density (and particulate morphology; cf. Figure 8.6) [188, 189, 191, 192] and the presence of a virtually erratic induction period at the beginning of the polymerizations [188, 192]. Recently, more kinetic data became available that supported a mechanism of simultaneous polymerization in the homogeneous solution and the heterogeneous precipitated polymer phase [148, 193–195].

Figure 8.6 Powder morphology of PVDF polymers produced in $scCO_2$ at (a) 50 °C (P = 280 bar, [VDF] = 3.125 mol l^{-1}, [DEPC] = 0.5 mol%, 2 h) and (b) 115 °C (P = 280 bar, [VDF] = 3.125 mol l^{-1}, [DEPC] = 0.5 mol%, 21.4 h) [189].

A surfactant-free polymerization of VDF in $scCO_2$ was initiated with γ-irradiation at 40 °C, below 250 bar. During the first, slow, phase of polymerization the rate of polymerization could be described in terms of a homogeneous solution polymerization model [196]. A partially successful dispersion polymerization of VDF in $scCO_2$ was reported on applying a poly(methyl vinyl ether-alt maleic anhydride) copolymer, grafted with fluorinated alcohols, as steric stabilizer at concentrations of 1–5 wt% relative to the monomer [197, 198]. However, only with very high concentrations of the monomer (10 mol l^{-1}) and very low conversions (5–14%) was it possible to stabilize the growing polymer particles. A dispersion polymerization of VDF in the presence of 5 wt% (with respect to monomer) of an ammonium carboxylate perfluoropolyether as stabilizer was described to yield spherical PVDF particles at 50 °C and 380 bar. The particle diameter increased with monomer conversion, and reached ~1000 nm at 78% conversion. Depending on the operating conditions, control was maintained up to conversions of 60–80% [199]. The authors claimed that the reaction data were in accordance with their model of the dispersion polymerization process [200], indicating that the polymerization kinetics became dominated by the interphase transport of the active radicals between the continuous phase and the polymer particles [201]. In spite of the tedious reaction kinetics of VDF polymerization in $scCO_2$, successful attempts to set up continuous polymerization reactor systems have been published [202–204] that may be the prerequisite for technical implementation. Morphology control of the precipitated PVDF is still an issue, since the precipitation polymerizations of PVDF from $scCO_2$ yields fluffy powders of low density and ill-controlled powder morphology [185–201]. It has been reported that strong stirring of the reactor during polymerization of VDF in $scCO_2$ at 40 °C and 350 bar, and also the addition of either polydimethylsiloxane monomethacrylate (PDMS-MA, $M_n \approx 1.000\,g\,mol^{-1}$) or FOMA can result in the formation of macroporous PVDF beads with diameters of several hundred micrometers. The procedure was effective at high monomer concentrations (6–8 mol l^{-1}) and well-adjusted, preferably low shear forces [205]. Focusing on PDMS-MA as a stabilizing additive (addition of up to 9 wt% PDMS-MA with respect to VDF) showed that PDMS-MA helps to control the morphology without affecting the microstructure and thermal properties of the resulting PVDF. Obviously PDMS-MA was incorporated in the PVDF to only a minor extent [206].

Perfluoroolefin Copolymers Perfluorinated homopolymers possess important materials properties, but they are expensive and not perfect: PTFE, for example, is difficult to process and exhibits only moderate mechanical strength due to its plasticity. Copolymerization allows the properties of polymers to be improved and tailored further for specific applications. Hence a series of perfluorinated copolymers have been introduced [119], in particular copolymers of TFE with hexafluoropropene (HFP), P[TFE-*co*-HFP] [207], ethylene, P[TFE-*co*-E] [208], 2,2-bis(trifluoromethyl)-4,5-difluoro-1,3dioxole (PDD), P[TFE-*co*-PDD) [209], and perfluorovinyl ether (PFE), P [TFE-*co*-PFE] [210]. Radical copolymerization of TFE or VDF with other fluoroalkenes [211–216] can yield thermoplastics [217, 218], elastomers [212–214, 219, 220], and thermoplastic elastomers [221]. Examples are alternating copolymers of VDF and hexafluoroisobutene (Allied Company, CMX) [222, 223] and VDF–methyl trifluoroacrylate [224]. Random copolymers of VDF with chlorotrifluoroethylene (CTFE), P(VDF-*co*-CTFE), have been commercialized as KelF. Copolymers of VDF with hexafluoropropene (HFP), P[VDF-*co*-HFP], are semicrystalline, flexible, and tough materials for HFP contents below ~20 mol% ("flexible PVDF"). At higher HFP contents, the compounds become melt processable elastomers [207, 225], insoluble in hydrocarbons, aromatic and chlorinated solvents, acids, and alkalis, but they can easily be dissolved in polar aprotic solvents such as DMSO or acetone [226]. The HFP content cannot exceed 50 mol%, since HFP cannot homopolymerize. The copolymers have attracted a great deal of attention as ferroelectric materials, since the polymer samples change their length in electric fields (="electrostriction"), which is an important feature for producing actuators. P[VDF-*co*-HFP] copolymers exhibit large electrostrictive responses exceeding 4% (0.84 MV m^{-1} for P[$VDF_{85mol\%}$-*co*-$HFP_{15mol\%}$]) [227–229]. P[$VDF_{81mol\%}$-*co*-$HFP_{19mol\%}$] copolymers with a molecular weight of 85 kg mol^{-1} are soluble in $scCO_2$ beyond 500 bar and temperatures between 0 and 230 °C. On increasing the molecular weight to 210 kg mol^{-1} at an HFP content of 22 mol%, the solubility region shifts to $P > 750$ bar and $T > 190$ °C [14, 230]. Precipitation-type copolymerizations of VDF and HFP in $scCO_2$ were investigated with diethyl peroxydicarbonate as initiator at 40–50 °C and 350 bar. The HFP content of the copolymers was varied between 1 and 37 mol% to yield copolymerization parameters $r_{VDF} = 5$, $r_{HFP} = 0$. The molecular weight distribution was multimodal in any case ($M_w/M_n = 2$–3), and the low molecular weight fraction – as supported by turbidimetric and fractionation experiments – was attributed to be formed in the homogeneous solution phase. The thermal and spectroscopic properties of the copolymers were close to those of commercial products [231, 232]. Continuous production in a stirred continuous tank reactor system was subsequently reported, using perfluorobutyl peroxide as the initiator, a CO_2 pressure between 207 and 400 bar and a temperature of 40 °C. The copolymerization ratios were similar to those previously reported ($r_{VDF} = 3.2$), but the polydispersity of the molecular weight distributions was strongly affected by the monomer concentration: At 1.8 mol l^{-1} monodisperse products were obtained, whereas the distributions showed an intensive high molecular weight shoulder at 5.9 mol l^{-1}. The observed first-order dependence of the rate and the degree of polymerization on the monomer concentration led to the conclusion that the continuous polymerization system was dominated by solution-phase polymerization [233].

P[TFE-*co*-PDD] (Teflon AF) copolymers are amorphous materials with well-tunable T_Gs, soluble in common fluorinated solvents. They combine optical transparency, low refractive index, high thermal and chemical stability, excellent electrical properties with low surface energies and high gas permeabilities [209]. P[TFE-*co*-PDD] copolymers have been prepared in $scCO_2$ at 35 °C and 138 bar, using bis(perfluoro-2-*n*-propoxypropionyl) peroxide as initiator. The PDD content of the copolymers was varied between 30 and 100 mol% to obtain copolymers with T_Gs ranging from 67 to 334 °C. On comparing commercial copolymers with the products obtained in $scCO_2$, no difference was found [234]. It has been reported that such CO_2-originated copolymers may have the potential to serve in key lithographic techniques, such as next-generation photolithography (193 nm, 157 nm, and immersion lithography) or soft lithography [235].

Binary copolymers of TFE and vinyl acetate (VAc) have been prepared in $scCO_2$ at 45 °C and pressures between 230 and 260 bar using diethyl peroxydicarbonate as radical initiator. The VAc content could be varied between 33 and 64 mol%, and the molecular weights (SEC values against PS) were 77 000–116 000 g mol^{-1} at polydispersities M_w/M_n around 1.8. The molecular weight increased with incorporation of VAc. In contrast to copolymers prepared by aqueous emulsion polymerization, the CO_2 based polymers were predominantly of linear structure. The effect was attributed to the inertness of CO_2 towards hydrogen abstraction reactions. The VAc units were partially hydrolyzed to yield TFE–VAc–vinyl alcohol (VAl) terpolymers P[TFE-*ter*-VAc-*ter*-VAl] [236, 237]. Terpolymers from TFE, vinyl acetate (VAc), and polydimethylsiloxane monomethacrylate ($PDMS_{1k}$-MA, 800–1000 g mol^{-1}), P[VDF-*ter*-VAc-*ter*-$PDMS_{1k}$-MA] have been prepared in $scCO_2$ at 65 °C and 350 bar. Since VDF does not readily copolymerize with methacrylates, the VAc was used as a "bridging" comonomer. The molecular weights of the terpolymers obtained were M_n = 29 000–173 000 g mol^{-1} at a polydispersity of M_w/M_n = 1.8–3.7, and the yields were up to 64%. The terpolymers containing 2–12 mol% $PDMS_{1k}$-MA and 57–33 mol% VAc were strongly hydrophobic (advancing contact angle against water 112°, receding contact angle 69°), and XPS measurements demonstrated the preferential segregation of the PDMS chains towards the air interface [238].

Fluoropolymers are rather expensive products that must justify their cost by superior performance. Because of their high material costs, slightly higher production costs as caused by a supercritical production technology may be tolerated by the market, provided that the materials' performance is increased with respect to products originating from standard technology. Political pressure, such as the ban of perfluorinated surfactants, in combination with demands for environmentally benign production schemes may help to shift the polymerization technology for fluoropolymers towards $scCO_2$-based processes.

8.2.1.3 Poly(Methyl Methacrylate)

Atactic poly(methyl methacrylate) (PMMA) as obtained by FRP is an amorphous thermoplastic with a T_G of 105 °C. At ambient temperature, PMMA is a glassy material with a refractive index $n_D^{20} = 1.490$ and very high transparency between 400 and 700 nm. It has a Young's modulus of 1800 MPa [239], a tensile strength of 62 MPa, and an elongation at break of 4% [240]. Due to this combination of mechanical and optical properties, together with its excellent melt processability, PMMA is a lightweight alternative to glass (Plexiglas, Perspex) for use as window-screen materials in architecture, in the aviation industry, in automotive taillights, in many consumer products (e.g. shower doors, decorative window mosaics, side glazing), and in advanced optical applications such as camera lenses, prisms, and fibers in fiber optics [240–242]. Bulk polymerization is used to prepare PMMA sheets or PPMA/MMA syrups, the latter being used to produce PMMA sheets in a batch process, or to make synthetic marble by mixing with inorganic compounds and subsequent curing [240]. In terms of green chemistry, these processes are hard to match since they work without any solvent and generate 100% of a highly pure polymer. However, because of the gel or Trommsdorf effect (= autoacceleration due to increased reaction mixture viscosity that reduces the rate of termination, while the rates of polymerization and initiation remain almost constant) FRP bulk polymerization is difficult to control and relatively expensive [243]. Hence the bulk of the PMMA produced is made by solution, suspension, and emulsion polymerization [240]. The polymerization of MMA has become the most investigated polymerization reaction in $scCO_2$ – at present it is *the* model *par excellence* for FRP in $scCO_2$. PMMA is insoluble in $scCO_2$ under conventional polymerization conditions ($P = 350$ bar, $T = 60$–$100\,°C$) [14], and early investigations revealed that precipitation polymerization resulted in low monomer conversions (10–40%), and also a PMMA with unfavorable molecular weights ($M_n = 77\,000$–$149\,000\ \mathrm{g\,mol^{-1}}$) and uncontrolled particle shapes and particle size distributions [244, 245].

The enthalpy of polymerization was measured as $-56.9 \pm 2.2\ \mathrm{kJ\,mol^{-1}}$, close to values obtained in conventional solvents [246]. High-pressure DSC was used to measure the polymerization kinetics under isothermal conditions at 56 bar. The first-order kinetics with respect to monomer was confirmed, and the apparent polymerization energy of activation was found to be $E_A(\mathrm{app}) = 51.6\ \mathrm{kJ\,mol^{-1}}$, which is 24% below the value reported for conventional MMA polymerizations ($\sim 68\ \mathrm{kJ\,mol^{-1}}$) [247, 248]. Other reaction kinetic methods employed were reaction calorimetry [249] and pulsed laser polymerization. At 180 bar, the propagation rate constant of MMA in liquid CO_2 and $scCO_2$ is described by $k_p(\mathrm{MMA}) = (5.2 \pm 3) \times 10^6\ \mathrm{l\,mol^{-1}\,s^{-1}} \times \exp[-(25\,400 \pm 1.200)/RT]$ in the temperature range 20–80 °C [250]. Reaction kinetic experiments revealed that in later stages of the polymerization, that is, at monomer conversions between 20 and 80%, the gel (= Trommsdorf) effect appeared in the monomer swollen particles [251]. The effects of pressure and temperature on the precipitation polymerization of MMA in $scCO_2$ were recently investigated, and the importance of phase equilibrium for the product yield and properties was confirmed [252]. Based on the reaction kinetic data, models have been proposed that allow one to simulate the course of precipitation and dispersion polymerizations of MMA in $scCO_2$ with satisfactory agreement between theory and experiment [252, 253]. The first free radical dispersion polymerization of MMA in $scCO_2$ was reported in 1994, using PFOMA as a steric stabilizer. The polymerization was initiated with either AIBN or a fluorinated AIBN derivative at 65 °C and 207 bar. Almost complete monomer conversion was achieved, the molecular weights increased to 190 000–325 000 $\mathrm{g\,mol^{-1}}$, and the PMMA precipitated in the form of spherical particles with a narrow diameter distribution between 1 and 3 μm. The key role of PFOA and its methacrylate analog PFOMA as stabilizers that allowed for control of the reaction rate and the reaction products particle morphology was consolidated by subsequent work [251]. As little as 0.24 wt% PFOMA [$M_n = (1.1$–$2) \times 10^4\ \mathrm{g\,mol^{-1}}$] with respect to the monomer were sufficient to obtain a stable latex of PMMA spheres in $scCO_2$. Up to 83% of the stabilizer could be removed from the precipitated polymer by extraction with CO_2, an operation which can easily be integrated in the polymerization and work-up scheme [8, 251]. The effect of different kinds of possible stabilizers on the dispersion polymerization of MMA in $scCO_2$ has been investigated. So far it seems that virtually any CO_2-soluble homo-, block, or copolymer bearing perfluorinated units as mentioned in Section 8.2.1.1 can be used. Polydimethylsiloxane-based block, graft, and copolymers have also been included in these investigations [8, 90]. P[MMA-*co*-HEMA-*co*-(HEMA-*graft*-HFPO)] copolymers were prepared by reaction of hexafluoropropene oxide (HFPO) with random copolymers from MMA and 2-hydroxyethyl methacrylate (HEMA), and investigated as stabilizers for the dispersion polymerization of MMA in $scCO_2$ [74]. In subsequent work, random copolymers of 1*H*,1*H*-perfluorooctyl methacrylate and 2-dimethylaminoethyl methacrylate, P [FOMA-*co*-DMAEMA], containing 34 wt% FOMA were also shown to act as stabilizers [255]. By variation of (i) the molecular weight of the CO_2-insoluble P[MMA-*co*-HEMA] backbone, (ii) the grafting density, and (iii) the molecular weight of the CO_2-soluble perfluorinated grafted side-chains, the effect of architecture and composition of a random graft copolymer was investigated. The molecular weight of the backbone was found to be critical, because with increasing length on the one hand the anchoring strength of the stabilizer on the PMMA particles grew, whereas on the other hand the solubility of the stabilizer in $scCO_2$ diminished. The general tendency was a decrease in particle diameter and diameter distribution with growing backbone length. At constant backbone length, an increase in grafting density enhanced the stability of the dispersions and reduced the particle diameter. The effect of the graft chain length depended on both of the other parameters; however, regions in the compositional space were found, where the increasing graft length acted in parallel with grafting density and backbone length [74]. The work demonstrated that a stabilizer must be adapted and optimized to the polymeric particles/solvent system to be stabilized. Hence general conclusions of experimental studies that are based on only one or two polymer homologs should be handled with great care. The effect of the non-perfluorinated end group on the anchoring strength and stabilization efficiency of monofunctional PFPE-X was investigated with $X = OH$, $O{-}CO{-}CH_3$, MA. While the low molecular weight end groups improved the anchoring strength as compared with the non-functionalized PFPE, block copolymers consisting of a PFPE block and a PMMA block ($PFPE_x$-*b*-$PMMA_y$, $x = 1400$–$4700\,g\,mol^{-1}$, $y = 1700$–$5600\,g\,mol^{-1}$) were found to be more effective. The best stabilizer was $PFPE_{1.75k}$-*b*-$PMMA_{2k}$; increasing the PMMA block beyond a degree of polymerization of about 20 resulted in lower molecular weights of the PMMA produced and loss of morphology control. The observed high yields, high molecular weights, and uniform particle sizes were attributed to an optimized balance between anchoring and solubility. It was outlined that the stabilizer does not require complete solubility from the initial stages of reaction; instead, it should be dispersed throughout the initial CO_2–MMA reaction mixture with the aid of MMA as a cosolvent [256]. A commercial carboxylic acid-terminated perfluoropolyether was also reported to act as stabilizer. Its performance was not as good as that of block copolymers; however, because of the reversible nature of its anchoring interaction (hydrogen bonds), the stabilizer could easily be extracted from the PMMA with CO_2 or organic solvents [257]. In addition to the already reviewed stabilizers [8, 90] P[FOA-*co*-(3-[tris(trimethylsilyloxy)silyl]propyl methacrylate)] [258], P[poly(ethylene glycol monomethacrylate)-*co*-FOMA] [259], monofunctionalized perfluoropolyether (PFPE) bearing OH, $-OCO{-}CH_3$, $-O{-}CO{-}C(CH_3){=}CH_2$ [256],

−COOR [260, 261], or −COOH groups [257], PFPE-*b*-PMMA block copolymer [255], perfluoroalkyl-terminated poly(ethylene glycol) $C_nF_{2n+1}C_mH_m$-*b*-PEO [262], random copolymers from FOMA and 3-hydroxypropyl methacrylate, P[FOMA-*co*-HPMA] [263], random copolymers from FOA and DMAEMA [264], alternating copolymers from maleic anhydride (MSA) and methyl vinyl ether (MVE) that were grafted with 1*H*,1*H*,2*H*,2*H*-perfluorooctan-1-ol ($C_8F_{17}C_2H_4$-OH) (= P[MVE-*alt*-MSA-*graft*-F8H2OH)] [265, 266], and FOMA–MMA block copolymer, PFOMA-*b*-PMMA [267], have been investigated.

Siloxane-based stabilizers are much cheaper then perfluoropolymers; however, polysiloxanes exhibit a lower solubility in $scCO_2$ than perfluoro side-chain copolymers [14]. Polydimethylsiloxane monomethacrylate (PDMS-MA) is a special auxiliary agent amongst the others, since this CO_2 soluble macromonomer is able to copolymerize with MMA to form random copolymers, P[MMA-*co*-(PDMS-MA)], bearing PDMS graft side-chains. In other words, the stabilizer polymer should be formed *in situ* during the polymerization process. First investigations showed that only small amounts of the PDMS-MA actually copolymerized, while the rest could be removed from the PMMA particles [268]. Due to the low anchoring capacity of the non-copolymerized PDMS-MA on PMMA, the latices were of low stability compared with PFOMA as stabilizer, and lower monomer conversions were found. However, with optimization of the reaction conditions, spherical PMMA particles with diameters from 1.1 to 5.7 μm and narrow diameter distributions were obtained [268, 269].

Reaction kinetic investigations on the early stages of the MMA–PDMS-MA (2 wt% with respect to MMA) stabilized dispersion polymerization in $scCO_2$ have been carried out using turbidimetric techniques [270, 271]. The continuous measurement of the number of growing particles and their turbidimetry-averaged diameter at 207 bar [270] or 330 bar, 60 °C [270] showed the particles to be nucleated through coagulation of PMMA molecules that were formed in the homogeneous phase. The particles precipitated upon exceeding diameters of 150–170 nm and continued to grow by collecting freshly formed polymer molecules from the homogeneous phase. At the end of the particle nucleation period, the particle concentration remained constant, because the already formed particles harvested the new PMMA macromolecules faster than the latter could agglomerate into new particles [272–274]. The turbidimetric results were found to be in semi-qualitative agreement with a previous model, developed to describe the polyvinylpyridine-stabilized dispersion polymerizations of styrene in alcoholic solvents [275]. Carrying out the dispersion polymerization of MMA–PDMS-MA systems in a reaction calorimeter and analyzing the molecular weight distribution of the resulting PMMA gave further insights. The heat released during the nucleation stage of the reaction was measured and permitted discussion on the polymerization locus during this stage. Scanning electron microscopy (SEM) analysis of the PMMA particles demonstrated that their surface area decreased with increasing PDMS-MA content (0–10 wt% with respect to the monomer), while simultaneously the molecular weight decreased due to the development of a low molecular weight PMMA species. By comparing the experimental results with those of modeling studies, radical exchange events between continuous and dispersed phase were held responsible for the observations. The pressure was found to affect the partitioning of the monomer between the two reaction phases [276–279]. The effect of the non-siloxane end group on the anchoring strength, and the stabilization efficiency of monofunctional PDMS-X was investigated with X = OH, O–CO–CH_3, MA. Whereas these low molecular weight end groups showed no improvement with respect to PDMS-MA, a block copolymer consisting of a PDMS block and an MMA–methyl acrylate (MA) copolymer block (PDMSb-P[$MMA_{1.1k}$-*co*-$MA_{0.5k}$]) was found to be very effective. The observed high yields, high molecular weights, and uniform particle sizes were attributed to an optimized balance between anchoring and solubility [280]. Similar results were claimed for the stabilization of polymerizing MMA–scCO_2 systems using trimethylsiloxy-terminated polydimethylsiloxane and benzoyl peroxide as initiator [281], and aminopropyl-terminated polydimethylsiloxane [282]. Also, a thiol-terminated PDMS (PDMS-SH) was applied at 300 bar, 65 °C that simultaneously acted as a chain transfer agent and as a stabilizer (Transtab), and allowed sub-micrometer-sized spherical PMMA particles to be generated [283]. When an amphipathic PDMS-P[methacrylic acid] stabilizer was used, the resulting PMMA particles could be redispersed in water subsequent to removal of the CO_2 [284]. A polydimethylsiloxane copolymer that contained initiator-active diazo groups in its main chain (P[PDMS-*alt*-N = N–]) was used in the dispersion polymerization of MMA in scCO_2 at 300 bar and 65 °C, simultaneously working as radical initiator and stabilizer. On heating, the PDMS compound expelled N_2 and the resulting radical-terminated PDMS blocks initiated the polymerization of MMA, hence *in situ* generating PDMS-*b*-PMMA- and PMMA-*b*-PDMS-*b*-PMMA-type stabilizers. The quality of the PMMA dispersions obtained was similar to that obtained for other PDMS-X-type stabilizers [285]. Another type of siloxane stabilizer is polydimethylsiloxane grafted with pyrrolidonecarboxylic acid side groups (Monasil PCA). With increasing stabilizer content, the PMMA particle diameters decreased (2–5.0 μm) and the size distribution became narrower. After isolation from CO_2, the PMMA particles were dispersible in aqueous buffer solutions containing a Pluronic (PEO-*b*-PPO-*b*-PEO) surfactant. The degree of dispersion and the dispersion stability could be enhanced considerably by means of ultrasound treatment [286]. Blocky structures are not necessarily required with siloxane-type stabilizers. Ring-opening metathesis polymerization was used to prepare poly[(bicyclo[2.2.1]hept-5-en-2-yl)triethoxysilane] of high molecular weight and low polydispersity. The macromolecule bears CO_2-soluble triethoxysilane side-chains and was successfully used as a stabilizer to prepare 2.7 μm sized PMMA particles [287]. Random copolymers from 3-tris(trimethoxysilyl)propyl methacrylate (TMOSiMA) and 2-dimethylaminoethyl methacrylate (DMAEMA), P[$TMOSiMA_{71mol\%}$-*co*-$DMAEMA_{29mol\%}$], were also found to be effective stabilizers in MMA polymerization, as demonstrated by high polymer yields (>93%) and the formation of micrometer-sized spherical PMMA particles [288]. The dispersion polymerization of MMA in scCO_2 has furthermore been investigated with respect to the size of the reaction vessels used. When working in reactors with diameters below 1 mm the control of the particle size distribution was gradually lost with decreasing reactor size, indicating that possibly turbulent flow is required for Paine-type [275] dispersion polymerization. It was concluded that the polymerization parameters must be modified and optimized for use in microreactors [289]. A special kind of stabilizer is perfluoroalkyl-modified exfoliated clay particles. Sodium montmorillonite was exfoliated by means of 1*H*,1*H*,2*H*,2*H*-perfluorododecylpyridinium iodide and subsequently used as a surfactant in the AIBN-initiated polymerization of MMA in $scCO_2$ at 65 °C, 241 bar. Although the modified clay was not soluble in CO_2, high molecular weight (449 000 g mol^{-1}) PMMA was obtained in 85% yield in the form of small, roundish particles about 10 μm in diameter that contained the clay particles at their surface and in their interior. Note that the dodecyl- and octadecylpyridinium-modified clays gave only 12% and 38% polymer yields, respectively. Compression-molded polymer–clay samples exhibited the typical PMMA–clay nanocomposite behavior such as enhanced T_G (PMMA 124 °C, PMMA–clay 132 °C) and increased storage modulus [$T = 26$ °C, G'(PMMA) = 1.28 GPa, G'(PMMA–clay) 1.88 GPa] [290, 291]. Analogous "pseudo-dispersion" polymerization of MMA was subsequently reported with polydimethylsiloxane-modified montmorillonites [292].

A continuous dispersion polymerization apparatus, based on a continuous stirred tank reactor, has been developed and tested with MMA in $scCO_2$ at 65 °C, 250 bar. The polymerization was initiated with AIBN and PDMS-MA was used as the stabilizer. The system produced PMMA microparticles that were similar in size and size distribution to particles prepared in batch reactions under similar conditions. The same held true for the molecular properties of the PMMA produced [293].

Controlled radical polymerization was also used to polymerize MMA in $scCO_2$. A dispersion polymerization was conducted under typical ATRP conditions using a bromo-terminated polydimethylsiloxane that simultaneously aced as the initiator and as a stabilizer ("inistab"). Microparticles consisting of a PMMA-*b*-PMMA block copolymer with low polydispersity ($M_w/M_n \approx 1.25$) were obtained [294]. Polymeric, CO_2-soluble macroligands for ATRP polymerization of MMA in $scCO_2$ have been prepared by copolymerization of FOA and 2-hydroxyethyl acrylate (HEA) to yield P[FOA-*co*-HEA]. The side-chain hydroxyl groups were esterified with acrylic acid to obtain P[FOA-*co*-(2-acryloylethyl acrylate)], and subsequently the side-chain acrylate groups were converted by means of Michael addition with $(C_2H_5)_2NCH_2CH_2NHCH_2CH_2N(C_2H_5)_2$ into P[FOA-*co*-{*N*,*N*-bis(*N*-diethylamino-2-ethyl)-3-propionylethyl acrylate}]. The resulting self-stabilizing ligands were successfully used in the copper-mediated ATRP reaction [295].

The RAFT method has also been employed to polymerize MMA in $scCO_2$, using the CO_2-soluble α-cyanobenzyl dithionaphthylate as RAFT agent and $PDMS_{10k}$-MA ($M_n = 10\,000$ g mol^{-1}) as stabilizer. At 65 °C, 276 bar the PMMA molecular weight could be controlled between 3000 and 63 000 g mol^{-1} at reasonable polydispersities of 1.09–1.45 and polymer yields up to 99%. The polymer particles were of spherical shape (diameter: 1–5 μm) and their diameter polydispersity was claimed to fall below that of PDMS-MA stabilized FRP products. It was pointed out as a critical condition that the RAFT agent and the RAFT oligomers require a sufficient degree of mobility in the precipitated PMMA phase. However, due to the known strong swelling of PMMA in $scCO_2$ under the selected polymerization conditions, the presumption was valid [296]. These findings were supported, and the control of the particles size

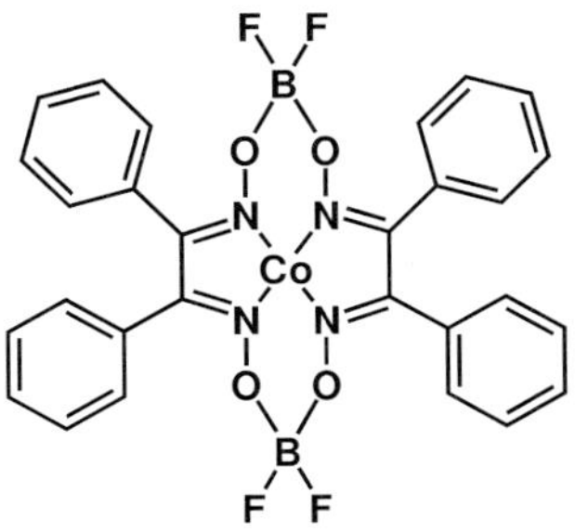

Scheme 8.4 Cobalamine complex used for the catalytic chain transfer polymerization of MMA in $scCO_2$.

distribution and macromolecular parameter became improved upon variation of the RAFT reagent [297]. At 50 °C, 150 bar the catalytic chain transfer polymerization of MMA in $scCO_2$ using a tetraphenylcobaltoxime–boron fluoride catalyst (Scheme 8.4) was shown to proceed faster than in liquid organic solvents, and the transfer constant ($k_{tr} \approx 10^7\,l\,mol^{-1}\,s^1$) led to the conclusion that the transfer step was diffusion controlled [298].

A dispersion catalytic chain transfer polymerization of MMA was achieved with the same cobaloxime as the transfer agent. The reaction was initiated with AIBN and stabilized with 5% (with respect to the monomer) of PDMS-MA to yield 91–93% PMMA particles. The polymer molecular weights ranged from 25 000 to 238 000 g mol^{-1} and the polydispersity was between 1.25 and 1.44. The activity of the chain transfer catalyst can be altered by the dispersion polymerization, and it was claimed that the technique may allow vinyl-terminated PMMA macromonomers to be generated [299].

"Immortal" polymerization of MMA was carried out in $scCO_2$ in the presence of 5,10,15,20-tetraphenylporphinatocobalt(II) and 5,10,15,20-tetra(pentafluorophenyl) porphinatocobalt(II) as the transfer agent. The polymers obtained were of low molecular weight, but also of low polydispersity. The chain transfer constant was measured as 1.3×10^3, which is comparable to those for liquid solvent systems [300].

8.2.1.4 **Polystyrene**

Polystyrene (PS) arising from FRP is an atactic, amorphous, thermoplastic material with a T_G of 100 °C. At ambient temperature, PS is a brittle, transparent glass with a density of 1.05 g cm^{-3}. The polymer has a Young's modulus of 3170 MPa and a tensile yield of 42 MPa with an elongation at break of 1.8%. Note that these mechanical data apply only for PS with a molecular weight exceeding about 150 000 g mol^{-1}; the lower molar mass homologs are too brittle for materials applications. General-purpose PS materials typically have molecular weights of 200 000–300 000 g mol^{-1} and can be processed and molded with great ease between 120 and 260 °C [301]. With a thermodynamic depolymerization (= ceiling) temperature of 310 °C [302], PS is thermally stable. At ambient temperature, PS is slowly degraded by air and exhibits photodegradation due to natural UV irradiation (i.e. shows low weatherability), but it is virtually resistant against biodegradation. PS is resistant against aqueous acids and bases, but readily dissolves at ambient temperature in aromatic liquids and semipolar solvents such as esters, ketones and halogenated hydrocarbons. At elevated temperatures, non polar solvents such as alkanes can also dissolve the polymer, but the liquids will damage PS surfaces with time at room temperature. The electrically nonconducting polymer is used for commodity articles such as tumblers, dining utensils, houseware, toys, camera parts, and audio tape cassettes. Because of its low thermal conductivity ($\lambda = 0.0015\,J\,s^{-1}\,K^{-1}\,cm^{-1}$) [303], PS is used as a thermal insulation material in the form of foamed ("expanded") PS in buildings and in consumer products. Expanded PS is extensively used for packaging purposes. PS is mainly produced by continuous solution polymerization with ethylbenzene as the solvent (2–30% ethylbenzene) and in minor quantities by suspension polymerization [304]. Since PS and modified PS polymers make up about 4% of the total polymer production [1], the replacement of the organic solvent-based production for "green" manufacture seems worthwhile. Pulsed laser spectroscopy was applied to measure the propagation rate constant k_P of the FRP of styrene in the homogeneous $scCO_2$ phase. At 65 °C and 180 bar, $k_P = 460 \pm 70\,l\,mol^{-1}\,s^{-1}$ was found, a value that is close to that for styrene bulk polymerization [99]. However, since PS rapidly precipitates from $scCO_2$ reaction mixtures, most investigations are directed at precipitation or dispersion polymerization reactions. At 58 °C and 100 bar, solvent mixtures consisting of THF–CO_2 or cyclohexene–CO_2 have been investigated in precipitation polymerizations. In these investigations, the effect of CO_2 as a non-solvent was intended to reduce the polydispersity of the resulting polymer molecular weight distribution [305]. In ethanol–CO_2 mixtures, the precipitation polymerization of styrene was limited to low yields (15–60%) of low molecular weight products ($M_n < 26\,000\,g\,mol^{-1}$) [306], indicating that efficient stabilization of the polymer particles is required. Pure PFOA was not a good stabilizer, probably because of its incompatibility with PS [8]. A series of PS-*co*-PFOA diblock copolymers were used to investigate the required composition of the stabilizer for successful styrene dispersion polymerization in $scCO_2$ [8], where the polystyrene block was intended as an "anchor" to the PS particle surface, whereas the PFOA block provided compatibility and solubility with the surrounding solvent. In the presence of the block stabilizers, up to 98% polymer yield in the form of a microparticulate free-flowing powder could be obtained. With a constant length of the PS block ($\langle X_n \rangle_{PS} = 37$), any variation of the PFOMA block length ($\langle X_n \rangle_{PFOMA} = 25$–$45$) had virtually no influence on the yield or the properties of the polymeric product. Increasing the length of the anchor resulted in smaller particles and narrower particle diameter distributions (optimum values: PS_{45}-*b*-$PFOA_{45}$ 98% yield, $M_n^{PS} = 22\,500\,g\,mol^{-1}$, $M_w/M_n = 3.1$, $d_P = 240$ nm, PSD = 1.3) [307]. Hence the anchor block must be of sufficient length to adhere to the particle surface. When PFOA with a single thiuram group was used that allowed PS blocks to be grafted on the PFOMA during the polymerization, micrometer-sized PS particles were obtained [8]. Two CO_2-insoluble PS_n-*b*-$PDMS_m$ ($n : m = 43 : 338$ and $96 : 878$, respectively) diblock stabilizers were compared in the dispersion polymerization of styrene in $scCO_2$. At 65 °C and 360 bar, the addition of ~7.5% (with respect to the monomer) stabilizer PS_{96}-*b*-$PDMS_{338}$ resulted in 95% monomer conversion ($M_n = 50\,000\,g\,mol^{-1}$, $M_w/M_n = 2.9$) and almost monodisperse spherical particles with 220 nm diameter, whereas in the presence of the stabilizer with long PDMS block, coagulated particles were found. Note that the polymerization results depended not only on the stabilizer, but also on the monomer concentration, temperature, and pressure [75, 308], which can make it difficult to compare the results of stabilizer experiments performed by different working groups.

PDMS-modified montmorillonite was used as a dispersing agent in the FRP of styrene, and analogous results to those reported for MMA were obtained. However, the distribution of the clay nanoparticles was found to differ from the MMA samples, because PS exhibited a different interaction parameter to clay than PMMA [292]. The AIBN-initiated alternating copolymerization between styrene and maleic anhydride (MAH) was investigated in $scCO_2$ at 50–80 °C and 205 bar. The molecular weights [$M_n = (2\text{–}8) \times 10^6\,\mathrm{g\,mol^{-1}}$, $M_w/M_n = 2.5\text{–}3.5$] of the P[Sty-*alt*-MAH] polymers obtained were 100–1000-fold higher than those of the copolymers obtained in butanone, DMF, and decalin. The effect was explained by the formation of Sty–MAH charge-transfer complexes that were more frequently formed in CO_2, because this weak Lewis acid reduced the electron density of the MAH C=C-bond by interaction with the carbonyl oxygen atoms of the MAH [309].

Polymerization of styrene in the presence of bifunctional monomers results in cross-linked polystyrene. Porous cross-linked PS microparticles are of importance for, for example, chromatographic stationary phases, and surface functionalized particles are of importance for solid-state (e.g. Merrifield) syntheses. Highly cross-linked P(Sty-*cross*-divinylbenzene) and P(Sty-*cross*-ethylene glycol dimethacrylate) microparticles with diameters between 1 and 5 μm have been prepared in $scCO_2$ at 65 °C and 306 bar. The diameters could be reduced to less than 500 nm in the presence of PMMA-*b*-PFOMA block copolymer stabilizers. Even surface-functionalized particles were obtained by addition of 4-vinylbenzyl chloride or pentafluorostyrene to the reaction mixture [310]. Block copolymers from styrene and methylphenylsiloxane (PS-*b*-PMPS) have also been used as stabilizers in the dispersion polymerization of styrene in $scCO_2$ [311].

In 2004, a RAFT polymerization of styrene in $scCO_2$ was reported to occur at 31 °C, 300 bar in the presence of cumyl dithiobenzoate as the controlling agent in the concentration range $3.5 \times 10^{-3}\text{–}2.1 \times 10^{-2}\,\mathrm{mol\,l^{-1}}$. At low styrene conversions (<15%), the molecular weight of the PS could be adjusted between 2700 and 40 000 $\mathrm{g\,mol^{-1}}$ with polydispersities M_w/M_n between 1.07 and 1.61. The typical behavior of a controlled polymerization, namely a linear dependence of M_n on the monomer conversion, narrow molecular weight distributions, and predictable molecular weights, was observed. The rate of the polymerization was slower in $scCO_2$ than in toluene by a factor of 1.5, but the molecular weight distributions were comparable in both cases [312].

The first CRP of styrene in $scCO_2$ by means of the NMP technique employed *N-tert*-butyl-*N*-(1-diethylphosphono-2,2-dimethylpropyl) nitroxide together with a polydimethylsiloxane-based azo initiator as "inistab" (=initiator + stabilizer). At 110 °C, the reaction proceeded smoothly to give high polymer yields, provided that sufficient amounts of inistab were present. The molecular weights increased linearly with conversion, indicating the controlled character of the polymerization, although the polydispersity of the products was not as low, and the agreement between theoretical and experimental molecular weight was not as good as for solvent-based NMP [313]. When a PDMS-*b*-PMMA stabilizer was used, the control was improved and the polydispersity values could be reduced to $M_w/M_n = 1.1$–1.4. A large excess of nitroxide reagent was required to achieve control, possibly because of a more favorable solubility of the nitroxide reagent in the CO_2 phase. Without stabilizer, the authors observed the onset of PS precipitation when the polymers degree of polymerization reached $\langle X_n \rangle = 28$ at 110 °C, 420 bar and 3.84 mol l^{-1} styrene concentration [314]. The precipitation polymerization in $scCO_2$ was found to proceed slower by a factor of 1.58 than the analogous bulk polymerization [315]. Subsequently the inistab principle was applied to NMP polymerization of styrene by synthesis and application of PDMS and PS blocks that were end capped with the transfer agent *N*-*tert*-*N*-butyl-*N*-[1-diethylphosphono(2,2-dimethylpropyl)] nitroxide. Also in these cases uniform spherical particles with narrow diameter distributions below 1 μm and PS with narrow molecular weight distributions were obtained [316]. In recent work, it was even possible to obtain molecular weight distributions from $scCO_2$ RAFT precipitation polymerization that were narrower than those obtained from RAFT reactions in toluene. The key issue was to increase the monomer concentration from 40% w/v (3.8 mol l^{-1}) to 70% w/v (6.7 mol l^{-1}). Since styrene acts as a cosolvent for PS in CO_2, the polymer precipitated around ~8% monomer conversion at an average degree of polymerization of $\langle X_n \rangle = 28$ at the lower and at 26% conversion, corresponding to $\langle X_n \langle = 114$ at the higher initial monomer concentration. Possibly at low conversions (and short chains) a considerably large fraction of growing radical chains had not reacted with the RAFT reagent and hence could not participate in the dynamic equilibrium between dormant and polymerization active species, which is a prerequisite for RAFT CRP. The cosolvency effect kept the growing PS chains sufficiently long in solution to permit almost complete functionalization of the chains with the RAFT agent. The hypothesis was further supported by the observation that a large excess of RAFT reagent was required with low styrene concentrations to achieve any control, and the fact that more reactive RAFT agents improved control further [317].

8.2.1.5 Other Vinyl Monomers

A series of other vinyl monomers have been polymerized in $scCO_2$, but with much less intensity than MMA and styrene. Table 8.3 gives an overview of radical homo- and copolymerizations that were published between 1999 and 2008. Below the table a subjective selection of some reactions are mentioned.

With a production of about 26×10^6 t in 2005, poly(vinyl chloride) (PVC) is the third most produced polymer after polyethylene and polypropylene. The polymer is cheap since it is used as the chlorine sink of the world's electrolytic sodium hydroxide production, but also exhibits good mechanical properties, chemical stability, and weathering resistance. Pure (="rigid") PVC has a tensile strength of 55 MPa (23 °C) and a tensile modulus of 2.7–3 GPa at 1% strain. The bulk of PVC is not used as the pure polymer, but contains 10–15% of additives such as plasticizers, heat stabilizers, lubricants, and fillers. The tensile strength of such "flexible PVC" grades is between 7.5 and 30 MPa with an elongation at break of 140–400% [366]. Rigid PVC is used as a

Table 8.3 Vinyl monomers polymerized in $scCO_2$(without MMA, styrene, and fluorinated monomers).[a]

Monomers	Reaction	Type	Stabilizer (additive)	Ref.
Acrylic acid	FRP	CP		[318–320]
Acrylic acid	FRP	CP	– (EtOH as cosolvent)	[321]
Acrylic acid	FRP	P		[322]
Acrylic acid (cross-linked)	FRP	P, CP	–	[323]
Acrylic acid (cross-linked: DVB)	FRP	P	– [(*S*)-propanolol]	[324]
Acrylic acid–acrylamide	FRP	P	–	[325]
Acrylic acid–FOA	FRP	D	PFOA	[326]
Acrylic acid–NIPA	FRP	P	–	[327, 328]
Acrylic acid–*N*-vinylpyrrolidone	FRP	P	–	327]
Acrylamide	FRP	E	–	[329, 330]
Acrylamide	FRP	P	– (Ethanol)	[331]
Acrylamide–*N,N*-MBAA	FRP	E	HFPO-$COONH_4$/PVA	[332]
N,N-Dimethylacrylamide	FRP	P	–	[333]
N-Ethylacrylamide	FRP	E	–	[334]
Acrylonitrile	FRP	D	PFOA	[335]
Acrylonitrile	FRP	P	–	[336–338]
Acrylonitrile	FRP	P	–	[339]
Acrylonitrile	FRP	D	P[Sty-*co*-AN]-*b*-P[FOMA]	[339]
Acrylonitrile–methyl acrylate	FRP	P	–	[340]
Acrylonitrile–2-chlorostyrene	FRP	P	–	[340]
Acrylonitrile–vinyl acetate	FRP	D	P[Sty-*co*-AN]-*b*-P[FOMA]	[341]
Isobornyl methacrylate	FRP	D	PDMS-MA, HFPO-$COONH_4$	[342]
Butyl acrylate–MMA	FRP	D	PDMS-MA, PFOA	[343]
Diethylene glycol dimethacrylate	FRP	D	HFPO-$COONH_4$	[344]
N,N-Dimethylaminoethyl MA	FRP	D	PDMS-MA, HFPO-$COONH_4$	[345]

4,4′-Divinylbiphenyl	FRP	P	–	[346]
Ethyl methacrylate–MMA	FRP	D	PDMS-MA	[347]
2-Ethylhexyl acrylate	FRP	D	PDMS-MA	[348]
Glycidyl-MA	FRP	D	PFOA	[349]
Glycidyl-MA	FRP	D	PDMS-MA + cross-linker	[350]
Glycidyl-MA–(MMA, BuMA), (DEAEMA, MA)	FRP	P	–	[351]
Glycidyl MA–MAA	FRP	P	–	[352]
Glycidyl-MA-2-hydroxyethyl-MA	FRP	P	–	[352]
Glycidyl-MA-Styrene	FRP	P	–	[352]
2-Hydroxyethyl-MA	FRP	D	PEO-*b*-P[F8H2A]	[353]
2-Hydroxyethyl-MA	FRP	D	PS-*b*-PFOA	[354]
2-Hydroxyethyl-MA	FRP	D	PFOA, P[FOA-*co*-MMA], P[DMS-*graft*-PCA]	[355]
NIPAM–MBAA	FRP	P	–	[356–358]
2-Methylene-1,3-dioxepane	FRP	P	–	[359]
2-Methylene-1,3-dioxepane–NVP	FRP	P	–	[359]
PEG MA	FRP	D	PDMS-MA, HFPO-$COONH_4$	[342]
PEG-di-MA–DEGDMA	FRP	D	HFPO-COOH	[360]
TMOSiMA	FRP	P	–	[361]
2,2,2-Trifluoromethyl-MA	FRP	P	–	[362]
Trimethylolpropane-tri-MA	FRP	P	–	[363]
Vinyl acetate	FRP	D	P[Sty-*co*-AN]-*b*-PFOMA	[364]
Vinyl chloride	FRP	P	–	[365]

[a]Reaction: FRP = free radical polymerization.

Type: CP = continuous precipitation polymerization, P = precipitation polymerization, D = dispersion polymerization, E = emulsion polymerization.

Monomers: AN = acrylonitrile, BuMA = butyl methacrylate, DEAEMA = *N,N*-dimethylaminoethyl methacrylate, MAA = methacrylic acid, MBAA = *N,N*-methylene bisacrylamide, VAc = vinyl acetate, MA = methacrylate, MeA = methyl acrylate, PEG = poly(ethylene glycol), TMOSiMA = 3-[tris(trimethylsilyloxy)silyl]propyl methacrylate.

Stabilizers: HFPO = poly(hexafluoropropene oxide), PCA = pyrrolidonecarboxylic acid, PDMS-MA = polydimethylsiloxane mono methacrylate, PFO(M)A = polyperfluorooctyl (meth)acrylate.

construction material, for example for window frames, gutters, pipes, and car seatbacks, in the domestic area. It is found in bottles, blister packing, curtain rails, drawer sides, and laminates. Flexible PVC is used for waterproof membranes (also in clothing), cable insulation, roof lining, greenhouses, shower curtains, leather cloth, hosepipes, and cling films. Medicinal uses include oxygen tents, bags and tubing for blood, and drips for infusion. About 80% of PVC is produced by free radical suspension polymerization of pressure-liquefied vinyl chloride in water, yielding particles with sizes between 100 and180 μm, and 12% is made by aqueous emulsion polymerization resulting in PVC latices with particle diameters controlled between 0.1 and 3 μm. Only 8% of all the PVC produced originates from solvent-free precipitation polymerization in liquid vinyl chloride [366]. Regarding the industrial importance of the polymer and its ease of preparation by means of FRP, it seems strange that the efforts to produce PVC in $scCO_2$ are limited to just a handful of publications. In 1968 and 1970 patents were filed describing the production of PVC by chemically- and gamma ray-induced FRP in $scCO_2$ [367, 368]. PVC was also shown to be formed by means of precipitation polymerization in $scCO_2$ between 55 and 70 °C and pressures from 102 to 272 bar. The yields obtained were low, between 20 and 76% [365]. Other publications mention PVC only in connection with CO_2 extraction, grafting with other monomers, or the preparation of semi-IPNs by polymerization monomers in CO_2-swollen PVC [369–372].

The dispersion polymerization of vinyl acetate and vinyl acetate–acrylonitrile mixtures was performed with P[Sty-*co*-AN]-*b*-PFOMA as a stabilizer in $scCO_2$ at 65 °C and unreported pressure. Depending on the length of the PFOMA block, the polymer yields ranged from 55 to 72%, and the polymer particle morphology changed from irregular to a spherical shape. With a molecular weight of 4500 g mol^{-1} for the P [Sty-*co*-AN] block, the PFOMA block molecular weight had to exceed 17 500 g mol^{-1} to obtain effective stabilizers [373].

"Molecular imprinting" involves the cross-linking dispersion polymerization of a functional monomer in the presence of a non-polymerizable template. The monomer should associate with the molecule because upon cross-linking polymerization the polymer network formed will closely adapt around the template, building an imaging shape around the molecule. When the template is removed, the remaining void may be able to recognize "its" molecule [374]. This recognition principle has been demonstrated to be of use to generate selective chromatographic stationary phases, catalysts ("artificial enzymes"), sensors, and functional filters [375, 376]. Molecular imprinting experiments have been performed in $scCO_2$ with the precipitation polymerization of acrylic acid, cross-linked with divinylbenzene, in the presence of (*S*)-propranolol. The precipitated P[AA-*cross* -DVB] particles were redispersed in organic solutions of (*R*)- or (*S*)-propranolol and the uptake of the molecules by the particles was measured. The particles were found to be selective for (*S*)-propranolol, because the cross-reactivity to the enantiomer was only about 5%. The absolute uptake capacity was comparable to that of conventionally prepared imprinted particles, and could be controlled by the amount of applied template [324]. A similar approach was reported for the generation of pharmaceutical (salicylic acid, acetylsalicylic acid) imprinted cross-linked particles from poly(ethylene glycol dimethacrylate) (P[MA–PEO–MA-*cross*-DEGDMA]) that were prepared by dispersion polymerization in $scCO_2$ with the help of a PFPE-COOH stabilizer [360].

Emulsion polymerization involves the polymerization of a monomer that is insoluble in the solvent (typically water) in the presence of a surfactant and a solvent-soluble initiator. According to the Harkin model, prior to polymerization the surfactant forms micelles and stabilizes large droplets of the monomer in the continuous phase, while the monomer distributes between the droplets, the continuous phase, and the micelles. The initiator decomposes in the continuous phase and the initiation step occurs in this phase. After a few additional steps, the growing polymer chains become insoluble and enter the monomer-swollen micelles that start to act as microreactors. Subsequently, the monomer becomes completely transferred from the droplets in the micelles that gradually transform into large monomer-swollen polymer-filled latex particles. The reaction ends when the monomer is consumed in the droplets and later in the latex particles. The whole process follows unique reaction kinetics (Smith–Edwards kinetics) that are different from the typical FRP kinetics, and characterized by an extended phase of constant rate of polymerization. In particular, the rate of polymerization and the number-average molecular weight are proportional to $[S]^{3/5}$, where $[S] =$ surfactant concentration [377].

Since nearly all liquid monomers are soluble in $scCO_2$, only an "inverse" emulsion polymerization can be implemented, where aqueous solutions of water-soluble but CO_2-insoluble monomers become dispersed in the CO_2 phase. The attempts were reviewed in 2005, showing that suitable monomers are scarce and that in future research a strong emphasis must be put on surfactant development. Inverse emulsion polymerizations were performed in the presence of perfluoropolyether-based surfactants; however, these compounds were considered to be too expensive for extensive technical use [378]. Acrylamide, *N*-vinylformamide, and vinylpyrrolidone have been polymerized in the presence of polyether-, PDMS-, and PFOA-glycosidic block polymers and modified sulfosuccinate surfactants; however, the rates of reaction and the molecular weights did not depend perfectly on the surfactant concentration as expected for true emulsion polymerization reactions. However, the dispersion of the water–monomer mixture was made possible by the surfactants, and also reasonable yields of high molecular weight polymers were obtained [378]. Note that the term "emulsion polymerization" is commonly used in the literature, even in cases where dispersion polymerizations were carried out.

In 2006, a successful microemulsion polymerization was reported to yield particles consisting of acryloxyethyltrimethylammonium chloride cross-linked with *N,N'*-methylenebisacrylamide. Aqueous solutions of both of the monomers were dispersed in $scCO_2$ by means of sodium bis[2-(perfluorohexyl)ethyl]phosphate to form water-in-CO_2 dispersions. After initiating the reaction with the CO_2-soluble initiator AIBN, spherical polymer particles ~20 nm in diameter were obtained [379]. In an analogous approach, 50–200 nm acrylamide particles were generated that became cross-linked subsequent to polymerization by an imidization reaction [329].

8.2.2
Metal-catalyzed Polymerizations

8.2.2.1 Polyolefins

Polyolefins, in particular polyethylene (PE), polypropylene (PP), and ethylene–propylene–diene (EPDM) elastomers represent the bulk (>50%) of all prepared polymer materials. Of the world polymer production in 2006 (245×10^6 t) about one-third (33.5%) was PE, and one-fifth (19.2%) PP [1]. Only low-density PE (LDPE), comprising 30% of all polyethylene polymers produced, is prepared by means of free radical polymerization in a solvent-free high-pressure (1000–3000 bar) gas-phase process in supercritical ethylene [380]. All other polyolefins are synthesized by means of coordinative catalyzed polymerizations, mostly by means of Phillips [381], Ziegler Natta [382], or metallocene catalysts [383]. It is an advantage of coordination catalysis that the stereoregularity, that is, the tacticity and consequently the crystallinity of the polymers, can be controlled very precisely. The mechanical and thermal properties of polymers that depend very sensitively on the polymer microstructure and morphology can hence be controlled to a great extent [384]. It is a disadvantage of the production process that, because of the extremely high reactivity of the catalyst systems towards air, water, protic, and polar impurities, the reactions must be carried out in aliphatic solvents. In these heterogeneous polymerizations, typically a polymer–catalyst slurry is produced in gasoline fractions that contains only about 20–30% solids [385]. Using highly flammable solvents or cosolvents, such as isobutane or propane, strongly increases the risk of accidents, while the unavoidable escape of solvent to the environment reduces the processes' environmental compatibility. Finally, the required recycling of the solvent and the drying of the polymer consume energy and make the processes more expensive. In spite of this enormous relevance, the level of research on coordinative polymerizations in $scCO_2$ is small compared with the efforts devoted to radical polymerization of vinyl monomers. Obviously the replacement of aliphatic hydrocarbon solvents with carbon dioxide will be of advantage; however, the technically well-established early transition metal catalyst systems cannot be used because of their high oxophilicity. Even the mild acidity of CO_2 will poison the catalysts [385]. The solubility and compatibility of a series of organometallic compounds with CO_2 have been tested [386]. Present research on the polymerization of olefins in $scCO_2$ is directed towards the development of late transition metal catalysts, since these compounds are known to tolerate CO_2 and other oxo functional groups [387–389]. A review on the palladium-based Brookhart catalyst (Scheme 8.5a) showed the performance of this catalyst in the homopolymerization of 1-hexene and ethylene and the copolymerization of ethylene and methyl acrylate. Poly(1-hexene) of molecular weight around 100 000 g mol^{-1} and a polydispersity $M_w/M_n = 1.5$–2 was obtained, and the catalyst achieved turnover-frequencies of 450–1000 h^{-1}. Polyethylene obtained under comparable conditions had molecular weights between 120 000 and 180 000 g mol^{-1} ($M_w/M_n = 1.8$–2.4) but exhibited extensive chain branching (about one branch per five monomer units) that resulted in sticky polymers ($T_G \approx -67\,°C$) with very low degrees of crystallinity of about 0.8–1.2%. It was pointed out that polymerization reactions are possible, but much work will be required to make the systems technically

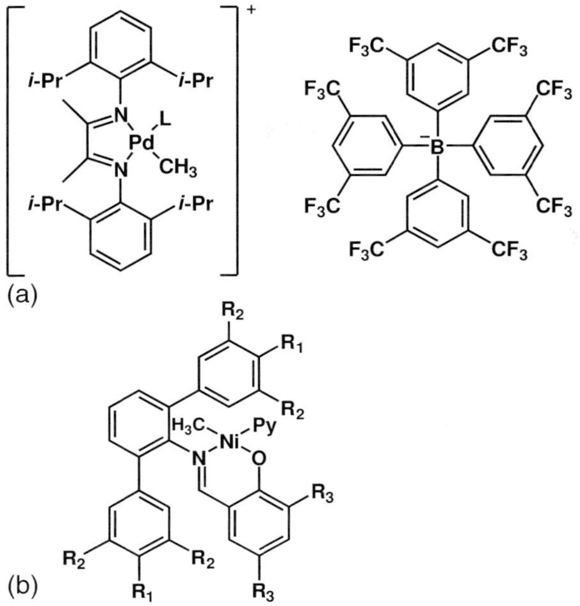

Scheme 8.5 (a) Brookhart catalyst precursor (L = OEt_2, CH_3CN) [390]; (b) neutral Ni(II)–salicylaldiminato complex (Py = pyridine) [77].

feasible [390]. At least the catalyst did not complex or react with CO_2, as evidenced by the high molecular weights and the fact that the same results could be obtained in comparative experiments carried out in CH_2Cl_2 as the solvent [78]. Neutral nickel(II) salicylaldiminato complexes (Scheme 8.5b) bearing solubilizing perfluorooctyl or trifluoromethyl substituents yielded polyethylenes with lower molecular weights (M_n = 10 000–66 000 g mol^{-1}) and larger polydispersities (M_w/M_n = 1.3–4.8), but the degree of branching could be reduced considerably to 1–2 branches per 50 monomer units. This success resulted in partially crystalline PE (degree of crystallinity 39–52%) that melted between 74 and 124 °C. Careful adjustment of the catalyst substituent pattern and the reaction conditions allowed the turnover numbers to be increased to 4900. The stability of the catalyst offered the possibility of incorporating perfluorinated surfactants in the reaction system; however, no definite conclusions about the effects of surfactants on polymer yield or morphology could be drawn [79].

It has been claimed that ethylene and norbornene can be copolymerized in $scCO_2$ by means of a (cyclooctadienyl)(chloro)(methyl)palladium complex, but the products were not sufficiently characterized [390].

When compared with the technically applied coordinative catalyst systems, the above-mentioned polymerizations must be considered as developments in their infancies. At present they are not satisfactory with respect to efficiency, turnover, reaction rates, and delivered product performance. Furthermore, the laboratory developments must be transferred to the technical scale. Although the 50 years' development advantage of Ziegler–Natta catalysts will be hard to catch up with, exactly this has to be attempted, simply because of the vital importance of polyolefin production for the polymer industry. In terms of green production, it makes little sense to replace specialty polymer productions – such as fluoropolymers – by an environmentally benign $scCO_2$ process, while ~50% of all polymers produced are still made in hydrocarbon solvent-based plants.

8.2.2.2 Other Metal-catalyzed Polymerizations

The formation of aliphatic polycarbonates (PC, $[X{-}O{-}CO{-}O]_n$) by metal-catalyzed copolymerization of carbon dioxide and oxiranes is of particular interest for two main reasons: first, polycarbonates represent a polymer class of high value, and second, this reaction is one of the least carbon dioxide-consuming reactions in polymer chemistry. Although it cannot be expected to reduce substantially the atmospheric CO_2 level by large-scale industrial PC production, an industrial process that not only uses an environmentally friendly solvent (CO_2) but also consumes a greenhouse gas deserves the title "green chemistry". The copolymerization of propene oxide or cyclohexene oxide with CO_2 in $scCO_2$ as solvent (cf. Scheme 8.6) has

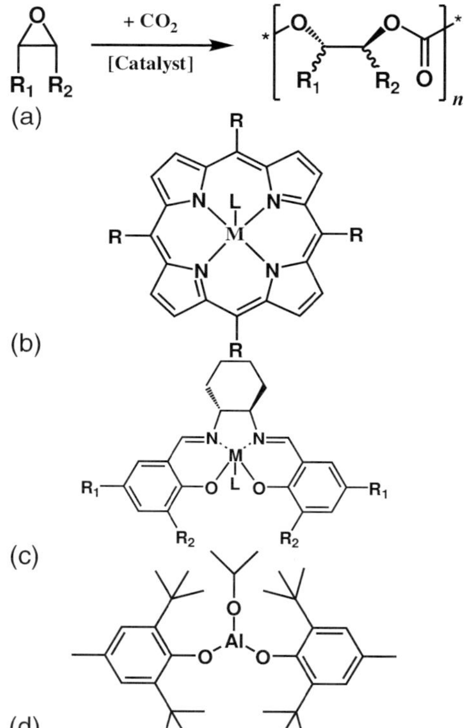

Scheme 8.6 (a) Copolymerization of epoxides and carbon dioxide and examples of (b) porphyrin (M = Al, Cr) and (c) salen (M = Co, Cr) complex catalysts. (d) Aluminum alkoxide catalyst causing the formation of polyether carbonate polymers.

recently been reviewed [391, 392]. Atactic poly(cyclohexyl carbonate) vitrifies at 115 °C, resulting in polymers with similar properties to polystyrene [393]. Note that atactic poly(propene carbonate) exhibits a T_G below 40 °C and depolymerizes below 250 °C into cyclic propene carbonate [394], impeding its use as a structural material. As mentioned in Section 8.1.3, propene oxide–CO_2 copolymers with a low degree of carbonate linkages (<25 mol%) belong to one of the rare non-fluorinated polymer classes that are soluble in $scCO_2$ under relatively mild conditions [29]. The possible application as a building block in stabilizers for dispersion polymerizations in $scCO_2$ has led to massive research activity since the discovery of its CO_2 solubility. Among the numerous publications on catalyst development, the zinc carboxylate, aluminum porphyrin, and salen complexes will be mentioned here.

In 1962, zinc complexes were the first catalysts employed for the alternating copolymerization of epoxides and carbon dioxide [395]. CO_2-insoluble zinc glutarates were applied in CO_2 to copolymerize CO_2 and propene oxide in low yields (10–20%) and molecular weights of about 10 000 g mol^{-1} [396]. CO_2-soluble zinc catalysts were obtained from zinc oxide and mono-(1*H*,1*H*,2*H*,2*H*-perfluorooctyl) maleate [397]. The latest development is mixed zinc glutarate–diethyl sulfinate complexes. Propylene oxide ($R_1 = CH_3$, $R_2 = H$ in Scheme 8.6a) could be copolymerized with carbon dioxide by means of an diethyl sulfinate–zinc complex [$Zn(SO_2Et)_2$] at 60 °C and 40 bar to yield poly(propene carbonate) with a high molecular weight of 500 000 g mol^{-1} and a catalyst productivity of 603 g of polymer per mole of zinc catalyst. When ethyl sulfinate-containing zinc glutarate catalysts were used, the molecular weights could be maintained between 80 000 and 178 000 g mol^{-1}; however, the catalyst productivity increased to 19 600 g of polymer per mole of zinc. In both cases the polymer yield exceeded 91% [398]. Mechanistic investigations revealed that the polymerization to proceeds via an insertion-type chain-growth mechanism embedded in a catalytic cycle. It is expected that the catalyst activity, selectivity, and productivity and the control of the molecular weight distributions of the polymeric products can be improved by finding the right ligand sphere around the catalytically active zinc atoms [391].

In 1978, the first example of a porphyrin-type catalyst (Scheme 8.6b) was presented to copolymerize cyclohexyl epoxide and CO_2. The molecular weights of the resulting polymers were relative low (typically <10 000 g mol^{-1}), but the reaction belongs to the class of "immortal" polymerizations [399]. This means that the catalyst can leave a growing chain which simultaneously transforms into a dormant species. The catalyst can start new polymer chains to grow, so that more polymer chains than catalyst sites are present in the system. The catalyst can also add to a dormant species and let it grow further. If this "chain hopping" of the catalyst is fast compared with the rate of polymerization, polymers with narrow molecular weight distributions are obtained. The chain hopping of aluminum tetraphenylporphyrin, (tpp)Al, is accelerated by protic substances, and even hydrochloric acid does not stop the polymerization reaction: Instead, (tpp)AlCl is formed that subsequently starts new chains [399]. The present state of the art is represented by a tpp manganese catalyst that allowed the polymerization of cyclohexylene oxide and CO_2 at 80 °C and 50 bar to a polymer with

99% carbonate linkages and a molecular weight of 6700 g mol^{-1} [400]. The technique was transferred to reaction in $scCO_2$ by employing tetrakis(pentafluorophenyl) porphyrinchromium(III) chloride as the catalyst. At 95–110 °C, up to 75% polymer yield was obtained. The material contained 90–97% carbonate linkages, but the molecular weight was only 3000 g mol^{-1} ($M_w/M_n < 1.4$) [401].

Aluminum alkoxides with bulky substituents (e.g. Scheme 8.6d) have been demonstrated to catalyze the copolymerization of propene oxide and CO_2 into copolymers that contain high fractions of ether linkages. At 60 °C and 80 bar CO_2 pressure, the formation of a polyether carbonate of $M_n = 5000$ g mol^{-1} with a broad molecular weight distribution ($M_w/M_n = 2.9$) was reported. The catalyst was not very active (turnover frequency = 2 h^{-1}) [29, 402] but worked to allow the generation of CO_2-soluble block units.

Cobalt and chromium Schiff base-type (salen) metal complexes (Scheme 8.6c) have been used since 2000 to catalyze the copolymerization of epoxides and CO_2 [391]. The special interest in the research arose from the fact that these catalyst systems allow the control of the stereochemistry of the polycarbonate chain. At 70 °C and 6.8 bar (*R,R*)-(*N,N'*-bis(3,5-di-tert-butylsalicylidene)-1,2-diaminopropane)cobalt pentafluorophenolate produced syndiotactic poly(cyclohexene carbonate) from cyclohexene with a turnover frequency of 400 h^{-1} [403]. At present the use of the metal salen catalysts is limited to propene and cyclohexene epoxides, and the control of molecular weight must be improved considerably. However, one can be optimistic since the mechanistic understanding has grown strongly during recent years [391], in particular by efficient cooperation of theory and experiment [404].

Polyketones of structural formula $[-CO-CHR-CH_2-]_n$ can be considered as alternating copolymers between carbon monoxide and vinyl monomers. The metal-catalyzed copolymerization of carbon monoxide and alkenes in conventional solvents is well established, and has been thoroughly reviewed [405–407]. The process was successfully transferred into $scCO_2$ as the solvent by using a cationic palladium catalyst with CO_2-soluble perfluoroalkyl chain substituents at its bipyridine ligands ($[Pd(CH_3)(NC-CH_3)$(4,4'-bis(1*H*,1*H*,2*H*,2*H*-perfluorodecyl)bipyridine)] $[B(3,5\text{-}(CF_3)_2C_6H_3)_4]$). At 37 °C and 250 bar, 4-*tert*-butylstyrene and carbon monoxide were copolymerized to yield highly syndiotactic alternating copolymers with molecular weights up to $M_n = 87\,000$ g mol^{-1} and relative narrow molecular weight distributions ($M_w/M_n = 1.2$) [408].

Phenylacetylene was polymerized in CO_2 by means of 2,5-norbornadienerhodium acetylacetonate to yield polyphenylacetylene. The polymerization was faster in CO_2 than in TFH or *n*-hexane. The catalyst produced a mixture of polymer chains with mainly *cis*-transoidal and *cis*-cisoidal microstructures, where the latter were insoluble in THF. When tris[4-(1,1,2,2-tetrahydroperfluorooctyl)phenyl]phosphane was added to the reaction mixture, the catalyst became CO_2 soluble due to ligand-exchange reactions. The soluble catalyst preferably produced *cis*-transoidal, that is, THF-soluble, polymers [409].

1,6-Heptadiyne monomers bearing a perfluoroalkyl ester chain connected to position 3 (e.g. 3,3,4,4,5,5,6,6,7,7,8,8,8-tridecafluorooctyidipropargyl acetate and

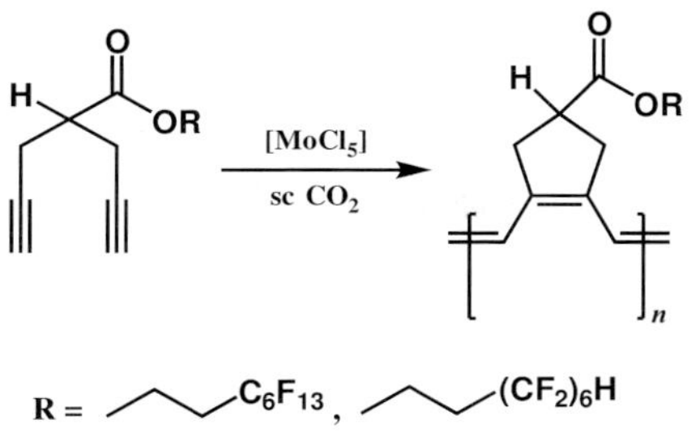

Scheme 8.7 Preparation of polymers with π-conjugated backbone.

3,3,4,4,5,5,6,6,7,7,8,8-dodecafluoroheptyldipropargyl acetate; see Scheme 8.7) were polymerized with $MoCl_5$ as a catalyst in $scCO_2$ at 40 °C and 345 bar. The polymer yields were between 29 and 69% and the molecular weights ranged from $M_n = 11\,000$ to $16\,000\ g\,mol^{-1}$ with large polydispersities M_w/M_n between 1.3 and 6.5. The formation of a polymer backbone with cyclopentenyl units was confirmed, and UV/VIS spectroscopy gave evidence of conjugated structure elements along the backbone [410]. The authors did not comment, however, on possible branching or cross-linking side reactions as implied by some high M_w/M_n values. A similar type of reaction was demonstrated to work with dimethyl dipropargylmalonate to yield a conjugated polymer that contained more than 95% cyclopentene units in the backbone [411].

Polyolefins containing alternating >C = C< double bonds and cycloalkane structures in the main chain can be produced by ring-opening metathetical polymerization (ROMP) of bicyclic olefin monomers (Scheme 8.8) [412]. The ROMP reaction is used industrially on a small scale for rapid injection molding (RIM), because very high rates of polymerization can be achieved with highly concentrated monomer mixtures [413–415]. Early attempts to perform ROMP reactions in $scCO_2$ used

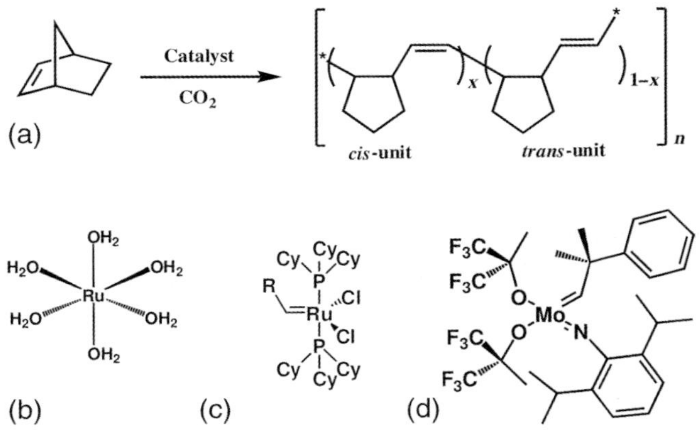

Scheme 8.8 Ring-opening metathetical polymerization (ROMP) of norbornene (a) and ROMP catalysts used to polymerize norbornene in $scCO_2$ (b). R = Ph, $Ph_2C = CH-$, Cy- = cyclohexyl.

norbornene (bicyclo[2.2.1]hept-2-ene) as monomer and the CO_2-insoluble ruthenium hexaaquo bis(*p*-toluenesulfonate) [$Ru(H_2O)_6(ptos)_2$; Scheme 8.8b] complex as the catalyst (Scheme 8.8a). At 65 °C and pressures of 60–345 bar, insoluble polynorbornene precipitated in yields between 30 and 76%. The molecular weight of the products ranged from 10 000 to 100 000 g mol^{-1} with a polydispersity of M_w/M_n 2–3.6. By addition of up to 16% methanol to the reaction mixture, the catalyst became soluble, and the polymerization rate increased by a factor of three. Simultaneously, the polymers microstructure changed from 83 mol% *cis*-addition (no MeOH) to 33 mol% *cis*-units at 16% MeOH content [416]. Grubbs- and Schrock-type carbene catalysts [417] exhibit higher activity in ROMP reactions, and both were tested in $scCO_2$. Whereas the ruthenium-based Grubbs-type catalysts were virtually insoluble in $scCO_2$, the molybdenum-based Schrock-type catalyst shown in Scheme 8.8d was partially soluble. At 25–45 °C and 100 bar, 94% of polynorbornene precipitated, exhibiting molecular weights up to 10^6 g mol^{-1}. The *cis*-unit contents depended on the CO_2 density and could be controlled from 66 mol% (ρ_{CO_2} = 0.57 g cm^{-3}) to 82 mol% (ρ_{CO_2} = 0.75 g cm^{-3}) [8]. Subsequent investigations confirmed these results and showed that the properties of the polymeric products closely resembled those of ROMP polynorbornenes obtained from conventional solvents [418, 419].

Semi-interpenetrating networks from polycyclooctene and poly(4-methyl-1-pentene) (PMP)have been prepared by swelling the PMP with $scCO_2$–cyclooctene mixtures and subsequent ROMP reaction that polymerized the cyclooctene in the swollen semicrystalline PMP phase [420].

8.2.3 Ionic Chain Polymerizations

8.2.3.1 Cationic Polymerizations

Cationic polymerizations are intrinsically difficult since carbocations and oxonium ions are highly reactive, giving raise to numerous side reactions, such as rearrangements and elimination and transfer reactions. At low temperatures and applying the right initiator systems, a series of monomers, including isobutene, branched alkenes, and vinyl ethers, can be polymerized without suppression of chain termination reactions [421]. Well-defined polymers can be obtained by means of living cationic polymerization methods that essentially work due to the reduction of the cation reactivity due to nucleophilic interactions [422, 423]. However, the high rates of polymerization and the tendency for side reactions make it advisable to work at low temperatures and to use non-nucleophilic solvents. Note that because of its Lewis base properties, CO_2 is a potential candidate for the attack of growing cationic species. Hence successful polymerizations can only be expected if the polymerization proceeds much faster than any side reactions with the solvent. The low polarity of CO_2 impedes the formation of free (=completely dissociated) ions, and prefers the presence of covalent species with polarized bonds, contact ion pairs, and a certain extent of solvent-separated contact ion pair species [421]. Although this drastically reduces the rate of polymerization, it also may help to reduce the side reactions between growing ions and CO_2.

The early efforts to use CO_2 as a solvent for cationic precipitation polymerizations have been summarized elsewhere. Mainly liquid CO_2 instead of $scCO_2$ was used ($T < 35\,^{\circ}C$), and the polymers of isobutene, isobutene–styrene, 3-methyl-1-butene, 4-methyl-1-pentene, formaldehyde, and isobutyl vinyl ether were obtained in poor yields and with limited molecular weights. More promising results were found with BF_3-initiated ring-opening reactions of cyclic ethers such as oxetanes bearing a perfluoroalkyl substituent {3-methyl-3′-[(1,1-dihydroheptafluorobutoxy)methyl]oxetane}. Due to the inherent solubility of the polyoxetane formed, the reaction proceeded under homogeneous conditions at $-10\,^{\circ}C$ and 290 bar and gave similar polymer yields (~70%) and comparable molecular weights ($M_n = 20\,000\,g\,mol^{-1}$, $M_w/M_n = 2$) to polymerizations carried out in CH_2Cl_2 [424]. A cationic dispersion polymerization of styrene succeeded at $15\,^{\circ}C$ and 330 bar after development of a suitable stabilizer consisting of a P[methyl vinyl ether]$_{17}$-*b*-P[2-(*n*-propyl)-*N*-perfluorooctylsulfonamido)ethyl vinyl ether]$_{28}$ block copolymer. Upon initiating with $TiCl_4$, PS yields up to 97% could be obtained in the presence of 4% stabilizer (with respect to styrene) within 4 h. The PS had a molecular weight of $18\,000\,g\,mol^{-1}$ and a polydispersity $M_w/M_n = 4$, also the particle sizes were widely distributed [425].

Ring-opening cationic polymerization of cyclo[tetra(dimethyl)siloxane] was possible at 110–125 °C and 150–200 bar by means of trifluoromethylsulfonic acid (TfOH) as the initiator. Polymer yields between 56 and 80% could be realized, and the molecular weight of the produced PDMS was adjustable between 121 000 and $168\,000\,g\,mol^{-1}$ with polydispersity $M_w/M_n = 1.5$–2.3. The CO_2 solubility of the PDMS was of importance for success, because at lower pressures, where the polymer had to precipitate, the yields dropped to 18% [426].

Carbon dioxide and ethyl vinyl ether have been co-oligomerized by means of a tris (acetylacetonato)aluminum-catalyzed cationic reaction. Only minor yields (<2%) could be obtained at 65 °C and 60 bar within a reaction time of 45 min. Increasing the temperature or pressure was of no advantage with respect to product formation. The oligomers obtained exhibited bimodal molecular weight distributions consisting of a major fraction of $M_n \approx 100$–150 g/mol ($\langle X_n \rangle \approx 2$), and a second minor fraction of $M_n \approx 220\,g\,mol^{-1}$ ($\langle X_n \rangle \approx 3$–$4$). The CO_2 content of the products could be adjusted between 12 and 48 mol% [427]. Although one hesitates to consider this reaction as "polymerization," the CO_2-consuming aspect is of importance. Future work will show if conversions and molecular weights can be driven towards the polymer regime.

Boron trifluoride dietherate-initiated cationic ring-opening polymerization of 2-substituted oxazolines was attempted at temperatures between 40 and 90 °C and pressures of 150–200 bar. The optimum conditions varied slightly between methyl-, ethyl-, and phenyloxazoline, but in any case monomer conversions up to 99% could be achieved. The molecular weight of the products was, however, restricted to $M_n = 1500$–$2500\,g\,mol^{-1}$ with polydispersities $M_w/M_n = 1.3$–1.5. MALDI-TOF investigations showed in any case the insertion of CO_2, giving rise to the formation of 10% (PhOx)–25% (MeOx) carboxylate end groups [428].

Monomers from renewable vegetable resources and polymerization reactions in $scCO_2$ are natural allies with respect to green chemistry. Purified soybean oil consists mainly of triglycerides with unsaturated chains. Since the average number of C=C double bonds per triglyceride exceeds 3–4, cross-linking polymerization should be possible. Oligomers with low molecular weights were produced by $BF_3{\cdot}(Et_2O)_2$-initiated cationic polymerization from soybean oil that were claimed to be useful as lubricants and hydraulic fluid additives [429].

8.2.3.2 Coordinative Anionic Polymerization

Because of their extreme nucleophilicity, carbanions will immediately react with carbon dioxide to form carboxylates. This reaction between CO_2 and, for example, Grignard reagents or lithium organic compounds is well known and of preparative importance [430]. It has-been tested, however, whether less reactive oxo-anions or "stabilized" anions can be used. The ring-opening polymerization of cyclic esters such as ε-caprolactone (CL) can be initiated by means of metal alkoxides or carboxylates (= "coordinative polymerization"). The reaction is characterized by an "activated chain" mechanism (Scheme 8.9) and proceeds via intermediate carboxylate species of low nucleophilicity. Most frequently aluminum dialkyl alkoxides (Scheme 8.9a) or bis(2-ethylhexyl) stannates (Scheme 8.9b) are used as the initiators, although zirconium and lanthanoid compounds have also been used. The key issue is obviously that a strongly Lewis acidic metal coordinates to the carbonyl group of the ester monomer and activates it for the attack of the relatively weak nucleophilic intermediates.

The coordinative polymerization of cyclic esters has been thoroughly reviewed [431], so here we can focus on the use of this techniques to prepare aliphatic polyesters in $scCO_2$.

Polycaprolactone Poly(ε-caprolactone) (PCL) is a semicrystalline polymer with a melting temperature of 57 °C and a T_G of −62 °C. The degree of crystallinity decreases with increasing molecular weight. Whereas at $M_n = 5000\,g\,mol^{-1}$ 80% of the material is crystalline, this fractions decreases to 45% at $M_n = 60\,000\,g\,mol^{-1}$. A 44 000 $g\,mol^{-1}$ PCL possesses a tensile strength of 16 MPa, a tensile modulus of 400 MPa, and breaks at 80% elongation. The polymer is soluble in chlorinated hydrocarbons, cyclohexanone, and 2-nitropropane, but insoluble in aliphatic hydrocarbons, diethyl ether, and alcohols [432]. It is biocompatible, non-toxic, biodegradable, and can be degraded enzymatically [433]. It is produced by anionic ring-opening polymerization of ε-caprolactone from organic solvents, and was proposed for use as degradable packaging material [432]. In 1999, the anionic polymerization of CL initiated by potassium *tert*-butyrate and also aluminum, yttrium and lanthanum isopropoxides was reported. It was shown that the polymerization proceeded without the incorporation of CO_2 in the polyester [426, 434]. Using dimethyltin dimethoxide $[Me_2Sn(OMe)_2]$ as initiator, a living polymerization was achieved at 40 °C and 210 bar. Up to molecular weights of 20 000 $g\,mol^{-1}$, low polydispersities and good agreements between theoretically expected and experimentally found molecular weights were observed. The rate of polymerization was slower in $scCO_2$ than in Freon-113 or toluene or in bulk. Whereas in the conventional reaction media the rate of polymerization was first order with respect to the initiator concentration, no such

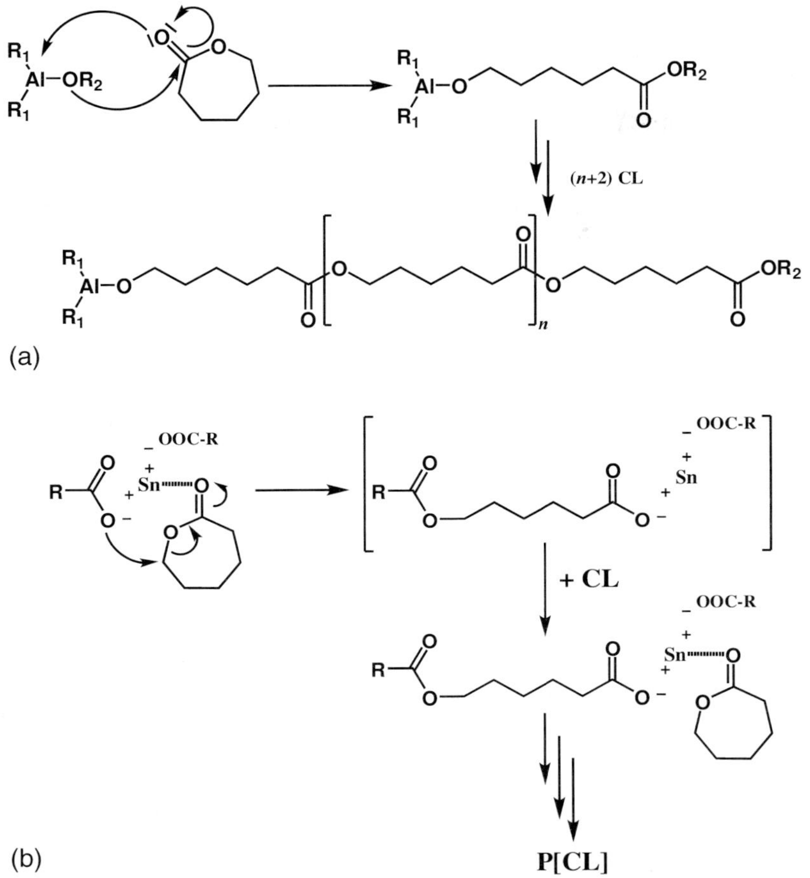

Scheme 8.9 Ring-opening coordinative polymerization of ε-caprolactone by means of (a) aluminum alkoxides and (b) tin carboxylates.

dependence was seen in $scCO_2$ [435]. It turned out that the reaction was slowed by a reversible side reaction of the tin initiator that formed carbonates which behaved as dormant species [436]. Subsequent investigations involving Y(O-*i*Pr)$_3$, La(O-*i*Pr)$_3$, and Al(O-*i*Pr)$_3$ showed that stronger ionic alkoxides undergo the carbonate-forming reactions, whereas less ionic compounds were inert against carbonate formation but also exhibited a lower polymerization activity [437]. The authors reviewed their studies with tin alkoxides in 2004 and further demonstrated the formation of PCL–clay composites by means of tin alkoxide-initiated polymerization of CL in the presence of montmorillonite. It was also possible to change the precipitation into a dispersion polymerization in the presence of a PCL_{175}-*b*-$PFOA_{75}$ diblock stabilizer. At 40 °C and 300bar, with 5% stabilizer (with respect to monomer), micrometer-sized PCL microspheres were prepared [438].

The polymerization of CL with bis(2-ethylhexyl) stannates ($SnOct_2$) was shown to proceed in $scCO_2$ by the same mechanism as in conventional solvents. At 80 °C and 235 bar, up to 94% monomer conversion could be achieved, resulting in PCL of $M_n = 68\,000\ g\ mol^{-1}$ ($M_w/M_n = 1.15$). The reaction was a living polymerization, as evidenced by the dependence of the molecular weight on the monomer conversion and the lack of termination reactions. In bulk the polymerization proceeded most fat, but the reaction rates were comparable in toluene, $scCO_2$, and THF [$k_{app}^{(THF)}$, $k_{app}^{(bulk)} = 8.7 \times 10^{-3}$; $k_{app}^{(CO_2)}$, $k_{app}^{(bulk)} = 9.7 \times 10^{-3}$; $k_{app}^{(toluene)}$, $k_{app}^{(bulk)} = 11.9 \times 10^{-3}$] [439].

The application temperature of caprolactone-based polyesters can be increased by copolymerization with lactic acid monomers [433]. This approach was checked in $scCO_2$ with mixtures of ε-caprolactone and (L,L)-lactide (LA) initiated with $Me_2Sn(OMe)_2$. Random copolymers were obtained, but the copolymerization parameters ($r_{LA} = 5.8$, $r_{CL} = 0.7$) demonstrated a strongly preferred incorporation of lactide units in the copolymer [440]. At 80 °C and 250 bar, CL and LA were copolymerized by means of $SnOct_2$ in the presence of a PCL-*b*-PFPE stabilizer. The reaction conditions of the dispersion polymerization were optimized with respect to yield, molecular weight, polydispersity, and powder morphology. It was demonstrated that CO_2 extracts monomer and stabilizer from the formed polymer, resulting in a polymeric product of high purity [441]. Note that the tin initiator is covalently bound to the polymer and cannot be extracted.

A series of branched pentablock copolymers $[PLA_{3-72}$-*b*-$PCL_{23}]_2$-*b*-PFPE-*b*-$[PLA_{3-72}$-*b*-$PCL_{23}]_2$ were prepared from a perfluoro polyether central block ($M_n^{PFPE} = 2.200\ g\ mol^{-1}$) bearing two hydroxy groups at each end. At 120 °C, four PCL arms were grown by means of $SnOct_2$ catalysis on the PFPE central block with a degree of polymerization $\langle X_n \rangle^{PCL} = 23$ per arm ($[PCL_{23}]_2$-*b*-PFPE–$[PCL_{23}]_2$). The synthesis was repeated in $scCO_2$ (85 °C, 300 bar), but only 70% monomer conversion was found (bulk polymerization: 99%). Subsequently, polylactide blocks were grown on the prepolymer in $scCO_2$ at 75 °C and 300 bar, forming PLA blocks with lengths $\langle X_n \rangle = 3$–72. The pentablock polymers were semicrystalline materials with melting temperatures between 43 °C ($[PLA_{72}$-*b*-$PCL_{23}]_4$PFPE) and 57 °C ($[PLA_6 b$-$PCL_{23}]_4$PFPE) [442].

Poly(Hydroxy Ester)s Aliphatic polyesters derived from hydroxycarbonic acids became a matter of interest because their easy hydrolysis, biocompatibility, and biodegradability. In particular, poly(α-hydroxy acid)s such as poly(glycolic acid) and poly(L-lactic acid), and also poly(β-hydroxyalkanoates), such as poly(β-hydroxybutyrate) and poly(β-propiolactone), and copolymers thereof have been investigated for use as biodegradable packaging material or as materials in biomedical applications such as degradable surgical suture materials or drug-delivery systems [432, 443, 444]. The poly(α-hydroxy acid)s can be prepared by metal-catalyzed ring-opening polymerizations of the acids' cyclic dimeric esters, namely glycolide (1,4-dioxane-2,5-dione) (GL) and L,L-lactide [(3S,6S)-3,6-dimethyl-1,4-dioxane-2,5-dione] (LA) (Scheme 8.10).

Copolymers of GA and LA have been prepared by coordinative ring-opening precipitation polymerization in $scCO_2$ using $SnOct_2$ as the catalyst. Because of the heterogeneous nature of the polymerizations, the molecular weights were limited to $M_n < 2200\ g\ mol^{-1}$ ($M_w/M_n = 1.4$) [445]. The copolymerizations were repeated in the

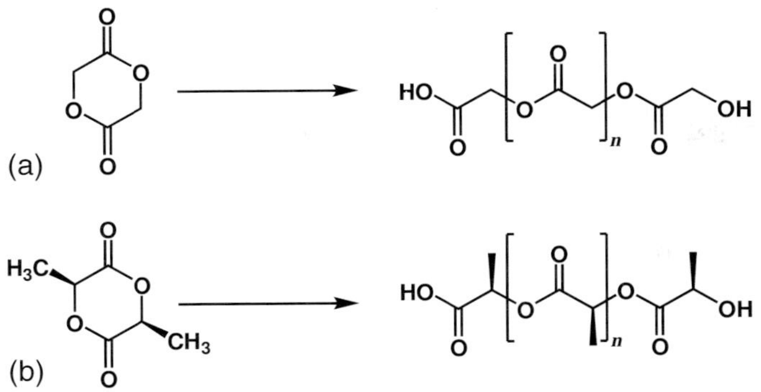

Scheme 8.10 Preparation of poly(glycolic acid) and poly(L-lactic acid) by ring-opening polymerization of glycolide (a) and L,L-lactide (b).

presence of 0.5 mol% PFOA (with respect to the monomer) as stabilizer. At 70 °C and 200 bar the molecular weight could be increased to 13 000 g mol^{-1} ($M_w/M_n = 2.3$) after 48 h of reaction. After 72 h, the final monomer conversion was 97% for LA and 60% for glycolide, demonstrating the different reactivities of the monomers [446].

A symmetric PCL_{12}–PFPE–PCL_{12} stabilizer was prepared from dihydroxy-terminated PFPE [447] in analogy with the $(PCL)_2$-*b*-PFPE-*b*-$(PCL)_2$ polymer [442]. At 80 °C and 241 bar, the homopolymerization of LA was performed in the presence of 10% triblock stabilizer. Monomer conversions between 88 and 92% were found, resulting in PLA of $M_n = 11\,600$–$14\,500$ g mol^{-1} ($M_w/M_n = 1.14$–1.38). The polymers formed fine powders consisting of irregular-shaped particles, but the stirring rate was found to control mainly the powder morphology. At low stirring rates (50 rpm), the particle diameter was around 45–50 µm, whereas on stirring at 300 rpm the particle diameters were reduced to 5–10 µm [447]. An oligomeric HO–PPO-*b*-PEO-*b*-PPO–OH Pluronic triblock copolymer ($M_n = 2700$ g mol^{-1}) was end-capped with acetyl chloride, and the diacetylated compound obtained was used as a stabilizer to polymerize GL in $scCO_2$ at 80 °C and 249 bar. The $SnOct_2$-catalyzed polymerization yielded up to 94% PGL with molecular weights below 4900 g mol^{-1} (M_w/M_n 2.2–2.4) and broad particle size distributions ranging from 10 to 300 µm with a maximum around 100 µm [448].

A series of new stabilizers have been prepared to control the polymerization of dimeric hydroxy acids in $scCO_2$, including PDMS-*b*-PLA, CH_3O–PEO-*b*-PLA, $C_9F_{19}CH_2$–PLA, $C_{10}H_{21}$–PLA, C_4H_9O–PPO-*b*-PLA [449], PFPE-*b*-PCL [450], PDMS-*b*-PAA, and PDMS-*b*-PMAA [451]. The properties of the polymeric products obtained with the different stabilizers did not vary substantially, although the powder morphologies may be affected. The last issue is difficult to compare, since the preparations were carried out under different reaction conditions and, in particular, stirring rates. Hence it cannot be judged to what extent the observed effects were caused by the stabilizer polymers.

The $SnOct_2$–decanol-initiated precipitation copolymerization of LA and GL in the presence of dexamethasone was used to prepare foamed microparticles containing the polymer-dispersed pharmaceutical. Polymerization in $scCO_2$ at 110–130 °C and 250 bar yielded P[LA-*co*-GA] with $M_n = 13\,000\ g\,mol^{-1}$ ($M_w/M_n = 1.8$). When the reaction mixture was rapidly expanded into air or toluene, foamed drug-loaded biodegradable microparticles were obtained that can possibly be exploited for long-term drug delivery systems [452]. Note that the presence of tin in the polymer may be counterintuitive with respect to applications in the body interior.

For this reason, it was tried to replace the tin catalysts with zinc(II) 2-ethylhexanoate ($ZnOct_2$) in the copolymerization of GL and LA in $scCO_2$. It was demonstrated that the $ZnOct_2$-catalyzed products were very similar to the polymers prepared with tin catalysts [453].

So far, it can be fairly stated that aliphatic polyesters of moderate molecular weights can be prepared in $scCO_2$ by coordinative precipitation and dispersion polymerization. Stabilizer polymers increase the yield and the molecular weight of the polymers; however, it may depend on the respective application whether these auxiliaries can be tolerated.

8.3 Conclusion

The long series of examples presented in Section 8.2 demonstrated that virtually any polymer can be made in liquid or supercritical carbon dioxide as solvent. CO_2 essentially behaves like a non-polar fluid of adjustable density and the polymers resulting from polymerizations in $scCO_2$ are similar to the products obtained from liquid solvents. As a rule of thumb, the control of yields, molecular weights, and powder morphology in dispersion polymerization are superior to those in simple precipitation polymerization reactions. Control becomes further improved when initiators or catalysts are made CO_2 soluble. The efficiency of dispersion polymerizations depends on the choice of the stabilizer, and it seems that the principles of stabilizer design for amorphous polymers have been found. However, the task of finding suitable stabilizers for dispersion polymerizations of crystalline polymers such as poly(vinylidene difluoride) has not yet been solved well. Another difficult area is the preparation of polyolefins in $scCO_2$, because polyethylene and polypropylene cannot be prepared in qualities comparable to those using conventional technology. Comparing the importance of technically prepared macromolecules with respect to their production volume (polyolefins $>$ PVC $\gg$ PS $\gg$ PMMA $\gg$ fluoropolymers) with the research efforts dedicated to polymer production in CO_2 (PMMA $>$ fluoropolymers $\gg$ PS $>$ polyolefins $\gg$ PVC), some mismatch must be acknowledged. Since polyolefins and PVC comprise more than 60% of all polymeric material production, a green chemistry that would like to offer alternative procedures to the industrial establishment should now focus on these materials. Obviously PVC that can be made by free radical polymerization may become the short-term target, whereas long-term research should be directed towards catalyst development to permit the green production of polyolefins in $scCO_2$.

References

1 Plastics Europe Deutschland, http://www.vke.de.

2 Kemmere, M. (2005) Supercritical carbon dioxide for sustainable polymer processes, in *Supercritical Carbon Dioxide in Polymer Reaction Engineering* (eds M.F. Kemmere and T. Meyer), Wiley-VCH Verlag GmbH, Weinheim. Chapter 1, pp. 1–14.

3 DeSimone, J.M. and Turmas, W. (eds) (2003) *Green Chemistry Using Liquid and Supercritical Carbon Dioxide*, Oxford University Press, Oxford.

4 Aresta, M. (ed.) (2003) *Carbon Dioxide Recovery and Utilization*, Kluwer, Dordrecht.

5 Young, J.L. and DeSimone, J.M. (2003) Synthesis and characterization of polymers: from polymeric micelles to step-growth polymerizations, in *Green Chemistry Using Liquid and Supercritical Carbon Dioxide* (eds J.M. DeSimone and W. Tumas), Oxford University Press, New York, pp. 149–163.

6 Villaroya, S., Thurecht, K.J., Heise, A. and Howdle, S.M. (2007) *Chemical Communications*, **37**, 3805–3813.

7 Tai, H., Popov, V.K., Shakesheff, K.M. and Howdle, S.M. (2007) *Biochemical Society Transactions*, **35**, 516–521.

8 Kendall, J.L., Canleas, D.A., Young, J.L. and DeSimone, J.M. (1999) *Chemical Reviews*, **99**, 543–563.

9 McHugh, M.A. and Krukonis, V.J. (1993) *Supercritical Fluid Extraction: Principles and Practice*, 2nd edn Butterworth-Heineman, Stoneham, MA.

10 Manke, C.W. and Gulari, E. (2003) Rheological properties of polymers modified with carbon dioxide, in *Green Chemistry Using Liquid and Supercritical Carbon Dioxide* (eds J.M. DeSimone and W. Turmas), Oxford University Press, Oxford, pp. 197–210.

11 Filardo, G., Galia, A. and Giaconia, A. (2003) Modification of polymers in supercritical carbon dioxide, in *Carbon Dioxide Recovery and Utilization* (ed. M. Aresta) Kluwer, Dordrecht, Chapter 8, pp. 197–210.

12 Fleming, O.S. and Kazarian, S.G. (2005) Polymer procesing with supercritical fluids, in *Supercritical Carbon Dioxide in Polymer Reaction Engineering* (eds M.F. Kemmere and T. Meyer), Wiley-VCH Verlag GmbH, Weinheim, Chapter 10, pp. 205–238.

13 NIST Web-Book, http://webbook.nist.gov/chemistry/formser.html.

14 (a) Rindfleisch, F., DiNoia, T. and McHugh, M.A. (1996) *Polymeric Materials Science and Engineering*, **74**, 178–179; (b) Rindfleisch, F., DiNoia, T.P. and McHugh, M.A. (1996) *The Journal of Physical Chemistry*, **100**, 15581–15587.

15 Hoefling, T.A., Newman, D.A., Enick, R.M. and Beckman, E.J. (1993) *Journal of Supercritical Fluids*, **6**, 165–171.

16 Bartle, K.D., Clifford, A.A., Jafar, S.A. and Shilstone, G.F. (1991) *Journal of Physical and Chemical Reference Data*, **20**, 713–756.

17 Tan, B., Woods, H.M., Licence, P., Howdle, S.M. and Cooper, A.I. (2005) *Macromolecules*, **38**, 1691–1698.

18 Kemmere, M.F. (2005) Supercritical carbon dioxide for sustainable polymer processes, in *Supercritical Carbon Dioxide in Reaction Engineering*, 1st edn (eds M.F. Kemmere and T. Meyer), Wiley-VCH Verlag GmbH, Weinheim, Chapter 1.

19 Ganapathy, H.S., Yuvaraj, H., Hwang, H.S., Kim, J.S., Choi, B.-C., Gal, Y.-S. and Lim, K.T. (2006) *Synthetic Metals*, **156**, 576–581.

20 Ahmed, T.S., DeSimone, J.M. and Roberts, G.W. (2004) *Chemical Engineering Science*, **59**, 5139–5144.

21 Saraf, M.K., Sylvain, G., Wojcinski, L.M. II, Charpentier, P.A., DeSimone, J.M. and Roberts, G.W. (2002) *Macromolecules*, **35**, 7976–7985.

22 Greiner, W. and Neise, L. (2004) *Thermodynamics and Statistical Mechanics*, 1st edn Springer, Berlin.

23 Lee, L.L. (1988) *Molecular Thermodynamics of Nonideal Fluids*, Butterworth, Stoneham, MA.

24 Prausnitz, J.M., Lichtenthaler, R.N. and de Azevedo, E.G. (1986) *Molecular Thermodynamics of Fluid-phase Equilibria*, 2nd edn Prentice-Hall, Englewood Cliffs, NJ.

25 Israelachvili, J. (1992) *Intermolecular and Surface Forces*, 2nd edn Academic Press, London.

26 Harris, J.E. (1970) *Journal of Physics B-Atomic Molecular and Optical Physics*, **3**, L150–L152.

27 Kazarian, S.G., Vincent, M.F., Bright, F.V., Liotta, C.L. and Eckert, C.A. (1996) *Journal of the American Chemical Society*, **118**, 1729–1736.

28 O'Shea, K.E., Kimse, K.M., Fox, M.A. and Johnston, K.P. (1991) *The Journal of Physical Chemistry*, **95**, 7863–7867.

29 Sarbu, T., Styranec, T. and Beckman, E.J. (2000) *Nature*, **405**, 165–168.

30 Patterson, D. (1982) *Polymer Engineering and Science*, **22**, 103–112.

31 Flory, P. (1953) *Principles of Polymer Chemistry*, Cornell University Press, Ithaca, NY, Chapter 12.

32 Somcynsky, T. (1982) *Polymer Engineering and Science*, **22**, 97–102.

33 Armeniades, C.D. and Baer, E. (1977) Transitions and relaxations in polymers, in *Introduction to Polymer Science and Technology* (eds H.S. Kaufman and J.J. Falcetta), John Wiley and Sons, Inc., New York, Chapter 6, pp. 239–300.

34 de Loos, T.W. (2005) Polymer thermodynamics, in *Handbook of Polymer Reaction Engineering*, vol. 1 (eds. T. Meyer and J. Keurentjes), Wiley-VCH Verlag GmbH, Weinheim, Chapter 2, pp. 17–56.

35 Burger, C. and Kreuzer, F.-H. (1996) Polysiloxanes and polymers containing siloxane groups, in *Silicon in Polymer Synthesis* (ed. H.R. Kricheldorf), Springer, Berlin, pp. 113–222.

36 Huang, Z., Shi, C., Xu, J., Kilic, S., Enick, R.M. and Beckman, E.J. (2000) *Macromolecules*, **33**, 5437–5442.

37 Guo, J., Andre, P., Adam, M., Panyukov, S., Rubinstein, M. and DeSimone, J.M. (2006) *Macromolecules*, **39**, 3427–3434.

38 Shi, C., Huang, Z., Kilic, S., Xu, J., Enick, R.M., Beckman, E.J., Carr, A.J., Mendelez, R.E. and Hamilton, A.D. (1999) *Science*, **286**, 1540–1543.

39 Hachisuka, H., Sato, T., Imai, T., Tsujita, Y., Takizawa, A. and Kinoshita, T. (1990) *Polymer Journal*, **22**, 77–79.

40 O'Neill, M.L., Cao, Q., Fang, M., Johnston, K.P., Wilkinson, S.P., Smith, C.D., Kerschner, J.L. and Jureller, S.H. (1998) *Industrial & Engineering Chemistry Research*, **37**, 3067–3079.

41 Meredith, J.C., Johnston, K.P., Seminario, J.M., Kazarian, S.G. and Eckert, C.A. (1996) *The Journal of Physical Chemistry*, **100**, 10837–10848.

42 Xiang, Y. and Kiran, E. (1995) *Polymer*, **36**, 4817–4826.

43 Dris, G. and Barton, S.W. (1996) *Polymeric Materials Science and Engineering*, **74**, 226–227.

44 Lora, M., Lim, J.S. and McHugh, M.A. (1999) *The Journal of Physical Chemistry*, **103**, 2818–2822.

45 Luna-Barcenas, G., Mawson, S., Takishima, S., DeSimone, J.M., Sanchez, I.C. and Johnston, K.O. (1998) *Fluid Phase Equilibria*, **146**, 325–337.

46 Hsiao, Y.-L., Maury, E.E., DeSimone, J.M., Mawson, S. and Johnston, K.P. (1995) *Macromolecules*, **28**, 8159–8166.

47 Mertdogan, C.A., Byun, H.-S., McHugh, M.A. and Tuminello, W.H. (1996) *Macromolecules*, **29**, 6548–6555.

48 DiNoia, T.P., McHugh, M.A., Cocchiaro, J.E. and Morris, J.B. (1997) *Waste Management*, **17**, 151–158.

49 DiNoia, T.P., Conway, S.E., Lim, J.S. and McHugh, M.A. (2000) *Macromolecules*, **33**, 6321–6329.

50 Lora, M., Lim, J.S. and McHugh, M.A. (1999) *The Journal of Physical Chemistry. B*, **103**, 2818–2822.

51 Baradie, B., Shoichet, M.S., Shen, Z., McHugh, M.A., Hong, L., Wang, Y., Johnson, J.K., Beckman, E.J. and Enick,

R.M. (2004) *Macromolecules*, **37**, 7799–7807.
52 Gerhart, L.J., Manke, C.W. and Gulari, E. (1997) *Journal of Polymer Science Part B-Polymer Physics*, **35**, 523–534.
53 von Solms, N., Nielsen, J.K., Hassager, O., Rubin, A., Dandekar, A.Y., Andersen, S.I. and Stenby, E.H. (2004) *Journal of Applied Polymer Science*, **91**, 1476–1488.
54 Tauer, K. (2003) Heterophase polymerizations, in *Encyclopedia of Polymer Science and Technology*, 3rd edn vol. 6, (eds H.F. Mark and J.I. Kroschwitz), Wiley–Interscience, Hoboken, NJ, pp. 410–527.
55 Barrett, K.E. (1975) *Dispersion Polymerization in Organic Media*, John Wiley & Sons Inc., New York.
56 Juba, M.R. (1979) *ACS Symposium Series*, **104**, 267–279.
57 Lovel, P. (1997) *Emulsion Polymerization and Emulsion Polymers*, John Wiley & Sons Inc., New York.
58 Schork, F.J., Luo, Y., Smulders, W., James, J.P., Butte, A. and Fontenot, K. (2005) *Advances in Polymer Science*, **175**, 129–255.
59 Xu, X.-J. and Gan, L.M. (2005) *Current Opinion in Colloid & Interface Science*, **10**, 239–244.
60 Sudol, E.D. (1997) *NATO ASI Ser. E Appl. Sci.*, **337**, 141–153.
61 Kawaguchi, S. and Ito, K. (2005) *Advances in Polymer Science*, **175**, 299–328.
62 Trommsdorff, E. (1954) *Macromolecular Chemistry*, **13**, 76–89.
63 Vivaldo-Lima, E., Wood, P.E., Hamielec, A.E. and Penlidis, A. (1997) *Industrial & Engineering Chemistry Research*, **36**, 939–965.
64 Allan, G. and Bevington, J.C. (1990) *Comprehensive Polymer Science*, vol. 4, Elsevier, Amsterdam.
65 Jacobasch, H.-J. and Freitag, K.-H. (1979) *ACTA Polymerica*, **30**, 453–469.
66 Consani, K.A. and Smith, R.D. (1990) *Journal of Supercritical Fluids*, **3**, 51–65.
67 Flory, P.J. and Kriegbaum, W.R. (1950) *Journal of Chemical Physics*, **18**, 1086–1094.
68 Smitham, J.B., Evans, R. and Napper, D.H. (1975) *Journal of the Chemical Society, Faraday Trans 1*, **71**, 285–297.
69 Croucher, M.D. and Hair, M.L. (1979) *The Journal of Physical Chemistry*, **83**, 1712–1717.
70 ONeill, M.L., Yates, M.Z., Harrison, K.L., Johnston, K.P., Canelas, D.A., Betts, D.E. and DeSimone, J.M. (1997) *Macromolecules*, **30**, 5050–5059.
71 Yates, M.Z., ONeill, M.L., Johnson, K.P., Webber, S., Canelas, D.A., Betts, D.E. and DeSimone, J.M. (1997) *Macromolecules*, **30**, 5060–5067.
72 Meredith, J.C. and Johnston, K.P. (1998) *Macromolecules*, **31**, 5508–5517.
73 Johnston, K.P., Da Rocha, S.R.P., Holmes, J.D., Jacobson, G.B., Lee, C.T. and Yates, M.Z. (2003) Interfacial phenomena with carbon dioxide soluble surfactants, in *Green Chemistry Using Liquid and Supercritical Carbon Dioxide* (eds J.M. DeSimone and W. Tumas), Oxford University Press, Oxford, Chapter 8, pp. 134–148.
74 Lepilleur, C. and Beckmann, E.J. (1997) *Macromolecules*, **30**, 745–756.
75 Canelas, D.A., Betts, D.E. and DeSimone, J.M. (1997) *Macromolecules*, **30**, 5673–5682.
76 Hyatt, J.A. (1984) *The Journal of Organic Chemistry*, **49**, 5097–5101.
77 Ahmed, T.S., DeSimone, J.M. and Roberts, G.W. (2004) *Science*, **59**, 5139–5144.
78 Carey, F.A. and Sundberg, R.J. (1983) *Advanced Organic Chemistry, Part B*, Plenum Press, New York, p. 260.
79 Bastero, A., Francio, G., Leitner, W. and Mecking, S. (2006) *Chemistry - A European Journal*, **12**, 6110–6116.
80 Arzoumanidis, G.G. and Peaches, C.PCT Int. Appl. WO 9300372 A1 19930107 (to Amoco Corp.), priority date1 July 1993, CAN 119:118060.
81 de Vries, T.J., Vorstman, M.A.G., Keurentjes, J.T.F. and Duchateau, R. 2000 *Chemical Communications*, 263–264.

82 Gorner, T., Dellachrief, J. and Perut, M. (1990) *Journal of Chromatography*, **514**, 309–316.

83 Rosselli, A.C., Boyer, D.S. and Houk, R.K. (1989) *Journal of Chromatography*, **465**, 11–15.

84 Leichter, E., Strode, J.T.B. III, Taylor, L.T. and Schweighardt, F.K. (1996) *Analytical Chemistry*, **68**, 894–898.

85 Raynie, D.E. and Delaney, T. (1994) *Journal of Chromatography Science*, **32**, 298–300.

86 Hsiao, Y.-L. and DeSimone, J.M. (1997) *Journal of Polymer Science Part A-Polymer Chemistry*, **35**, 2009–2013.

87 Elias, H.-G. (1999) *Macromolecules, Vol. 1, Structure and Syntheses*, 6th edn Wiley-VCH Verlag GmbH, Weinheim.

88 Davidson, T.A. and DeSimone, J.M. (1999) Polymerization in dense carbon dioxide, in *Chemical Synthesis Using Supercritical Fluids* (eds P.G. Jessop and W. Leitner), Wiley-VCH Verlag GmbH, Weinheim, Chapter 4. 5, pp. 297–325.

89 Young, J.L. and DeSimone, J.M. (2000) *Pure and Applied Chemistry*, **72**, 1357–1363.

90 Wells, S.L. and DeSimone, J.M. (2001) *Angewandte Chemie-International Edition*, **40**, 518–527.

91 Kemmere, M.F. and Meyer, T. (eds) (2005) *Supercritical Carbon Dioxide in Polymer Reaction Engineering*, 1st edn Wiley-VCH Verlag GmbH, Weinheim.

92 Ghosh, S., Bhattacharjee, C. and Mukhopadhyay, M. (2005) *International Chemical Engineering A*, **47**, 224–234.

93 Villaroya, S., Thurecht, K.J., Heise, A. and Howdle, S.M. 2007 *Chemical Communications*, 3805–3813.

94 Darensbourg, D. (2007) *Chemical Reviews*, **107**, 2388–2410.

95 Odian, G. (1991) *Principles of Polymerization*, 3rd edn John Wiley & Sons Inc., New York, Chapter 3.

96 Forbes, M.D.E. and Yashiro, H. (2007) *Macromolecules*, **40**, 1460–1465.

97a Charpentier, P.A., DeSimone, J.M. and Roberts, G.W. (2000) *Chemical Engineering Science*, **55**, 5341–5349.

97b Bilgin, N., Baysal, C. and Menceloglu, Y.Z. (2005) *Journal of Polymer Science Part A-Polymer Chemistry*, **43**, 5312–5322.

98 Xu, W.Z., Li, X. and Charpentier, P.A. (2007) *Polymer*, **48**, 1219–1228.

99 van Herk, A.M., Manders, B.G., Canelas, D.A., Quadir, M. and DeSimone, J.M. (1997) *Macromolecules*, **30**, 4780–4782.

100 Ryan, J., Erkey, C. and Shaw, M. (1997) *Polym. Prepr.*, **38**, 428–429.

101 Beuermann, S., Buback, M., Schmaltz, C. and Kuchta, F.-D. (1998) *Macromolecular Chemistry and Physics*, **199**, 1209–1216.

102 Beuermann, S. and Buback, M. (2005) Kinetics of free-radical polymerization in homogeneous phase of supercritical carbon dioxide in *Supercritical Carbon Dioxide* (eds. M.F. Kemmere and T. Meyer), Wiley-VCH Verlag GmbH, Weinheim, pp. 55–80.

103 Beuermann, S. and Buback, M. (2005) Kinetics of free-radical polymerization in homogeneous phase of supercritical carbon dioxide, in *Supercritical Carbon Dioxide in Polymer Reaction Engineering* (eds M.F. Kemmere and T. Meyer), Wiley-VCH Verlag GmbH, Weinheim, Chapter 4, pp. 55–80.

104 Percec, V. and Tirrell, D.A. (2000) *Journal of Polymer Science Part A-Polymer Chemistry*, **38**, 1710–1738, and comments.

105 Beuermann, S., Buback, M., Kuchta, F.-D. and Schmaltz, C. (2005) *Macromolecular Chemistry and Physics*, **205**, 876–883.

106 Beuermann, S. and Nelke, D. (2003) *Macromolecular Chemistry and Physics*, **204**, 460–470.

107 Beuermann, S., Buback, M., Isemer, C., Lacik, I. and Wahl, A. (2002) *Macromolecules*, **35**, 3866–3869.

108 Beuermann, S., Buback, M. and Nelke, D. (2001) *Macromolecules*, **34**, 6637–6640.

109 Matyaszewski, K. and Braunecker, W.A. (2007) Radical polymerization, in *Macromolecular Engineering*, vol. 1 (eds K.

Matyaszewski, Y. Gnanou and L. Leibler), Wiley-VCH Verlag GmbH, Weinheim, Chapter 5, pp. 161–216.

110 Matyaszewski, K., Gnanou, Y.nad Leibler, L. (eds) (2007) *Macromolecular Engineering*, vol. 2, Wiley-VCH Verlag GmbH, Weinheim.

111 Sciannamea, V., Jerome, R. and Detrembleur, C. (2008) *Chemical Reviews*, **108**, 1104–1126.

112 Chevalier, C., Guerret, O. and Gnanou, Y. (2006) Nitroxide-mediated living/controlled free radical polymerization, in *Living and Controlled Polymerization: Synthesis, Characterization and Properties of the Respective Polymers and Copolymers* (ed. J. Jagur-Grodzinski), Nova Science Publishers, Hauppauge, NY, pp. 51–63.

113 Matyaszewski, K. and Xia, J. (2001) *Chemical Reviews*, **101**, 2921–2990.

114 Matyaszewski, K. and Davies, T.P. (2002) *Handbook of Radical Polymerization*, Wiley-Interscience, Hoboken, NJ.

115 Barner-Kowollik, C. (ed.) (2008) *Handbook of RAFT Polymerization*, Wiley-VCH Verlag GmbH, Weinheim.

116 Moad, G., Rizzardo, E. and Thang, S.H. (2008) *Polymer*, **49**, 1079–1131.

117 Inoue, S. (2000) *Journal of Polymer Science Part A-Polymer Chemistry*, **38**, 2861–2871.

118 Endres, A. and Maas, G. (2000) *Chemie in Unserer Zeit*, **34**, 82–393.

119 Israelachvili, J. (1991) *Intermolecular and Surface Forces*, 2nd edn, Academic Press, Amsterdam-Boston-Heidelberg-London-New York-Oxford-Paris-San Diego-San Francisco-Singapore-Sydney-Tokyo.

120 Amedouri, B. and Botevi, B. (2004) *Well Architectured Fluoropolymers: Synthesis, Properties and Applications*, Elsevier, Amsterdam.

121 (a) Madorskaya, L.Ya., Loginova, N.N., Shadrina, N.E., Kleshcheva, M.S. and Panshin, Yu.A. (1980) *Vysokomol. Soedin., Ser. B: Krat. Soobsh.*, **22**, 904–906; (b) Wilson, C.W. and Santee, E.R. Jr. (1965) *Journal of Polymer Science C*, **8**, 97–112.

122 Scheirs, J. (ed.) (1997) *Modern Fluoropolymers: High Performance Polymers for Diverse Applications*, John Wiley & Sons, Ltd, Chichester, Chapters 9, 22 and 26.

123 Rouette, H.K. (1995) *Lexikon für Textilveredelung*, Laumann-Verlag, Dülmen.

124 Wood, C.D., Yarbrough, J.C., Roberts, G. and DeSimone, J.M. (2005) Production of fluoropolymers in supercritical carbon dioxide, in *Supercritical Carbon Dioxide in Polymer Reaction Engineering* (eds M.F. Kemmere and T. Meyer), Wiley-VCH Verlag GmbH, Weinheim, Chapter 9, pp. 189–204.

125 DeSimone, J.M., Guan, Z. and Eisbernd, C.S. (1992) *Science*, **257**, 945–947.

126 Guan, Z., Combes, J.R., Menceloglu, Y.Z. and DeSimone, J.M. (1993) *Macromolecules*, **26**, 2663–2669.

127 Guan, Z., Combes, J.R., Elsbernd, C.S. and DeSimone, J.M. (1993) *Polym. Prepr.*, **34**, 446–447.

128 Ehrlich, P. (1993) *Chemtracts: Organic Chemistry*, **6**, 92–94.

129 Kapellen, K.K., Mistele, C.D. and DeSimone, J.M. (1996) *Macromolecules*, **29**, 495–496.

130 Hwang, H.S., Yuvaraj, H., Kim, W.S., Lee, W.-K., Gal, Y.-S. and Lim, K.T. (2008) *Journal of Polymer Science Part A-Polymer Chemistry*, **46**, 1365–1375.

131 Ye, W. and DeSimone, J.M. (2000) *Industrial & Engineering Chemistry Research*, **39**, 4564–4566.

132 Ye, W., Wells, S. and Desimone, J.M. (2001) *Journal of Polymer Science Part A-Polymer Chemistry*, **39**, 3841–3849.

133 Yoshida, E. and Imamura, H. (2007) *Colloid and Polymer Science*, **285**, 1463–1470.

134 Ganapathy, H.S., Hwang, S.S., Lee, M.Y., Jeong, Y.T., Gal, Y.S. and Lim, K.T. (2008) *Journal of Materials Science*, **43**, 2300–2306.

135 Martinez, H.J., Green, J.W., Blanda, M.T., Cassidy, P.E. and Fitch, J.W. (1999) *Polym. Prepr.*, **40** (1), 99–100.

136 Lacroix-Desmazes, P., Andre, P., Desimone, J.M., Ruzette, A.-V. and Boutevin, B. (2004) *Journal of Polymer Science Part A-Polymer Chemistry*, **42**, 3537–3552.

137 Yong, T.-M., Hems, W.P., van Nunen, J.L.M., Andrew, A.B., Steinke, J.H.G., Taylor, P.L., Segal, J.A. and Griffin, D.A. (1997) *Chemical Communications*, **18**, 1811–1812.

138 Yoshida, E. and Imamura, H. (2007) *Colloid and Polymer Science*, **285**, 1463–1470.

139 Xia, J., Johnson, T., Gaynor, S.G., Matyjaszewski, K. and DeSimone, J.M. (1999) *Macromolecules*, **32**, 4802–4805.

140 Environmental Protection Agency. Draft Risk Assessment of Potential Human Health Effects Associated with PFOA and its Salts, Report EPA-SAB-06-006, EPA, Washington, DC, (2006).

141 Lau, F.S., Wesson, J.P. and Wunderlich, B. (1984) *Macromolecules*, **17**, 1102–1104.

142 McLane, D.I. (1970) *Tetrafluoroethylene Polymers* in *Encyclopedia of Polymer Science and Technology*, vol. 13, John Wiley & Sons Inc., New York, pp. 623–654.

143 Renfrew, M.M. and Lewis, E.E. (1946) *Industrial & Engineering Chemistry*, **38**, 870–877.

144 Kerbow, D.L. and Sperati, C.A. (1999) Chapter V, Physical constants of fluoropolymers, in *Polymer Handbook*, vol. 1, (eds J. Brandrup, E.H. Immergut and E.A. Gulke), Wiley-Interscience, Hoboken, NJ, pp. V31–V41.

145 Gangal, S.V. (1985) Polytetrafluoroethylene, homopolymer of tetrafluroethylene, in *Encyclopedia of Polymer Science and Engineering*, vol. 16, (eds H.F. Mark, N.M. Bikales and C.G. Overberger), John Wiley & Sons, Inc., New York, pp. 577.

146 Kennedy, K.A., Roberts, G.W. and DeSimone, J.M. (2005) *Advances in Polymer Science*, **175**, 329–346.

147 Bramer, D. J. V., Shiflett, M. B. and Yokozeki, A.(to E. I. du Pont de Nemours), US Patent 5 345 013, priority date 9 June (1994), CAN 122:16167.

148 Kemmere, M.F. and Meyer, T. (2005) *Supercritical Carbon Dioxide in Polymer Reaction Engineering*, Wiley-VCH Verlag GmbH, Weinheim, Chapter 6.

149 Beginn, U., Moeller, M., Pladis, P. and Kiparissides, C. (2004) *DECHEMA Monographs*, **138**, 381–386.

150 Quintero-Ortega, I.A., Vivaldo-Lima, E., Gupta, R.B., Luna-Barcenas, G. and Penlidis, A. (2007) *J. Macromol. Sci. A: Pure and Applied Chemistry*, **44**, 205–213.

151 Romack, T.J., Combes, J.R. and DeSimone, J.M. (1995) *Macromolecules*, **28**, 1724–1726.

152 DuPont (2002) *DuPont Introduces Fluoropolymers Made with Supercritical CO_2 Technology*, DuPont, Wilmington, DE, press release,

153 Kemmere, M.F. and Meyer, T. (eds) (2005) *Supercritical Carbon Dioxide in Polymer Reaction Engineering*, Wiley-VCH Verlag GmbH, Weinheim, Chapter 9.

154 Giaconia, A., Scialdone, O., Apostolo, M., Filardo, G. and Galia, A. (2007) *Journal of Polymer Science Part A-Polymer Chemistry*, **46**, 257–266.

155 (a) Humphrey, J. S. and Amin-Sanayei, R. (2003) Vol. 4, Vinylidene fluoride polymers, in *Encyclopedia of Polymer Science and Engineering*, 3rd edn (eds. H. F. Mark and J. I. Kroschwitz), Wiley Interscience Hoboken, USA, pp. 510–533.
(b) Seiler, D. A. (1997) PVDF in the Chemical Process Industry in *Modern Fluoropolymers: High Performance Polymers for Diverse Applications*, 1st edn, (ed. J. Scheirs) *Modern Fluoropolymers: High Performance Polymers for Diverse Applications*, 1st edn, John Wiley & Sons, Inc., New York, Chapter 25.

156 Lovinger, A.J. (1981) *Developments in Crystalline Polymers I*, Applied Science, London. 195.

157 (a) Leshchenko, S.S., Karpov, V.L. and Kargin, V.A. (1959) *Vysokomolekulyarnye Soedineniya Seriya B*, **1**, 1538–1548;

(b) Görlitz, M., Minke, R., Trautvetter, W. and Weisberger, G. (1973) *Angewandte Makromolekulare Chemie*, **29/30**, 137–162; (c) Weinhold, S., Litt, M.H. and Lando, J.B. (1979) *Journal of Polymer Science, Polymer Chemistry Edition*, **17**, 585–589.

158 Lando, J.B., Olf, H.G. and Peterlin, A. (1966) *Journal of Polymer Science, Part A1*, **4**, 941–951.

159 Kawai, H. (1969) *Japanese Journal of Applied Physics*, **8**, 975–976.

160 Ueberschlag, P. (2001) *Sensor Review*, **21**, 118–125.

161 (a) Bergman, J.G., McFee, J.H. and Crane, G.R. (1971) *Applied Physics Letters*, **18**, 203–205; (b) Fedosov, S.N. and von Seggern, H. (2008) *Journal of Applied Physiology*, **103**, 014105/1–014105/8.

162 Noeckel, D. (1983) *Swiss Chemistry*, **5**, 27–32.

163 Onishi, T. and Takemoto, T.(to Plantex Ltd.), Japanese Patent JP 2001289574, priority date 4 October (2001), CAN 135:305577.

164 Pelrine, R. E., Kornbluh, R. D., Stanford, S. E., Pei, Q., Heydt, R., Eckerle, J. S. and Heim, J. R., US Patent 7,064,472, B2 (to: SRI International), priority date 20.06. 2006, CAN 145:148046.

165 Tokarsky, E. W., Uy, W. C. and Young, R. T.(to E. I. du Pont de Nemours), US Patent 6 841 243, priority date 25 March (2005), CAN 140:272275.

166 Jozokos, M. A. and Globus, Y. I.(to Alphagary Corporation), US Patent Application 2007-078209, priority date 7 August (2007) CAN 146:380956.

167 Robinson, D. and Seiler, D. A. (1993) Potential Applications for Kynar Flex PVDF in the nuclear industry. in Proceedings of the American Glovebox Society National Conference, Linings, Coatings and Materials, Seattle, WA, 16–19 August 1993, Section 3C, pp. 10–14.

168 Governal, R.A. (1994) *Semiconductor International*, **17**, 176–178.

169 Jan, D., Jeon, J.S. and Raghavan, S. (1994) *Journal of Adhesion Science and Technology*, **8**, 1157–1168.

170 Scheirs, J., Burks, S. and Locaspi, A. (1995) *Trends in Polymer Science*, **3**, 74–82.

171 Hiatt, W.C., Vitzthum, G.H., Wagener, K.B. and Gerlach, K. (1985) *ACS Symposium Series*, **269**, 229–244.

172 Bottino, A., Capannelli, G. and Munari, S. (1987) *Chimica Oggi-Chemistry Today*, **4**, 11–15.

173 Brandwein, H. and Aranha-Creado, H. (2000) *Developmental Biology*, **102**, 157–163.

174 Allegrezza, A. E. Jr and Burke, E. T.(to Millipore), European Patent Application EP 245863, priority date 19 November (1987), CAN 108:133028.

175 Reid, G.E., Rasmussen, R.K., Dorow, D.S. and Simpson, R.J. (1998) *Electrophoresis*, **19**, 946–955.

176 Arora, P. and Zhang, Z. (2004) *Chemical Reviews*, **104**, 4419–4462.

177 Micheron, F. (1986) *Macromol. Chem. Macromol. Symp. Ser.*, **1**, 173–178.

178 Lang, S.B. and Muensit, S. (2006) *Applied Physics A-Materials Science & Processing*, **85**, 125–134.

179 Bearman, K.R., Blackmore, D.C., Carter, T.J.N., Colin, F., Ross, S.A. and Wright, J.D. (2004) *Springer Series on Chemical Sensors and Biosensors*, **1**, 203–226.

180 Lin, B. and Giurgiutiu, V. (2005) *Proceedings of SPIE*, **5765**, 1033–1044.

181 McCain, G. H., Semancik, J. R. and Dietrich, J. J. (to Diamond Alkali), French Patent 1 530 119, priority date 21 June (1968), CAN 69:403923.

182 Madorskaya, L.Ya., Loginova, N.N., Shadrina, N.E., Kleshcheva, M.S. and Panshin, Yu.A. (1980) *Vysokomol. Soedin., Ser. B: Krat. Soobsh.*, **22**, 904–906.

183 Pantani, R., Speranza, V., Besana, G. and Titomanlio, G. (2003) *Journal of Applied Polymer Science*, **89**, 3396–3403.

184 Doll, W.W. and Lando, J.B. (1970) *Journal of Applied Polymer Science*, **14**, 1767–1773.

185 Asahi Glass, French Patent 1 570 233, priority date 6 June (1969) CAN 72:22140.

186 Beuermann, S. and Imran-ul-haq, M. (2007) *Macromolecular Symposia*, **259**, 210–217.

187 DeSimone, J.M., Wojcinski, L.M., Kennedy, K.A., Zannoni, L., Saraf, M., Charpentier, P. and Roberts, G.W. (2001) *Polymeric Materials Science and Engineering*, **84**, 137–137.

188 Charpentier, P.A., Kennedy, K., DeSimone, J.M. and Roberts, G.W. (1999) *Macromolecules*, **32**, 5973–5975.

189 Ellmann, J.Polymerization and reactive coating in supercritical carbon dioxide, PhD thesis, University of Ulm, (2001).

190 Najjar, R. (2006) *Polymerization Studies of Vinylidene Difluoride in Supercritical Carbon Dioxide*, 1st edn G. Mainz, Aachen.

191 Liu, J., Tai, H. and Howdle, S.M. (2005) *Polymer*, **46**, 1467–1472.

192 Ahmed, T.S., DeSimone, J.M. and Roberts, G.W. (2004) *Chemical Engineering Science*, **59**, 5139–5144.

193 Saraf, M.K., Sylvain, G., Wojcinski, L.M. II, Charpentier, P.A., DeSimone, J.M. and Roberts, G.W. (2002) *Macromolecules*, **35**, 7976–7985.

194 Mueller, P.A., Storti, G., Morbidelli, M. and Apostolo, M. (2004) *DECHEMA Monograph*, **138**, 367–374.

195 Bonavoglia, B., Storti, G. and Morbidelli, M. (2006) *Industrial & Engineering Chemistry Research*, **45**, 3335–3342.

196 Liu, J., Tai, H. and Howdle, S.M. (2005) *Polymer*, **46**, 1467–1472.

197 Galia, A., Caputo, G., Spadaro, G. and Filardo, G. (2002) *Industrial & Engineering Chemistry Research*, **41**, 5934–5940.

198 Tai, H., Wang, W. and Howdle, S.M. (2005) *Macromolecules*, **38**, 1542–1545.

199 Tai, H., Wang, W. and Howdle, S.M. (2005) *Polymer*, **46**, 1062610636.

200 Mueller, P.A., Storti, G., Morbidelli, M., Costa, I., Galia, A., Scialdone, O. and Filardo, G. (2006) *Macromolecules*, **39**, 6483–6488.

201 Mueller, P.A., Storti, G., Morbidelli, M., Apostolo, M. and Martin, R. (2005) *Macromolecules*, **38**, 7150–7163.

202 Galia, A., Giaconia, A., Scialdone, O., Apostolo, M. and Filardo, G. (2006) *Journal of Polymer Science Part A-Polymer Chemistry*, **44**, 2406–2418.

203 Ahmed, T.S., DeSimone, J.M. and Roberts, G.W. (2004) *Science*, **59**, 5139–5144.

204 Royer, J. R. and Roberts, G. W.(to North Carolina State University), US Patent Application 2003-072690, priority date 17 April (2003) CAN 138:304727.

205 Charpentier, P.A., Kennedy, K.A., DeSimone, J.M. and Roberts, G.W. (1999) *Macromolecules*, **32**, 5973–5975.

206 Tai, H., Liu, J. and Howdle, S.M. (2005) *European Polymer Journal*, **41**, 2544–2551.

207 Tai, H., Wang, W., Martin, R., Liu, J., Lester, E., Licence, P., Woods, H.M. and Howdle, S.M. (2005) *Macromolecules*, **38**, 355–363.

208 Gangal, V. (2003) Perfluorinated polymers, perfluorinated ethylene–propene copolymers, in *Encyclopedia of Polymer Science and Technology*, 3rd edn vol. 3, (eds H.F. Mark and J.I. Kroschwitz), Wiley-Interscience, Hoboken, NJ, pp. 364–378.

209 Gangal, V. (2003) Perfluorinated polymers, perfluorinated ethylene–ethylene copolymers, in *Encyclopedia of Polymer Science and Technology*, 3rd edn vol. 3, (eds H.F. Mark and J.I. Kroschwitz), Wiley-Interscience, Hoboken, NJ, pp. 402–418.

210 Gangal, V. (2003) Perfluorinated polymers, perfluorinated ethylene–perfluorodioxole copolymers, in *Encyclopedia of Polymer Science and Technology*, 3rd edn vol. 3, (eds H.F. Mark and J.I. Kroschwitz), Wiley-Interscience, Hoboken, NJ, pp. 418–422.

211 Gangal, V. (2003) perfluorinated polymers, perfluorinated ethylene–perfluorovinyl ether copolymers, in *Encyclopedia of Polymer Science and Technology*, 3rd edn vol. **3**, (eds H.F. Mark and J.I. Kroschwitz), Wiley-Interscience, Hoboken, NJ, pp. 422–435.

212 Carlson, D.P. and Schmiegel, W. (2001) Fluoropolymers, in *Industrial Polymers Handbook*, Vol. 1, (ed. E.S. Wilks), Wiley-VCH Verlag GmbH, Weinheim, pp. 506–564.

213 Novitskaya, S.P., Nudelman, Z.N. and Dontsov, A.A. (1989) *Fluoroelastomers. (Ftorelastomery)*, Khimiya, Moscow.

214 Ameduri, B., Boutevin, B. and Kostav, G. (2002) *Macromolecular Chemistry and Physics*, **203**, 1763–1778.

215 Ameduri, B., Boutevin, B. and Kostav, G. (2001) *Progress in Polymer Science*, **26**, 105–187.

216 Souzy, R., Ameduri, B. and Boutevin, B. (2004) *Macromolecular Chemistry and Physics*, **205**, 476–485.

217 Arcella, V. and Ferro, R. (1997) Fluorocarbon elastomers, in *Modern Fluoropolymers* (ed. J. Scheirs), John Wiley & Sons, Ltd, Chichester, Chapter 2, pp. 71–90.

218 Tournut, C. (1994) *Macromolecular Symposia*, **82**, 99–109.

219 Tournut, C. (1997) Thermoplastic copolymers of vinylidene fluoride, in *Modern Fluoropolymers* (ed. J. Scheirs), John Wiley & Sons, Inc., Chichester, pp. 577–596.

220 Logothetis, A.L. (1989) *Progress in Polymer Science*, **14**, 251–296.

221 Cook, D. and Lynn, M. (1990) *RAPRA Reviews of Reproduction*, **3**, 32/1–32/112.

222 Hull, D.D.E., Johnson, B.V., Rodricks, J.P., Staley, J.B. (1997) THV fluoroplastic, in *Modern Fluoropolymers* (ed. J. Scheirs), John Wiley & Sons, Ltd, Chichester, pp. 257–270.

223 Minhas, P.S. and Petrucelli, F. (1977) *Plastics Engineering*, **33**, 60–63.

224 Allied, CMX Technical Brochure, Allied, (1992) Allied-Signal, Morristown, NJ, USA.

225 Souzy, R., Guiot, J., Ameduri, B., Boutevin, B. and Paleta, O. (2003) *Macromolecules*, **36**, 9390–9395.

226 Worm, A.T. and Grootaert, W. (2005) Fluorocarbon elastomers, in *Encyclopedia of Polymer Science and Technology*, 3rd edn vol. 2, (eds H.F. Mark and J.I. Kroschwitz), Wiley-Interscience, Hoboken, NJ, pp. 577–590.

227 McCarthy, T.F. (1998) *Journal of Applied Polymer Science*, **70**, 2211–2225.

228 Lu, X., Schirokauer, A. and Scheinbeim, J.I. (2000) *IEEE Trans. Ultrason. Ferroelectr. Freq. Control*, **47**, 1291–1291.

229 Scheinbeim, J.I. and Gao, Q. (2001) *Proceedings SPIE*, **4329**, 131–132.

230 Jayasuriya, A.C., Schirokauer, A. and Scheinbeim, J.I. (2001) *Journal of Polymer Science Part B-Polymer Physics*, **39**, 2793–2801.

231 Mertdogan, C.A., DiNoia, T.P. and McHugh, M.A. (1997) *Macromolecules*, **30**, 7511–7515.

232 Tai, H., Wang, W. and Howdle, S.M. (2005) *Macromolecules*, **38**, 9135–9142.

233 Beginn, U., Najjar, R., Ellmann, J., Vinokur, R., Martin, R. and Möller, M. (2006) *Journal of Polymer Science Part A-Polymer Chemistry*, **44**, 1299–1316.

234 Ahmed, T.S., DeSimone, J.M. and Roberts, G.W. (2007) *Macromolecules*, **40**, 9322–9331.

235 Michel, U., Resnick, P., Kipp, B. and DeSimone, J.M. (2003) *Macromolecules*, **36**, 7107–7113.

236 Wood, C.D., Michel, U., Rolland, J.P. and DeSimone, J.M. (2004) *Journal of Fluorine Chemistry*, **125**, 1671–1676.

237 Lousenberg, R.D. and Shoichet, M.S. (2000) *Macromolecules*, **33**, 1682–1685.

238 Baradie, B. and Shoichet, M.S. (2002) *Macromolecules*, **35**, 3569–3575.

239 Baradie, B. and Shoichet, M.S. (2005) *Macromolecules*, **38**, 5560–5568.

240 Schreyer, G. (1972) *Konstruieren mit Kunststoffen*, vol. 1, Carl Hanser, Munich.

241 Slone, R.V. (2005) Methacrylic ester polymers, in *Encyclopedia of Polymer Science and Technology*, 3rd edn vol. 3, (eds H.F. Mark and J.I. Kroschwitz), Wiley-Interscience, Hoboken, NJ, pp. 249–277.

242 Emslie, C. (1988) *Journal of Materials Science*, **23**, 2281–2293.

243 Kaetsu, I., Yoshida, K. and Okubo, O. (1979) *Applied Radiation and Isotopes*, **30**, 209–212.

244 Louie, B.M., Carratt, G.M. and Soong, D.S. (1985) *Journal of Applied Polymer Science*, **30**, 3985–4012.

245 Cooper, A.L. and DeSimone, J.M. (1996) *Polymeric Materials Science and Engineering*, **74**, 262–263.

246 DeSimone, J.M., Maury, E.E., Menceloglu, Y.Z., McClain, J.B., Romack, T.J. and Combes, J.R. (1994) *Science*, **265**, 356–359.

247 Fortini, S. and Meyer, T. (2004) *DECHEMA Monographs*, **138**, 361365.

248 Lu, S., Zhang, Z., Nawaby, A.V. and Day, M. (2003) *Annual Technical Conference Society Plastics Engineering*, **61**, 1794–1798.

249 Lu, S., Zhang, Z., Nawaby, A.V. and Day, M. (2004) *Journal of Applied Polymer Science*, **93**, 1236–1239.

250 Wang, W., Griffiths, R.M.T., Giles, M.R., Williams, P. and Howdle, S.M. (2003) *European Polymer Journal*, **39**, 423–428.

251 Quadir, M.A., DeSimone, J.M., Van Herk, A. and German, A.L. (1998) *Macromolecules*, **31**, 6481–6485.

252 Hsiao, Y.-L., Maury, E.E., DeSimone, J.M., Mawson, S.M. and Johnston, K.P. (1995) *Macromolecules*, **28**, 8159–8166.

253 Wang, Z., Yang, Y.J., Dong, Q.Z. and Hu, C.P. (2008) *Journal of Applied Polymer Science*, **110**, 468–474.

254 Rosell, A., Storti, G. and Morbidelli, M. (2001) *DECHEMA Monographs*, **137**, 467–474.

255 Chatzidoukas, C., Pladis, P. and Kiparissides, C. (2003) *Industrial & Engineering Chemistry Research*, **42**, 743–751.

256 Hwang, H.S., Gal, Y.-S., Johnston, K.P. and Lim, K.T. (2006) *Macromolecular Rapid Communications*, **27**, 121–125.

257 Woods, H., Nouvel, C., Licence, P., Irvine, D.J. and Howdle, S.M. (2005) *Macromolecules*, **38**, 3271–3282.

258 Rosell, A., Storti, G., Morbidelli, M., Bratton, D. and Howdle, S.M. (2004) *Macromolecules*, **37**, 2996–3004.

259 Deniz, S., Baran, N., Akgun, M., Uzun, I., Nimet, D. and Dincer, S. (2005) *Polymer International*, **54**, 1660–1668.

260 Hwang, H.S., Lee, W.-K., Hong, S.-S., Jin, S.-H. and Lim, K.T. (2007) *Journal of Supercritical Fluids*, **39**, 409–415.

261 Wang, W., Naylor, A. and Howdle, S.M. (2003) *Macromolecules*, **36**, 5424–5427.

262 Tai, H., Popov, V. K., Shakesheff, K. M., and Howdle, S. M. (2007) *Biochemical Society Transactions*, **35**, 516–521.

263 Galia, A., Pierro, P. and Filardo, G. (2004) *Journal of Supercritical Fluids*, **32**, 255–263.

264 Ding, L. and Olesik, S.V. (2003) *Macromolecules*, **36**, 4779–4785.

265 Hwang, H.S., Yuvaraj, H., Kim, W.S., Lee, W.-K., Gal, Y.S. and Lim, K.T. (2008) *Journal of Polymer Science Part A-Polymer Chemistry*, **46**, 13651375.

266 Giles, M.R., OConnor, S.J., Hay, J.N., Winder, R.J. and Howdle, S.M. (2000) *Macromolecules*, **33**, 1996–1999.

267 Giles, M.R., Griffiths, R.M.T., Aguiar-Ricardo, A., Silva, M.M.C.G. and Howdle, S.M. (2001) *Macromolecules*, **34**, 20–25.

268 Hems, W.P., Yong, T.-M., van Nunen, J.L., Cooper, A.I., Holmes, A.B. and Griffin, D.A. (1999) *Journal of Materials Chemistry*, **9**, 14031407.

269 Shaffer, K.A., Jones, T.A., Canelas, D.A., DeSimone, J.M. and Wilkinson, S.P. (1996) *Macromolecules*, **29**, 2704–2706.

270 Giles, M.R., Hay, J.N., Howdle, S.M. and Winder, R.J. (2000) *Polymer*, **41**, 6715–6721.

271 (a) ONeill, M.L., Yates, M.Z., Johnston, K.P., Smith, C.D. and Wilkinson, S.P. (1996) *Macromolecules*, **31**, 2838–2847; (b) ONeill, M.L., Yates, M.Z., Johnston, K.P., Smith, C.D. and Wilkinson, S.P. (1996) *Macromolecules*, **31**, 2848–2856.

272 Fehrenbacher, U., Muth, O., Hirth, T. and Ballauff, M. (2000) *Macromolecular Chemistry and Physics*, **201**, 1532–1539.

273 Fehrenbacher, U., Ballauff, M., Muth, O. and Hirth, T. (2001) *Applied Organometallic Chemistry*, **15**, 613–616.

274 Fehrenbacher, U. and Ballauff, M. (2002) *Macromolecules*, **35**, 3653–3661.

275 Ballauff, M., Fehrenbacher, U. and Hirth, T. (2003) *Chemie Ingenieur Technik*, **75**, 1638–1642.

276 (a) Paine, A.J., Wayne, L. and McNulty, J. (1990) *Macromolecules*, **23**, 3104–3109; (b) Paine, A.J. (1990) *Macromolecules*, **23**, 3109–3117.

277 Mueller, P.A., Storti, G. and Morbidelli, M. (2005) *Chemical Engineering Science*, **60**, 377–397.

278 Mueller, P.A., Storti, G., Morbidelli, M., Mantelis, C.A., Charalampos, A. and Meyer, T. (2007) *Macromolecular Symposia*, **259**, 218–225.

279 Mantelis, C.A. and Meyer, T. (2008) *AICHE Journal*, **54**, 529–536.

280 Mantelis, C.A., Barbey, R., Fortini, S. and Meyer, T. (2007) *Macromolecules Reaction Engineering*, **1**, 78–85.

281 Ganapathy, H.S., Hwang, H.S., Jeong, Y.T. and Lim, K.T. (2006) *Studies in Surface Science and Catalysis*, **159**, 797–800.

282 Fujii, S., Minami, H. and Okubo, M. (2005) *Colloid and Polymer Science*, **284**, 327–333.

283 Okubo, M., Fujii, S. and Minami, H. (2004) *Progress Colloid and Polymer Science*, **124**, 121–125.

284 Fujii, S., Minami, H. and Okubo, M. (2004) *Colloid and Polymer Science*, **282**, 569–574.

285 Yates, M.Z., Li, H., Shim, J.J., Maniar, S., Johnston, K.P., Lim, K.T. and Webber, S. (1999) *Macromolecules*, **32**, 1018–1026.

286 Okubo, M., Fujii, S., Maenaka, H. and Minami, H. (2002) *Colloid and Polymer Science*, **280**, 183–187.

287 Park, J.-Y. and Shim, J.-J. (2003) *Journal of Supercritical Fluids*, **27**, 297–307.

288 Giles, M.R. and Howdle, S.M. (2001) *European Polymer Journal*, **37**, 1347–1351.

289 Hwang, H.S. and Lim, K.T. (2006) *Macromolecular Rapid Communications*, **27**, 722–726.

290 Yang, J., Wang, W., Sazio, P.J. and Howdle, S.M. (2007) *European Polymer Journal*, **43**, 663–667.

291 Zhao, Q. and Samulski, E.T. (2005) *PMSE Prepr.*, **93**, 892–893.

292 Zhao, Q. and Samulski, E.T. (2005) *Macromolecules*, **38**, 7967–7971.

293 Zhao, Q. and Samulski, E.T. (2006) *Polymer*, **47**, 663–671.

294 Giaconia, A., Filardo, G., Scialdone, O. and Galia, A. (2006) *Journal of Polymer Science Part A-Polymer Chemistry*, **44**, 4122–4135.

295 Minami, H., Kagawa, Y., Kuwahara, S., Shigematsu, J., Fujii, S. and Okubo, M. (2004) *Desig. Mon. Polym.*, **7**, 553–562.

296 Grignard, B., Jerome, C., Calberg, C., Jerome, R. and Detrembleur, C. (2008) *European Polymer Journal*, **44**, 861–871.

297 Thurecht, K.J., Gregory, A.M., Wang, W. and Howdle, S.M. (2007) *Macromolecules*, **40**, 2965–2967.

298 Gregory, A.M., Thurecht, K.J. and Howdle, S.M. (2008) *Macromolecules*, **41**, 1215–1222.

299 Forster, D.J., Heuts, J.P.A., Lucien, F.P. and Davis, T.P. (1999) *Macromolecules*, **32**, 5514–5518.

300 Wang, W., Irvine, D.J. and Howdle, S.M. (2005) *Industrial & Engineering Chemistry Research*, **44**, 8654–8658.

301 Mang, S.A., Dokolas, P. and Andrew, A.B. (1999) *Organic Letters*, **1**, 125–127.

302 Priddy, D. (2005) Styrene polymers, in *Encyclopedia of Polymer Science and Technology*, 3rd edn vol. 4, (eds H.F. Mark, and J.I. Kroschwitz), Wiley-Interscience, Hoboken, NJ, pp. 247–336.

303 Odian, G. (1991) *Principles of Polymerization*, 3rd edn John Wiley & Sons Inc., New York, Chapter 3-9c-1, p. 285.

304 Pasquino, A.M. and Pilsworth, M.N. Jr. (1964) *Journal of Polymer Science Part C-Polymer Letters*, **2**, 253–255.

305 Odian, G. (1991) *Principles of Polymerization*, 3rd edn John Wiley & Sons Inc., New York. Chapter 3-13b-2, pp. 306–308.

306 Liu, J., Han, B., Liu, Z., Wang, J. and Huo, Q. (2001) *Journal of Supercritical Fluids*, **20**, 171–176.

307 Mingotaud, A.-F., Begue, G., Cansell, F. and Gnanou, Y. (2001) *Macromolecular Chemistry and Physics*, **202**, 2857–2863.

308 Canelas, D.A., Betts, D.E. and DeSimone, J.M. (1996) *Macromolecules*, **29**, 2818–2821.

309 Baran, N., Deniz, S., Akguen, M., Uzun, I. N., and Dincer, S. (2005) *European Polymer Journal*, **41**, 1159–1167.

310 Qiu, G.-M., Zhu, B.-K., Ku, Y.-Y. and Geckeler, K.E. (2006) *Macromolecules*, **39**, 3231–3237.

311 Cooper, A.I., Hems, W.P. and Holmes, A.B. (1999) *Macromolecules*, **32**, 2156–2166.

312 Jadhav, A.V., Patwardhan, S.V., Ahn, H.W., Selby, C.E., Stuart, J.O., Zhang, X.K., Kuo, C.M., Sheerin, E.A., Vaia, R.A., Smith, S.D. and Clarson, S.J. (2007) *ACS Symposium Series*, **964**, 116–133.

313 Arita, T., Beuermann, S., Buback, M. and Vana, P. (2004) *e-Polymers*, **03**, 1–14.

314 Ryan, J., Aldabbagh, F., Zetterlund, P.B. and Okubo, M. (2005) *Polymer*, **46**, 9769–9777.

315 McHale, R., Aldabbagh, F., Zetterlund, P.B., Minami, H. and Okubo, M. (2006) *Macromolecules*, **39**, 6853–6860.

316 McHale, R., Aldabbagh, F., Zetterlund, P.B. and Okubo, M. (2007) *Macromolecular Chemistry and Physics*, **208**, 1813–1822.

317 McHale, R., Aldabbagh, F., Zetterlund, P.B. and Okubo, M. (2006) *Macromolecular Rapid Communications*, **27**, 1465–1471.

318 Aldabbagh, F., Zetterlund, P.B. and Okubo, M. (2008) *Macromolecules*, **41**, 2732–2734.

319 Liu, T., DeSimone, J.M. and Roberts, G.W. (2004) *Polym. Prepr.*, **45**, (1), 510–511.

320 Liu, T., Desimone, J.M. and Roberts, G.W. (2005) *Journal of Polymer Science Part A-Polymer Chemistry*, **43**, 2546–2555.

321 Liu, T., Garner, P., DeSimone, J.M., Roberts, R.G. and George, G.D. (2006) *Macromolecules*, **39**, 6489–6494.

322 Liu, T., DeSimone, J.M. and Roberts, G.W. (2006) *Chemical Engineering Science*, **61**, 3129–3139.

323 Xu, Q., Han, B. and Yan, H. (2000) *Polymer*, **42**, 1369–1373.

324 Liu, T., DeSimone, J.M. and Roberts, G.W. (2006) *Polymer*, **47**, 4276–4281.

325 Ye, L., Yoshimatsu, K., Kolodziej, D., Da Cruz, F.J. and Dey, E.S. (2006) *Journal of Applied Polymer Science*, **102**, 2863–2867.

326 Ma, L., Zhang, L., Le, J.-C., Yang, J.-C. and Xie, X.-M. (2002) *Journal of Applied Polymer Science*, **86**, 2272–2278.

327 Hu, H., He, T., Feng, J., Chen, M. and Cheng, R. (2002) *Polymer*, **43**, 6357–6361.

328 Cao, L. and Chen, L. (2006) *Polymer Bulletin*, **57**, 651–659.

329 Cao, L., Chen, L. and Lai, W. (2007) *Journal of Polymer Science Part A-Polymer Chemistry*, **45**, 955–962.

330 Ohde, H., Wai, C.M. and Rodriguez, J.M. (2007) *Colloid and Polymer Science*, **285**, 475–478.

331 Adamsky, F.A. and Beckman, E.J. (1994) *Macromolecules*, **27**, 312–14.

332 Zhang, B., Chen, M. and Liu, W. (2004) *Polym. Prepr.*, **45**, (1), 512–513.

333 Butler, R., Davies, C.M. and Cooper, A.I. (2001) *Advanced Materials*, **13**, 1459–1463.

334 Galia, A., Muratore, A. and Filardo, G. (2003) *Industrial & Engineering Chemistry Research*, **42**, 448–455.

335 Ye, W. and DeSimone, J.M. (2005) *Macromolecules*, **38**, 2180–2190.

336 Shiho, H. and DeSimone, J.M. (2000) *Macromolecules*, **33**, 15651569.

337 Teng, X.-R., Hu, X.-C. and Shao, H.-L. (2002) *Polymer Journal*, **34**, 534–538.

338 Teng, X.-R. (2003) *Journal of Applied Polymer Science*, **88**, 1393–1398.

339 Okubo, M., Fujii, S., Maenaka, M., Hiroshi, M. and Minami, H. (2003) *Colloid and Polymer Science*, **281**, 964–972.

340 Wang, Z., Yang, Y.J., Dong, Q., Liu, T. and Hu, C.P. (2006) *Polymer*, **47**, 7670–7679.

341 Yeo, S.-D. and Erdogan, E. (2004) *Macromolecules*, **37**, 8239–8248.

342 Yang, Y., Dong, Q., Wang, Z., Shen, C., Huang, Z., Zhu, H., Liu, T. and Hu, C.P. (2006) *Journal of Applied Polymer Science*, **102**, 5640–5648.

343 Giles, M., Griffiths, R.M.T., Irvine, D.J. and Howdle, S.M. (2003) *European Polymer Journal*, **39**, 1785–1790.

344 Wang, D. and DeSimone, J.M. (1999) *Polymeric Materials Science and Engineering*, **80**, 526–527.

345 Casimiro, T., Banet-Osuna, A.N., Ramos, A.M., da Ponte, M.N. and Aguiar-Ricardo, A. (2005) *European Polymer Journal*, **41**, 19471953.

346 Wang, W., Giles, M.R., Bratton, D., Irvine, D.J., Armes, S.P., Weaver, J.V. and Howdle, S.M. (2003) *Polymer*, **44**, 3803–3809.

347 Okubo, M., Fujii, S., Maenaka, M. and Minami, H. (2002) *Colloid and Polymer Science*, **280**, 1084–1090.

348 Giles, M., Hay, J.N. and Howdle, S.M. (2000) *Macromolecular Rapid Communications*, **21**, 1019–1023.

349 Shim, J.-J., Yates, M.Z. and Johnston, K.P. (1999) *Industrial & Engineering Chemistry Research*, **38**, 3655–3662.

350 Shiho, H. and DeSimone, J.M. (2001) *Macromolecules*, **34**, 11981203.

351 Wang, W., Griffiths, R.M.T., Naylor, A., Giles, M.R., Irvine, I.J. and Howdle, S.M. (2002) *Polymer*, **43**, 6653–6659.

352 Matsuyama, K., Mishima, K., Keiji, T., Ryugen, R. and Tomokage, H. (2003) *Journal of Chemical Engineering of Japan*, **36**, 516–521.

353 Zhang, D., Mishima, K., Matsuyama, K., Zhou, L. and Zhang, R. (2007) *Journal of Applied Polymer Science*, **103**, 2425–2431.

354 Ma, Z. and Lacroix-Desmazes, P. (2004) *Polymer*, **45**, 6789–6797.

355 Shiho, H. and DeSimone, J.M. (2000) *Journal of Polymer Science Part A-Polymer Chemistry*, **38**, 3783–3790.

356 Oh, K.S., Bae, W. and Kim, H. (2008) *European Polymer Journal*, **44**, 415–425.

357 Cao, L., Chen, L., Chen, X., Zuo, L. and Li, Z. (2006) *Polymer*, **47**, 4588–4595.

358 Cao, L., Chen, L., Jiao, J., Zhang, S. and Gao, W. (2007) *Colloid and Polymer Science*, **285**, 1229–1236.

359 Cao, L., Chen, L., Cui, P. and Wang, J. (2008) *Journal of Applied Polymer Science*, **108**, 3843–3850.

360 Kwon, S., Lee, K., Bae, W. and Kim, H. (2008) *Polymer Journal*, **40**, 332–338.

361 Duarte, A.R.C., Casimiro, T., Aguiar-Ricardo, A., Simplicio, A.L. and Darte, C.M.M. (2006) *Journal of Supercritical Fluids*, **39**, 102–106.

362 Shiho, H. and Desimone, J.M. (2000) *Journal of Polymer Science Part A-Polymer Chemistry*, **38**, 3100–3105.

363 Kwon, S., Bae, W. and Kim, H. (2004) *Korean Journal of Chemical Engineering*, **21**, 910–914.

364 Wood, C. and Cooper, A.I. (2001) *Macromolecules*, **34**, 5–8.

365 Ma, Z. and Lacroix-Desmazes, P. (2004) *Polymer*, **45**, 6789–6797.

366 Li, G., Johnston, K.P., Zhou, H., Venumbaka, S.R. and Cassidy, P. (2001) *Polym. Prepr.*, **42**, (2), 817–817.

367 Allsopp, M.W. and Vianello, G. (2005) Vinyl chloride polymers, in *Encyclopedia of Polymer Science and Technology*, 3rd edn vol. 8, (eds H.F. Mark and J.I. Kroschwitz), Wiley-Interscience, Hoboken, NJ, pp. 473–476.

368 Sumitomo Chemical, French Patent 1 524 533, priority date 5 October (1968), CAN 71:3903.

369 Fukui, K., Kagiya, T., Yokota, H., Toriuchi, Y. and Kuniyoshi, F.(to Sumitomo Chemical), Japanese Patent JP 45025305, priority date 21 August (1970) CAN 74:3997.

370 Lee, S., Kwak, S. and Azzam, F.US Patent 5 663 237, priority date 2 September (1997) CAN 127:221178.££.

371 Muth, O., Hirth, T. and Vogel, H. (2000) *Journal of Supercritical Fluids*, **17**, 65–72.

372 Li, D. and Han, B. (2000) *Macromolecules*, **33**, 4555–4560.

373 Trivedi, A.H., Kwak, S. and Lee, S. (2001) *Polymer Engineering and Science*, **41**, 1923–1937.

374 Yajun, W., Zi, G., Yan, L., Liu, T., Hu, C. and Dong, Q. (2006) *Journal of Applied Polymer Science*, **102**, 1146–1151.

375 Wulff, G. and Sarhan, A. (1972) *Angewandte Chemie (International Edition in English)*, **11**, 341.

376 Komiyama, M., Takeushi, T., Mukawa, T. and Asanuma, H. (2003) *Molecular Imprinting – From Fundamentals to Applications*, 1st edn Wiley-VCH Verlag GmbH, Weinheim.

377 Li, W. and Li, S. (2007) *Advances in Polymer Science*, **206**, 191–211.

378 Lovell, P.A. and El-Aasser, M.S. (eds) (1997) *Emulsion Polymerization and Emulsion Polymers*, John Wiley & Sons, Ltd, Chichester.

379 Beckman, E.J. (2005) Inverse emulsion polymerization in carbon dioxide, in *Supercritical Carbon Dioxide in Polymer Reaction Engineering* (eds M.F. Kemmere and T. Meyer), Wiley-VCH Verlag GmbH, Weinheim, Chapter 7, pp. 139–156.

380 Ye, W.-J., Keiper, J.S. and DeSimone, J.M. (2006) *Chinese Journal of Polymer Science*, **24**, 95–101.

381 Peacock, A.I. (2000) *Handbook of Polyethylene: Structures: Properties, and Applications*, Marcel Dekker, New York, Chapter 3.

382 Groppo, E., Lamberti, C., Bordiga, S., Spoto, G. and Zecchina, A. (2005) *Chemical Reviews*, **105**, 115–183.

383 Zucchini, U., DallOcco, T. and Resconi, L. (1993) *Indian Journal of Technology*, **31**, 247–262.

384 Kaminsky, W. (2001) *Advances in Catalysis*, **46**, 89–159.

385 Brintzinger, H.H., Fischer, D., Mülhaupt, R., Rieger, B. and Waymouth, R.M. (1995) *Angewandte Chemie (International Edition in English)*, **34**, 1143–1170.

386 Cecchin, G., Morini, G. and Piemontesi, F. (2007) Ziegler–Natta catalysts, in *Kirk–Othmer Encyclopedia of Chemical Technology* (eds A. Seidel, R. E. Kirk and D. Othmer), 5th edn, vol. 26, Wiley-VCH Verlag GmbH, Weinheim, pp. 502–554.

387 Kreher, U., Sebastian, S. and Walther, D. (1998) *Zeitschrift Fur Anorganische und Allgemeine Chemie*, **624**, 602–612.

388 Boffa, L.S. and Novak, B.M. 2000 *Chemical Reviews*, **100**, 1479–1494.

389 Ittel, S.D., Johnson, L.K. and Brookhart, M. (2000) *Chemical Reviews*, **100**, 1169–1204.

390 Rieger, B., Baughm, L.S., Kacker, S. and Striegler, S. (eds) (2003) *Late Transition Metal Polymerization Catalysis*, Wiley-VCH Verlag GmbH, Weinheim.

391 Kemmere, M., de Vries, T.J. and Keurentjes, J. (2005) Catalytic polymerizations of olefins in supercritical carbon dioxide, in *Supercritical Carbon Dioxide in Polymer Reaction Engineering* (eds M.F. Kemmere and T. Meyer), Wiley-VCH Verlag GmbH, Weinheim, Chapter 8, pp. 157–187.

392 Coates, G.W. and Moore, D.R. (2004) *Angewandte Chemie-International Edition*, **43**, 6618–6639.

393 Darensbourg, D.J. (2007) *Chemical Reviews*, **107**, 2388–2410.

394 Koning, C., Wildeson, J., Parton, R., Plum, B., Steeman, P. and Darensbourg, D.J. (2001) *Polymer*, **42**, 3995–4004.

395 Peng, S.M., An, Y., Chen, C., Fei, B., Zhuang, Y. and Dong, L. (2003) *Polymer Degradation and Stability*, **80**, 141–147.

396 Inoue, S., Tsuruta, T. and Furukawa, J. (1962) *Macromolecular Chemistry*, **53**, 215–218.

397 Darensbourg, D.J., Stafford, N.W. and Katsuaro, T. (1995) *Journal of Molecular Catalysis A-Chemical*, **104**, L1–L4.

398 Costello, C.A., Berluche, E., Han, S.J., Sysyn, D.A., Super, M.S. and Beckman, E.J. (1996) *Polym. Prepr.*, **74**, (1), 430.

399 Eberhardt, R., Allmendinger, M., Zintl, M., Troll, C., Luinstra, G.A. and Rieger, B. (2004) *Macromolecular Chemistry and Physics*, **205**, 4247.

400 Inoue, S. (1978) *Journal of Polymer Science Part A-Polymer Chemistry*, **38**, 2861–2871.

401 Sugimoto, H., Ohshima, H. and Inoue, S. (2003) *Journal of Polymer Science Part A-Polymer Chemistry*, **41**, 3549–3555.

402 Mang, S., Cooper, A.I., Colclough, M.E., Chauhan, N. and Holmes, A.B. (2000) *Macromolecules*, **33**, 303–308.

403 Sarbu, T., Styranec, T.J. and Beckman, E.J. (2000) *Industrial & Engineering Chemistry Research*, **39**, 4678–4683.

404 Cohen, C.T., Thomas, C.M., Peretti, K.L., Lobkovsky, E.B. and Coates, G.W. 2006 *Dalton Transactions*, 237–249.

405 Luinstra, G.A., Haas, G.R., Molnar, F., Bernhart, V., Eberhart, R. and Rieger, B. (2005) *Chemistry - A European Journal*, **11**, 6298–6314.

406 Belov, G.P. and Novikova, E.V. (2004) *Russian Chemical Reviews*, **73**, 267–291.

407 Nozaki, K. (2005) *Polymer Journal*, **37**, 871–876.

408 Garcia Suarez, E.J., Godard, C., Ruiz, A. and Claver, C. (2007) *European Journal of Inorganic Chemistry*, (18), 2582–2593.

409 Gimenez-Pedros, M., Tortosa-Estorach, C., Bastero, A., Masdeu-Bulto, A.M., Solinas, M. and Leitner, W. (2006) *Green Chemistry*, **8**, 875–877.

410 Hori, H., Six, C. and Leitner, W. (1999) *Macromolecules*, **32**, 3178–3182.

411 Naidu, B.V.K., Oh, B.-H., Nam, D.-H., Hwang, C.-K., Jin, S.-H., Lim, K.-T., Hwang, H.-S., Lee, J.W. and Gal, Y.-S. (2006) *Journal of Polymer Science Part A-Polymer Chemistry*, **44**, 1555–1560.

412 Pack, J.W., Hur, Y.-J., Kim, H. and Lee, Y.-W. 2005 *Chemical Communications*, 5208–5210.

413 Buchmeiser, M.R. (ed.) (2005) *Metathesis Polymerization*, vol. 176, Advances in Polymer Science, Springer, Berlin.

414 Khosravi, E. (2002) *NATO Science Series II*, **56**, 105–115.

415 Chuah, H.H., Ellison, R.H., Dale, D.L. Jr. and Scardino, B.M. (1997) *Journal of Polymer Science Part A-Polymer Chemistry*, **35**, 3049–3063.

416 Hine, P.J., Leejarkpai, T., Khosravi, E., Duckett, R.A. and Feast, W.J. (2001) *Polymer*, **42**, 9413–9422.

417 Mistele, C.D., Thorp, H.H. and DeSimone, J.M. (1996) *J. Macromol. Sci. A: Polym. Chem.*, **33**, 953–960.

418 Grubbs, R.H. (2003) *Handbook of Metathesis*1st edn vol. 1, Wiley-VCH Verlag GmbH, Weinheim.

419 Fuerstner, A., Ackermann, L., Beck, K., Hori, H., Koch, D., Langemann, K., Liebl, M., Six, C. and Leitner, W. (2001) *Journal of the American Chemical Society*, **123**, 9000–9006.

420 Hu, X., Blanda, X.M.T., Venumbaka, S.R. and Cassidy, P.E. (2005) *Polymers for Advanced Technologies*, **16**, 146–149.

421 Cao, C. and McCarthy, T.J. (2001) *Polymeric Materials Science and Engineering*, **84**, 47–48.

422 Matyaszewki, K. (ed.) (1996) *Cationic Polymerizations*, Marcel Dekker, New York.

423 Higashimura, T., Sawamoto, M. and Miyamoto, M. (1984) *Macromolecules*, **17**, 265–268.

424 Sawamoto, M. (1996) Controlled polymer synthesis by cationic polymerization, in *Cationic Polymerizations* (ed. K. Matyaszewki), Marcel Dekker, New York, Chapter 5, pp. 381–436.

425 Clark, M.R. and DeSimone, J.M. (1995) *Macromolecules*, **28**, 3002–3004.

426 Clark, M.R., Kendeall, J.L. and DeSimone, J.M. (1997) *Macromolecules*, **30**, 6011–6014.

427 Mingotaud, A.-F., Dargelas, F. and Cansell, F. (2000) *Macromolecular Symposia*, **153**, 77–86.

428 Yokoyama, C., Kawase, Y., Shibasaki-Kitakawa, N. and Smith, R.L. Jr. (2003) *Journal of Applied Polymer Science*, **89**, 3167–3174.

429 Veiga de Macedo, C., Soares da Silva, M., Casimiro, T., Cabrita, E.J. and Aguiar-Ricardo, A. (2007) *Green Chemistry*, **9**, 948–953.

430 Liu, Z., Sharma, B.K. and Brajendra, S.Z. (2007) *Biomacromolecules*, **8**, 233–239.

431 Carey, F.A. and Sundberg, R.J. (1983) *Advanced Organic Chemistry*, 2nd edn Plenum Press, New York, Chapter 6.

432 Penczek, S., Duda, A., Kubisa, P. and Slomkowski, S. (2007) Ionic and coordination ring-opening polymerization, in *Macromolecular Engineering*, vol. 1, (eds K. Matyaszewski, Y. Gnanou and L. Leibler), Wiley-VCH

Verlag GmbH, Weinheim, Chapter 5, pp. 161–216.

433 Kumar, N., Ezra, A., Ehrenfroind, T., Krasko, M.Y. and Domb, A.J. (2005) Biodegradable polymers, in *Encyclopedia of Polymer Science and Technology*, 3rd edn vol. 5, (eds H.F. Mark and J.I. Kroschwitz), Wiley-Interscience, Hoboken, NJ, pp. 263–285.

434 Albertsson, A.-C. and Varma, I.K. (2002) Aliphatic polyesters, in *Biopolymers*, vol. 4, (eds Y. Doi and Y. Steinbüchel), Wiley-VCH Verlag GmbH, Weinheim, Chapter 2, pp. 26–52.

435 Mingotaud, A.F., Cansell, F., Gilbert, N. and Soum, A. (1999) *Polymer Journal (Tokyo)*, **31**, 406–410.

436 Stassin, F., Halleux, O. and Jerome, R. (2001) *Macromolecules*, **34**, 775–781.

437 Stassin, F. and Jerome, R. 2003 *Chemical Communications*, 232–233.

438 Bergeot, V., Tassaing, T., Besnard, M., Cansell, F. and Mingotaud, A.-F. (2004) *Journal of Supercritical Fluids*, **28**, 249–261.

439 Lecomte, Ph., Stassin, F. and Jerome, R. (2004) *Macromolecular Symposia*, **215**, 325–338.

440 Bratton, D., Brown, M. and Howdle, S.M. (2005) *Macromolecules*, **38**, 1190–1195.

441 Stassin, F. and Jerome, R. (2000) *Journal of Polymer Science Part A-Polymer Chemistry*, **43**, 2777–2789.

442 Saner, B., Bilgin, N., Nalan, A., Namuslu, A., Piskin, E. and Mencelogiu, Y.Z. (2006) *Polymeric Materials Science and Engineering Prepr.*, **95**, 413–414.

443 Saner, B., Menceloglu, Y.Z. and Oncu, N.B. (2007) *High Performance Polymers*, **19**, 649–664.

444 Doi, Y. and Steinbüchel, Y. (eds) (2002) *Biopolymers*, vols 2–4, Wiley-VCH Verlag GmbH, Weinheim.

445 Philip, S., Keshavarz, T. and Roy, I. (2007) *Journal of Chemical Technology and Biotechnology*, **82**, 233–247.

446 Hile, D.H. and Pishko, M.V. (1999) *Macromolecular Rapid Communications*, **20**, 511–514.

447 Hile, D.H. and Pishko, M.V. (2001) *Journal of Polymer Science Part A-Polymer Chemistry*, **39**, 562–570.

448 Bratton, D., Brown, M. and Howdle, S.M. (2003) *Macromolecules*, **36**, 5908–5911.

449 Bratton, D., Brown, M. and Howdle, S.M. 2004 *Chemical Communications*, 808–809.

450 Hwang, H.S., Park, E.J., Jeong, Y.T., Heo, H. and Lim, K.T. (2004) *Studies in Surface Science and Catalysis*, **153**, 389–392.

451 Bratton, D., Brown, M. and Howdle, S.M. (2005) *Journal of Polymer Science Part A-Polymer Chemistry*, **43**, 6573–6585.

452 Ganapathy, H.S., Hwang, H.S., Jeong, Y.T., Lee, W.-K. and Lim, K.T. (2006) *European Polymer Journal*, **43**, 119–126.

453 Asandei, A.D., Erkey, C., Burgess, D.J., Saquing, C., Sa-ha, G. and Zolnik, B.S. (2005) *Materials Research Society Symposium Proceedings*, **845**, 243–248.

454 Mazarro, R., de Lucas, A., Gracia, I. and Rodriguez, J.F. (2008) *Journal of Biomedical Materials Research*, **85B** (1), 196–203.

9
Synthesis of Nanomaterials

Zhimin Liu and Buxing Han

9.1
Introduction

Research on nanomaterials is a very promising and rapidly growing field. Supercritical fluids (SCFs) are attractive media for the synthesis of nanomaterials because they have unique properties, such as high diffusivity, low viscosity, and near-zero surface tension. Furthermore, their physical properties can be tuned continuously by changing the pressure and/or temperature. Accordingly, SCFs were successfully used in the syntheses and processing of materials, and many nanomaterials which are difficult to synthesize by conventional methods were successfully prepared using SCF technology. In addition, the synthetic processes can contribute to sustainable materials development, especially when supercritical CO_2 ($scCO_2$) or supercritical water (scH_2O) is used. It is not surprising that this area has attracted much attention in recent years. In this chapter, we focus on recent research on the chemical synthesis of nanomaterials related to SCFs, and mainly on metal or semiconductor nanocrystals, metal oxide nanoparticles, carbon nanomaterials, and some nanocomoposites. Approaches for the preparation of nanomaterials using SCFs by physical routes are not included here, although they are also very interesting and promising.

9.2
Metal and Semiconductor Nanocrystals

9.2.1
Direct Synthesis of Nanocrystals in SCFs

Direct synthesis of nanocrystals in SCFs refers to the formation of nanocrystals via the direct chemical conversion of precursors in SCFs. Based on the characteristics of the SCFs used, different routes have been developed to synthesize various nanocrystals.

Handbook of Green Chemistry, Volume 4: Supercritical Solvents. Edited by Walter Leitner and Philip G. Jessop

ISBN: 978-3-527-32590-0

9.2.1.1 Synthesis in $scCO_2$

$scCO_2$ provides a potentially useful medium for nanocrystal synthesis as it is non-toxic, non-flammable, and environmentally benign. For the synthesis of nanoparticles in $scCO_2$, precursors and capping agents that are soluble in $scCO_2$ are needed. To meet these requirements, some special precursors have been designed or selected, which mainly involve fluorinated organometallic compounds, such as Ag(1,1,1,5,5,5-hexafluoropentane-2,4-dione)(tetraglyme) [Ag(hfpd)(tetraglyme)] [1], palladium(II) hexafluoroacetylacetonate [$Pd(hfac)_2$], triphenylphosphinegold(I) perfluorooctanoate (TPAuFO) [2], and hydrocarbon-based or oxygenated hydrocarbon-based precursor complexes including dimethyl(cyclooctadiene)platinum(II) [(cod)Pt-Me_2] [3], bis (2,2,6,6-tetramethyl-3,5-heptanedionato)copper(II) [$Cu(tmhd)_2$], bis(cyclopentadienyl)nickel [$NiCp_2$] [4], and silver acetylacetonate [Ag(acac)] [5]. These compounds can be readily reduced to the metallic form in $scCO_2$. The capping agents for the nanocrystals are mainly CO_2-philic fluorinated stabilizing ligands such as fluoroalkanes, fluoroethers, polymeric fluoroethers, fluoroacrylates, siloxanes, and branched hydrocarbons. For example, perfluorodecanethiol ($C_8F_{17}C_2H_4SH$) was used successfully to stabilize nanocrystals in $scCO_2$ [6]. Two more families of fluorinated surfactants were found to disperse nanocrystals in CO_2 more effectively than the fluoroalkanes: perfluoropolyether (PFPE)- and fluorooctyl methacrylate (FOMA)-based ligands. Perfluoropolyether-based thiols {$F[CF(CF_3)CF_2O]_3CF(CF_3)CONH(CH_2)_2SH$} disperse nanocrystals in CO_2 at moderate pressures, allowing the dispersion of small gold nanocrystals, about 2.0 nm in diameter, in liquid CO_2 at the vapor pressure [7]. However, the most effective thiol to date is a fluorooctyl methacrylate-based thiol [$CF_3(CF_2)_5CH_2CH_2OCOCH_2CH_2SH$], which can be used to disperse larger gold nanocrystals of about 3.5 nm diameter in liquid CO_2 at the vapor pressure [7].

Johnston and co-workers explored the synthesis and dispersibility of nanocrystals (e.g. Ag, Pt) in $scCO_2$ [8]. They showed that silver nanocrystals capped with partially fluorinated ligands (1*H*,1*H*,2*H*,2*H*-perfluorodecanethiol) were redispersible in $scCO_2$. Subsequently, they synthesized Ag, Ir, and Pt nanocrystals ranging from 2.0 to 12 nm in diameter in CO_2 using hydrogen to reduce organometallic precursors in the presence of fluorinated ligands [9]. The nanocrystals capped by 1*H*,1*H*,2*H*,2*H*-perfluorooctanethiol can be redispersed in acetone and fluorinated solvents. However, they cannot be redispersed in neat $scCO_2$. In another study, perfluorodecanethiol was used to stabilize Ag nanocrystals synthesized via reducing Ag(acac) with hydrogen in $scCO_2$ [10]. The influence of pressure, stabilizer, and precursor concentration on the formation and dispersion of Ag nanoparticles was investigated. At a thiol to precursor ratio of $\sim$2.5 and a temperature of 80 °C, the size of the Ag nanoparticles synthesized was found to decrease with increase in pressure from 4.0 ± 2.1 nm at 20.7 MPa to 1.7 ± 1.1 nm at 25.9 MPa. Above 25.9 MPa, the average size of the Ag nanoparticles remained virtually constant with smaller standard deviations. The CO_2 density used during synthesis controls the particle size and polydispersity. At high solvent densities ($P > 25$ MPa, $T = 80$ °C), the ligands provide a strong steric barrier that maintains small particles with a diameter of 2.0 nm. At lower solvent densities ($P < 25$ MPa, $T = 80$ °C), the osmotic repulsions between capping ligands are weak, resulting in 4.0 nm diameter nanocrystals with higher polydispersity. At early stages in the growth process, metal core coagulation competes with ligand adsorption. Favorable factors for steric stabilization, such as long ligands and high solvent density, quench nanocrystal growth at relatively low ligand binding densities, which leads to smaller nanocrystals. Under poor solvent conditions, particles grow to larger sizes before the coverage of capping ligand is sufficient to prevent coagulation of metal particles. Perfluorodecanethiol-coated Ag nanocrystals, synthesized in either good or poor solvent conditions, can be readily redispersed in acetone, fluorinated solvents, and $scCO_2$ (at high density). The precursor concentration, thiol to precursor ratio, and reaction time do not affect the nanocrystal size appreciably in the range studied, although these parameters do affect the polydispersity. The authors proposed that the stabilizing ligands must interact with the solvent to provide protection against irreversible aggregation and exhibit a binding strength weak enough to allow particle growth, yet strong enough to arrest particle growth in the nanometer size range.

Esumi *et al.* [11] synthesized Au nanoparticles of 1 nm in diameter via the reduction of triphenylphosphinegold(I) perfluorooctanoate with dimethylamineborane in $scCO_2$. A high dispersion stability of the Au nanoparticles in $scCO_2$ can be obtained due to binding both triphenylphosphine and fluorocarbon ligands on the surface of the Au nanoparticles. Fan *et al.* [12] selected silver bis(3,5,5-trimethyl-1-hexyl)sulfosuccinate (Ag-AOT-TMH), a highly soluble non-fluorinated complex, as precursor, and perfluorooctanethiol as stabilizing ligand for the synthesis of Ag nanoparticles. Particles 1–6 nm in diameter were produced in $scCO_2$ solution containing 0.06 wt % Ag-AOT-TMH and 0.5 wt% perfluorooctanethiol at 40 °C and 28.6 MPa using $NaBH_4$ as the reducing agent. Attempts to produce Ag nanoparticles from Ag-AOT-TMH in a completely non-fluorous system were unsuccessful because effective stabilization could not be attained with silicone-, PEG-, or hydrocarbon-based thiols.

9.2.1.2 Synthesis in Supercritical Organic Solvents

The synthesis of some semiconductor (e.g. Si and Ge) nanocrystals requires significantly elevated temperatures to achieve core crystallinity. Supercritical alkanes with high critical temperatures were shown to be effective solvents for reactions beyond 400 °C by pressurizing to 300 bar. Holmes *et al.* reported the synthesis of crystalline Si nanocrystals in supercritical solvents at temperatures above 400 °C using organosilane precursors [13]. By thermally degrading the Si precursor, diphenylsilane, in a supercritical hexane–octanol mixture at 500 °C and 34.5 MPa, Si nanocrystals with relatively high monodispersity which are sterically stabilized and with diameter from 15 to 4.0 nm could be obtained in significant quantities. In this method, supercritical octanol serves as an effective capping ligand for Si nanocrystals. It binds to the Si nanocrystal surface through an alkoxide linkage and provides steric stabilization through the hydrocarbon chain. The as-synthesized Si nanocrystals exhibit significant size-dependent optical properties due to quantum confinement effects. Additionally, the smallest nanocrystals produced, 1.5 nm in diameter, exhibit previously unobserved discrete optical transitions. Although the above route produced high-quality Si nanocrystals and nanowires, the authors found that the product yields were relatively low, being limited by the coproduction of organosilane byproducts. To alleviate organosilane byproduct formation, they explored trisilane as an alternative Si precursor, and succeeded in the synthesis of colloidal submicrometer diameter amorphous silicon (a-Si) particles via the direct decomposition of trisilane in supercritical hexane at temperatures ranging from 400 to 500 °C and at pressures up to 34.5 MPa [14]. It was discovered that trisilane reactions produced only a-Si colloids with average diameters ranging between 60 and 450 nm, depending on the reaction conditions, with very high yields of >90% conversion of trisilane to colloidal a-Si product. The colloidal product was of high quality: the particles were not aggregated and exhibited a spherical shape, and they had relatively narrow size distributions with standard deviations about the mean diameter as low as about 10%. Moreover, the thermal annealing of the a-Si particles after isolation from the reactor resulted in crystallization to diamond structure silicon at 650 °C.

Similarly, crystalline Ge nanocrystals were synthesized via decomposition of Ge precursors (e.g. diphenylgermane and tetraethylgermane) in supercritical hexane and octanol at 400–550 °C and 20.7 MPa in a continuous flow reactor [15]. The average nanocrystal diameter could be changed over a wide range from ~2 to ~70 nm by varying the reaction temperature and precursor concentration. Nanocrystals with relatively high monodispersity could be produced, with standard deviations about the mean diameter as low as ~10%. UV–visible absorbance and photoluminescence (PL) spectra of Ge nanocrystals in the 3–4 nm diameter size range exhibited optical absorbance and PL spectra blue shifted by ~1.7 eV relative to the bandgap of bulk Ge, with quantum yields up to 6.6%.

To synthesize semiconductor nanowires, a supercritical fluid–liquid–solid (SFLS) approach has been developed, and some semiconductors including Si [16, 17], Ge [18, 19], and GaAs [20] were successfully prepared. In this approach, sterically stabilized metal nanocrystals are input as seed particles that direct wire growth at temperatures that exceed the metal/semiconductor eutectic temperature. The temperature must be sufficiently high to degrade the molecular precursor and to induce nanowire formation. Therefore, high critical temperature solvents (e.g. hexane, cyclohexane) are generally needed. Holmes *et al.* [16] adapted the SFLS method to prepare monodisperse Si nanowires. In their method monodisperse alkanethiol-capped Au nanocrystals of 2.5 nm in diameter were used as seeds to direct one-dimensional Si crystallization via the decomposition of diphenylsilane in supercritical hexane at 500 °C and 20.0 or 27.0 MPa. The crystal structure of the nanowires produced using this SFLS growth process could be controlled by changing the reaction pressure. At 20.0 MPa, Si nanowires grew preferentially along the (100) plane, and along the (110) direction at 27.0 MPa, as illustrated in Figure 9.1.

Molecularly tethered Au nanocrystals with an average diameter of 7 nm were used to grow single-crystal Si nanowires on a substrate through an SFLS mechanism [17]. Si wires were obtained by degrading diphenylsilane in cyclohexane heated and pressurized well above its critical point (T_c = 281 °C; P_c = 4.1 MPa) and the Si/Au eutectic temperature (363 °C). Temperatures greater than 450 °C are needed to promote rapid crystallization of straight Si nanowires. The nanowire diameter reflects the Au nanocrystal diameter, ranging from 5 to 30 nm, with lengths of

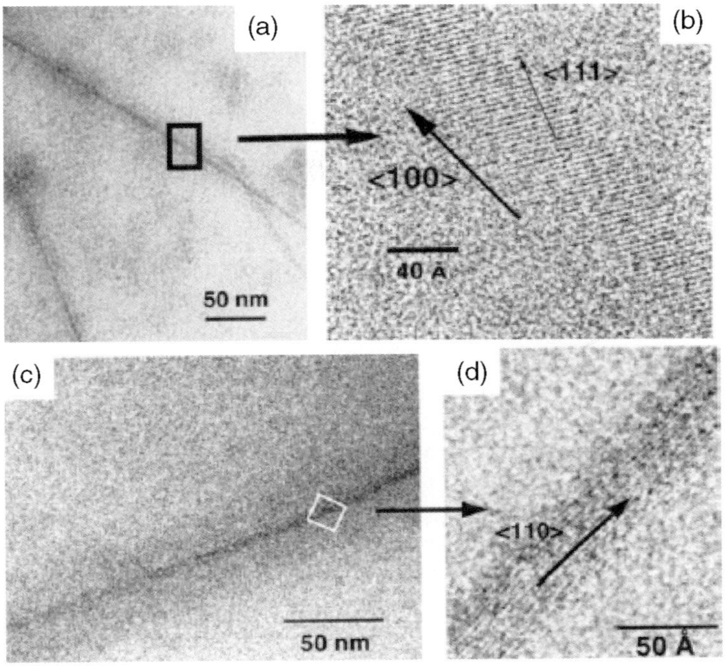

Figure 9.1 Transmission electron microscopy (TEM) images of Si nanowires synthesized at 500 °C in hexane at 20.0 MPa (a, b) and 27.0 MPa (c, d).

several micrometers. Both batch and flow reactors were used. In the flow reactor, the diphenylsilane concentration can be controlled at the optimum diphenylsilane concentration favoring straight wire growth without leading to homogeneous particle production. Scanning electron microscopy (SEM) images of Si nanowires grown at a series of temperatures, reactor residence times, and Si precursor concentrations revealed that the wire growth kinetics influence the nanowire morphology significantly and could be controlled effectively using a flow reactor. A kinetic analysis of the process qualitatively explains the dependence of the nanowire morphology on the reaction conditions.

9.2.1.3 Synthesis in Supercritical Water (scH_2O)

The dielectric constant of scH_2O is very low, which makes it dissolve organics rather than inorganics (see Chapter 13). Moreover, in contrast to organic solvents, scH_2O is stable even at high temperatures. Therefore, it is an excellent solvent for the synthesis of many nanocrystals. By using organic capping ligands that are miscible with scH_2O to arrest the particle growth, coalescence can be prevented. Ziegler *et al.* [21] employed an alkanethiol as a capping ligand to control particle formation of Cu in scH_2O. Cu nanocrystals with an average diameter of 7.0 nm were synthesized at 400 °C and 20.0 MPa using copper nitrate as the precursor in the presence of hexanethiol. On increasing the temperature to 425 and 450 °C, the average particle size was increased to 7.8 and 9.2 nm, respectively. These nanocrystals were spherical, highly crystalline and protected from agglomeration by the thiol passivating layer. The authors proposed a mechanism for sterically stabilized nanocrystal growth in SCW that described competing pathways of hydrolysis to large oxidized copper particles versus exchange of thiol ligands and arrested growth by thiols to produce small monodisperse Cu nanoparticles.

Sue *et al.* [22] used nickel formate as a precursor to synthesize Ni particles directly in scH_2O at 400 °C and 30 MPa. In this method, formate anion thermally decomposed to produce H_2, which formed a homogeneous mixture with scH_2O. In this environment, Ni^{2+} was easily reduced to Ni, and Ni crystals precipitated. High-crystallinity Ni particles with sizes in the sub-micrometer range were obtained.

9.2.2 Synthesis of Nanomaterials in SCF-based Microemulsions

Reverse micelles formed in SCFs can function like "nanoreactors" for a wide range of chemical syntheses in SCFs, which have been exploited for the solution-phase synthesis of metallic and semiconductor nanoparticles. Surfactant-stabilized aqueous cores dispersed in a continuous supercritical phase provide a unique medium for exploring nanocrystal synthesis in supercritical solvents. Supercritical alkanes and $scCO_2$ have been used as a continuous phase of microemulsions, leading to two families of water-in-SCF microemulsions. For the water-in-supercritical alkane microemulsion systems, sodium bis(2-ethylhexyl)sulfosuccinate (AOT) is generally used as the surfactant, and fluorinated surfactants are used for the formation of water-in-CO_2 microemulsions. These systems have been applied in the synthesis of nanomaterials, including metal nanocrystals, metal oxide nanoparticles, and semiconductor crystals.

9.2.2.1 Water-in-Supercritical Alkane Microemulsion

Roberts and co-workers [23–25] used water-in-alkane microemulsions to synthesize metal (e.g. Ag, Cu) nanoparticles. Cu particles of <20 nm were produced by the reduction of copper ions from copper bis(2-ethylhexyl)sulfosuccinate [$Cu(AOT)_2$], incorporated within AOT reverse micelles by the addition of hydrazine reducing agent into the compressed propane and supercritical ethane (scC_2H_6), along with small amounts of isooctane cosolvent. Particle growth rates in the compressed fluids were compared with those in normal liquid solvents to study the effects of the compressed fluids on the growth behavior of these nanoparticles within the reverse micelles. It was demonstrated that the growth behavior is different from that in normal liquid solvents for both Cu and Ag. The particle growth rate in AOT reverse micelles in scC_2H_6 is faster than that in normal liquid solvents, partly because of intermediate physical and transport properties of the supercritical solvent (e.g. increased mass transfer, decreased viscosity and density, lower dielectric constant), which results in increased collision frequencies and exchange rates between micelles. In addition, increasing the water content leads to a faster reaction.

The UV–visible spectrum of these nanosized Cu particles is sensitive to particle size, and hence time-resolved spectral measurements can be used to monitor the particle growth process.

The thermophysical properties of compressed liquids and SCFs can be tuned through adjustments in temperature and pressure, which affords the unique ability to control particle synthesis within the reverse micelle systems by adjusting the strength of the interaction between the surfactant tails and the bulk solvent. In reverse micelle reaction media, the role of the anionic surfactant is twofold: creating a thermodynamically stable microemulsion system consisting of the continuous oil phase and the reverse micelles containing water and metal ions, and acting as a stabilizing ligand for the synthesized nanoparticles. It was demonstrated that an increase in pressure or a decrease in temperature results in slightly larger particles. For the synthesis of Cu nanoparticles, the median particle diameter shifted from 5.4 to 9.0 nm with an increase in pressure from 24.1 to 34.5 MPa in compressed propane. Synthesis in compressed ethane and scC_2H_6 was achieved at a pressure of 51.7 MPa and could be slightly adjusted with temperature and with the addition of a liquid alkane cosolvent [26].

9.2.2.2 **Water-in-$scCO_2$ Microemulsions**

Reverse micelles formed in liquid and $scCO_2$ allow highly polar or polarizable compounds to be dispersed in this non-polar fluid. CO_2-philic surfactants (e.g. fluorinated surfactants) are required to form such reverse micelles in CO_2 with a water core in which CO_2-phobic polar species (e.g. metal salts) can dissolve. Several groups have utilized water-in-CO_2 microemulsions as nanoreactors for synthesizing various nanomaterials, such as metal nanocrystals, metal oxide nanoparticles, and semiconductor nanocrystals. Ji *et al.* [27] reported the synthesis of Ag nanoparticles in a water-in-CO_2 microemulsion by chemical reduction of silver ions in the water cores with the reducing agent sodium triacetoxyborohydride. In this work, AOT together with perfluoropolyether phosphate ether (PFPE-PO_4) were used to form water-in-CO_2 microemulsions. Although neither of the two surfactants is soluble in $scCO_2$, their mixtures with appropriate ratios can form stable microemulsions with water cores in $scCO_2$. These mixed surfactant systems have been used to synthesize various nanoparticles. For example, Ohde *et al.* [28] reported the synthesis of nanometer-sized Ag and Cu metal particles by chemical reduction of Ag^+ and Cu^{2+} ions solubilized in water-in-$scCO_2$ microemulsions formed using the mixture of AOT and PFPE-PO_4 as surfactant. The concentrations of AOT and PFPE-PO_4 were 12.8 and 25.3 mM, respectively, and the water to surfactant molar ratio was 12. Sodium cyanoborohydride ($NaBH_3CN$) and *N,N,N′,N′*-tetramethyl-*p*-phenylenediamine (TMPD), which are soluble in $scCO_2$, were used as reducing agents for synthesizing these metal nanoparticles. Formation of the metal nanoparticles was monitored spectroscopically using a high-pressure fiber-optic reactor equipped with a CCD array UV–visible spectrometer. Ag and Cu nanoparticles synthesized in the microemulsions showed characteristic surface plasmon resonance absorption bands centered at 400 and 557 nm, respectively. Diffusion and distribution of the oxidized form of the reducing agent between the micellar cores and $scCO_2$ appeared to be the rate-determining step for the formation of the Ag nanoparticles in this system. Using the same water-in-$scCO_2$ microemulsion system, they also synthesized semiconductor nanoparticles via mixing two water-in-$scCO_2$ microemulsions with the same water to surfactant molar ratio (*W*) and containing different ions in the water cores [29, 30]. By collision, exchange of ions between the reverse micelles takes place, leading to the formation of nanoparticles in the water core. For example, CdS and ZnS nanoparticles were synthesized by mixing two water-in-CO_2 microemulsions containing S^{2-} ions and Cd^{2+} or Zn^{2+} ions separately in the water cores. The size of the nanoparticles formed in the microemulsions depended on *W*. At $W = 6$ and 12, the bandgaps of CdS were calculated as 3.50 and 3.15 eV, corresponding to particle radii of 1.4 and 1.7 nm, respectively. The ZnS particles synthesized at $W = 12$ showed a bandgap of 4.26 eV, corresponding to a mean particle radius of 1.6 nm. This microemulsion-plus-microemulsion approach can be used to synthesize a variety of nanoparticles in $scCO_2$ using water-soluble reagents as starting materials. Moreover, the synthesis of semiconductor nanoparticles in $scCO_2$ offers several advantages over the conventional water-in-oil microemulsion approach, including fast reaction, rapid separation, and easy removal of solvent from the nanoparticles. Since the stability of the CO_2 microemulsion depends on the density of the solvent, one can change the temperature and pressure of the system to break down the microemulsion to deposit small nanoparticles on substrates placed directly in the SCF system.

Holmes *et al.* [31] used ammonium carboxylate perfluoropolyether (PFPE-$COONH_4$)-stabilized water-in-CO_2 microemulsion with Cd^{2+} to synthesize CdS particles. An aqueous solution of Na_2S was injected directly into the CO_2 phase, forming CdS particles in the water core. Using this method, the *W* value of the microemulsion changes as the Na_2S solution is introduced into the CO_2 phase. In this case, the particle radii of approximately 0.9 and 1.8 nm, respectively, were consistent with the microemulsion droplet diameters determined by small-angle neutron scattering (SANS). In this method, the binding of the surfactant molecules to the particle surface is sufficient to enable the particles collected to be redispersed in ethanol without coalescence.

Liu *et al.* [32, 33] employed the commonly used surfactant AOT with 2,2,3,3,4,4,5,5-octafluoro-1-pentanol (F-pentanol) as a cosurfactant to prepare water-in-$scCO_2$ microemulsions, which were used to synthesize Ag and AgI nanocrystals. Not only do the nanometer-sized aqueous domains in the microemulsion cores act as nanoreactors, but also the surfactant interfacial monolayer helps the stabilization of the metal and semiconductor nanoparticles. TEM results showed that Ag and AgI nanocrystals with average diameters of about 6.0 nm were formed.

Fernandez and Wai investigated the effects of density and *W* of water-in-$scCO_2$ microemulsions on the synthesis of nanoparticles [34]. Monodisperse CdS and ZnS nanoparticles were synthesized by chemical reaction of cadmium or zinc nitrate with NaS, using two water-in-$scCO_2$ microemulsions as nanoreactors, followed by protection with a fluorinated thiol stabilizer. The size and size distribution of CdS and ZnS nanoparticles can be tuned over a wide range of values by adjusting the density of the fluid phase. The average size of the ZnS nanoparticles decreases linearly from approximately 9.1 to 1.9 nm with increase in fluid density from 0.86 to 0.99 $g\,cm^{-3}$

at a *W* value of 10. At $W = 6$, the particle size can be tuned from 7.0 to 1.5 nm in the same density range. In the case of CdS nanocrystals, the size varied from 7.1 to 2.0 nm with $W = 10$ and from 4.0 to 1.3 nm with $W = 6$, in the same density range. Their subsequent work showed that the size of Ag nanoparticles synthesized by the $scCO_2$ microemulsion-templating method can be tuned continuously over a large range of values by density variation of the fluid phase [35]. Moreover, this approach gave consistent results for two different values of *W* and for two different concentrations of surfactant tested. The variation of the size of the nanocrystals with the density of the solvent follows a linear relationship. The $scCO_2$ microemulsion method represents a simple approach using a density-tunable solvent for synthesizing size-controlled semiconductor nanoparticles.

9.2.2.3 Recovery of Nanoparticles from Reverse Micelles Using $scCO_2$

Synthesis of nanoparticles in reverse micelles formed in apolar solvents is a commonly used method, and recovery of the nanoparticles from the reverse micelles is one of the main steps. In conventional methods, the surfactant is usually precipitated together with the nanoparticles. Ideally, the synthesized nanoparticles should be precipitated while the surfactant remains in solution. Zhang and co-workers proposed a method to recover nanoparticles from reverse micelles using compressed CO_2 [36, 37]. In this method, CO_2 is added to a reverse micellar solution with synthesized nanoparticles to precipitate the particles. Using this method, ZnS in AOT reverse micelles in isooctane and Ag nanoparticles in AOT reverse micelles in isooctane with tetraethylene glycol dodecyl ether ($C_{12}E_4$) as a cosurfactant were recovered. The unique feature of the method is that all of the ZnS and Ag particles in the reverse micelles can be recovered, while the surfactant and cosurfactant remain in the isooctane continuous phase.

9.3 Metal Oxide Nanoparticles

9.3.1 Supercritical Hydrothermal Synthesis

In recent years, high-temperature SCF approaches for producing nanoscale materials have been conducted in near-critical or supercritical water (scH_2O). scH_2O allows highly elevated temperatures to be used that cannot be sustained by most conventional solvents, which is a useful property for the synthesis of many metal and semiconductor materials. The high temperature is particularly helpful in the crystallization of the materials. The low dielectric constant of scH_2O makes the solvation characteristics much more like those of a hydrocarbon solvent, which has led to considerable efforts to use scH_2O to hydrolyze metal salts to form metal oxide particulates. Adschiri *et al.* showed that metal nitrates can decompose to form a wide variety of metal oxide crystals in scH_2O [38, 39]. Poliakoff and co-workers also demonstrated that metal–organic precursors can be used to obtain metal oxide

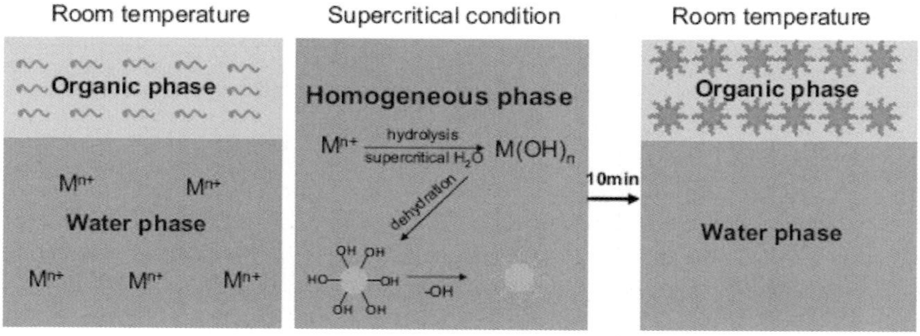

Figure 9.2 The strategy for the synthesis of metal oxide nanocrystals in an organic ligand-assisted supercritical hydrothermal process.

nanocrystals through a similar mechanism [40–42]. Both of these groups reported the ability to control the oxidation state of metal nanoparticles through the addition of reductants or oxidants to the reaction chamber during particle growth, and the particle size, morphology, and oxidation state of metal nanoparticles produced in scH_2O can be readily controlled.

Nanoparticles are apt to aggregate due to their high surface energy, hence it is more problematic to collect nanoparticles. This problem has been overcome by the surface modification of nanoparticles in scH_2O. Adschiri and colleagues developed a green and facile strategy to synthesize metal oxide nanocrystals stabilized by organic ligands based on the SC hydrothermal synthesis process [43–45]. The organic ligands used can be hydrophilic and hydrophobic compounds, which are soluble in scH_2O to form a homogeneous phase while they are not soluble in water at room temperature. This allows the stabilized nanoparticles to be dispersed in aqueous and organic solvents. Figure 9.2 illustrates schematically a strategy to synthesize metal oxides using a hydrothermal synthesis process in supercritical media mediated by organic ligands [44]. This strategy depends on (1) sub-decananometer single-crystal formation in a supercritical hydrothermal process; (2) the miscibility of the organic ligand molecules with high-temperature water; and (3) controlled nanocrystal growth from the selective reaction of organic ligand molecules with specific inorganic crystal surfaces. Importantly, the use of water, instead of an organic solvent, provides a green chemistry route to the preparation of nanometer-sized building blocks for advanced materials and devices. This method has the following features: the particle size and crystal phase can be tuned by changing the concentration of the modifier and the resultant nanoparticles can be dissolved in aqueous and organic solvents depending on the modifier used.

In a typical procedure for the synthesis of CeO_2 nanocrystals, the cerium precursor was first prepared by mixing $Ce(NO_3)_3$ solution with NaOH solution, and then transferred into a high-pressure vessel [44]. To modify the surface of the nanocrystals and induce their anisotropic growth, an appropriate amount of decanoic acid was also loaded into the reactor vessel. Subsequently, a hydrothermal reaction was performed in the reactor at 400 °C for 10 min, resulting in monodispersible organic

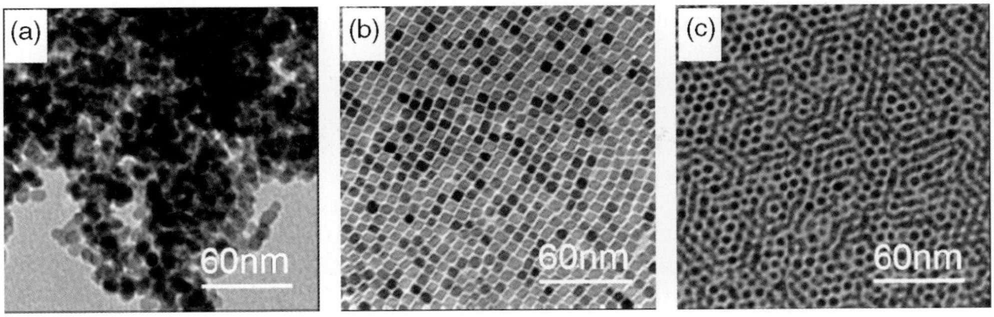

Figure 9.3 TEM and HRTEM images of the synthesized ceria nanocrystals. The molar ratios of decanoic acid to the ceria precursor was (a) 0, (b) 6 : 1, and (c) 24 : 1.

ligand-modified CeO_2 nanocrystals. These CeO_2 nanoparticles can easily dissolve in an organic solvent and can self-assemble into close-packed superlattices on carbon substrates. Figure 9.3 shows the TEM and high-resolution TEM (HRTEM) images of CeO_2 prepared with different decanoic acid to ceria precursor ratios. It is obvious that organic ligand molecules had a pronounced effect on the morphology of the nanocrystals. Without modifier, the CeO_2 nanoparticles obtained were randomly aggregated nanoparticles with an average size of about 7 nm, as shown in Figure 9.3a, and the shape of the naked CeO_2 nanocrystals was a truncated octahedron. As decanoic acid was introduced into the supercritical hydrothermal reaction system, monodisperse CeO_2 nanoparticles with uniform size were obtained, which can be self-assembled on the amorphous carbon grid, as illustrated in Figure 9.3b and c. With a molar ratio of decanoic acid to CeO_2 precursor of 6 : 1, the particle size was about 6 nm and the shape of the CeO_2 nanocrystals was cubic (Figure 9.3b), while the nanocrystal size decreased to about 5 nm (Figure 9.3c) and the shape was a truncated octahedron enclosed by the {111} and {200} planes as the ratio of decanoic acid to CeO_2 precursor reached 24 : 1. The gap between the particles on the TEM image is around double the length of the organic modifier, which suggests sufficient and uniform surface modification.

The method described above can be extended to the synthesis of many other metal oxides and oxide hybrids [46–49]. By changing organic modifiers, the particle morphology and crystal structure can be tuned. For example, highly dispersible boehmite (AlOOH) nanoparticles were synthesized in scH_2O by using $CH_3(CH_2)_4CHO$ and $CH_3(CH_2)_5NH_2$ as modifier reagents [47]. The reagents bind chemically to the surface of the AlOOH nanoparticles, and the surface modification affects crystal growth, reduces the particle size, and changes the morphology of the particles. 1-Hexanal was also used to modify the surface of TiO_2 nanocrystals during a supercritical hydrothermal synthesis process, which enabled the synthesized TiO_2 nanocrystals to be perfectly dispersed in isooctane because of the hydrophobic nature of the surface modifier [48]. Furthermore, the above one-pot route is also effective for the synthesis of hybrid materials. For the preparation of a $CoAl_2O_4$ hybrid, a mixed sol of $Co(OH)_2$ and $Al(OH)_3$ from the reactions of sulfates ($CoSO_4$, Al_2SO_4) with NaOH was used as precursor and oleic acid or decanoic acid as surface modifier of nanoparticles [45]. The hydrothermal synthesis was performed at 400 °C and 38 MPa for 10 min, resulting in organic ligand-modified nanocrystals which can be extracted from the product mixtures with hexane. The resultant nanoparticles have an average particle size of less than 8 nm with well-stabilized single nanocrystals. The organic ligand capping helps to diminish the scattering effect of the blue nano-pigment, resulting in a transparent cobalt blue nano-pigment without any post-heat treatment. In a subsequent study, the synthesis and *in situ* organic surface modification of cobalt aluminate nanoparticles were carried out to change the hydrophilic surface of pigment nanoparticles to a hydrophobic surface by using organic reagents such as hexanoic acid and 1-hexylamine along with the respective metal hydroxide as starting materials at supercritical temperatures and 30–40 MPa [49].

To produce metal oxide nanocrystals on a large scale, a flow-through supercritical water (FT-scH_2O) method for continuous hydrothermal synthesis was established for forming micro- or nanosized metal oxide fine particles in scH_2O [50, 51]. In this method, a flow-type tubular reactor was used. An aqueous solution of the metal salt was mixed with preheated water to increase the solution temperature rapidly to the supercritical state. Because of the high reaction rate of hydrothermal synthesis and low solubility of metal oxides, extremely high supersaturation is attained just after the mixing point. The hydrothermal reaction rate and metal oxide solubility can be varied widely since the reaction solvent properties are strongly dependent on the thermodynamic conditions in supercritical regions. Therefore, the FT-scH_2O method can be used to change the size, morphology, and crystal structure of many types of particles. Due to the unusual properties of scH_2O, this FT-scH_2O method has great potential for producing nanosized metal oxide fine particles. There are a few reports that demonstrate the controllability of the particle size on the basis of metal oxide solubility and supersaturation [52, 53]. Sue *et al.* [54]. studied the experimental conditions for producing metal oxide nanoparticles under 10 nm in diameter by the FT-scH_2O method. Hydrothermal synthesis of metal oxide nanoparticles (e.g. AlOOH/Al_2O_3, CuO, Fe_2O_3, NiO, ZrO_2) from metal nitrate aqueous solution was carried out at 400 °C and pressures ranging from 25 to 37.5 MPa. It was demonstrated that supersaturation should be set to higher than 10^4 to obtain particles under 10 nm in diameter at 400 °C. High supersaturation reduces the crystallinity, possibly because high supersaturation leads to the inclusion of water molecules during the formation of particles.

9.3.2
Direct Sol–Gel Synthesis in scCO_2

The sol–gel process is a commonly used method to prepare metal oxides in aqueous solutions catalyzed by acid or base. Recently, the non-aqueous sol–gel process was performed in scCO_2 to synthesize metal oxides. In this process, alkoxide is used as precursor and organic acids as condensation agents. One of the attractive features associated with this method is that the oxide nanomaterials can be readily and reproducibly prepared on a large scale in an autoclave reactor, and pure oxide with crystalline phase can be obtained after calcination. Another feature is the environmentally friendly synthesis conditions. Furthermore, the size and the morphology of the oxide nanomaterials can be tuned by varying the concentration of the precursor and the organic acid used.

Sui *et al.* [55] reported the synthesis of TiO_2 nanofibers by the esterification and condensation of titanium alkoxides using acetic acid as the polymerization agent in $scCO_2$ from 40 to 70 °C and at different pressures. The diameters of the TiO_2 nanofibers ranged from 9 to 100 nm. N_2 physisorption and powder diffraction patterns showed that the nanofibers exhibit relatively high surface areas of up to $400\,m^2\,g^{-1}$ and anatase and/or rutile nanocrystallites were formed after calcination. In subsequent work, TiO_2 colloids were prepared in $scCO_2$ via polycondensation of titanium isopropoxide (TIP) and titanium butoxide (TBO) with acetic acid (HAc), respectively [56]. The authors explored the effects of the type of precursor and the precursor concentration on the morphology of the resultant TiO_2 nanostructures. It was shown that the morphology of the resultant TiO_2 colloids can be tailored by changing the concentration of the starting materials or by changing the type of precursor. Increasing the ratio of HAc to titanium alkoxide results in a morphology change from spheres to fibers when using TBO as the precursor, while increasing the ratio results in a larger fiber diameter and the morphology changes from curled to straight fiber when using TIP as the precursor. Fiber formation is enhanced by a higher HAc/Ti ratio and the use of the TIP monomer. Fourier transform infrared (FTIR) spectroscopic analysis indicated that the formation of nanofibers is favored by a titanium hexamer that leads to one-dimensional condensation, whereas nanospheres are favored by a hexamer that permits three-dimensional condensation.

Porous ZrO_2 aerogels with nanostructure were also synthesized via the direct sol–gel route in $scCO_2$, which involved the coordination and polycondensation of a zirconium alkoxide using acetic acid, followed by $scCO_2$ drying and calcination [57]. Either a translucent or opaque monolith was obtained, both of which had high surface areas (up to $399\,m^2\,g^{-1}$) and porosity. Electron microscopy results showed that the translucent monolithic ZrO_2 had a well-defined mesoporous structure, whereas the opaque monolith, formed using added alcohol as a cosolvent, was composed of loosely compacted nanospherical particles with a diameter of ca. 20 nm. X-ray diffraction (XRD) results indicated that the ZrO_2 exhibited tetragonal and/or monoclinic phases after calcination at 400 and 500 °C. *In situ* IR spectroscopy results showed the formation of a zirconium acetate coordination complex at the initial stage of the polycondensation, followed by further condensation of the complex into macromolecules.

Some oxides, including SiO_2, TiO_2, and ZrO_2, can be produced in $scCO_2$ via nonhydrolytic acylation/deacylation of alkoxides by anhydrides on the basis of the equation

$$M(OR)_n + \frac{n}{2}(R'C{=}O)_2O \rightarrow MO_{\frac{n}{2}} + nR'COOR \qquad (9.1)$$

where M represents Si, Ti, Zr, and so on, and R and R′ denote alkyl chains. Because organic anhydrides and metal alkoxides are soluble in $scCO_2$, the route shown in

Equation 9.1 is suitable for preparing metal oxide powders in $scCO_2$ via a homogeneous reaction. The reaction rate at room temperature of thermal acylation/deacylation is lower than that of a hydrolytic route [58], allowing for more uniform particle sizes. For the synthesis of TiO_2 powders using this method, the precursor, TIP, reacted with acetic anhydride or trifluoroacetic anhydride, respectively, affording TiO_2 powders with organic groups on their surfaces together with the formation of isopropyl acetate or trifluoroacetate. A small amount of titanium tetrachloride was added as a Lewis acid catalyst to promote the condensation. In a typical experiment, suitable amounts of TIP, acetic anhydride, and titanium(IV) tetrachloride were added by syringe to the high-pressure reaction cylinder. Liquid CO_2 was then pumped into the vessel via a manual high-pressure generator. The mixture was stirred for a period at room temperature, then heated to 110 °C and stirred at this temperature for 20 h, resulting in white TiO_2 powder. The as-prepared TiO_2 in $scCO_2$ was amorphous, and became crystalline after calcination at 500 °C. The BET surface area of the TiO_2 powder was about 80 $m^2\,g^{-1}$, somewhat higher than that of TiO_2 prepared by hydrolysis of titanium isopropoxide in $scCO_2$ (65 $m^2\,g^{-1}$) [59].

A method for producing crystalline nanosized metal oxides using a seed to enhance crystallization (SSEC) process has been developed. This method is a modification of the sol–gel process performed in SCFs, in which the seeding material acts as heterogeneous seeds for nucleus formation and facilitates the formation of crystalline nuclei at low temperature. Furthermore, the seeding material can cause the reactants to be uniformly distributed on the seeding material, resulting in the coating of the seeding materials by nanoparticles with a very narrow size distribution. Polymers, ceramics, metal fibers, and natural materials can be used as seeding materials. For the synthesis of TiO_2 nanocrystals, TIP was used as precursor, and five different seeding materials, hydrophilic polypropene, hydrophobic polypropene, ceramic, metal fiber and natural fiber, were adopted. In all cases, crystalline TiO_2 of size less than 10 nm was obtained in the temperature range 96–175 °C. The crystallization temperature is lower by 100–250 °C than traditional sol–gel processes, which can be attributed to the fact that the rate of nucleation in supercritical fluids is significantly higher than the crystal growth rate [60].

9.3.3 Synthesis Using Water-in-CO_2 Microemulsions

Water-in-CO_2 microemulsions have also been employed in the synthesis of metal oxides [61]. For instance, titania powders were synthesized via hydrolysis of TIP in water-in-CO_2 microemulsions. A commercial anionic fluorosurfactant, DuPont Zonyl FSP, was employed to form a stable water-in-CO_2 microemulsion. Injection of TIP into the microemulsion resulted in precipitation of spherical titania particles, and free-flowing white titania powders were isolated in 65–70% yield by slow isothermal depressurization. The titania powders had broad particle size distributions (20–800 nm) and specific surface areas in the range 100–500 $m^2\,g^{-1}$. Calcination of a surfactant-free titania powder at 300 °C in air decreased the specific surface area from ~300 to 65 $m^2\,g^{-1}$ and increased the mean cylindrical pore diameter from

2.6 to 4.9 nm, consistent with the collapse of micropores. Titania nanoparticle synthesis via TIP hydrolysis in $scCO_2$ was attempted using water-in-CO_2 microemulsions formed with PFPE-NH_4. However, injection of TIP into PFPE-NH_4-stabilized microemulsions produced TiO_2 particles with sizes of 0.3–2 μm. The use of other CO_2-soluble titanium(IV) alkoxides gave qualitatively similar results.

9.4 Carbon Nanomaterials

9.4.1 Carbon Nanotubes (CNTs)

Vohs *et al.* [62] presented a low-temperature route to prepare CNTs from the catalytic decomposition of CCl_4 over iron-encapsulated polypropylenimine (PPI) catalyst in $scCO_2$. The Fe-encapsulated dendrimer catalyst was prepared by encapsulating various concentrations of Fe^{3+} ions inside a fourth-generation (G_4) PPI dendrimer in ethanol solution, followed by the reduction of the iron cations with $NaBH_4$. The dried Fe@dendrimer powder was placed in a vial with a volume of 15 ml, surrounded with a beaker (50 ml) that contained 5 ml of CCl_4. The reactor was charged with 8.3 MPa of CO_2, which increased to 27.6 MPa after reaching a final temperature of 175 °C. After 24 h, CNTs were formed. The average diameter of the resultant nanotubes was 20–25 nm, much smaller than those reported through other low-temperature methods [63]. Some branched and bent nanotube morphologies were observed, probably resulting from a more complex growth mechanism involving the Fe@PPI-G_4 catalysts used where encapsulated and peripheral iron atoms would exhibit different catalytic rates for nanotube growth. The authors proposed multiple roles of $scCO_2$ for nanotube formation. First, it serves to purge the reactor of oxygen and water, which would result in the formation of oxychloride species from the decomposition of CCl_4, rather than graphitic carbon. Second, the high pressure exerted by the CO_2 assists the decomposition of CCl_4, supported by the fact that no nanotube growth was observed with an analogous experimental setup in the absence of CO_2. Hence this medium is needed to deliver the intermediate species to the encapsulated iron surface for nucleation and growth of CNTs. It should be noted that the temperature and pressure used in this work for CNT growth are far below those for decomposition of the G_4-PPI dendrimer itself, suggesting the presence of nanotube dendritic linkages.

Lee *et al.* [64] reported the synthesis of multiwalled carbon nanotube (MWNT) nanofilaments in supercritical toluene, catalyzed by ferrocene or nanocrystals of Fe or FePt. In this process, toluene serves as both the carbon source for carbon nanostructure growth and the reaction solvent. Under the synthesis conditions producing the highest quality nanotubes (600 °C, ~12.4 MPa), toluene degrades catalytically at the metal particle surfaces, with only minimum homogeneous toluene degradation. MWNTs ranging from 10 to 50 nm in outer diameter were produced along with carbon nanofilaments. The morphology of the nanotubes depends on the catalyst. Larger particles produce solid nanofilaments and smaller particles yield MWNTs. In contrast to vapor-phase synthetic routes, the supercritical solvent provides a high precursor concentration and a homogeneous reaction environment with dispersed catalyst particles. The supercritical fluid approach to synthesizing CNTs has potential for much higher throughput relative to the heterogeneous gas-phase approaches. Toluene is chemically stable up to 650 °C, but metallocenes (e.g. ferrocene) catalyze the decomposition of toluene to carbon and promote nanotube formation at lower temperatures. Therefore, carbon formation occurs only at the catalyst. In the SCF, the carbon reactant and dispersed metal catalyst concentrations can be orders of magnitude higher than those possible in vapor-phase processes. In Lee *et al.*'s work, toluene had a conversion to carbonaceous product of less than 1%, and 2% of the carbon product was nanotubes. In subsequent work [65], the authors used a continuous flow reactor to synthesize CNTs. MWNTs with wall thicknesses of 5–20 nm were synthesized using metallocenes such as cobaltocene, nickelocene, and ferrocene or cobalt or iron nanocrystals as catalysts in supercritical toluene at temperatures ranging from 600 to 645 °C at 8.3 MPa. The SEM images of the MWCNTs obtained under different conditions are shown in Figure 9.4. Relatively

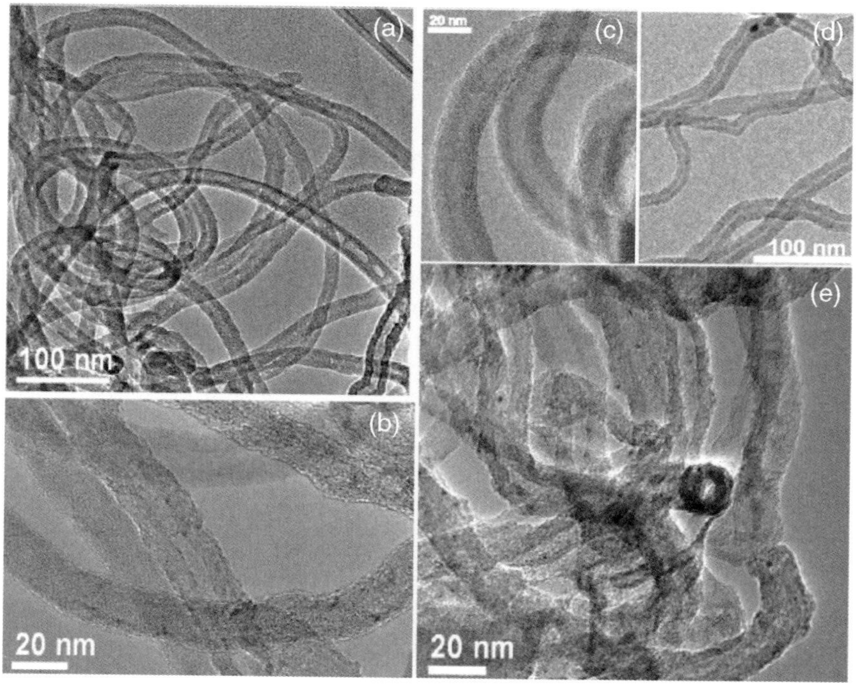

Figure 9.4 TEM images of MWNTs obtained from reactions carried out under the following conditions: (a) 0.6 mM cobaltocene, 3.7 mM ethanol, and 0.3 mM deionized (DI) H_2O at 640 °C; (b, d) 8.2 mM cobaltocene, 3.7 mM ethanol, and 0.2 mM DI-H_2O at 640 °C with a flow rate of 1.5 ml min^{-1}; (c) 8.2 mM cobaltocene, 3.7 mM ethanol, and 0.2 mM DI-H_2O at 640 °C; and (e) 8.2 mM cobaltocene, 3.7 mM ethanol, and 0.04 mM DI-H_2O at 640 °C.

high yields of up to 4% conversion of toluene to MWNTs were reached using cobaltocene as a catalyst in supercritical toluene with the addition of ethanol (30 vol.%) and water (0.75 vol.%). Nickelocene, ferrocene, cobaltocene, Co and Fe nanocrystals all work as catalysts, but cobaltocene gives the highest yields and purity, followed by nickelocene. Relative to these catalysts, ferrocene is not very effective. Water was found to be a critical additive, preventing amorphous carbon and carbon filament formation effectively. The addition of ethanol increases the yield by almost an order of magnitude relative to pure toluene.

Li *et al.* [66] presented a continuous flow SCF method for growing CNTs with selectivity of MWNTs up to 80% of the carbonaceous materials and in high yield. In this method, scCO was used as the carbon source for generating the carbon nanotubes. The diameter of the MWNTs synthesized at 750 °C in the pressure range 5.17–10.34 MPa ranged from 10 to 20 nm, with a length of several tens of micrometers. The SCF-grown nanotubes also exhibited field-emission characteristics similar to CVD-grown MWNTs. These results suggest that SCF-grown nanotubes have potential for the large-scale fabrication of carbon nanotube-based devices.

9.4.2
Carbon Nanocages

Li *et al.* [67] reported a method for preparing carbon nanocages (CNCs) in $scCO_2$ via *p*-xylene decomposition over an MgO-supported Co/Mo catalyst. The reaction temperature played a key role in the formation of the CNCs. At a reaction pressure of 10.34 MPa, amorphous carbon was formed at 600 °C, whereas cage-like carbons were produced when the temperature was increased to 650 °C. Nanocages with diameters between 10 and 60 nm can be synthesized at temperatures between 650 and 750 °C. HRTEM images revealed that the walls of the carbon cages prepared at 700 °C were very thin and were composed of a curved carbon structure consisting of 5–10 carbon layers. Similar cage-like carbon was produced with different *p*-xylene to CO_2 molar ratios ranging from 0.13 to 1.3%. The surface area and pore volume of the nanocages depended on the reaction temperature and pressure employed. At fixed pressure, the surface area of the CNCs decreased with increase in temperature. For example, the BET surface areas of CNCs prepared at 650, 700, and 750 °C and 10.34 MPa were 1240, 698, and 680 $m^2\,g^{-1}$, respectively. The highest total pore volume of the CNCs prepared at 650 °C was up to 5.84 $cm^3\,g^{-1}$. The surface area and microporosity of the CNCs increased with pressure. This is possibly because the decomposition rate of the carbon precursor depends on temperature and pressure.

9.5
Nanocomposites

In recent years, significant progress has been made in fabricating various nanocomposites using supercritical media, and some new routes have been developed. An outstanding feature of SCFs as the composite formation medium is to control the morphology through adjustable solvent properties. Composite formation using $scCO_2$ and scH_2O can be carried out more sustainably and in this way reduce the environmental impact. To date, many nanocomposites including polymer-based composites and porous material-supported composites have been fabricated via chemical reactions in SCFs. Here, a brief survey of the application of SCFs in the chemical synthesis of nanocomposites is given.

9.5.1 Synthesis of Polymer-based Composites

The development of supercritical technology provides a new opportunity for the synthesis of polymer-based nanocomposites. The most widely used SCFs for the synthesis of this kind of nanomaterials are $scCO_2$ and $scCO_2$-based solutions. Although $scCO_2$ cannot dissolve polymers, it can swell most polymers and has considerable solubility in polymers, which reduces the glass transition temperature of the polymers. Several strategies have been developed for the synthesis of this kind of nanomaterials, among which the *in situ* chemical conversion of precursors in an $scCO_2$-swollen polymer matrix is a widely used protocol. In a typical procedure, polymer substrate is first swollen by $scCO_2$ solution, and the precursors dissolved in $scCO_2$ are impregnated into the swollen polymer, which convert to nanoparticles or nanophases to distribute in the polymer substrate, forming polymer-based composites. The precursors used are fairly soluble in $scCO_2$, and include organic vinyl monomers and organometallic compounds. According to the type of precursor, different polymer-based composites including polymer–polymer, metal–polymer, and inorganic oxide–polymer composites were fabricated.

Watkins and McCarthy [68, 69] reported an approach to synthesize polymer–polymer composites via the polymerization of monomers in an $scCO_2$-swollen polymer substrate. In this method, the vinyl monomer and initiator dissolve in $scCO_2$ and infuse into an $scCO_2$-swollen polymer substrate, followed by polymerization of monomer in the polymer substrate, generating polymer–polymer nanocomposites. This procedure involves a soaking period at a temperature at which the half-life of the initiator is hundreds of hours and a polymerization period at a temperature at which the initiator half-life is several hours. Using this route, different polymer–polymer composites were fabricated [70–73]. Generally, the newly formed polymer phase possesses nanoscale size and distributes uniformly through the bulk polymer substrate, and the resultant composites display enhanced mechanical properties. For example, styrene was trapped inside a polychlorotrifluoroethylene (PCTFE) matrix using $scCO_2$, and PCTFE–polystyrene (PS) composites were obtained after free radical polymerization of the styrene [69]. The PS content and distribution in the blends can be controlled by adjusting the concentration of styrene in the SCF or by tuning the soaking time. The PS existed as discrete phase-segregated regions throughout the PCTFE film.

Watkins and McCarthy also prepared metal–polymer composites with the aid of $scCO_2$ [74]. They used [$PtMe_2$(cod)] as a metal precursor and poly(4-methylpent-1-ene) (PMP) as a polymer substrate. The precursor was impregnated into the PMP films at 80 °C and 15.5 MPa. The Pt nanoparticles with a maximum particle size of approximately 15 nm were dispersed uniformly in the polymer substrate. Some other metal–polymer nanocomposites, including Ag–polyimide[75, 76], Ag–poly(ether ether ketone) [77], and Ag–poly(styrene–divinylbenzene) [78], have also been prepared through similar procedures by different researchers. Yoda *et al.* employed $[Pt^{II}(acac)_2]$ and $[Pd^{II}(acac)_2]$ as precursors, and impregnated them into polyimide films using $scCO_2$ [79]. The operating conditions and the type of the polyimide films are important factors for the morphology and the metal loading content of the resulting composites. The Pt or Pd content in polyimide films depends significantly on the impregnation time and temperature.

Inorganic oxide–polymer hybrids have attracted much attention due to their potential applications in many fields, such as optical, electronic, and magnetic devices, fine ceramics, catalysts, pigments, and sensors. These hybrids with nano-sized inorganic oxides in a polymer matrix can combine the properties of the components or have synergistic effects. Silica–poly(styrene-*co*-allyl alcohol) (PSAA) composites were prepared by infusing $SiCl_4$ into the PSAA using $scCO_2$ as a carrier [80]. The infused $SiCl_4$ reacted with the alcohol groups present along the polymer backbones to form Si–O–CH_2PSAA linkages and silica cross-linked composites were formed, as shown by FTIR analysis. SEM images showed that the silica was distributed homogeneously throughout the PSAA matrix and the silica particle size was in the range 50–200 nm. Thermal analysis indicated that the formation of the reinforcing silica network throughout the organic substrates improved the thermal stability of the polymer substrate and also its solvent resistance. Sun *et al.* [81] developed a two-step method to prepare silica–polypropylene (PP) nanocomposites. In this method, tetraethyl orthosilicate (TEOS) is first impregnated into a PP matrix using $scCO_2$ as a swelling agent for PP and also a carrier for TEOS, and then the TEOS-intercalated PP is treated in acidic water, resulting in hydrolysis/condensation reaction of TEOS confined in the polymer network. The uptake of the TEOS is an important parameter affecting the morphology and properties of the resulting composites. The TEOS uptake in the PP matrix can be tuned by changing soaking time, temperature, CO_2 density, and concentration of the TEOS in the fluid phase during the soaking process. The resulting composites showed unusual microstructures. For example, fine silica nanoparticles distributed uniformly in the PP matrix without macroscopic phase segregation, and the polymer molecules entangled with the silica network. This special structure led to a significant increase in Young's modulus and tensile strength relative to the original PP. Similarly, inorganic precursors of TEOS and titanium isopropoxide (TIP) mixture were impregnated into a PP matrix simultaneously with the aid of $scCO_2$, and SiO_2–TiO_2–PP composites were produced via the hydrolysis and condensation reaction of the precursors confined in a polymer network [82]. Wang *et al.* [83] presented a simple and efficient route to prepare metal oxide–polymer composites in a CO_2-based supercritical solution. The resultant composites had structures in which the metal oxide nanoparticles were not only decorated on the outer surface of the polymer substrate but were also intercalated into the shell and further into the inner cavity of the hollow polymer spheres.

Some researchers have reported the use of $scCO_2$ in the preparation of polymer–clay nanocomposites [84, 85]. Zerda *et al.* [84] presented a synthetic route to produce nanocomposites with significantly high concentrations of clay. At high levels of clay (>20 wt%), the viscosity of the mixture is high and is overcome by using $scCO_2$ as a reaction medium. Homogeneous dispersion of monomer, initiation, and subsequent polymerization all occur with a significantly reduced viscosity in this medium. The resulting nanocomposites have a homogeneous morphology even if the intercalated nanocomposites contain 40 wt% of clay.

9.5.2
Decoration of Nanoparticles on Carbon Nanotubes

Since the discovery and successful large-scale production of CNTs, this field has attracted much attention in both fundamental and applied science [86, 87]. The preparation of CNT-based nanocomposites usually involves introducing a functional component to the surface of CNTs. However, the surface of CNTs is rather inert, and it is not easily wetted by liquids with high surface tension. Therefore, it is difficult to form a continuous and homogeneous coating layer without surface treatment. So far, functionalization of CNTs with a variety of metals, metal compounds, organics, and polymers has been achieved by different strategies, including chemical vapor deposition, wet chemical processing, capillary action, the arc discharge technique, electrodeless deposition, and the SCF deposition (SCFD) technique. SCFD shows some unique advantages for the synthesis of various CNT-based composites.

Wai and co-workers deposited Pd [88, 89], Rh [90], Ru [90] and Pt [91] uniformly on CNTs through hydrogen reductions of metal–β-diketone complexes in $scCO_2$. The functions of the resulting composites were also studied. Using $[M(hfa)_2{\cdot}H_2O]$ (M = Pd, Ni, Cu; hfa = hexafluoroacetylacetonate) as precursors, metal nanoparticles can be deposited on CNTs via the reduction of these precursors in $scCO_2$ [92]. Most of the metal nanoparticles with sizes of 4–8 nm were well dispersed on the surface of CNTs and some were intercalated into the inner cavities of the CNT with a filling percentage of about 10%. Using $Pt(acac)_2$ and $Ru(acac)_2$ mixture as precursors, PtRu alloy nanoparticles have been decorated on CNTs via the hydrogen reduction of the precursors in $scCO_2$–methanol solution at 200 °C [93]. The PtRu nanoparticles showed a good distribution on the surface of CNTs, and the average particle size was about 5–10 nm. Energy-dispersive X-ray spectroscopy (EDS) indicated that the contents of Pt and Ru in the composites were identical with those of the precursors. PtRu–CNT catalysts exhibited high activity for methanol oxidation, which resulted from the high surface area of the CNTs and the nanostructure of PtRu particles.

Bayrakceken *et al.* [94] investigated the adsorption isotherm of the precursor $[PtMe_2(cod)]$ on CNTs in $scCO_2$. It was shown that the isotherm obeys the Langmuir model. Heating the CNTs with adsorbed precursor produced Pt–CNT composites, in which Pt particles were uniformly distributed on CNTs and the particle size was about 2 nm. Since the thermal reduction of $[PtMe_2(cod)]$ occurred at the CNT surface alone,

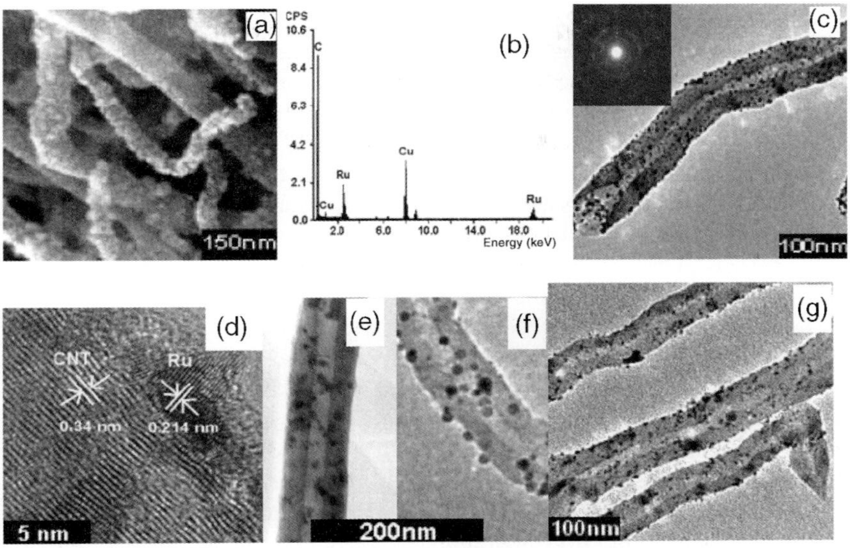

Figure 9.5 (a) SEM image, (b) EDS spectrum, (c) TEM image, and (d) HRTEM image of the Ru-CNT composites prepared with an $RuCl_3{\cdot}H_2O$ to CNT weight ratio of 1 : 1 at 400 °C. TEM images of the composite prepared with different $RuCl_3{\cdot}H_2O$ to CNT ratios at different temperatures: (e) 1 : 2, 400 °C, (f) 1 : 2, 450 °C; (g) TEM image of the composite (prepared with the initial weight ratio of $RuCl_3{\cdot}H_2O$ to CNT of 1 : 1 at 400 °C) used four times for benzene hydrogenation.

higher uniformity of the particle size was achieved. Furthermore, the Pt loading density can be controlled thermodynamically based on the adsorption isotherm.

Sun *et al.* [95] prepared Ru–CNT composites by heating an aqueous solution of $RuCl_3{\cdot}H_2O$ with dispersed CNTs. Figure 9.5 shows SEM and TEM images and the EDS spectrum of the Ru–CNT nanocomposites, which were fabricated by treating CNTs and $RuCl_3$ in scH_2O at 400 or 450 °C. From EDS analysis (Figure 9.5b), it can be deduced that $RuCl_3{\cdot}H_2O$ was converted into metallic Ru in scH_2O in the presence of CNTs. The SEM and TEM images show that the Ru nanoparticles are uniformly decorated on the outer surface of the CNTs (Figure 9.5c, e, f, and g) under all experimental conditions. It seems that some particles were also inserted into CNTs (Figure 9.5c). This might result from the unusual properties of scH_2O, including low viscosity, high diffusivity, and near-zero surface tension, which allows the precursor dissolved in scH_2O to be delivered into the cavities of CNTs, permitting deposition of Ru in the internal channels of CNTs. The Ru nanoparticles are crystalline, as confirmed by HRTEM analysis (Figure 9.5d). The initial weight ratio of $RuCl_3{\cdot}H_2O$ to CNTs and the reaction temperature are two important factors for the tuning of the particle loading content and particle size. Increasing the weight ratio of $RuCl_3{\cdot}H_2O$ to CNTs leads to an increase in the amount of Ru nanoparticles loaded on the CNT (Figure 9.5c vs e). The Ru particles prepared at 450 °C are larger than those obtained at 400 °C (Figure 9.5e and f). The Ru–CNT nanocomposites display good catalytic activity for the hydrogenation of benzene, and can be reused several times without obvious loss of catalytic activity, which may be due to their stable structure (Figure 9.5g). It is worth mentioning that this method is simple and clean and can produce Ru–CNT composites on a large scale. It also can be extended to synthesize other noble metal-CNT composites.

Methanol can dissolve some inorganic salts, such as H_2PtCl_6, $HAuCl_4$, and $RuCl_3$, and reduce them under suitable conditions. Supercritical methanol ($scCH_3OH$) can be used as a medium for the synthesis of metal–CNT composites. In these cases methanol acts as the solvent and as the reducing agent for some metal precursors. Using a mixture of H_2PtCl_6 and $RuCl_3$ as metal precursor, bimetallic PtRu nanoparticles were attached to CNTs in $scCH_3OH$ [96]. Uniform and well-dispersed crystalline nanoparticles were decorated on the surfaces of the CNTs and their average particle size was about 2–3 nm. It was shown that Pt and Ru nanoparticles can be produced simultaneously during the reaction process, and both can be deposited on the surfaces of CNTs. Moreover, some PtRu alloy nanoparticles were formed and deposited on CNTs.

A CO_2-based microemulsion route to deposit metal nanoparticles on CNTs at room temperature was reported by Wai and co-workers [97]. In this method, metallic nanoparticles are first synthesized by hydrogen reduction of metal ions dissolved in the water core of a water-in-CO_2 microemulsion, and then the microemulsion solution containing metal nanoparticles is put in contact with the CNTs. As a consequence, metal nanoparticles are able to deposit homogeneously on the surfaces of the CNTs. In that study, the microemulsion was stabilized by sodium bis(2-ethylhexyl)sulfosuccinate (AOT) in liquid CO_2 with hexane as modifier. The particle diameters were in the range 2–10 nm. Approximately 70% of the metal ions present initially can be deposited as metal nanoparticles on the CNT surfaces. Bimetallic nanoparticles can also be deposited on CNTsurfaces by this method using a metal salt mixture in the water cores of the microemulsion. The as-synthesized CNT-supported PdRh bimetallic nanoparticles are very effective catalysts for the hydrogenation of aromatic compounds.

Attachment of metal oxides on CNTs is another important functionalization of CNTs. SCFs also show unique advantages for the synthesis of this kind of composite. $scCO_2$ cannot dissolve metal nitrates, whereas ethanol has strong solvent power for such salts. Therefore, $scCO_2$–ethanol solution can dissolve some metal nitrates. Using $scCO_2$-ethanol mixtures as the reaction media, a simple route was developed to deposit metal oxides on CNTs via the decomposition of metal nitrates in $scCO_2$-ethanol solution at relatively low temperatures. In a general procedure, nitrate solution in ethanol and CNTs are loaded into a high-pressure cell. The vessel is placed in a water-bath with a constant temperature of 35 °C, and CO_2 is then charged into the cell up to a desired pressure sufficient to bring the solution in the vessel to a homogeneous state. Subsequently, the cell is moved into an oven with a desired temperature in the range 100–200 °C, at which the nitrate in the $scCO_2$–ethanol solution decomposes. Some metal oxide–CNT nanocomposites, including Co_3O_4–CNT [98], Eu_2O_3–CNT [99], Al_2O_3–CNT [100], Fe_2O_3–CNT [101], and ZrO_2–CNTs[102], were synthesized using this method. Generally, the metal oxides are decorated on the outer surface of the CNTs in the form of nanoparticles with a uniform size, and some nanoparticles can also be inserted into the inner cavity. Some resultant metal oxide–CNT nanocomposites possess special structures which are difficult to achieve. For example, Co_3O_4 spheres are beaded on CNTs and crystalline Eu_2O_3 film can be coated on CNTs, and amorphous Al_2O_3 and Fe_2O_3 films wrapped around CNTs have been made. Obvious advantages of this technique are its flexibility, simplicity, and large-scale production of the composites. In addition, this route does not require tedious surface modification of the CNTs. The differences in the microstructures of the resultant metal oxide–CNT composites may originate from the differences in the interactions between the precursors and the CNTs used. The resultant metal oxide–CNT composites have promising applications. For instance, the Co_3O_4–CNT nanocomposites were utilized to fabricate a Schottky-junction diode. The measured current–voltage profile exhibits clear rectifying behavior and no reverse bias breakdown is observed up to a measured voltage of 2.5 V. [98] The Eu_2O_3–CNT composites can be used as red light-emitting nanomaterials since their photoluminescence spectrum has two striking excitation peaks at 590 and 615 nm [99]. The Al_2O_3–CNT nanocomposites can be used to prepare a field-effect transistor (FET), which has distinct p-type characteristics [100]. The chemiluminescent sensor made of the ZrO_2–CNT nanocomposites is stable and very sensitive and selective to ethanol [102].

In addition, polymer–CNT composites have also been fabricated via SCF-mediated routes [103, 104]. SCFs demonstrate unusual advantages for the fabrication of this kind of material. Polymers can coat the outer surface of CNTs and can also be impregnated inside CNTs and grafted on to single-walled CNTs [105].

9.5.3
Deposition of Nanoparticles on Porous Supports

"Porous substrate-supported nanocomposites" are porous materials loaded with nanoparticles in their pores or channels. This kind of material is of great importance due to the wide applications as catalysts, sensors, and optical and electronic devices. Using porous materials with a uniform pore size and high surface area as hosts, the nanoparticles can be supported on well-defined matrices. So far, for example, mesoporous silica, carbon molecular sieves (CMS), and natural clays have been employed as supports of nanoparticles. The unique properties of SCFs make them excellent media for the deposition of nanoparticles on these substrates.

C_nFSM-16 are micro- and mesoporous silica materials with uniform pore size and high surface areas, where n ($n = 8$, 10, 12, and 16) is the carbon number of the alkyl chain of the surfactant template used for the porous material synthesis. Wakayama *et al.* [106] used a two-step approach to deposit Pt nanoparticles on C_nFSM-16 ($n = 8$, 10, 12, and 16) matrices. To prepare the composites, $Pt(acac)_2$ is first carried into the pores of C_nFSM-16 by $scCO_2$ modified with acetone at 150 °C and 32 MPa, followed by H_2 reduction at 400 °C after releasing CO_2. The Pt particles are highly dispersed in the C_nFSM-16 supports. The average sizes of the Pt nanoparticles incorporated in C_nFSM-16 with $n = 8$, 10, 12, and 16) are 1.5, 1.7, 2.3 and 2.9 nm, respectively, which are slightly smaller than the pore diameters of the corresponding C_nFSM-16. This implies that the particles mainly exist inside the pores of C_nFSM-16.

Erkey and co-workers successfully supported Pt and Ru nanoparticles with a particle size of 1 nm and a metal content up to 40 wt% on a wide range of porous materials, including carbon aerogel (CA), carbon black (CB), silica aerogel (SA), silica (SiO_2), and γ-alumina (Al_2O_3) [107–111]. The organometallic precursors used were [$PtMe_2$(cod)], [Ru(acac)$_3$], and [Ru(cod)(tmhd)$_2$], which are soluble in $scCO_2$. In a typical experiment to deposit Pt nanocrystals on porous silica, the porous substrate is exposed to a solution of CO_2 and [$PtMe_2$(cod)] at 80 °C and 27.6 MPa. After depressurization, the impregnated precursor is reduced to elemental Pt by heat treatment in a nitrogen atmosphere. HRTEM images show a good distribution of highly dispersed Pt particles throughout the bulk of the support used. The TEM observations reveal that Pt particles are uniformly dispersed on the substrate with particle sizes ranging from 1.2 to 6.4 nm and a narrow particle size distribution. They explored the effects of the substrates on the size and size distribution of the supported Pt nanoparticles. It was shown that Pt nanoparticles on SiO_2 supports are larger and have broader size distributions than Pt nanoparticles on carbon and alumina supports at a similar metal content. These observations indicate that particle formation and growth are mainly governed by the strength of the interaction between the organometallic precursors and the support. Erkey and co-workers synthesized Pt–CA nanocomposites with a Pt loading of 40 wt% using a supercritical deposition method [107, 109, 112]. As a precursor, [$PtMe_2$(cod)] is first impregnated into porous CA in $scCO_2$. The impregnated aerogels are converted to Pt–CA composites by heat treatment at temperatures ranging from 300 to 1000 °C in the presence of nitrogen gas. Increasing the reduction temperature increases both the crystal size and the polydispersity. The crystallite size is 1 nm at low reduction temperatures and the particles are single crystalline. Howdle and co-workers exploited the synthesis of Ag [78] and Pd nanoparticles [113] in silica aerogel (SA) via a supercritical route. Chatterjee *et al.* [114] reported an approach to deposit Au nanoparticles in the channels of mesoporous material MCM-48 in $scCO_2$ using a hydrogen reduction technique.

Sun *et al.* [115] described the monolayer and double-layer coating of TiO_2 on the inner surfaces of SBA-15 via a surface sol–gel process in $scCO_2$. In this method, SBA-15 with a high level of silanol groups is used as the substrate. TIP dissolved in $scCO_2$ is passed through SBA-15 and then grafted on to SBA-15 via the reaction of TIP with the surface silanol groups, forming a coating layer on the inner surface of the SBA-15. Calcination at 550 °C generates TiO_2-coated SBA-15 composites. The TiO_2-loaded samples show similar nitrogen sorption isotherms to the original SBA-15 material used. However, the pore size reduction of about 1.0 nm for each modification cycle implies the deposition of a TiO_2 film of about 0.5 nm thickness. In this study, the silanol groups on SBA-15 serve as active sites for titanium grafting and are consumed during the reaction process. The high concentration of hydroxy groups on the silica surface is required to obtain a high surface coverage with titanium species. Mokaya and co-workers investigated the use of $scCO_2$ for post-synthesis alumination of MCM-41 to form mesoporous aluminosilicates [116]. The processing procedure involves soaking the silica substrate in $scCO_2$–aluminum isopropoxide solution and calcining the soaked silica at 600 °C for several hours. Tatsuda and co-workers developed a method for the preparation of TiO_2–carbon composites which involves the penetration of titanium tetraisopropoxide (TTIP) dissolved in $scCO_2$ into activated carbon (AC) and the conversion of TTIP into TiO_2 nanoparticles inside AC [117–119]. TiO_2–AC composites were also prepared in supercritical 2-propanol, in which tetraisopropyl titanate (TPT) is soaked in AC pores and converted into anatase nanoparticles [120]. The supercritically treated TiO_2–AC composites, particularly at 300 and 350 °C, have much higher formaldehyde-decomposing ability than a non-composite comprising a simple mixture of AC and TiO_2 granules. This indicates that the supercritical treatment can be effective for preparing photocatalytic composites that have a synergetic effect of adsorption and photocatalytic decomposition of formaldehyde for environmental clean-up.

9.5.4
Some Other Nanocomposites

SCFs have also been utilized in the synthesis of other inorganic–inorganic nanocomposites. For example, through hydrogen reduction of metal precursors in $scCO_2$ [121], Cu and Pd nanocrystals were deposited on SiO_2 nanowires to form different types of nanostructured materials, including nanocrystal–nanowire, spherical aggregation-nanowire, and shell-nanowire composites, and "mesoporous" metals supported by the framework of nanowires. This SCF deposition technique is an attractive approach for modifying nanowires because of its generality and simplicity. The modified nanowires can be useful as catalysts and for further fabrication of multifunctional composites. By using a simple low-temperature SCF-assisted method, crystalline multilayer Eu_2O_3 sheaths were coated uniformly on Ga_2O_3 nanoribbons [122]. The resulting core–sheath coaxial heterostructures, with their various possible functionalities, represent an important class of nanoscale building blocks for optoelectronic applications.

Natural clays such as montmorillonite (MMT) and vermiculite (VMT) possess layered structures and high surface areas, and can be used as catalyst supports and polymer fillers. Yoda and co-workers developed a strategy for the fabrication of oxide-pillared clay [123, 124]. The clay is first modified using organic ions (e.g. octadecyltrimethylammonium ion, $C_{18}TMA$) to control the interlayer spacing and hydrophobic nature, and then alkoxides dissolved in $scCO_2$ are intercalated into the interlayers of the modified clay, followed by hydrolysis with interlayer-adsorbed water, resulting in mesoporous oxide-pillared clay composites after calcination at high temperature.

9.6
Conclusion

Many methods for synthesizing nanomaterials using SCFs have been proposed and numerous nanomaterials have been synthesized, including those which are difficult to obtain by other approaches. There is no doubt that SCF technology will be used more widely in the synthesis of various kinds of nanomaterials with some unique advantages. Like most other technologies, SCF technology also has its inherent drawbacks, such as the need to perform syntheses and other operations at high pressure. However, with well-designed processes the advantages of this technology can effectively compensate the disadvantages. With this in mind, controllable syntheses of nanostructures using SCFs will certainly have a bright future, and it can be envisioned that many related fundamental and technological problems might also be tackled effectively with SCF technology.

References

1 McLeod, M.C., Gale, W.F. and Roberts, C.B. (2004) *Langmuir*, **20** (17), 7078–7082.
2 Esumi, K., Sarashina, S. and Yoshimura, T. (2004) *Langmuir*, **20** (13), 5189–5191.
3 Watkins, J.J., Blackburn, J.M. and McCarthy, T.J. (1999) *Chemistry of Materials*, **11** (2), 213–215.
4 Blackburn, J.M., Long, D.P., Cabanas, A. and Watkins, J.J. (2001) *Science*, **294** (5540), 141–145.
5 Shah, P.S., Husain, S., Johnston, K.P. and Korgel, B.A. (2002) *The Journal of Physical Chemistry. B*, **106**, 12178–12185.
6 Shah, P.S., Holmes, J.D., Doty, R.C., Johnston, K.P. and Korgel, B.A. (2000) *Journal of the American Chemical Society*, **122**, 4245–4246.
7 Shah, P.S., Novick, B.J., Hwang, H.S., Lim, K.T., Carbonell, R.G., Johnston, K.P. and Korgel, B.A. (2003) *Nano Letters*, **3**, 1671–1675.
8 Shah, P.S., Holmes, J.D., Doty, R.C., Johnston, K.P. and Korgel, B.A. (2000) *Journal of the American Chemical Society*, **122**, 4245–4246.
9 Shah, P.S., Husain, S., Johnston, K.P. and Korgel, B.A. (2001) *The Journal of Physical Chemistry. B*, **105**, 9433–9440.
10 Shah, P.S., Shabbir, H., Johnston, K.P. and Korgel, B.A. (2002) *The Journal of Physical ChemistryB*, **106**, 12178–12185.
11 Esumi, K., Sarashina, S. and Yoshimura, T. (2004) *Langmuir*, **20**, 5189–5191.
12 Fan, X., McLeod, M.C., Enick, R.M. and Roberts, C.B. (2006) *Industrial & Engineering Chemistry Research*, **45**, 3343–3347.
13 Holmes, J.D., Ziegler, K.J., Doty, R.C., Pell, L.E., Johnston, K.P. and Korgel, B.A. (2001) *Journal of the American Chemical Society*, **123**, 3743–3748.
14 Pell, L.E., Schricker, A.D., Mikulec, F.V. and Korgel, B.A. (2004) *Langmuir*, **20**, 6546–6548.
15 Lu, X., Ziegler, K.J., Ghezelbash, A., Johnston, K.P. and Korgel, B.A. (2004) *Nano Letters*, **4**, 969–974.
16 Holmes, J.D., Johnston, K.P., Doty, R.C. and Korgel, B.A. (2000) *Science*, **287**, 1471–1473.
17 Lu, X., Hanrath, T., Johnston, K.P. and Korgel, B.A. (2003) *Nano Letters*, **3**, 93–99.
18 Hanrath, T. and Korgel, B.A. (2002) *Journal of the American Chemical Society*, **124**, 1424–1429.
19 Hanrath, T. and Korgel, B.A. (2003) *Advanced Materials*, **15**, 437–440.
20 Davidson, F.M., Schricker, A.D., Wiacek, R. and Korgel, B.A. (2004) *Advanced Materials*, **16**, 646–649.
21 Ziegler, K.J., Doty, R.C., Johnston, K.P. and Korgel, B.A. (2001) *Journal of the American Chemical Society*, **123**, 7797–7803.
22 Sue, K., Suzuki, A., Suzuki, M., Arai, K., Hakuta, Y., Hayashi, H. and Hiaki, T. (2006) *Industrial & Engineering Chemistry Research*, **45**, 623–626.
23 Cason, J.P. and Roberts, C.B. (2000) *The Journal of Physical Chemistry. B*, **104**, 1217–1221.

24 Cason, J.P., Khambaswadkar, K. and Roberts, C.B. (2000) *Industrial & Engineering Chemistry Research*, **39**, 4749–4755.

25 Cason, J.P., Miller, M.E., Thompson, J.B. and Roberts, C.B. (2001) *The Journal of Physical Chemistry. B*, **105**, 2297–2302.

26 Kitchens, C.L. and Roberts, C.B. (2004) *Industrial & Engineering Chemistry Research*, **43**, 6070–6081.

27 Ji, M., Chen, X., Wai, C.M. and Fulton, J.L. (1999) *Journal of the American Chemical Society*, **121**, 2631–2632.

28 Ohde, H., Hunt, F. and Wai, C.M. (2001) *Chemistry of Materials*, **13**, 4130–4135.

29 Ohde, H., Rodriguez, J.M., Yea, X.R. and Wai, C.M. 2000 *Chemical Communications*, 2353–2354.

30 Ohde, H., Ohde, M., Bailey, F., Kim, H. and Wai, C.M. (2002) *Nano Letters*, **2**, 721–724.

31 Holmes, J.D., Bhargava, P.A., Korgel, B.A. and Johnston, K.P. (1999) *Langmuir*, **15**, 6613–6615.

32 Liu, J.C., Raveendran, P., Shervania, Z. and Ikushima, Y. 2004 *Chemical Communications*, 2582–2583.

33 Liu, J.C., Raveendran, P., Shervani, Z., Ikushima, Y. and Hakuta, Y. (2005) *Chemistry - A European Journal*, **11**, 1854–1860.

34 Fernandez, C.A. and Wai, C.M. (2007) *Chemistry - A European Journal*, **13**, 5838–5844.

35 Fernandez, C.A. and Wai, C.M. (2006) *small*, **2**, 1266–1269.

36 Zhang, J.L., Han, B.X., Liu, J.C., Zhang, X.G., He, J., Liu, Z.M., Jiang, T. and Yang, G.Y. (2002) *Chemistry - A European Journal*, **8**, 3879–3883.

37 Zhang, J.L., Han, B.X., Liu, J.C., Zhang, X.G., Liu, Z.M. and He, J. (2001) *Chemical Communications*, 2724–2725.

38 Adschiri, T., Kanazawa, K. and Arai, K. (1992) *Journal of the American Ceramic Society*, **75**, 1019–1022.

39 Adschiri, T., Kanazawa, K. and Arai, K. (1992) *Journal of the American Ceramic Society*, **75**, 2615–2618.

40 Cabanas, A., Darr, J.A., Lester, E. and Poliakoff, M. 2000 *Chemical Communications*, 901–902.

41 Cabanas, A., Darr, J.A., Lester, E. and Poliakoff, M. (2001) *Journal of Materials Chemistry*, **11**, 561–568.

42 Cabanas, A. and Poliakoff, M. (2001) *Journal of Materials Chemistry*, **11**, 1408–1416.

43 Adschiri, T. (2007) *Chemistry Letters*, **36**, 1188–1193.

44 Zhang, J., Ohara, S., Umetsu, M., Naka, T., Hatakeyama, Y. and Adschiri, T. (2007) *Advanced Materials*, **19**, 203–206.

45 Rangappa, D., Naka, T., Kondo, A., Ishii, M., Kobayashi, T. and Adschiri, T. (2007) *Journal of the American Chemical Society*, **129**, 11061–11066.

46 Mousavand, T., Ohara, S., Umetsu, M., Zhang, J., Takami, S., Naka, T. and Adschiri, T. (2007) *Journal of Supercritical Fluids*, **40**, 397–401.

47 Mousavand, T., Zhang, J., Ohara, S., Umetsu, M., Naka, T. and Adschiri, T. (2007) *Journal of Nanoparticle Research*, **6**, 1067–1071.

48 Kaneko, K., Inoke, K., Freitag, B., Hungria, A.B., Midgley, P.A., Hansen, T.W., Zhang, J., Ohara, S. and Adschiri, T. (2007) *Nano Letters*, **7**, 421–425.

49 Rangappa, D., Ohara, S., Naka, T., Kondo, A., Ishii, M. and Adschiri, T. (2007) *Journal of Materials Chemistry*, **17**, 4426–4429.

50 Sue, K., Murata, K., Kimura, K. and Arai, K. (2003) *Green Chemistry*, **5**, 659–662.

51 Mousavand, T., Takami, S., Umetsu, M., Ohara, S. and Adschiri, T. (2006) *Journal of Materials Science*, **41**, 1445–1448.

52 Adschiri, T., Hakuta, Y., Sue, K. and Arai, K. (2001) *Journal of Nanoparticle Research*, **3**, 227–235.

53 Reverchon, E. and Adami, R. (2005) *Journal of Supercritical Fluids*, **37**, 1–22.

54 Sue, K., Suzuki, M., Arai, K., Ohashi, T., Ura, H., Matsui, K., Hakuta, Y., Hayashi, H., Watanabed, M. and Hiaki, T. (2006) *Green Chemistry*, **8**, 634–638.

55 Sui, R., Rizkalla, A.S. and Charpentier, P.A. (2005) *Langmuir*, **21**, 6150–6153.

56 Sui, R., Rizkalla, A.S. and Charpentier, P.A. (2006) *The Journal of Physical Chemistry. B*, **110**, 16212–16218.

57 Sui, R., Rizkalla, A.S. and Charpentier, P.A. (2006) *Langmuir*, **22**, 4390–4396.

58 Arnal, P., Corriu, R.J.P., Leclercq, D., Mutin, P.H. and Vioux, A. (1996) *Journal of Materials Chemistry*, **6**, 1925–1932.

59 Stallings, W.E. and Lamb, H.H. (2003) *Langmuir*, **19**, 2989–2994.

60 Jensen, H., Joensen, K.D., Iversen, S.B. and Søgaard, E.G. (2006) *Industrial & Engineering Chemistry Research*, **45**, 3348–3353.

61 Stallings, W.E. and Lamb, H.H. (2003) *Langmuir*, **19**, 2989–2994.

62 Vohs, J.K., Brege, J.J., Raymond, J.E., Brown, A.E., Williams, G.L. and Fahlman, B.D. (2004) *Journal of the American Chemical Society*, **126**, 9936–9937.

63 Jiang, Y., Wu, Y., Zhang, S., Xu, C., Yu, W., Xie, Y. and Qian, Y. (2000) *Journal of the American Chemical Society*, **122**, 12383–12384.

64 Lee, D.C., Mikulec, F.V. and Korgel, B.A. (2004) *Journal of the American Chemical Society*, **126**, 4951–4957.

65 Smith, D.K., Lee, D.C. and Korgel, B.A. (2006) *Chemistry of Materials*, **18**, 3356–3364.

66 Li, Z., Andzane, J., Erts, D., Tobin, J.M., Wang, K., Morris, M.A., Attard, G. and Holmes, J.D. (2007) *Advanced Materials*, **19**, 3043–3046.

67 Li, Z., Jaroniec, M., Papakonstantinou, P., Tobin, J.M., Vohrer, U., Kumar, S., Attard, G. and Holmes, J.D. (2007) *Chemistry of Materials*, **19**, 3349–3354.

68 Watkins, J.J. and McCarthy, T.J. (1994) *Macromolecules*, **27**, 4845–4847.

69 Watkins, J.J. and McCarthy, T.J. (1995) *Macromolecules*, **28**, 4067–4074.

70 Kung, E., Lesser, A.J. and McCarthy, T.J. (1998) *Macromolecules*, **31**, 4160–4165.

71 Zhang, J.X., Busby, A.J., Roberts, C.J., Chen, X.Y., Davies, M.C., Tendler, S.J.B. and Howdle, S.M. (2002) *Macromolecules*, **35**, 8869–8874.

72 Busby, A.J., Zhang, J.X., Roberts, C.J., Lester, E. and Howdel, S.M. (2005) *Advanced Materials*, **17**, 364–369.

73 Abbett, K.F., Teja, A.S., Kowalik, J. and Tolbert, L. (2003) *Macromolecules*, **36**, 3015–3019.

74 Watkins, J.J. and McCarthy, T.J. (1995) *Chemistry of Materials*, **7**, 1991–1993.

75 Rosolovsky, J., Boggess, R.K., Rubira, A.F., Taylor, L.T., Stoakley, D.M. and Clair, A.K.St. (1997) *Journal of Materials Research*, **12**, 3127–3133.

76 Boggess, R.K., Taylor, L.T., Stoakley, D.M. and Clair, A.K.St. (1997) *Journal of Applied Polymer Science*, **64**, 1309–1317.

77 Nazem, N., Taylor, L.T. and Rubira, A.F. (2002) *Journal of Supercritical Fluids*, **23**, 43–57.

78 Morley, K.S., Marr, P.C., Webb, P.B., Berry, A.R., Allison, F.J., Moldovan, G., Brown, P.D. and Howdle, S.M. (2002) *Journal of Materials Chemistry*, **12**, 1898–1905.

79 Yoda, S., Hasegawa, A., Suda, H., Uchimaru, Y., Haraya, K., Tsuji, T. and Otake, K. (2004) *Chemistry of Materials*, **16**, 2363–2368.

80 Green, J.W., Rubal, M.J., Osman, B.M., Welsch, R.L., Cassidy, P.E., Fitch, J.W. and Blanda, M.T. (2000) *Polymers for Advanced Technologies*, **11**, 820–825.

81 Sun, D.H., Zhang, R., Liu, Z.M., Huang, Y., Wang, Y., He, J., Han, B.X. and Yang, G.Y. (2005) *Macromolecules*, **38**, 5617–5624.

82 Sun, D.H., Huang, Y., Han, B.X. and Yang, G.Y. (2006) *Langmuir*, **22**, 4793–4798.

83 Wang, J.Q., Zhang, C.L., Liu, Z.M., Ding, K.L. and Yang, Z.Z. (2006) *Macromolecular Rapid Communications*, **27**, 787–792.

84 Zerda, A.S., Caskey, T.C. and Lesser, A.J. (2003) *Macromolecules*, **36**, 1603–1608.

85 Yan, C., Ma, L. and Yang, J.C. (2005) *Journal of Applied Polymer Science*, **98**, 22–28.

86 Iijima, S. (1991) *Nature*, **354**, 56–58.

87 Dai, H.J., Hafner, J.H., Rinzler, A.G., Colbert, D.T. and Smalley, R.E. (1996) *Nature*, **384**, 147–150.

88 Ye, X.R., Lin, Y.H. and Wai, C.M. 2003 *Chemical Communications*, 642–643.

89 Ye, X.R., Lin, Y.H., Wai, C.M., Talbot, J.B. and Jin, S.H. (2005) *Journal of Nanoscience and Nanotechnology*, **5**, 964–969.

90 Ye, X.R., Lin, Y.H., Wang, C.M., Engelhard, M.H., Wang, Y. and Wai, C.M. (2004) *Journal of Materials Chemistry*, **14**, 908–913.

91 Lin, Y.H., Cui, X.L., Wang, C. and Wai, C.M. (2005) *The Journal of Physical Chemistry. B*, **109**, 14410–14415.

92 Ye, X.R., Lin, Y.H., Wang, C. and Wai, C.M. (2003) *Advanced Materials*, **15**, 316–319.

93 Lin, Y.H., Cui, X.L., Yen, C.H. and Wai, C.M. (2005) *Langmuir*, **21**, 11474–11479.

94 Bayrakceken, A., Kitkamthorn, U., Aindowb, M. and Erkey, C. (2007) *Scripta Materialia*, **56**, 101–103.

95 Sun, Z.Y., Liu, Z.M., Han, B.X., Wang, Y., Du, J.M., Xie, Z.L. and Han, G.J. (2005) *Advanced Materials*, **17**, 928–930.

96 Sun, Z.Y., Fu, L., Liu, Z.M., Han, B.X., Liu, Y.Q. and Du, J.M. (2006) *Journal of Nanoscience and Nanotechnology*, **6**, 691–697.

97 Wang, J.S.F., Pan, H.B. and Wai, C.M. (2006) *Journal of Nanoscience and Nanotechnology*, **6**, 2025–2030.

98 Fu, L., Liu, Z.M., Liu, Y.Q., Han, B.X., Hu, P.A., Cao, L.C. and Zhu, D.B. (2005) *Advanced Materials*, **17**, 217–220.

99 Fu, L., Liu, Z.M., Liu, Y.Q., Han, B.X., Wang, J.Q., Hu, P.A., Gao, L.C. and Zhu, D.B. (2004) *Advanced Materials*, **16**, 350–352.

100 Fu, L., Liu, Y.Q., Liu, Z.M., Han, B.X., Cao, L.C., Wei, D.C., Yu, G. and Zhu, D.B. (2006) *Advanced Materials*, **18**, 181–184.

101 Sun, Z.Y., Yuan, H.Q., Liu, Z.M., Han, B.X. and Zhang, X.R. (2005) *Advanced Materials*, **17**, 2993–2996.

102 Sun, Z.Y., Zhang, X.R., Na, N., Liu, Z.M., Han, B.X. and An, G.M. (2006) *The Journal of Physical Chemistry. B*, **110**, 13410–13414.

103 Liu, Z.M., Dai, X.H., Xu, J., Han, B., Zhang, J.L., Wang, Y., Huang, Y. and Yang, G.Y. (2004) *Carbon*, **42**, 458–460.

104 Steinmetz, J., Kwon, S., Lee, H.J., Abou-Hamad, E., Almairac, R., Goze-Bac, C., Kim, H. and Park, Y.W. (2006) *Chemical Physics Letters*, **431**, 139–144.

105 Yue, B.H., Wang, Y.B., Huang, C.Y., Pfeffer, R. and Iqbal, Z. (2007) *Journal of Nanoscience and Nanotechnology*, **7**, 994–1000.

106 Wakayama, H., Setoyama, H. and Fukushima, Y. (2003) *Advanced Materials*, **15**, 742–745.

107 Saquing, C.D., Cheng, T.T., Aindow, M. and Erkey, C. (2004) *The Journal of Physical Chemistry. B*, **108**, 7716–7722.

108 Zhang, Y., Kang, D.F., Saquing, C., Aindow, M. and Erkey, C. (2005) *Industrial & Engineering Chemistry Research*, **44**, 4161–4164.

109 Zhang, Y., Kang, D.F., Aindow, M. and Erkey, C. (2005) *The Journal of Physical Chemistry. B*, **109**, 2617–2624.

110 Zhang, Y. and Erkey, C. (2005) *Industrial & Engineering Chemistry Research*, **44**, 5312–5317.

111 Haji, S., Zhang, Y., Kang, D.F., Aindow, M. and Erkey, C. (2005) *Catalysis Today*, **99**, 365–373.

112 Saquing, C.D., Kang, D., Aindow, M. and Erkey, C. (2005) *Microporous and Mesoporous Materials*, **80**, 11–23.

113 Morley, K.S., Licence, P., Marr, P.C., Hyde, J.R., Brown, P.D., Mokaya, R., Xia, Y. and Howdle, S.M. (2004) *Journal of Materials Chemistry*, **14**, 1212–1217.

114 Chatterjee, M., Ikushima, Y., Hakuta, Y. and Kawanami, H. (2006) *Advanced Synthesis and Catalysis*, **348**, 1580–1590.

115 Sun, D.H., Liu, Z.M., He, J., Han, B.X., Zhang, J.L. and Huang, Y. (2005) *Microporous and Mesoporous Materials*, **80**, 165–171.

116 O'Neil, A.S., Mokaya, R. and Poliakoff, M. (2002) *Journal of the American Chemical Society*, **124**, 10636–10637.

117 Wakayama, H., Itahara, H., Tatsuda, N., Inagaki, S. and Fukushima, Y. (2001) *Chemistry of Materials*, **13**, 2392–2396.

118 Wakayama, H., Inagaki, S. and Fukushima, Y. (2002) *Journal of the American Ceramic Society*, **85**, 161–164.

119 Tatsuda, N., Itahara, H., Setoyama, N. and Fukushima, Y. (2004) *Journal of Materials Chemistry*, **14**, 3440–3443.

120 Huang, B. and Saka, S. (2003) *Journal of Wood Science*, **49**, 79–85.

121 Ye, X.R., Zhang, H.F., Lin, Y.H., Wang, L.S. and Wai, C.M. (2004) *Journal of Nanoscience and Nanotechnology*, **4**, 82–85.

122 Fu, L., Liu, Z.M., Liu, Y.Q., Han, B.X., Wang, J.Q., Hu, P.A., Cao, L.C. and Zhu, D.B. (2004) *The Journal of Physical Chemistry. B*, **108**, 13074–13078.

123 Yoda, S., Sakurai, Y., Endo, A., Miyata, T., Otake, K., Yanagishita, H. and Tsuchiya, T. (2002) *Chemical Communications*, 1526–1527.

124 Yoda, S., Nagashima, Y., Endo, A., Miyata, T., Yanagishita, H., Otake, K. and Tsuchiya, T. (2005) *Advanced Materials*, **17**, 367–370.

10
Photochemical and Photo-induced Reactions in Supercritical Fluid Solvents

James M. Tanko

10.1
Introduction

10.1.1
"Solvent" Properties of Supercritical Fluids

There are several potential advantages that may be realized with the use of supercritical fluids (SCFs) as solvents for chemical reactions from the standpoint of reactivity and selectivity. As many of the examples discussed in this chapter illustrate, the unique features of SCFs can be exploited to control the behavior (i.e. kinetics and selectivity) of many chemical processes in a way not possible with conventional liquid solvents.

Changes in reaction rates arising from direct effects of temperature and pressure on the kinetics of a reaction are governed by transition-state theory, and the same considerations pertain to reactions both in SCF media and in conventional solvents [1–4]. However, a unique feature of SCFs is that solvent properties, such as polarity (dielectric constant), viscosity, and solubility parameter, vary with temperature and pressure, and changes in these properties may alter reaction rates. The polarity of the reaction medium will exert an effect on the rate of a chemical reaction if the polarities of the reactants and transition state are different. Solvent viscosity will exert its influence on reactions that are diffusion controlled or on reactions in which cage effects are important. Hence control of these solvent properties, via manipulation of temperature and pressure, provides a way of adjusting the kinetics of a chemical process unique to the SCF medium [1–3]. As many of the studies cited here demonstrate, with an SCF solvent it is possible to study solvent effects on reaction rates *without varying the molecular functionality of the solvent.*

However, there are still *additional* factors which may affect reactivity. Numerous studies have shown that, for SCFs in the compressible region of the phase diagram, the local solvent density about a solute is often enhanced relative to the bulk solvent density [3, 5]. The term "solvent–solute clustering" has been coined to describe this phenomenon. Because of enhanced local solvent density, the rotational and

Handbook of Green Chemistry, Volume 4: Supercritical Solvents. Edited by Walter Leitner and Philip G. Jessop

ISBN: 978-3-527-32590-0

translational motion of a solute may be restricted (attributable to increased local viscosity). In this scenario, reaction rate and selectivity will be perturbed only when the reactions are extremely rapid (diffusion-controlled), or if cage effects are important – as is often the case with photochemical reactions. There is also evidence that, in some cases, "solute–solute clustering" may be important [6–12], and conceivably reaction rates could be affected because of locally higher concentrations. In general, clustering is most important near the critical point, and there is considerable interest in what effect this phenomenon has on reaction rates and selectivities[13, 14] (see below). An excellent review of clustering and solvation in supercritical fluids appeared in 1999 [15].

10.1.2
Scope of This Chapter

This chapter summarizes the literature through 2007 pertaining to photochemical and photo-initiated organic reactions in SCF solvents. Photochemical reactions involving organometallic compounds in SCF solvents have been extensively studied by Poliakoff and co-workers [16, 17]. In this chapter, the emphasis is on reaction chemistry in which stable products are formed and isolated, and how the unique features of the SCF medium influence reaction yields, rates, and/or selectivities.

What are *not* discussed at length are photophysical phenomena in SCF solvents (e.g. fluorescence quenching, triplet–triplet annihilation, charge transfer, and exiplex formation), which have been extensively used to probe SCF properties, in general, and have been especially informative regarding the existence of clusters (solvent–solute and solute–solute) and their effect on reactivity. Absorption and fluorescence spectroscopy (both steady-state [9, 18–35] and time-resolved) [11, 36–50], vibrational spectroscopy [51–56], pulse radiolysis [57, 58], and electron paramagnetic resonance (EPR) [59–62] have all been utilized in this regard. The interested reader is directed to the references provided above for more information on these topics.

10.1.3
Experimental Considerations

Photochemical reactions conducted under supercritical conditions require high-pressure reaction vessels equipped with a "window" which permits light to enter, and the necessary hardware/plumbing to generate high pressures and maintain constant temperature. A typical reaction vessel [63] used for photochemical reactions conducted in supercritical carbon dioxide ($scCO_2$) (Figures 10.1 and 10.2) is fabricated from a strong (inert) alloy such as stainless steel or Hastelloy, and is equipped with a sapphire window. Sapphire is especially suited for high-pressure work and is optically transparent in the region 150–6000 nm. CaF_2 and quartz have also been used. Provisions for stirring may also be included; in many cases, a simple magnetic stir bar suffices. Temperature control is typically achieved through the use of a resistive heater, thermocouple, and a temperature controller. Utilizing such a reactor, pressures up to 70–100 MPa can be achieved.

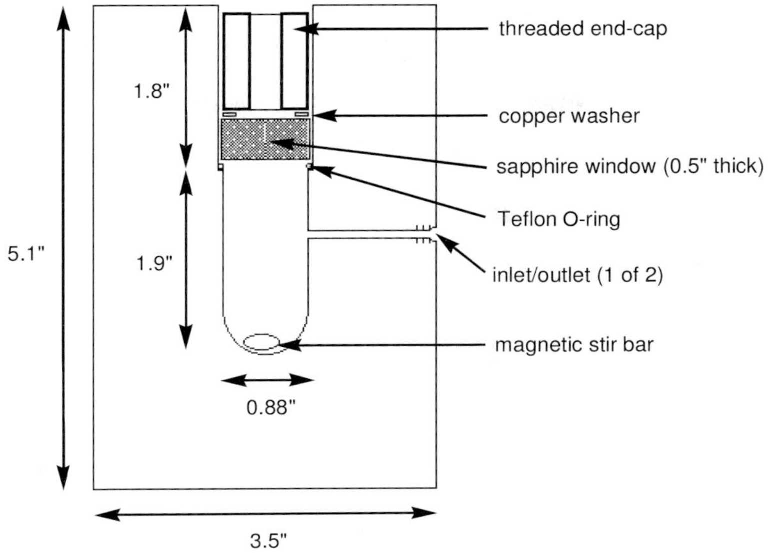

Figure 10.1 Cross-section of $scCO_2$ reactor.

A complete system for generating $scCO_2$ (Figure 10.3) [63] requires a device to generate high pressures (compressor, high-pressure piston, or HPLC pump), a pressure transducer, and the necessary plumbing. Often, a high-pressure release valve (rupture disk) is used to ensure that pressures in the reactor do not exceed specification. Because many photochemical and free radical reactions require the exclusion of oxygen, provisions can be made for purging the system with an inert gas

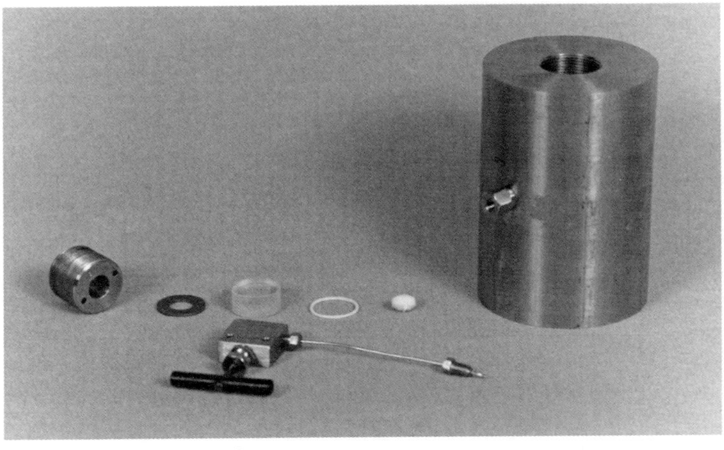

Figure 10.2 Photograph of $scCO_2$ reactor.

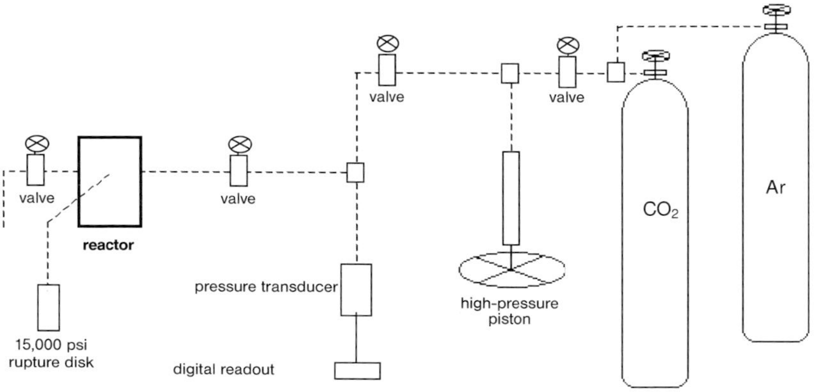

Figure 10.3 Schematic diagram of apparatus for generating $scCO_2$.

such as argon. Solid and liquid samples can be added to the reactor under an argon backflush; volatile liquids or gases can be introduced in glass vials which rupture when the reactor is pressurized.

Generally, light sources typically used for photochemical reactions in conventional solvents (e.g. medium- or high-pressure mercury lamps) are used for reactions in SCF solvents. Because many popular SCFs used for photochemical experiments do not absorb UV–visible light (e.g. CO_2, CHF_3, low molecular weight alkanes), and because these experiments are usually conducted at low substrate concentrations, heating of the reactor and the concomitant increase in pressure are generally not a problem. Occasionally, heating may occur because of the heat given off by the lamp itself, but separating the lamp and reactor and providing ample ventilation can avoid this.

After irradiation, the reactor must be vented so as to bring the system to ambient pressure. Products that are solids often precipitate from solution as the pressure is lowered, and are readily recovered. Separation of volatile compounds from an SCF is often more challenging, and usually entails "bubbling" the contents of the reactor into an organic solvent. Another strategy, applicable to SCFs that are not gases at room temperature (e.g. H_2O), is to cool the reactor to room temperature where the system is no longer pressurized.

Two procedures that illustrate some of the experimental protocols associated with these experiments are highlighted below:

The photodimerization of isophorone in $scCO_2$ [64] (see below) was accomplished first by pressurizing (to ~0.4 MPa) and depressurizing the reactor with CO_2 several times to purge the system of air. The reagent was introduced into the reactor via a syringe under a positive pressure of CO_2. Subsequently, the reactor was sealed, brought to the desired temperature and pressure, and irradiated with a 450 W medium-pressure mercury lamp. At completion of the reaction, the reactor contents were bubbled into methylene chloride, the reactor and lines were rinsed with solvent, and the combined solutions were analyzed by gas chromatography (GC).

For the chlorination of cyclohexane in $scCO_2$, the following procedure was followed [65, 66]. The appropriate volume of cyclohexane was placed in a 1 ml ampoule. The ampoule was degassed by several freeze–pump–thaw cycles (freezing to $-198\,°C$, evacuating to less than 0.01 mbar, and warming to room temperature), sealed under vacuum, and placed in the reactor. A second sealed ampoule containing the appropriate amount of Cl_2 (similarly degassed) was added to the reactor. The reactor was then sealed, covered with aluminum foil (to prevent premature initiation of the reaction via action of ambient laboratory light), and brought to 40 °C (the desired reaction temperature). Following several argon purges, the reactor was pressurized with CO_2 and allowed to equilibrate at 40 °C for several minutes. The aluminum foil was removed and the reactor was illuminated as with a 450 W mercury arc lamp. Following illumination, the contents of the reactor were bubbled slowly into hexanes cooled to 0 °C. An internal standard was added, and direct analyses by GC were performed to assess product yields.

10.2 Photochemical Reactions in Supercritical Fluid Solvents

10.2.1 Geometric Isomerization

Aida and Squires examined the photoisomerization of (*E*)-stilbene (Equation 10.1) in a conventional organic solvent (cyclohexane) and $scCO_2$. This system was selected for study because the solvent effect on the isomerization was already documented; increased viscosity facilitates the $E \rightarrow Z$ conversion [67].

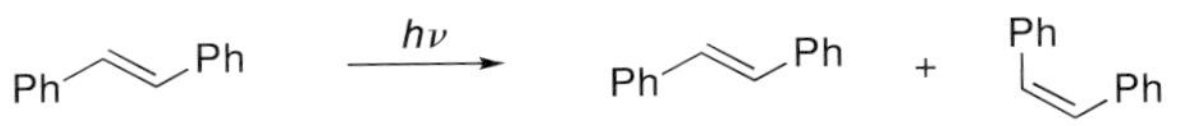

(10.1)

For *liquid* CO_2 at 25 °C, a change in pressure from 8.3 to 21.4 MPa (corresponding to a change in viscosity from ~0.07 to 0.1 cP) changes the $Z{:}E$ ratio from 5.5 to 6.8. For $scCO_2$, where an analogous pressure variation changes the viscosity from 0.02 to 0.08 cP, the effect is more dramatic, with the $Z{:}E$ ratio changing from 1.4 to 7.0 [67]. This study provided one of the first examples of how the outcome of a photochemical reaction can be altered by varying the solvent properties of the SCF via manipulation of pressure.

10.2.2 Photodimerization

In 1989, Fox and co-workers [64]. studied the [2 + 2] photodimerization of isophorone (Equation 10.2) in $scCO_2$ (38 °C) and $scCHF_3$ (34.5 °C). Three dimers were produced, a head-to-head dimer (**H-H**$_{anti}$), and two diastereomeric head-to-tail dimers (**H-T**$_{anti}$ and **H-T**$_{syn}$). In conventional solvents, Chapman *et al.* found that more polar solvents favor production of the more polar product: the ratio **H-H**$_{anti}$:**H-T**$_{total}$ was 1 : 4 in cyclohexane compared with 4 : 1 in methanol [68].

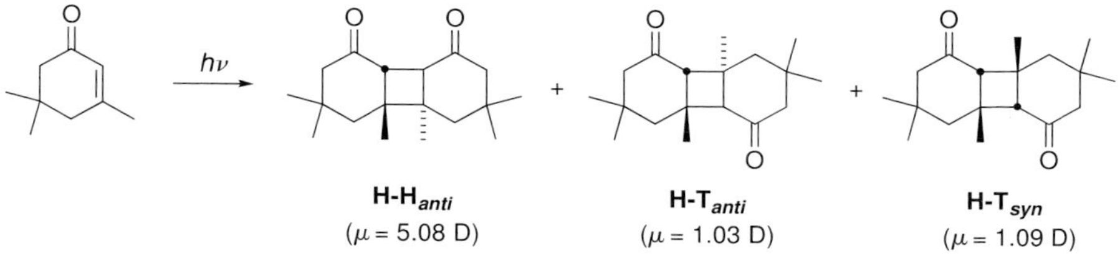

(10.2)

Analogous results were observed in SCF solvents: The more polar product (**H-H**$_{anti}$) was a major product in the more polar solvent CHF_3 (where the **H-H:H-T**$_{total}$ ratio varied from 0.75 to 1.0 with increasing pressure) and only a minor product in CO_2 (in which the **H-H:H-T**$_{total}$ ratio was essentially 0.10, independent of pressure) [64]. These observations are explicable on the basis that over the range of pressures examined, the dielectric constant (a measure of solvent polarity) varies more for CHF_3 (from 2.5 to 8.4) than it does for CO_2 (from 1.34 to 1.54).

An unexpected result of this study was that for the head-to-tail dimers, the *anti*:*syn* ratio varied with pressure (Figure 10.4). In conventional solvents, both are formed in approximately equal amounts. The authors suggested that "differential solvent reorganization" was responsible (i.e. that more desolvation must occur to form the *syn* isomer compared with the *anti* isomer, Scheme 10.1) [64]. Thus, at higher pressures (higher solvent densities), solvent reorganization was important thereby favoring the anti isomer.

Sun and co-workers reported that in $scCO_2$, the quantum yield for anthracene dimerization (Equation 10.3) was (a) 10 times greater in $scCO_2$ than in conventional liquid solvents at comparable anthracene concentrations, and (b) pressure dependent, with the yield *decreasing* at higher pressures [69]. The key step in the dimerization

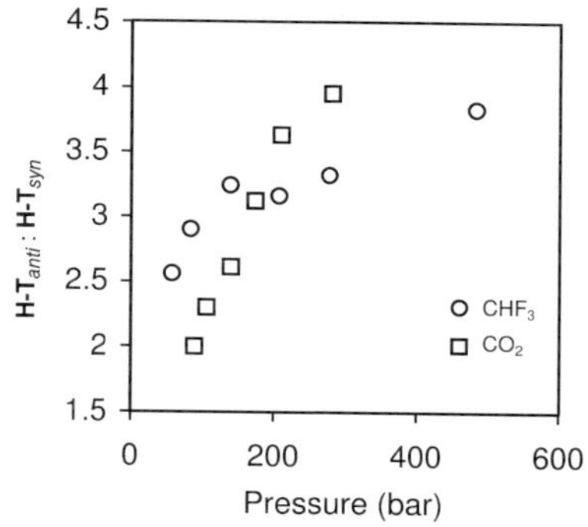

Figure 10.4 *Anti*:*syn* ratio for the "head-to-tail" dimers formed in the dimerization of isophorone in $scCHF_3$ (34.5 °C) and $scCO_2$ (38 °C) as a function of pressure. Data taken from [72].

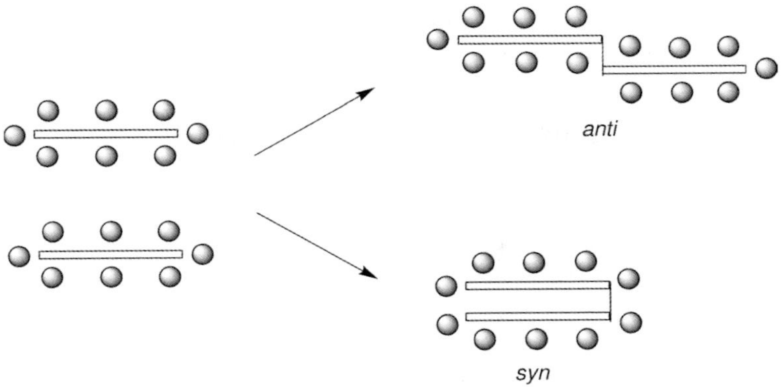

Scheme 10.1

process involves the formation of an eximer via diffusion-controlled reaction of anthracene in its ground and singlet excited states, A and $^1A^*$, respectively (Equation 10.4). In conventional liquid solvents, the rate constant for a diffusion-controlled reaction is on the order of $10^{10}\ M^{-1}\ s^{-1}$. However, because of the higher diffusivity of the SCF medium, the limit for diffusion control is higher ($\sim 10^{11}\ M^{-1}\ s^{-1}$). Thus, because the quantum yield for anthracene dimerization is directly proportional to k_{dim}, a 10-fold increase in efficiency is achieved in SCF media. The pressure effect arises because the CO_2 viscosity increases with increase in pressure [69].

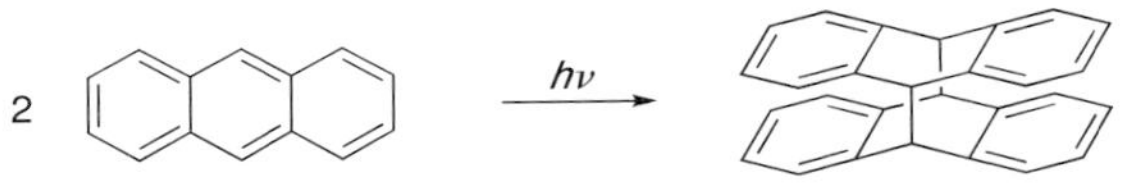

(10.3)

$$^1A^* + A \xrightarrow{k_{dim}} {}^1(AA)^* \qquad (10.4)$$

10.2.3
Carbonyl Photochemistry

In 1992, Kraus and Kirihara reported that acylhydroquinones could be synthesized photochemically from the corresponding aldehyde and quinone (Scheme 10.2) [70], and in 2001 the same group successfully transferred this chemistry into $scCO_2$ solvent [71]. Irradiation of the quinone generates the triplet state, which abstracts the relatively weak C–H bond of the formyl group, generating radical pair **3**. Radical–radical coupling forms **4**, which, after enolization, yields product **5**. This chemistry provides an environmentally benign alternative to the Friedel–Crafts acylation reaction because it eliminates the use of acid chlorides, strong Lewis acids, and benzene as a solvent for the synthesis of acylhydroquinones.

The Norrish Type I photo-cleavage represents a classic process in organic photochemistry, and has been extensively studied in conventional solvents. In 1991, Fox and co-workers extended this reaction to SCF solvent: in an attempt to probe for cage

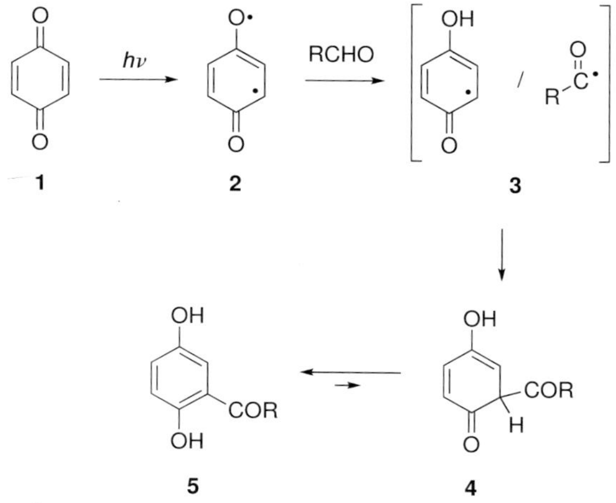

Scheme 10.2

effects and possibly *enhanced* cage effects attributable to solvent–solute clustering, the photolysis of an unsymmetrical dibenzyl ketone was examined in $scCO_2$ and scC_2H_6 [72]. Dibenzyl ketone photolysis had been shown to lead to cage effects in conventional solvents. The rationale behind this experiment is outlined in Scheme 10.3. Photolysis of an unsymmetrical dibenzyl ketone, A(C=O)B, leads to the formation of two benzyl radicals and carbon monoxide in the solvent cage, depicted as $[A^\bullet\ CO\ {}^\bullet B]_{cage}$. In-cage coupling is expected to yield exclusively the cross-coupling product A–B, whereas cage escape will lead to all possible coupling products, A–A, A–B, and B–B, in a statistical ratio of 1 : 2 : 1.

Over a pressure range from just above P_c to 30.0 MPa, this reaction yielded only a *statistical* distribution of products [72]. Hence no cage effect (enhanced or otherwise) was observed for this reaction.

This problem was later discussed by Chateauneuf and co-workers, who examined the decarbonylation of the phenylacetyl radical ($PhCH_2C^\bullet{=}O \rightarrow PhCH_2^\bullet + C{=}O$)

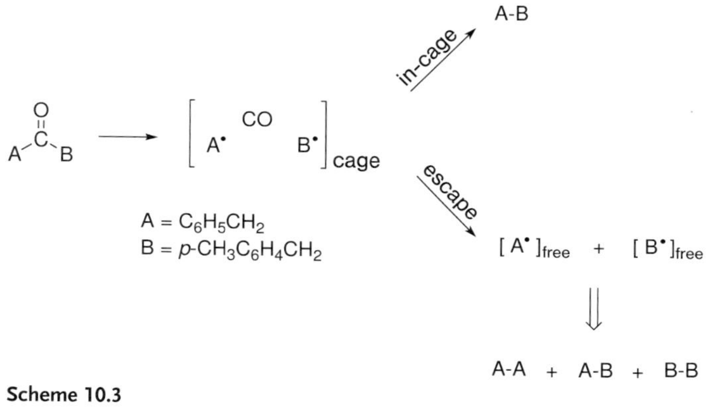

Scheme 10.3

by laser flash photolysis [44]. These workers again found no evidence for a cage effect (enhanced or otherwise) in SCF solvent. Moreover, to explain the absence of a cage effect in these reactions, they went on to suggest that the integrity of the cage is maintained for only a few picoseconds, whereas decarbonylation occurs in the time regime of a few hundred nanoseconds (i.e. the cage disintegrates long before in-cage coupling can occur) [44].

In order to address this issue, a process which involves a much shorter-lived radical pair needed to be examined. Toward this end, Weedon and co-workers examined the photo-Fries rearrangement of naphthyl acetate (Scheme 10.4) in $scCO_2$ at 35 and 46 °C [73]. Photolysis of **6** leads to caged-pair [**7**, **8**]; reaction in-cage yields the photo-Fries products, 2- or 4-acetylnaphthol (**9**). On the other hand, cage escape, followed by hydrogen abstraction (2-propanol was present as a hydrogen atom donor), leads to α-naphthol (**10**).

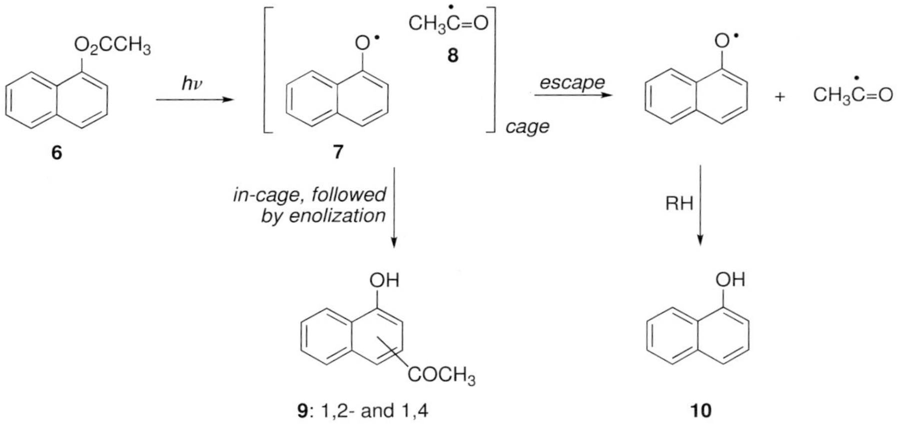

Scheme 10.4

A plot of the product ratio **9** : **6** as a function of pressure is presented in Figure 10.5. This plot exhibits a dramatic spike at pressures near the critical pressure, which the authors attributed to the onset of solvent–solute clustering; disintegration of the caged pair is inhibited because the viscosity at the molecular level is much greater than the bulk viscosity.

Tanko and Pacut examined the behavior of *geminate* and *diffusive* caged radical pairs, both generated by the photolysis of dicumyl ketone in $scCO_2$ [74]. A *geminate* caged radical pair arises when the two radicals are generated simultaneously from a common precursor, such as caged pair **12** generated from photolysis of dicumyl ketone (**11**, Scheme 10.5). Geminate caged pair **12** partitions between two pathways: cage escape (k_{esc}, the magnitude of which is viscosity dependent) and in-cage hydrogen abstraction (k_H). Because distinct products arise from each of these two competing pathways, the rate constant ratio k_{esc}/k_H is readily determined, and provides a measure of solvent viscosity at the molecular level. In these experiments, the magnitude of the cage effect was found to be greater than expected based on

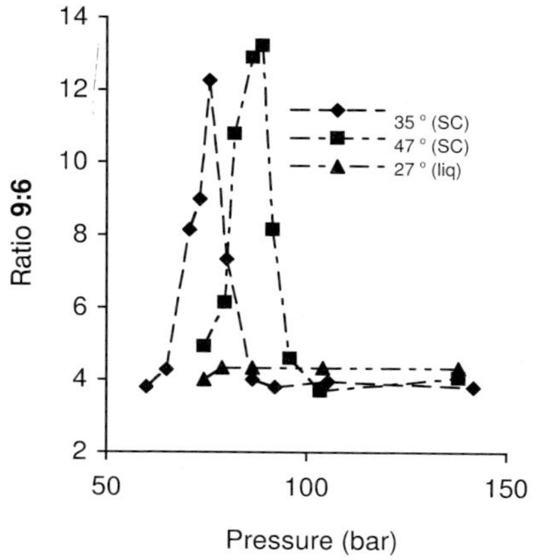

Figure 10.5 Ratio of products produced from photolysis of α-naphthyl acetate in CO_2. Data taken from [81].

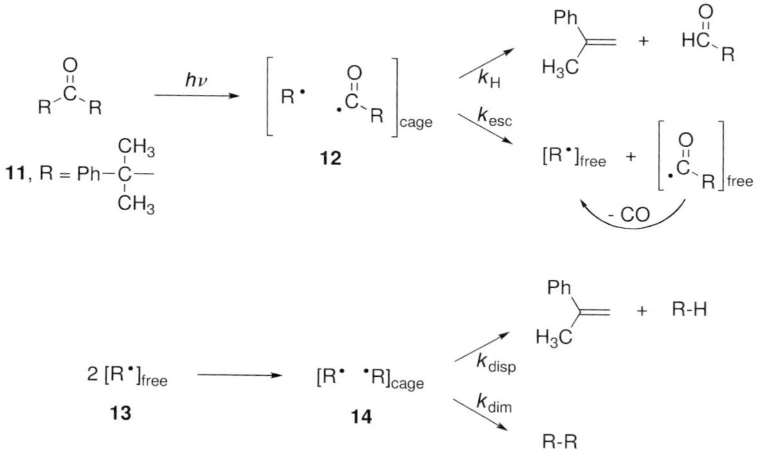

Scheme 10.5

extrapolations from conventional solvents [74]. These extrapolations were based on the relationship between the diffusion coefficient (D) and viscosity as described by the Stokes–Einstein equation, which may overestimate D and hence underestimate the cage effect in $scCO_2$.

In this system, a diffusive caged radical pair (**14**) is also formed by two cumyl radicals diffusing together. Radical pair **14** also partitions between two pathways: dimerization (k_{dim}) and disproportionation (k_{disp}). The rate constant ratio k_{dim}/k_{disp} is also viscosity dependent, and easily measured by product yields. The rate constant k_{dim} decreases with increasing viscosity because the two radicals must rotate and align in order for bond formation to occur. In contrast, the geometric constraints for disproportionation are less rigid, and k_{disp} is less sensitive to viscosity. Again, the magnitude of the cage effect was found to be greater than expected. For both the geminate and diffusive caged pairs in this system, the magnitude of the cage effect was found to increase near the critical pressure, possibly the result of enhanced local viscosity attributable to solvent–solute clustering [74].

10.2.4
Photosensitization and Photo-induced Electron Transfer

Using $scCO_2$ as both a solvent and reactant, Chateauneuf *et al.* [75] reported the synthesis of 9,10-dihydroanthracene-9-carboxylic acid from anthracene via photo-induced electron transfer. Irradiation of anthracene to its triplet state in the presence of a good electron donor such as *N,N*-dimethylaniline leads to the anthracene radical anion. In the presence of a hydrogen atom donor such as 2-propanol, the carboxylic acid is formed in 57% yield, presumably by the mechanism depicted in Scheme 10.6. Consistent with the electrophilic nature of CO_2, trapping by anthracene radical anion is best viewed as nucleophile–electrophile coupling, rather than radical addition to CO_2.

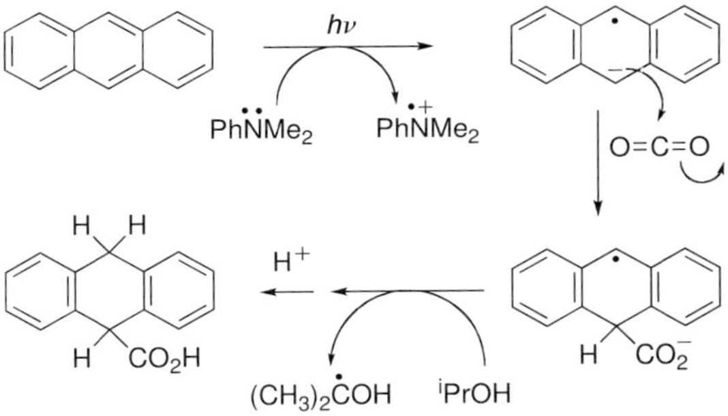

Scheme 10.6

Using a chiral sensitizer, Inoue and co-workers [76–78] found that photoaddition of alcohols to 1,1-diphenylpropene was enantioselective in $scCO_2$ (Equation 10.5). The enantiomeric excess (*ee*) exhibited a jump as the pressure was increased from below to above the critical pressure, the magnitude of which also varied with the structure of ROH. For example, at 8 and 18 MPa, the observed *ee* with MeOH was 7 and 21%, respectively. With 2-propanol at these same pressures, *ee*s of 21 and 42% were observed, respectively. This peculiar behavior near the critical pressure was attributed to clustering of ROH around the sensitizer. In a related study, these authors also reported an anomalous pressure dependence on the enantioselectivity of the photo-sensitized $Z \rightarrow E$ isomerization of cyclooctene in $scCO_2$, which was attributed to CO_2 clustering near the critical pressure [79].

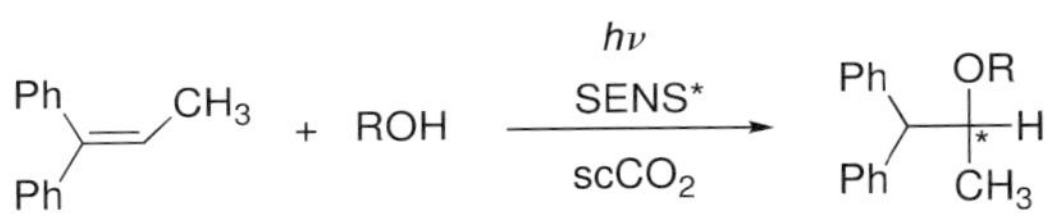

(10.5)

10.2.5
Photo-oxidation Reactions

Koda and co-workers reported the photo-induced (KrF laser, 248 nm) oxidations of ethylene [80] and benzene [81] in $scCO_2$ (Equations 10.6 and 10.7). In the context of product distribution and yields, no unusual behavior was noted near the critical point.

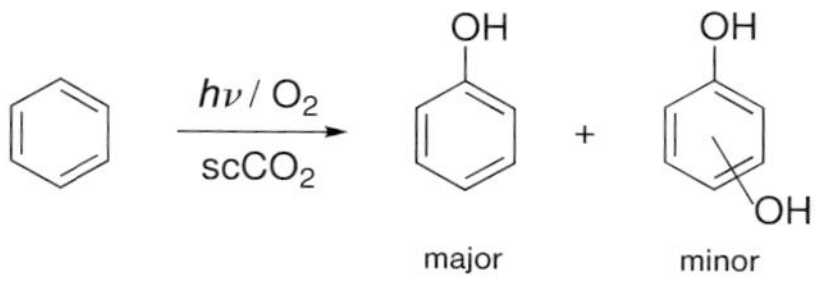

(10.6)

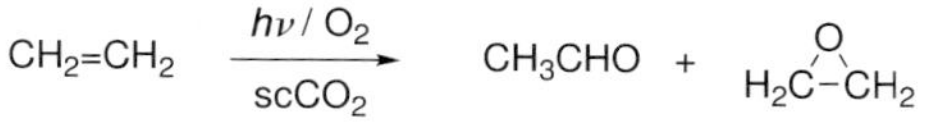

(10.7)

10.3
Photo-initiated Radical Chain Reactions in Supercritical Fluid Solvents

10.3.1
Free Radical Brominations of Alkyl Aromatics in Supercritical Carbon Dioxide

In 1994, Tanko and Blackert reported that the free radical bromination of alkylaromatics (e.g. toluene) could be carried out in $scCO_2$ [82]. This reaction is photo-initiated, and proceeds via the chain process outlined in Scheme 10.7 [83].

$$Br_2 \xrightarrow{h\nu} 2\,Br^\bullet \quad \textit{initiation}$$

$$Br^\bullet + PhCH_3 \longrightarrow HBr + PhCH_2^\bullet$$
$$PhCH_2^\bullet + Br_2 \longrightarrow PhCH_2Br + Br^\bullet \quad \textit{propagation}$$

$$PhCH_3 + Br_2 \longrightarrow PhCH_2Br + HBr \quad \textit{overall}$$

Scheme 10.7

Reaction yields were analogous to those observed using conventional solvents (e.g. CCl_4). Via competition experiments, the relative reactivity of the "secondary" hydrogens of ethylbenzene versus the "primary" hydrogens of toluene on a *per hydrogen* basis, $r(2^\circ/1^\circ)$, were assessed. Within experimental error, the selectivity did not vary over a pressure range of 7.5–42.3 MPa [$r(2^\circ/1^\circ) = 30 \pm 2$] at 40 °C [82]. In retrospect, this result is reasonable because (a) the hydrogen abstraction step is insensitive to solvent polarity effects, and is well below the diffusion-controlled limit so that viscosity effects are unimportant, and (b) the difference in the volume of activation for hydrogen abstraction from toluene versus ethylbenzene is small ($\sim$4.8 $cm^3\,mol^{-1}$), so that over the range of pressures examined the selectivity change would be of the same magnitude as experimental error [63, 82].

The observed selectivity in $scCO_2$ is nearly identical with that found in conventional organic solvents: $r(2°/1°) = 35 \pm 1$, 34 ± 1, and 29 ± 1 for CCl_4, Freon 113, and CH_2Cl_2, respectively. at 40 °C [63, 82]. These results confirm the role of $Br^{\bullet}$ as the chain carrier in these experiments, as depicted in Scheme 10.7, and suggest that $Br^{\bullet}$ selectivity is not altered by complexation to CO_2. It is noteworthy that $Br^{\bullet}$ *does* form a complex with CS_2 (which is isoelectronic with CO_2) and that this complex does exhibit enhanced selectivities in hydrogen atom abstractions [84].

With molecular bromine (Br_2) as the brominating agent, a small amount of *p*-bromotoluene is formed, arising from the competing electrophilic aromatic substitution (EAS) process. However, with the use of *N*-bromosuccinimide (NBS) as the brominating agent in direct analogy with the classical Ziegler reaction (Equation 10.8), the EAS side-product is completely eliminated. Reaction yields and selectivities are identical with those observed in CCl_4, the solvent most widely used for the Ziegler reaction [63, 82].

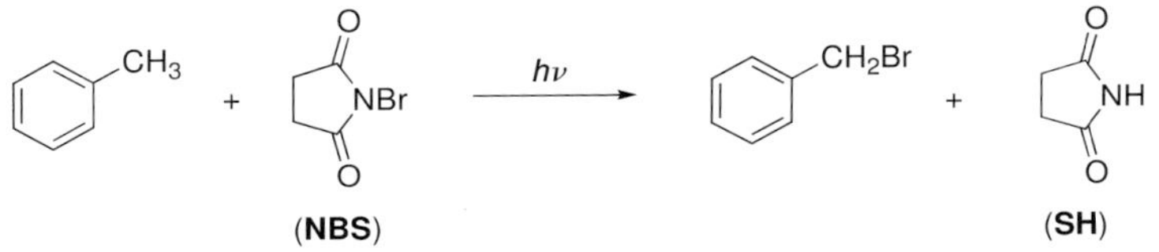

(10.8)

Competition experiments (ethylbenzene versus toluene) confirm the role of $Br^{\bullet}$ as chain carrier in the Ziegler bromination in $scCO_2$ [63, 82]. The role of NBS in this reaction is to maintain a low, steady-state concentration of Br_2, by scavenging HBr as it is produced during the course of the reaction (Equation 10.9).

$$\text{NBS} + \text{HBr} \longrightarrow \text{SH} + Br_2 \qquad (10.9)$$

10.3.2
Free Radical Chlorination of Alkanes in Supercritical Fluid Solvents

The free radical chlorination of alkanes is a classic procedure for the functionalization of alkanes. Many of the details of this reaction have been well understood for more than half a century [85]. In the laboratory, this reaction is initiated by action of visible light, with product formation occurring via the propagation steps outlined in Scheme 10.8: Chlorine atom abstracts hydrogen from the alkane yielding an alkyl radical and HCl. The alkyl radical subsequently reacts with molecular chlorine yielding the product alkyl chloride and regenerating chlorine atom.

$$Cl_2 \xrightarrow{h\nu} 2\,Cl^{\bullet} \quad \text{initiation}$$
$$RH + Cl^{\bullet} \longrightarrow R^{\bullet} + HCl$$
$$R^{\bullet} + Cl_2 \longrightarrow RCl + Cl^{\bullet} \quad \text{propagation}$$
$$RH + Cl_2 \longrightarrow RCl + HCl \quad \text{overall}$$

Scheme 10.8

A chlorine atom is a highly reactive species and exhibits low selectivity in hydrogen abstractions. In solution, the preference decreases in the order tertiary C−H (4.2) > secondary C−H (3.6) > primary C−H (1.0), on a *per hydrogen* basis (25 °C) [83]. Absolute rate constants for hydrogen abstraction are slightly below the diffusion-controlled limit [86].

The chlorine atom cage effect, first discovered by Skell and Baxter in 1983 [87], has been the subject of numerous investigations [88, 89]. Put briefly, for the chlorine atom abstraction step in the free radical chlorination of an alkane (RH_2), the geminate RHCl−Cl• caged pair is partitioned between three pathways (Scheme 10.9): diffusion apart (k_{diff}), abstraction of hydrogen from RH_2 comprising the cage walls (k_{RH_2}), and a second *in-cage* abstraction of hydrogen from the alkyl chloride (k_{RHCl}). Although the k_{diff} and k_{RH_2} steps result in the formation of monochloride (RHCl), the k_{RHCl} step results in the formation of polychlorides. In conventional solvents, the ratio of mono- to polychlorinated products (M/P) has been shown to depend on solvent viscosity [90].

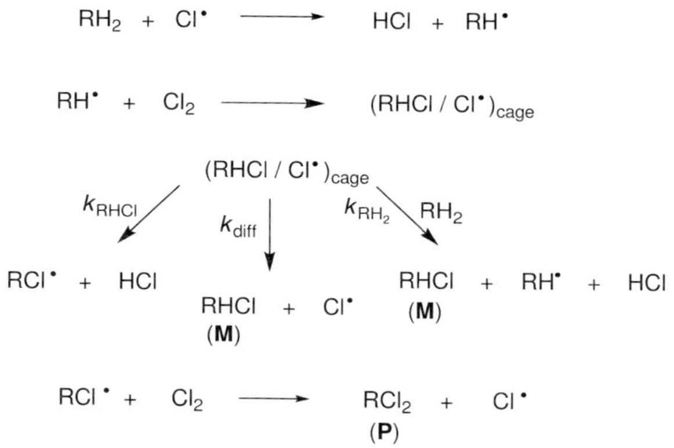

Scheme 10.9

Tanko and co-workers utilized the chlorine atom cage effect as a highly sensitive probe for studying the effect of SCF viscosity and the possible role of solvent clusters on cage lifetimes and reactivity [65, 66]. These experiments were conducted in $scCO_2$ (40 °C at various pressures), with parallel experiments in conventional solvents and in the gas phase.

Cage effects are typically quantified in terms of the Noyes model, which predicts that the efficiency of cage escape should vary linearly with the inverse of viscosity ($1/\eta$) [91]. In Figure 10.6, the ratio of mono- to polychlorides (M/P) observed in the chlorination of 2,3-dimethylbutane, neopentane, and cyclohexane is plotted as a function of $1/\eta$ for the experiments conducted in $scCO_2$ *and* in conventional solvents. Overall, these plots are linear over a range of viscosities spanning 1.7 orders of magnitude (from conventional solvents to $scCO_2$) and provide no indication of an enhanced cage effect (unusually low observed M/P ratio) near the critical pressure. It is also worth noting that *the best straight line through the solution-phase results successfully predicts the SCF phase results* [65, 66].

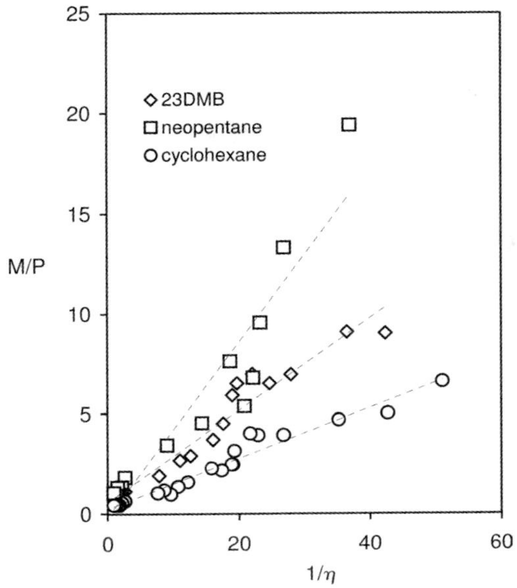

Figure 10.6 Ratio of mono- to polychlorides produced in the free radical chlorination of 2,3-dimethylbutane (23DMB), neopentane, and cyclohexane in conventional and supercritical fluid solvents as a function of inverse viscosity at 40 °C. Data taken from [73] and [74].

Based on these observations, there was no indication of an enhanced cage effect near the critical point in $scCO_2$ solvent. The magnitude of the cage effect observed in $scCO_2$ at all pressures examined is well within what is expected based on extrapolations from conventional solvents. It was suggested that for instances where enhanced cage effects have been observed attributable to solvent–solute clustering, this enhancement may be unique to the specific systems studied [65, 66].

The experiments with 2,3-dimethylbutane (23DMB) provided insight into the extent that $Cl^•$ selectivity varies as a function of pressure. In the gas phase (40 °C), the relative reactivity of the tertiary and primary hydrogens of 23DMB [$r(3°/1°)$] is 3.97. In the condensed phase (neat 23DMB, 40 °C), $r(3°/1°) = 3.27$. In $scCO_2$, $r(3°/1°)$ varies with pressure and falls between the gas- and liquid-phase values. The fact that $r(3°/1°)$ is so close to the solution and gas-phase values suggests that $Cl^•$ selectivity is not altered by complexation to CO_2 [65].

The slight variation in $r(3°/1°)$ can be explained as follows. The rate constants for primary, secondary, or tertiary hydrogen abstractions by $Cl^•$ from alkanes are nearly diffusion controlled in conventional solvents. Consequently, the intrinsic selectivity of $Cl^•$ is diminished in conventional solvents because of the onset of diffusion control. In the gas phase, selectivity is slightly higher because the barrier imposed by diffusion is eliminated. The viscosity of a supercritical fluid (a) lies between those of conventional liquid solvent and the gas phase and (b) varies with pressure. Because of the low viscosity of supercritical fluids, bimolecular rate constants greater than the

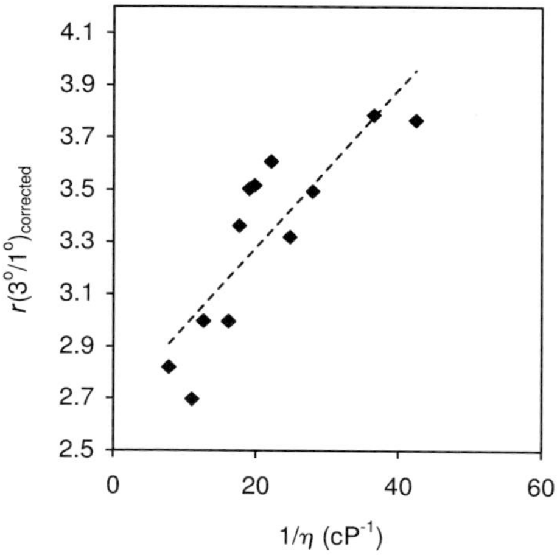

Figure 10.7 Chlorine atom selectivity in $scCO_2$ solvent at 40 °C. Data taken from [73].

$10^{10}\,M^{-1}\,s^{-1}$ liquid-phase diffusion-controlled limit can be realized in SCFs and, as a consequence, enhanced selectivity is achieved. Consistent with this interpretation is the observation that the plot of $r(3^\circ/1^\circ)$ versus inverse viscosity is approximately linear (Figure 10.7) [65].

10.4 Conclusion

The examples discussed here demonstrate that the unique nature of SCFs provides a means of "dialing up" the selectivity of a chemical process in a manner which is simply impossible using conventional solvents, that is by manipulation of temperature and pressure. Reaction rates may vary with temperature and pressure depending on the magnitude of the activation energy and activation volume, respectively, for reactions conducted in *both* SCF and conventional solvents. What is unique about SCF solvents is that the actual nature of the solvent (polarity, viscosity, etc.) *also* varies with temperature and pressure. In addition, numerous studies have demonstrated surprisingly large variations in selectivity with pressure (density) near the critical pressure – generally attributed to clustering. However, from a practical perspective (chemical manufacturing and synthesis), large variations in selectivities resulting from small changes in pressure are not likely to be useful. Rather, it is the fact that selectivities tend to be either constant, or at least predictable, at pressures sufficiently above the critical pressure that will make SCFs desirable for chemical manufacturing purposes. SCF solvents such as CO_2 and H_2O are especially attractive as they are "environmentally benign" alternatives to a number of classical solvents that pose hazards to either health or the environment. Coupled with the tunable properties of a supercritical fluid, these solvents emerge not only as *viable* alternatives to conventional organic solvents, but in some cases at least, also as *superior* alternatives. Finally, SCF solvents are superb tools for probing solvent effects in chemical processes, as it is possible to vary (via manipulation of pressure) pertinent solvent properties (e.g. viscosity and polarity) without changing the molecular functionality of the solvent.

Acknowledgment

Support from the National Science Foundation (CHE-0548129) during the writing of this chapter is acknowledged and appreciated.

References

1 Kim, S. and Johnston, K.P. (1987) *Supercritical Fluids* (eds T.G. Squires and M.E. Paulaitis), American Chemical Society, Washington, DC. p. 42.
2 Kim, S. and Johnston, K.P. (1988) *Chemical Engineering Communications*, **63**, 49.
3 Johnston, K.P. and Haynes, C. (1987) *AICHE Journal*, **33**, 2017.
4 Wu, B.C., Klein, M.T. and Sandler, S.I. (1991) *Industrial & Engineering Chemistry Research*, **30**, 822.
5 Johnston, K.P., McFann, G.J., Peck, D.G. and Lemert, R.M. (1989) *Fluid Phase Equilibria*, **52**, 337.
6 Bunker, C.E. and Sun, Y.-P. (1995) *Journal of the American Chemical Society*, **117**, 10865.
7 Bunker, C.E., Sun, Y.-P. and Gord, J.R. (1997) *Journal of Physical Chemistry A*, **101**, 9233.
8 Debenedetti, P.C., Patsche, I.B. and Mohamed, R.S. (1989) *Fluid Phase Equilibria*, **52**, 347.
9 Ellington, J.B., Park, K.M. and Brennecke, J.F. (1994) *Industrial & Engineering Chemistry Research*, **33**, 965.
10 Rhodes, T.A. and Fox, M.A. (1996) *The Journal of Physical Chemistry*, **100**, 17931.
11 Roberts, C.B., Zhang, J., Chateauneuf, J.E. and Brennecke, J.F. (1995) *Journal of the American Chemical Society*, **117**, 6553.
12 Zhang, J., Roek, D.P., Chateauneuf, J.E. and Brennecke, J.F. (1997) *Journal of the American Chemical Society*, **119**, 9980.
13 Randolph, T.W., O'Brien, J.A. and Ganapathy, S. (1994) *The Journal of Physical Chemistry*, **98**, 4173.
14 Tucker, S.C. and Maddox, M.W. (1998) *The Journal of Physical Chemistry. B*, **102**, 2437.
15 Kajimoto, O. (1999) *Chemical Reviews*, **99**, 355.
16 Poliakoff, M. and Turner, J.J. (1995) in *Molecular Cryospectroscopy* (eds R.J.H. Clark and R.E. Hester), John Wiley and Sons, Inc., New York, p. 275.
17 Darr, J.A. and Poliakoff, M. (1999) *Chemical Reviews*, **99**, 495.
18 Brennecke, J.F., Tomasko, D.L., Peshkin, J. and Eckert, C.A. (1990) *Industrial & Engineering Chemistry Research*, **29**, 1682.
19 Deye, J.F., Berger, T.A. and Anderson, A.G. (1990) *Analytical Chemistry*, **62**, 615.
20 Ikushima, Y., Saito, N. and Arai, M. (1992) *The Journal of Physical Chemistry*, **96**, 2293.
21 Kajimoto, O., Futakami, M., Kobayashi, T. and Yamasaki, R. (1988) *The Journal of Physical Chemistry*, **92**, 1347.
22 Kim, S. and Johnston, K.P. (1987) *Industrial & Engineering Chemistry Research*, **26**, 1206.
23 Lemert, R.M. and DeSimone, J.M. (1991) *Journal of Supercritical Fluids*, **4**, 186.
24 Okada, T., Kobayashi, Y., Yamasa, H. and Mataga, N. (1986) *Chemical Physics Letters*, **128**, 583.
25 Rollins, H.W., Dabestani, R. and Sun, Y.-P. (1997) *Chemical Physics Letters*, **268**, 187.

26 Sigman, M.E., Lindley, S.M. and Leffler, J.E. (1985) *Journal of the American Chemical Society*, **107**, 1471.

27 Sun, Y.-P., Bennett, G., Johnston, K.P. and Fox, M.A. (1992) *The Journal of Physical Chemistry*, **96**, 10001.

28 Sun, Y..-P. and Bunker, C.E. (1995) *Berichte Der Bunsen-Gesellschaft-Physical Chemistry*, **99**, 976.

29 Sun, Y.-P., Bunker, C.E. and Hamilton, N.B. (1993) *Chemical Physics Letters*, **210**, 111.

30 Sun, Y.-P. and Fox, M.A. (1993) *Journal of the American Chemical Society*, **115**, 747.

31 Sun, Y.-P., Fox, M.A. and Johnston, K.P. (1992) *Journal of the American Chemical Society*, **114**, 1187.

32 Takahashi, K., Abe, K., Sawamura, S. and Jonah, C.D. (1998) *Chemical Physics Letters*, **282**, 361.

33 Yonker, C.R., Frye, S.L., Kalkwarf, D.R. and Smith, R.D. (1986) *The Journal of Physical Chemistry*, **90**, 3022.

34 Yonker, C.R. and Smith, R.D. (1988) *The Journal of Physical Chemistry*, **92**, 2374.

35 Adams, J.E. (1998) *The Journal of Physical Chemistry. B*, **102**, 7455.

36 Anderton, R.M. and Kauffman, J.F. (1995) *The Journal of Physical Chemistry*, **99**, 13759.

37 Betts, T.A., Zagrobelny, J. and Bright, F.V. (1992) *Journal of the American Chemical Society*, **114**, 8163.

38 Gehrke, C., Schroeder, J., Schwarzer, D., Troe, J. and Voss, F. (1990) *The Journal of Physical Chemistry*, **92**, 4805.

39 Heitz, M.P. and Bright, F.V. (1996) *The Journal of Physical Chemistry*, **100**, 6889.

40 Ji, Q., Lloyd, C.R., Eyring, E.M. and van Eldik, R. (1997) *The Journal of Physical Chemistry A*, **101**, 243.

41 Kajimoto, O., Sekiguchi, K., Nayuki, T. and Kobayashi, T. (1997) *Berichte Der Bunsen-Gesellschaft-Physical Chemistry*, **101**, 600.

42 Roberts, C.B., Chateauneuf, J.E. and Brennecke, J.F. (1992) *Journal of the American Chemical Society*, **114**, 8455.

43 Roberts, C.B., Zhang, J., Brennecke, J.F. and Chateauneuf, J.E. (1993) *The Journal of Physical Chemistry*, **97**, 5618.

44 Roberts, C.B., Zhang, J., Chateauneuf, J.E. and Brennecke, J.F. (1993) *Journal of the American Chemical Society*, **115**, 9576.

45 Zagrobelny, J., Betts, T.A. and Bright, F.V. (1992) *Journal of the American Chemical Society*, **114**, 5249.

46 Grimm, C., Kling, M., Schroeder, J. and Troe, J.Z.J. (2004) *Israel Journal of Chemistry*, **43**, 305.

47 Serpa, C., Gomes, P.J.S., Arnaut, L.G., Formosinho, S.J., Pina, J. and Seixas de Melo, J. (2006) *Chemistry - A European Journal*, **12**, 5014.

48 Aizawa, T., Kanakubo, M., Ikushima, Y. and Smith, J.R.L. (2004) *Fluid Phase Equilibria*, **219**, 37.

49 Takahashi, K., Sawamura, S., Dimitrijevic, N.M., Bartels, D.M. and Johan, C.D. (2002) *Journal of Physical Chemistry A*, **106**, 108.

50 Nunes, R.M.D., Arnaut, L.G., Solntsev, K.M., Tolbert, L.M. and Formosinho, S.J. (2005) *Journal of the American Chemical Society*, **127**, 11890.

51 Akimoto, S. and Kajimoto, O. (1993) *Chemical Physics Letters*, **209**, 263.

52 Hegarty, J.N.M., McGarvey, J.J., Bell, S.E.J. and Al-Obaidi, A.H.R. (1996) *The Journal of Physical Chemistry*, **100**, 15704.

53 Sun, X..-Z., George, M.W., Kazarian, S.G., Nikiforov, S.M. and Poliakoff, M. (1996) *Journal of the American Chemical Society*, **118**, 10525.

54 Urdahl, R.S., Rector, K.D., Myers, D.J., Davis, P.H. and Fayer, M.D. (1996) *The Journal of Physical Chemistry*, **105**, 8973.

55 Arakcheev, V.G., Bagratashvilli, V.N., Valeev, A.A., Gordiyenko, V.M., Kireev, V.V., Morozov, V.B., Olenin, A.N. and Lomonosova, M.V. (2003) *Journal of Raman Spectroscopy*, **34**, 952.

56 Lalanne, P., Tassaing, T., Danten, Y., Cansell, F., Tucker, S.C. and Besnard, M. (2004) *Journal of Physical Chemistry A*, **108**, 2617.

57 Takahashi, K. and Jonah, C.D. (1997) *Chemical Physics Letters*, **264**, 297.

58 Dimitrijevic, N.M., Takahashi, K., Bartels, D.M., Jonah, C.D. and Trifunac, A.D. (2008) *Journal of Physical Chemistry A*, **104**, 568.

59 Ganapathy, S., Carlier, C., Randolph, T.W. and O'Brien, J.A. (1996) *Industrial & Engineering Chemistry Research*, **35**, 19.

60 Randolph, T.W. and Carlier, C. (1992) *The Journal of Physical Chemistry*, **96**, 5146.

61 Batchelor, S.N. (1998) *The Journal of Physical Chemistry. B*, **120**, 615.

62 Tachikawa, T., Akiyama, K., Yokoyama, C. and Tero-Kubota, S. (2003) *Chemical Physics Letters*, **376**, 350.

63 Tanko, J.M., Blackert, J.F. and Sadeghipour, M., (1994) in *Benign by Design. Alternative Synthetic Design for Pollution Prevention* (eds P.T. Anastas and C.A. Farris), American Chemical Society, Washington, DC, p. 98.

64 Hrnjez, B.J., Mehta, A.J., Fox, M.A. and Johnston, K.P. (1989) *Journal of the American Chemical Society*, **111**, 2662.

65 Fletcher, B., Suleman, N.K. and Tanko, J.M. (1998) *Journal of the American Chemical Society*, **120**, 11839.

66 Tanko, J.M., Suleman, N.K. and Fletcher, B. (1996) *Journal of the American Chemical Society*, **118**, 11958.

67 Aida, T. and Squires, T.G., (1987) in *Supercritical Fluids: Chemical and Engineering Principles and Applications* (ed. M.E. Paulaitis), American Chemical Society, Washington, DC. p. 58.

68 Chapman, O.L., Nelson, P.J., King, R.W., Trecker, D.J. and Griswold, A. (1967) *Rec. Chem. Prog.*, **28**, 167.

69 Bunker, C.E., Rollins, H.W., Gord, J.R. and Sun, Y.-P. (1997) *The Journal of Organic Chemistry*, **62**, 7324.

70 Kraus, G.A. and Kirihara, M. (1992) *The Journal of Organic Chemistry*, **57**, 3256.

71 Pacut, R., Grimm, M.L., Kraus, G.A. and Tanko, J.M. (2001) *Tetrahedron Letters*, **42**, 1415.

72 O'Shea, K.E., Combes, J.R., Fox, M.A. and Johnston, K.P. (1991) *Photochemistry and Photobiology*, **54**, 571.

73 Andrew, D., Islet, B.T.D., Margaritis, A. and Weedon, A.C. (1995) *Journal of the American Chemical Society*, **117**, 6132.

74 Tanko, J.M. and Pacut, R. (2001) *Journal of the American Chemical Society*, **1123**, 5703.

75 Chateauneuf, J.E., Zhang, J., Foote, J., Brink, J. and Perkovic, M.W. (2002) *Advances in Environmental Research*, **6**, 487.

76 Nishiyama, Y., Kaneda, M., Saito, R., Mori, T., Wada, T. and Inoue, Y. (2004) *Journal of the American Chemical Society*, **126**, 6568.

77 Nishiyama, Y., Wada, T., Mori, T. and Inoue, Y. (2007) *Chemistry Letters*, **36**, 1488.

78 Nishiyama, Y., Masayuki, K., Asaoka, S., Saito, R., Mori, T., Wada, T. and Inoue, Y. (2007) *Journal of Physical Chemistry A*, **111**, 13432.

79 Saito, R., Kaneda, M., Wada, T., Katoh, A. and Inoue, Y. (2002) *Chemistry Letters*, **31**, 860.

80 Koda, S., Ebukuro, T., Otomo, J., Tsuruno, T. and Oshima, Y. (1998) *Journal of Photochemistry and Photobiology A: Chemistry*, **115**, 7.

81 Kawahata, T., Otomo, J., Oshima, Y. and Koda, S. (1998) *Journal of Supercritical Fluids*, **13**, 197.

82 Tanko, J.M. and Blackert, J.F. (1994) *Science*, **263**, 203.

83 Russell, G.A. (1958) *Journal of the American Chemical Society*, **80**, 4997.

84 Sadeghipour, M., Brewer, K. and Tanko, J.M. (1997) *The Journal of Organic Chemistry*, **62**, 4185.

85 Ingold, K.U., Lusztyk, J. and Rayner, K.D. (1990) *Accounts of Chemical Research*, **23**, 219.

86 Bunce, N.J., Ingold, K.U., Landers, J.P., Lusztyk, J. and Scaiano, J.C. (1985) *Journal of the American Chemical Society*, **107**, 5464.

87 Skell, P.S. and Baxter, I.H.N. (1985) *Journal of the American Chemical Society*, **107**, 2823.

88 Raner, K.D., Lusztyk, J. and Ingold, K.U. (1988) *Journal of the American Chemical Society*, **110**, 3519.

89 Tanko, J.M. and Anderson, F.E. (1988) *Journal of the American Chemical Society*, **110**, 3525.

90 Tanner, D.D., Oumar-Mahamat, H., Meintzer, C.P., Tsai, E.C., Lu, T.T. and Yang, D. (1991) *Journal of the American Chemical Society*, **113**, 5397.

91 Koenig, T. and Fischer, H. (1973) in *Free Radicals*, vol. 1 (ed. J.K. Kochi), John Wiley & Sons, Inc., New York. p. 157.

11
Electrochemical Reactions

Patricia Ann Mabrouk

11.1
Introduction

Supercritical fluids (SCFs) represent a very intriguing medium in which to study redox-based reaction chemistry. Merely by changing pressure and temperature, the dielectric constant and therefore the solvent polarity and solvating power of the supercritical fluid can be varied at will. SCFs exhibit gas-like diffusivity compared with normal liquids that potentially translates into enhanced mass transport to/from the electrode surface, which can be exploited to probe reactions that occur on a shorter time-scale than in normal liquids. In addition, these solvents have relatively large potential windows, which increases the number of redox reactions that can be examined in them.

The field is relatively new: Silvestri *et al.* [1] reported the first electrolysis in the supercritical "phase" in 1981. In 1984, Bard and co-workers initiated a series of efforts to explore the potential of near-critical and supercritical solvents in electrochemistry [2–7]. They investigated a number of classical electrochemical analytes in a wide range of relatively polar supercritical solvents, including water, acetonitrile, and ammonia, that caught the attention of many in the community who subsequently initiated their own efforts, many of which are summarized in this chapter. The literature is reviewed topically from the perspective of an electrochemist examining the methods, analytes, electrolytes, solvents, and applications.

11.2
Electrochemical Methods

To date, investigators have exclusively used controlled potential methods and the bulk of the work has focused on cyclic voltammetry (CV), although chronoamperometry (CA) [5–8] and chronocoulometry (CC) [2] have also been utilized with conventional electrochemical instrumentation.

Handbook of Green Chemistry, Volume 4: Supercritical Solvents. Edited by Walter Leitner and Philip G. Jessop

ISBN: 978-3-527-32590-0

11.3
Analytes

Table 11.1 provides an overview of the analytes that have been investigated under supercritical conditions so far. The majority of studies have focused on reagents that are stable and that undergo a well-defined redox reaction, which has been demonstrated to be well behaved from an electrochemical standpoint under a wide range of experimental conditions including solvent, pressure, and temperature in order to determine whether or not electrochemical techniques can be employed in near-critical and supercritical solvents.

The electrochemistry of ferrocene or derivatives of ferrocene has most frequently been reported. Ferrocene is a standard reference compound in electrochemical investigations as it exhibits reversible voltammetry for the $Fe^{3+/2+}$ couple at a wide range of electrode substrates both in aqueous [30] and non-aqueous [31] media. Ferrocene is the IUPAC-recommended non-aqueous potential reference standard [32]. All of these characteristics make ferrocene a logical choice for electrochemical investigations in supercritical solvents. However, as Table 11.1 clearly shows, a large variety of analytes have been dissolved in and investigated in a wide array of supercritical solvents. The wide array of analytes suggests that in principle any redox system can be studied in this medium with a judicious selection of solvent (see below).

Table 11.1 Analytes investigated by electrochemistry in supercritical media.

Supercritical solvent	Analyte	References
Acetonitrile	Decamethylferrocene, osmium(II) tris(bipyridyl), methyl viologen	[4]
Ammonia	*m*-Chloronitrobenzene, pyrazine, quinoxaline, phenazine, solvated electrons	[2, 6]
Carbon dioxide	Ferrocene, anthracene, *p*-benzoquinone, ruthenium(II) tris(bipyridyl), phenol, *N,N,N′,N′*-tetramethyl-*p*-phenylenediamine, bis(tetradodecylammonium)nickel maleonitrile	[8–21]
Chlorodifluoromethane	Cobaltocene, ferrocene, 2,5-dichlorohydroquinone, *p*-methoxyphenol, copper bis(diethyldithiocarbamate), ruthenium(II) bis(bipyridyl)	[22, 23]
Difluoromethane	Ferrocene, ferrocenecarboxylic acid	[24–26]
Dinitrogen oxide	Ferrocene	[12]
Hydrogen chloride	Iodine/iodide	[1]
Sulfur dioxide	Iron(II) tris(bipyridyl)	[7]
1,1,1,2-Tetrafluoroethane	Cesium(I) (18-crown-6) tetrafluoroborate	[24]
Trifluoromethane	Cobaltocene, decamethylferrocinium hexafluorophosphate	[27, 28]
Water	Bromine/bromide, iodine/iodide, Cu(0)/Cu(I)/Cu(II)	[3, 5, 29]

11.4
Electrolytes

The majority of studies have applied supercritical carbon dioxide ($scCO_2$) and supercritical hydrofluorocarbons. Consequently, in this section we focus on electrolytes used in these solvents. In the case of $scCO_2$, its poor solvating power has made identification of suitable electrolytes particularly challenging. The normally employed array of non-aqueous electrolytes are not soluble in $scCO_2$.

DiMaso *et al.* [10] attempted to use acetonitrile as a polar modifier in an effort to circumvent the poor solvating power of carbon dioxide and use standard non-aqueous electrolytes. Abbott and Harper [33] attempted to use a hydrophobic electrolyte tetrakis(decyl)ammonium tetraphenylborate in $scCO_2$. However, the voltammetry reported was poorly defined. No later work was ever reported using either strategy. Identification of suitable electrolytes remains a key issue for electrochemistry in $scCO_2$. The poor solvent power of $scCO_2$ has led researchers to investigate alternative strategies to facilitate charge transport, such as ionically modified electrodes [11–15] and microemulsions [20, 21] (see below).

The situation is much rosier for hydrofluorocarbons, which exhibit higher dielectric constants and therefore better solvating power than carbon dioxide. Traditional non-aqueous electrolytes including tetrabutylammonium tetrafluoroborate ($TBABF_4$) [24, 26, 27], tetrabutylammonium perchlorate ($TBAClO_4$) [24], and tetrabutylammonium hexafluorophosphate ($TBAPF_6$) [28] have been used successfully in supercritical difluoromethane, chlorodifluoromethane, trifluoromethane, and 1,1,1,2-tetrafluoroethane.

11.5
Electrochemical Cell and Supercritical Fluid Delivery System

Researchers typically construct their own electrochemical cell and supercritical fluid delivery system. Most groups have used stainless steel [2, 5, 6, 9, 11, 13, 20, 22, 23, 34] for their electrochemical cell body. Corrosion of cell materials was found to be a significant problem in supercritical water [5, 29], sulfur dioxide [7], and ammonia [2, 6]. Consequently, high nickel alloy Iconel 600 was in the end used with ammonia and an alumina tube was used with supercritical water [5, 29] and sulfur dioxide [7]. Electrochemical cells have been exclusively single compartment. Cell volumes have ranged from 6 [22] to 96 ml [35] but typically average about 10 ml [9, 11, 23, 26, 28], as there is not much to be gained by using a large cell volume with the microelectrodes that are typically used in SCFs (see below). Several groups have modified their electrochemical cells with sapphire [22, 23, 28] or quartz [11, 13, 14] windows to allow visual inspection of the contents and to facilitate *in situ* spectroelectrochemical interrogation by UV–visible and/or fluorescence methods. Electrochemical cells are usually wrapped with heating tape or placed in an oven in order to heat them. The cell temperature is measured with an appropriate thermocouple inserted directly into the electrochemical cell.

11.6 Electrodes

11.6.1 Working Electrode

Microelectrodes [36, 37] are often used in resistive media such as non-aqueous solvents. Basically, these are approximately micron-diameter wires often heat sealed in glass capillaries. Microelectrodes are advantageous in that due to the nano- or picoampere currents used, solution resistance typically produces a negligible change in potential, often referred to as ohmic distortion (*iR* drop). Hence they can be used in the absence of added electrolyte or in relatively non-polar (highly resistive) solvents. In addition, mass transport at microelectrodes is enhanced due to convergent diffusion. Consequently, increased signal-to-noise ratio is also a benefit as the reduced current diminishes the likelihood that significant amounts of electrogenerated contaminants will be produced. Pt disk microelectrodes sealed in glass have most often been used by investigators [9–11, 13, 14, 16, 17, 20–25, 27–29, 33, 38], although carbon [11, 27] and tungsten [2] fibers have also been utilized.

Both macro- and microelectrodes (electrodes with micrometer dimensions) have been investigated with supercritical solvents. Macroelectrodes were used in the earliest electrochemical studies [1–3, 5] that fortuitously focused on relatively polar supercritical solvents (acetonitrile, ammonia, and water) and which were carried out just about when microelectrodes came on the scene. Sizeable *iR* drops were observed even in those systems.

11.6.2 Reference Electrode

In order to relate measured redox potentials to each other, an arbitrary choice must be made for a standard reference electrode. The standard hydrogen electrode (SHE), for which the half-reaction is the two-electron reduction of 2 mol of protons to produce 1 mol of hydrogen gas, serves this function in traditional solvents. The potential for SHE is defined to be 0 V at all temperatures in all solvents.

The design of a reference electrode that will provide thermodynamically meaningful results is not trivial due the elevated pressures and temperatures at which the SCFs are typically generated. It is also not clear what conditions would provide a meaningful "standard" state for the wide range of pressures and temperatures relevant to the various supercritical solvents. Ideally, the reference would be placed inside the cell in close proximity to the working electrode. If the SHE were used, a constant pressure of hydrogen would have to be maintained without contaminating the working electrode. Of course, an alternative is to place the reference external to the cell and use a salt bridge. However, this would likely produce a significant junction potential and potential error of unknown and possibly varying magnitude as a function of pressure and temperature.

In view of these challenges, quasi-reference electrodes (QREs) have typically been employed. These are merely metal wires which, when inserted into the test solution, assume relatively stable and reproducible, albeit unknown, potentials. Using QREs, relative potentials and the relative shift in potentials can be determined. However, it must be stressed that since the absolute potential that the QRE assumes is unknown and dependent on the experimental conditions such as the solvent (in this case, temperature and pressure) and electrolyte, it is difficult to reference potentials to the SHE, the absolute reference standard. Under these circumstances, if the experimental objective is the characterization of novel redox processes, an internal redox standard such as ferrocinium/ferrocene can be added to the test solution. Of course, this contaminates the test solution and may be inconvenient in terms of study of the test system.

The Ag QRE has been used extensively in SCFs in carbon dioxide [9, 11, 35], hydrofluorocarbons including difluoromethane, trifluoromethane, and 1,1,1,2-tetrafluoroethane [24, 25, 28], and ammonia [2]. The Pt QRE has also frequently been used in $scCO_2$ [13, 14, 17, 20] and tungsten has been used in near-critical and supercritical ammonia [6] and chlorodifluoromethane [22]. Several groups have used the stainless-steel cell body as both the quasi-reference electrode and counter electrode in $scCO_2$ [10, 16, 39, 40] and in chlorodifluoromethane [22]. The idea here is essentially the same as that of the "wire" QRE in that the electrochemical cell itself assumes a stable, albeit unknown, potential.

As most studies used microelectrodes, a two-electrode configuration [17, 22] – working and reference electrodes (no counter electrode) – is most frequently used.

11.7 Solvents

Table 11.2 lists the supercritical solvents that have been investigated to date as media for electrochemistry. Carbon dioxide, which has a relatively mild critical temperature and pressure (Table 11.2), is relatively non-polar and highly resistive. In general, the more polar the liquid, the higher is its critical temperature. The relatively high critical temperature and pressure of many potentially attractive supercritical solvents such as hydrogen chloride, acetonitrile, and water present experimental challenges. Because of the high pressures and temperatures, stainless steel is attractive as a material for the cell body. However, the polar supercritical solvents such as hydrogen chloride, ammonia, and water are corrosive to stainless steel and also reactive with many analytes and electrolytes. On the other hand, although non-polar supercritical solvents such as carbon dioxide are compatible with a wide range of metals and can safely be contained in cells made from stainless steel, their low permittivity means that many potentially attractive redox couples and electrolytes in these solvents are insoluble. Furthermore, the increased solution resistance in non-polar supercritical solvents produces uncompensated ohmic distortion that may shift the half-wave potential, which limits their reliability/value.

Table 11.2 Critical temperatures and pressures of SCFs that have been used in electrochemistry.[a]

Supercritical solvent	Critical temperature (°C)	Critical pressure (bar)
Acetonitrile [41]	274.8	48.3
Ammonia	132.4	113.2
Carbon dioxide	31.1	73.8
Chlorodifluoromethane [22]	96.2	49.7
Difluoromethane	78.1	57.8
Dinitrogen oxide	36.4	72.5
Hydrogen chloride	51.5	82.6
Sulfur dioxide [4]	158	78.8
1,1,1,2-Tetrafluoroethane [42]	101.0	40.55
Trifluoromethane [28]	26.2	48.58
Water	374.0	220.6

[a]Supercritical temperature and pressure data are taken from Chapter 3 unless noted otherwise.

Bard and co-workers investigated the use of polar SCFs, including supercritical ammonia [2, 5], acetonitrile [41], and sulfur dioxide [4] and near-critical and supercritical water [3, 5]. In this series of studies, well-defined, reversible voltammetry was obtained for a wide range of analytes. Ammonia [2, 6] was shown to be a useful solvent in which to study unstable intermediates, radical anions, and so on. The ability to reduce the radical anion of phenazine to produce the dianion and observe a well-defined, reversible second redox wave for this species demonstrates the potential of this protic solvent in electrochemical studies. The diffusion coefficient was determined by CC as a function of temperature and pressure for *m*-chloronitrobenzene in supercritical ammonia. The diffusion coefficient was higher than that in solution, consistent with the expectation of enhanced mass transport due to the comparatively lower viscosity of the supercritical ammonia. The experimentally derived diffusion coefficient compared favorably with the theoretical value calculated using the Stokes–Einstein equation. Nonetheless, work in supercritical ammonia was found to be challenging. A convection-distorted cyclic voltammetric peak shape was observed at a scan rate of $500\,\mathrm{mV\,s^{-1}}$ at temperatures at or below 100 °C and at short measurement times. Contributions due to convection increased as the temperature was raised. Overall, this series of thoughtful studies whetted the appetite of many electrochemists with the promise of improved reaction kinetics, increased mass transport rates, and wide potential windows with low background currents, potentially allowing the electrochemist to probe novel reactions and affect reaction thermodynamics.

11.7.1 Supercritical Carbon Dioxide

Electrochemists recognized, however, that the high critical temperatures and pressures and corrosivity of polar supercritical solvents such as ammonia and water presented serious experimental challenges. In the chromatographic community, growing interest in supercritical fluid extraction (SFE) and supercritical fluid chromatography (SFC) [43], largely practiced using carbon dioxide, shifted researchers' focus to $scCO_2$. The relatively low critical temperature and critical pressure of carbon dioxide (Table 11.2), together with its non-toxicity, non-flammability, and relatively low cost, made it at least initially comparatively attractive as a possible electrochemical solvent.

The earliest reports [1, 9] suggested that voltammetry in $scCO_2$ was not possible even using microelectrodes. $scCO_2$ was simply too resistive to support charge transport even when microelectrodes were used. Taking a cue from Bard and coworkers [2, 3, 5], Philips *et al.* [9] added water (~0.64 M), a common polar modifier, to the $scCO_2$ to increase the conductivity and solvating power of the $scCO_2$. Under these conditions, distorted voltammetry was obtained for ferrocene, which was complicated by a significant ohmic contribution reflected in shifts in the half-wave potential with temperature and pressure and a peak on the reduction wave attributed to precipitation of the ferrocinium cation on the electrode surface. Addition of low concentrations (~0.01 M) of tetrahexylammonium hexafluorophosphate as supporting electrolyte improved the voltammetry, reducing the iR drop and diminishing the cathodic ferrocinium precipitation wave. Subsequently, Niehaus *et al.* [11] reported the first successful direct voltammetry for ferrocene at a carbon microdisk in $scCO_2$. Again, in the absence of added moisture, no voltammetry was obtained. However upon addition of a small amount of moisture (~55 mM), oxidation of ferrocene was observed at 8.96 MPa and 80 °C in $scCO_2$. Reduction of ferrocinium to ferrocene, however, was complicated by the precipitation of ferrocene on the electrode surface. Water, however, is not an ideal modifier for voltammetry as it significantly decreases the otherwise normally wide potential limits for electrochemical detection exhibited by supercritical solvents such as carbon dioxide. Addition of a molten salt, tetrahexylammonium hexafluorophosphate, appeared to yield classical steady-state voltammetry for the oxidation of ferrocene.

11.7.1.1 **Electrode Modification**

Niehaus *et al.* [11] capitalized on the relative insolubility of the molten salt, tetrahexylammonium hexafluorophosphate, and applied it to their microdisk electrodes as a permeable, conductive liquid film to facilitate mass transport of their analyte, ferrocene, between phases ($scCO_2$ and molten salt film) and facilitate its voltammetry. Under these conditions, well-defined oxidative voltammetry exhibiting minimal ohmic distortion was obtained. Subsequently, they demonstrated that analytes dissolved in the supercritical solvent partition into the film of molten salt on the working electrode [12]. Irreversibility of the electrochemistry was determined to be due to instability of the analyte in the molten salt film.

In 1993, Sullenberger and Michael [13] investigated the use of microelectrodes coated with the polyelectrolyte Nafion in $scCO_2$. They found that a polar modifier, specifically water, was useful in obtaining voltammetry in $scCO_2$. Since water significantly decreases the effective potential window for electrochemistry and supercritical fluid-based separations do not typically use water-modified solvents, they subsequently investigated the use of electrodes coated with a thin film of poly (ethylene oxide) (PEO) containing lithium triflate [14, 15]. Incorporation of ruthenium tris(bipyridyl) bis(hexafluorophosphate) into the PEO films was found to swell the PEO films and further enhance their electrical conductivity. As was the case with the molten salt films, the limiting factor in their utility proved to be the ability of the analyte to partition quickly between the supercritical fluid and the polymer film.

11.7.1.2 **Hydrophobic Electrolytes**

Abbott and Harper [19] discovered that tetraalkylammonium tetraarylborate electrolytes can be dissolved in useful concentrations (~0.03 M) in $scCO_2$ and investigated their utility as supporting electrolytes in electrochemical investigations in $scCO_2$. Unfortunately, the voltammetry reported for bis(tetradodecylammonium)nickel maleonitrile in $scCO_2$ containing tetrakis(decyl)ammonium tetraphenylborate was poorly defined and likely complicated by both adsorption and significant ohmic distortion. No follow-up work was reported.

11.7.1.3 **Water-in-Carbon Dioxide Microemulsions**

Since water-in-carbon dioxide microemulsions [44, 45] had been shown to be useful in a wide range of applications using $scCO_2$, Ohde *et al.* [20] investigated the use of water-in-$scCO_2$ microemulsions as a strategy in facilitating redox processes in $scCO_2$. They used a perfluoropolyether phosphate and sodium bis(2-ethylhexyl)sulfosuccinate to prepare stable microemulsions in $scCO_2$. Well-defined voltammetry was obtained for two model systems, ferrocene and *N,N,N′,N′*-tetramethyl-*p*-phenylenediamine. Voltammetry in both cases was reversible, although some ohmic distortion was evident. Calculated diffusion coefficients demonstrated that diffusion in the aqueous core of the micelles was the rate-limiting step. Subsequently, Lee *et al.* [21] used the ionic surfactant ammonium perfluoropolyether carboxylate to create water-in-carbon dioxide microemulsions. They examined the voltammetry of ~100 μM ferrocene, which is soluble in carbon dioxide but relatively insoluble in water, trimethyl(ferrocenylmethyl)ammonium trifluoromethanesulfonate, a water-soluble derivative of ferrocene, and potassium iodide at a Pt microdisk electrode. Well-defined voltammetry was obtained for all three systems. Experimentally derived diffusion coefficients again supported the view that the analytes are in the aqueous core of the micelles. Based on these two reports, the use of water-in-carbon dioxide microemulsions does not appear to suffer from problems of adsorption or product precipitation and analytes may be quantitated.

11.7.2
Hydrofluorocarbon Supercritical Solvents

Due to the aforementioned challenges presented by $scCO_2$, interest shifted to hydrofluorocarbon solvents such as chlorodifluoromethane [22], trifluoromethane [27, 28, 35], and 1,1,1,2-tetrafluoroethane [34]. Hydrofluorocarbon solvents are more polar than carbon dioxide and therefore solvate polar analytes better, eliminating the need for solvent modifiers. Furthermore, standard non-aqueous electrolytes such as $TBAPF_6$ and $TBABF_4$ are reasonably soluble and electrically conductive in these solvents. Finally, hydrofluorocarbons have mild critical temperatures and pressures (see Table 11.2), which makes them reasonably easy to work with in the research laboratory.

For example, Olsen and Tallman [22, 23] were able to prepare homogeneous solutions of supercritical chlorodifluoromethane containing 67 μM ferrocene and 8 mM $TBABF_4$. Voltammetry was well defined, reversible, and showed modest ohmic distortion compared with what had previously been reported in $scCO_2$. In addition, peak currents were noted to be lower than expected and to decrease upon repeated potential cycling.

Subsequently, Olsen and Tallman [23, 27] investigated the voltammetry of ferrocene in sub- and supercritical chlorodifluoromethane containing millimolar concentrations of $TBABF_4$ as the supporting electrolyte at Pt microelectrodes. Ferrocene exhibited reversible voltammetry in supercritical chlorodifluoromethane. The reversibility makes ferrocene potentially useful as an internal potential reference standard in supercritical chlorodifluoromethane. In the same study, they presented some promising preliminary work in trifluoromethane. Well-defined voltammetry was exhibited for the ferrocinium/ferrocene redox couple in 10 mM $TBABF_4$ in supercritical trifluoromethane. The experimentally derived diffusion coefficient for ferrocene in supercritical chlorodifluoromethane was higher than that in liquid chlorodifluoromethane, consistent with expectations based on the significantly lower viscosity of the supercritical solvent.

Subsequently, Abbott's group initiated a series of thoughtful electrochemical studies [19, 24–26, 46] using hydrofluorocarbon SCFs. This represents some of the most promising work since Bard and co-workers opened the field of supercritical fluid electrochemistry a decade earlier. In 1998, Abbott *et al.* [24] demonstrated the potential superiority/utility of hydrofluorocarbons, specifically difluoromethane and 1,1,1,2-tetrafluoroethane, as liquid and supercritical solvents for electrochemical processes. Due to the increased polarity of these solvents, standard non-aqueous electrolytes including $TBABF_4$ and $TBAClO_4$ were shown to dissolve readily in these solvents; for example, 0.1 M solutions of $TBAClO_4$ in 1,1,1,2-tetrafluoroethane were prepared at 25 °C and 1.0 MPa. Both solvents were shown to have wide potential windows – ~6 V (9 V in the case of $TBAClO_4$ in 1,1,1,2-tetrafluoroethane), which expands the range of redox processes that could potentially be probed or exploited electrochemically in synthetic or analytical applications. The authors illustrated this by probing the oxidation of saturated xenon in liquid 1,1,1,2-tetrafluoroethane containing $TBAClO_4$ using CV. An oxidation wave at + 3.5 V was attributed to the oxidation of xenon.

Abbott and Eardley [46] investigated the structure of the double layer produced in liquid and supercritical difluoromethane containing $TBABF_4$. Several peaks in the double-layer capacitance–potential plot at high pressure were found to dependent on potential. Their potential dependence was interpreted as evidence of electrolyte ion adsorption and reorientation of the polar solvent molecules adjacent to the electrode surface due to the change in the surface charge produced by varying the electrode potential. They found that as the pressure decreases, the thickness of the diffuse layer decreases and eventually collapses at pressures near the critical pressure. The collapse was attributed to the decreased viscosity and accompanying increase in thermal motion of the SCF near the critical point. A subsequent microbalance study in the same laboratory [26] of supercritical difluoromethane containing the electrolyte $TBABF_4$ demonstrated that the addition of electrolyte to the SCF produces significant changes in the structure of the SCF. The lower than predicted peak currents typically obtained in supercritical media were attributed to an increase in order within the fluid due to long-range electrostatic interactions that ultimately is reflected in an increase in solvent viscosity. The net effect is that the mass transport characteristics of the electrolyte–supercritical solvent system end up being closer to those of a liquid than of an SCF.

Goldfarb and Corti [28] analyzed homogeneous solutions of trifluoromethane containing 5×10^{-5} M decamethylferrocene. Well-defined voltammetry was obtained even in the absence of electrolyte. Addition of 5.1×10^{-4} M $TBAPF_6$ decreased the ohmic distortion significantly. However, strong ion association of the electrolyte at low densities of supercritical trifluoromethane was found to affect the limiting current for the analyte.

Abbott and Durling [25] investigated variations in the redox potential of ferrocene, ferrocenecarboxylic acid, and tetrabutylammoniumferrocene carboxylate as a function of pressure in supercritical difluoromethane containing 0.02 M $TBABF_4$ as supporting electrolyte. Voltammetry of all three analytes was reversible. Since there is essentially no volume change in these compounds upon oxidation, variation in the half-wave potential as a function of pressure provides information on solvation. While the redox potential of ferrocene was found to be relatively unaffected by changes in pressure, the redox potential of tetrabutylammoniumferrocene carboxylate and, to a lesser extent, that of ferrocenecarboxylic acid were perturbed. This suggests that ionic dissociation and solvation become more pronounced in supercritical difluoromethane at higher pressures for ionic species.

Abbott and Eardley [34] electrochemically reduced carbon dioxide in a liquid and supercritical mixture of 1,1,1,2-tetrafluoroethane (polar modifier) and carbon dioxide containing 20 mM $TBABF_4$ as supporting electrolyte. In aqueous solution, electroreduction of carbon dioxide is slow due to its poor solubility in water, which of course limits its mass transport to the electrode. In addition, the faradaic reaction efficiency is low in water due to hydrogen evolution. Findings in the SCF were very promising: Reduction kinetics and the faradaic efficiency for the electroreduction of carbon dioxide at Pt to produce oxalate were shown to be significantly enhanced in the supercritical mixture due to the high concentration of carbon dioxide produced at the electrode surface.

Exploring the potential for electrochemistry as a detection mode for SCF separations, Olsen and Tallman, using CV measurements, prepared a plot of the limiting current for ferrocene versus concentration and demonstrated good linearity (correlation coefficient 0.996) between 18 and 168 μM ferrocene at 115 °C and 9.0 MPa in supercritical chlorodifluoromethane. The sensitivity derived from the slope was a respectable 62 pA μM^{-1}. This suggests that at least in principle, electrochemical detection should be a sensitive mode of detection when coupled with SFC.

11.8 Applications

11.8.1 Electrochemical Synthesis in Supercritical Solvents

In 1987, Dombro *et al.* [47] demonstrated the feasibility of performing a synthetic-scale electrolysis in an SCF in the case of the electrochemical synthesis of dimethyl carbonate in $scCO_2$ containing methanol and tetrabutylammonium bromide as a supporting electrolyte and source of bromide ions. Significant current efficiencies, between 40 and 110%, were reported for the 3–12 h process even though a sizable ohmic drop ($\sim$11 V) was present.

Mabrouk and co-workers [39, 40] successfully synthesized two conducting polymers, polypyrrole (PPy) and polyaniline, potentiostatically at indium tin oxide (ITO)-coated glass electrodes in $scCO_2$ containing a small amount of acetonitrile (modifier) and *n*-tetrabutylammonium phosphate (electrolyte) from pyrrole or aniline hydrochloride, respectively. The conducting polymers synthesized electrochemically in $scCO_2$ were found to exhibit high conductivity [$4.4 \pm 2.0\ \text{S cm}^{-1}$ ($n = 16$) for $scCO_2$-synthesized PPy/ITO] – comparable to that produced potentiostatically in aqueous or non-aqueous (acetonitrile) media. Furthermore, the new method produced a fairly flat polymer film with raised nodules. The surface morphology of the conducting polymer films was distinctly different from that obtained in aqueous or non-aqueous media and may be potentially attractive in realizing optical applications of conducting polymers such as dielectric coatings.

Atobe *et al.* [35] extended this work to supercritical trifluoromethane and synthesized polypyrrole and polythiophene. They did not report any electrical conductivity for their films, which were apparently very thin. Reported surface morphology for the polypyrrole films was unremarkable, suggesting that the unique characteristics reported by Mabrouk and co-workers [39, 40] may require synthesis using a more non-polar SCF. More recently, Atobe *et al.* [48] reported the electrochemical synthesis of conducting polymer nanoparticles in supercritical trifluoromethane by using a nanoporous alumina membrane (Whatman Anodisc) coated with a thin film of Pt as a template.

11.8.2 Electrochemical Detection in Supercritical Solvents

SFE [49] and SFC [43, 50] has been used in a wide array of applications in the pharmaceutical, agrochemical, and petrochemical industries in the separation and analysis of, for example, pharmaceutical (chiral) compounds, food products, and polymers (high molecular weight). Packed columns and to a lesser extent capillary columns are often used. Consequently, the ideal detector for SFC is inexpensive, universal, and sensitive. Carbon dioxide continues to be the most frequently used solvent in SCF chromatographic separations. Due to the non-polarity of $scCO_2$, discussed earlier, polar modifiers such as methanol are typically added to dissolve and elute polar analytes. UV–visible, light scattering, and flame ionization detection (FID)

(GC) are most frequently, used though fluorescence and Fourier transform infrared (FT-IR) spectroscopy have also been successfully interfaced with SFC instruments.

Electrochemical methods are generally very sensitive and therefore attractive as detection methods in liquid chromatography [51, 52]. In principle, the enhanced mass transport in supercritical solvents mentioned earlier in this chapter should increase the analytical sensitivity for electrochemical detection, making this approach even more attractive as a detection technique for SFC in spite of the resistivity of SCFs. An obvious consideration in using electrochemical detection with SFC is whether or not an electrolyte is required. Addition of a supporting electrolyte to the SCF mobile phase would potentially complicate the coupling of the SFC instrument to the electrochemical detector, potentially clogging the outlet restrictor on the column if sufficient electrolyte precipitated.

In 1990 Di Maso *et al.* [10] first demonstrated the feasibility of on-line electrochemical detection for SFC using $scCO_2$ containing $TBABF_4$. No difficulties were encountered due to the precipitation of electrolyte at the restrictor, even though this had been perceived to be a potentially significant problem.

Almquist *et al.* [16] recognized that the use of electrolyte in an SCF separation would in the long term be impractical and limit the use of electrochemical detection, and instead investigated the use of a water-saturated column as a practical means of modifying the carbon dioxide and facilitating charge transport such that electrochemical detection would be possible. They demonstrated this, in principle, for the separation and detection of a series of ferrocenyl derivatives (ferrocene, acetylferrocene, and benzoylferrocene) although the detector sensitivity decreased with time due to precipitation of the ferrocinium on the working electrode, a problem noted earlier by other investigators using water as a modifier in $scCO_2$.

In 1995–96, Dressman and co-workers [17, 18] published two studies demonstrating the feasibility of on-line electrochemical detection for SFC. Both studies used a microelectrode coated with a conductive poly(ethylene oxide) (PEO) (MW 600 000) film containing ruthenium(II) tris(2,2-bipyridyl)bis(hexafluorophosphate) as a possible means of avoiding the introduction of electrolyte into the mobile phase. In the first study, ferrocene, *p*-benzoquinone, hydroquinone, and anthracene were run singly and in combination through an SFC column in $scCO_2$ both with and without modifiers (3% acetonitrile or 3% methanol). Voltammetric detection was employed and the quality of the results was compared side-by-side with those for post-column FID. For the admittedly somewhat idealized case reported at least, electrochemical detection proved possible in unmodified carbon dioxide. Overall, good sensitivity was demonstrated both in the absence and presence of polar modifier. Not surprisingly, the sensitivity was dependent on the ability of the redox active analyte to partition into the PEO film on the working electrode, with higher sensitivity reported for analytes that partition effectively between the supercritical solvent and the PEO film. The performance of the electrochemical detector was also shown to be unaffected by the modifier (acetonitrile).

In the second study, Dressman *et al.* [18] evaluated on-line voltammetric detection quantitatively for SFC by using voltammetric detection and FID simultaneously following the separation of a series of phenols and polyaromatic hydrocarbons by

SFC using neat and 1% v/v methanol-modified carbon dioxide as a mobile phase. Voltammetric detection exhibited good selectivity, nanogram detection limits, and acceptable linearity over two concentration decades.

Since 1998, four reports [8, 38, 53, 54] have appeared using amperometric detection following SFE- or SFC-based separations. All used $scCO_2$ as a mobile phase. Three [8, 38, 53] introduced a polar modifier (water, trimethylamine, or methanol). Toniolo *et al.* [54] instead investigated the use of porous electrodes (Pt) in unmodified $scCO_2$. In each of the studies, electrochemical detection was shown to exhibit good selectivity, comparable if not better sensitivity, nano- to picogram detection limits, and good long-term stability compared with frequently used methods of detection, specifically UV–visible detection and FID.

11.9
Conclusion and Outlook

To date, the body of work on electrochemistry in SCFs remains limited. Much of the work has focused on identifying the experimental conditions under which reversible voltammetry can be obtained for well-characterized, robust, reversible systems. Significant emphasis has also been placed on demonstrating the feasibility of using electrochemical detection for SCF separations, principally in carbon dioxide. That said, given the wide variety of systems that have been studied (inorganic, organic, organometallic, polymeric, etc.) to date, it is clear that, in principle, any redox active analyte should be amenable to electrochemical study in SCFs. Of course, key considerations would include analyte and electrolyte solubility – especially challenging in supercritical solvents with low permittivity such as carbon dioxide – and stability. Hydrofluorocarbons appear to be the most promising solvents due to their moderate critical temperatures and pressures, solvent power, and large potential windows. Gratifyingly, a number of more recent studies have begun to probe the fundamental nature of supercritical solvents and reactions that take place in them, to employ electrochemical detection in analysis, and exploit the unique properties of SCFs in the synthesis of novel materials including nanoparticles. This suggests that the future is bright indeed for applications involving electrochemistry in SCFs, such as probing reactions that previously could not be probed in traditional solvents.

References

1 Silvestri, G., Gambino, S., Filardo, G., Cuccia, C. and Guarino, E. (1981) *Angewandte Chemie International Edition in English*, **20**, 101.

2 Crooks, R.M., Fan, F.-R.F. and Bard, A.J. (1984) *Journal of the American Chemical Society*, **106**, 6851.

3 McDonald, A.C., Fan, F.-R.F. and Bard, A.J. (1986) *Journal of Physical Chemistry*, **90**, 196.

4 Cabrera, C.R. and Bard, A.J. (1989) *Journal of Electroanalytical Chemistry*, **273**, 147.

5 Flarsheim, W.M., Tsou, Y.-M., Trachtenerg, I., Johnston, K.P. and Bard,

A.J. (1986) *Journal of Physical Chemistry*, **90**, 3857.
6 Crooks, R.M. and Bard, A.J. (1987) *Journal of Physical Chemistry*, **91**, 1274.
7 Cabrera, C.R., Garcia, E. and Bard, A.J. (1989) *Journal of Electroanalytical Chemistry*, **260**, 457.
8 Palenzuela, B., Rodriquez-Amaro, R., Rios, A. and Valcarcel, M. (2002) *Electroanalysis*, **14**, 1427.
9 Philips, M.E., Deakin, M.R., Novotny, M.V. and Wightman, R.M. (1987) *Journal of Physical Chemistry*, **91**, 3934.
10 DiMaso, M., Purdy, W.C. and McClintock, S.A. (1990) *Journal of Chromatography*, **519**, 256.
11 Niehaus, D., Philips, M., Michael, A. and Wightman, R.M. (1989) *Journal of Physical Chemistry*, **93**, 6232.
12 Niehaus, D.E., Wightman, R.M. and Flowers, P.A. (1991) *Analytical Chemistry*, **63**, 1728.
13 Sullenberger, E.F. and Michael, A.C. (1993) *Analytical Chemistry*, **65**, 3417.
14 Sullenberger, E.F. and Michael, A.C. (1993) *Analytical Chemistry*, **65**, 2304.
15 Sullenberger, E.F., Dressman, S.F. and Michael, A.C. (1994) *Journal of Physical Chemistry*, **98**, 5347.
16 Almquist, S.R., Nyholm, L. and Markides, K.E. (1994) *Journal of Microcolumn Separations*, **6**, 495.
17 Dressman, S.F. and Michael, A.C. (1995) *Analytical Chemistry*, **67**, 1339.
18 Dressman, S.F., Simeone, A.M. and Michael, A.C. (1996) *Analytical Chemistry*, **68**, 3121.
19 Abbott, A.P. and Harper, J.C. (1999) *Physical Chemistry Chemical Physics*, **1**, 839.
20 Ohde, H., Hunt, F., Kihara, S. and Wai, C.M. (2000) *Analytical Chemistry*, **72**, 4738.
21 Lee, D., Hutchinson, J.C., DeSimone, J.M. and Murray, R.W. (2001) *Journal of the American Chemical Society*, **123**, 8406.
22 Olsen, S.A. and Tallman, D.E. (1994) *Analytical Chemistry*, **66**, 503.
23 Olsen, S.A. (1995) PhD thesis, North Dakota State University (Fargo).
24 Abbott, A.P., Eardley, C.A., Harper, J.C. and Hope, E.G. (1998) *Journal of Electroanalytical Chemistry*, **457**, 1.
25 Abbott, A.P. and Durling, N.E. (2001) *Physical Chemistry Chemical Physics*, **3**, 579.
26 Abbott, A.P., Hope, E.G. and Palmer, D.J. (2005) *Analytical Chemistry*, **77**, 6702.
27 Olsen, S.A. and Tallman, D.E. (1996) *Analytical Chemistry*, **68**, 2054.
28 Goldfarb, D.L. and Corti, H.R. (2000) *Electrochemistry Communications*, **2**, 663.
29 Flarsheim, W.M., Bard, A.J. and Johnston, K.P. (1989) *Journal of Physical Chemistry*, **93**, 4234.
30 Bond, A.M., McLennan, E.A., Stojanic, R.S. and Thomas, F.G. (1987) *Analytical Chemistry*, **59**, 2853.
31 Gagne, R.R., Koval, C.A. and Lisensky, G.C. (1980) *Inorganic Chemistry*, **19**, 2854.
32 Gritzner, G. and Kuta, J. (1984) *Pure and Applied Chemistry*, **56**, 461.
33 Abbott, A.P. and Harper, J.C. (1996) *Journal of the Chemical Society, Faraday Transactions*, **92**, 3895.
34 Abbott, A.P. and Eardley, C.A. (2000) *Journal of Physical Chemistry B*, **104**, 775.
35 Atobe, M., Ohsuka, H. and Fuchigami, T. (2004) *Chemistry Letters*, **33**, 618.
36 Howell, J.O. and Wightman, R.M. (1984) *Analytical Chemistry*, **56**, 524.
37 Wightman, R.M. (1988) *Science*, **240**, 415.
38 Senorans, F.J., Markides, K.E. and Nyholm, L. (1999) *Journal of Microcolumn Separations*, **11**, 385.
39 Badlani, R.N., Mayer, J.L., Anderson, P.E. and Mabrouk, P.A. (2002) *Polymer Preprints*, **43**, 938.
40 Anderson, P.E., Badlani, R.N., Mayer, J. and Mabrouk, P.A. (2002) *Journal of the American Chemical Society*, **124**, 10284.
41 Crooks, R.M. and Bard, A.J. (1988) *Journal of Electroanalytical Chemistry*, **243**, 117.
42 Abbott, A.P. and Eardley, C.A. (1998) *Journal of Physical Chemistry B*, **102**, 8574.
43 Lee, M.L. and Markides, K.E. (1987) *Science*, **235**, 1342.
44 Johnston, K.P., Harrison, K.L., Clarke, M.J., Howdle, S.M., Heitz, M.P., Bright,

F.V., Carlier, C. and Randolph, T.W. (1996) *Science*, **271**, 624.

45 Clarke, M.J., Harrison, K.L., Johnston, K.P. and Howdle, S.M. (1997) *Journal of the American Chemical Society*, **119**, 6399.

46 Abbott, A.P. and Eardley, C.A. (1999) *Journal of Physical Chemistry B*, **103**, 6157.

47 Dombro, R.A., Jr., Prentice, G.A. and McHugh, M.A. (1987) *Journal of the Electrochemical Society*, **135**, 2219.

48 Atobe, M., Iizuka, S., Fuchigami, T. and Yamamoto, H. (2007) *Chemistry Letters*, **36**, 1448.

49 Turner, C. (2006) *ACS Symposium Series*, **926**, 189.

50 Rouessac, F. (2000) *Chemical Analysis*, John Wiley & Sons, Ltd., Chichester.

51 LaCourse, W.R. (1997) *Pulsed Electrochemical Detection in High-Performance Liquid Chromatography*, John Wiley & Sons, Inc., New York.

52 Flanagan, R.J., Perrett, D. and Whelpton, R. (2005) *Electrochemical Detection in HPLC: Analysis of Drugs and Poisons*, Royal Society of Chemistry, Cambridge.

53 Lang, Q., Cheng, I.F., Wai, C.M., Paszczynski, A., Crawford, R.L., Barnes, B., Anderson, T.J., Wells, R., Corti, G., Allenbach, L., Erwin, D.P., Assefi, T. and Mojarradi, M. (2002) *Analytical Biochemistry*, **301**, 225.

54 Toniolo, R., Comisso, N., Schiavon, G. and Bontempelli, G. (2004) *Analytical Chemistry*, **76**, 2133.

12
Coupling Reactions and Separation in Tunable Fluids: Phase Transfer-Catalysis and Acid-catalyzed Reactions

Pamela Pollet, Jason P. Hallett, Charles A. Eckert, and Charles L. Liotta

12.1
Introduction

The implementation of the principles of green chemistry in catalytic processes is a major challenge that requires innovative approaches. Phase transfer catalysis (PTC) eliminates the need for high-boiling and non-recyclable solvents such as dimethyl sulfoxide (DMSO). Phase transfer catalysts, however, can be difficult and expensive to separate and recycle, resulting in huge amount of contaminated water wastes [1]. Similarly, Brønsted and Lewis acids are the most widely used catalysts in industry, and they produce more than 1×10^8 metric tons per year of products [2]. However, acid-catalyzed industrial processes often suffer from large amounts of waste, toxicity, corrosion, and difficulty of separation. For example, in the Friedel–Crafts acylation of methyl benzoate with acetic anhydride, approximately 20 kg of $AlCl_3$ are used per kilogram of product produced [3].

In this chapter, we show CO_2-enhanced extraction to be an efficient tool to separate phase transfer catalysts with some water solubility, minimizing the production of contaminated wash water by up to 95%. Next, we demonstrate the potential of alkylcarbonic acids (from alcohols + CO_2) and near-critical water to form *in situ* reversible acid catalysts requiring no downstream neutralization, remediation, or regeneration [4]. The common theme of these three processes is to maintain, if not improve, the selectivities and performances of current processes while minimizing environmental impact and cost, fusing green chemistry with good economics.

12.2
Phase Transfer Catalysis

12.2.1
Background

The reactions of organic substrates with inorganic salts are particularly challenging because they require either a dipolar, aprotic solvent or a biphasic mixture of solvents

Handbook of Green Chemistry, Volume 4: Supercritical Solvents. Edited by Walter Leitner and Philip G. Jessop

ISBN: 978-3-527-32590-0

combined with a phase transfer catalyst [5]. The former strategy is not particularly desirable because dipolar, aprotic solvents are relatively expensive and difficult to remove from the products. For example, DMSO, dimethylformamide (DMF), and hexamethylphosphoramide (HMPA) have boiling points of 189, 153, and 235 °C, respectively. In the latter strategy, phase transfer catalysts are often difficult to separate and recycle. However, the separation of the phase transfer catalyst by extraction, distillation, adsorption, or binding to an insoluble support is known [5].

12.2.2
Phase Transfer Catalysis Quaternary Ammonium Salt-catalyzed Reactions

The first published example of PTC in a supercritical fluid (SCF) was the nucleophilic displacement of benzyl chloride with potassium bromide in supercritical CO_2 ($scCO_2$) with 5 mol% acetone in the presence of tetraheptylammonium bromide (THAB) [6]. Chandler *et al.* later reported that a highly reactive phase, the omega phase, was formed at the surface of the salt in the reaction of benzyl chloride with potassium cyanide in $scCO_2$ in the presence of the phase transfer catalyst tetraheptylammonium chloride [7]. Their results suggested that the reaction occurs in this catalyst-rich phase (Figure 12.1). This pioneering work opened new avenues in reacting a soluble substrate like an organic electrophile with an insoluble inorganic nucleophile in SCF.

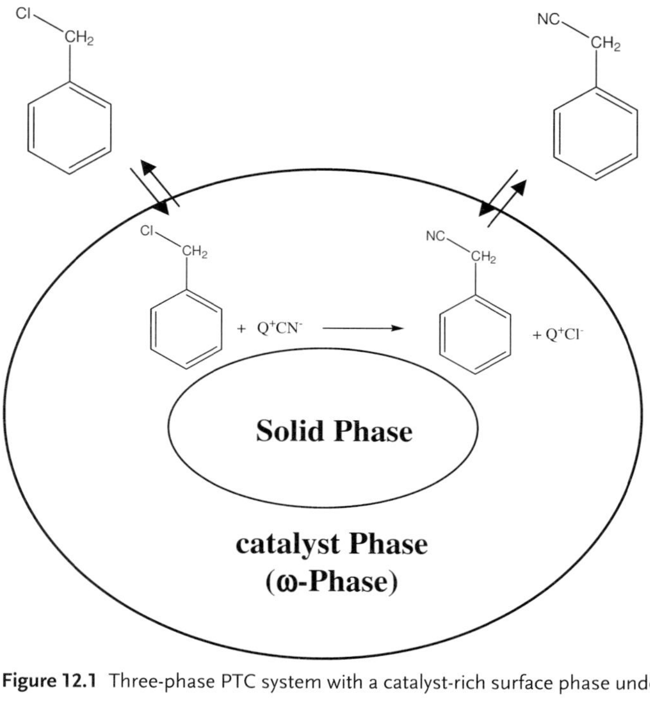

Figure 12.1 Three-phase PTC system with a catalyst-rich surface phase under dynamic conditions.

12.2.3
PTC Separation and Recycling Using CO_2

In commercial processes, phase transfer catalysts are commonly separated by extraction. Unfortunately, the complete removal of the phase transfer catalyst from the organic phase requires many wash cycles, using massive quantities of water per kilogram of product. In addition, recycling of the catalyst is often precluded by the energy-intensive evaporation of these large amounts of water to concentrate the catalyst.

Since CO_2 is non-polar and a very weak solvent, the addition of a significant quantity of CO_2 can reduce the ability of polar organic solvents to dissolve ions and polar organic solutes. Xie *et al.* investigated the effect of CO_2 on the partitioning behavior of two lipophilic phase transfer catalysts [benzyltriethylammonium bromide (BTEAB) and tetrabutylammonium picrate (TBAP)] between organic and aqueous phases [8]. The distribution coefficients of the benzyltriethylammonium cation at different CO_2 pressures are shown in Figure 12.2. The biphasic system comprised acetonitrile and an aqueous solution with 20 wt% NaCl. At pressures less than 6 MPa, the dissolved CO_2 drove BTEAB into the aqueous phase, altering the distribution coefficient by about 200-fold. Another finding of practical interest was that the acetonitrile concentration in the aqueous phase decreased fivefold with the addition of CO_2, mitigating potential solvent losses.

TBAP is an analog of tetrabutylammonium bromide (a common industrial phase transfer catalyst) that could be conveniently assayed by UV–visible spectroscopy. The distribution coefficients of TBAP as a function of CO_2 pressure are shown in Figure 12.3 for the case of three organic solvents: butyl acetate, methyl isobutyl ketone (MIBK), and methylene chloride. In each case, the distribution coefficients were increased upon addition of less than 6 MPa of CO_2. With methylene chloride, for example, the distribution coefficient was increased about 60-fold.

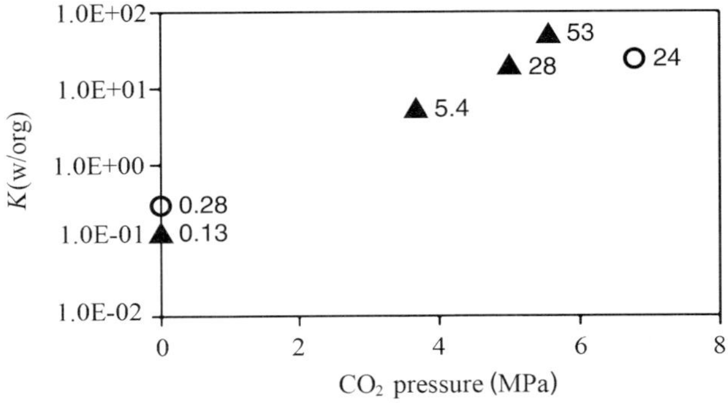

Figure 12.2 BTEAB distribution coefficient as a function of CO_2 pressure at room temperature (23–25 °C). Data (○): 50 g NaCl solution (20 wt%), 20 ml acetonitrile, 0.3 g (0.0011 mol) BTEAB. Data (▲): 11.5 g NaCl solution (20 wt%), 20 ml acetonitrile, 0.0068 g (2.5×10^{-5} mol) BTEAB.

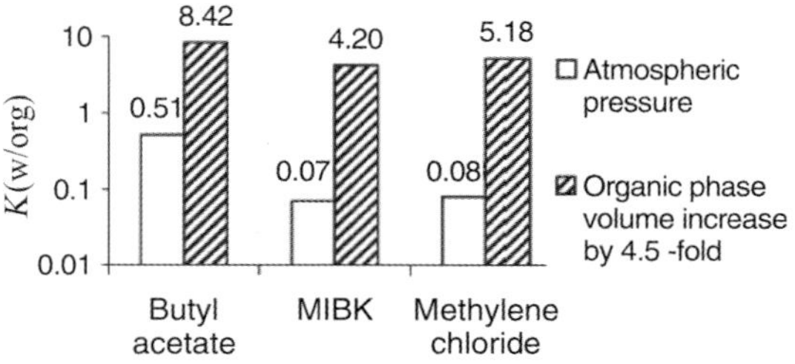

Figure 12.3 TBAP distribution coefficient as a function of CO_2 pressure at room temperature (23–25 °C). Organic solvents used: butyl acetate, methyl isobutyl ketone, and methylene chloride. 20 ml 8.770 × 10^{-5} M TBAP aqueous solution, 20 ml organic solvent. For the three solvents used, CO_2 pressures are less than 6 MPa to achieve a 4.5-fold volume expansion of the organic phase.

Table 12.1 Comparison of the water usage of a conventional aqueous extraction at atmospheric pressure and a CO_2-enhanced aqueous extraction with 5 MPa CO_2 pressure for equal volume batch and countercurrent modes: the calculation is based on the data for the CH_2Cl_2–H_2O system in Figure 12.3 and concentration reduction by 100-fold in 1 l of organic solvent.

Pressure	Water usage (l)	
	Equal volume batch	Countercurrent
Atmospheric pressure	60	20–30
5 MPa CO_2 pressure	2.7	0.3–0.5

These large changes in the distribution coefficients of the phase transfer catalysts in immiscible aqueous–organic systems can result in substantial reductions in the amount of wash water required for their removal. The water usages of a conventional aqueous extraction at atmospheric pressure and an aqueous extraction enhanced with 5 MPa CO_2 pressure are compared in Table 12.1 for equal-volume batch and countercurrent modes. Reductions in wash water volumes of more than 95% are possible. These results demonstrate that CO_2-enhanced extraction is an efficient tool to separate lipophilic catalysts such as benzyltriethylammonium bromide and tetrabutylammonium picrate.

12.3 Near-critical Water

12.3.1 Definition

Near-critical water (NCW) is defined as water that has been heated to a temperature range of 200–300 °C where its properties have begun to differ significantly from those of ambient water. NCW has been called a variety of names, including "high-temperature water," "hot water," "subcritical water," and "near-subcritical water." In this chapter, NCW will be used for consistency. All liquids exhibit dilation (a rapid decrease in density) as they approach their critical points. In the NCW region, water is not yet an SCF, but a significant change in properties such as a density decline has already begun to occur. Hence the properties of NCW are intermediate between those of ambient and supercritical water. Supercritical water (scH_2O) has been the subject of much investigation, primarily for waste remediation by scH_2O oxidation. This oxidation process typically operates at 400–500 °C and pressures of 20–50 MPa, where water provides an excellent medium for the destruction of chemicals, but is much less useful for synthesis. Moreover, the insolubility of salts in this medium has caused clogging difficulties in process applications. In contrast, NCW at 200–300 °C and pressures less than 10 MPa is a much better solvent and vastly superior for making chemicals, and it has the advantage of doing this at lower temperatures and pressures than scH_2O.

12.3.2 Properties

There are several important differences in the physiochemical properties of NCW compared with ambient water. The most important are the changes in the structure of water due to the reduction in (exothermic) hydrogen bonding with increase in temperature. Many experimental and theoretical studies have attempted to quantify the structure of water at elevated (mostly supercritical) temperatures [9]. These studies show that water loses ~55–60% of its hydrogen bonding network as the temperature is increased from 25 to 300 °C, with a corresponding reduction in molecular ordering [10]. The ordered hydrogen bonding network gives rise to the fundamental characteristics of ambient water, and its breakdown contributes to important changes in the properties of NCW, such as the density, dielectric constant, and dissociation constant.

This breakdown of hydrogen bonding in NCW causes a variety of changes in the physical properties of water; the most apparent physical property change is that of density. Following the saturation vapor pressure of water in Figure 12.4, the density decreases from 1 g cm^{-3} at 25 °C to about 0.75 g cm^{-3} at 300 °C [11]. In contrast, the density of scH_2O just above the critical point is around 0.1 g cm^{-3}. Naturally, the density of water is easily tuned by adjusting the temperature and pressure of the system, and since the compressibility becomes infinite at the critical point, in this region the tuning becomes more sensitive. For example, increasing the system pressure from saturation to 100 MPa at 200 °C increases the density of water by only 10% ($\rho = 0.85$ g cm^{-3} at 4 MPa, $\rho = 0.95$ g cm^{-3} at 100 MPa). However, at 300 °C, an increase of nearly 25% occurs ($\rho = 0.7$ g cm^{-3} at 7 MPa, $\rho = 0.9$ g cm^{-3} at 100 MPa) and nearly 50% at 350 °C ($\rho = 0.55$ g cm^{-3} at 15 MPa, $\rho = 0.8$ g cm^{-3} at 100 MPa) [11]. The decrease in density with temperature is also a contributing factor to other property changes of NCW, such as the dielectric constant and dissociation constant [12]. It is also a major factor in improved transport in NCW, as a reduction in

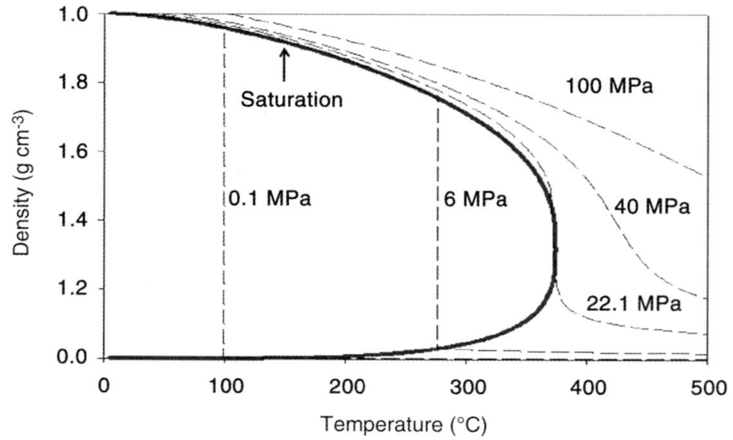

Figure 12.4 Density of water as a function of temperature from 0 to 500 °C at various pressures. Correlation from NIST steam tables [10].

density corresponds to an increase in diffusion [13]. The higher density of NCW compared with scH_2O also contributes to the predominance of ionic as opposed to free radical mechanisms in NCW [14]. Although for NCW the pressures required for significant density effects are too large to be of practical interest, such density changes can be used in research to give insights into the nature of molecular events.

In addition, the dielectric constant of NCW is greatly reduced when compared with water at ambient conditions, as shown in Figure 12.5. Uematsu and Franck correlated the dielectric constant of water as a function of both temperature and density [12]. For example, at 300 °C and saturation pressure, the dielectric constant of water is approximately 20, a nearly 75% reduction from the value of 78 at ambient conditions.

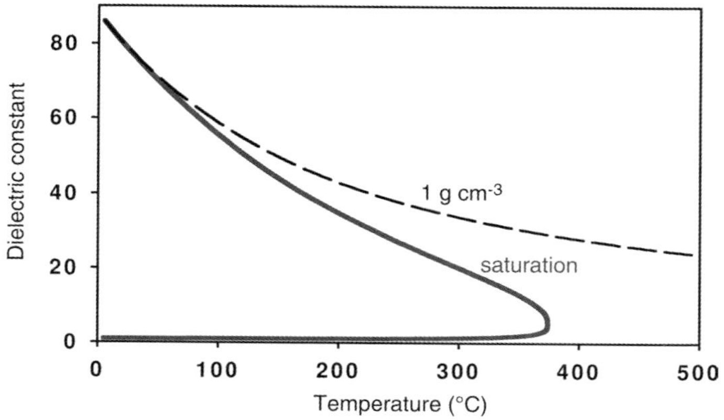

Figure 12.5 Dielectric constant of water as a function of temperature from 0 to 500 °C. Correlation from Uematsu and Franck [12].

This dielectric constant most closely corresponds to a moderately polar solvent such as acetone (dielectric constant = 21.4 at 25 °C). This reduction in dielectric constant, resulting from a major decrease in the hydrogen bonding relative to ambient water, greatly enhances the solubility of non-polar organic species in NCW. Increasing the density can increase the dielectric constant of NCW, but hundreds of MPa pressures are required. For example, increasing the density at 300 °C from saturation conditions ($\rho \approx 0.75\ g\,cm^{-3}$) to the density of ambient water ($\rho = 1\ g\,cm^{-3}$) nearly doubles the dielectric constant to approximately 35.

The reduced dielectric constant of NCW, while increasing the solubility of non-polar organics, reduces the solubility of inorganic salts. Therefore, for some applications of NCW as a reaction medium, one must strike a careful balance to achieve the desired solubility of ionic and non-ionic species. Although pressure is a potential variable, adjustments in temperature are generally used, and this ability to tune properties offers important advantages for NCW in reaction processes.

A major advantage of using NCW as a reaction solvent is the wealth of opportunities related to the increase in dissociation constant. The K_W value of water increases by as much as three orders of magnitude, from 10^{-14} at 25 °C to nearly 10^{-11} at around 250 °C (at saturation density) [15], where a maximum occurs, as displayed in Figure 12.6. This maximum can be increased by about another order of magnitude (and shifted to higher temperature) by increasing the density of the solvent – but again this requires hundreds of MPa pressure and is predominantly of academic interest. Increased temperature renders dissociation more favorable, but the decrease in dielectric constant reduces the ability of the medium to solvate the resulting ions. This is also why the maximum can be increased by increasing density, as the

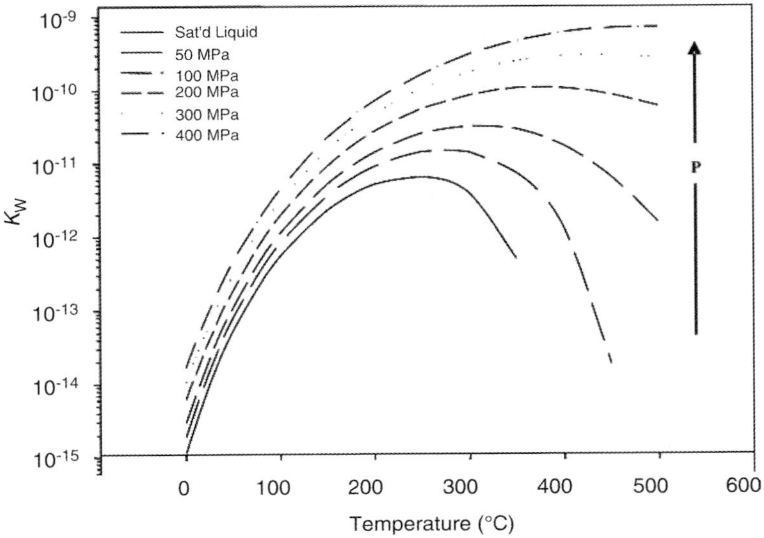

Figure 12.6 Dissociation constant of water as a function of temperature from 0 to 600 °C at various pressures. Correlation from Marshall and Franck [15].

dielectric constant also increases. The reason for interest in the high dissociation constant of NCW is that it provides a means of performing both acid and base catalysis in NCW without adding mineral acids or bases. By simply heating the solution, the dissociation of water increases, providing an enormous increase in the concentration of hydronium and hydroxide ions, resulting in a catalytic medium for the chemical transformation. Cooling the mixture restores the ambient ion concentrations; NCW is a self-neutralizing catalytic medium. This is important for ease of separations (no catalyst to be recovered), elimination of processing steps (no neutralization required), and waste reduction (no salt disposal).

To perform chemical transformations on organic substrates in NCW, the medium must be able to dissolve those substrates in sufficient quantities for homogeneous reaction to take place. Highly augmented solubility is achieved for even non-polar organic substrates at NCW temperatures, while maintaining the ability to dissolve salts. Moreover, it also creates a facile separation by simply cooling the post-reaction mixture for a simple decantation of products to replace more difficult forms of product recovery. The solubility of organic species in NCW varies greatly with the functionality of the organic molecule.

The upper critical solution temperature (UCST) is the temperature above which the organic substrate and water are miscible in all proportions (at the bubble pressure of the mixture). Polar organics, such as acetonitrile, have very low UCSTs (−1 °C for acetonitrile) [16]. Functionalized organics have much higher UCSTs, for example acetophenone (228 °C), 1-octanol (278 °C), and anisole (291 °C) [17]. Hydrocarbons can have much higher values, as evidenced by benzene (305 °C) toluene (310 °C) and *n*-hexane (355 °C) [18]. The UCST is an important measure, but it does not represent the entire solubility phenomenon. Benzene is completely miscible at 305 °C, despite having an ambient solubility in water of 500 ppm [19]. *n*-Hexane solubility increases by almost five orders of magnitude from ambient water to NCW [20]. Structural effects are not pronounced, as the UCST for 1-hexanol (221 °C) and 2-hexanol (230 °C) are fairly similar [21]. Most phase equilibrium models for the solubility of organics in water at elevated temperatures are extremely poor [22]; hence data acquisition at elevated temperature is important.

12.3.3
Friedel–Crafts Chemistry in NCW

Friedel–Crafts alkylation reactions are useful for attaching carbon functionalities to aromatic ring systems. They are usually conducted in the presence of a Lewis acid such as $AlCl_3$ and BF_3 or protic acids such as H_2SO_4, HF, and H_3PO_4, which must be subsequently neutralized and separated from the product. By employing NCW to replace the required acid catalyst, we eliminate the need for expensive base neutralization, catalyst regeneration, and disposal of salt byproducts.

The reaction of phenol with *tert*-butanol in water at 250, 275, and 300 °C to produce 2-*tert*-butylphenol and 4-*tert*-butylphenol (Equation 12.1) has been reported by Chandler *et al.* [23]. The product yields are shown as a function of time at each temperature in Figures 12.7–12.9. When the products were subjected to NCW at 275 °C, phenol was

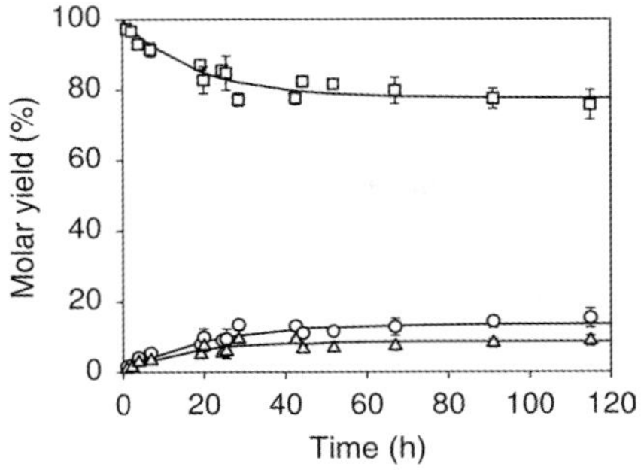

Figure 12.7 Mole fraction product yields as a function of time for the reaction of phenol with *tert*-butanol in water at 250 °C and 17.2 MPa: (□) phenol; (○) 2-*tert*-butylphenol; (△) 4-*tert*-butylphenol.

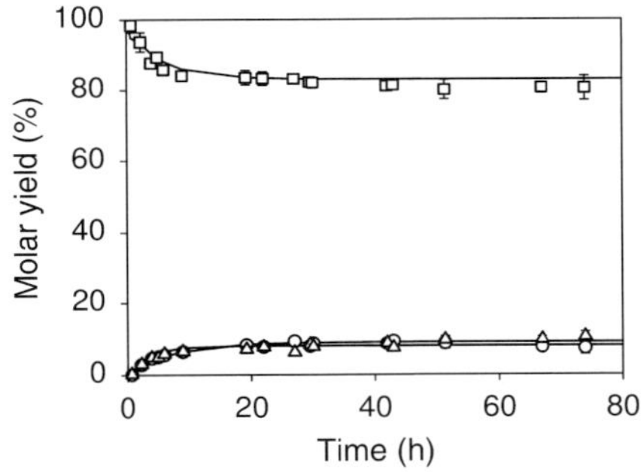

Figure 12.8 Mole fraction product yields as a function of time for the reaction of phenol with *tert*-butanol in water at 275 °C and 17.2 MPa: (□) phenol; (○) 2-*tert*-butylphenol; (△) 4-*tert*-butylphenol.

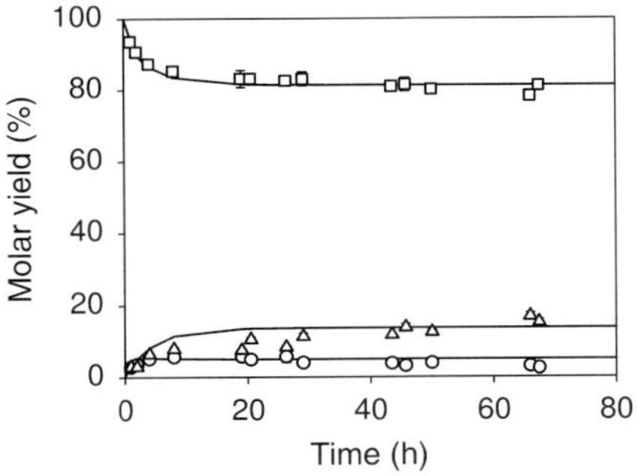

Figure 12.9 Mole fraction product yields as a function of time for the reaction of phenol with *tert*-butanol in water at 300 °C and 17.2 MPa: (□) phenol; (○) 2-*tert*-butylphenol; (△) 4-*tert*-butylphenol.

produced, indicating that the reaction is reversible. Small quantities (5%) of 2,4-di-*tert*-butylphenol are also formed. The reaction kinetics were described using a simple reaction network involving two reversible, first-order reactions (Equation 12.2).

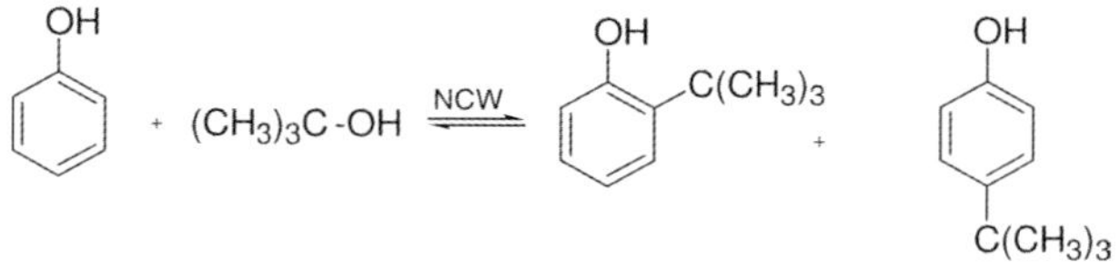

(12.1)

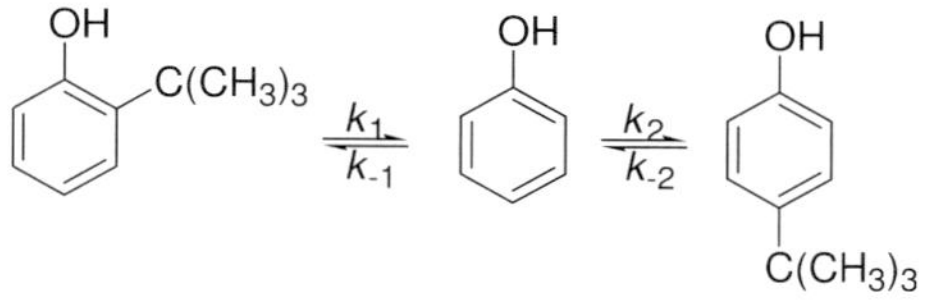

(12.2)

Xu *et al.* reported that under neutral conditions in water in the temperature range 225–320 °C, *tert*-butanol undergoes rapid dehydration to form isobutene [24]. The alkylation can occur through the incipient tertiary carbonium ion formed from either the *tert*-butanol or the isobutene (Scheme 12.1). However, because the *tert*-butanol was present in large excess, the reaction of phenol to form both 2-*tert*-butylphenol and 4-*tert*-butylphenol was assumed to be pseudo-first order in phenol. Additionally, 2,4-di-*tert*-butylphenol was produced in very low concentrations and was not included in the proposed reaction network.

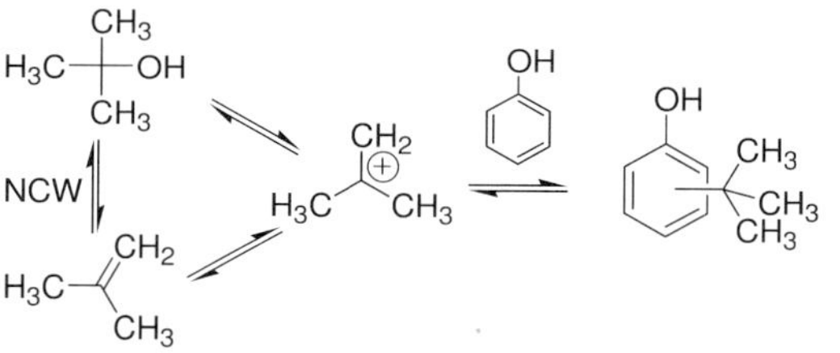

Scheme 12.1 Reaction sequence of the Friedel–Crafts alkylation of phenol from *tert*-butanol or isobutene.

Table 12.2 summarizes the pseudo-first order rate constants for each of the temperatures studied. Activation energies were calculated from Arrhenius plots with values reported in Figures 12.10 and 12.11. Based on the differences between

Table 12.2 Pseudo-first-order rate constants for the alkylation reaction of phenol with *tert*-butanol in NCW.

Temperature (°C)	$k_1 \times 10^5$ (s^{-1})	$k_{-1} \times 10^5$ (s^{-1})	$k_2 \times 10^5$ (s^{-1})	$k_{-2} \times 10^5$ (s^{-1})
250	0.19	1.1	0.17	1.4
275	0.61	6.4	0.36	3.3
300	2.1	35	0.92	5.4

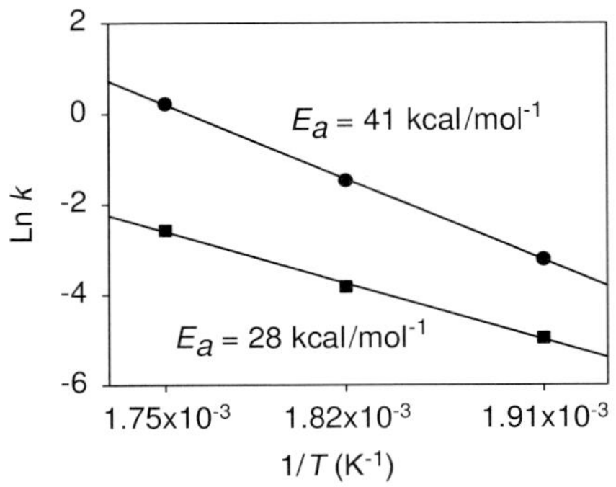

Figure 12.10 Arrhenius plot for the forward and reverse rate constants for the alkylation of phenol to form 2-*tert*-butylphenol in NCW: (■) k_1; (●) k_{-1}.

the forward and reverse reaction activation energies, the heat of reaction of the alkylation of phenol to form 2-*tert*-butylphenol was calculated to be $-54.4\,\mathrm{kJ\,mol^{-1}}$ and the heat of the forward reaction of phenol to form 4-*tert*-butylphenol was calculated to be $+16.7\,\mathrm{kJ\,mol^{-1}}$. The difference in the heats of reaction between the formation of the *ortho*-substituted product and the *para*-substituted product is large, and for comparison, the heats of reaction were calculated from heats of formation of the reactants and products. The calculated heat of reaction for the formation of the *ortho*-isomer was reported to be $-20.9\,\mathrm{kJ\,mol^{-1}}$. The corresponding heat of formation of the *para*-isomer was estimated to be approximately the same value; hence the calculated heat of reaction was not consistent with those determined experimentally. However, these calculated values were for a gas-phase reaction at 25 °C, compared with the NCW reactions that are in highly non-ideal aqueous solutions at high temperatures. Table 12.2 shows that the rate of formation of 2-*tert*-butylphenol was faster than the rate of formation of 4-*tert*-butylphenol at all reaction temperatures. In addition, the activation energies and heats of

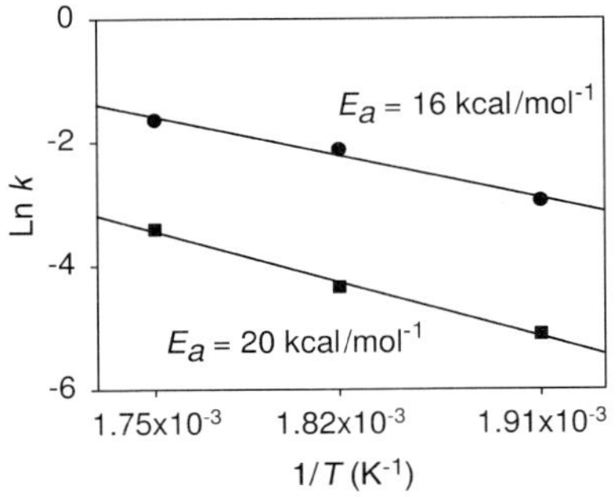

Figure 12.11 Arrhenius plot for the forward and reverse rate constants for the alkylation of phenol to form 4-*tert*-butylphenol in NCW: (■) k_2; (●) k_{-2}.

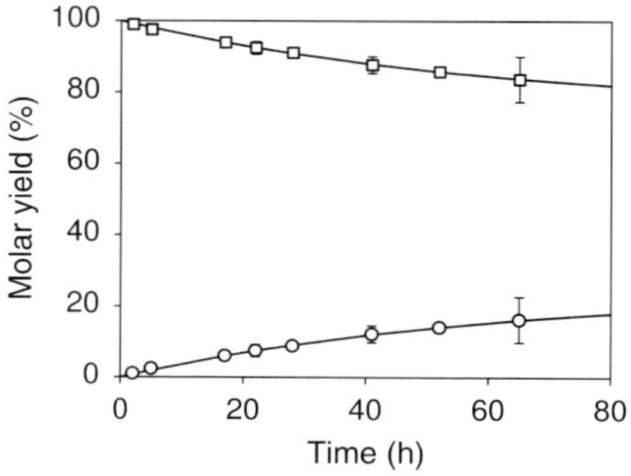

Figure 12.12 Mole fraction product yields as a function of time for the reaction of *p*-cresol with *tert*-butanol in water at 250 °C and 17.2 MPa: (□) *p*-cresol; (○) 2-*tert*-butyl–4-methylphenol.

reaction revealed that the formation of 2-*tert*-butylphenol was exothermic, and thus the equilibrium concentration of the 2-isomer actually decreased with increasing temperature. Conversely, the formation of 4-*tert*-butylphenol was found to be endothermic and the equilibrium concentration increased with increasing temperature.

We have also reported the reaction of *p*-cresol with *tert*-butanol to form 2-*tert*-butyl-4-methylphenol (Equation 12.3) at 250, 275, and 300 °C, and the product yields are shown as a function of time in Figures 12.12–12.14 [25]. The 2-*tert*-butyl-4-methylphenol converts back to *p*-cresol when subjected to NCW at 275 °C, indicating that the reaction was reversible. The pseudo-first order rate constants, summarized in Table 12.3, increased with increase in temperature, and Figure 12.15 shows the effect of temperature on the forward and reverse rate constants along with the activation energies. For the forward reaction, the heat of reaction was calculated to be $-25.1\,kJ\,mol^{-1}$. The corresponding value calculated from heats of formation was $-12.5\,kJ\,mol^{-1}$. As in the previous case, this latter value is for a gas-phase

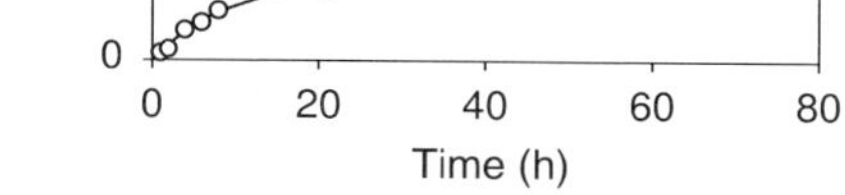

Figure 12.13 Mole fraction product yields as a function of time for the reaction of *p*-cresol with *tert*-butanol in water at 275 °C and 17.2 MPa: (□) *p*-cresol; (○) 2-*tert*-butyl–4-methylphenol.

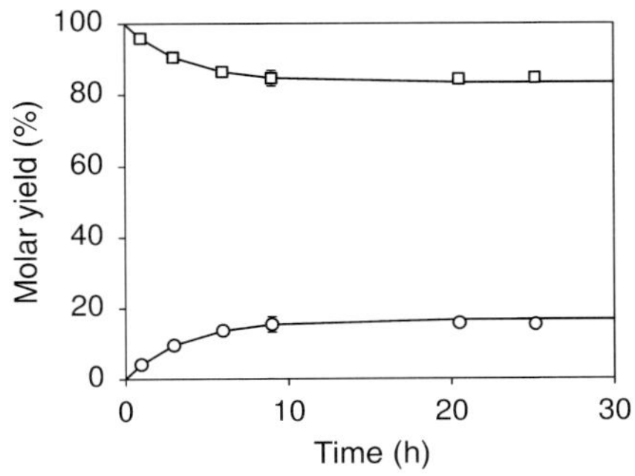

Figure 12.14 Mole fraction product yields as a function of time for the reaction of *p*-cresol with *tert*-butanol in water at 300 °C and 17.2 MPa: (□) *p*-cresol; (○) 2-*tert*-butyl–4-methylphenol.

reaction at 25 °C. Finally, the reaction of *tert*-butanol with *p*-cresol reached equilibrium at approximately 1 h, which is much faster than the corresponding reaction with phenol.

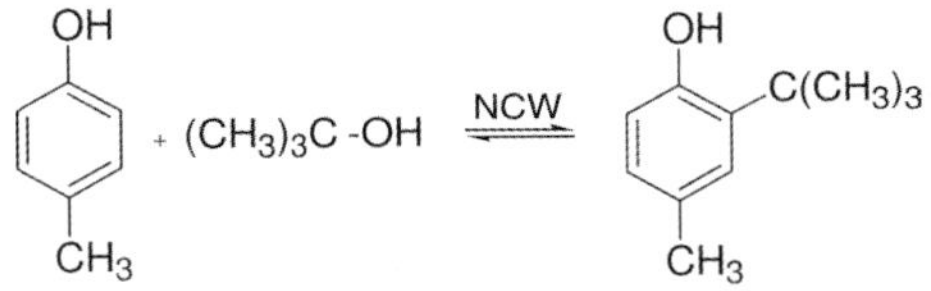

(12.3)

The reactions of phenol with 2-propanol and 1-propanol are slow, and the major products with 2-propanol are 2-isopropylphenol and 2,6-diisopropylphenol, as shown in Figure 12.16. These results are in contrast to the reaction of phenol with *tert*-butanol, where the second *tert*-butyl group went to the 4-position of the phenol, probably due to steric effects. Reaction of phenol with 1-propanol proceeded even more slowly than the isopropyl alcohol. The yield of products was less than 5% yield over a time period of 144 h, giving 2-isopropylphenol plus very small amounts of 2,6-diisopropylphenol and 2-*n*-propylphenol. Clearly, the incipient primary carbonium ion rearranged to the more stable secondary carbonium ion.

Table 12.3 Pseudo-first-order rate constants for the alkylation reaction of *p*-cresol with *tert*-butanol in NCW.

Temperature (°C)	$k_1 \times 10^5$ (s^{-1})	$k_{-1} \times 10^5$ (s^{-1})
250	0.11	0.35
275	0.48	1.8
300	1.3	6.3

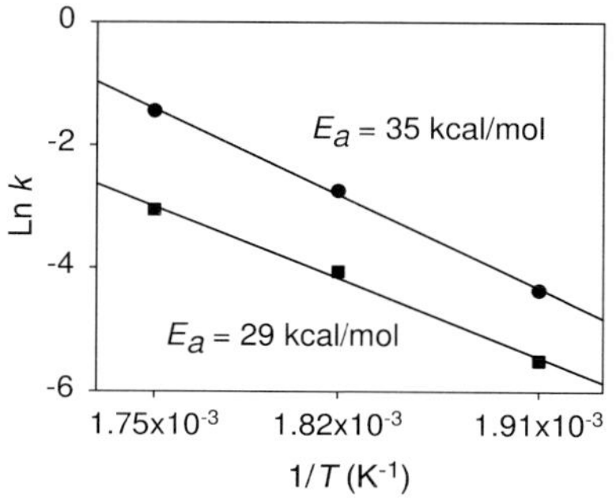

Figure 12.15 Arrhenius plot for the forward and reverse rate constants for the alkylation of *p*-cresol to form 2-*tert*-butyl-4-methylphenol in NCW: (■) k_1; (●) k_{-1}.

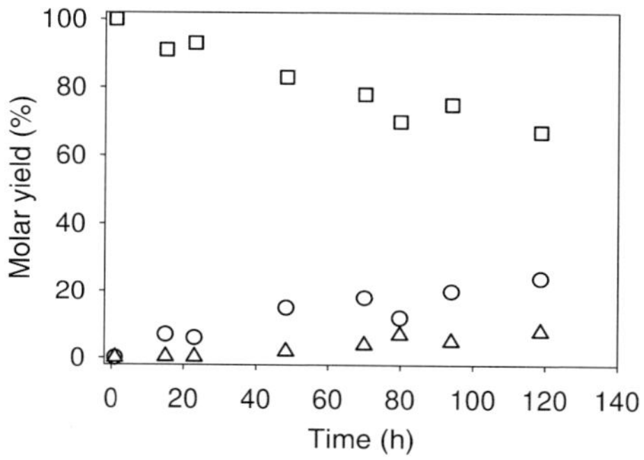

Figure 12.16 Mole fraction product yields as a function of time for the reaction of phenol with 2-propanol in water at 275 °C: (□) phenol; (○) 2-isopropylphenol; (△) 2,6-diisopropylphenol.

12.4
Alkylcarbonic Acids

12.4.1
Probing Alkylcarbonic Acids – Alkylcarbonic Acids with Diazodiphenylmethane (DDM)

The reversible reaction of carbon dioxide with water to form carbonic acid is well known; analogously, CO_2 reacts reversibly with alcohols to form alkylcarbonic acids (Scheme 12.2). This provides *in situ* acid formation of catalysts which can be neutralized readily by the removal of carbon dioxide. Much of our recent research has used CO_2 with organic solvents to form gas-expanded liquids [26]. For example, we verified the presence of alkylcarbonic acids in CO_2-expanded alcohols using diazodiphenylmethane (DDM) as a reactive probe to trap the carbonic acid species [27]. Product analysis confirmed that DDM was reacting with an alkylcarbonic acid species and not just the protic alcohol. The rates of reaction of a series of

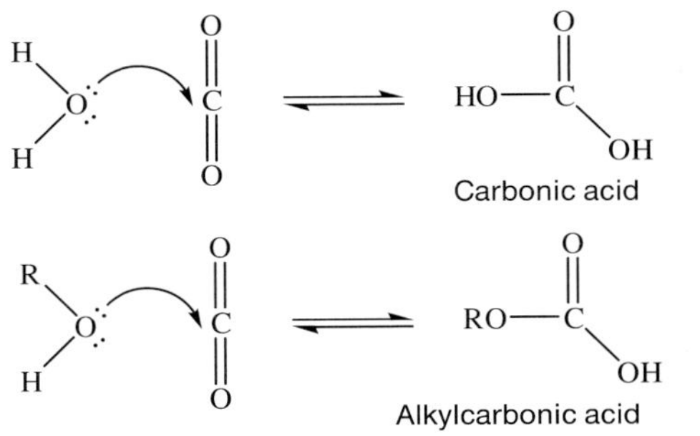

Scheme 12.2 Formation of carbonic acid and alkylcarbonic acid.

alcohols with DDM indicate that the relative reaction rates follow the order primary secondary > tertiary. To characterize further these types of acids, the relative reaction rates of DDM were measured in both methanol and ethanol with 20 mole CO_2. The rate in methanol is roughly 2.8 times that in ethanol, suggesting that the expanded methanol solution is more acidic than the expanded ethanol. However, the rates are influenced by both the actual strength of the acids (pK_a) and the prior equilibria to form the alkylcarbonic acids (concentration). The two equilibria involved in the formation and dissociation of alkylcarbonic acids are shown in Scheme 12.3. The addition of CO_2 to the system will affect both of these equilibria, as increasing the amount of CO_2 will drive the formation of the alkylcarbonic acid but simultaneously decrease the polarity of the solvent mixture and thereby decrease acid dissociation.

To compare the rates of the DDM reaction with alkylcarbonic acids with the rate with carbonic acid, West *et al.* used acetone as a diluent to maintain a single liquid phase with constant CO_2 concentration [28]. They measured kinetics at 40 °C using a solution containing 60 mol% acetone, 20 mol% CO_2, and 20 mol% ROH, where R = alkyl or hydrogen (Figure 12.17). The rates of the reactions run in alcohols follow a logical progression with the fastest rate in methanol and the slowest in *tert*-butanol. This relationship is consistent with steric and electronic factors affecting the equilibrium and dissociation of the alkylcarbonic acids. The corresponding rate using water (carbonic acid) was found to be slower than with either methanol or ethanol.

Weikel *et al.* dissolved DDM in 60 mol% acetone, 20 mol% alcohol, and 20 mol% CO_2 in a pressure vessel in order to compare the relative reaction rates of DDM with various alkylcarbonic acids formed *in situ* [29]. This study included methanol,

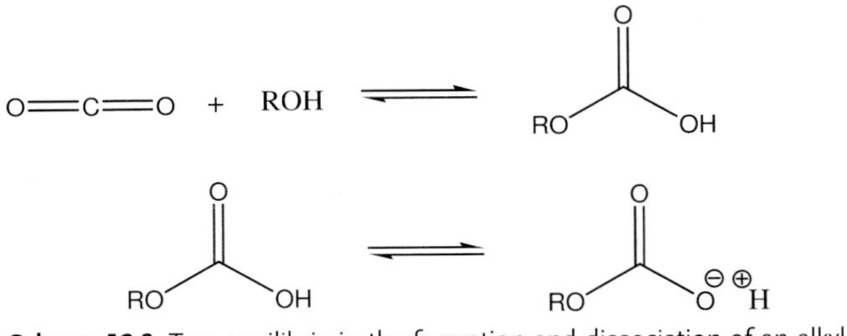

Scheme 12.3 Two equilibria in the formation and dissociation of an alkylcarbonic acid.

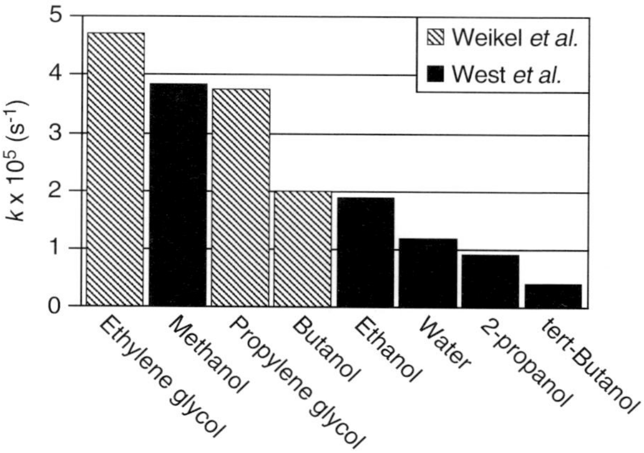

Figure 12.17 Comparison of the effect of various alcohols, diols, and water on the pseudo-first-order reaction rate constant (k, s^{-1}) of DDM (with corresponding carbonic acid) in 60 mol% acetone–20 mol% ROH–20 mol% CO_2 at 40 °C.

ethylene glycol, propylene glycol, benzyl alcohol, 4-nitrobenzyl alcohol, 4-chlorobenzyl alcohol, and 4-methoxybenzyl alcohol. The fastest rate was for ethylene glycol, whereas methanol and propylene glycol lead to similar rates (Figure 12.17). The increased rate with ethylene glycol relative to ethanol was attributed to the adjacent alcohol group stabilizing the alkylcarbonic acid through intramolecular hydrogen bonding. The authors also reported experiments without the acetone diluent, using instead pure methanol to investigate CO_2 effects on the rates of the DDM reaction (Figure 12.18). The presence of a maximum (at 6 MPa CO_2) is consistent with the hypothesis that the DDM rate is affected by both the concentration of the carbonic acid (which increases with CO_2 pressure) and by the rate of proton transfer to the DDM (which decreases with CO_2 pressure as the dielectric constant of the mixture decreases). This second effect was confirmed through dielectric constant measurements. Electronic effects were studied using a series of substituted benzyl alcohols: the

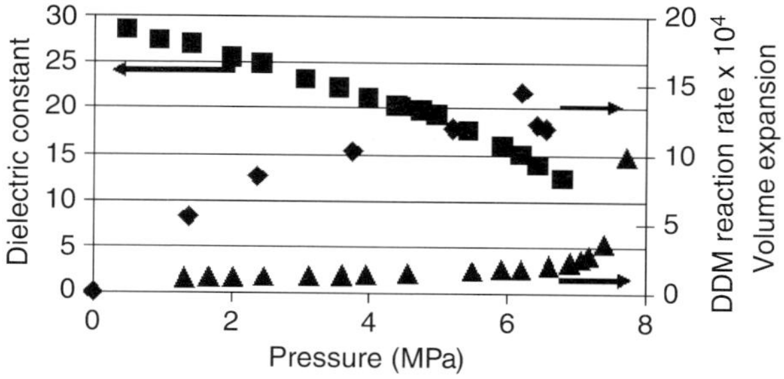

Figure 12.18 Pseudo-first-order rate constant (k, s^{-1}) for the reaction of DDM with methylcarbonic acid (diamonds), dielectric constant of MeOH–CO_2 (squares) and volume expansion of MeOH–CO_2 (triangles) at 40 °C versus pressure.

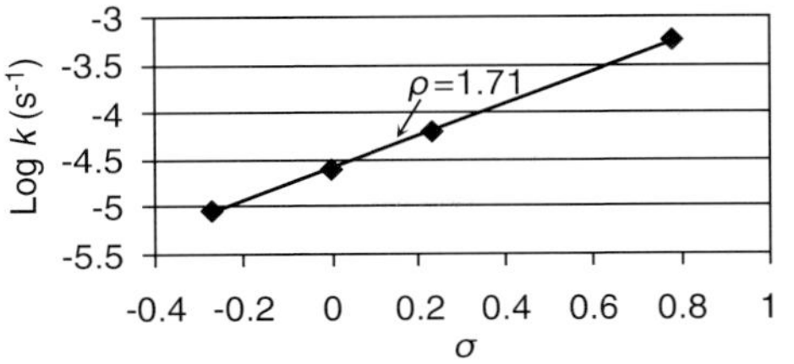

Figure 12.19 Hammett plot for pseudo-first-order reaction rates (k, s^{-1}) for the reaction of substituted benzylalkylcarbonic acids and DDM in acetone at 40 °C.

excellent linear Hammett correlation (Figure 12.19), with a slope ρ of 1.71, indicates the reaction is sensitive to electronic effects. Since the fastest rate occurred with the highly electron-withdrawing nitro substitution, this means that the second step in Scheme 12.3 is most affected by electronic effects. The first step is a nucleophilic attack on the carbon in CO_2, which should be increased by electron-donating groups and conversely decreased by electron-withdrawing groups. On the other hand, the second step of proton dissociation should be increased by electron-withdrawing groups and decreased by electron-donating groups. The fact that benzoic acid, which must only undergo the second step (proton dissociation) has a similar ρ value, is consistent with the hypothesis that the acid formation step for alkylcarbonic acids is not limiting. The Hammett plot also shows a two orders of magnitude difference between the most electron-withdrawing and the most electron-donating groups. Electron-withdrawing substituents enhance the proton transfer step in the reaction with DDM and electron-donating groups decrease the rate of proton transfer. These results are consistent with the proton transfer step being rate-determining.

12.4.2
Reactions Using Alkylcarbonic Acids

12.4.2.1 Ketal Formation

Acetals and ketals are commonly used to protect aldehyde and ketone functionalities from basic media. Acetal groups are normally formed by reacting the carbonyl-containing substrate with an excess of alcohol in the presence of a strong acid catalyst. Xie *et al.* employed the alkylcarbonic acids of methanol and ethylene glycol, two of the most common protection agents, to form the dimethyl ketal and cyclic ethylene ketal of cyclohexanone [30]. Initial rates under pseudo-first-order conditions for the ketal formation reaction between methanol and cyclohexanone in methanol and CO_2-expanded methanol at 25–50 °C are shown in Figure 12.20, where the catalytic effect of CO_2 addition is clear. Also, when CO_2 was replaced by ethane, the rate of ketal formation was similar to the (very slow) rate in pure alcohol (with no added CO_2). This result negates the possibility that simple pressure or viscosity reduction was responsible for the increased rate in CO_2-expanded methanol [31].

Figure 12.20 shows maxima in the rate constants at each temperature, probably due to the tradeoff between methylcarbonic acid formation and the rate of the proton

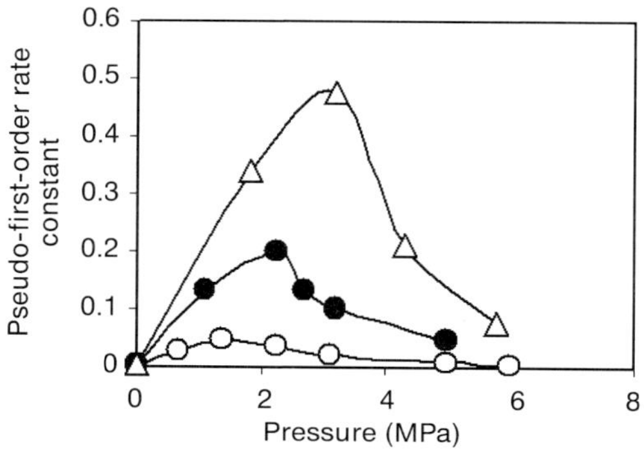

Figure 12.20 Pseudo-first-order rate constants of cyclohexanone acetal formation in CO_2-expanded methanol at various CO_2 pressures and 25 °C (open circles), 40 °C (filled circles) and 50 °C (triangles).

transfer step when CO_2 concentration is increased. The shifting of the maximum from about 2 MPa at 25 °C to 3.5 MPa at 50 °C strengthens this argument, as more pressure would be required to maintain a similar CO_2 concentration in the liquid phase at higher temperatures. Equation-of-state modeling revealed that the mole fractions of CO_2 at the various maxima are similar, although not identical [32]. The catalytic effects observed in the CO_2–ethylene glycol system demonstrate that the alkylcarbonic acid species can exist even at low CO_2 concentrations (~2% at 4 MPa) [33].

12.4.2.2 **Formation of Diazonium Salts**

Weikel *et al.* reported the formation of diazonium intermediates using methylcarbonic acid as a reversible acid catalyst. [34] The diazonium species were coupled with electron-rich aromatics such as *N,N*-dimethylaniline or reacted with potassium iodide in a Sandmeyer-type sequence (Scheme 12.4). The reaction of aniline, sodium

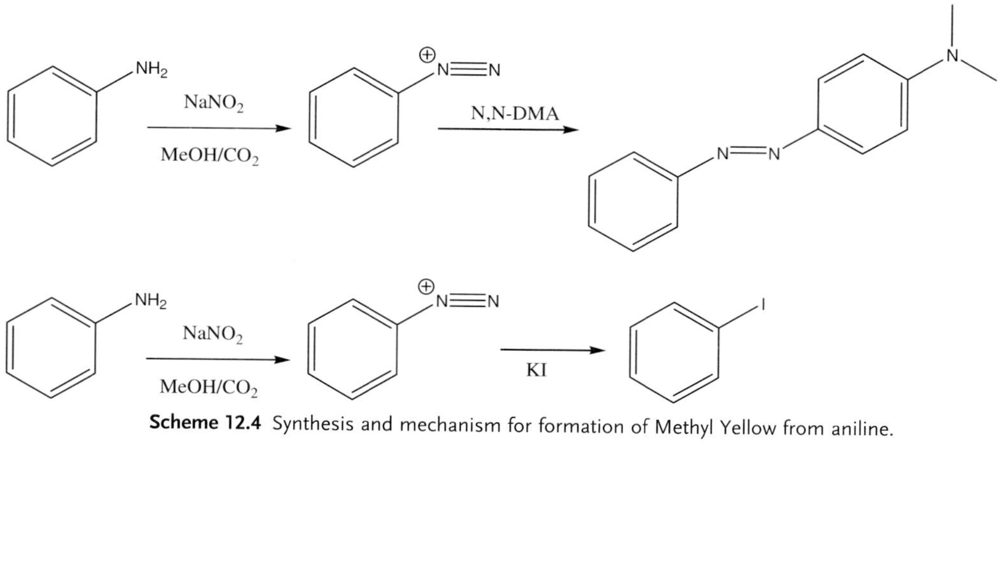

Scheme 12.4 Synthesis and mechanism for formation of Methyl Yellow from aniline.

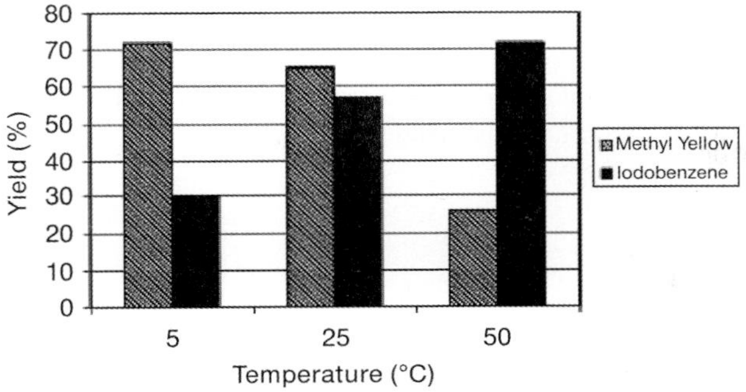

Figure 12.21 Effect of temperature on yield with 2.2 equiv. of $NaNO_2$. The pressures varied from $P = 1$ MPa (5 °C, 0.54 mol of CO_2) to $P = 4.7$ MPa (50 °C, 1.25 mol of CO_2).

nitrite, *N,N*-dimethylaniline, and CO_2 in methanol forms Methyl Yellow, an industrial dye. The most successful reaction conditions were at 5 °C with excess nitrite salt and a high CO_2 loading (Figure 12.21). The reaction was run for 24 h and produced a 97% conversion of aniline with an average yield of 72% of Methyl Yellow. The major by-product was benzene, formed from the elimination of molecular nitrogen from the diazonium intermediate followed by hydrogen abstraction from the solvent. Higher CO_2 loadings gave improved yields up to 4.7 MPa; apparently, under these conditions, the CO_2 concentration did not sufficiently lower the dielectric constant of the solvent to hinder proton transfer and the formation of the diazonium intermediate.

The most effective conditions for the synthesis of iodobenzene were similar to those for the coupling reaction except for the temperature. The best yield was an average of 72% at a high nitrite loading (2.2-fold excess) and CO_2 loading (1.25 mol) and at higher temperature (50 °C). The effect of nitrite and CO_2 loading was not as pronounced for iodobenzene as it was for Methyl Yellow. However, for iodobenzene, an increase in temperature had a significant impact on yield, as seen in Figure 12.21. This could be an effect of both an increased reaction rate and improved potassium iodide solubility. Again, the only observed by-product for the substitution reactions was benzene.

12.5 Conclusion

As environmental concerns become more pressing and as waste disposal becomes more costly, ever greater benefits will be derived from more benign processes. CO_2-enhanced aqueous extraction has been shown to be both efficient and environmentally benign for the separation of BTEAB and TBAP, thus making this technique attractive for sustainable processes involving PTC.

In situ-generated acids require no neutralization and eliminate the waste resulting from acid neutralization in processes such as Friedel–Crafts alkylation and acylation, acetal formation, and formation and reaction of diazonium salts. The use of NCW and alkylcarbonic acids provides not only environmental benefits from waste reduction but also economic processing advantages (higher selectivities, simplified separations), fostering the implementation of sustainable technologies.

References

1 (a) Starks, C., Liotta, C. and Halpern, M. (1994) *Phase-transfer Catalysis: Fundamentals, Applications, and Industrial Perspectives*, Chapman & Hall, New York; (b) Halpern, M. and Grinstein, R. (1999) Choosing a phase-transfer catalyst to enhance reactivity and catalyst separation, Part 1, in *Pilot Plants and Scale-up of Chemical Processes* (ed. W. Hoyle), Royal Society of Chemistry, Cambridge, **236**, 30–39.

2 Corma, A. (1997) *Current Opinion in Solid State & Materials Science*, **2**, 63–75.

3 Dartt, C.B. and Davis, M.E. (1994) *Industrial & Engineering Chemistry Research*, **33**, 2887–2899.

4 Eckert, C.A., Liotta, C.L., Kitchens, C.L. and Hallett, J.P. (2005) *Pharmaceutical Manufacturing*, 38–45.

5 Starks, C.M., Liotta, C.L. and Halpern, M. (1994) *Phase-Transfer Catalysis: Fundamentals, Applications, and Industrial Perspectives*, Chapman & Hall, New York, p. 688.

6 Boatright, D.L., Suleiman, D., Liotta, C.L. and Eckert, C.A. (1994) Solid-supercritcal fluid phase transfer catalyst, presented at the ACS 207th National Meeting, San Diego, CA, 15 March; Dillow, A.K., Yun, S.L.J., Boatright, D.L., Suleiman, D., Liotta, C.L. and Eckert, C.A. (1998) *Industrial & Engineering Chemistry Research* **37**, 3252.

7 Chandler, K., Culp, C.W., Lamb, D.R., Liotta, C.L. and Eckert, C.A. (1998) *Industrial & Engineering Chemistry Research*, **37**, 3252–3259.

8 Xie, X., Brown, J.S., Joseph, P.J., Liotta, C.L. and Eckert, C.A. (2002) *Chemical Communications*, 1156–1157.

9 Akiya, N. and Savage, P.E. (2002) *Chemical Reviews*, **102** (8), 2725–2750.

10 Hoffmann, M. and Conradi, M.S. (1997) *Journal of the American Chemical Society*, **119** (16), 3811–3817.

11 Harvey, A.H. and Klein, S.A. (1996) *NIST/ASME Steam Properties*, NIST Standard Reference Database 10, Version 2.01 (1996), Washington, DC.

12 Uematsu, M. and Franck, E.U. (1980) *The Journal of Physical Chemistry Ref. Data*, **9**, 1291–1306.

13 Lamb, W.J., Hoffman, G.A. and Jonas, J. (1981) *Journal of Chemical Physics*, **74** (12), 6875–6880.

14 Antal, M.J. Jr., Brittain, A., DeAlmeida, C., Ramayya, S. and Roy, J.C. (1987) *ACS Symposium Series*, 329, 77–86.

15 Marshall, W.L. and Franck, E.U. (1981) *The Journal of Physical Chemistry Ref. Data*, **10** (2), 295–304.

16 Szydlowski, J. and Szykula, M. (1999) *Fluid Phase Equilibria*, **154** (1), 79–87.

17 Brown, J.S., Hallett, J.P., Bush, D. and Eckert, C.A. (2000) *Journal of Chemical and Engineering Data*, **45**, 846–850.

18 Connolly, J.F. (1966) *Journal of Chemical and Engineering Data*, **11**, 13–16.

19 Sherman, S.R., Trampe, D.B., Bush, D.M., Schiller, M., Eckert, C.A., Dallas, A.J. *et al.* (1996) *Industrial & Engineering Chemistry Research*, **35** (4), 1044–1058.

20 Connolly, J.F. (1966) *Journal of Chemical and Engineering Data*, **11**, 13–16.

21 Hallett, J.P. (2002) Enhanced recovery of homogeneous catalysts through manipulation of phase behavior, Thesis, Georgia Institute of Technology, Atlanta, GA.

22 Hooper, H.H., Michel, S. and Prausnitz, J.M. (1988) *Industrial & Engineering Chemistry Research*, **27** (11), 2182–2187.

23 Chandler, K., Deng, F., Dillow, A.K., Liotta, C.L. and Eckert, C.A. (1997) *Industrial & Engineering Chemistry Research*, **36** (12), 5175–5179; Chandler, K., Liotta, C.L. and Eckert, C.A. (1998) *AICHE Journal*, **44** (9), 2080–2087.

24 Xu, X., Antal, M.J. Jr. and Anderson, D.G.M. (1997) *Industrial & Engineering Chemistry Research*, **36** (1), 23–41.

25 Chandler, K., Deng, F., Dillow, A.K., Liotta, C.L. and Eckert, C.A. (1997) *Industrial & Engineering Chemistry Research*, **36** (12), 5175–5179; Chandler, K., Liotta, C.L. and Eckert, C.A. (1998) *AICHE Journal*, **44** (9), 2080–2087.

26 West, K.N., Wheeler, C., McCarney, J.P., Griffith, K.N., Bush, D., Liotta, C.L. and Eckert, C.A. (2001) *The Journal of Physical Chemistry A*, **105**, 3947–3948; Weikel, R.R., Hallett, J.P., Levitin, G.R., Liotta, C.L. and Eckert, C.A. (2006) *Topics in Catalysis*, **37**, 75–80; Xie, X., Liotta, C.L. and Eckert, C.A. (2004) *Industrial & Engineering Chemistry Research*, **43**, 2605–2609; Weikel, R.R., Hallett, J.P., Liotta, C.L. and Eckert, C.A. (2007) *Industrial & Engineering Chemistry Research*, **46**, 5252–5257; Hallett, J.P., Kitchens, C.L., Hernandez, R., Liotta, C.L. and Eckert, C.A. (2006) *Accounts of Chemical Research*, **39**, 531–538; Eckert, C.A., Liotta, C.L., Bush, D., Brown, J.S. and Hallett, J.P. (2004) *The Journal of Physical Chemistry. B*, **108**, 18108–18118.

27 West, K.N., Wheeler, C., McCarney, J.P., Griffith, K.N., Bush, D., Liotta, C.L. and Eckert, C.A. (2001) *The Journal of Physical Chemistry A*, **105**, 3947–3948.

28 West, K.N., Wheeler, C., McCarney, J.P., Griffith, K.N., Bush, D., Liotta, C.L. and Eckert, C.A. (2001) *Journal of Physical Chemistry A*, **105**, 3947–3948.

29 Weikel, R.R., Hallett, J.P., Levitin, G.R., Liotta, C.L. and Eckert, C.A. (2006) *Topics in Catalysis*, **37**, 75–80.

30 Xie, X., Liotta, C.L. and Eckert, C.A. (2004) *Industrial & Engineering Chemistry Research*, **43**, 2605–2609.

31 Frank, M.J.W., Kuipers, J.A.M. and van Swaaij, W.P.M. (1996) *Journal of Chemical and Engineering Data*, **41**, 297–302.

32 Xie, X., Liotta, C.L. and Eckert, C.A. (2004) *Industrial & Engineering Chemistry Research*, **43**, 2605–2609.

33 Jou, F.Y., Deshmukh, R.D., Otto, F.D. and Mather, A.E. (1990) *Chemical Engineering Communications*, **87**, 223–231.

34 Weikel, R.R., Hallett, J.P., Liotta, C.L. and Eckert, C.A. (2007) *Industrial & Engineering Chemistry Research*, **46**, 5252–5257.

13
Chemistry in Near- and Supercritical Water

Andrea Kruse and G. Herbert Vogel

13.1
Introduction

Water is a cheap and environmentally friendly solvent which is neither toxic nor burnable. The critical point of water is fairly high: $T_c = 374\,°C$, $p_c = 22.1$ MPa. In spite of this and the experimental challenges associated with these extreme conditions, a wide range of reactions in near- and supercritical water (NSCW) have been studied [1–5]. The reasons are the extraordinary properties, changing with the process parameters temperature and pressure (density). NSCW is a tunable solvent, much more than any other supercritical fluid. Here a short overview about reactions in NSCW is given with a focus on those which are expected to be interesting for technical applications.

To reduce the number of citations, only reviews and papers with extensive reference lists are mentioned. Therefore, not only the publication cited but also the literature included therein should be taken into consideration by the reader.

13.2
Properties

The reaction rate of chemical reactions depends on the temperature, pressure, the presence of a catalyst, and the properties of the solvent such as density, dielectric constant, heat capacity, viscosity, and ionic product in the case of protic solvents. The distinctness of NSCW is the variability of the solvent properties that open up the opportunity to influence the chemical kinetics and therefore the selectivity of single reactions of a reaction network (Figure 13.1). In fact, the influence of water is in some cases so strong that water can be regarded as a catalyst (see below).

The physicochemical properties of water depend strongly on the temperature. At elevated temperatures but below the critical point, the ionic product is higher than under ambient conditions (Figure 13.2). This temperature range is chosen for acid- or

Handbook of Green Chemistry, Volume 4: Supercritical Solvents. Edited by Walter Leitner and Philip G. Jessop

ISBN: 978-3-527-32590-0

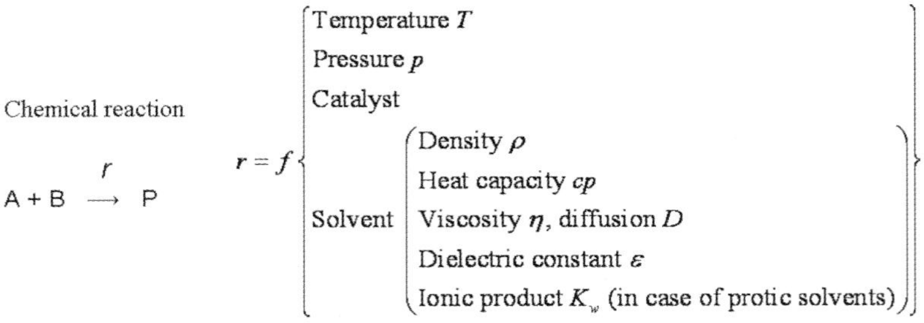

$$r = k_0 \cdot \exp\left(-\frac{E_a}{RT}\right) \cdot \exp\left(\frac{V^{\#} \cdot (p - p_0)}{RT}\right) \cdot C_A \cdot C_B$$

Figure 13.1 Parameters influencing chemical reaction kinetics: r = reaction rate; E_a = activation energy = f(catalyst); $V^{\neq}$ = activation volume = f(solvent); C = molarity = $f(p, T)$.

base-catalyzed reactions, because these reactions often occur under these conditions without any acid or base addition. Usually, it is assumed that the reactions occur because of the high ionic product ($pK_W = 11$, Figure 13.2). On the other hand, recent studies show that often the reaction mechanism is different from that under ambient conditions and in many cases it is not the H^+ or the OH^- ion which attack the reacting molecule but the water molecule. This is a result of the different microstructure of water rendering a single water molecule more mobile and therefore more active because of the decreased strength of the hydrogen bond network [1, 6].

In the supercritical state, the solvent properties of water resemble those of non-polar substances even though the solvent still consists of polar molecules. This

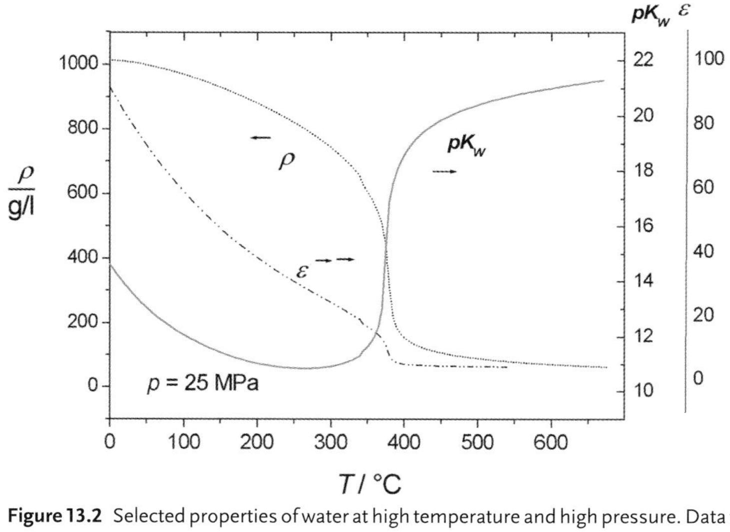

Figure 13.2 Selected properties of water at high temperature and high pressure. Data taken from [7].

unique combination should allow reactions that are impossible in any other solvent. The low dielectric constant of supercritical water (SCW) makes it a powerful solvent for compounds, which are purely soluble in ambient water. On the other hand, the solubility of salts decreases drastically near the critical point. The complete miscibility with gases and also the high solubility of organic compounds and high transport velocity in this medium make SCW a superior solvent for reactions of organic compounds with gases. This is the reason why oxidations are intensively studied in SCW. A high solubility together with fast diffusion is very interesting for reactions in the presence of heterogeneous catalysts. These properties are needed in order to avoid coking by removing coke precursors. Unfortunately, the stability of heterogeneous catalysts is still an issue; often the stabilities change drastically and/or the catalyst's activities are strongly altered by severe corrosion. Below the critical point, the solubility for salts is still high and protective layers of metals are dissolved, and therefore the corrosive attack of water on pure metal is possible. On the other hand, the drastic change in the solubility of salts can be used to precipitate compounds with very special properties. Therefore, it is not surprising that NSCW is a superior solvent for inorganic reactions and fine particle formation [8, 9].

In summary, it can be stated that at high temperatures and pressures water is a turntable solvent: sub- and supercritical water show different properties that are both unique and can be used to control chemical reactions.

An example of a reaction that only occurs in NSCW in this way is the Cannizzaro reaction, for example of benzaldehyde (Scheme 13.1). Usually, this reaction is observed only in the presence of strong bases. In NSCW, this reaction is observed without the addition of bases or acids. The extraordinary properties of NSCW seem to open up the opportunity for a new mechanism for this reaction[1, 4].

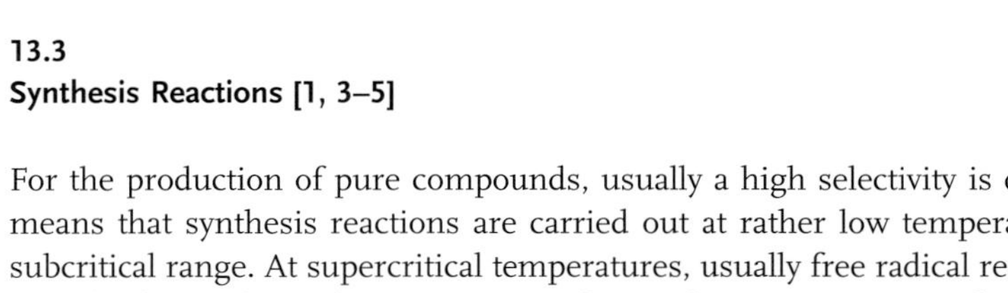

Scheme 13.1 Cannizzaro reaction of benzaldehyde [1, 4].

13.3
Synthesis Reactions [1, 3–5]

For the production of pure compounds, usually a high selectivity is desired. This means that synthesis reactions are carried out at rather low temperatures in the subcritical range. At supercritical temperatures, usually free radical reactions dominate leading to low selectivities except for gas formation ([10]; see below).

Of course, hydration and hydrolysis reactions in water are of special interest. Here, water simultaneously is the solvent and a reactant. The high ionic product and/or the increased mobility of single molecules may help to reach high reaction rates.

13.3.1
Hydrations

The addition of water to double bonds is found only with small yields and/or at low temperatures for thermodynamic reasons [11, 12]; with alkynes the yields are better at relatively low temperature [11]. Usually acid catalysts are necessary.

The reactivity of polar bonds is much higher. By addition of water, nitriles react to give amides and further to acids via hydrolysis with high selectivities and without additives (Scheme 13.2). Under ambient conditions, the addition of strong acids or bases is necessary. In a similar way, imines form ketones [4]. Kinetic investigations, for example for acetonitrile and benzonitrile, have been carried out, which show that the hydration rates of both are very similar in the range between 350 and 400 °C [13]. Of special technical interest is the reaction of ε-aminocapronitrile to ε-caprolactam, which is a stable end product up to 380 °C. The overall selectivity is approximately 80% [4].

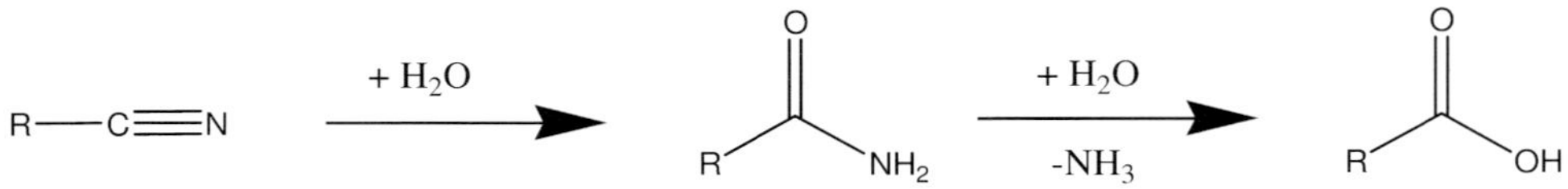

Scheme 13.2 Hydration of a nitrile to an amide and hydrolysis of the amide to an carboxylic acid [4].

13.3.2
Hydrolysis

Reactions investigated include the hydrolysis of amines, amides, esters, ethers, acetals, alkyl metal halides, anhydrides, nitro compounds, and silanes. Often acid catalysts such as CO_2, H_3PO_4, or CH_3COOH are added to reach high yields.

13.3.2.1 Esters

The hydrolysis of esters is of technical interest and therefore many different esters, such as acetates, phthalates, natural fats, and others, have been investigated [1, 4, 5]. A detailed investigation of the hydrolysis of ethyl acetate (tubular reactor, 23–30 MPa, 250–450 °C, 5.4–230 s) [14] without catalyst addition shows a lower activation energy under subcritical conditions than under supercritical conditions, indicating two different reaction mechanisms. For the subcritical region, nucleophilic attack on a protonated ester is assumed to be the rate-determining step. The formation of a protonated ester is favored in the subcritical region (Scheme 13.3). At 350 °C and a reaction time of 150 s, almost complete conversion to acetic acid and ethanol is observed. The ester group R shows a significant effect: the rate of hydrolysis rises in the sequence benzyl > methyl > ethyl > *n*-butyl. The equilibrium conversions are 88% for methyl acetate, 98% for ethyl acetate, 60% for *n*-butyl acetate, and 44% for benzyl acetate [4].

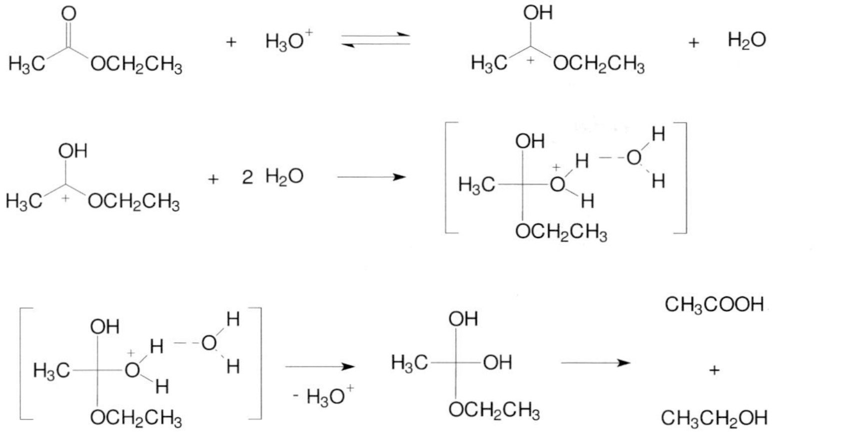

Scheme 13.3 Hydrolysis of esters in the subcritical region (as suggested in [14]: $A_{AC}2$ mechanism).

13.3.2.2 Ethers

Examples for ethers investigated are methoxynaphthalenes, dibenzyl ether, and anisoles [1, 4, 5]. Also in the conversion of biomass hydrolysis, reaction steps are very important: Cellulose and starch are rapidly hydrolyzed to glucose in NSCW. Therefore, a solid biomass is converted to soluble intermediates in a short period of time [2].

13.3.2.3 Amides

The hydrolysis of acetamide to acetic acid is almost complete (300–400 °C); only very small amounts of acetonitrile are formed. Benzamide hydrolyzes completely to benzoic acid. An interesting result is that in the temperature range 300–400 °C, the hydrolysis of acetamide is faster than that of benzamide, but the hydrolysis of benzamide is faster above 400 °C.

13.3.3 Dehydrations

It seems to be surprising to carry out dehydration reactions in water. In fact, for example, the equilibrium between simple alcohols and the corresponding unsaturated compounds formed via water elimination is shifted to the side of alkenes with increase in temperature. Therefore, at the relatively high temperature of NSCW, many alcohols react to give alkenes. The water is also often assumed to be the source of H^+ ions to catalyze this reaction. The water elimination reaction of *tert*-butanol, ethanol, glycerol, glycol, fructose, lactic acid, propylene glycol, polypropylene glycol, phenyl alcohols, and cyclohexanol, often in the presence of a mineral acid, have been studied in NSCW. In studies concerning dehydration of biomass-derived polyols (1,2- and 1,3-propanediol, 1,2-butanediol, glycerol and *m*-erythritol) the catalytic effect of different electrolytes was investigated (see below

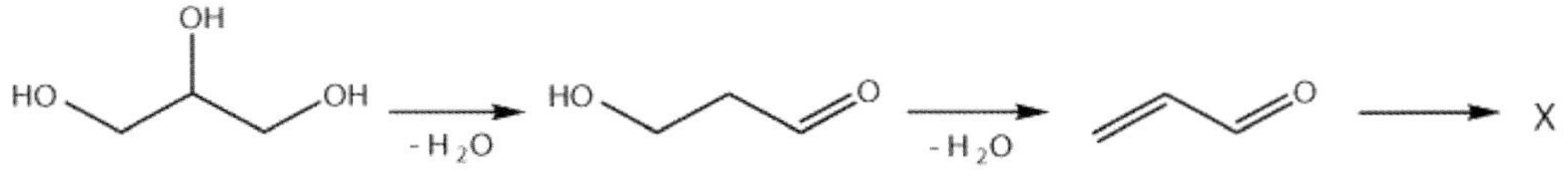

Scheme 13.4 Water elimination from glycerol [15, 16].

and Fig. 13.3). Water elimination from 1,2-propanediol, for example, was promoted by $ZnSO_4$, leading to a yield above 80% at 360 °C, but was suppressed by Na_2SO_4 under the same conditions [15]. The dehydration of glycerol leads to the formation of acrolein with a selectivity up to nearly 60% [15, 16]; acrolein shows consecutive reactions (Scheme 13.4). This observation of the effects of salts opens up new opportunities to influence reactions in NSCW.

13.3.4 Condensations

Other water elimination reactions include the formation of ethers by condensation. The dehydration of 1,4-butanediol in sub- and supercritical water leads selectively to the formation of tetrahydrofuran [4]. Different examples of the Friedel–Crafts alkylation or acylation have also been investigated (Scheme 13.5) [1].

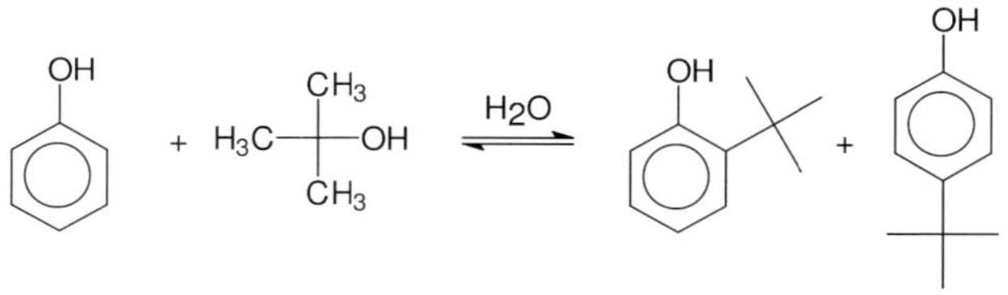

Scheme 13.5 Friedel–Crafts alkylation [1].

The aldol (Scheme 13.6), Claisen, and Dieckmann condensations like the Cannizzaro reaction (Scheme 13.1), are typically catalyzed by strong bases under ambient conditions. In NSCW, these reactions occur in high yield also in the absence of basic catalysts. In addition, the reverse aldol conversion, called aldol splitting here, occurs without the presence of a catalyst. This reaction is necessary to split the rings of glucose and fructose and is therefore a key reaction for the conversion of biomass in NSCW (Figure 13.4).

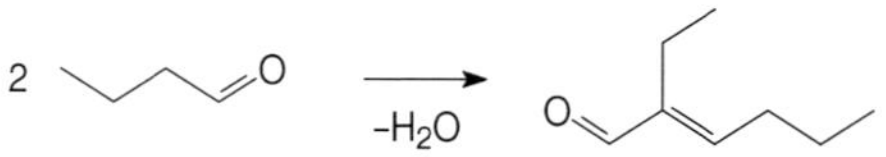

Scheme 13.6 Aldol condensation [1].

13.3.5 Diels–Alder Reactions

Diels–Alder reactions in water as the solvent and under ambient conditions have been investigated for a long time (Scheme 13.7). In contrast to the low solubility under these conditions, most of the possible dienophiles and dienes are completely soluble in SCW. Additionally, the reaction should be accelerated by high pressure. On the other hand, the hydrophobic effect in Diels–Alder reactions should vanish in SCW. However, many different reactions of this type have been investigated [1–5].

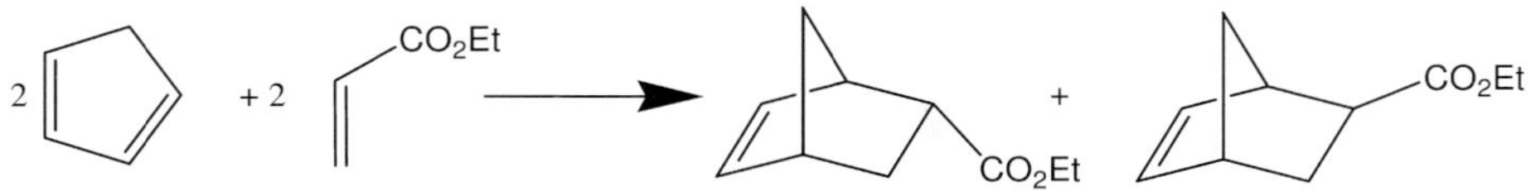

Scheme 13.7 Diels–Alder reaction.

13.3.6 Rearrangements

The focus of studies was to investigate whether rearrangements needing acid catalysts under ambient conditions would proceed without catalysts in regions of high ionic product in near-critical water. This was found for the pinacol–pinacolone rearrangement (Scheme 13.8a) and the Beckmann rearrangement (Scheme 13.8b).

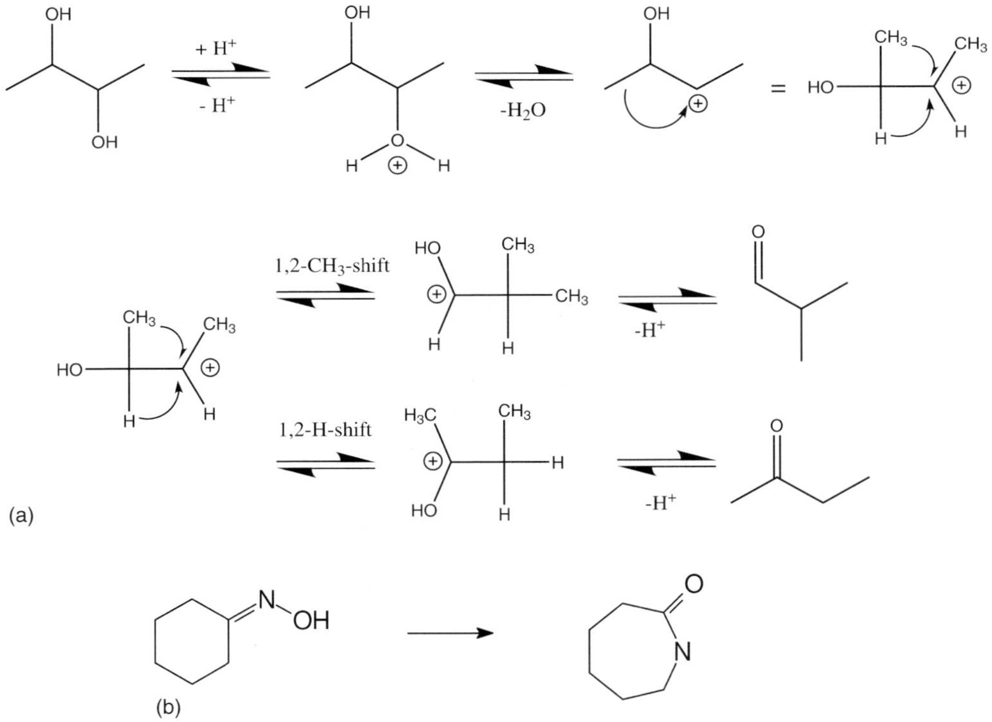

Scheme 13.8 (a) Pinacol–pinacolone rearrangement with 2,3-butanediol; (b) Beckmann rearrangement with ε-caprolactam [1].

In both cases high selectivities are reached. Other rearrangements studied include the formation of methylcyclopentene from cyclohexanol or cyclohexene in the presence of $SnCl_2$ and the carvone–carvacro, Claisen, and Rupe rearrangements [1–5] and recently the benzyl rearrangement [17].

13.3.7
Partial Oxidations

The high solubility of organic compounds and gases in SCW and also its inertness should make it an ideal solvent for partial oxidations. A selective conversion of alkanes to alcohols or aldehydes would be very interesting with a view to technical applications. An example is the partial oxidation of cyclohexane with oxygen (Scheme 13.9). The products identified are cyclohexene, cyclohexanol, cyclohexanone, carboxylic acids, CO, and CO_2 [4].

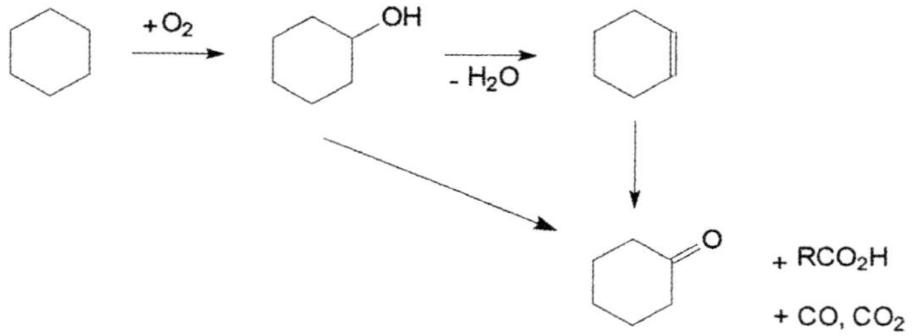

Scheme 13.9 Partial oxidation of cyclohexane in SCW [4].

With increasing temperature from 300 to 400 °C, the selectivity with respect to useful products (-ene, -ol, -one) rises to 30% while the conversion of oxygen is complete. The direct conversion of methane to methanol is of special interest. The highest selectivity for methanol was 39% at 420 °C, 35 MPa and a reaction time of 23 s. The associated consumption of oxygen was 13% [4].

Unfortunately, in all these investigations of partial oxidations of alkenes, the yields have so far been too low and the stability of the heterogeneous catalysts, if they were used, is still unsatisfactory (e.g. [18]). Much higher yields are found for the oxidation of alkylarenes to aldehydes, ketones, and organic acids in the presence of, for example, $MnBr_2$ as the catalyst. In the case of *p*-xylene, a yield higher than 80% (300 °C, $MnBr_2$, 5–15 min) was achieved in a batch reactor [2, 19].

13.3.8
Reductions

In the subcritical range, the reductions of nitroarenes to the corresponding amines and quinolines and also of azides to amines with zinc have been reported. HCO_2Na with Pt/C as catalyst reduced alkynes to alkanes. In SCW, the hydrogenation of naphthalenes was carried out by using CO as precursor of active hydrogen formed via the water-gas shift reaction (Equation 13.1)[1–5].

$$CO + H_2O \rightarrow CO_2 + H_2 \qquad (13.1)$$

13.3.9 Organometallic Reactions

SCW may be used as a "non-polar" solvent for organometallic-catalyzed reactions. The advantage of carrying out these types of reactions in SCW may be the easy separation after cooling. Reactions studied are Heck coupling, cyclotrimerization of alkynes (Scheme 13.10), Glaser coupling, and hydroformylation [1].

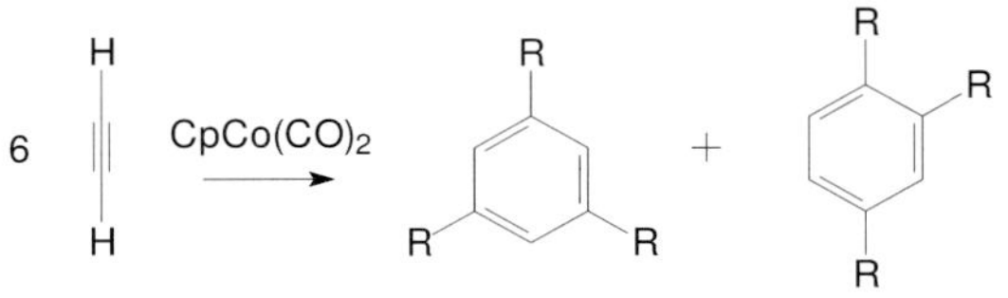

Scheme 13.10 Cyclotrimerization of alkynes.

13.4 Biomass Conversion

Today, biomass is the only renewable carbon source and is therefore an interesting feedstock for the production of platform chemicals and fuels. A disadvantage of biomass is that the land available for production is limited. To achieve the highest possible benefit of the available amount of biomass, the application of the biorefinery concept [20] is useful. Here, different products are produced in order to achieve a complete conversion and maximum profit out of the limited resources. Such a biorefinery may produce platform chemicals for the chemical industry combined with fuel or energy. The variety of the solvent properties of NSCW is assumed to be an advantageous tool to achieve the high selectivities.

13.4.1 Platform Chemicals

The use of biomass as feedstock has the disadvantage that biomass is not a pure compound. The main components are cellulose, hemicellulose, lignin, proteins, and oil/fats in varying proportions. These different ingredients of biomass may influence reactions with each other and, in addition, variations of compositions make it difficult to achieve high yields of single platform chemicals.

13.4.1.1 Carbohydrates

As a consequence of the varying composition of biomass, studies to produce platform chemicals from biomass usually start with glucose or fructose. Both are low-cost products from biomass.

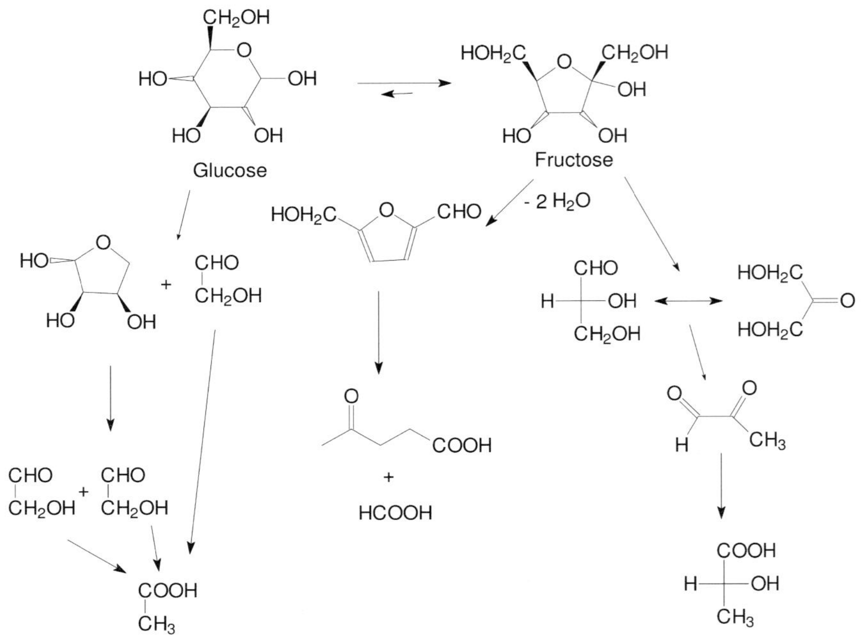

Scheme 13.11 Simplified reaction scheme for the synthesis of platform chemicals from glucose or fructose.

Glucose isomerizes to fructose in water (Scheme 13.11). The elimination of water from fructose leads to the formation of 5-hydroxymethylfurfural (5-HMF). From the technical point of view, the formation of 5-HMF from fructose is of special interest, because it can be oxidized to furandicarboxylic acid, which can be used for polyester production similarly to terephthalic acid. In the temperature range 250–350 °C and 25 MPa, 5-HMF was produced from fructose with a selectivity of roughly 30% at 50% conversion [4] (see also [21]). This reaction is also part of the reaction network of biomass degradation (see below).

An HMF yield of ~65% (mol/mol) from fructose can be achieved in the presence of H_3PO_4 at 240 °C. For erythrose up to 50% ($g\,g^{-1}$) yield at 400 °C and for glycol aldehyde up to 64% ($g\,g^{-1}$) at 450 °C from glucose are found. Usually lactic acid is one of many different compounds formed during the conversion of glucose or fructose. The additions of small quantities of $ZnSO_4$ increases the yield substantially to 42 and 48% ($g\,g^{-1}$) for glucose and fructose, respectively [22].

The partial oxidation of cellulose was studied with a view to the production of acetic acid. Here the yield was 16% (carbon base).

Usually, carbohydrates and often also the corresponding degradation products are "over-functionalized", which means that there are too many OH-groups with nearly the same reactivity in a single molecule. Here the idea of combining the biochemical production of carboxylic acids with a selective dehydration catalyzed by Zn salts is very promising. In the same way glycerol, a by-product of biodiesel production,

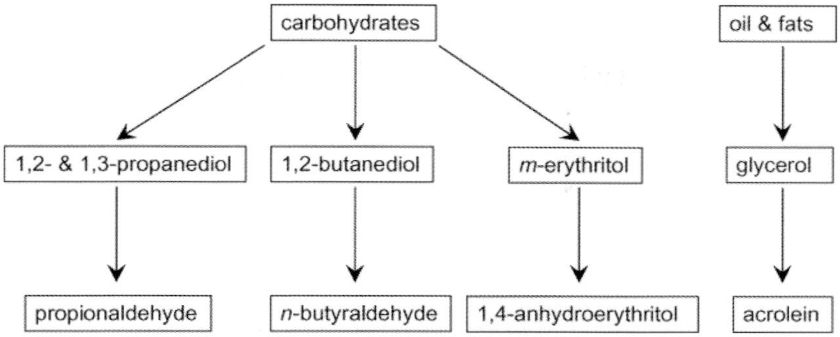

Figure 13.3 Products of the selective dehydration of renewable raw materials [15].

can be converted to acrolein, which is an interesting platform chemical (see Section 13.3.4 [15], Scheme 13.4).

13.4.1.2 **Lignin**

The idea of producing phenols from lignin is not new [23], but has become increasingly attractive in recent years, especially in view of carrying it out in NSCW. It is desirable to use not only the carbohydrate part of biomass but also the lignin component for the transformation into basic chemicals. Furthermore, the increased knowledge of the properties of NSCW could be used advantageously in the area of lignin transformation. The hope was that the aggressiveness of NSCW concerning polar bonds and the high solvent power of SCW (here usually at around 400 °C) would reduce the “re-polymerization”, which is the main challenge in this process. The first aspect should lead to the fast splitting of lignin and the second to solvation of the degradation products suppressing “re-polymerization”. On the other hand, yields found in different studies have been below 40% (based on carbon) so far. In some studies, it is assumed that the polymers are formed via a Friedel–Crafts reaction and that it is useful to carry out the reaction in water–phenol or water–cresol mixtures. Here, the phenol or cresol works as a “catching agent,” forming compounds with two aromatic rings (Scheme 13.12). In this way, cross-linking of phenols with residual lignin is avoided [24]. However, further studies are necessary.

13.4.1.3 **Proteins**

The hydrolysis of proteins or protein-containing biomass [25, 26] to obtain amino acids and the consecutive decarboxylation of amino acids to amines has been investigated with a view to the production of platform chemicals from biomass. The decarboxylation competes with the hydrolysis to carboxylic acid and ammonia. In spite of this, the reaction of glycine, for example, leads to a yield of around 84% of ethylamine (Scheme 13.13) [27].

In addition, a novel protein-based, biodegradable plastic has been produced by the conversion of serum albumin of bovine blood in subcritical water [28].

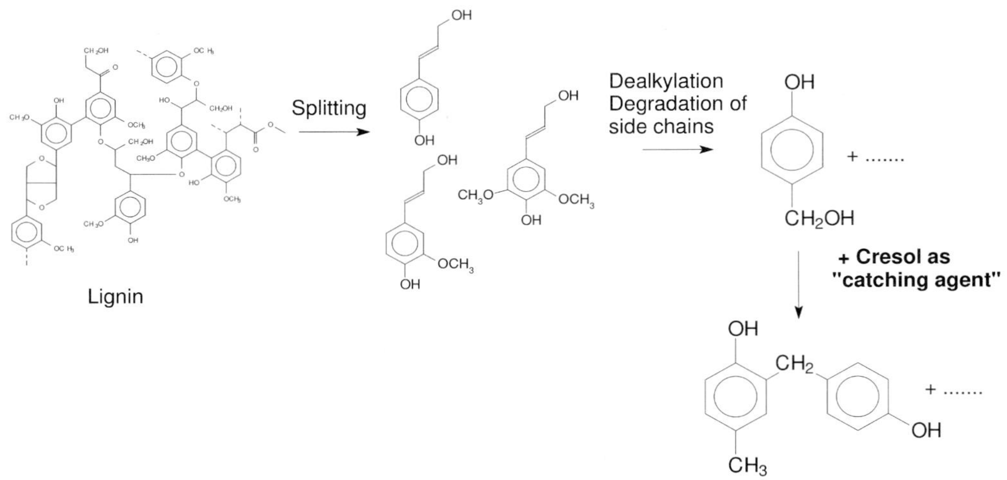

Scheme 13.12 Degradation of lignin in NSCW with a "catching agent".

13.4.2
Oil, Gases, Coke

Most of the biomass which is not used at present is the so-called wet biomass with a water content of >50% (g g^{-1}). The natural water content of biomass is 80–95% (g g^{-1}). The advantage of hydrothermal processes in NSCW to produce oil, gases, and coke is that such wet biomass can be used without expensive drying. The special properties of the reaction medium permit higher selectivities and significantly lower temperatures compared with the corresponding "dry" processes. All the compounds shown in Scheme 13.11 are intermediates in the pathway from biomass to gases.

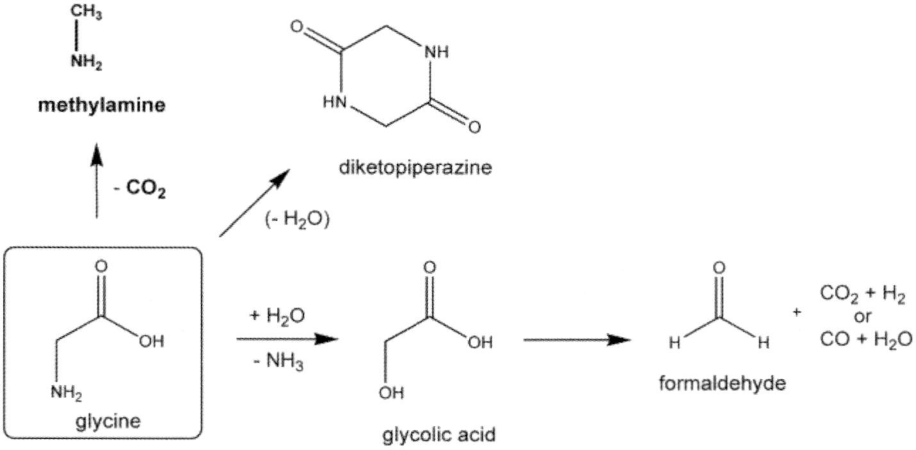

Scheme 13.13 Reaction network of hydrothermal glycine decomposition. Main products are indicated in bold [27]

If the starting material is biomass, usually high yields for single components are not possible. Instead, a mixture of many different compounds is found as oil. The word “oil” is commonly used but may be misleading: Usually these mixtures start to flow above 80 °C. This liquefaction, also often called “hydrothermal upgrading,” which is the name of a process developed originally by Shell, is usually carried out in the temperature range 300–400 °C [29]. The advantages of this process are that the oil produced has

- a relatively high heating value, higher than that of pyrolysis oil from “dry” processes;
- a lower reaction temperature than the pyrolysis of dry biomass;
- with no salts included, which is of special interest for the conversion of biomass with high salt content such as algae.

For some applications, the high viscosity of the oil produced is a disadvantage. The most important reason standing against technical applications is the large amount of wastewater produced, which is contaminated with many different organic compounds. To reduce the amount of organics in the water phase and to find applications for it are a current subject of research (see, e.g., [30]).

Equilibrium calculations predict a complete conversion of biomass to gases also at such a low temperature. Here methane and at higher temperature hydrogen are the preferred burnable gases [31]. In both cases, the corresponding amount of CO_2 is also formed.

The conversion of biomass to methane at 200–400 °C needs hydrogenation catalysts such as Ni or Pt [32, 33]. This process has also been successfully demonstrated with chemical wastes. The relative high reactivity of biomass in NSCW permits methane formation, which is not possible in the thermochemical dry process [32, 33].

The formation of hydrogen at 600–700 °C is supported by the presence of alkali metal salts, increasing the rate of the water gas shift reaction [2]. These salts are natural ingredients of biomass and usually there is no need to add them. It was observed that not only was the gas composition changed in the presence of alkali metal salts, but also the concentration of intermediates was influenced. Typical intermediates were found in studies with model compounds such as glucose and with biomass. As key compounds, they indicate changes in the reaction network, for example caused by salts. Figure 13.4 shows a simplified reaction network based on key compounds. The presence of salts decreases the formation of 5-HMF and suppresses the polymerization of small intermediates. The latter is assumed to be the consequence of the formation of “active hydrogen” formed via the water gas shift reaction, which is catalyzed by the salts. Probably this “active hydrogen” reacts with reactive compounds, which then lose their ability to polymerize. Formic acid and formates are assumed to be intermediates of the water gas shift reaction and are known to be effective hydrogenation agents. This effect of the water gas shift reaction is enhanced in reactors with strong back-mixing. Here, this late consecutive reaction is also able to proceed with early intermediates. In addition, other components of biomass influence the biomass gasification. The presence of proteins decreases the

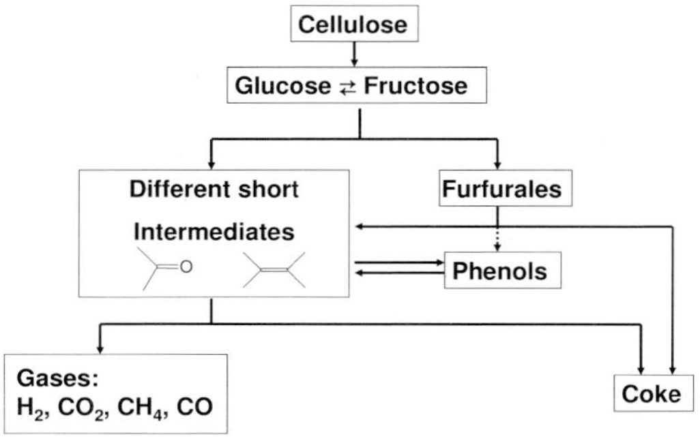

Figure 13.4 Simplified reaction scheme for biomass gasification in NSCW [2].

gas yield, especially at short reaction times. It was shown that this is probably a consequence of the heterocyclic compounds formed from amines, which are degradation products of proteins and carbohydrates. These nitrogen-containing compounds are able to form relatively stable free radicals that act as free radical scavengers [34]. The gas formation reactions are mainly free radical reactions inhibited by these compounds until they are consumed [2].

For this high-temperature hydrogen formation, charcoal has been used as a catalyst to increase the gas yield [31, 32]. The conversion of real biomass to hydrogen was successfully demonstrated on a larger scale with a throughput of 100 kg h^{-1} biomass [35].

In the studies mentioned for the production of hydrogen, high temperatures were chosen for thermodynamic reasons. At low temperatures methane and at higher temperatures hydrogen dominate as burnable gases in the equilibrium. On the other hand, the biomass gasification is a reaction that is highly kinetically driven. Therefore, some groups have succeeded in producing high yields of hydrogen at much lower temperatures in the presence of a noble metal catalyst such as Pd or Pt [36]. This process is often called "aqueous (phase) reforming" and so far it has been applied only with model compounds, because the heterogeneous catalysts are poisoned by the use of real biomass. The production of high-quality coke in the form of carbon microspheres from biomass of model compounds is usually carried out at 200 °C or higher (400, 500 °C) [37, 38].

13.5 Supercritical Water Oxidation (SCWO)

SCWO was developed to destroy chemical wastes associated with the production of heat. Two different processes have been and are still being investigated:

1. Catalytic SCWO (CSCWO) usually under near- or subcritical conditions in the presence of a catalyst. The challenge here is to achieve high stability of the catalyst [19, 39].
2. SCWO without a catalyst at around 600 °C [40–43].

In both cases, nearly complete conversion, usually higher than 99%, is achieved. Many model compounds and real wastes have been investigated. In various studies, the SCWO of simple compounds has been described by kinetic models of elementary reactions originating from gas-phase kinetics or by lumped kinetic models consisting also of many differential equations [3, 44]. The experimental results are usually sufficiently described by the first models, but not the pressure dependence, which is more complex [45]. The relatively high cost of SCWO compared with other incineration processes prevents its widespread commercialization.

13.6 Inorganic Compounds in NSCW

The special properties of NSCW and the change in the solution behavior of inorganic materials near the critical point create both opportunities and challenges. The high solubility below and the low solubility above the critical point make water at high temperature and pressure an excellent medium for the formation of inorganic particles, but also cause problems such as corrosion in preheaters and unwanted salt deposition under supercritical conditions.

13.6.1 Particle Formation

The special properties of NSCW permit the controlled precipitation of different inorganic compounds (e.g. hydroxyapatite, CdS, PbS, ZnS, TiO_2, ZnO, $LiCoO_2$, α-$NiFe_2O_4$, noble metal particles such as Au, Ag, Pt, etc.) and composite materials, mainly to produce nanoparticles (e.g. [8, 9]).

13.6.2 Corrosion

Especially in studies of SCWO, corrosion was and is a challenge, especially in the presence of Cl^- ions and oxidizing agents. Not only Cl^- but also other heteroatoms such as sulfur and phosphorus are converted to the corresponding acids, here H_2SO_4 and H_3PO_4, which are very corrosive. Bonded nitrogen leads mainly to the formation of N_2 and smaller amounts of N_2O. In many cases, if the pressure is not too high, but not always, the strongest corrosion is observed in the preheater or at the beginning of the reactor, where the temperature is near but below the critical point. Here the solubility of salts, also of those that usually form protective layers on metal surfaces, is fairly high. Without the protective layers, metals are usually oxidized very rapidly in this aggressive and oxygen-containing environment. This effect is enhanced because of the relatively high dissociation of acids and bases under these conditions. With increased pressure and therefore higher density, the region of high corrosion rates shifts to higher temperatures, corresponding to the dependence of the water properties on density. The use of a base for neutralization may lead to an increased salt problem (see below).

Intensive corrosion studies help in finding the best material for a specific feedstock. It can be concluded that for every feedstock a suitable material, but not a single material for all types of feedstock, can be found [46–52].

In addition, heterogeneous catalysts may show severe corrosion, leading to very short lifetimes. Intensive studies of catalytic hydrothermal gasification to produce methane show that only ruthenium is both an active and stable catalyst material. Stable support materials for the catalyst include monoclinic zirconia, rutile, titania, and carbon [33].

Studies of the stability of potential oxidation catalysts, such as Cu, Ag, Ni, Pd, Ru, and Co oxides, show that all materials except Ni were stable in pure NSCW (300–500 °C, 25–50 MPa). In the presence of oxygen, all materials were more or less unstable [18]. Also in CSCWO, the choice of a suitable catalyst is a challenge [19, 39].

13.6.3 Unwanted Salt Precipitation and Salt Plugging

The solubility of salts above the critical point and especially at relatively low pressures is very low. Therefore, under these conditions salts precipitate, leading to plugging of the reactor. One way to overcome both challenges of SCWO, corrosion and salt plugging, is to use the so-called transpiring wall reactor [53, 54], which is a double-wall concept. The outer wall is the autoclave. Inside, contact of the reactants, injected into the center of the reactor, with the inner wall of the reactor is avoided by a steady water flow rising through this porous inner wall. Therefore, no contact with the wall means no corrosion and no salt deposition.

13.6.4 Poisoning of Heterogeneous Catalysts

Poisoning by sulfur is a challenge in NSCW as in all other media [55]. However, the regeneration of sulfur-poisoned Ru/TiO_2 by subcritical water has been reported [56]. In investigations of the catalytic gasification of municipal and animal waste, poisoning by inorganic compounds of Ca, Mg, and P was reported [55].

13.7 Conclusion

The special properties of NSCW and their variability open up a wide range of opportunities for chemical reactions. Many reactions are being studied, including degradation reactions such as biomass gasification and total oxidation of aqueous wastes. Obviously, the properties of NSCW influence chemical reactions. On the other hand, the relatively high investment costs of the high-pressure equipment will hinder the widespread technical application of NSCW. Of special interest are reactions which occur only in this medium. Examples are:

1. The biomass conversion processes occurring at much lower temperatures than in the corresponding dry processes. Thermochemical methane formation is possible only in NSCW.
2. The Cannizzaro reaction and others which proceed in NSCW without the addition of bases.
3. The opportunity to influence the chemical reaction by salts opens up new opportunities to control reactions.

In some cases, processes in NSCW show clear advantages, such as the possibility of converting wet biomass without drying or of avoiding separation processes after reaction, because the product is relatively less soluble in ambient water.

In some cases, the fact that high solubility occurs at high pressures and temperatures and not under ambient conditions opens up the opportunity of an easy separation of the products, which is interesting because avoiding separation processes saves a lot of money.

13.8 Future Trends

As a result of the higher prices and the restricted supply of fossil fuels, biomass as a resource is of increasing interest. The possibility of producing gases, oil, and coke from wet biomass, without drying and at lower temperatures than in the "dry" processes, is therefore very interesting. In addition, the new processes to form platform chemicals in NSCW are very promising: The special properties of the reaction medium potentially combined with the remarkable effect of salts open up the opportunity for the production of platform chemicals originating from biomass. The amount of renewable primary products is limited by the farmland available which is not used for food or feed production. Therefore, it is necessary to maximize the yield of valuable products from biomass. Therefore, the development of hydrothermal biorefineries combining, for example, food production with reactions in NSCW to produce platform chemicals, fuel, or energy, seems to make sense.

References

1 Kruse, A. and Dinjus, E. (2007) *Journal of Supercritical Fluids*, **39** (3), 362–380.

2 Kruse, A. and Dinjus, E. (2007) *Journal of Supercritical Fluids*, **41** (3), 361–379.

3 Watanabe, M., Sato, T., Inomata, H., Lee Smith, R. Jr., Arai, K., Kruse, A. and Dinjus, E. (2004) *Chemical Reviews*, **104** (12), 5803–5821.

4 Bröll, D., Kaul, C., Krämer, A., Krammer, D., Richter, T., Jung, M., Vogel, H. and Zehner, P. (1999) *Angewandte Chemie International Edition*, **38** (20), 2998–3014.

5 Savage, P.E. (1999) *Chemical Reviews*, **99**, 603–621.

6 Hunter, S.E. and Savage, P.E. (2004) *Chemical Engineering Science*, **59** (22–23), 4903–4909.

7 Meyer, C. A., McClintock, R. B., Silvestri, G. J. and Spencer, R. C. Jr (1992) Steam Tables – Thermodynamic and Transport

Properties of Steam, computer program, ASME, Version 6.
8 Byrappa, K. and Adschiri, T. (2007) *Progress in Crystal Growth and Characterization of Materials*, **53** (2), 117–166.
9 Mousavand, T., Takami, S., Umetsu, M., Ohara, S. and Aschiri, T. (2006) *Journal of Materials Science*, **41** (5), 1445–1448.
10 Bühler, W., Dinjus, E., Ederer, H.J., Kruse, A. and Mas, C. (2002) *Journal of Supercritical Fluids*, **22** (1), 37–53.
11 An, J., Bagnell, I., Cabelwski, T., Strauss, C.R. and Trainor, R.W. (1997) *Journal of Organic Chemistry*, **62**, 2505–2511.
12 Tomita, K., Koda, S. and Oshima, Y. (2002) *Industrial and Engineering Chemistry Research*, **41**, 3341–3344.
13 Kramer, A., Mittelstadt, S. and Vogel, H. (1999) *Chemical Engineering and Technology*, **22** (6), 494–500.
14 Klein, M.T., Torry, L.A., Wu, B.C. and Townsend, S.H. (1990) *Journal of Supercritical Fluids*, **3**, 222–227.
15 Lehr, V., Sarlea, M., Ott, L. and Vogel, H. (2007) *Catalysis Today*, **121** (1–2), 121–129.
16 Ott, L., Bicker, M. and Vogel, H. (2006) *Green Chemistry*, **8** (2), 214–220.
17 Comisar, C.M. and Savage, P.E. (2007) *Industrial and Engineering Chemistry Research*, **46** (6), 1690–1695.
18 Kaul, C., Vogel, H. and Exner, H.E. (1999) *Materialwissenschaft und Werkstofftechnik*, **30** (6), 326–331.
19 Savage, P., Dunn, J. and Yu, J. (2006) *Combustion Science and Technology*, **178** (1–3), 443–465.
20 Kamm, B., Gruber, P. and Kamm, M. (2006) *Industrial Processes and Products: Status Quo and Future Directions*, Wiley-VCH Verlag GmbH, Weinheim.
21 Aida, T.M., Sato, Y., Watanabe, M., Tajima, K., Nonaka, T., Hattori, H. and Arai, K. (2007) *Journal of Supercritical Fluids*, **40** (3), 381–388.
22 Bicker, M., Endres, S., Ott, L. and Vogel, H. (2005) *Journal of Molecular Catalysis A: Chemical*, **239** (1–2), 151–157.
23 von Wacek, A. (1938) *Holz als Roh- und Werkstoff*, **1** (14), 543–548.
24 Matsumura, Y., Sasaki, M., Okuda, K., Takami, S., Ohara, S., Umetsu, M. and Aschiri, T. (2006) *Combustion Science and Technology*, **178** (1), 509–536.
25 Lamoolphak, W., Goto, M., Sasaki, M., Suphantharika, M., Muangnapoh, C., Prummuag, C. and Shotipuk, A. (2006) *Journal of Hazardous Materials*, **137** (3), 1643–1648.
26 Yoshida, H., Takahashi, Y. and Terashima, M. (2004) *Journal of Chemical Engineering of Japan*, **36** (4), 441–448.
27 Klingler, D., Berg, J. and Vogel, H. (2007) *Journal of Supercritical Fluids*, **43** (1), 112–119.
28 Abdelmoez, W. and Yoshida, H. (2006) *AIChE Journal*, **52** (7), 2607–2617.
29 Goudriaan, F. and Peferoen, D.G.R. (1990) *Chemical Engineering Science*, **45** (8), 2729–2734.
30 Watanabe, M., Bayer, F. and Kruse, A. (2006) *Carbohydrate Research*, **341** (18), 2891–2900.
31 Antal, M.J., Allen, S.G., Schulman, D., Xu, X.D. and Divilio, R.J. (2000) *Industrial and Engineering Chemistry Research*, **39** (11), 4040–4053.
32 Matsumura, Y., Minowa, T., Potic, B., Kersten, S.R.A., Prins, W., van Swaaij, W.P.M., van de Beld, B., Elliott, D.C., Neuenschwander, G.G., Kruse, A. and Antal, M.J. (2005) *Biomass and Bioenergy*, **29** (4), 269–292.
33 Elliott, D.C., Hart, T.R. and Neuenschwander, G.G. (2006) *Industrial and Engineering Chemistry Research*, **45** (11), 3776–3781.
34 Kruse, A., Maniam, P. and Spieler, F. (2007) *Industrial and Engineering Chemistry Research*, **46** (1), 87–96.
35 Boukis, N., Diem, V., Galla, U. and Dinjus, E. (2006) *Combustion Science and Technology*, **178**, 467–485.
36 Huber, G.W., Shabaker, J.W., Evans, S.T. and Dumesic, J.A. (2006) *Applied Catalysis B: Environmental*, **62** (3–4), 226–235.
37 Mi, Y., Hu, W., Dan, Y. and Liu, Y., *Materials Letters*, **62** (8–9), 1194–1196.

38 Titirici, M.M., Thomas, A., Yu, S.H., Muller, J.O. and Antonietti, M. (2007) *Chemistry of Materials*, **19** (17), 4205–4212.

39 Ding, Z.Y., Frisch, M.A., Li, L. and Gloyna, E.F. (1996) *Industrial and Engineering Chemistry Research*, **35** (10), 3257–3279.

40 Veriansyah, B. and Kim, J.D. (2007) *Journal of Environmental Sciences*, **19** (5), 513–522.

41 Bermejo, M.D. and Cocero, M.J. (2006) *AIChE Journal*, **52** (11), 3933–3951.

42 Oshima, Y., Hayashi, R. and Yamamoto, K. (2006) *Environmental Sciences: an International Journal of Environmental Physiology and Toxicology*, **13** (4), 213–218.

43 Abeln, J., Kluth, M., Petrich, G. and Schmieder, H. (2001) *High Pressure Research*, **20** (1–6), 537–545.

44 Belkacemi, K., Larachi, F. and Sayari, A. (2000) *Journal of Catalysis*, **193** (2), 224–237.

45 Henrikson, J.T., Grice, C.R. and Savage, P.E. (2006) *Journal of Physical Chemistry A*, **110** (10), 3627–3632.

46 Boukis, N., Claussen, N., Ebert, K., Janssen, R. and Schacht, M. (1997) *Journal of the European Ceramic Society*, **17** (1), 71–76.

47 Schacht, M., Boukis, N., Dinjus, E., Ebert, K., Janssen, R., Meschke, F. and Claussen, N. (1998) *Journal of the European Ceramic Society*, **18** (16), 2373–2376.

48 Friedrich, C., Kritzer, P., Boukis, N., Franz, G. and Dinjus, E. (1999) *Journal of Materials Science*, **34** (13), 3137–3141.

49 Kritzer, P., Boukis, N. and Dinjus, E. (2000) *Corrosion*, **56**, 1093–1104.

50 Schacht, M., Boukis, N. and Dinjus, E. (2000) *Journal of Materials Science*, **35**, 6251–6258.

51 Lee, H.C., Son, S.H., Hwang, K.Y. and Lee, C.H. (2006) *Industrial and Engineering Chemistry Research*, **45** (10), 3412–3419.

52 Teysseyre, S. and Was, G.S. (2006) *Corrosion*, **62** (12), 1100–1116.

53 Bermejo, M.D. and Cocero, M.J. (2006) *Journal Of Hazardous Materials*, **137** (2), 965–971.

54 Bermejo, M.D., Fdez-Polanco, F. and Cocero, M.J. (2006) *Journal of Supercritical Fluids*, **39** (1), 70–79.

55 Ro, K.S., Cantrell, K., Elliott, D. and Hunt, P.G. (2007) *Industrial and Engineering Chemistry Research*, **46** (26), 8839–8845.

56 Osada, M., Hiyoshi, N., Sato, O., Arai, K. and Shirai, M., *Energy and Fuels*, **22** (2), 845–849.

Index

a
abstractions
– hydrogen 413
accelerated diffusion 262
acetal groups 451
acetate
– α-naphthyl 408
– vinyl 338
acetonitrile 273
acetylation, enantioselective 290
acid-catalyzed processes 219
acid-catalyzed reactions 452
acid-sensitive functional groups 222
acids
– alkylcarbonic 448, 449
– atropic 138
– benzoic 162
– carbonic 221, 449
– carboxylic 460
– decanoic 379
– dimeric hydroxy 351
– fatty 23
– *in situ* generation 120
– levulinic 195
– natural carbonic, $scCO_2$ 9
– peroxocarbonic 147
– poly(glycolic acid) 351
– poly(L-lactic acid) 351
– reactor cleaning 58
– solid 149
– tiglic 138
active hydrogen 469
acylation 233, 435
acylhydroquinones 405
addition
– cycloaddition 221
– photo- 225
– propargylic alcohols and amines 205
adjustable solvating power 8
adsorption isotherm 388
aerogels 381
agglomeration, particles 311
aggregated nanoparticles 379
AISI, material number 36
alcohols
– benzyl 162
– enantioselective acylation 233
– propargylic 205
alcoholysis, lipase-catalyzed 284
aldehydes
– cinnam- 195
– sacrificial 206
aliphatic polycarbonates 342
alkaline phosphatase 230
alkanes
– free radical chlorination 411
– iso- 262
– supercritical 371, 374
alkene metathesis, Grubbs catalyst 148
alkoxides 344, 381
alkyl aromatics 410
alkylation 255, 266
– Friedel–Crafts 422
alkylcarbonic acid 448
– carbonic acid 449
allylic amines 123
aluminum alkoxides 344
amides, hydrolysis 461
amidine-functionalized phosphine ligand 128
amination, aromatic 211
amines 205
– allylic 123
– carbamate protection 212
– secondary 124
– tertiary 124

Handbook of Green Chemistry, Volume 4: Supercritical Solvents. Edited by Walter Leitner and Philip G. Jessop

ISBN: 978-3-527-32590-0

ammonia, synthesis 20
ammonium salt-catalyzed reactions 436
amorphous films 391
amphipathic polymers 318
analytes, electrochemical 420
anchoring strength 328
anhydride, maleic 193, 194
anhydrous media 283
aniline 452
anionic polymerization, coordinative 348
anthracene 227
anthraquinones, functionalized 258
anti-inflammatory drug, non-steroidal 138
anti:syn ratio 404
AOT, *see bis*(2-ethylhexyl)sulfosuccinate
applications
– electrochemical reactions 429
– emulsions 174
– industrial 16
– SCFs 9
aqueous (phase) reforming 470
aqueous–SCF biphasic systems 159
aromatic amination 211
aromatics, alkyl 410
Arrhenius plot 445, 448
artificial enzymes 338
asymmetric binary mixtures 88
asymmetric hydroformylation, styrene 203
asymmetric hydrogenation 198
– atropic and tiglic acid 138
atactic PMMA 326
atropic acid 138
attenuated total reflection mode (ATR) 56
Aufbau reaction 14
autoclave 18, 53
automated supercritical flow reactors 69
autotuning procedures 48
aza-Diels–Alder reactions 223

b
Bacillus circulans 231
Bacillus megaterium 289
back-pressure regulator 69, 72, 166, 296
backbone, π-conjugated 345
Baeyer–Villiger oxidation, ketones 208
base-catalyzed processes 218
batch processes 41
– dosage 60
Baylis–Hillman reaction 218
Beckmann rearrangement 463
benzaldehyde 284, 459
benzene, diffusion coefficients 115
benzoic acid 162
benzyl alcohol 162
bifunctional monomers 334
binary mixtures, asymmetric 88
binary phase diagrams 86
biocatalysis 150
biodiesel 287
biomass 465, 470
biotransformations 229
biphasic catalysis 23
biphasic reactions, emulsions 172
biphasic systems
– aqueous–SCF 159
– classification 101
– IL-scCO_2 294
– liquid–SCF 101, 159
– phase behavior 102
– polymer–SCF 167
bis(2-ethylhexyl)sulfosuccinate (AOT) 374, 375
bond migration, double 210
bonding, hydrogen 117, 439
boron trifluoride 347
branched pentablock copolymers 350
bromination, free radical 410
bromine, molecular 411
Brookhart catalyst precursor 341
Brownian movement, thermal 312
bubbling 402
bulk density 3
bursting discs 53
butene, supercritical 154

c
caffeine 162
cage effects 412
caged radical pairs 407
Cagniard de LaTour 10, 11, 16
Candida antarctica 283, 290
Cannizzaro reaction 459, 473
canon de fusil 11
capping agents 370
capping ligands, organic 373
caprolactone 350
– ε- 348
carbamates 212, 219, 285
carbohydrates, biomass conversion 465
carbon molecular sieves (CMS) 391
carbon nanocages 385
carbon nanomaterials 383
carbon nanotubes (CNT) 383, 390
– multiwalled 383, 384
carbonic acid 221
– alkylcarbonic acid 449
– natural, scCO_2 9
carbonyl photochemistry 405

carbonylation 139
– free-radical 229
– Pauson–Khand 142
carboxylation 234
carboxylic acid 460
catalysis
– bio- 150
– biphasic 23
– chemisorption of gases in liquids 120
– electro- 149
– enzymatic 281, 294
– expanded liquid phases 101
– heterogeneous catalysis 18, 243
– multiphase 101
– phase transfer 435
catalysts
– acids 219
– ammonium salt 436
– bases 218
– Brookhart precursor 341
– catalyst-rich surface phase 437
– cobalt 142
– colloids 163
– Grubbs 123
– heterogeneous 472
– heterogenized catalyst systems 141
– immobilization 263
– lipases 282, 284, 286
– lipophilic phase transfer 437
– mesoporous 248
– metallocene 340
– metals 214, 342
– noble metal 258
– Novozym 290
– palladium 127, 208
– Phillips 340
– pore effectiveness factor 250
– porphyrin-type 343
– product/catalyst separation 154
– rhodium 204, 263
– ROMP 345
– ruthenium-based Grubbs-type 346
– solid acids 149
– stabilization 155
– switchable solvents 124
– Wilkinson's 170
– zeolite 262
– Ziegler–Natta 341
– zinc complexes 343
cationic polymerizations 346
cell
– electrochemical 421
– variable-volume view 42
ceria nanocrystals 379
chain hopping 343
chain polymerizations, ionic 346
chain reactions, radical 410
chemical fixation, CO_2 19
chemical functionality of solutes 94
chemical reactions 16
chemical strains 34
chemical synthesis, SCFs in 6
chemisorption, gases in liquids 120
chemistry
– electrochemical reactions 243
– green chemistry 243
– macromolecular 304
– NCW and scH_2O 457
chemoenzymatic reactor 297
chemoselectivity, enzymatic catalysis 281
chiral sensitizers 409
chlorides, mono-/poly- 413
chlorinated solvents 58
chlorination 403, 411
chromatography 14, 86
cinnamaldehyde 195
– solventless hydrogenation 136
citral 191
citronellol 232
classification, biphasic systems 101
clays, natural 391, 393
cleaning, reactors 57, 67, 69
cleavage, photo- 405
clustering, solvent–solute/solute–solute 8, 399
CMS, *see* carbon molecular sieves
CNT, *see* carbon nanotubes
CO_2
– CO_2-expanded liquids (CXLs) 103, 169
– CO_2-induced liquefaction 137
– electroreduction 428
– phase diagram 2
– physical properties 306
– quadrupole moment 83
– self-diffusivity 79, 80
– supercritical, $scCO_2$ 4
– thermal conductivity 84
– water-in-CO_2 microemulsions 374
co-reductant, sacrificial 146
coating 392
– free enzymes 292
cobalamine complex 332
cobalt-catalyzed cycloaddition 142
coexistence curve, liquid–vapor 77
coil expansion factor, hydrodynamic 310
coke 266
– biomass conversion 468
colloid-catalyzed reduction 166

competition experiments 411
complexation of polar substrates 285
complexes
– cobalamine 332
– monophos 201
– Schiff base-type metal 344
– solid-supported metal 199, 200
– zinc 343
compressed gases 5
– multiphase catalysis 101
compressed liquids, thermophysical properties 375
compressibility 247
compression 33
compressors, diaphragm or reciprocating 34
computational approaches 190
condensation 66, 462
conductivity, thermal 84
configurational entropy 308
continuous flow processes 154, 166
– dosage 61
– solventless hydroformylation 172
continuous flow reactors 44
continuous green enzyme reactor 296
continuously stirred tank reactors (CSTRs) 45
controlled pore glasses (CPGs) 249
"controlled" radical polymerization (CRP) 316, 331
controller, mass flow 61, 65
conversion
– biomass 465
– heterogeneous catalysis 252
coordinative anionic polymerization 348
copolymers 324, 350
Coriolis mass flow controller 62, 63
corresponding states, law of 1
corrosion 33, 35, 471
– risks 67
cosolvent tuning 8
coupling reactions 435
– Heck reactions 127, 211
– palladium-mediated 208
cracking 270
– stress corrosion 35
o-cresol, oxybromination 122
p-cresol 446
critical end-points 89, 92
critical line, mixtures 88
critical opalescence 247
critical point 11
critical pressure/temperature 1
critical solution temperature, upper 442
critical temperature, low 62, 63
cross-linked enzyme crystals 293
cross-linked PDMS 170
crossover pressure 97
crystallization
– solutes 130
– SSEC process 382
– tunable 130
crystallization curve 321
crystals
– cross-linked enzyme 293
– nanocrystals 293
CSTRs, *see* continuously stirred tank reactors
cutinase 291
CXLs, *see* CO_2-expanded liquids
cyclic voltammetry (CV) 419
cycloaddition 221
– cobalt-catalyzed 142
– 1,3-dipolar 224
cyclohexane
– chlorination 403
– homogeneous oxidation 144
– partial oxidation 464
cyclohexanone 162
cyclopropanations, enantioselectivity 217

d

damping effects, substituents 95
DDM, *see* diazodiphenylmethane
deactivation, enzymes 291
decanoic acid 379
decomposition, hydrothermal glycine 468
decoration, nanoparticles 388
degradation, lignin 468
dehydration 461
– selective 467
dehydrogenation 253, 260
delivery system, SCF 421
dendrimers 209, 383
denitrogenation 225
density
– density–pressure plane 78
– expanded liquid phases 109
– fluctuations 247
– local augmentation 3
– SCFs 3
– solvents 304
– water 439, 440
deposition, nanoparticles on porous supports 390
depressurization 67
depressurized gas streams 66
desorption 250
detection, electrochemical 429, 430
diaphragm compressors 34

diazodiphenylmethane (DDM) 448, 450, 451
diazonium salts 451
dielectric constant 81, 284
– NCW 440
– water 82, 83, 440
Diels–Alder reactions 221, 462
diesel, bio- 287
diethyl ether 11
– volumetric expansion 105
diffusibility, $scCO_2$ 58
diffusion, accelerated 262
diffusive caged radical pairs 407
diffusivity (diffusion coefficient)
– benzene 115
– effective 270
– expanded liquid phases 114
– *n*-pentane 249
– SCFs 7
– self-diffusivity 79
– solutes 307
difluoromethane 427
digester 10
dimeric hydroxy acids 351
dimerization
– isophorone 404
– photo- 403
dimmers, head-to-tail 404
dimethyl sulfoxide 435
1,3-dipolar cycloaddition 195
direct synthesis
– nanocrystals 369
– sol–gel 380
directional interactions 308
discharging devices, gases 34
dispersion polymerization 320
– stabilizer design 310
disproportionation 268
dissociation constant, water 441
dissolution, CO_2 168
dosage
– batch processes 60
– continuous flow processes 61
– gases 58
– liquids 63
– solids 64
dosing unit 63
double bond migration 210
double layers 427
– coating 392
drug precursor 138
drugs, anti-inflammatory 138
dye, hydrophilic 132
dynamic kinetic resolution 297

e
edible oils 23
effective diffusivity 270
effectiveness factor 250
– pore 250
EFL
– enhanced fluidity liquids 103
electrocatalysis 149
electrochemical cell 421, 429
electrochemical detection
– on-line 430
electrochemical reactions 419
– applications 429
– ferrocene 420
– solvents 423
electrochemical synthesis 429
electrodes 422
– modification 425
– standard hydrogen 422
electrolytes 421
– hydrophobic 426
electron transfer
– photo-induced 409
electron-withdrawing groups 451
electroreduction 428
– CO_2 428
"electrostriction" 325
elimination
– water 462
ELP, *see* expanded liquid phases
emulsion polymerization 157
emulsions
– applications 174
– biphasic reactions 172
– inverse 172
– ionic liquid-in-SCF 173
– microemulsions 174
– SCF-in-water 173
enantioselectivity 216
– acetylation 290
– acylation 233
– cyclopropanations 217
– enzymes 290
– hydrogenation 138
– photoisomerization 227
end-points
– upper/lower critical 89, 92
energy
– Gibbs free 306
– intermolecular potential 307
– surface 308
engineering
– process and production 32
enhanced fluidity liquids (EFL) 103

entrapment
– enzymes 292
entropy change– mixing 308
environmental benefit– SCF 189
enzymatic catalysis 281
enzymatic
– IL-scCO_2 biphacis systems 294
– near-critical/supercritical fluids 284
– SCF 283
enzymatic esterification 171
enzymatic reactions 285
– deactivation processes 291
enzymes
– alkaline phosphatase 230
– artificial 338
– carboxylation of pyrrole 234
– coating 292
– cross-linked crystals 293
– cutinase 291
– enantioselectivity 290
– entrapment 292
– esterases 285
– lipases 282, 285, 286, 291
– non-aqueous environments 281
– reactors 287, 288, 296
– stabilized 292
epoxidation
– sharpless 207
equilibrium
– vapor–liquid 105
equilibrium reactions 121
era of polymers 303
esterases 285
esterification 232
– enzymatic 171
esters 460
ethane 108, 350, 460
– octacosane–ethane system 90
ethanol
– interfacial tension 113
– volumetric expansion 105
ethers 11, 461
– diethyl 105
ethylbenzene 72
ethylcyclohexane 72
ethylene 108
ethylene glycol 450
EtOAc 105
evaporation 66
expanded liquid phases (ELP) 101, 102
– density 109
– diffusivity 114
– gas solubility 118
– hydrogenation reactions 135
– interfacial tension 112
– melting point 111
– oxidation reactions 143
– polarity 115
– viscosity 110
expansion
– volumetric 105–107
explosion prevention
– oxygen-containing systems 142
extraction
– in SCFs 452
extractive phase 295

f
fats
– valorization 285
fatty acids 23
ferrocene 420, 427
ferroelectric materials 325
films
– amorphous 391
fine chemicals
– high-value 9
Fischer–Tropsch synthesis 254, 260
fittings 45
– National Pipe Thread 41
– tube 47
fixation
– chemical 339
fixed-bed reactor 259
flammable compounds 60
flammable SCFs 189
Flory–Huggins interaction parameter 311
flow controller
– mass 61, 65
flow-through scH_2O method 380
flowmeter
– variable-area 66
fluctuations
– density 247
fluidity
– enhanced fluidity liquids (EFL) 103
– pure 81
– supercritical fluids 1
– thermophysical properties 59
– tunable 130
fluorides
– poly(vinylidene fluoride) 324
fluorinated solids 134
fluorinated solvents 325
fluorinated surfactants 375
fluoroform 93
fluoroolefin polymers 319
fluoropolymers

– side-chain 317
Fourier transform (FT) IR measurements 56
free enzymes
– coating 292
free-radical bromination 410
free-radical carbonylation 229
free-radical chlorination–alkanes 411
Friedel–Crafts acylation 420, 435, 442
Friedel–Crafts chemistry 442
fructose 466
FTIR, *see* Fourier transform (FT) IR measurements
fumehood 33, 71
functional groups
– acid-sensitive 222
functionality
– chemical 94
– molecular 399
functionalization 334
– alkanes 411
– amidine-functionalized phosphine ligand 128
– anthraquinones 258
– carbon nanotubes 390

g
gas chromatography (GC) 54
gas-expanded liquids (GXLs) 101, 103, 243
– conversions in 271
– heterogeneous catalysis 244
“gas-like” properties 54
gas shift reaction
– reverse water 257
gas solubility
– expanded liquid phases 118
gas supply 445
gases
– biomass conversion 468
– chemisorption 120
– compressed 5, 101
– depressurized streams 66
– discharging devices 34
– dosage 58
– greenhouse 214
– head-gas 4
– ideal 93, 307
gasification
– biomass 470
geminate caged radical pairs 407
general reactor design 41
in situ generation
– acids 120
geometric isomerization 403
Geotrichum candidum 289
germanium nanoparticles 372
Gibbs free energy 306
glasses– controlled pore 249
glucose 466
glycerol 462
glycine decomposition
– hydrothermal 468
glycol
– ethylene 450
gold nanocrystals 371
– molecularly tethered 372
gravimetric dosage 60
green chemistry
– 12 principles 243
green enzyme reactor
– continuous 296
greenhouse gas emission 7
groups
– acetal 451
– electron-withdrawing 451
– functional 222
Grubbs catalysts
– alkene metathesis 148
– ruthenium-based 346
GXLs
– gas-expanded liquids 243

h
Haber–Bosch process 20, 243
Hammett correlation 451
Harkin model 339
hazards
– organic solvents 109
– SCFs 5
head-gas 4
head-to-tail dimmers 404
heat transfer effects 245
heated solvents– reactor cleaning 58
heating
– reactor 109
Heck reactions 209
– coupling 211
– Pd-catalyzed 127
Henry reaction 219
Henry’s law 102, 104
heterogeneous
– catalysis 18, 243, 244
– catalyst poising 472
– conversions 252
– polymerizations 310
heterogenized catalyst systems 141
2-hexanone 162
high-pressure liquid chromatography 69
high-pressure methods and equipment 31

high-pressure NMR investigations 42
high-pressure reactors 452
– windows 400
high-pressure regime
– dosage of gases 58
high-pressure stirred tank reactor 288
high-pressure systems
– phase behavior 86
high-value fine chemicals 9
highly fluorinated solids 134
Hildebrand parameter 3, 284
history
– SCFs 9
homogeneous hydrogenation 136, 170
homogeneous mixture
– isotropic 307
homogeneous oxidation
– cyclohexane 144
homopolymers
– perfluorinated 324
HPLC, *see* high-pressure liquid chromatography
hydrations 460
hydroaminomethylation 123
– rhodium-catalyzed 204
hydrocarbons
– cracking 270
– para-critical 14
hydrodynamic coil expansion factor 310
hydrofluorocarbon supercritical solvents 426
hydroformylation 202, 254, 262
– asymmetric 204
– continuous flow 172
– 1-octene 139, 141
hydrogen
– abstractions 413
– active 469
– bonding 117, 439
– solubility 272
– standard electrode 422
hydrogenation 190, 253
– asymmetric 138, 198
– enantioselective 138
– expanded liquid phases 135
– heterogeneous catalysis 256
– homogeneous 136, 170
– selective 198
– solventless 136, 137
– supercritical reactor 71
hydrolysis 460
hydrophilic dye 132
hydrophobicity
– electrolytes 426
– surface 265
hydrothermal glycine decomposition 468
hydrothermal synthesis 377, 380
hydrothermal synthesis–supercritical 377
hydrothermal upgrading 469
hydroxy acids
– dimeric 351

i
ideal gas 93, 307
IL, *see* ionic liquids
imidazolium IL–scCO_2 mixtures 127
imidization reaction 339
imines 201
immiscibility
– unintentional 151
immobilization
– catalysts 263
– 1,3-regiospecific lipase 291
"immortal" polymerization 332
imprinting
– molecular 338
in situ generation
– acids 120
indole derivatives 94
induced phase change 112
industrial applications
– SCFs 16, 58
industrial production
– 2-propanol 151
infrared 56
inline IR measurements 56
inorganic compounds
– in NCW and scH_2O 471
inorganic–inorganic nanocomposites 393
inorganic oxide–polymer hybrids 387
inorganic salts 390
inorganic SCF 85
interfacial tension
– expanded liquid phases 112
intermolecular potential energy 307
inverse emulsions
– water-in-SCF 172
iodobenzene 453
ionic chain polymerizations 346
ionic liquids (IL) 106, 293, 294
– crystallization of solutes 130
– IL-scCO_2 biphasic systems 294
– IL-in-SCF emulsions 173
– IL–SCF biphasic systems 163
– imidazolium 115, 164
– supported ionic liquid phase 140
– water-immiscible 294
IR measurements 56, 126
irregular-shaped particles 351

isoalkanes 262
isomerization 256, 269
– geometric 403
– photo- 227
isomers
– *ortho-* 445
isophorone 193
isopropoxide
– titanium 381
isotherm
– adsorption 388
– solubility 97
isotropic homogeneous mixture 307

j
Joule–Thomson effect 67

k
ketal formation 451
ketones 202, 208
kinetic products
– removal 156
kinetic resolution
– dynamic 297

l
laboratory meters 66
laurate
– vinyl 296
laws and equations
– Gibbs free energy 306
– Henry's law 102, 104
– law of corresponding states 1
– partial molar volume 248
– Stokes–Einstein equation 408
– van der Waals equation 87
– virial equation 61
LCEP, *see* lower critical end-points
levulinic acid 195
lifetime/stability enhancement 250
ligands
– organic capping 373
– phosphine 128
light organic solvents 109
lignin 467
limonene 135, 192
lipase-catalyzed reactions 282, 286
– alcoholysis 284
lipases 282, 285
– 1,3-regiospecific 291
lipophilic phase transfer catalysts 437
liquefaction
– CO_2-induced 137
liquefaction point 120
liquid–liquid demixing line 321
liquid organic solvents 251
liquid polymers 167
– CO_2-expanded 118, 119
liquid–SCF biphasic systems 101, 159
– ionic 163
liquid–vapor coexistence curve 77
liquids
– chemisorption 120
– CO_2-expanded 103
– compressed 375
– dosage 62
– enhanced fluidity liquids (EFL) 103
– expanded liquid phases (ELP) 102
– gas-expanded 101, 243
– gas-expanded liquids 101
– ionic 104, 111
– ionic liquids 293
– non-volatile 160
– polymers 106
LOC values 60
local density augmentation 3
low critical temperatures 84, 425
low-polarity organic compounds 55
lower critical end-points (UCEP) 89, 92
Lurgi technology 21

m
macromolecular chemistry 304
macromolecules
– solubility 306
magnetic drive
– packless 52
maleic anhydride 193
manometer 47
mass and heat transfer effects 245
mass flow controller (MFC) 61
mass flow meter 62
materials
– sealing 39
– templated 158
"MegaMethanol" 21
melting point
– depression 90, 137
– expanded liquid phases 111
– naphthalene 112
"memory" phenomenon 283
meniscus 2
mesoporous catalysts 248
mesoporous silica 391
metal catalysts
– cobalt 142
– palladium 127
– rhodium catalysts 128

– supported noble metal catalysts 258
metal-catalyzed polymerizations 340
metal-catalyzed processes 214
metal complexes
– Schiff base-type 344
– solid-supported 200
metal nanocrystals 369
metal oxide nanoparticles 377
metal–polymer composites 386
metallocene catalysts 340
metathesis
– alkene 148
– ring-closing 215
metathetical polymerization
– ring-opening 345
metering pumps 63
Methyl Yellow 452
MFC, *see* mass flow controller
micelles
– reverse 375
microemulsions
– SCF-based 374
– water-in-CO_2 426
– water-in-oil 376
microstructured reactors (MSRs) 45
migration
– double bond 210
miscibility
– tunable 134
mixing entropy change 308
mixtures
– asymmetric binary 88
– critical line 88
– isotropic homogeneous 307
model
– Harkin 339
– Noyes 412
modulus
– Thiele 270
– Young's 332, 387
molar volume
– partial 248
molecular bromine 411
molecular functionality 399
molecular imprinting 338
molecular sieves
– carbon 391
molecularly tethered gold nanocrystals 372
monochlorides 413
monodisperse nanoparticles 376
monolayer coating 392
monomers
– bifunctional 334
– vinyl 335, 336
monophos complexes 201
montmorillonite 334
– sodium 331
Mucor miehei 291
multicatalytic processes 296
multicomponent systems 245
multienzymatic processes 298
multiphase
– compressed gases 101
multiwalled carbon nanotubes (MWNT) 383, 384

n
nanocages
– carbon 385
nanocomposites 385
– inorganic–inorganic 393
nanocrystals 369
– ceria 379
nanomaterials 369, 383
nanoparticles
– decoration on nanotubes 388
– deposition on porous supports 391
– metal oxide 377
– monodisperse 376
– palladium 205
– platinum 392
– recovery 377
– surface energy 378
nanoreactors 376
nanotubes
– carbon 383
– multiwalled 383, 384
nanowires 372, 373
naphthalene 93
– hydrogenation 197
– melting point 112
α-naphthyl acetate 408
National Pipe Thread fittings 41
natural carbonic acid
– $scCO_2$ 9
natural clays 391, 393
NCW, *see* near-critical water
near-critical region 2, 14, 243
near-critical water (NCW) 438
– chemistry in 457
– dielectric constant 440
– Friedel–Crafts chemistry 442
– inorganic compounds 471
– synthesis 459
neoteric solvents 284
nitrile 460
nitroarenes 122

nitrobenzene
– volumetric expansion 105
NMR investigations
– high-pressure 42
noble metal catalysts 264
– supported 258
non-steroidal anti-inflammatory drug 138
non-volatile liquids 160
nonanal 139
norbornene 345
Norrish type I photo-cleavage 405
Novozym 230
Noyes model 412
NSCW
– near-critical water, scH_2O 457
nucleation 329
nucleophilicity 348

o
O-ring seal design 41
octacosane–ethane system 90
oils
– biomass conversion 468
– edible 23
– valorization 285
– water-in-oil microemulsions 376
olefins
– fluoroolefin polymers 319
– perfluoroolefin copolymers 324
– polyolefins 340
on-line electrochemical detection 430
opalescence
– critical 247
optical windows 53
optimization
– self- 48
organic capping ligands 373
organic chemistry
– synthetic 189
organic compounds
– low-polarity 55
– unsaturated 257
organic salts
– CO_2-induced liquefaction 137
organic SCF 86
organic solvent mixtures
– phase behavior 133
organic solvents
– hazardous 109
– light 109
– liquid 251
– reactor cleaning 57
– supercritical 371
organic synthesis
– solid-phase 142
organometallic reactions 465
ortho-isomer 445
orthosilicate
– tetraethyl 387
overoxidized products 263
overpressure
– peak 32
oxidation 205, 254
– Baeyer–Villiger 208
– expanded liquid phases 143
– heterogeneous catalysis 263
– homogeneous 144
– partial 272, 464
– photo- 410
– radical 145
– scH_2O 22, 470
– Wacker 213
oxide–polymer hybrids
– inorganic 387
oxybromination
– *o*-cresol 122
oxyfuntionalization 265
oxygen-containing systems
– explosion prevention 59

p
packed-bed enzyme reactor 288
packless magnetic drive 52
palladium-catalyzed Heck reactions 127
palladium-mediated coupling reactions 208
palladium nanoparticles 205
para-critical, (i.e. near-critical) hydrocarbons 14
partial molar volume 314
partial oxidation 272, 464
particle agglomeration 311
particle formation, in NCW and scH_2O 471
partition coefficients 162, 284
Pauson–Khand carbonylations 142
Pauson–Khand reaction 214
PDMS, cross-linked 170
PE, *see* polyethylene
peak overpressure 32
PEG, *see* poly(ethylene glycol)
pentablock copolymers, branched 350
n-pentane
– diffusivity 249
perfluorinated homopolymers 324
perfluorinated surfactants 341
perfluoroolefin copolymers 324
peroxocarbonic acid 147
Perspex 326
PFRs, *see* plug flow reactors

PGSS process 14
phase behavior
– biphasic systems 102
– IL-scCO_2 biphacis systems 295
– organic solvent mixtures 133
– SCF 96
– water–SCF mixtures 161
phase change, induced 112
phase diagrams
– binary 86
– CO_2 2
– five classes 88
– pure substances 78
phase separation, tunable 131
phase transfer catalysis (PTC) 435, 437
– lipophilic catalysts 437
– separation and recycling 437
phase transfer reactions 435
phenol 443, 448
– 4-tertbutyl phenol 198
2-phenylethanol 171
Phillips catalysts 340
phosphatase, alkaline 230
phosphine ligand, amidine-functionalized 128
photoaddition 225
photochemical reactions 224, 399
photochemistry, carbonyl 405
photocleavage, Norrish Type I 405
photodimerization 403
photoinduced reactions 399
– electron transfer 409
– radical chain reactions 410
photoinitiator 316
photoisomerization 227
photooxidation 410
photosensitization 409
physical properties
– CO_2 306
– SCFs 7, 77
physical strains 34
physicochemical properties, water 457
picrate, tetrabutylammonium 437, 438
piezoelectric polymers 322
pinacol–pinacolone rearrangement 463
pinene 191
platform chemicals 465
platinum nanoparticles 392
Plexiglas 326
plug flow reactors (PFRs) 45, 51
plugging, salt 472
PMMA, *see* poly(methyl methacrylate)
poising, heterogeneous catalysts 472
polar SCFs 424
polar solids 93
polar substrates, complexation 285
polarity
– expanded liquid phases 115
– relative 117
polarity–composition diagram 115
polarity parameter 116
polyacrylates 309
polyacrylonitrile 313
polycaprolactone 348
polycarbonates, aliphatic 342
polychlorides 413
polycondensation reactions 314
polydispersity 332
polyesters, caprolactone-based 350
poly(ethylene glycol) (PEG) 297
– PEG-stabilized Pd nanoparticles 205
polyethylene (PE) 340
poly(glycolic acid) 351
poly(hydroxy ester)s 350
polyketones 344
poly(L-lactic acid) 351
polymer-based nanocomposites 386
polymer–SCF biphasic systems 170
polymerization
– cationic 346
– “controlled” radical 316, 331
– coordinative anionic 348
– dispersion 310, 320
– emulsion 157
– heterogeneous 310
– “immortal” 332
– in scCO_2 303, 315
– ionic chain 346
– metal-catalyzed 340, 342
– radical 315
– radical-initiated 126
– ring-opening metathetical 345
polymers
– amphipathic 318
– copolymers 324
– fluoro- 317
– fluoroolefin 319
– inorganic oxide–polymer hybrids 387
– liquid 106, 160, 168
– metal–polymer composites 386
– π-conjugated backbone 345
– piezoelectric 322
– polymer–polymer composites 386
– semicrystalline thermoplastic 322
– solubility 168
poly(methyl methacrylate) (PMMA) 326
polyolefins 340
polypropylene (PP) 340
polystyrene (PS) 332

polytetrafluoroethylene (PTFE) 319
poly(vinyl chloride) (PVC) 335
poly(vinylidene fluoride) 322, 335
poppet 65
pore effectiveness factor 250
porous solids 244, 246
porous supports, deposition of nanoparticles 391
porphyrin-type catalyst 343
post-reaction separation 152
potential energy, intermolecular 307
potential windows 419
powder morphology 323
precipitation
– salts 472
– tunable 131
precipitation polymerization 310
pressure
– adjustment 61
– critical 1
– crossover 97
– density–pressure plane 78
– electrochemistry 424
– reactors 65
– reduced 1
– supercritical biocatalysis 289
– supply 64
pressure effects 96
pressure regulator
– back- 69, 72, 166
pressure relief valves (PRVs) 53, 402
pressure sensors 47, 49
pressure transmitters 47
pressure vessels 5, 41
12 principles of green chemistry 243
process and production engineering 32
processes
– acid-catalyzed 219
– aqueous (phase) reforming 470
– base-catalyzed 218
– batch processes 60
– biomass conversion 465
– continuous flow 61, 154, 166
– design 24, 77
– enzyme deactivation 291
– Haber–Bosch 20, 243
– hydrothermal upgrading 469
– metal-catalyzed 214
– multi-enzymatic 298
– multicatalytic 296
– PGSS 14
– RESS 14
– sequential reaction–separation 130
– slurry-phase 260
– SSEC 382
product/catalyst separation 154
propagation constants 317
2-propanol
– industrial production 151
propanol
– volumetric expansion 105
propargylic alcohols 205
propene
– hydrogenation 190
protection strategies
– temporary 120
proteins 467
protonation 125
PS, *see* polystyrene
pseudo-first-order rate constants 444, 447, 452
PTC, *see* phase transfer catalysis
PTFE, *see* polytetrafluoroethylene
pumps
– metering 63
pure fluids
– viscosity 81
pure substances
– phase diagram 78
PVC, *see* poly(vinyl chloride)
pyrolysis 469
pyrone formation 214
pyrrole 234

q
quadrupole moment
– CO_2 83
quasi-reference electrodes (QREs) 423

r
radical chain reactions
– photo-induced 410
radical-initiated polymerization
– styrene 126
radical oxidation
– steel-promoted 145
radical polymerization 315
radical reactions 228
– transfer 316
radicals
– caged pairs 407
– free-radical 407
rate constants
– pseudo-first order 444, 447, 452
rate enhancement 247, 251
raw materials
– renewable 467
reaction–separation processes

– sequential 130
reactions
– acetylation 290
– acid-catalyzed 435
– acylation 233
– alkylation 255, 266
– alkylcarbonic acids 451
– amination 210
– ammonium salt-catalyzed 436
– asymmetric hydrogenation 198
– Aufbau 14
– Baylis–Hillman 211
– biphasic 172
– Cannizzaro 459, 473
– carbonylation 139
– carboxylation 234
– chlorination 403
– condensation 66, 462
– coupling 208, 435
– crystallization 130
– cycloaddition 142, 221
– cyclopropanation 217
– dehydration 461
– dehydrogenation 253, 260
– denitrogenation 225
– Diels–Alder 221, 463
– dimerization 404
– disproportionation 268
– electrochemical– electrochemical 419
– electroreduction 428
– enzymatic 285
– enzymatic esterification 171
– equilibrium 121
– esterification 232
– Fischer–Tropsch synthesis 254, 260
– free radical bromination 410
– free radical chlorination 411
– Friedel–Crafts acylation 435
– Friedel–Crafts alkylation 220, 442
– geometric isomerization 403
– glycine decomposition 468
– Heck 127, 209
– Henry 219
– hydration 157
– hydroaminomethylation 123, 204
– hydroformylation 139, 141, 172, 254, 262
– hydrogen abstraction 413
– hydrogenation 135
– hydrolysis 460
– imidization 339
– in SCFs 4, 16
– isomerization 256, 269
– lipase-catalyzed 282, 286
– metathesis 148
– nucleation 329
– organometallic 465
– oxidation 205
– oxybromination 122
– Pauson–Khand 209, 214
– phase transfer 435
– photochemical 224, 399
– photodimerization 403
– photoinduced 399, 409
– photosensitization 409
– polycondensation 314
– polymerization 303
– post-reaction separation 152
– precipitation 130
– protonation 125
– radical 228, 316, 410
– rearrangements 463
– reverse water gas shift 257
– ring-closing metathesis 215
– sharpless epoxidation 207
– supercritical biocatalysis 289
– Suzuki 209
– transesterification 231
– tunable phase separation 131
– tunable precipitation/crystallization 130
reactive ketones 202
reactors
– chemoenzymatic 297
– cleaning 57, 69
– continuous flow 44
– continuous green enzyme 296
– continuously stirred tank reactors 45
– design 41, 251
– enzyme 287
– fixed-bed 259
– heating 48
– high-pressure 34, 42
– hydroformylation 202
– microstructured 45
– nano- 376
– packed-bed enzyme 288
– plug flow 45, 51
– pressure 65
– $scCO_2$ 401
– stirred tank reactor 41
– supercritical hydrogenation 70
rearrangements 463
reciprocating compressors 34
reciprocating pumps 63
recirculation 288
recovery of nanoparticles 377
recycling
– PTC 437
reduced pressure/temperature 1

reduction
– colloid-catalyzed 163
– in NCW and scH_2O 464
reference electrode 422
– quasi- 422
reforming
– aqueous (phase) 470
regioisomers 139
regioselectivity
– enzymatic catalysis 281
1,3-regiospecific lipase
– immobilized 291
relative polarity 117
relative volume expansion 107
renewable raw materials 347, 467
resistance temperature detectors 48
RESS process 14
reverse micelles 375, 377
reverse water gas shift (RWGS) reaction 257
rhodium catalysts 122, 263
ring-closing metathesis 215
ring-opening metathetical polymerization (ROMP) 345
ring systems
– aromatic 442
ROMP, *see* ring-opening metathetical polymerization
rupture disk 401
ruthenium-based Grubbs-type catalysts 346

s
sacrificial aldehyde 206
sacrificial co-reductant 145
safety valves 53
safety warnings 58
salts
– ammonium 436
– diazonium 452
– inorganic 390
– organic 137
– plugging 472
– unwanted precipitation 472
sapphire 44
$scCO_2$ 4, 7
– diffusibility 58
– electrochemical reactions 424
– free radical brominations 410
– IL-$scCO_2$ biphacis systems 294
– imidazolium IL–$scCO_2$ mixtures 165
– nanomaterials synthesis 370
– polymerization in 303
– polymerization in 315
– reactor 401
– water-in-$scCO_2$ microemulsions 375, 382
SCF, *see* supercritical fluids
SCF-in-water emulsions 173
Schiff base-type metal complexes 344
scH_2O (SCW) 7, 85
– chemistry in 457
– flow-through method 380
– inorganic compounds 471
– nanomaterials synthesis 373
– oxidation 22, 470
– synthesis 459
– water-in-$scCO_2$ microemulsions 375, 382
SCWO, *see* supercritical water oxidation
sealing materials 35, 39, 41
secondary amines 124
seed to enhance crystallization (SSEC) process 382
selectivity
– benzaldehyde 264
– control 135
– dehydration 467
– hydrogenation 198
– shape-selectivity effects 267
– tuning 248
self-diffusivity
– CO_2 79, 80
self-optimization 48
semiconductor nanocrystals 369
semicrystalline thermoplastic polymer 322
sensitizers 227
– chiral 409
– photo- 409
sensors
– pressure 48, 49
separation 435
– post-reaction 152
– product/catalyst 154
– PTC 437
sequential reaction–separation processes 130
shape-selectivity effects 267
sharpless epoxidation 207
shift reaction
– reverse water gas 257
shut-off devices 33
side-chain fluoropolymers 317
sieves
– molecular 391
silica
– mesoporous 391
silicon nanoparticles 371
siloxane-based stabilizers 329
SILP, *see* supported ionic liquid phase
silver nanocrystals 370
six-component mixture 140
six-port valve 54

slurry-phase process 260
sodium montmorillonite 331
sol–gel synthesis
– direct 380
solid acids 149
solid-phase organic synthesis (SPOS) 142
solid supports 200, 292
solids
– dosage 64
– fluorinated 134
– polar 93
– porous 244, 246
solubility
– biphasic systems 102, 104
– experimental data 305
– gases 118
– hydrogen 272
– isotherm 97
– macromolecules 306
– polymers 168
– SCF 92
– structure–solubility relationship 95
– temperature and pressure effects 96
solutes
– chemical functionality 94
– clustering 8
– clustering 400
– crystallization 130
– diffusivity 307
solution temperature
– upper critical 442
solvating power
– adjustable 8
solventless continuous flow
hydroformylation 172
solventless hydrogenation
– cinnamaldehyde 136
solvents
– chlorinated 58
– density 304
– electrochemical reactions 423
– fluorinated 325
– hazardous organic 109
– heated 58
– light organic 109
– liquid organic 251
– neoteric 284
– organic 57
– organic solvent mixtures 133
– relative volume expansion 107
– SCF 92, 399
– SCF discovery 9
– selection 159
– solvent–solute clustering 399
– supercritical 426
– supercritical organic 371
– switchable 124
– tunable 4
– viscosity 399
– volumetric expansion 107
sound level 32
stability enhancement 250
stabilization
– catalysts 155
– steric 311, 313, 371
stabilized enzymes 292
stabilizers
– design 310
– for CO_2 314
– siloxane-based 329
standard hydrogen electrode 422
statistical distribution of products 406
steel-promoted radical oxidation 145
stereoselectivity– enzymatic catalysis 281
steric stabilization 311, 313, 371
stirred tank reactor (STR) 41
– continuously 45
– high-pressure 288
stirrer types 51
Stokes–Einstein equation 408
storage vessel 61
streams
– depressurized gas 66
stress corrosion cracking 35
structure–solubility relationship 95
styrene
– asymmetric hydroformylation 204
– homogeneous hydrogenation 170
– radical-initiated polymerization 126
substituents
– damping effect 95
substrates
– polar 285
sulfosuccinate
– bis(2-ethylhexyl) 374, 375
sulfoxide
– dimethyl 435
supercritical alkanes 371, 374
supercritical biocatalysis 289, 292
supercritical butene 154
supercritical CO_2, *see* $scCO_2$
supercritical difluoromethane 427
supercritical flow reactors
– automated 69
supercritical fluids (SCF) 1
– aqueous–SCF biphasic systems 159
– biotransformations 229
– biphasic systems 101, 159, 163

– chromatography in 452
– conversions in 256
– delivery system 421
– density 3
– environmental benefit 189
– enzymatic catalysis 283
– extraction in 14
– hazards 5
– history and applications of 9
– industrial applications 16, 20
– inorganic 85
– ionic liquid-in-SCF emulsions 173
– macromolecular chemistry 304
– microemulsions 374
– nanomaterials synthesis 370
– organic 86
– phase behavior 86
– photochemical and photo-induced reactions 399
– physical properties 7, 77
– polar 424
– polymer–SCF biphasic systems 167
– reactions 4, 16
– solubility 92
– solvents 92, 399
– synthetic organic chemistry 189
– thermophysical properties 375
– viscosity 413
– water-in-SCF inverse emulsions 172
– water–SCF mixtures 161
supercritical hydrogenation reactor 70
supercritical hydrothermal synthesis 377
supercritical organic solvents 371
supercritical solvents
– hydrofluorocarbon 426
supercritical water oxidation (SCWO) 22, 470
supercritical water (SCW), *see* scH_2O
supply pressure 138
supported ionic liquid phase (SILP) 140
supported noble metal catalysts 258
supports
– porous 391
– solid 292
surface area-to-volume ratio 44
surface energy
– nanoparticles 378
surface-functionalized particles 334
surface hydrophobicity 265
surface phase
– catalyst-rich 436
surfactants
– AOT 374, 375
– fluorinated 375
– perfluorinated 341
Suzuki reactions 209
swelling 252
switchable solvents 124
synthesis
– ammonia 20
– chemisorption of gases in liquids 120
– direct sol–gel 380
– electrochemical 429
– Fischer–Tropsch 254, 260
– hydrothermal 377
– in NCW and scH_2O 459
– nanomaterials 369
– SCFs in 6
– solid-phase organic 142
– supercritical hydrothermal 377
synthetic organic chemistry in SCF 189
syringe pumps 63

t

tank reactors
– stirred tank reactor 288
TBAP, *see* tetrabutylammonium picrate
Teflon FEP 321
temperature
– critical 1
– detectors 48
– electrochemistry 424
– low critical 62
– reduced 1
– supercritical biocatalysis 289
– upper critical solution 442
temperature effects 96
templated materials 158
temporary protection strategies 120
tensile strength 46
tension– interfacial 12
4-tertbutyl phenol 198
tertiary amines 124
tethered gold nanocrystals
– molecularly 372
tetrabutylammonium picrate (TBAP) 437, 438
tetrabutylammonium tetrafluoroborate ($TBABF_4$) 421, 427, 428
tetraethyl orthosilicate (TEOS) 387
tetrafluoroborate
– tetrabutylammonium 421, 427, 428
thermal Brownian movement 312
thermal conductivity 84
– CO_2 84
thermal mass flow meter 62
thermocouples 48
thermophysical properties
– compressed liquids and SCF 375
– fluid systems 59

thermoplastic polymer
– semicrystalline 322
Thiele modulus 270
tiglic acid 1, 138
titanium alkoxides 381
titanium isopropoxide (TIP) 381
toluene
– volumetric expansion 104
toroidal design 43
total reflection mode
– attenuated 56
total volume measurement 66
toxicity
– SCFs 5
transesterification 231
transfer catalysis
– phase transfer catalysis 435
transfer effects
– mass and heat 245
transfer reactions
– phase 435
– radical 316
trifluoride
– boron 347
triple point 247
tube fittings 47
tubes 45
tunable
– crystallization 130
– fluids 435
– miscibility 134
– phase separation 131
– precipitation 130
– solvents 4

u
UCEP, *see* upper critical end-points
UCST, *see* upper critical solution temperature
unintentional immiscibility 151
units, dosing 62
unsaturated organic compounds 257
unwanted salt precipitation 472
upgrading, hydrothermal 469
upper critical end-points (UCEP) 89
upper critical solution temperature (UCST) 92, 442

v
valorization, oils and fats 285
valves 47
– pressure relief 33, 53
– safety 53
– six-port 54
van der Waals equation 87
van der Waals potential 313
vapor
– liquid–vapor coexistence curve 77
– vapor–liquid equilibrium 105
variable-area flowmeter 66
variable-volume view cells 42
vegetable resources, renewable 348
vessels
– pressure 41
– pressurized 5
– storage 61
view cells, variable-volume 42
vinyl acetate 338
vinyl carbamate 219
vinyl laurate 296
vinyl monomers 335, 336
virial coefficient 310
virial equation 61
viscosity 79
– CO_2 81
– expanded liquid phases 110
– pure fluids 81
– SCF 7, 413
– solvents 399
VLE, vapor 105
voltammetry 419, 427
volume
– expansion 272
– partial molar 248
– surface area-to-volume ratio 44
– total 66
volumetric expansion 104, 106
– CO_2 dissolution 168
– solvents 107

w
Wacker oxidation 213
water
– aqueous–SCF biphasic systems 159
– density 439, 440
– dielectric constant 82, 83, 440
– dissociation constant 441
– elimination from glycerol 462
– near-critical 438
– physicochemical properties 457
– reverse water gas shift reaction 257
– SCF-in-water emulsions 173
– supercritical, scH_2O 7
– supercritical oxidation 22, 470
– usage 438
– volumetric expansion 104
– water-immiscible ILs 294
– water-in-CO_2 microemulsions 426
– water-in-oil microemulsions 376

– water-in-SCF inverse emulsions 172
water–SCF mixtures, phase behavior 161
wet meters 66
Wilkinson's catalyst, homogeneous
  hydrogenation 170
windows
– high-pressure reaction vessels 400
– optical 53
working electrode 422
working phase 295

x
xanthene 266

y
Young's modulus 332, 387

z
zeolite catalysts 266
Ziegler–Natta catalysts 340
zinc complexes 343